Lecture Notes in Computer Science 16444

Founding Editors

Gerhard Goos
Juris Hartmanis

The series Lecture Notes in Computer Science (LNCS), including its subseries Lecture Notes in Artificial Intelligence (LNAI) and Lecture Notes in Bioinformatics (LNBI), has established itself as a medium for the publication of new developments in computer science and information technology research, teaching, and education.

LNCS enjoys close cooperation with the computer science R & D community, the series counts many renowned academics among its volume editors and paper authors, and collaborates with prestigious societies. Its mission is to serve this international community by providing an invaluable service, mainly focused on the publication of conference and workshop proceedings and postproceedings. LNCS commenced publication in 1973.

Emilio Di Giacomo · Debajyoti Mondal

Editors

WALCOM: Algorithms and Computation

20th International Conference and Workshops on Algorithms
and Computation, WALCOM 2026
Perugia, Italy, March 4–6, 2026
Proceedings

 Springer

Editors
Emilio Di Giacomo 🆔
Università degli Studi di Perugia
Perugia, Italy

Debajyoti Mondal 🆔
University of Saskatchewan
Saskatoon, SK, Canada

ISSN 0302-9743 ISSN 1611-3349 (electronic)
Lecture Notes in Computer Science
ISBN 978-981-95-7126-0 ISBN 978-981-95-7127-7 (eBook)
https://doi.org/10.1007/978-981-95-7127-7

This Springer imprint is published by the registered company Springer Nature Singapore Pte Ltd.
The registered company address is: 152 Beach Road, #21-01/04 Gateway East, Singapore 189721, Singapore

If disposing of this product, please recycle the paper.

Preface

This volume contains the papers presented at the 20th International Conference and Workshops on Algorithms and Computation (WALCOM 2026), which took place March 4–6, 2026 at the University of Perugia in Perugia, Italy. Notably, this conference marked WALCOM's debut in Europe. WALCOM provides an international forum in the areas of algorithms and computation where established researchers and students can meet, discuss their work, present their results, and engage in scientific collaborations.

The conference received 108 submissions from 38 countries, which highlights WALCOM's broad international scope. Each submission underwent a rigorous single-blind review process, with at least three reviews provided by members of the Program Committee and external referees. A total of 38 papers were accepted for presentation, corresponding to an acceptance rate of 35%. Selected high-quality papers were invited for submission to special issues of the Journal of Graph Algorithms and Applications (JGAA) and Theoretical Computer Science (TCS).

The conference program featured three invited talks from renowned researchers in the field. Paolo Ferragina (Scuola Superiore Sant'Anna, Pisa) spoke on *"Data Compression and Indexing in the Era of Machine Learning"*. Petra Mutzel (University of Bonn) presented *"Graph Similarity: Concepts, Methods, and Challenges"*. Bart Jansen (Eindhoven University of Technology) discussed *"Sparsification and Kernelization for Constraint Satisfaction Problems through the Lens of Non-Redundancy"*. The abstracts of these talks are included in this volume.

The Program Committee conferred the Best Paper Award on M. Chimani and M. H. Wagner for their paper *"Computing Beyond-Planar Crossing Numbers via Forbidden Crossing Patterns"*. The Best Student Paper Award was presented to T. Gima, Y. Kobayashi, Y. Otachi, and T. Sato for their paper *"Forcing a Unique Minimum Spanning Tree and a Unique Shortest Path"*.

The success of WALCOM 2026 was made possible through the contributions of many individuals and organizations. We thank all the authors who submitted their papers. We extend our gratitude to the Steering Committee for their guidance, and the Program Committee members and the external reviewers for their careful evaluation of the submissions and for assembling a strong and engaging program. Our warmest thanks go to the Organizing Committee, chaired by Carla Binucci, Luca Grilli, and Alessandra Tappini of the University of Perugia, for their dedication and effort in ensuring the smooth organization of the conference. We are grateful to our sponsors, Sileon and Tom Sawyer Software, whose financial support was instrumental in the success of the event. Finally, we express our gratitude to Springer for publishing these proceedings in their distinguished LNCS series.

March 2026

Emilio Di Giacomo
Debajyoti Mondal

Organization

WALCOM Steering Commitee

Tamal Dey	Purdue University, USA
Seok-Hee Hong	University of Sydney, Australia
Giuseppe Liotta	University of Perugia, Italy
Petra Mutzel	University of Bonn, Germany
Shin-ichi Nakano	Gunma University, Japan
Subhas Chandra Nandy	Indian Statistical Institute, Kolkata, India
Md. Saidur Rahman	Bangladesh University of Engineering and Technology, Bangladesh
Ryuhei Uehara	Japan Advanced Institute of Science and Technology, Japan

WALCOM Organizing Committee

Carla Binucci (Co-chair)	University of Perugia, Italy
Emilio Di Giacomo	University of Perugia, Italy
Walter Didimo	University of Perugia, Italy
Luca Grilli (Co-chair)	University of Perugia, Italy
Fabrizio Grosso	University of Perugia, Italy
Giuseppe Liotta	University of Perugia, Italy
Fabrizio Montecchiani	University of Perugia, Italy
Giacomo Ortali	University of Perugia, Italy
Daniel Perz	University of Perugia, Italy
Tommaso Piselli	University of Perugia, Italy
Alessandra Tappini (Co-chair)	University of Perugia, Italy

Program Commitee

Akanksha Agrawal	Indian Institute of Technology Madras, India
Reyan Ahmed	University of Arizona, USA
Hugo A. Akitaya	University of Massachusetts Lowell, USA
Michael A. Bekos	University of Ioannina, Greece
Carla Binucci	University of Perugia, Italy
Hans L. Bodlaender	Utrecht University, the Netherlands

Tiziana Calamoneri	"Sapienza" University of Rome, Italy
Marek Chrobak	University of California, Riverside, USA
Sabine Cornelsen	University of Konstanz, Germany
Minati De	Indian Institute of Technology Delhi, India
Emilio Di Giacomo (Co-chair)	University of Perugia, Italy
Stephane Durocher	University of Manitoba, Canada
Zachary Friggstad	University of Alberta, Canada
Konstantinos Georgiou	Toronto Metropolitan University, Canada
Thekla Hamm	Eindhoven University of Technology, the Netherlands
Meng He	Dalhousie University, Canada
Philipp Kindermann	Trier University, Germany
Debajyoti Mondal (Co-chair)	University of Saskatchewan, Canada
Shin-ichi Nakano	Gunma University, Japan
Subhas Chandra Nandy	Indian Statistical Institute, Kolkata, India
Rahnuma Islam Nishat	Brock University, Canada
Martin Nöllenburg	TU Wien, Austria
Fahad Panolan	University of Leeds, UK
Md. Saidur Rahman	Bangladesh University of Engineering and Technology, Bangladesh
Morteza Saghafian	Institute of Science and Technology Austria, Austria
Toshiki Saitoh	Kyushu Institute of Technology, Japan
Christiane Schmidt	Linköping University, Sweden
Ryuhei Uehara	Japan Advanced Institute of Science and Technology, Japan
Mingyu Xiao	University of Electronic Science and Technology of China, China
Meirav Zehavi	Ben-Gurion University of the Negev, Israel

External Reviewers

Acharyya, Ankush	Bhowmick, Partha
Ahmed, Shareef	Biniaz, Ahmad
Alipour, Sharareh	Bose, Prosenjit
An, Shinwoo	Brötzner, Anna
Asarin, Eugene	Carayol, Arnaud
Azgor, Ruhul	Ceccherini-Silberstein, Tulio
Bai, Tian	Chakraborty, Sourav
Banbara, Mutsunori	Chattopadhyay, Amit
Banik, Aritra	Chen, Jingbang
Bataa, Magsarjav	Chlebikova, Janka

Corò, Federico
Cáceres, Manuel
Das, Gautam Kumar
Das, Sandip
Das, Syamantak
Depian, Thomas
DeWolfe, Ryan
Dey, Sanjana
Didimo, Walter
Dobler, Alexander
Dong, Rui
Fernau, Henning
Fertin, Guillaume
Firbas, Alexander
Fujisawa, Jun
Fujiwara, Hiroshi
Förster, Henry
Ghosh, Anirban
Goetze, Miriam
Gorla, Daniele
Grilli, Luca
Grosso, Fabrizio
Habib, Siam
Hanaka, Tesshu
Hasan, Md. Manzurul
Herschlag, Gregory
Hesterberg, Adam
Hoang, Duc A.
Huang, Haoqiang
Ito, Takehiro
Jabal Ameli, Afrouz
Jacob, Dalu
Jain, Pallavi
Jana, Satyabrata
Janzen, Jacob
Jeffery, Stacey
Jeż, Artur
Jin, Ce
Kawahara, Jun
Keikha, Vahideh
Kimmel, Shelby
Kiya, Hironori
Kiyomi, Masashi
Klostermeyer, William
Koechlin, Florent

Kolay, Sudeshna
Kosub, Sven
Krithika, R.
Kuhn, Fabian
Kurita, Kazuhiro
Liu, Hsiang-Hsuan
Liu, Jiamou
Liu, Yuxi
Majumdar, Diptapriyo
Mandal, Ratnadip
Mann, Kevin
Mashghdoust, Amirhossein
Matsui, Tomomi
Michoel, Tom
Mishra, Gopinath
Misra, Pranabendu
Montecchiani, Fabrizio
Moore, Ben
Mouawad, Amer
Mumey, Brendan
Muttakin, Md. Nurul
Müller, Haiko
N. R. Aravind
Nanoti, Saraswati Girish
Ntasiou, Aikaterini Maria
Okamoto, Yoshio
Otachi, Yota
Paesani, Giacomo
Pandit, Supantha
Pareek, Akash
Paul, Kaustav
Paul, Subhabrata
Pavlidi, Maria Eleni
Peng, Junqiang
Perz, Daniel
Peters, Tom
Pighizzini, Giovanni
Plätz, Lukas
Polishchuk, Valentin
Pradhan, Dinabandhu
Richter, Marena
Rieck, Christian
Rizzo, Nicola
Sadhu, Sanjib
Sahu, Abhishek

Salvo, Ivano	Uno, Yushi
Scheffer, Christian	van der Zanden, Tom
Schoenbach, Gabe	van Renssen, André
Sinaimeri, Blerina	Varshney, Sakshi
Singh, Satyam	Verbeek, Kevin
Sivalingam, Krishna	Verma, Shaily
Sorge, Manuel	Wasa, Kunihiro
Stock, Frederick	Wu, Kaiyu
Strozecki, Yann	Xu, Chao
Sultana, Shaheena	Yamanaka, Katsuhisa
Takenaga, Yasuhiko	Yanhaona, Muhammad Nur
Terziadis, Soeren	Zhou, Yi
Toth, Csaba	Zhu, Binhai

Sponsors

Abstracts of the Invited Talks

Data Compression and Indexing in the Era of Machine Learning

Paolo Ferragina

Scuola Superiore Sant'Anna, Pisa, Italy

Abstract. Key-value stores, vector and graph DBs, as well as search engines are posing a continuously growing need to efficiently store, retrieve and analyze massive datasets under the many and different requirements posed by data types, users, devices and applications. Such a new level of complexity cannot be properly handled by just known techniques, so academic and industrial researchers started recently to devise new compression and indexing schemes that integrate classic approaches with various kinds of advanced techniques drawn from computational geometry, combinatorial pattern matching and machine learning. In this talk, I'll survey some of those "combinations", discuss their theoretical and experimental performance, and highlight their limits and some recent results that are pointing out new challenges worthy of future research.

Graph Similarity: Concepts, Methods, and Challenges

Petra Mutzel

Universität Bonn, Germany

Abstract. Graph similarity plays a central role in network analysis, pattern recognition, and algorithmic data science. This talk provides a unified perspective on key concepts, methodological approaches, and computational challenges inherent in comparing complex graph structures. We will discuss structural approaches, covering both distance-based techniques (e.g., graph edit distance) and isomorphism-based methods via subgraph search, alongside embedding-based techniques derived from the Weisfeiler–Leman refinement. A central theme of the talk is how algorithm engineering can bridge the gap between theoretical difficulty and practical applicability, enabling similarity computations on large and complex graph data.

Sparsification and Kernelization for Constraint Satisfaction Problems through the Lens of Non-redundancy

Bart Jansen

Eindhoven University of Technology, the Netherlands

Abstract. Sparsifying a graph to obtain a smaller representation is a popular tool to speed up computations involving flows and cuts. In the field of parameterized complexity, kernelization is a popular method to build fixed-parameter tractable algorithms by applying reduction rules that reduce the input size without changing the YES/NO answer. While sparsification and kernelization share the common theme of data reduction, these two research areas developed independently since the lossy techniques used for sparsification typically fail in the exact setting of kernelization. However, recent work has shown that in the context of constraint satisfaction problems, which extends the setting of graph cuts, there is a unifying concept of non-redundancy that explains the extent to which data reduction is possible. This talk surveys the most powerful ideas for sparsification and kernelization for constraint satisfaction problems, presents the concept of non-redundancy as a unifying perspective, and showcases recent progress on analyzing non-redundancy via algebraic techniques.

(Based on joint work with Joshua Brakensiek, Venkatesan Guruswami, Victor Lagerkvist, and Magnus Wahlström.)

Induced subgraphs and dominating sets

Complexity

Games and Graph Reconfiguration

Shortest Paths and Minimum Spanning Trees

Geometric Problems

Enumeration Problems

Graph Drawings and Embeddings

Computing Beyond-Planar Crossing Numbers via Forbidden Crossing Patterns

Markus Chimani[iD] and Mirko H. Wagner[✉][iD]

Theoretical Computer Science, Osnabrück University, Osnabrück, Germany
{markus.chimani,mirko.wagner}@uos.de

Abstract. Beyond-planarity concepts (e.g., k-planarity or fan-planarity) apply certain restrictions on the allowed patterns of crossings in drawings. It is natural to ask for the minimum number of crossings over all so-restricted drawings for any given beyond-planarity concept. Unfortunately, for most studied beyond-planarity concepts, already recognizing such graphs is NP-complete.

We augment exact crossing number formulations, based on integer linear programs (ILPs), to be able to forbid crossing patterns, and discuss necessary algorithmic consequences. To this end, we first adapt crossing patterns to the context of abstract topological graphs, and propose further extensions that allow for provable polyhedral strengthenings over natural ad-hoc constraint classes. From the users' perspective, they only need to provide such patterns (for known or novel beyond-planarity concepts). Our framework then automatically establishes the required model and separates the necessary constraints dynamically, without the need for additional case-specific implementations. If the computation is successful, our framework solves the membership problem, outputs a crossing-minimal feasible drawing w.r.t. the crossing pattern, and also generates a proof of membership/optimality that can be externally verified.

In a short exploratory study, we see that the additional beyond-planar constraints do not drastically impair the practical performance and that our extended patterns are beneficial. Furthermore, we identify algorithmic subproblems (namely primal heuristics and preprocessing routines) that are in need of the ability to soundly tackle beyond-planar scenarios.

1 Introduction

Traditionally, much graph drawing research focused on planar graphs and, if the graphs were not planar, on drawing them with as few crossings as possible. Beyond-planarity concepts (prominent examples include k-planarity and fan-planarity) are ideas to bridge the divide between these two strategies [21,24]: By restricting the patterns of crossings that are allowed to occur in drawings, one establishes graph classes (and drawing paradigms) that are richer than planar graphs. Unfortunately, while planarity can be tested in linear time [6,20,25], already testing membership is NP-complete for most established beyond-planarity concepts [3–5,23,28]. Given a graph, perhaps even known to

E. Di Giacomo and D. Mondal (Eds.): WALCOM 2026, LNCS 16444, pp. 3–18, 2026.
https://doi.org/10.1007/978-981-95-7127-7_1

lie in a specific beyond-planar graph class, it is natural to ask for a feasible drawing (w.r.t. the considered drawing paradigm) that minimizes the number of crossings. We know that such *beyond-planar crossing numbers* can drastically deviate from the traditional crossing number (called *normal* in the following), sometimes being larger by a factor linear (or even quadratic) in the number of vertices [9, 15, 34]. Recently, it was shown that beyond-planar crossing numbers are FPT in the number of crossings for beyond-planarity concepts that can be expressed by (well-defined) *forbidden crossing patterns* [32, 33]. Alas, these algorithms do not seem applicable in practice, and there are effectively no known practical algorithmic approaches to find suitable drawings for virtually any such concepts.

In the context of the (NP-complete [22]) normal crossing number, there are integer linear programs (ILPs) that solve the problem in practice on not-too-large not-too-dense graphs [7, 13, 16]. These implementations are in particular also able to extract externally verifiable proofs [19]. This functionality is also exposed via a free web-service and for years now used by several different international research groups to check conjectures during research or establish otherwise tedious and uninteresting base cases for inductive proofs on infinite graph classes.

Our Contribution. We augment these ILP approaches to several beyond-planarity concepts. We base our approach on the forbidden crossing patters by Münch and Rutter [32]. However while their definition is sufficient in the context of manual proofs, we reiterate on the precise details to make them applicable in the context of our mathematical models in Sect. 3.

In Sect. 4, we can then describe our system: It not only tackles the crossing number within some specific beyond-planarity concepts, but allows the user to define arbitrary forbidden crossing patterns as part of the input in a short, natural encoding. Based on this, the implementation automatically separates necessary constraints dynamically during the computation and also outputs a formal, externally verifiable proof for the resulting crossing number (provided the computation was successful). Our framework is resilient enough to also consider graphs that do not belong to the considered beyond-planar graph class; it implicitly solves the recognition problem itself. Even if the precise optimal number of crossings may be unattainable within the allowed time frame, it may still affirm (or refute) membership.

We close the paper with a short practical feasibility study in Sect. 5. The main focus in this paper is the development of the framework and its theoretical foundations, *not* a thorough, hypothesis-driven experimental study on various beyond-planarity concepts. At the current state of research, we first require exploratory tests to pinpoint the bottlenecks and cruxes for further in-depth development, and show the potential use of this line of research.

2 Preliminaries

Graph Definitions. Let $G = (V, E)$ be a graph. *Subdividing* an edge $\{v, w\} \in E$ means to replace it by a *chain* (path with at least one edge and degree-2 inner vertices) between v and w; we may call these new path's edges *subedges*. A *subdivided graph* (or *subdivision*) $\tilde{G} = (V, S, \tilde{E})$ of G is graph $(V \cup S, \tilde{E})$ arising from G by subdividing its edges; the new degree-2 vertices (*subdivision vertices*, as opposed to the *real vertices* V) are collected in set S. $\tilde{G}$ induces a mapping $\sigma_{\tilde{G}} \colon \tilde{E} \to E$ from subedges to their respective original edges (we may omit the subscript if this is clear from the context). As a generalization of subdivided graphs, we define a *subdivision-graph* as any graph $D = (R_D, S_D, E_D)$ consisting of distinct *real* vertices R_D and *subdivision* vertices S_D, the latter of degree one or two, even without an underlying graph G.

Kuratowski's theorem [30] states that a graph G is planar if and only if it does not contain a subdivison of a K_5 or $K_{3,3}$ as a subgraph; these are called *Kuratowski subdivisions*.

A *drawing* of a graph is a mapping of vertices to distinct points in the plane, and of edges to simple Jordan curves connecting the respective end points, not traversing through other vertex points. A *crossing* is an interior point common to two edge curves. An interior point common to k edges would naturally be counted as $\binom{k}{2}$ crossings. A drawing is *simple* if any two edge curves share at most one point (either a common end point or a crossing), and no three edge curves share a common interior point. The normal crossing number (minimizing the number of crossings over all possible drawings) is always achieved by a simple drawing. This does not necessarily hold for other beyond-planar crossing numbers.

An *abstract topological graph (AT-graph)* $(G, \mathcal{C})$ is a graph $G = (V, E)$ together with a set $\mathcal{C}$ of edge pairs that are supposed to cross. The *simple realizability problem* asks whether there exists a simple drawing of G such that precisely the crossings of $\mathcal{C}$ occur. If each edge occurs in at most one pair of $\mathcal{C}$, the problem can be answered in linear time by replacing the crossings via degree-4 dummy nodes and testing planarity. However, if edges are crossed by several other edges, it is hard to find or rule-out a feasible order of these crossings along the respective edges. This problem is indeed NP-complete [29], even in surprisingly small configurations [31]. Let $(G, \mathcal{C})$ be an AT-graph and $\mathcal{C}' \subseteq \mathcal{C}$ a subset of its crossing pairs. Then $(G', \mathcal{C}')$ is an *AT-subgraph* if G' is the subgraph of G induced by the edges considered in $\mathcal{C}'$. We say $(\tilde{G}, \tilde{\mathcal{C}})$ is an *AT-subdivision* of $(G, \mathcal{C})$ if $\tilde{G}$ is a subdivision of G, $|\mathcal{C}| = |\tilde{\mathcal{C}}|$, and for each $\{e, f\} \in \mathcal{C}$ there is precisely one $\{\tilde{e}, \tilde{f}\} \in \tilde{\mathcal{C}}$ with $e = \sigma(\tilde{e})$ and $f = \sigma(\tilde{f})$.

Crossing Number ILPs. We chiefly summarize the known ILP models for the normal crossing number, discussing only the tidbits necessary within this paper. For the general gist of how to solve ILPs via branch-and-cut-and-price, please see [35]. Let $G = (V, E)$ be the (non-planar) graph in question; we ask for its (normal) crossing number $\mathrm{cr}(G)$.

Models. Recall that $\mathrm{cr}(G)$ is always attained by a *simple* drawing. The core idea is to have $\mathcal{X} := \big\{ \{e, f\} : e, f \in E, e \cap f = \emptyset \big\}$ be all non-adjacent edge pairs in G and to consider a variable set $\boldsymbol{x} = \big\{ x_{\{e,f\}} \in \{0, 1\} \big\}_{\{e,f\} \in \mathcal{X}}$. Variable $x_{\{e,f\}}$ shall be 1 if and only if e crosses f in the solution. However, these variables do not suffice due to the mentioned NP-hard simple realizability problem, and we need to encode the order of crossings along edges as well.

Table 1. Simplified view on the ILP models for the normal crossing number. For model $\mathbb{S}$, let $G_\mathbb{S} = (V, S, E_\mathbb{S})$ be a sufficient subdivision of G with $\mathcal{X}_\mathbb{S} := \big\{\{\tilde{e}, \tilde{f}\} : \tilde{e}, \tilde{f} \in E_\mathbb{S}, \{\sigma_{G_\mathbb{S}}(\tilde{e}), \sigma_{G_\mathbb{S}}(\tilde{f})\} \in \mathcal{X}\big\}$.

subdivision-based model ($\mathbb{S}$)		ordering-based model ($\mathbb{O}$)	
$\min \displaystyle\sum_{\{\tilde{e},\tilde{f}\}\in\mathcal{X}_\mathbb{S}} \tilde{x}_{\{\tilde{e},\tilde{f}\}}$		$\min \displaystyle\sum_{\{e,f\}\in\mathcal{X}} x_{\{e,f\}}$	
$\displaystyle\sum_{\tilde{f}\,:\,\{\tilde{e},\tilde{f}\}\in\mathcal{X}_\mathbb{S}} \tilde{x}_{\{\tilde{e},\tilde{f}\}} \leq 1$	$\forall \tilde{e} \in E_\mathbb{S}$	$\text{OrderConstraints}(y_{e,.,.})$	$\forall e \in E$
		$\text{Kuratowski}(K, x, y)$	$\forall K \in \mathcal{K}_\mathbb{O}$
$\text{Kuratowski}(K, \tilde{x})$	$\forall K \in \mathcal{K}_\mathbb{S}$	$x_{\{e,f_1\}} \geq y_{e,f_1,f_2} \leq x_{\{e,f_2\}}$	$\forall \{e,f_1\}, \{e,f_2\} \in \mathcal{X}, f_1 \neq f_2$
$\tilde{x}_{\{\tilde{e},\tilde{f}\}} \in \{0,1\}$	$\forall \{\tilde{e}, \tilde{f}\} \in \mathcal{X}_\mathbb{S}$	$x_{\{e,f\}} \in \{0,1\}$	$\forall \{e,f\} \in \mathcal{X}$
		$y_{e,f_1,f_2} \in \{0,1\}$	$\forall \{e,f_1\}, \{e,f_2\} \in \mathcal{X}, f_1 \neq f_2$

There are two distinct approaches to achieve this, see Table 1: In the *subdivision-based* [7] model (denoted by $\mathbb{S}$), each edge is subdivided several times, yielding a subdivision $G_\mathbb{S}$ (over which, formally, we establish the variables $\tilde{x}$ instead of x). If we subdivide the edges into long enough chains, we can restrict the number of crossings per subedge to 1 (the *simplicity constraint*) without effectively restricting the crossing number of underlying G but yielding a trivial realizability instance. In the *ordering-based* model [16] (denoted by $\mathbb{O}$), we orient all edges arbitrarily and add binary variables y_{e,f_1,f_2}, for all $e, f_1, f_2 \in E$ with $\{e, f_1\}, \{e, f_2\} \in \mathcal{X}$, $f_1 \neq f_2$, which shall be 1 if and only if e is (along its orientation) crossed by f_1 before it is crossed by f_2. For each e, we can then enforce a linear order (via the standard 3-cycle constraints known from the linear order polytope) over the edges crossing it.

Given any of the two models, the workhorse of both are the *Kuratowski constraints* (which are known to yield facets of the corresponding crossing number polytope [8]). Let $\tilde{x}'$ (x' and y') be some, not necessarily feasible, binary variable assignment for model $\mathbb{S}$ (model $\mathbb{O}$, respectively). A *partial planarization* results from $\tilde{G}$ (G) by adding degree-4 dummy vertices according to $\tilde{x}'$ (x' and y', respectively). For model $\mathbb{S}$ (model $\mathbb{O}$), let $\mathcal{K}_\mathbb{S}$ ($\mathcal{K}_\mathbb{O}$) denote the set of *all* Kuratowski subdivisions over *all* possible partial planarizations. Let $K \in \mathcal{K}_\mathbb{S}$. Avoiding notational details, a Kuratowski constraint then enforces at least one crossing on K, as long as all variables $\tilde{x}_{\{\tilde{e},\tilde{f}\}}$ with $\tilde{x}'_{\{\tilde{e},\tilde{f}\}} = 1$ are 1. (And analogously for $K \in \mathcal{K}_\mathbb{O}$ w.r.t. x', y').

Solving the Models. Crucial initial ingredients to solve either of the above models in practice are (a) a strong preprocessing routine [10], and (b) strong primal heuristics to quickly compute (hopefully even optimal) upper bounds [11,14].

As the number of Kuratowski subdivisions (and subsequent constraints) is exponential, we cannot consider them all explicitly, but have to use a (heuristic) separation routine: We "round" a fractional solution (setting only variables above a threshold – e.g., 0.7 or 0.99 – to 1), compute the corresponding partial planarization, run a planarity test which yields a Kuratowski subdivision as a witness to its negative answer (an augmented algorithm [17] even finds multiple in a single run), and add the corresponding constraint (if it is indeed violated). As this routine always finds a violated constraint in an infeasible integral solution, the overall process is guaranteed to terminate with an optimum solution.

In both models, the number of variables, although polynomial, is a bottleneck in practice and we deploy a form of column generation. In contrast to the standard method of (algebraic) *pricing*, the crossing number framework uses a beneficial, purely combinatorial detection paradigm, which is easiest to describe in the context of model $\mathbb{S}$: assume no edge is initially subdivided, and there are no simplicity constraints. As long as there are no multiple crossings over a single edge, the realization problem is still easily tractable. Only when a current solution requires two crossings over a single edge e, we may then subdivide e into two subedges, add the simplicity constraint for the first subedge, and decrease this subedge's cost by a small ε. The second subedge may later still obtain more than one crossings, triggering another subdivision step. While this simple method suffices for the proof extraction method [19] (see below), some stronger, more fine-tuned column generation routines are deployed to find the optimal solution faster in the first place [13,16].

Proof System. The full proof system [19] runs as follows: First, a computation with the model $\mathbb{O}$ and all available speed-up techniques is started, as this is typically faster than model $\mathbb{S}$ (but requires a column generation that is harder to explain and verify). With the optimal solution (P1) for best-possible bounding, we start a computation with model $\mathbb{S}$, with a column generation simplified as described above. From this, we can extract a case distinction according to the branch-tree (P2), the necessary number of subdivisions for each individual edge (P3), and a sufficient subset $\mathcal{K}_{\mathrm{active}}$ of Kuratowski subdivisions (P4), and write the data in a JSON-file, together with (P1). An external program (in Java instead of C++) can read this file, check that (P2) captures all cases, and run an independent (possibly slow but exact) LP solver for each case using (P3), including the simplicity constraints for all but the "last" subedges, and (P4). It only needs to verify that solutions cannot be better than (P1) while all "last" subedges have a crossing sum of at most 1 (without explicit simplicity constraint). If all tests succeed, the optimality of (P1) is soundly established.

Beyond-Planarity Concepts and Combinatorial Crossing Patterns. Many beyond-planarity concepts can be characterized by one or more forbidden crossing patterns. To show FPT results for various beyond-planarity concepts in [32], this notion was formalized as follows (where the moniker "combinatorial" was later added in [33]):

Definition 1 (Combinatorial Crossing Pattern [32]). *A combinatorial crossing pattern is a graph $P = (V_P, E_P)$ with $V_P = R \cup S \cup C$ such that (i) each vertex in S has degree 1; (ii) each vertex in C represents a crossing, has degree 4, and its incidences are partitioned into two sets of size 2; and (iii) each vertex in $R \cup S$ is adjacent to at least one vertex of C.*

The idea is to forbid that a drawing of a graph G has a substructure resembling a given pattern; thereby, vertices S are used to easily model that edges

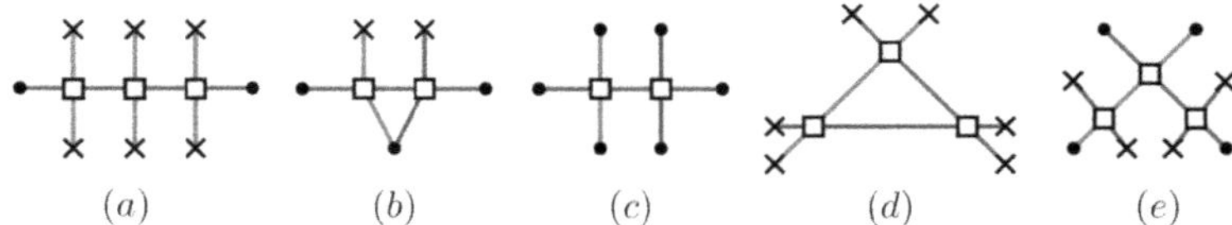

Fig. 1. Combinatorial crossing patterns from [32]: R, S, and C vertices are circles, crosses and squares, respectively. (a) k-planar (here: k=2), (b) fan-crossing-free, (c) adjacency-crossing, (d) quasi-planar, (e) min-1-planar.—While (c) was proposed for fan-crossing, to the best of our knowledge, its absence in a drawing only guarantees that that drawing is *adjacency-crossing*; in the former, one also needs to forbid an edge to cross all three edges of a triangle.

in the pattern may only be part of an edge in G. Figure 1 shows the patterns introduced in [32] (see Sect. 3 for the formal definition of the concepts).

3   Basic and Extended Abstract Crossing Patterns

Abstract Crossing Patterns. We argue that the order of the crossings on an edge in a crossing pattern is seldom important and if it is irrelevant, then it is beneficial to not restrict it. In fact, in all known crossing patterns the crossing order is indeed irrelevant. By forcing specific crossing orders as part of a pattern (e.g. for k-quasi-planarity with $k \geq 4$ or for min-k-planarity with $k \geq 2$), we need $\Omega(k^2)$ formally different patterns (one for each possible order). By avoiding specific orders, we reduce them into a single unifying pattern. Indeed, also the proofs in [32] implicitly use the automorphism groups in the combinatorial crossing patterns; we prefer them to be explicit. We thus propose the following variation of the above crossing patterns, using the language of AT-graphs:

Definition 2 (Basic Abstract Crossing Pattern). *A basic abstract crossing pattern $P = (G_P, \mathcal{C}_P)$ is an AT-graph such that (1) G_P is a subdivision-graph and (2) each edge of G_P is part of at least one crossing in $\mathcal{C}_P$.*

Definition 3 (Containment). *Let $(G, \mathcal{C})$ be an AT-graph and P a basic abstract crossing pattern. $(G, \mathcal{C})$ contains P if there is an AT-subdivision of G that contains an AT-subgraph isomorphic to an AT-subdivison of P.*

A drawing satisfying some beyond-planarity concept must not contain any of its corresponding forbidden basic abstract crossing patterns. In the language of combinatorial crossing patterns, one basic abstract crossing pattern P gives rise to a set CCP(P) of combinatorial crossing patterns (one for each possible planarization of P). Containment of P in $(\tilde{G}, \mathcal{C})$ is then equivalent to containing (in the sense of [32]) some element of CCP(P).

In the context of ILPs, it is often beneficial to have constraints that rule out undesired solutions early on, even when (most) variables attain fractional values in the LP relaxation. To this end, it can be helpful to combine several basic constraints into a single more powerful constraint. We extend the basic abstract crossing patterns with some integer ϱ_P and relax the containment semantics: A pattern is contained in an AT-graph already if a subpattern with at least ϱ_P crossings is contained.

Definition 4 (Extended Abstract Crossing Pattern). *An extended abstract crossing pattern $P = (G_P, \mathcal{C}_P, \varrho_P)$ is a basic abstract crossing pattern $(G_P, \mathcal{C}_P)$ and an integer $\varrho_P \leq |\mathcal{C}_P|$ such that, for each $r \geq \varrho_P$, any two AT-subgraphs induced by r crossings (called* subpatterns*) are isomorphic basic abstract crossing patterns.*

Definition 5 (Extended Containment). *Let $(G, \mathcal{C})$ be an AT-graph and P an extended abstract crossing pattern. Then $(G, \mathcal{C})$ contains P, if it contains any subpattern of P.*

Abstract Crossing Patterns for Beyond-Planarity Concepts. For all beyond-planarity concepts considered in [32], as well as grid-free graphs, we can establish basic and often also extended abstract crossing patterns. Our basic patterns are naturally direct translations of the original combinatorial crossing patterns. We start each considered concept with a chief definition. See Fig. 2 for visualizations of the patterns.

k-planar: *Any edge is crossed at most k times.*

A basic abstract crossing pattern $P_k^{\mathsf{B}} = (G_{P_k^{\mathsf{B}}}, \mathcal{C}_{P_k^{\mathsf{B}}})$ (B for "basic") for k-planar holds a pattern graph $G_{P_k^{\mathsf{B}}}$ with $k + 2$ edges which form a matching; since it is not important whether any of these edges are adjacent (in non-simple drawings they may even be subedges of a common edge), all its vertices are subdivison vertices. We forbid that the first edge e_0 is crossed by all of the other $k+1$ edges. Formally, let $[k + 1] := \{0, 1, ..., k + 1\}$ and we have

$$P_k^{\mathsf{B}} = ((\, R_{P_k^{\mathsf{B}}}, \qquad S_{P_k^{\mathsf{B}}}, \qquad\qquad E_{P_k^{\mathsf{B}}} \qquad\qquad), \qquad \mathcal{C}_{P_k^{\mathsf{B}}} \qquad)$$
$$= ((\; \emptyset,\; \{s_i, s_{i+2k+2}\}_{i \in [k+1]},\; \{e_i = \{s_i, s_{i+2k+2}\}\}_{i \in [k+1]}\;),\; \{\{e_0, e_i\}\}_{i \in [k+1]}).$$

We extend these patterns in two ways. In the first extension, we forbid that a single edge may be crossed by any subset of at least $k + 1$ edges, out of arbitrarily many more edges. We thus establish the extended abstract crossing pattern $P_k^{\mathsf{M}} := (G_{P_\ell^{\mathsf{B}}}, \mathcal{C}_{P_\ell^{\mathsf{B}}}, k+1)$, where we may choose $\ell > k$ arbitrarily large, even tending to infinity. The second extension is specific to 1-planarity: clearly, there can be at most one crossing between any three edges e_0, e_1, e_2. We yield an extended pattern $P_1^{\mathsf{0}} := (G_{P_1^{\mathsf{B}}}, \mathcal{C}_{P_1^{\mathsf{B}}} \cup \{\{e_1, e_2\}\}, 2)$ by adding the additional crossing to $\mathcal{C}_{P_1^{\mathsf{B}}}$, and forbidding any subset of size at least 2 (out of all three crossings) to exist.

In the following, we reuse these extension methods in other scenarios as well, where "M" is used when considering more edges than necessary, "0" when conceptually closing a cycle.

quasi-planar: *There are no pairwise crossings between any three edges.*

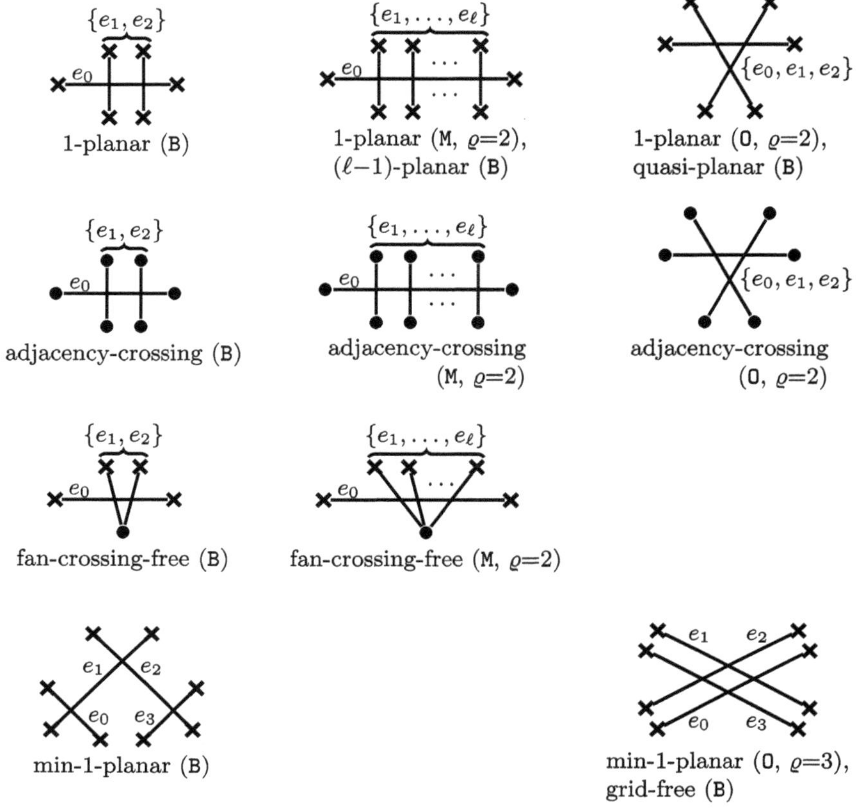

Fig. 2. Visualizations of abstract crossing patterns. Real vertices are filled circles, subdivision vertices are crosses. For easy comprehensibility, the figures show crossings to represent the edge pairs; they do not imply any order.

For quasi-planar, we construct a basic crossing pattern that consists of three independent edges and crossings between all three possible pairs. Note that the only difference between this pattern (only forbidding all 3 crossings to exist simultaneously) and the 0-extension for 1-planarity is that the latter already forbids 2 crossings on the same structure.

adjacency-crossing: *Two edges crossing the same edge must share a vertex.*

For adjacency-crossing we construct the basic and the two extended patterns identical to the 1-planar case, but require all vertices to be real instead of subdivided vertices.

fan-crossing-free: *No edge can cross a fan, i.e., two edges that share a vertex.*

We can extend the natural basic abstract pattern (which follows from the combinatorial crossing pattern), using method M, i.e., consider an arbitrary number of edges adjacent to the previously-degree-2 real vertex, while setting the size requirement $\varrho_P = 2$.

min-1-**planar:**   *In each crossing, one of the involved edges is crossed only once.*

For min-1-planar, the basic abstract crossing pattern is four independent edges $e_0, \ldots, e_3$ on 8 subdivision vertices and crossing pairs $\{e_i, e_{i+1}\}$ for $0 \leq i < 2$. We can extend this pattern following the O-paradigm by additionally listing the crossing pair $\{e_3, e_0\}$, but forbidding any planarization realizing at least $\varrho_P = 3$ of these crossings.

grid-**free:**   *No pair of edges can cross another pair of edges.*

The basic abstract crossing pattern uses the graph and crossings of the extended min-1-planar pattern. It thus forbids to have all these four crossings simultaneously.

4   Integrating Crossing Patterns Into the ILP Models

We are now ready to discuss how to efficiently integrate basic and extended abstract crossing patterns into the ILP framework. As they give rise to an exponential number of constraints, we have to devise a sufficient and efficient separation routine. In particular, we aim at a system that is not restricted and specific to the beyond-planarity patterns above described. Our goal is to deduce the constraints and their effective separation (as well as their form in generated proofs) purely from the pattern description. This shall allow the use of the framework for any new desired patterns, without additional implementation efforts. In line with much of current research, we restrict our implementation to simple drawings.

From Patterns to Constraints.   Let $P = (G_P, \mathcal{C}_P)$ be a basic abstract crossing pattern, and G the input graph to the ILP. We need to forbid selecting crossing pairs $\mathcal{C}^*$ such that the AT-graph $(G, \mathcal{C}^*)$ contains P.

We start with the simpler – but arguably the only practically relevant – case that P contains no degree-2 subdivision vertices. Let $\tilde{G}$ be any subdivision of G that contains a subgraph $\tilde{G}' \subseteq \tilde{G}$ isomorphic to G_P, and let $\varphi \colon E(G_P) \to E(\tilde{G}')$ be the corresponding mapping. This naturally gives rise to the *basic crossing pattern constraint*

$$\sum_{\{e_P, f_P\} \in \mathcal{C}_P} x_{\{\sigma_{\tilde{G}}(\varphi(e_P)), \sigma_{\tilde{G}}(\varphi(f_P))\}} \leq |\mathcal{C}_P| - 1.$$

We can use this constraint directly in model $\mathbb{O}$. For model $\mathbb{S}$, these x-variables do not exist; we can still use the constraint by substituting any variable $x_{\{e,f\}}$ over an original edge pair by summing over the corresponding subedges in $G_{\mathbb{S}}$. Formally, let $\psi_{\mathbb{S}}(e, f) := \{\{\tilde{e}, \tilde{f}\} : \sigma_{G_{\mathbb{S}}}(\tilde{e}) = e, \ \sigma_{G_{\mathbb{S}}}(\tilde{f}) = f\}$ and use the substitution $x_{\{e,f\}} = \sum_{\{\tilde{e}, \tilde{f}\} \in \psi_{\mathbb{S}}(e,f)} \tilde{x}_{\{\tilde{e}, \tilde{f}\}}$.

If P is not a basic but an extended crossing pattern, it additionally holds some value ϱ_P. We construct the same constraint above (for $\mathcal{C}_P$ or any subset of $\mathcal{C}_P$ with $r \geq \varrho_P$ elements), but set the right-hand side to $\varrho_P - 1$, independent of r and $|\mathcal{C}_P|$.

It remains to discuss the case that G_P contains subdivision vertices of degree 2; call them S_2. S_2 forces some order on certain crossing pairs, for the pattern to be contained. We start by establishing the constraint as above, suppressing S_2 (i.e., we consider the edge to go through uninterruptedly). Then, one such vertex $s \in S_2$ after another, let e_1, e_2 be the two incident edges to s and E_1 (E_2) the edges that form crossing pairs in $\mathcal{C}_P$ with e_1 (e_2, respectively). Both e_1 and e_2 (through φ and $\sigma_{\tilde{G}}$) correspond to a common original edge e in G. Consider model $\mathbb{O}$, and assume that e is oriented such that E_1 comes before E_2. On the left-hand side of the constraints we add all variables y_{e,f_1,f_2} with $f_1 \in E_1$ and $f_2 \in E_2$; we increase the right-hand side of the constraint by the number of so-added variables. Finally – to further strengthen the constraints – we may remove all $x_{\{e,f\}}$ variables from the left-hand side for which at least one $y_{e,f,\cdot}$ or $y_{e,\cdot,f}$ variable is also in the constraint, and decrease the right-hand side by the number of so-removed variables. Analogous operations allow us to tackle degree-2 subdivision vertices in the $\mathbb{S}$ model as well.

We can formally show the (at least theoretical) benefit of extended abstract crossing patterns over basic crossing patterns. A minimizing ILP model $\mathbb{M}_{\text{strong}}$ is called *polyhedrally stronger* than another model $\mathbb{M}_{\text{weak}}$, if (1) $\mathbb{M}_{\text{strong}}$'s LP-relaxation always yields a solution no less than $\mathbb{M}_{\text{weak}}$'s LP-relaxation (i.e., it establishes an at least as good lower bound for the ILP solution), and (2) there is an instance where this bound is strictly stronger.

Theorem 1. *The above extended abstract crossing patterns always lead to strictly stronger polyhedral models than using only the corresponding basic patterns.*

Proof (Sketch). For the ease of exposition, consider patterns without subdivision vertices of degree 2. In all our constraints for any fixed beyond-planarity concept, the difference between a basic and its extended constraint is that the former holds a subset of the variables of the latter on its left-hand side, with an identical right-hand side (observe that ϱ is always the number of crossing pairs in the original basic pattern). Thus any solution satisfying all basic constraints by construction also satisfies all extended constraints, establishing (1).

Let us describe the key idea to establish (2). Consider some constraint for a basic pattern P; it sums $|\mathcal{C}_P|$ variables on the left-hand side, with a right-hand side of $|\mathcal{C}_P| - 1$. Thus setting all involved variables to the fractional value $\frac{|\mathcal{C}_P|-1}{|\mathcal{C}_P|}$ satisfies all basic constraints. However, the corresponding extended constraints (summing over more than $|\mathcal{C}_P|$ so-set variables) are violated, and a more expensive solution to the LP-relaxation is required. By essentially embedding individual crossing patterns into a larger planar graph structure, it is relatively straight-forward to find actual graphs in which these weak solutions occur, even when considering all interactions with Kuratowski constraints. □

Separating Pattern Constraints. As for Kuratowski constraints, we propose a heuristic separation routine (sufficient to guarantee optimal ILP solutions) by considering the crossings arising from rounding the current fractional LP solution. Let $P = (G_P, \mathcal{C}_P)$ be a basic abstract crossing pattern and lexicographically sort the edges primarily in decreasing order in the number of occurrences in $\mathcal{C}_P$, and secondarily in decreasing order in the number of real vertex incidencies. Matching edges in this order, we deploy a backtracking algorithm that tries to identify a suitable edge with sufficiently many crossings in the rounded solution, consistent with the previously identified edges. Thereby, it is relatively straight-forward to detect automorphisms within the pattern (and thus avoid symmetric matches). The edge order is chosen such that (at least heuristically) we first identify the rarest-to-successfully-match edges and thus minimize the number of backtracking steps.

Let $P = (G_P, \mathcal{C}_P, \varrho_P)$ be an extended pattern. We choose any of the isomorphic subpatterns with ϱ_P crossings, and separate it as a basic crossing pattern as above. Then, we iteratively pick a not-yet-considered crossing from $\mathcal{C}_P$ (all choices are equivalent) and try to add it to the already identified subpattern; this may include matching new edges. The crossing may not exist in the rounded solution, we only have to check that the matched edges are consistent with G_P. We grow our matched pattern until we fail to grow for the first time.

Each identified pattern gives rise to a constraint, which, if it is also violated in the fractional solution, is added to the set of active constraints. By the use of our backtracking approach, we can identify several constraints in a single run.

Subdivision Bound in Model $\mathbb{S}$.   When solving model $\mathbb{S}$, we require an upper bound β on the number of subdivisions per edge; an unnecessarily high upper bound leads to large subdivisions and variable spaces. The total number of crossings in any feasible drawing (according to the crossing pattern at hand) yields an upper bound, but there is a severe lack of heuristics to attain drawings (see below). Thus we have to resort to more crude bounds. In simple drawings, no edge is crossed more than once per other edge, yielding a trivial bound $\beta = |E| - 1$. For k-planar drawings, we have a natural bound of $\beta = k$ subedges per edge. In simple adjacency-crossing and fan-crossing-free drawings, the maximum degree and the number of vertices yield bounds, respectively.

Primal Heuristics and Preprocessing. For the normal crossing number, primal heuristics are a cornerstone to the approach's practicality, as they typically quickly identify the crossing number for which to prove optimality. While there are no known specific heuristics for any of the considered beyond-planarity concepts (and strong concepts like planarization approaches are intrinsically hard in such scenarios [27]), we can still use a normal crossing number heuristic and check (using the algorithm used for separation) if the solution satisfies the beyond-planarity by chance.

In particular on "real-world" graphs, the *non-planar core reduction* [10] is an established preprocessing routine that (sometimes drastically) improves the computation time for the normal crossing number. We cannot soundly use this strategy in our scenario.

Proof System. Our framework also extends the proof generation for the normal crossing number to the pattern-restricted use case. Again, we (1) first compute an optimum solution using either model and then (2) start a second computation (with model $\mathbb{S}$) to prove that there is no solution with strictly fewer crossings. Step (1) improves both the bounding behavior in the branch-tree as well as (potentially) the subdivision bound β. Additionally to the proof outputs described in Sect. 2, we also output the list of machine-verifiable crossing pattern constraints that were generated in the process. In case of non-membership, (1) shows infeasibility and (2) extracts a corresponding proof.

5   Feasibility Study

In the following, we report on a short feasibility study, where we want to explore three questions:

Q1. Are our formulations feasible in practice?
Q2. Do extended crossing patterns yield better practical performance than basic ones and, if so, to which extent?
Q3. What are bottlenecks and important-to-tackle subproblems?

Due to space constraints, we can only give a very high-level overview on the main findings herein. Still, we feel that it provides the necessary context to judge our contribution.

Our C++ implementation uses OGDF [12], ABACUS 3.0.1 [26], and CPLEX 20.10. We aim to make the software freely available, also as part of the crossing number web compute service [18]. We consider two benchmark sets: The *Rome* graphs [2] and graphs collected by the crossing number web compute service [18] (*WebCompute*). We consider the 16 beyond-planar crossing patterns described in Sect. 3, as well as the normal crossing number (i.e., no pattern). We use a time limit of 10 min per run. All instances, results, proofs, and the independent proof verifier can be found at [1].

Q1. Practical Viability. For several concepts – namely 3-planar, quasi-planar, grid-free – their success rates are nearly on par with the normal crossing number (90% for *Rome* and 96–99% for *WebCompute*), see Table 2. These are the most "relaxed" patterns, and many optimal crossing number solutions do not violate them anyhow. Other concepts, in particular 1-planar (76%, 93%) and fan-crossing-free (85%, 66%) and (to a lesser extent) adjacency-crossing seem more complicated. There, the heuristic and optimal solutions for the normal crossing number are typically invalid. Still, our experiments show that the framework is capable of resolving a significant number of instances for all patterns.

Q2. Benefit of Extended Patterns. Interestingly, M-extended patterns yield only a slight advantage, if any. O-extended patterns have no significant impact at all. It remains open, whether the extensions would become beneficial for larger or harder instances, which are out of reach with our current hard- and software.

Q3. Bottlenecks. Investigating the behavior sketched for *Q1* further, we see clearly, that whenever the initial primal heuristic finds the optimal solution, the ILP has a good chance of proving it; if the primal heuristic (which ignores crossing patterns) fails to yield strong upper bounds, this directly results in

Table 2. Success rates (in %).

		WebCompute				Rome			
		𝕆		𝕊		𝕆		𝕊	
	with heuristics?:	✗	✓	✗	✓	✗	✓	✗	✓
1-planar	B	56	57	**88**	**88**	56	58	**70**	**70**
	M	59	59	(88)	(88)	62	61	(70)	(70)
	O	56	57	**88**	**88**	57	57	69	**70**
2-planar	B	88	89	84	**91**	81	**82**	76	**82**
	M	89	90	(84)	(91)	81	**82**	(76)	(82)
3-planar	B	96	**97**	84	**97**	87	**90**	73	89
	M	96	**97**	(84)	(97)	87	**90**	(76)	(89)
adjacency-cr.	B	91	92	87	**94**	65	65	**71**	**71**
	M	92	93	87	**94**	68	68	**71**	**71**
	O	91	92	86	**94**	65	66	**71**	**71**
fan-cr.-free	B	63	64	62	**66**	83	**85**	69	81
	M	63	65	62	**66**	84	**85**	69	82
min-1-planar	B	75	**76**	71	74	77	**78**	64	71
	O	75	**77**	71	75	77	**78**	63	72
quasi-planar	B	97	**99**	83	**99**	88	**90**	66	88
grid-free	B	95	**96**	82	92	88	**90**	66	88
normal	-	97	**99**	83	98	88	**91**	66	89
	with preprocessing	98	**99**	89	98	92	**93**	74	92

weak overall performance. Similarly, we see that the models achieve a much higher success rate for the normal crossing number when first preprocessing the instances using the non-planar core reduction [10]; however, this routine is not sound w.r.t. beyond-planar patterns, and thus cannot be used in these contexts.

Further Insights. Being able to compute beyond-planar crossing numbers, some practical questions come into reach. From theory, we know that a beyond-planar crossing number can be much larger than the normal crossing number [9,15,34]. But how much do they deviate in practice? It turns out that for the vast number of solved *Rome* instances, the 1-planar crossing number is identical (about 77%) or at least very similar to the normal crossing number (at most larger by 2). Only about 0.2% of the *Rome* graphs are found to be not 1-planar, see Fig. 3. However, as roughly a quarter of the *Rome* instances remain unsolved w.r.t. 1-planarity, we can formally only conclude that we at least never observe large deviations in the crossing numbers as of now.

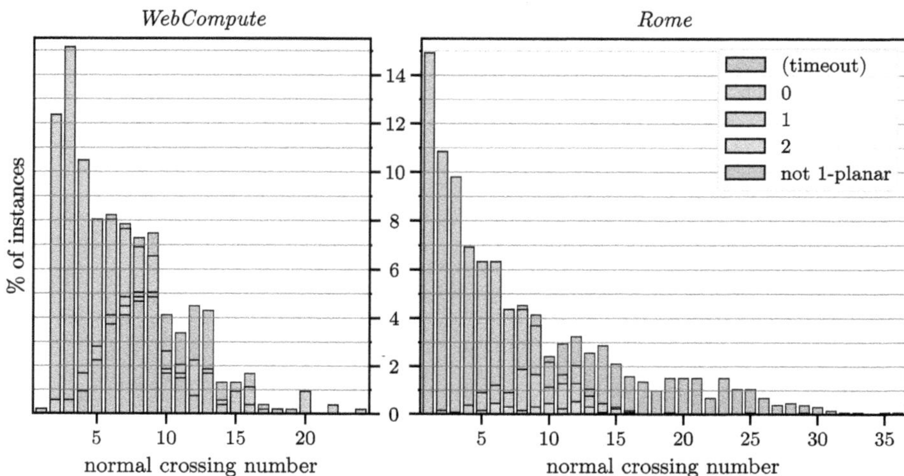

Fig. 3. Percentage of instances in the respective benchmark set with the specified difference between the normal and the 1-planar crossing number.

Conclusion. Overall our models are able to solve relevant instances in practice. We also identified two main algorithmic bottlenecks of the approach as interesting research questions on their own: preprocessing routines and constructive heuristics for beyond-planar crossing numbers.

References

1. Webpage with instances, results, proofs, and the independent proof verifier (2025). https://tcs.uos.de/research/cr
2. Di Battista, G., Garg, A., Liotta, G., Tamassia, R., Tassinari, E., Vargiu, F.: An experimental comparison of four graph drawing algorithms. Comput. Geom. **7**(5–6), 303–325 (1997)
3. Bekos, M.A., Cornelsen, S., Grilli, L., Hong, S., Kaufmann, M.: On the recognition of fan-planar and maximal outer-fan-planar graphs. Algorithmica **79**, 401–427 (2017)
4. Binucci, C., et al.: Min-k-planar drawings of graphs. J. Graph Algorithms Appl. **28**(2), 1–35 (2024)
5. Binucci, C., et al.: Fan-planarity: properties and complexity. Theor. Comput. Sci. **589**, 76–86 (2015)
6. Boyer, J.M., Myrvold, W.J.: On the cutting edge: simplified o(n) planarity by edge addition. J. Graph Algorithms Appl. **8**(3), 241–273 (2004)
7. Buchheim, C., et al.: A branch-and-cut approach to the crossing number problem. Discret. Optim. **5**(2), 373–388 (2008)
8. Chimani, M.: Facets in the crossing number polytope. SIAM J. Discret. Math. **25**(1), 95–111 (2011)
9. Chimani, M., Donzelmann, N., Kloster, T., Koch, M., Völlering, J., Wagner, M.H.: Crossing numbers of beyond planar graphs re-revisited: a framework approach. In: Proc. GD 2024, LIPIcs 320, pp. 1–17 (2024)

10. Chimani, M., Gutwenger, C.: Non-planar core reduction of graphs. Discret. Math. **309**(7), 1838–1855 (2009)
11. Chimani, M., Gutwenger, C.: Advances in the planarization method: effective multiple edge insertions. J. Graph Algorithms Appl. **16**(3), 729–757 (2012)
12. Chimani, M., Gutwenger, C., Jünger, M., Klau, G.W., Klein, K., Mutzel, P.: The open graph drawing framework (OGDF). In: Handbook on Graph Drawing and Visualization, pp. 543–569. Chapman and Hall/CRC (2013). http://www.ogdf.net/
13. Chimani, M., Gutwenger, C., Mutzel, P.: Experiments on exact crossing minimization using column generation. ACM J. Exp. Algorithmics **14**, 3–4 (2009)
14. Chimani, M., Ilsen, M., Wiedera, T.: Star-struck by fixed embeddings: modern crossing number heuristics. In: Purchase, H.C., Rutter, I. (eds.) GD 2021. LNCS, vol. 12868, pp. 41–56. Springer, Cham (2021). https://doi.org/10.1007/978-3-030-92931-2_3
15. Chimani, M., Kindermann, P., Montecchiani, F., Valtr, P.: Crossing numbers of beyond-planar graphs. Theor. Comput. Sci. **898**, 44–49 (2022)
16. Chimani, M., Mutzel, P., Bomze, I.: A new approach to exact crossing minimization. In: Halperin, D., Mehlhorn, K. (eds.) ESA 2008. LNCS, vol. 5193, pp. 284–296. Springer, Heidelberg (2008). https://doi.org/10.1007/978-3-540-87744-8_24
17. Chimani, M., Mutzel, P., Schmidt, J.M.: Efficient extraction of multiple Kuratowski subdivisions. In: Proc. GD 2007, LNCS 4875, pp. 159–170 (2007)
18. Chimani, M., Wiedera, T.: Crossing number web compute (2016). http://crossings.uos.de
19. Chimani, M., Wiedera, T.: An ILP-based proof system for the crossing number problem. In: Proc. ESA 2016, LIPIcs 57, pp. 1–13 (2016)
20. de Fraysseix, H., Ossona de Mendez, P., Rosenstiehl, P.: Trémaux trees and planarity. Int. J. Found. Comput. Sci. **17**(5), 1017–1030 (2006)
21. Didimo, W., Liotta, G., Montecchiani, F.: A survey on graph drawing beyond planarity. ACM Comput. Surv. **52**(1), 1–37 (2019)
22. Garey, M.R., Johnson, D.S.: Crossing number is NP-complete. SIAM J. Algebraic Disc. Meth. **4**(3), 312–316 (1983)
23. Grigoriev, A., Bodlaender, H.L.: Algorithms for graphs embeddable with few crossings per edge. Algorithmica **49**(1), 1–11 (2007)
24. Hong, S., Tokuyama, T. (ed.): Beyond Planar Graphs. Communications of NII Shonan Meetings. Springer (2020)
25. Hopcroft, J.E., Tarjan, R.E.: Efficient planarity testing. J. ACM **21**(4), 549–568 (1974)
26. Jünger, M., Thienel, S.: The ABACUS system for branch-and-cut-and-price algorithms in integer programming and combinatorial optimization. Softw. Pract. Exp. **30**(11), 1325–1352 (2000)
27. Katheder, J., Kindermann, P., Klute, F., Parada, I., Rutter, I.: On k-plane insertion into plane drawings. In: Proc. GD 2024, LIPIcs 320, pp. 1–11 (2024)
28. Korzhik, V.P., Mohar, B.: Minimal obstructions for 1-immersions and hardness of 1-planarity testing. J. Graph Theory **72**(1), 30–71 (2013)
29. Kratochvíl, J.: String graphs. II. recognizing string graphs is NP-hard. J. Comb. Theory B **52**(1), 67–78 (1991)
30. Kuratowski, K.: Sur le probleme des courbes gauches en topologie. Fundam. Math. **15**(1), 271–283 (1930)
31. Da Lozzo, G., Didimo, W., Montecchiani, F., Münch, M., Patrignani, M., Rutter, I.: Simple realizability of abstract topological graphs. In: Proc. ISAAC 2024, LIPIcs 322, pp. 1–15 (2024)

32. Münch, M., Rutter, I.: Parameterized algorithms for beyond-planar crossing numbers. In: Proc. GD 2024, LIPIcs 320, pp. 1–16 (2024)
33. Münch, M., Rutter, I.: Parameterized algorithms for crossing number with forbidden topological crossing patterns. In: Proc. EuroCG 2025 (2025)
34. van Beusekom, N., Parada, I., Speckmann, B.: Crossing numbers of beyond-planar graphs revisited. J. Graph Algorithms Appl. **26**(1), 149–170 (2022)
35. Wolsey, L.A.: Integer programming. Discr. Math. Optim. (1998)

Minimum-Weight Outerplane Laman Graphs

Oswin Aichholzer[1] , Yuya Higashikawa[2], Joachim Orthaber[1] ,
Daniel Perz[3(✉)] , Birgit Vogtenhuber[1] , and Alexandra Weinberger[4]

[1] Institute of Algorithms and Theory, Graz University of Technology, Graz, Austria
`{oswin.aichholzer,orthaber,birgit.vogtenhuber}@tugraz.at`
[2] Graduate School of Information Science, University of Hyogo, Kobe, Japan
`higashikawa@gsis.u-hyogo.ac.jp`
[3] University of Perugia, Perugia, Italy
`daniel.perz@unipg.it`
[4] FernUniversität in Hagen, Hagen, Germany
`alexandra.weinberger@fernuni-hagen.de`

Abstract. Given an outerplane drawing $\mathcal{D}$ with n vertices, a $\mathcal{D}$-*constrained maximum outerplane drawing* is an outerplane drawing that contains $\mathcal{D}$ and has the maximum number of edges among all such drawings. We show that (1) any such $\mathcal{D}$-constrained maximum outerplane drawing has at least $n + 1$ edges, and (2) this bound is best possible. If $\mathcal{D}$ is a plane spanning path, we present an $O(n^3)$-time algorithm to compute a $\mathcal{D}$-constrained maximum outerplane drawing that minimizes its weight, that is, the sum of the Euclidean length of the edges of its drawing. For the unweighted setting, our results imply the following: It can be decided in $O(n^3)$ time whether a given plane spanning path admits a polygon such that every edge of the path is on the polygon or an internal edge. Further, we present for several minimum-weight structures, like minimum spanning trees, points sets such that they are not subdrawings of the minimum-weight outerplane Laman graph of that point set.

1 Introduction

A *Laman graph* with n vertices is a graph that has exactly $2n - 3$ edges and, for each subset of $k \leq n$ vertices, the induced subgraph has at most $2k - 3$ edges. Laman graphs are central to rigidity theory: Drawn as straight-line graphs in the plane, which is similar to being realized as a two-dimensional generic *bar-joint framework*, they form a family of sparse graphs which describe *minimally rigid* systems of rods and joints in the plane [5,9].

Laman graphs appear in a wide range of applications such as mechanical statics and its design as linkages, CAD system design, the analysis of protein

We thank Rosna Paul for fruitful discussions. Y.H. is partially supported by JSPS KAKENHI Grant Number 23H03350. J.O. and A.W. were partially supported by the Austrian Science Fund (FWF) grant W1230. D.P. is supported in part by MUR PRIN Proj. 2022ME9Z78 - "NextGRAAL: Next-generation algorithms for constrained GRAph visuALization".

flexibility, and sensor network localization [14,16]. There, sensors are modeled as vertices and the links or measured distances between them as edges. Often, sensor positions are fixed by environmental constraints or deployment strategies. If the communication graph satisfies the Laman conditions for rigidity, then the relative positions of the sensors can be uniquely determined (up to Euclidean transformations). In practical applications, plane (that is, crossing-free) Laman graphs are particularly important. For example, in the area of architecture or mechanical engineering, it is often important to avoid intersections between bars causing design constraints.

On the theoretical side there has been a lot of research about planar Laman graphs. It has been shown that it can be decided in polynomial time whether a given graph is a planar Laman graph [12]. Further results include the generation of planar Laman graphs [3] and properties of them [8]. Note that every triangulation of a polygon is a planar Laman graph. Another notable subclass of planar Laman graphs are pointed pseudo-triangulations [6,15]. (A pseudo-triangle is a polygon with three convex vertices. A pseudo-triangulation is a tiling of the plane where every face except the unbounded face is a pseudo-triangle, and it is pointed if every vertex is incident to an angle greater than π.) For a survey about (pointed) pseudo-triangulations we refer to [13].

In this paper, we study outerplane Laman graphs and maximum outerplane drawings. A straight-line drawing is *outerplane* if it is plane and every vertex is incident to the outer (unbounded) face. We call an outerplane drawing of a Laman graph an *outerplane Laman graph*. We particularly focus on minimum-weight outerplane Laman graphs, and the positions of the vertices are fixed, where the weight of a drawing is defined as the sum of the Euclidean length of all its edges.

We say a set S of points in the plane *admits* a straight-line drawing of a given graph G if the vertices of G are represented by points in S and consequently the edges of G are straight-line segments between some of those points.

A (drawing of a) graph is *maximal* w.r.t. some property if it is impossible to add an edge without violating this property. A (drawing of a) graph is *maximum* w.r.t. some property if amongst all (drawings of) graphs that fulfill the property, this one has the most edges. A *maximum outerplane drawing* (or short MOD) on a point set S is an outerplane drawing of some graph G on S such that any other graph that admits an outerplane drawing on S has at most as many edges as G. Without additional restrictions, every maximum outerplane drawing $\mathcal{D}$ on n points in general position has exactly $2n-3$ edges since we can obtain a MOD by taking a plane Hamiltonian cycle of the point set and triangulating the interior of the cycle. Thus, outerplane Laman graphs are also maximal outerplane drawings. However the converse is not true, since there exist maximal outerplane drawings that are quite different from outerplane Laman graphs; for an example see Fig. 1. We discuss some of the differences and subtleties of maximal and maximum graphs in Sect. 2. Further note that if we have a certain substructure given, then a MOD might have fewer than $2n-3$ edges. Hence, outerplane Laman graphs do not exist for every given outerplane subdrawing of them.

1.1   Minimum-Weight and Constrained Graphs

A minimum-weight Laman graph is one that not only satisfies the necessary conditions for rigidity but also minimizes a cost function—be it related to material usage, weight, or construction cost. In architectural design, the practical challenge is to create a stable structure with the fewest and lightest possible elements. We therefore next consider the structure of minimum-weight graphs.

Since a maximal outerplane drawing is not necessarily an outerplane Laman graph, we cannot guarantee that arbitrarily extending outerplane drawings (while keeping the outerplane condition) will yield outerplane Laman graphs. To find minimum-weight outerplane Laman graphs, a greedy approach to iteratively add the shortest non-crossing edge does not suffice, since there exist minimum-weight outerplane Laman graphs that do not contain the shortest edge; see Sect. 4. In general, outerplane Laman graphs are a distinguished structure and there are no meaningful containment relations with any of the well-known minimum-weight structures (like minimum-weight spanning trees, triangulations, etc.); see again Sect. 4. The following question arises and remains open.

Open Problem 1 *What is the complexity of computing a minimum-weight outerplane Laman graph on a fixed point set?*

We remark that even when requiring the Laman graph to be plane (rather than outerplane), the complexity remains open. However, when the planarity condition is dropped altogether, minimum-weight Laman graphs can be computed in polynomial time [10].

Instead of computing a minimum-weight outerplane Laman graph from scratch in the fixed point set, one could also compute it with a partial drawing already given. Given an outerplane drawing $\mathcal{D}$, a $\mathcal{D}$-constrained outerplane Laman graph is an outerplane Laman graph that contains $\mathcal{D}$. Problems with such constrains have been considered for several different structures, a famous example being constrained Delaunay triangulations [2]. Adding constrains might change the complexity of a problem and even whether a structure can always be computed. For example, every point set admits a plane Hamiltonian cycle, but deciding for a point set and a given plane matching on that point set whether there is a plane Hamiltonian cycle containing that matching is NP-hard [11]. Hence, for the topic of outerplane Laman graphs, the following question arises.

Open Problem 2 *Given an outerplane drawing $\mathcal{D}$ on a fixed point set S, what is the complexity of computing a minimum-weight $\mathcal{D}$-constrained outerplane Laman graph on S if it exists?*

If the given drawing $\mathcal{D}$ is a simple polygon, then the classical works of Gilbert [4] and Klincsek [7] on minimum-weight triangulations imply the following result, as in this setting the minimum-weight triangulation and the minimum-weight outerplane Laman graph are identical.

Observation 1 *The minimum-weight outerplane Laman graph of a simple polygon with n vertices can be computed in $\mathcal{O}(n^3)$ time.*

1.2   Results and Outline

We start, in Sect. 2, by considering maximal and maximum outerplane drawings. We derive tight bounds for their number of edges and provide conditions for the relation between (minimum-weight) maximal outerplane drawings and Laman graphs.

In Sect. 3, we concentrate on the path-constrained setting, that is, the constraining outerplane drawing $\mathcal{D}$ is a path. Given a spanning path $\mathcal{P}$, we give a polynomial time algorithm that finds a minimum-weight $\mathcal{P}$-constrained MOD, and hence a $\mathcal{P}$-constrained outerplane Laman graph if it exists. The result is of independent interest since it implies that it can be decided in polynomial time whether a circumscribing polygon of a path exists. This is in contrast to families of cycles where the analogous decision question is known to be NP-hard [1, Theorem 4.7].

A common strategy to tackle problems like Problem 1 is to find a certain substructure. Unfortunately, we show in Sect. 4, that the minimum-weight outerplane Laman graph might not contain the shortest edge, minimum-weight matching, or minimum-weight Hamiltonian cycle. Further we show that the minimum-weight outerplane Laman graph might not be contained in the Delaunay triangulation or the minimum-weight triangulation.

2   Maximal Outerplane Drawings

In this section, we consider the relation between maximal and maximum outerplane drawings. Recall from the introduction that every MOD $\mathcal{D}$ on n points has exactly $2n - 3$ edges. For maximal plane drawings (instead of outerplane), it is well-known that they triangulate the whole given point set.

Observation 2 *Every maximal plane drawing on a point set S is a triangulation of S.*

In particular, every maximal plane drawing is also maximum plane. The other direction that every maximum plane drawing is maximal plane holds trivially.

Maximal outerplane drawings behave differently though. The direction that every MOD is maximal outerplane is still trivial. However, the other direction is not true in general. In fact, we show that maximal outerplane drawings might have surprisingly few edges.

Proposition 1. *Every maximal outerplane drawing on $n \geq 4$ points contains at least $n + 1$ edges. This bound is tight.*

Proof. Let $\mathcal{D}$ be an outerplane drawing. As long as $\mathcal{D}$ is not connected, we can add at least one more edge without inducing a crossing or violating outerplanarity. Indeed, by Observation 2 we can find an edge e connecting two different components such that $\mathcal{D} + e$ is still plane. As e only introduces a cut in the outer face of $\mathcal{D}$, $\mathcal{D} + e$ is still outerplane. This implies that every maximal outerplane drawing contains at least $n - 1$ edges (a spanning tree).

Furthermore, again by Observation 2 and an area subset argument, for every plane drawing $\mathcal{D}$ that is not yet a triangulation but where each inner face is a triangle, we can find an edge e in the outer face such that $\mathcal{D} + e$ is still plane and the newly introduced inner face is a triangle as well.

Hence, when $\mathcal{D}$ is connected and outerplane with less than $n + 1$ edges, it is either a spanning tree ($n - 1$ edges) or it has exactly one cycle with empty interior (n edges). In the first case, since two triangles cannot enclose a point and $n \geq 4$, we can add at least two more edges by the above argument without destroying outerplanarity. In the second case, if the cycle is a triangle we can add at least one more edge by the same argument; otherwise, we can triangulate the interior of that cycle, adding at least one more edge as well. Thus, every maximal outerplane drawing on $n \geq 4$ points contains at least $n + 1$ edges.

To see the tightness, we consider the drawing in Fig. 1. It is a plane tree plus two edges that together form a maximal outerplane drawing. And the construction can easily be generalized to every $n \geq 4$ by adding the appropriate number of points and stacked edges on top. $\qquad\square$

Fig. 1. A plane tree (left) and a plane path (right) on specific point sets, that allow only two more edges (red) to be added without violating outerplanarity. (Color figure online)

For an outerplane Laman graph, which is a triangulated polygon, we can observe that no chord of that triangulation can be used in a plane Hamiltonian cycle. Thus, every outerplane Laman graph contains exactly one Hamiltonian cycle (the boundary of the polygon). Contrarily, maximal outerplane drawings might not contain any Hamiltonian cycles, as can be seen in Fig. 1. In fact, we show that the existence of a Hamiltonian cycle characterizes whether a maximal outerplane drawing is maximum or not.

Proposition 2. *A maximal outerplane drawing $\mathcal{D}$ on n points in the plane is an outerplane Laman graph if and only if $\mathcal{D}$ contains a Hamiltonian cycle.*

Proof. If $\mathcal{D}$ is maximum outerplane then the boundary of the outer face forms a Hamiltonian cycle. On the other hand, if $\mathcal{D}$ contains a Hamiltonian cycle $\mathcal{C}$, then $\mathcal{C}$ must be plane because $\mathcal{D}$ is outerplane. Hence, to be maximal outerplane, the interior of $\mathcal{C}$ must be triangulated. But then $\mathcal{D}$ has at least $2n - 3$ edges and therefore is maximum outerplane. $\qquad\square$

With this characterization, we are well-equipped to distinguish general maximal outerplane drawings from outerplane Laman graphs. The next natural question, given our focus on minimum-weight structures, is how this relation is affected by minimum-weight conditions. That is, how do minimum-weight maximal outerplane drawings relate to minimum-weight MODs? In particular, is a minimum-weight maximal outerplane drawing also maximum outerplane for every point set S in the plane? We answer this in the negative; see Fig. 2.

Proposition 3. *There exist point sets in the plane for which no minimum-weight maximal outerplane drawing is an outerplane Laman graph.*

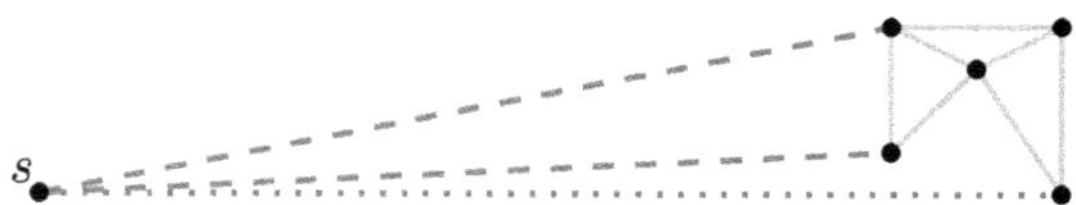

Fig. 2. A minimum-weight maximal outerplane drawing (gray, continuous edges and blue, dotted edge) different from the minimum-weight maximum outerplane drawing (which would include the red, dashed edges instead of the blue, dotted one). (Color figure online)

Proof. Figure 2 shows an example of such a point set. Since s is far away from the rest of the points, any minimum-weight maximal outerplane drawing will only use one edge incident to s. Hence, it does not contain any Hamiltonian cycle and thus, by Proposition 2, cannot be maximum outerplane.   □

3   Path-Constrained Maximum Outerplane Drawings

For a spanning path $\mathcal{P}$, we study in this section $\mathcal{P}$-constrained maximum outerplane drawings. We show that we can compute a minimum-weight $\mathcal{P}$-constrained MOD in $O(n^3)$ time. This implies that in $O(n^3)$ time we can decide whether a $\mathcal{P}$-constrained outerplane Laman graph exists by checking whether the computed drawing has $2n - 3$ edges and compute a minimum-weight one if it does.

The key idea for a polynomial time algorithm is to construct a weighted directed acyclic graph (weighted DAG for short) based on the instance with a source and a sink. Note that the maximum number of edges in an outerplane drawing depends on the number of points that have two angles on the unbounded face. We use as weights for arcs a pair (x, y). The purpose of x is to indicate whether a vertex has two angles on the unbounded face. The purpose of y is, if the arc corresponds to a potential edge on the unbounded face of a $\mathcal{P}$-constrained MOD, to state the weight of the minimum-weight triangulation (or short MWT) of the region enclosed by $\mathcal{P}$ and the corresponding edge. We remark that we do not count the weight of the edges of $\mathcal{P}$ in a MWT of a region.

Before we construct the DAG, we give some definitions and observations. Let us associate the integers $1, 2, \ldots, n$ with the vertices in S in the order along $\mathcal{P}$. We say an edge of a drawing $\mathcal{D}$ is a boundary edge if it is on the unbounded face. For two vertices $i, j \in S$ with $i < j$, let $\mathcal{P}_{ij}$ denote a subpath of $\mathcal{P}$ consisting of consecutive vertices $i, i+1, \ldots, j$. Observe that if the edge ij does not cross $\mathcal{P}$, then $\mathcal{P} \cup ij$ encloses a region distinct from the outer face with boundary $\mathcal{P}_{ij} \cup ij$. We denote this region enclosed by $\mathcal{P} \cup ij$ by R_{ij}. Let us call R_{ij} a *right-region* (resp. *left-region*) if consecutive vertices $i, i+1, \ldots, j$ are arranged in clockwise (resp. counterclockwise) order around R_{ij}. Note that this means that any left-region is interior disjoint from any right-region.

Consider a curve $\mathcal{C}$ connecting vertex 1 and n, which does not cross $\mathcal{P}$ or any of the regions R_{ij}. Note that $\mathcal{C}$ exists, since no region encloses vertex 1 or n due to outerplanarity. Now $\mathcal{P}$ together with $\mathcal{C}$ encloses a region $R_\mathcal{C}$ and has an unbounded face $\overline{R}_\mathcal{C}$. Similar as before we say $R_\mathcal{C}$ is the *right side* (resp. *left side*) if the vertices of $\mathcal{P}$ are encountered in clockwise (resp. counterclockwise) order. If $R_\mathcal{C}$ is the right side (resp. left side), we refer to $\overline{R}_\mathcal{C}$ as left side (resp. right side). Now given a $\mathcal{P}$-constrained outerplane drawing $\mathcal{D}$, note that $\mathcal{C}$ does not cross any edge of $\mathcal{D}$ by definition. We say a vertex i with $2 \leq i \leq n-1$ is seen from both sides if by adding $\mathcal{C}$ to $\mathcal{D}$ it is on incident to both faces that are incident to $\mathcal{C}$.

We now study which edges might be in a MOD. First note that any edge starting at one side of $\mathcal{P}$ and ending at the other side will enclose either vertex 1 or vertex n and thus cannot be part of an outerplane drawing containing $\mathcal{P}$. Hence, each potential boundary edge can be assigned a unique side of $\mathcal{P}$, "left" or "right" in the direction of $\mathcal{P}$, oriented from 1 to n. Secondly, if two edges $e = ij, f = i'j' \notin \mathcal{P}$ are boundary edges of some $\mathcal{P}$-constrained outerplane drawing with $i < i'$, then $j' > j$ (otherwise f is not a boundary edge) and $i' \geq j-1$ since $j-1$ would not be on the unbounded face otherwise. If $i' = j-1$, then e and f must lie on different sides of $\mathcal{P}$ since e and f would cross otherwise.

We construct the directed acyclic graph $\mathsf{DAG}(S, \mathcal{P})$ as follows (see Fig. 3 for an example): For every vertex i we have three nodes v_i^L, v_i^R and $v_i^\mathcal{P}$. The nodes v_i^L, $1 \leq i \leq n$, are associated with the left side, the nodes v_i^R are associated with the right side. The node $v_i^\mathcal{P}$ indicates whether i is seen from both sides in an outerplane drawing. For every $1 \leq i \leq n$, we add directed arcs $\{v_i^\mathcal{P}, v_i^L\}, \{v_i^\mathcal{P}, v_i^R\}$ with weight $(0,0)$. For every $1 \leq i < n$, we add $\{v_i^\mathcal{P}, v_{i+1}^\mathcal{P}\}$ with weight $(1,0)$ and $\{v_i^L, v_{i+1}^\mathcal{P}\}$ and $\{v_i^R, v_{i+1}^\mathcal{P}\}$ with weight $(1,0)$. For each edge ij with $i < j$ such that adding ij to $\mathcal{P}$ would still be outerplane, we add the arc $\{v_i^L, v_{j-1}^R\}$ if R_{ij} is a left-region and otherwise the arc $\{v_i^R, v_{j-1}^L\}$. In both cases the weight of the added arc is $(0, w_{ij})$ where w_{ij} is the weight of the MWT of R_{ij} minus the edges of $\mathcal{P}$ in R_{ij}. Finally, we add a source node s, a sink node t, and add the arc $\{s, v_1^\mathcal{P}\}$ with weight $(1,0)$ and $\{v_n^\mathcal{P}, t\}$ with weight $(0,0)$. See Fig. 3 for an example of a path $\mathcal{P}$ and its corresponding graph $\mathsf{DAG}(S, \mathcal{P})$.

An *st-path* in $\mathsf{DAG}(S, \mathcal{P})$ is a path from s to t. The weight of *st*-path are compared lexicographically. In other words an *st*-path with weight (x, y) is smaller than an *st*-path with weight (x', y') if and only if $x < x'$ or $x = x'$ and $y < y'$.

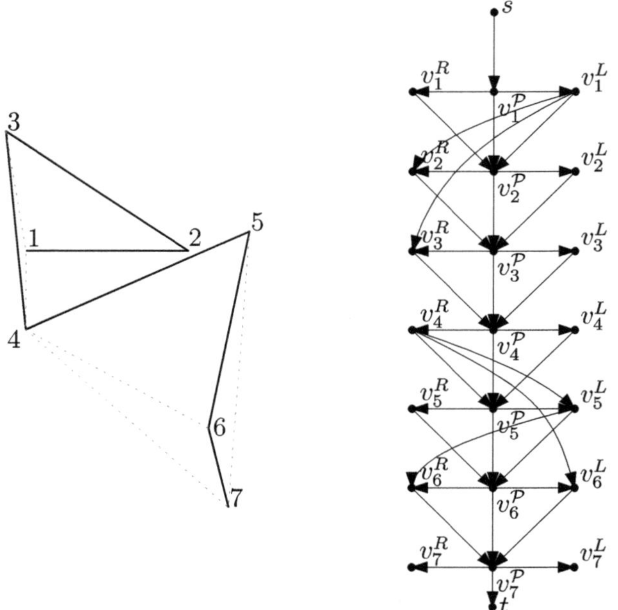

Fig. 3. Left: A path on a point set. The potential edges are dotted. Right: The corresponding graph $\mathsf{DAG}(\mathsf{S}, \mathcal{P})$. The edges 24, 25 and 35 are not considered because adding them would not preserve outerplanarity.

A *mintriangulated* $\mathcal{P}$-constrained outerplane drawing is a drawing where each region bounded by a boundary edge and the path $\mathcal{P}$ contains the MWT of that region. Note that a minimum-weight $\mathcal{P}$-constrained MOD is also a mintriangulated $\mathcal{P}$-constrained outerplane drawing with the largest number of edges and the minimum weight of its edges. In the next two lemmata we show that every st-path corresponds to a mintriangulated outerplane drawing and vice versa.

Lemma 1. *Every st-path in $\mathsf{DAG}(\mathsf{S}, \mathcal{P})$ with weight (x, y) corresponds to a mintriangulated $\mathcal{P}$-constrained outerplane drawing with weight y where x points are seen from both sides.*

Every mintriangulated $\mathcal{P}$-constrained outerplane drawing with weight y where x points are seen from both sides corresponds to an st-path in $\mathsf{DAG}(\mathsf{S}, \mathcal{P})$ with weight (x, y).

Proof. Let π be an st-path in $\mathsf{DAG}(\mathsf{S}, \mathcal{P})$ with weight (x, y). We construct a unique triangulated $\mathcal{P}$-constrained outerplane drawing $\mathcal{D}$ depending on π with weight y where x vertices are seen from both sides.

For every arc $\{v_i^L, v_{j-1}^R\}$ or $\{v_i^R, v_{j-1}^L\}$ in π we draw the edge ij and take the MWT of R_{ij}. The weight of the arc is $(0, w)$. Note that, by definition, w is exactly the weight of the MWT of R_{ij}. Let $\mathcal{D}$ be the drawing obtained by adding all these edges. Each of the edges $\{v_i^P, v_i^L\}$, $\{v_i^P, v_i^P\}$, $\{v_i^P, v_i^R\}$ has weight $(0,0)$ or $(1,0)$. So y is exactly the sum of all MWT of R_{ij}.

We have to show that $\mathcal{D}$ is an outerplane drawing. By definition, any arc $\{v_i^L, v_{j-1}^R\}$ or $\{v_i^R, v_{j-1}^L\}$ corresponds to an edge ij which added to $\mathcal{P}$ would not violate outerplanarity. Hence, no edge ij crosses $\mathcal{P}$. Let $ij, i'j'$ be two edges in $\mathcal{D}$ with $i < i' < j$. If two edges ij and $i'j'$ cross, then both edges are on the same side of $\mathcal{P}$, say to the left side. If ij and $i'j'$ are in π, then $\{v_i^L, v_{j-1}^R\}$ and $\{v_{i'}^L, v_{j'-1}^R\}$ are in π with $i < i' < j$. Note that the neighbors of v_{j-1}^R are v_j^P and v_h^L with $h > j$. Hence, π does not visit any node $v_{i'}^L$ with $i < i' < j$. So any two edges defined by $\{v_i^L, v_{j-1}^R\}$ or $\{v_i^R, v_{j-1}^L\}$ in π do not cross.

Hence, we constructed a mintriangulated $\mathcal{P}$-constrained drawing with weight y.

We still have to show that x is the number of vertices seen from both sides in $\mathcal{D}$. We show that $v_i^P \in \pi$ if and only if i is seen from both sides. Note that by construction and for any i, any st-path either contains v_i^P or an arc $\{v_h^L, v_j^R\}$ or $\{v_h^R, v_j^L\}$ with $h < i \le j$. If i is seen from both sides, then there is no arc $\{v_h^L, v_j^R\}$ or $\{v_h^R, v_j^L\}$ in π with $h < i \le j$. Then $v_i^P \in \pi$ since π is a st-path. On the other hand, if i is not seen from both sides, then there is an arc $\{v_h^L, v_j^R\}$ or $\{v_h^R, v_j^L\}$ in π with $h < i \le j$. So the number of vertices seen from both sides is the number of nodes v^P in π. Any arc with endpoint v^P has weight $(1,0)$. Hence, x is the number of nodes $v^P \in \pi$, which is also the number of vertices seen from both sides. Therefore, x is exactly the number of vertices seen from both sides.

For the other direction we construct the path π depending on the given drawing $\mathcal{D}$. For every boundary edge ij we take the arc $\{v_i^L, v_{j-1}^R\}$ or $\{v_i^R, v_{j-1}^L\}$ depending on whether ij describes a left or a right region. By similar arguments as before we get that y is the sum of all MWT of R_{ij} for every boundary edge ij, which is not in $\mathcal{P}$.

It remains to show that these arcs can be extended to a path. Given a boundary edge ij and another boundary edge $i'j'$ such that $i < j, i' < j'$ and there exists no other boundary edge ab with $i < a < i'$. If $i' = j - 1$, then the corresponding arcs in $\mathsf{DAG}(\mathsf{S}, \mathcal{P})$ are consecutive in π since ij and $i'j'$ are on different sides. Note that, if we have $i < i'$, then we also have $i' > j - 1$. If $i' > j - 1$, then all vertices between j and i' are seen from both sides. Hence, we add the edges $\{v_{j-1}^R, v_j^P\}$ and $\{v_k^P, v_{k+1}^P\}$, for $j \le k < i'$. If $R_{i'j'}$ is a right-region, then we add $\{v_{i'}^P, v_{i'}^R\}$. Otherwise, we add $\{v_{i'}^P, v_{i'}^L\}$. This also means that we have an arc with weight $(1,0)$ for every vertex that is seen from both sides. So we constructed a path π in $\mathsf{DAG}(\mathsf{S}, \mathcal{P})$ depending on $\mathcal{D}$ such that π has weight (x,y) and $\mathcal{D}$ has weight y and x vertices are seen from both sides. $\square$

Lemma 2. *The st-path in* $\mathsf{DAG}(\mathsf{S}, \mathcal{P})$ *with the smallest weight corresponds to the minimum-weight $\mathcal{P}$-constrained* MOD $\mathcal{D}$.

Proof. Let π be the st-path in $\mathsf{DAG}(\mathsf{S}, \mathcal{P})$ with the smallest weight. By Lemma 1, every MOD corresponds to an st-path in $\mathsf{DAG}(\mathsf{S}, \mathcal{P})$. Furthermore, every MOD minimizes the number of vertices seen from both sides. Hence, the st-path in $\mathsf{DAG}(\mathsf{S}, \mathcal{P})$ with the smallest weight corresponds to a MOD. Note that all MODs $\mathcal{D}_a$ with weight (x_a, y_a) have the same value x_a. Hence, the st-path in $\mathsf{DAG}(\mathsf{S}, \mathcal{P})$ with the smallest weight corresponds to the minimum-weight $\mathcal{P}$-constrained MOD $\mathcal{D}$. $\square$

We now have all ingredients to show our main result.

Theorem 1. *Given a set of n vertices $S \subset \mathbb{R}^2$ together with their pairwise Euclidean distances and a path $\mathcal{P}$ spanning S, a minimum-weight $\mathcal{P}$-constrained MOD on S can be computed in $\mathcal{O}(n^3)$ time.*

Proof. By Lemma 1, the minimum-weight $\mathcal{P}$-constrained MOD of S corresponds to a mintriangulated $\mathcal{P}$-constrained outerplane drawing. This drawing consists of (1) the edges of $\mathcal{P}$, (2) some boundary edges that form together with $\mathcal{P}$ a set of regions and each region is bounded by exactly one of those edges and $\mathcal{P}$, and (3) the MWTs of these regions. To compute the minimum-weight $\mathcal{P}$-constrained MOD of S, we need to determine a left weight and a right weight for each edge ij with $i < j$, that bounds a left-region or a right-region R_{ji}, which we need for the construction of our $\mathsf{DAG}(\mathsf{S}, \mathcal{P})$. This weight is of the length of the MWT of R_{ij} minus the length of the subpath $\mathcal{P}_{ij} := i, \ldots, j$.

To this end, we first check whether $P \cup ij$ is plane and whether both endpoints of $\mathcal{P}$ lie in the unbounded face of $P \cup ij$. If one of these checks fails, we set both weights of the edge ij to $+\infty$. If $j = i + 1$ we set both weights to the Euclidean distance $\|ij\|$. Otherwise, we further determine whether R_{ij} is a left-region or a right-region. Then we set the according left (or right) weight to the Euclidean distance $\|ij\|$ and the other weight to $+\infty$. This can easily be done in $O(n)$ time per edge and hence in $O(n^3)$ time for all edges.

To compute the MWTs of all potential left-regions, we apply the classic $O(n^3)$ dynamic programming minimum-weight triangulation algorithm for convex polygons to the "polygon" formed by $P \cup 1n$, with the base-edge $1n$ having weight 0 and the before-determined initial left weights of the edges. The result are two $n \times n$ matrices. One contains, for each edge ij with $i < j$, the weight of the MWT of the left region R_{ij}, including the weight of its boundary (or $+\infty$ if R_{ij} does not form a left-region, or the length of ij if $j = i + 1$). The other one contains for each edge ij with $j \neq i + 1$, the vertex k of the triangle ijk in the according triangulation.

We do the same for all potential right-regions with the before-determined initial right weights of the edges.

Lastly, we compute, again for each edge ij with $i < j$, the Euclidean length of the subpath $\mathcal{P}_{ij}$ of $\mathcal{P}$ and subtract the resulting value from the weights of the left-region and right-region MWTs from the previous step. Again, this step can easily be done in $O(n)$ time per edge and hence in $O(n^3)$ time for all edges.

After this preparation, we can now construct $\mathsf{DAG}(\mathsf{S}, \mathcal{P})$, where the edges corresponding to left-regions and right-regions, respectively, are exactly those for which the entry in the weight-matrix from the MWT step is finite. Since

$\mathsf{DAG}(\mathsf{S}, \mathcal{P})$ has $O(n)$ nodes and $O(n^2)$ arcs, each of which can be added in constant time with the help of the preprocessing data, $\mathsf{DAG}(\mathsf{S}, \mathcal{P})$ can be constructed in $\mathcal{O}(n^2)$ time.

We apply Dijkstra's algorithm with start node s to $\mathsf{DAG}(\mathsf{S}, \mathcal{P})$ to obtain a shortest paths tree $T(\mathsf{DAG}(\mathsf{S}, \mathcal{P}))$ rooted at s in $O(n^2)$ time. Finally, we extract the shortest path from s to t in $\mathsf{DAG}(\mathsf{S}, \mathcal{P})$ from $T(\mathsf{DAG}(\mathsf{S}, \mathcal{P}))$, and compute in $O(n)$ time the minimum-weight MOD of S containing $\mathcal{P}$ as well as its total weight using the information from the minimum-weight triangulation step.   □

Applying Theorem 1 to $\mathcal{P}$ and checking whether the obtained $\mathcal{P}$-constrained MOD has $2n - 3$ edges yields the following corollary.

Corollary 1. *Given a set of n vertices $S \subset \mathbb{R}^2$ together with their pairwise Euclidean distances and a path $\mathcal{P}$ spanning S, the existence of a minimum-weight $\mathcal{P}$-constrained outerplane Laman graph on S can be decided in $\mathcal{O}(n^3)$ time. In the affirmative case, such a Laman graph is also computed.*

Note that our algorithm first maximizes the number of edges, and then minimizes the weight of the edges. Hence, we also obtain the following corollary which might be of independent interest.

Corollary 2. *Given a set of n vertices $S \subset \mathbb{R}^2$ and a path $\mathcal{P}$ spanning S, it can be decided in $\mathcal{O}(n^3)$ time whether there exists a plane spanning cycle $\mathcal{C}$ of S such that every edge of $\mathcal{P}$ is either on the boundary or in the interior of $\mathcal{C}$.*

4   Relation to Other Minimum-Weight Structures

To gain more insight into minimum-weight outerplane Laman graphs, we investigate their relation to well-studied minimum-weight structures. We show for the best known minimum-weight structures that there are no containment-relations. While this means that the "easy way" to derive results (via better known structures) is blocked, it also ensures that minimum-weight outerplane Laman graphs are indeed its own and special drawing class and makes them more interesting to study. For proofs of all claims made in this section we refer to the full version.

The probably best studied minimum-weight structures are minimum-weight Hamiltonian cycles, minimum-weight spanning trees, and minimum-weight triangulations. It seems natural to ask whether all minimum-weight outerplane Laman graph contain a minimum-weight Hamiltonian cycle or minimum-weight spanning tree. Both questions can be answered in the negative. Any minimum-weight outerplane Laman graph in the point set in Fig. 4 does not only not contain any minimum-weight Hamiltonian cycle, but also no minimum-weight Hamiltonian path or minimum-weight perfect matching. In the point set as constructed in Fig. 5, the unique minimum-weight outerplane Laman graph does not even contain the shortest edge. We remark that this is a strong contrast to the minimum-weight triangulation [4].

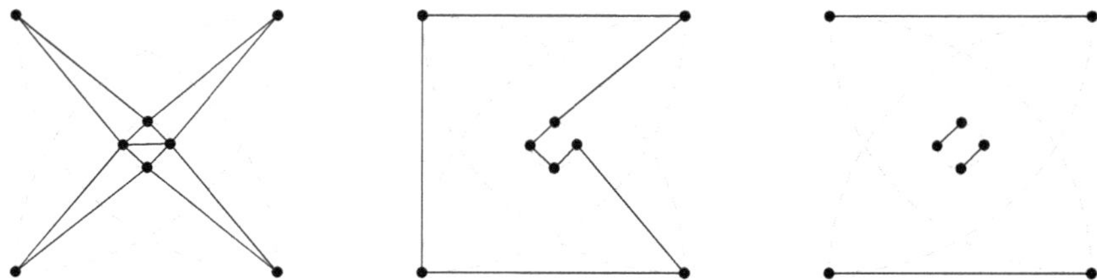

Fig. 4. A point set where no minimum-weight outerplane Laman graphs contains any minimum-weight Hamiltonian cycle or path nor any minimum-weight perfect matching. Left: A minimum-weight outerplane Laman graph. Middle: A minimum-weight Hamiltonian cycle. Right: A minimum-weight perfect matching.

When considering the other direction and looking for drawings that contain minimum-weight outerplane Laman graphs, triangulations seem strong candidates. However, also for this case we can show that there are no containment-relations with the best-known triangulations. The unique minimum-weight triangulation of the point set in Fig. 6(a) coincides with the Greedy triangulation of that point set, and this triangulation does not contain any outerplane Laman graph. Similarly, the point set in Fig. 7 has a unique Delaunay triangulation which does not contain any minimum-weight outerplane Laman graph on this point set. Of possibly independent interest, we can observe that the point sets given in Figs. 6(a) and 7(a) are point sets were no minimum-weight triangulation or Greedy triangulation contains any Hamiltonian cycle, and where no minimum-weight triangulation contains a minimum-weight spanning path, respectively.

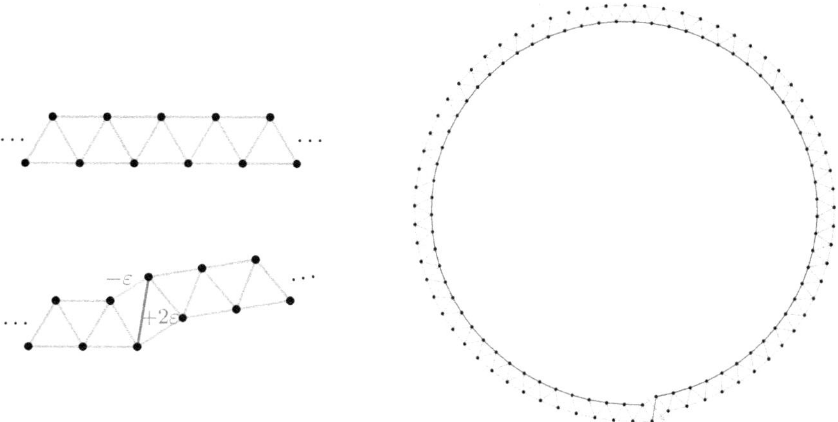

Fig. 5. Constructing a point set where no minimum-weight outerplane Laman graph contains the shortest edge. Left top: Starting configuration: A triangulated strip with uniform edge lengths d. Left bottom: Increasing the length of one diagonal edge by 2ε (red) while decreasing the length of a neighboring horizontal edge by ε (turquoise, this will be the shortest edge). Right: Bending the strip to a circle such that all the gray edges still have length d and the lengths of the blue edges are between (not including) $d - \varepsilon$ and d. (Color figure online)

Fig. 6. A point set where the minimum-weight triangulation coincides with the Greedy triangulation and does not contain any outerplane Laman graph. Left: Minimum-weight and Greedy triangulation. Right: A minimum-weight outerplane Laman graph.

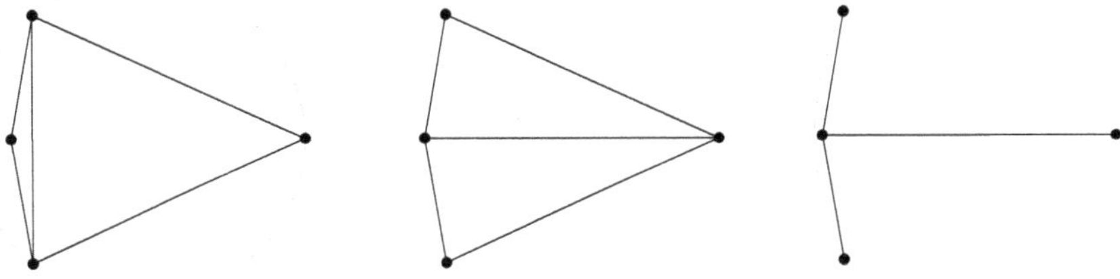

Fig. 7. A point set where the minimum-weight outerplane Laman graph is not contained in the Delaunay triangulation. Left: The unique minimum-weight outerplane Laman graph. Middle: The unique Delaunay triangulation. Right: The unique minimum-weight spanning tree.

5   Conclusion

In this work we studied minimum-weight path-constrained outerplane Laman graphs and minimum-weight path-constrained maximum outerplane drawings. Our next step is to consider the tree-constrained setting and to investigate the computational complexity of obtaining minimum-weight tree-constrained outerplane Laman graphs. The setting might also be interesting for other outerplane subdrawings. As also the complexity of computing minimum-weight plane Laman graphs on a given point set is open, all questions are also exciting for plane Laman graphs rather than outerplane Laman graphs. Another interesting variant would be to investigate these complexities for pointed pseudo-triangulations.

References

1. Akitaya, H.A., Korman, M., Korten, O., Rudoy, M., Souvaine, D.L., Tóth, C.D.: Circumscribing polygons and polygonizations for disjoint line segments. Discret. Comput. Geom. **68**(1), 218–254 (2022). https://doi.org/10.1007/S00454-021-00355-8
2. Chew, L.P.: Constrained Delaunay triangulations. Algorithmica **4**(1), 97–108 (1989). https://doi.org/10.1007/BF01553881
3. Fekete, Z., Jordán, T., Whiteley, W.: An inductive construction for plane Laman graphs via vertex splitting. In: Albers, S., Radzik, T. (eds.) ESA 2004. LNCS, vol. 3221, pp. 299–310. Springer, Heidelberg (2004). https://doi.org/10.1007/978-3-540-30140-0_28
4. Gilbert, P.D.: New results on planar triangulations. Master's thesis, University of Illinois (1979). https://apps.dtic.mil/sti/tr/pdf/ADA085016.pdf
5. Graver, J. E., Servatius, B., Servatius, H.: Combinatorial Rigidity, volume 2 of Graduate Studies in Mathematics (1993)
6. Haas, R., et al.: Planar minimally rigid graphs and pseudo-triangulations. In: Proceedings of the 19th Annual Symposium on Computational geometry (SoCG 2003), pp. 154–163 (2003). https://doi.org/10.1145/777792.777817
7. Klincsek, G.T.: Minimal triangulations of polygonal domains. In: Annals of Discrete Mathematics, vol. 9, pp. 121–123 (1980). https://doi.org/10.1016/S0167-5060(08)70044-X
8. Kobourov, S., Ueckerdt, T., Verbeek, K.: Combinatorial and geometric properties of planar Laman graphs. In: Proceedings of the 24th Annual ACM-SIAM Symposium on Discrete Algorithms (SODA 2013), pp. 1668–1678 (2013). https://doi.org/10.1137/1.9781611973105.120
9. Laman, G.: On graphs and rigidity of plane skeletal structures. J. Eng. Math. **4**(4), 331–340 (1970). https://doi.org/10.1007/BF01534980
10. Lee, A., Streinu, I.: Pebble game algorithms and sparse graphs. Discret. Math. **308**(8), 1425–1437 (2008). https://doi.org/10.1016/j.disc.2007.07.104
11. Rappaport, D.: Computing simple circuits from a set of line segments is NP-complete. In: Proceedings of the 3rd Annual Symposium on Computational geometry (SoCG 1987), pp. 322–330 (1987). https://doi.org/10.1145/41958.41993
12. Rollin, J., Schlipf, L., Schulz, A.: Recognizing planar Laman graphs. In: Proceedings of the 27th Annual European Symposium on Algorithms (ESA 2019), pp. 79–1 (2019). https://doi.org/10.4230/LIPIcs.ESA.2019.79
13. Rote, G., Santos, F., Streinu, I.: Pseudo-triangulations-a survey. arXiv preprint: math/0612672 (2006). https://arxiv.org/abs/math/0612672
14. Servatius, B.: The geometry of frameworks: rigidity, mechanisms and CAD. MAA Notes, pp. 81–87 (2000)
15. Streinu, I.: Pseudo-triangulations, rigidity and motion planning. Discret. Comput. Geom. **34**(4), 587 (2005). https://doi.org/10.1007/s00454-005-1184-0
16. Thorpe, M.F., Duxbury, P.M.: Rigidity Theory and Applications. Fundamental Materials Research (1999). https://doi.org/10.1007/b115749

Minimizing Vertical Length in Linked Bar Charts

Steven van den Broek[1], Marc van Kreveld[2], Wouter Meulemans[1],
and Arjen Simons[1(✉)]

[1] TU Eindhoven, Eindhoven, The Netherlands
{s.w.v.d.broek,w.meulemans,a.simons1}@tue.nl
[2] Utrecht University, Utrecht, The Netherlands
m.j.vankreveld@uu.nl

Abstract. A linked bar chart is the augmentation of a traditional bar chart where each bar is partitioned into blocks and pairs of blocks are linked using orthogonal lines that pass over intermediate bars. The order of the blocks readily influences the legibility of the links. We study the algorithmic problem of minimizing the vertical length of these links, for a fixed bar order. The main challenge lies with *dependent* links, whose vertical link length cannot be optimized independently per bar. We show that, if the dependent links form a forest, the problem can be solved in $O(nm)$ time, for n bars and m links. If the dependent links between non-adjacent bars form a forest, the problem admits an $O(n^4 m)$-time algorithm. Finally, we show that the general case is fixed-parameter tractable in the maximum number of links that are connected to one bar.

Keywords: Graph drawing · bar chart · length minimization · dynamic programming · fixed-parameter tractability

1 Introduction

Bar charts are a ubiquitous tool for visualizing scalar values across categories. Stacked bar charts, in particular, allow different quantities to be aggregated in a single column. In their traditional form, they primarily show *single-category* values: values that are uniquely attributable to a specific category. In many settings, however, certain quantities are not uniquely attributable to a single category but instead relate to multiple categories.

Such *cross-category* values may arise in different forms. They may represent shared quantities, for example, in a bar chart encoding total communication per account: the communication between two accounts is present in both bars. They may also represent pairwise uncertainties, that is, quantities that may belong to one of two categories. For example, in election poll results, groups of voters may hesitate between two political parties, or in the analysis of pollution, factories near country borders may contribute to the pollution of either country, though the exact distribution may not be known.

E. Di Giacomo and D. Mondal (Eds.): WALCOM 2026, LNCS 16444, pp. 33–48, 2026.
https://doi.org/10.1007/978-981-95-7127-7_3

As standard bar charts cannot directly visualize such cross-category values, *linked bar charts* were recently introduced [2]: the single- and cross-category values partition each bar into *blocks* of appropriate height, such that the total bar height reflects the aggregate value of the category. Each cross-category value is then visualized through a *link*: a polyline between the blocks that passes over intermediate bars (Fig. 1). We refer to blocks as *unlinked* or *linked* blocks, for single-category and cross-category values respectively, that is, depending on whether the block is linked to another.

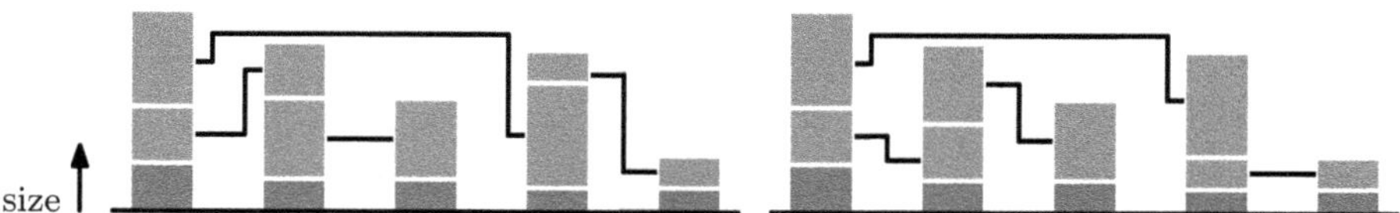

Fig. 1. Two linked bar charts [2] that show the same data using different vertical orderings, with cross-category scalar values between linked blocks (pink) and single-category values drawn as unlinked blocks (gray).

Note that linked bar charts effectively show a weighted graph: the bars are vertices, weighted by their single-category values, and linked blocks are edges, weighted by their cross-category values. The quality of the resulting drawing readily depends on the order of the bars, as well as the order in which the blocks are stacked on top of each other. Arising from (orthogonal) graph-drawing literature, there are various natural measures [10], such as the number of crossings [7,15,16], the length of the links [17,18] and the number of bends [3,12,15,17]. Vertical distance between elements has also been considered as a quality measure for other visualizations, such as storylines [11] and parallel coordinate plots [9].

For quality measures that rely solely on the bar order without considering the stacking order of the blocks, the problem is effectively that of drawing the (unweighted) graph with a one-page book embedding. The outerplanar graphs are exactly the family of graphs that can be drawn without edge crossings in this style [1]. For such graphs, minimizing total and maximum (horizontal) edge length, cutwidth, or bandwidth is polynomial-time solvable [13]. These problems are NP-hard for general graphs [14].

In contrast, the stacking order of the blocks adds a new dimension to this classic graph-drawing problem, which remains unstudied at this time.

Contributions. We study the following problem: given a weighted graph with fixed vertex order, we aim to compute a linked bar chart—a stacking of blocks in each bar—that minimizes the total vertical link length, while avoiding crossings between links that share an endpoint. In Sect. 2, we introduce our definitions. We distinguish between *dependent* and *independent* links, indicating whether the optimal position of a linked block depends on the position of the block it is linked to. In Sect. 3, we describe an $O(nm)$-time algorithm, for cases with n bars and m links, where the subgraph of dependent links is a forest. If the subgraph of dependent links between non-adjacent bars is a forest, the problem can be solved in $O(n^4 m)$ time (Sect. 4). The problem is fixed-parameter-tractable, parameterized by the maximum degree of a bar (Sect. 5). This algorithm still runs in polynomial time, if only the degree of the subgraph of dependent links is bounded by a constant. Omitted proofs can be found in the full version [6].

2   Definitions, Notation and Observations

Our input is a weighted graph $G = (V, E, w)$, where V represents a sequence of n vertices $v_1, \ldots, v_n$ in fixed order and $E \subseteq V^2$ is a set of m edges. The nonnegative weight function $w : V \cup E \to \mathbb{R}^+$ assigns weights both to vertices (single-category values) and edges (cross-category values). The *span* of an edge is the subsequence of vertices in V between its endpoints, including these endpoints. The *intermediate* vertices of an edge refer to its span minus its endpoints.

Bars, Blocks and Links. To visualize G, we draw the vertices in V as a sequence of unit-width *bars* $B_1, \cdots, B_n$, arranged on a horizontal baseline with a spacing of one unit between adjacent bars. Note that the horizontal position of the bars is fixed by their order in the sequence $B_1, \ldots, B_n$. Each vertex v_i corresponds to bar B_i for all $1 \le i \le n$; we use bar to refer to the vertex in V as well as its visual representation, as there is a one-to-one mapping between the two.

An edge e induces a *linked block* in each of the two bars it connects: a rectangle of unit width and height $w(e)$. A vertex v induces an *unlinked block* of height $w(v)$ in its corresponding bar. The height $h(b)$ of a block b is is denoted by $h(b)$.

We draw an edge e as an orthogonal *link* that connects the two linked blocks corresponding to e; see also Fig. 1. A link connects the centers of two blocks; we refer to these centers as the endpoints of the link. If at least one endpoint is placed higher than all intermediate bars of e, then the link generally has two bends; a link may be horizontal in some cases and have zero bends. Otherwise, the link is drawn using four bends. Links always pass over all intermediate bars. We assume that between two consecutive bars, there is enough horizontal space for the links to run between those bars, and ignore this issue from now on.

Stacking Blocks. As the bar order is fixed, the horizontal length of links is as well. To minimize total vertical length, we stack the blocks within a bar appropriately while avoiding crossings between links that connect to the same bar. A stacking of blocks corresponds to an order on the incident edges of a bar. Unlinked blocks are placed at the baseline, below the stacking of linked blocks.

Consider a bar B_j. All incident edges $\{B_i, B_j\}$ to an earlier bar B_i go towards the left from B_j. We denote this ordered sequence of edges by $L_j = \{l_1, \ldots, \}$, sorted in decreasing order of index of the other bar. Similarly, the rightward edges are denoted by $R_j = \{r_1, \dots, \}$—the edges $\{B_j, B_k\}$ with $k > j$—in increasing order of index of the other bar. To avoid crossings between links that originate from a common bar, the stacking order must contain both L_j and R_j as a subsequence. That is, the stacking order must be some merge of these two ordered sequences. Any merge is valid, but they incur different vertical link lengths, as shown in Fig. 1. For ease of notation, we identify an edge e in L_j or R_j with the corresponding block in bar B_j.

Merging left- and rightward subsequences in this way implies that the number of crossings in our visualization equals that in the crossings in the one-page book drawing of G that respects the fixed vertex order. Two links intersect if and only if their spans intersect in at least two bars and neither is a subset of the other.

Consider some linked block b in a bar B_j, corresponding to vertex v_j, for edge $l_i \in L_j$. Its vertical center point is determined by its height, the height of the unlinked block, and all blocks prior to it in L_j—the order of the blocks in L_j is fixed—and also by all blocks in R_j that are chosen to be below it in the stacking order. Hence, the center coordinate of a block is influenced only by the number of blocks in the other sequence occurring before it. With k blocks of R_j below b, the vertical center of b is given by $y(b, k) = \frac{1}{2}h(b) + w(v_j) + \sum_{x=1}^{i-1} h(l_x) + \sum_{x=1}^{k} h(r_x)$. As the y-position of a block is uniquely determined by k, we also refer to k as the position of the block. The situation is symmetrical for a block in R_j.

For the remainder, we do not explicitly treat unlinked blocks: their fixed effect is readily captured via the y-function defined above. Hence, we refer to linked blocks simply as blocks.

Link Types. Consider an edge $e = \{B_i, B_j\}$ with $i < j$. Its linked blocks b and b' must be in R_i and L_j. Let $\uparrow b = y(b, |L_i|)$ and $\downarrow b = y(b, 0)$ be the highest and lowest possible y-coordinate for b its center and analogously define $\uparrow b' = y(b', |R_j|)$ and $\downarrow b' = y(b', 0)$. Let H denote the height of the tallest intermediate bar.

We call a link between block b and b' *independent*, if its vertical length can be minimized by placing its blocks closer to a fixed target t that does not depend on the relative position of b and b'. That is, its vertical length can be expressed as $|t - y| + |t - y'|$, for given center coordinates y for b and y' for b', for some fixed target t. In the following cases, links are independent (see Fig. 2, left):

1. If there is an intermediate bar that is higher than the highest coordinate for one of the blocks, the link must always go up to H from the lower block: the target is H. Formally, if $H \geq \uparrow b$ or $H \geq \uparrow b'$, then $t = H$.
2. Otherwise, if the intervals are interior disjoint, the direction of the vertical piece is always the same: we can use as a target the lowest position of the higher interval. Formally, if $H < \min\{\uparrow b, \uparrow b'\}$ and $\downarrow b \geq \uparrow b'$, then $t = \downarrow b$; analogous for $\uparrow b \leq \downarrow b'$.
3. Otherwise, if one of the intervals is a singleton, we use the fixed value as a target. Formally, if $\uparrow b = \downarrow b$ (that is, $L_i = \emptyset$), then $t = \uparrow b$; analogous for $\uparrow b' = \downarrow b'$ (that is, $R_j = \emptyset$).

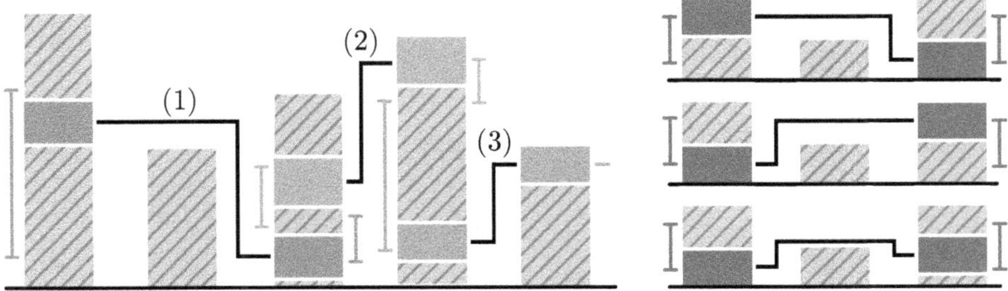

Fig. 2. Schematics of links, as determined by the interval of possible center coordinates (drawn ranges). Left: the three cases for independent links. Right: three possible placements for a single dependent link.

If none of the cases hold, the link is *dependent*: its vertical length depends on the relative position of the two blocks, and we cannot generally assign a static target (see Fig. 2, right). Dependent links come in two variants: *adjacent dependent links* and *non-adjacent dependent links*, depending on whether the two linked bars are consecutive.

We abbreviate the various link types as IL (independent link), DL (dependent link), ADL (adjacent dependent link) and NADL (non-adjacent dependent link). We refer to edges and blocks also via these types: for example, a dependent edge is an edge with a dependent link.

Each of the link types induces a subgraph of G. In particular, our results rely on the structure of the subgraph of dependent links.

Lemma 1. *The subgraph of dependent links is a plane one-page book embedding for the given bar order, and thus is outerplanar.*

Computing with Links. By their definition, the contribution of independent links to the total vertical link length can be decomposed into a *cost* for placing each of its blocks, being the distance of its placement to the target of the link. For dependent links, their *cost*—contribution to the total vertical link length—can be computed only when knowing the placement of both its blocks. We use λ to denote these costs. That is, an independent block b placed at position i has cost $\lambda(b,i) = |t - y(b,i)|$, where t denotes its target. For a dependent link $e = (b, b')$, the cost depends on the largest intermediate bar with height H: if both endpoints lie below H, it is the sum of the two vertical segments; otherwise, it is the absolute difference of the endpoints (see Fig. 2, right). Hence, letting i and j denote the positions of the endpoints, its cost is computed as:

$$\lambda(e,i,j) = \begin{cases} |H - y(b,i)| + |H - y(b',j)| & \text{if } H > y(b,i) \text{ and } H > y(b',j) \\ |y(b,i) - y(b',j)| & \text{otherwise} \end{cases}$$

We assume that we know, for each link, its type and the maximum height of its intermediate bars, such that the cost of a placed independent block and of a placed dependent link can be computed in $O(1)$ time. By the lemma below, we can precompute such information efficiently, and the running time of this precomputation step is $\leq$ dominated by all our other algorithms.

Lemma 2. *Given $G = (V, E, w)$ as adjacency list with only V in order, we can compute the L- and R-sets for all bars in $O(n + m)$ time, and augment the bars and links in $O(\min\{n^2, n + m \log n\})$ time, such that the type of a link and λ can be computed in $O(1)$ time.*

Finally, we observe that any interactions in how blocks are stacked in a bar arise from dependent links. Thus, we can minimize the cost for each connected component in the dependent-link subgraph separately, and sum their results to minimize the total vertical link length. The algorithms presented in the next sections hence look at ways of further decomposing the dependent-link subgraph into smaller parts, by parameterizing dependencies between these parts; that is, by choosing ways of placing the dependent links between different parts.

3   The Dependent Links Form a Forest

When the subgraph of DLs is a forest, we can minimize the vertical link length efficiently, by using the trees to determine an order in which to process the bars. Note that this case generalizes charts in which the bars are sorted by height.

Theorem 1. *Given $G = (V, E, w)$ where the subgraph of dependent links is a forest, minimizing the vertical link length takes $O(nm)$ time.*

Observation 1. *If the bars have a single local maximum in height, the dependent links form a collection of paths.*

Proof. An NADL requires that all intermediate bars are strictly lower than its endpoints: with a single local maximum, there are only ILs and ADLs.   □

Before we prove Theorem 1, we turn our focus to a simpler case, where there are no dependent edges.

Lemma 3. *Given $G = (V, E, w)$ without dependent edges, minimizing the vertical link length takes $O(nm)$ time.*

Proof (sketch). Without dependent links, each bar can be solved in isolation, via dynamic programming. Let $D(p, q)$ denote the minimum cost of stacking blocks the first p and q blocks from L and R, respectively. Since l_p or r_q must be the top block of this subproblem, D can be expressed recursively as:

$$D(p, q) = \min\{D(p - 1, q) + \lambda(l_p, q), D(p, q - 1) + \lambda(l_q, p)\}$$

The final solution is $D(|L|, |R|)$. A single bar of degree d_i takes $O(|L||R|) = O(d_i^2)$ time to solve. Since $d_i \leq n$, the total running time, for all bars, is $O(nm)$.   □

In the remainder of this section, we sketch the proof of Theorem 1. First, note that the trees in the forest can be optimized independently, as there are no dependent links between them. We describe below the algorithm for a single tree T. Note that bars without dependent links form trees with a single vertex and their dynamic program reduces to that of Lemma 3.

Setup. We root tree T at an arbitrary bar; each bar B in T is now the root of a subtree T_B. A dependent block in B has an *associated subtree*: subtree $T_{B'}$ when B is part of link $\{B, B'\}$. Without loss of generality, let $l_{p*} \in L$ be the block in B that corresponds to the dependent link that connects to the parent of B in T. We refer to this link as the *parent link*.

Let $\downarrow_B(p, q)$ denote the first $p < p^*$ and q blocks from L and R, respectively. Let $\Downarrow_B(p, q)$ denote the blocks in $\downarrow_B(p, q)$ together with all blocks in subtrees associated with dependent blocks in $\downarrow_B(p, q)$. We define $\uparrow_B(p, q)$ and $\Uparrow_B(p, q)$ symmetrically, for $p > p^*$. We define the cost of a set of blocks (such as $\uparrow_B(p, q)$ and $\Uparrow_B(p, q)$) as the cost of all independent blocks plus the cost of all dependent links whose endpoint blocks are both in the set.

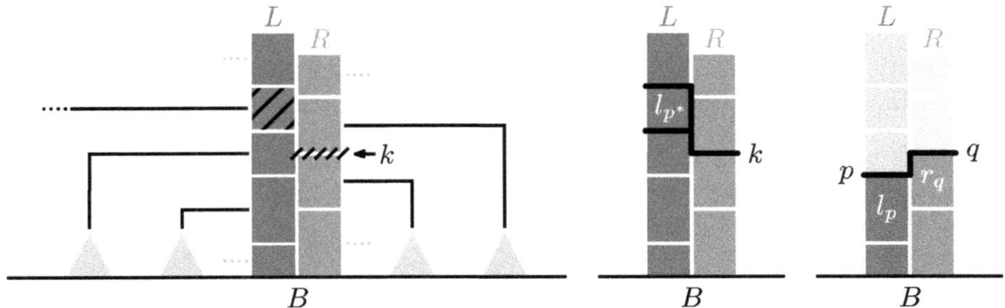

Fig. 3. Left: $P(B, k)$ is the minimum total cost of all blocks in B and its child subtrees, except that of block l_{p*} (hatched), when l_{p*} is placed with k blocks of R under it. Grey dots are independent links. Middle: block l_{p*} connects B to its parent in T. When placing k blocks of R under l_{p*}, B is split (black line) into two parts that can be stacked independently. Right: $D_\downarrow(p, q)$ is the minimum cost of the first p and q blocks from L and R, respectively.

Splitting the Cost. We process T using a post-order traversal to compute for each bar B the minimum cost $P(B, k)$ of the blocks in subtree T_B, given that block l_{p*} has k blocks of R below it (Fig. 3, left). To compute the $P(B, k)$ values efficiently, we observe that placing l_{p*} at position k splits B into two parts: $\downarrow_B(p^* - 1, k)$ and $\uparrow_B(p^* + 1, k + 1)$ (Fig. 3, middle). Let $D_\downarrow(p, q)$ and $D_\uparrow(p, q)$ denote the minimum cost of $\Downarrow_B(p, q)$ and $\Uparrow_B(p, q)$ respectively (Fig. 3, right). Then $P(B, k) = D_\downarrow(p^* - 1, k) + D_\uparrow(p^* + 1, k + 1)$; the only bar without a parent link is the root B_r, which is simply computed as $P(B_r) = D_\downarrow(|L_r|, |R_r|)$, reflecting the optimal cost for the entire tree.

The D tables can be computed using a dynamic program akin to that of Lemma 3. Specifically, $D_\downarrow$ follows the same recursion, and $D_\uparrow$ requires a simple inversion, where we consider which of the two options of the considered sets is the lowest block instead of the highest. Additionally, we need to explicitly consider the dependent links in these blocks. Consider a dependent link e between blocks $\{b, b'\}$; assume $b \in R$, the other case is analogous. Its cost is $\lambda(e, i, j)$ when placing b at $y(b, i)$ and b' at $y(b', j)$. Hence, the cost of placing block b at i becomes $\min_j \lambda(e, i, j) + P(B', j)$, where B' is the associated child bar of B in T.

Proving the Theorem. We compute both D tables in $O(d^2)$ for a bar of degree d and the total running time is $O(\sum d^2 + \sum d \cdot d')$, which sums to $O(nm)$ time.

4   The Non-adjacent Dependent Links Form a Forest

We take our previous results one step further by requiring that only the subgraph of the NADLs is a forest. That is, there may be ADLs in addition to this forest. Note that this case is also implied by a specific structure on the total bar heights.

Theorem 2. *Given $G = (V, E, w)$ where the subgraph of non-adjacent dependent links is a forest, minimizing the vertical link length takes $O(n^4 m)$ time.*

Observation 2. *If the bars have a single local minimum in height, the non-adjacent dependent links form a forest.*

Proof (sketch). The NADLs of a bar are either all in L or all in R. A cycle of NADLs must intersect itself, contradicting planarity (Lemma 1).   □

To prove Theorem 2, we develop an extended dynamic program, akin to that of Theorem 1. We first extend the forest of NADLs to a forest F containing all NADLs and a maximal set of ADLs; that is, ADLs are added until any further addition would create a cycle. As F has no DLs between different trees, we can optimize each tree independently. In the remainder, we sketch the algorithm for a single such tree $T \in F$.

Setup. We root T at its leftmost bar and denote the subtree at a bar B with T_B. As in Theorem 1, we associate to a block in B the subtree $T_{B'}$ of its child bar B'. The dependent links from B to its children form subsequences of L and R that we denote by $L^D = (l_1^D, l_2^D, \ldots)$ and $R^D = (r_1^D, r_2^D, \ldots)$. See the left of Fig. 4 for an illustration of these links and upcoming definitions.

We observe that a subtree has at most one leftward and one rightward ADL to a bar that is not in the subtree, because any bar in the range defined by the left- and rightmost bars in the subtree must be part of the subtree, and cannot have any dependent links to bars outside of this range, due to planarity.

An additional result of planarity and choice of the root is that the parent link of a bar B is the highest dependent link in its L- or R-set. Below, we assume that it is a leftward link; the other case is analogous.

Now, T_B and the parent of B have the following order in the linked bar chart, from left to right: the parent bar of B, then the subtrees associated with $(\ldots, l_2^D, l_1^D)$, then bar B, and lastly the subtrees associated with $(r_1^D, r_2^D, \ldots)$. An ADL can connect only bars that are adjacent in the linked bar chart. Let $\ldots, a_{\leftarrow,2}, a_{\leftarrow,1}, a_{\rightarrow,1}, a_{\rightarrow,2}, \ldots$ denote the ADLs that are not in F and have one endpoint in a child subtree of B and one endpoint in some other bar outside that subtree, in left-to-right order. The arrow in the subscript denotes on which side of B the link lies.

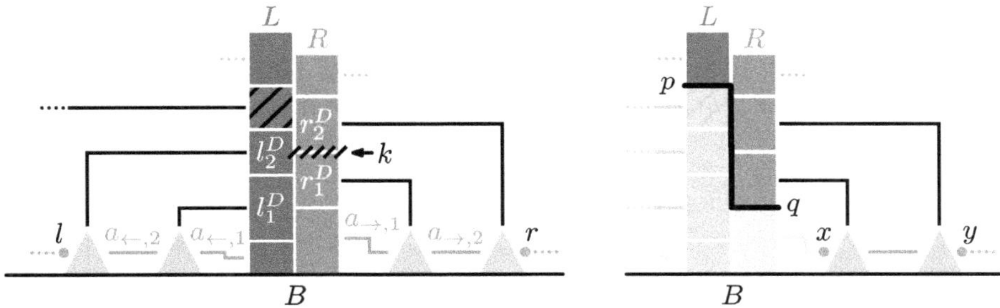

Fig. 4. Left: illustration of $P(B,k,l,r)$ and ADLs $a_{\leftarrow,1}, a_{\leftarrow,2}, \ldots$. The parent link (hatched) is excluded. Right: illustration of $D_\uparrow(p,q,x,y)$. There cannot be a leftward DL to a child of B.

As in the proof of Theorem 1, the cost of a set of blocks is the cost associated with all independent blocks and the cost of all dependent links whose endpoint blocks are both in the set. Consider the subtree T_B rooted at bar B. We aim to develop a dynamic program to compute the cost of blocks in T_B. There are, at most, three dependent links that have only one endpoint in T_B: the parent link, a leftward ADL at the leftmost bar in T_B, and a rightward ADL at the rightmost bar in T_B. By parameterizing their position by k, l and r, we derive a dynamic program $P(B,k,l,r)$: the minimum total cost for subtree T_B, with the three dependent links positioned according to the three parameters. As in Theorem 1, the root has no such parameters and $P(B)$ is the optimal solution of the entire tree.

Splitting the Cost. Our approach to computing $P(B,k,l,r)$ efficiently is structurally the same as that of Theorem 1. Again, let l_{p^*} denote the block in L that connects to the parent of B, and define $\Downarrow_B(p,q)$ and $\Uparrow_B(p,q)$. In addition to the parent link, there are at most two dependent links that have only one endpoint in

$\Downarrow_B(p,q)$: a leftward ADL at the leftmost bar in $\Downarrow_B(p,q)$, and a rightward ADL at the rightmost bar. Let $D_\downarrow(p,q,x,y)$ denote the minimum cost of $\Downarrow_B(p,q)$ when the at most two ADLs with only one endpoint in $\Downarrow_B(p,q)$ are positioned according to x and y (Fig. 5, left). We symmetrically define $D_\uparrow(p,q,x,y)$ (Fig. 4, right), noting that there cannot be dependent links above parent link l_{p^*} in L, by the choice of the root. Value $P(B,k,l,r)$ can be derived by combining the minimum costs of the two parts of B and aligning the at most one ADL that has an endpoint in each of the two parts. More precisely, the value is equal to $\min_{a,b}(D_\downarrow(p^*-1,k,l,a) + D_\uparrow(p^*+1,k+1,b,r))$.

Computing Each Part. We can define $D_\downarrow(p,q,x,y)$ recursively by observing that either l_p or r_q must be the top block. Without loss of generality, assume l_p is the top block; the other case is symmetric. Furthermore, assume w.l.o.g. that every DL l_i^D has a corresponding ADL $a_{\leftarrow,i}$; if in the following such an ADL does not exist, simply replace its cost by zero. If l_p is independent, then we need only add the cost of l_p to $D_\downarrow(p-1,q,x,y)$, which is easily determined in constant time. If l_p is dependent, then the previous leftmost ADL $a_{\leftarrow,i}$ with only one endpoint in $\Downarrow_B(p-1,q)$ now lies completely within $\Downarrow_B(p,q)$. Given that $a_{\leftarrow,i+1}$ is positioned at x, we need to determine two values: the minimum cost of all blocks in $\Downarrow_B(p-1,q)$ plus the cost of $a_{\leftarrow,i}$, and the minimum cost of the subtree $T_{B'}$ associated with l_p plus the cost of l_p. These two parts need to agree on where they place the left endpoint χ of $a_{\leftarrow,i}$. More precisely, these costs are:

$$\min_{x'}(D_\downarrow(p-1,q,x',y) + \lambda(a_{\leftarrow,i},\chi,x')), \text{ and} \tag{1}$$

$$\min_{k'}(P(B',k',x,\chi) + \lambda(l_p,k',q)), \text{ where } l_p \text{ connects } B \text{ to child bar } B'. \tag{2}$$

Cost $D_\downarrow(p,q,x,y)$ is then the minimum over χ of the sum of these costs. See the right of Fig. 5 for an illustration. The dynamic program for $D_\uparrow$ is similar.

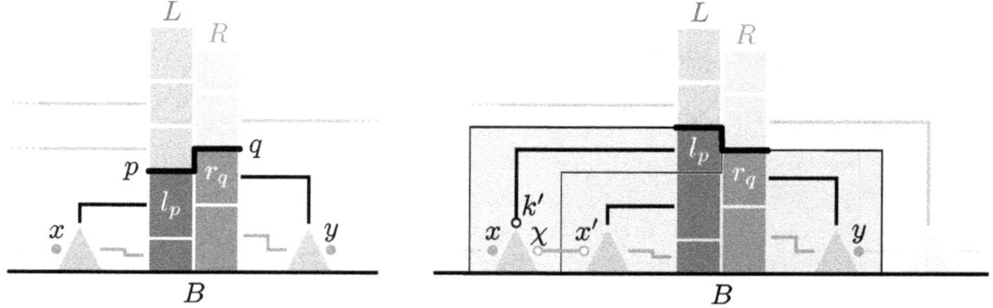

Fig. 5. Left: illustration of $D_\downarrow(p,q,x,y)$. Right: the two subproblems for computing $D_\downarrow(p,q,x,y)$ when dependent block l_p is on top, with Cost 1 (purple) and Cost 2 (orange).

Proving the Theorem. To compute the minimum vertical link length for G, we process each tree in F via a post-order traversal. Consider a bar B and assume all P values of child bars have been computed and stored. Let d be the degree of B and Δ be the maximum degree of a bar in graph G. First, we compute $D_\downarrow$ efficiently for increasing values of p and q, spending $O(\Delta^3 d^2)$ time per bar to determine $D_\downarrow$. After treating $D_\uparrow$ analogously, we compute each of the $O(\Delta^2 d)$ entries of P in $O(\Delta^2)$ time, and thus in $O(\Delta^4 d)$ time per bar. As processing a single bar takes $O(\Delta^4 d)$ time, the total running time becomes $O(\sum \Delta^4 d) = O(n^4 m)$.

5   General Case

We now investigate the case of a general graph G, that is, without placing further assumptions on the structure of the dependent links. Although the complexity of the problem remains open, a simple reduction from 3-Partition shows that a generalized version—in which bars have multiple unlinked blocks that can be ordered arbitrarily—is strongly NP-hard.

Theorem 3. *Minimizing vertical link length for a fixed bar order is strongly NP-hard, when vertices have multiple unlinked blocks and these may be ordered arbitrarily throughout the linked blocks.*

In the remainder, we prove Theorem 4, using two structural parameters:

Δ: the maximum degree of a bar in the full graph G;
δ: the maximum degree of a bar in the subgraph of dependent links.

As $\delta \leq \Delta \leq n - 1$, Theorem 4 implies that our problem is fixed-parameter tractable in Δ, and admits a polynomial-time algorithm when only $\delta = O(1)$.

Theorem 4. *Given $G = (V, E, w)$, the vertical link length can be minimized in $O(\Delta^{3\delta} \delta n)$ time.*

Our algorithm uses a rooted nice tree decomposition. These specific decompositions are generally useful for presenting dynamic programs, and are standard concepts in literature; see, for example, [5]. We briefly present their main aspects below, before presenting our results.

Rooted Nice Tree Decompositions. Let $G' = (V, E')$ denote the subgraph of dependent links. A *rooted tree decomposition* of G' is a rooted tree T in which each node[1] u of T has an associated *bag* X_u of bars from V, each link from E' has at least one bag containing both its endpoints, and every bar of V occurs only in the bags of a nonempty, connected subtree of T. Such a decomposition is *nice* if the root bag is empty and each node u is one of the following types:

Leaf: u has no children and $X_u = \emptyset$.

---

[1] To avoid ambiguity, we use nodes to refer to the vertices of T.

Forget: u has one child v with $X_u \subset X_v$ and $|X_u| = |X_v| - 1$.
Introduce: u has one child v with $X_u \supset X_v$ and $|X_u| = |X_v| + 1$.
Join: u has two children v and w with $X_u = X_v = X_w$.

Due to its outerplanarity (Lemma 1), the treewidth of G'—the minimal size of the largest bag in such a decomposition, minus one—is at most 2 [4]. Thus, we may compute a nice rooted tree decomposition of $O(n)$ nodes with at most 3 bars per bag in $O(n)$ time [8]. In the remainder, we omit the qualifications nice and rooted, and denote this tree decomposition as a triple (T, X, r).

Dynamic Program. To minimize the vertical link length in a linked bar chart of G, we run a dynamic program as a post-order traversal on the tree decomposition (T, X, r) of G'. Given a node $u \in T$, we denote the set of nodes in the subtree rooted at u as T_u. We say that a bar is in the *past* of u if it is contained in $\bigcup_{v \in T_u} X_v \setminus X_u$, that is, in the bag of a descendant of u, but not in that of u itself. The past of u is then the set of all bars in the past of u; see Fig. 6.

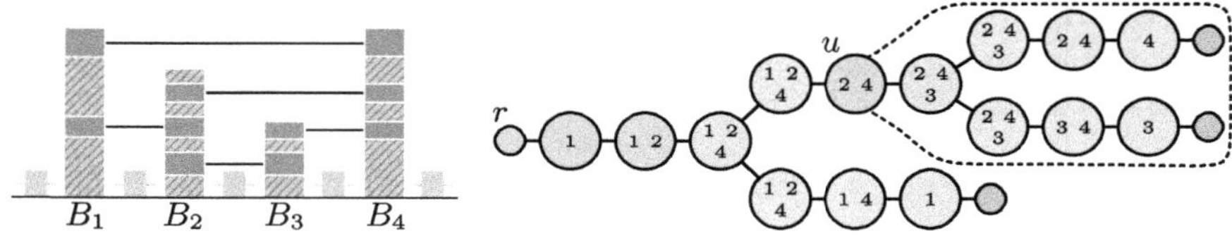

Fig. 6. Left: schematic of a linked bar chart, highlighting the subgraph G' of dependent links on bars B_1, B_2, B_3, B_4. Gray blocks in between these bars represent sequences of bars with only independent blocks. Hatched regions in gray bars indicate the remaining independent blocks. Right: a tree decomposition of G', with numbers referring to bar indices. Each node has a type: forget (orange), join (green), introduce (blue), leaf (red). At forget node u, $X_u = \{B_2, B_4\}$ and bar B_3 is forgotten. As B_3 is the only bar in subtree T_u (dashed outline) that is not in X_u, the past of u is precisely $\{B_3\}$. (Color figure online)

For a node $u \in T$, our dynamic program $P^*(u, \ldots)$ represents the minimum cost of all independent blocks in the bars of the past of u and of all dependent links with at least one of its endpoints in the past of u. The additional parameters of $P^*(u, \ldots)$ are the way in which the endpoints of the dependent links in each of the bars in X_u are positioned: we refer to this as the *state* of a bar. For a node with a bag of three bars, we symbolically represent them using S_1, S_2, S_3; for fewer bars in its bag, we number accordingly.

The root, having an empty bag, has no further parameters, and all bars are in its past: thus $P^*(r)$ represents the minimal vertical link length.

Lemma 4. *A node in (T, X, r) has at most $O(\Delta^{3\delta})$ states.*

Although we symbolically represent these states with at most three parameters, we must observe that these are δ-dimensional objects and, as such, reading an entry of $P^*(u,\ldots)$ takes $O(\delta)$ time.

We compute $P^*(u,\ldots)$ by treating each node type separately, as shown by Lemma 5 and Lemma 6 below. In particular, one node takes $O(\Delta^{3\delta}\delta)$ time. As there are $O(n)$ nodes to process, the total time required to compute $P^*(r)$ is $O(\Delta^{3\delta}\delta n)$, which proves Theorem 4.

Lemma 5. *Processing a join, introduce, or leaf node takes $O(\Delta^{3\delta}\delta)$ time.*

Proof (sketch). A join node u has two children with identical bags and disjoint pasts. Therefore, the cost of $P^*(u,\ldots)$ is the sum of the costs of the children.

An introduce node u has precisely one bar in its bag that is not present in the bag of its child: neither this bar nor any of its dependent links are in the past of u. Therefore, the cost for $P^*(u,\ldots)$ is identical to that of its child node, for the states of the bars they have in common.

Looking up the values of a child in $O(\delta)$ time for all $O(\Delta^{3\delta})$ states of $P^*(u,\ldots)$ takes $O(\Delta^{3\delta}\delta)$ time for join and introduce nodes.

A leaf node u has an empty bag X_u and past: $P^*(u)$ is trivially 0. $\square$

Lemma 6. *Processing a forget node takes $O(\Delta^{3\delta}\delta)$ time.*

Proof (sketch). In a forget node u, some bar B_f has become part of the past: it is not in X_u, but it is in the bag X_v of the child node v. The cost of its past includes both its independent blocks and its dependent links to the bars in X_v. We iterate over all possible states $\mathcal{S}_f$ of B_f: together with the states prescribed for the remaining bars, this determines the cost of the dependent links and prescribes between which dependent links each independent block is to be placed.

Assume X_u has two bars, B_1 and B_2. We can then express our dynamic program as $P^*(u, S_1, S_2) = \min_{S \in \mathcal{S}_f} P^*(v, S_1, S_2, S) + IL(S) + DL(S, S_1) + DL(S, S_2)$, where $IL(S)$ indicates the minimal cost of the independent blocks of B_f given the state S of its dependent links, and $DL(S, S_i)$ the cost of the dependent link between the bars of which the states are given.

If a link e between B_f and B_i exists, $DL(S, S_i) = \lambda(e, p, q)$ where p, q are the placements of the blocks of e prescribed by S and S_i; otherwise, $DL(S, S_i) = 0$. This is trivially computed in $O(\delta)$ time.

To compute $IL(S)$, we observe that the placement of dependent blocks prescribed by S, partitions the independent blocks into sets: one set between each pair of consecutively placed blocks, one before the first and one after the last dependent blocks. Each set contains exactly a consecutive subset, of L_f and R_f, characterized through indices i, p for L_f and j, q for R_f: $\{l_i, \ldots, l_p\} \cup \{r_j, \ldots, r_q\}$.

Each such set can be ordered independently, and the cost of such a set does not depend on the state S either, beyond it determining which sets to consider. So, we can precompute, once per forget node, the costs of such sets.

We precompute the costs via a dynamic program $D_{i,j}(p, q)$, the minimum cost of stacking $\{l_i, \ldots, l_p\} \subseteq L_f$ and $\{r_j, \ldots, r_q\} \subseteq R_f$; this excludes the cost of these lower blocks with fixed height. $D_{i,j}$ can be computed similarly to Lemma 3 (see Fig. 7). We only need values of p and q such that the corresponding sets do not include dependent links, and only need $D_{i,j}$ where at least i or j is the first link after a dependent link, or overall first link, in L_f or R_f, respectively. For $D_{i,j}$ with i being such first link, denote p_i being the largest value such that $\{l_i, \ldots, l_{pi}\}$ contains no dependent links. These subsequences are disjoint for all such i, and are

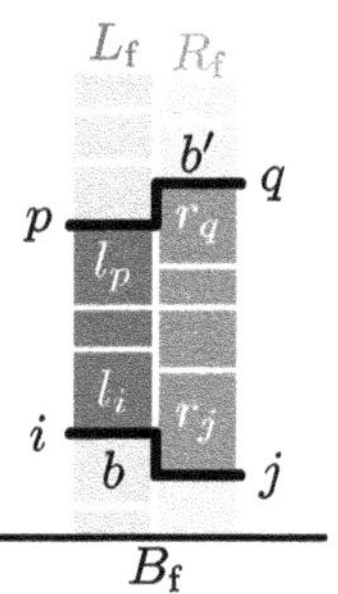

Fig. 7. $D_{i,j}(p, q)$ is the minimum cost of $\{l_i, \ldots, l_p\} \cup \{r_j, \ldots, r_q\}$ between dependent blocks b and b'.

partitioned by the dependent links: $\sum_i (p_i - i + 1) = \Delta - \delta$. With $|R_f|$ positions for j, and a dynamic program that takes $O(\sum_i (p - i + 1)|R_f|^2) = O(\Delta^3)$ time to compute, the total time for these instances of $D_{i,j}$ is $O(\sum_i (p - i + 1)|R_f|^2) = O(\Delta^3)$. The case with j just after a dependent link analogously takes $O(\Delta^3)$ time, resulting in a total preprocessing time of $O(\Delta^3)$ time for a forget node.

With $D_{i,j}$ available, we compute $IL(S)$, after ordering the placements of dependent links in $O(\delta)$ time, traversing all $O(\delta)$ consecutive pairs and looking up the minimum cost of intermediate independent blocks using D in $O(1)$ time.

Processing one state of $P^*(u, \ldots)$ takes $O(\Delta^\delta \delta)$ time. As a forget node has at most two bars, it has $O(\Delta^{2\delta})$ states. Thus, the total time to process a forget node is $O(\Delta^{3\delta}\delta + \Delta^3) = O(\Delta^{3\delta}\delta)$. $\qquad\square$

6 Conclusion

We studied algorithmically minimizing vertical link length for linked bar charts. This problem provides a new dimension to one-page book embeddings, by considering how the blocks (vertex and edge weights) at each bar (vertex) are to be stacked. For a fixed bar order, we described polynomial-time algorithms for special cases on the subgraph of dependent links. The general case is fixed-parameter tractable, parameterized by the maximum degree of a bar; the algorithm is polynomial when the degree in the dependent subgraph is bounded by a constant.

Future Work. The primary open problem is deciding whether minimizing vertical link length for a fixed bar order is NP-hard. Though we show a generalized version to be NP-hard, this reduction relies strongly on multiplicity of vertex weights. Our FPT results imply that, if a reduction is to be shown, this must rely on an instance with bars having many dependent links, and that is well connected to avoid being close to a tree.

Furthermore, we assumed a fixed bar order: what are the algorithmic possibilities for optimizing linked bar charts when we can control both the bar order as well as the stacking order of the blocks?

Generalizations. There are several natural generalizations to our problem.

Asymmetric links: The original design assumes that the blocks on both ends of a link are of equal height. However, one could also use a design where these sizes are not the same, for example, to communicate an expected distribution. We observe that our algorithms do not rely on linked blocks being the same size, and as such, generalize to solve this case.

Other measures: Beyond vertical link length, other quality measures could be considered. Our results straightforwardly apply when the measure permits decomposing its cost for independent links and computing the cost of a link takes $O(1)$ given its placed endpoints; for example, bend minimization.

Directed graphs and multigraphs: The original design assumes an undirected graph, with at most one link between two bars. A natural generalization would be to consider directed or multiple links between two bars. In the context of shared quantities, it may be relevant to show different sizes, for example, to indicate communication on different platforms, or to differentiate between sending and receiving messages. In context of uncertainty visualization, multiple links may be relevant to show different (uncertain) sizes; for example, multiple groups of indecisive voters, or multiple factories that may cause pollution in two countries. This opens up new questions, depending on whether the multiple links between two bars can be reordered, and whether they should be placed as one contiguous chunk within a bar.

Hypergraphs: A natural extension is also to go beyond regular graphs and instead consider a hypergraph to describe the cross-category values, linking multiple bars at once; representing, for example, communication in a group chat. Again, this generalization gives rise to new design questions that need to be answered before the algorithmic problem can be studied. The foremost question is how to represent a single link between three or more bars.

Acknowledgments. A. Simons and W. Meulemans are (partially) supported by the Dutch Research Council (NWO) under project number VI.Vidi.223.137. Research on the topic of this paper was initiated at the 8th Workshop on Applied Geometric Algorithms (AGA 2024) in Otterlo, The Netherlands.

References

1. Bernhart, F., Kainen, P.C.: The book thickness of a graph. J. Comb. Theory, Ser. B **27**(3), 320–331 (1979)
2. van Beusekom, N., et al.: Visualizing set sizes with dependent and independent uncertainties. In: Abstracts of the 17th IEEE Pacific Visualization Symposium (PacificVis, poster) (2024)
3. Bläsius, T., Rutter, I., Wagner, D.: Optimal orthogonal graph drawing with convex bend costs. ACM Trans. Algorithms **12**(3), 1–32 (2016)
4. Bodlaender, H.L.: A partial k-arboretum of graphs with bounded treewidth. Theoret. Comput. Sci. **209**(1), 1–45 (1998)

5. Bodlaender, H.L., Bonsma, P., Lokshtanov, D.: The fine details of fast dynamic programming over tree decompositions. In: Gutin, G., Szeider, S. (eds.) IPEC 2013. LNCS, vol. 8246, pp. 41–53. Springer, Cham (2013). https://doi.org/10.1007/978-3-319-03898-8_5

6. van den Broek, S., van Kreveld, M., Meulemans, W., Simons, A.: Minimizing vertical length in linked bar charts. CoRR (ArXiv.org), 2511.16812v1 (2025)

7. Buchheim, C., Chimani, M., Gutwenger, C., Jünger, M., Mutzel, P.: Crossings and planarization. In: Tamassia, R. (ed.) Handbook of Graph Drawing and Visualization, pp. 43–85. Chapman and Hall/CRC (2013)

8. Cygan, M., et al.: Parameterized Algorithms. Springer, Cham (2015)

9. Dasgupta, A., Kosara, R.: Pargnostics: screen-space metrics for parallel coordinates. IEEE Trans. Visual Comput. Graph. 16(6), 1017–1026 (2010)

10. Di Battista, G., Eades, P., Tamassia, R., Tollis, I.G.: Graph Drawing: Algorithms for the Visualization of Graphs. Prentice Hall, Hoboken (1999)

11. Dobler, A., Hegemann, T., Nöllenburg, M., Wolff, A.: Optimizing wiggle in storylines. CoRR (ArXiv.org), 2508.19802v1 (2025). to appear at GD 2025

12. Formann, M., et al.: Drawing graphs in the plane with high resolution. SIAM J. Comput. 22(5), 1035–1052 (1993)

13. Frederickson, G.N., Hambrusch, S.E.: Planar linear arrangements of outerplanar graphs. IEEE Trans. Circuits Syst. 35(3), 323–333 (1988)

14. Garey, M.R., Johnson, D.S.: Computers and Intractability; A Guide to the Theory of NP-Completeness. W. H. Freeman & Co., USA (1990)

15. Purchase, H.C., Cohen, R.F., James, M.: Validating graph drawing aesthetics. In: Proceedings of Graph Drawing (GD), LNCS 1027, pp. 435–446 (1996)

16. Radermacher, M., Reichard, K., Rutter, I., Wagner, D.: A geometric heuristic for rectilinear crossing minimization. In: Proceedings of the Twentieth Workshop on Algorithm Engineering and Experiments (ALENEX), pp. 129–138 (2018)

17. Tamassia, R.: On embedding a graph in the grid with the minimum number of bends. SIAM J. Comput. 16(3), 421–444 (1987)

18. Tamassia, R., Di Battista, G., Batini, C.: Automatic graph drawing and readability of diagrams. IEEE Trans. Syst. Man Cybern. 18(1), 61–79 (1988)

On Compaction and Realizability of Almost Convex Octilinear Representations

Henry Foerster[1,2], Giacomo Ortali[3]([envelope]), and Lena Schlipf[2]

[1] School of Computation, Information and Technology, TU Munich,
Heilbronn, Germany
`henry.foerster@tum.de`
[2] Wilhelm-Schickard-Institut für Informatik, Universität Tübingen,
Tübingen, Germany
`lena.schlipf@uni-tuebingen.de`
[3] Department of Engineering, University of Perugia, Perugia, Italy
`giacomo.ortali@unipg.it`

Abstract. Octilinear graph drawings are a standard paradigm extending the orthogonal graph drawing style by two additional slopes (± 1). We are interested in two constrained drawing problems where the input specifies a so-called representation, that is: a planar embedding; the angles occurring between adjacent edges; the bends along each edge. In ORTHOGONAL REALIZABILITY one is asked to compute any orthogonal drawing satisfying the constraints, while in ORTHOGONAL COMPACTION the goal is to find such a drawing using minimum area. While ORTHOGONAL REALIZABILITY can be solved in linear time, ORTHOGONAL COMPACTION is NP-hard even if the graph is a cycle. In contrast, already OCTILINEAR REALIZABILITY is known to be NP-hard. In this paper we investigate the OCTILINEAR REALIZABILITY and OCTILINEAR COMPACTION problems. We prove that OCTILINEAR REALIZABILITY remains NP-hard if at most one face is not convex or if each interior face has at most 8 reflex corners. We also strengthen the hardness proof of OCTILINEAR COMPACTION, showing that OCTILINEAR COMPACTION does not admit a PTAS even if the representation has no reflex corner except at most 4 incident to the external face. On the positive side, we prove that when the input graph is biconnected OCTILINEAR REALIZABILITY is FPT in the number of reflex corners and for OCTILINEAR COMPACTION we describe an XP algorithm on the number of edges represented with a ± 1 slope segment (i.e., the diagonals), again when the representation has no reflex corner except at most 4 incident to the external face.

Keywords: octilinear graph drawing · parameterized complexity · approximation

---

This work started at the Bertinoro Workshop on Graph Drawing 2024. We thank the organizers for organizing this event and the other participants for useful discussions.

1 Introduction

Octilinear (graph) drawings have been the standard visualization paradigm [13, 16,21] used in metro maps since the first map of the London Underground was designed by Henry Beck in 1933 [1][1]. It extends the popular orthogonal (graph) drawing style by two additional slopes (± 1) which can be useful in various diagramming applications such as UML [19] or BPMN [18]. Similar to studies on orthogonal drawings, research efforts on octilinear drawings have aimed to compute drawings with few bends per edge in small area [4,5].

Due to the similarities to the orthogonal drawing model, one may be tempted to adopt techniques that have been successfully employed in the context of orthogonal drawing. However, many problems related to octilinear drawing turn out to be more difficult than the corresponding ones for the orthogonal model. For instance, it is NP-hard to compute an octilinear drawing with the fewest total number of bends if the embedding is fixed [17] while the corresponding problem for orthogonal drawings is polynomial time solvable [22].

We are interested in two constrained drawing problems where the input specifies not only a planar embedding but a so-called representation, that is, a planar embedding and the angles occurring between adjacent edges, the bends along each edge, and the angles each bend forms. In the REALIZABILITY problem one is then asked to compute any drawing satisfying the constraints whereas in the COMPACTION problem the goal is to find such a drawing using minimum area. ORTHOGONAL REALIZABILITY can be trivially solved in polynomial time [7,22], but ORTHOGONAL COMPACTION is NP-hard [20] even if the graph is a cycle [11]. In contrast, already OCTILINEAR REALIZABILITY is known to be NP-hard [3]. The NP-hardness of ORTHOGONAL COMPACTION implies the NP-hardness of OCTILINEAR COMPACTION.

ORTHOGONAL COMPACTION is efficiently solvable under certain constraints [6,15]. In the past few years, these concepts have resurfaced in the graph drawing community [2,10] culminating in an FPT algorithm [8] parameterized by the number of so-called *kitty corners*. An angle is inflex, flat, or reflex when it is less than, equal to, or greater than 180°, respectively. Intuitively, kitty corners are pairs of reflex corners that "point" towards each other and are a measure for non-convexity. More in general, recently various parametrized algorithms have been presented for other variants of orthogonal drawings (see for example [8,9,12]), but as far as we know no parametrized approach has been investigated for octilinear drawings.

The rest of the paper is structured as follows. In Sect. 2 we give preliminary definitions. In Sect. 3 we strengthen the hardness results known for both OCTILINEAR REALIZABILITY and OCTILINEAR COMPACTION problems. In particular, in Sect. 3.2 we prove that OCTILINEAR COMPACTION admits no $\frac{9}{4}$-approximation even when the representation has no reflex corner except at most 4 incident to the external face; in Sect. 3.3 we show that OCTILINEAR REALIZABILITY is

---

[1] See also [23] for Beck's visualization.

para-NP-hard[2] when parameterized by the number of faces with reflex corners or by the maximum number of reflex corners per face. In Sect. 4 we present two parametrized algorithms: in Sect. 4.2 we show that when the input graph is biconnected OCTILINEAR REALIZABILITY is FPT parameterized by the number of reflex corners and in Sect. 4.3 we show that OCTILINEAR COMPACTION admits an XP algorithm parametrized by the number of edges of the representation represented with segments of slope ± 1 (i.e., the diagonals), again in the case where the representation has no reflex corner (except at most 4 incident to the external face).

2 Preliminaries

In an octilinear drawing of a graph $G = (V, E)$ each vertex $v \in V$ is represented by a point in $\mathbb{R}^2$ and each edge $e \in E$ is drawn as a sequence of horizontal, vertical and diagonal (with slope ± 1) segments. In this paper we only investigate simple planar drawings, hence from now on we omit the word "simple" and the word "planar". Observe that a necessary condition for a vertex to admit an octilinear drawing is having maximum degree at most 8.

Two important equivalence classes of octilinear drawings are *embeddings* and *octilinear representations*. An embedding contains all (octilinear) graph drawings with the same set of faces and the same external face. An *(abstract) octilinear representation* contains all octilinear drawings with the same embedding that have the same angles between consecutive edges around each vertex and the same sequence of bends along an oriented version of each edge (e.g., from lower index to higher index endpoint). If a drawing fullfills these constraints, we say that it *realizes* the representation. Since bends are predetermined, we can replace each bend by a vertex of degree 2 and assume our representations have no bends. We now define formally OCTILINEAR REALIZABILITY and COMPACTION. When $\mathcal{R} = \emptyset$, it means that there exists there exists no drawing realizing $\mathcal{R}$.

OCTILINEAR REALIZABILITY: *Given an octilinear representation $\mathcal{R}$, decide if $\mathcal{R} \neq \emptyset$.*

OCTILINEAR COMPACTION: *Given a realizable octilinear representation $\mathcal{R}$,* report a drawing $\Gamma \in \mathcal{R}$ on the integer grid such that the area of Γ, i.e., the area of the smallest axis-parallel rectangle containing Γ, is at most the area of Γ' for each $\Gamma' \in \mathcal{R}$ on the integer grid.

In the rest of the paper, unless otherwise stated, we denote by G a graph and by $\mathcal{R}$ an octilinear representation of G. For the evaluation of the area in the OCTILINEAR COMPACTION problem, we have to make the use of the grid comparable. Observe that we are employing a standard model in graph drawing, that is, we require that vertices must be placed at integer coordinates. Moreover, also observe that for the input of OCTILINEAR COMPACTION, we require that $\mathcal{R} \neq \emptyset$ so to separate the computational complexity of the actual compaction from the NP-hardness of the necessary realizability. Note that even in

---

[2] Para-NP-hardness means that the problem remains NP-hard for a bounded value of a parameter.

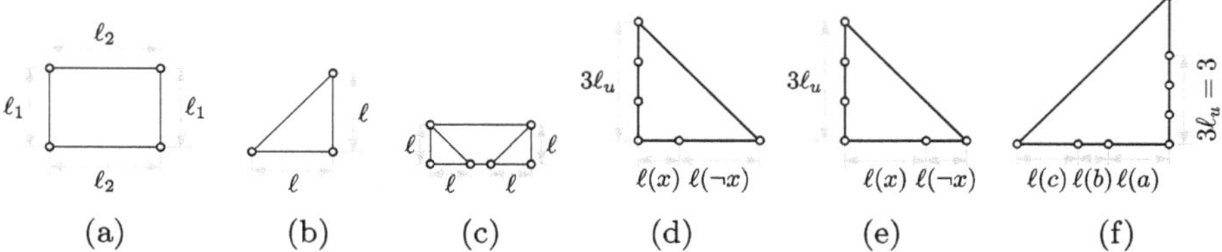

Fig. 1. Gadgets used in the reduction from 3-SAT to Octilinear Realizability [3] for $\ell_u = 1$: (a) Propagation, (b) rerouting, (c) copy, (d)–(e) variable and (f) clause gadget. (d) and (e) show $x = \top$ and $x = \bot$, respectively.

this restricted setting, we show APX-hardness in Sect. 3.2. We aim to apply the notion of limited non-convexity to investigate the parameterized complexity of Octilinear Realizability and Octilinear Compaction problems. Our results concern a set of more relaxed parameters that are related to the number of reflex corners. We consider four parameters: 1. ω is the total number of reflex corners in the entire representation; 2. φ is the maximum number of faces that contain at least one reflex corner; 3. κ is the maximum number of reflex corners per face; 4. δ is the number *diagonals* (edges having slope ± 1) of the entire representation.

3   Hardness Results

3.1   Review of the NP-Hardness Construction of Bekos Et Al. [3]

In both Sects. 3.2 and 3.3, we extend the NP-hardness reduction from 3-SAT to Octilinear Realizability and Compaction by Bekos et al. [3] to obtain new results. Here, we briefly describe the approach of [3]. As a central concept, information is encoded in the length ℓ_u of specific edges of the drawing. We first describe the proof for the easier case where $\ell_u = 1$. For reason of space, the approach to generalize to any value of ℓ_u will be only sketched in this section and described in details in the full version of the paper.

Case $\ell_u = 1$: The gadgets. There are three types of gadgets besides the clause and variable gadgets in [3]. The *propagation gadget* is a rectangular face, where the information can be propagated from either side to its opposite side. See Fig. 1a. The *rerouting gadget* is a triangular face which allows to propagate from a vertical to a horizontal edge or vice-versa. See Fig. 1b. The *copy gadget* are two triangular faces connected with two straight-line edges. This gadget copies the length of an edge to three other edges letting information related to a variable reach all the clauses that contain the variable. See Fig. 1c.

It remains to discuss how variables and clauses are encoded. The variable gadget for a variable x consists of a single triangular shaped face as shown in Figs. 1d and 1e, so that the left side of the face is formed by three edges of length ℓ_u while the bottom side of the face consists of two edges. One of those is representing the literal x while the other edge represents the literal $\neg x$. As a

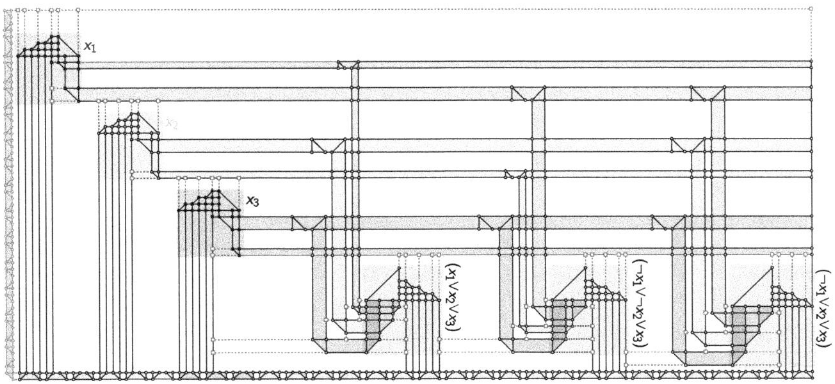

Fig. 2. Example of the reduction described in [3] with formula $(x_1 \vee x_2 \vee x_3) \wedge (\neg x_1 \vee \neg x_2 \vee x_3) \wedge (\neg x_1 \vee x_2 \vee x_3)$ when $\ell_u = 1$ and the satisfying assignment $x_1 = \bot$, $x_2 = x_3 = \top$ with our modifications (red and blue) as described in the proof of Theorem 1.

result, it holds that $\ell(x) + \ell(\neg x) = 3\ell_u$. It is easy to see that $\ell(x), \ell(\neg x) \in \{1, 2\}$ on the integer grid. We assume that the literal whose corresponding edge has length 2 to be true and check with the clause gadget in Fig. 1f. For clause $(a \vee b \vee c)$ at least one among the three literals a, b, and c is true, since the right side of the face has height more than three times the unit edge length $\ell_u = 1$. See Fig. 1f. See Fig. 2 for an example of the reduction.

Case $\ell_u = 1$: Routing information. Refer to Fig. 2. The first $3n$ copy gadgets at the bottom side connect to the n variable gadgets. The variable gadget of variable x_i appears above the variable gadget of variable x_{i+1}. The output literals of each variable are rerouted using rerouting gadgets so that the variable length is propagated towards the right via a series of propagation and copy gadgets, to which we refer as a *literal path* (shaded with three different colors in Fig. 2 and all the other figures). Note that if $\ell_u = 1$, exactly one literal of each variable is true. For each clause there are seven copy gadgets at the bottom side of the drawing and a clause gadget; the last three copies of the unit length of the associated copy gadgets are propagated via propagation and rerouting gadgets to the clause gadget. The clause gadget receives the information from the corresponding literal paths via the right copied length of a copy gadget on the literal path. We bound the upper and right side of the drawing by inserting a vertex at the top right corner. We connect this vertex with the topmost copy gadget horizontally and a path through the ends of each variable path and the rightmost copy gadget vertically. For reason of space the general case $\ell_u \geq 1$ is in the full version of the paper.

3.2 Approximability of Octilinear Compaction

We investigate the parameterized complexity of OCTILINEAR COMPACTION for $\omega = 4$ (the outer face contains at least 3 reflex corners in any octilinear drawing).

Theorem 1. *There is no $\frac{9}{4}$-approximation for* OCTILINEAR COMPACTION *even if $\omega = 4$ unless P=NP.*

Proof (Sketch). We adapt the NP-hardness construction by Bekos et al. [3] for the special case $\ell_u = 1$. Namely, we add a sequence of copy gadgets to the left side of the construction (red shaded gadgets in Fig. 2) which makes the area of the drawing dependent on ℓ_u. In addition, we add some extra (subdivision) vertices and edges to make all faces convex (see blue parts in Fig. 2). See For reason of space details and computation of the approximation ratio are in the full version of the paper.   □

Corollary 1. *Unless P=NP, there is no PTAS for* OCTILINEAR COMPACTION.

3.3 Para-NP-Hardness of OCTILINEAR REALIZABILITY

We shift our attention to the less restrictive parameters φ, the number of non-convex faces, and κ, the maximum number of reflex corners per face, and we prove para-NP-hardness even for OCTILINEAR REALIZABILITY:

Theorem 2. OCTILINEAR REALIZABILITY *remains NP-hard for $\varphi = 1$. Also, it remains NP-hard for $\kappa = 8$.*

Proof (Sketch). We adapt the NP-hardness proof by Bekos et al. [3] for the case $\ell_u \geq 1$. For $\varphi = 1$, we achieve that only the outer face contains reflex corners. Then, for $\kappa = 8$, we subdivide the outer face so that each gadget requiring reflex corners is contained in its own face. For reason of space details are in the full version of the paper.   □

4 Parametrized Results

In this section we show that OCTILINEAR REALIZABILITY and OCTILINEAR COMPACTION can be efficiently solved when there are either few reflex angles or diagonal edges, respectively. Namely, in Sect. 4.2 we prove that there exists an FPT algorithm on parameter ω for OCTILINEAR REALIZABILITY and in Sect. 4.3 we describe an XP algorithm on parameter δ for OCTILINEAR COMPACTION. Before that, in Sect. 4.1, we describe a flow procedure that will be used in the other two sections to prove the respective results.

4.1   Flow Algorithm for the Convex Case

Assume that in $\mathcal{R}$ all inner faces are convex and that the outer face contains no inflex angle (that is, the polygon bounding the outer face is convex as well). In this case, a realizing drawing exists if and only if for each face the sum of widths of the edges forming the top boundary is equal to the sum of the widths of the edges forming the bottom boundary and the same holds for the left and the right boundary. The reason for that is that one can use the obtained edge lengths to construct the drawing after fixing an arbitrary vertex to an arbitrary point in the plane. In any such solution, for each face, the edges forming the facial cycle are indeed realized as a crossing-free convex polygon. Thus, the resulting drawing correctly realizes the given planar embedding and it is therefore planar.

We can model this behavior using two auxiliary flow networks similar to techniques used in orthogonal graph drawing [7,22]. Namely, $G^{\rightarrow}$ (blue graph in Fig. 3) contains a node for each internal face and two nodes s and t for the outer face. In addition, arc (f_1, f_2) exists if there is an edge e in $\mathcal{R}$ so that f_1 occurs to the left of e and f_2 to the right (for the outer face all outgoing and incoming edges are attached to s and t, resp.). It is easy to see that any valid flow from s to t with positive value along each arc describes an assignment of vertical lengths to the edges of $\mathcal{R}$ satisfying the height constraint stated above. Similarly, we can define $G^{\downarrow}$ for the width constraint (red graph in Fig. 3). These two network flows can be described as a linear program. Observe that diagonal edges induce arcs in the red and in the blue graph.

To be more precise, we create a linear program formulation of the flow as follows. We associate each arc $a = (f_1, f_2) \in E(G^{\rightarrow}) \cup E(G^{\downarrow})$ with a variable x_a that must have strictly positive values, i.e., $x_a \geq 1$. Then, for each face f that is not the outer face we create two constraints that ensure flow conservation in both $G^{\rightarrow}$ and $G^{\downarrow}$, namely, $\sum_{a=(f,f')\in E(G^{\rightarrow})} x_a = \sum_{a=(f',f)\in E(G^{\rightarrow})} x_a$ and $\sum_{a=(f,f')\in E(G^{\downarrow})} x_a = \sum_{a=(f',f)\in E(G^{\downarrow})} x_a$.

Diagonal segments create an arc in both $G^{\rightarrow}$ and $G^{\downarrow}$ (orange highlights in Fig. 3). Thus, we must additionally require the flows along these two arcs to be the same as the height and the width of a diagonal at slope ± 1 are the same. These additional constraints are easily included in our linear program; e.g., we can simply use the same variable for both arcs.

The resulting linear program can be solved using polynomial time algorithms which yields a rational solution that can be encoded with a polynomial number of bits [14] that can then be scaled the to an integer solution in polynomial time.

We first state the following lemma, which is an interesting result by itself and that we use as the main tool in our FPT algorithm and XP algorithms. The lemma is directly implied by the described linear program modelling the flow.

Lemma 1. OCTILINEAR REALIZABILITY *is in P when there is no reflex angle in any internal face and no inflex angle in the outer face in* $\mathcal{R}$.

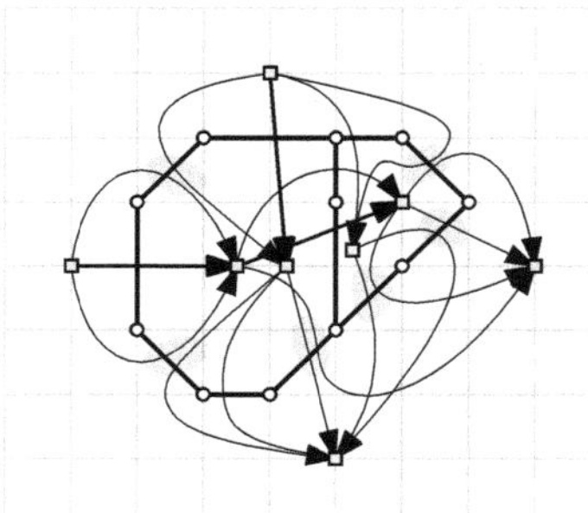

Fig. 3. A representation $\mathcal{R}$ and auxiliary flow networks $G^{\rightarrow}$ (blue) and $G^{\downarrow}$ (red). The value of flow along all thicker arcs is 2, the value of flow along all other arcs is 1. The positions on the grid corresponds to the solution yielded.

4.2 An FPT Algorithm for Octilinear Realizability

Given Lemma 1, our goal is now to extend it for fixed values of ω. In order to do that, we need further notation and intermediate results. Recall that $\mathcal{R}$ denotes an octilinear representation of a graph G. In this section we assume that G is biconnected. We describe how to enrich G and $\mathcal{R}$ so that each face containing at least one reflex angle has size $O(\omega)$. These enrichments are denoted by the *shadow graph* and *representation*. The key property of a shadow representation $\mathcal{R}'$ is that such a representation is realizable if and only if $\mathcal{R}$ is realizable (Lemma 2). Thus, we can work on the shadow graph and its shadow representation in order to obtain an FPT algorithm.

Shadow Graph and Representation. We begin by defining the *shadow graph G'* and an octilinear representation $\mathcal{R}'$ of G', called the *shadow representation*, which are obtained from the input graph G and its input octilinear representation $\mathcal{R}$, respectively. Initially, we set $G' = G$ and $\mathcal{R}' = \mathcal{R}$. During the construction of the shadow graph and representation, we analyze each face f and enrich the graph considering the vertices incident to f and their angles inside f.

Consider first the internal faces. For each internal face f of G, consider each vertex v incident to f and add a vertex v' if the angle of v in f is either strictly convex or reflex. We say that v' is the *shadow vertex* of v. We order the shadow vertices following the order of the corresponding vertices and add a cycle connecting them. Let C_f denote this cycle. See the thicker red cycle in Fig. 4b.

The newly created face f' bounded by C_f is called the *shadow face* of f. The angle inside f' of shadow vertex $v' \in C_f$ in G' is the same as the angle inside f of the vertex v in G. Finally, we use horizontal and vertical edges and, if necessary, subdivision vertices in order to connect C_f to the border of f in G'. We do this as follows. Consider vertex v, its shadow vertex v', and the angle α at v inside f in $\mathcal{R}$, which is the same as the angle of v' in $\mathcal{R}'$. Consider the angle α_f at v'

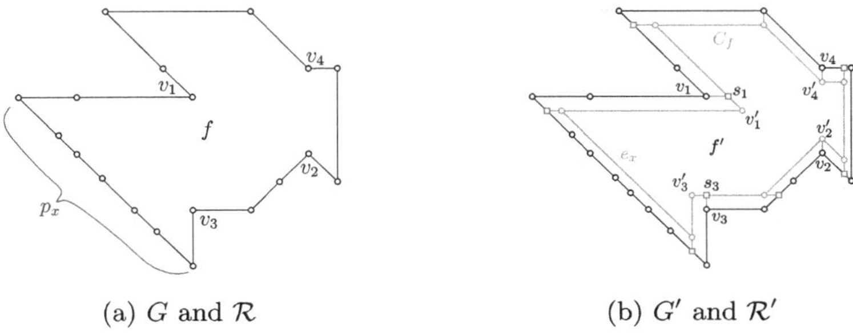

(a) G and $\mathcal{R}$ (b) G' and $\mathcal{R}'$

Fig. 4. The construction of G' and $\mathcal{R}'$.

in the face between the boundary of f and C_f. Note that either α is reflex and α_f is inflex or vice versa. We describe the construction for the case where α is reflex, the other case can be handled similarly. We proceed with a case analysis, refer to Fig. 4 where Fig. 4a depicts f in G also showing the configurations of the angles described in $\mathcal{R}$, while Fig. 4b shows the following construction.

If $\alpha = \frac{7}{4}\pi$, v is incident to either a horizontal or a vertical edge on the border of f. Consider the first case, the second can be treated similarly. We consider the diagonal edge incident to v' and we subdivide it with a vertex s. Then we add a horizontal edge from v to this subdivision vertex, fixing the angles accordingly. See v_1 in Fig. 4b. If $\alpha = \frac{3}{2}\pi$, v is either incident to two diagonal or to a horizontal and a vertical edge. If v is incident to two diagonal edges, one of its horizontal ports or one of its vertical ports is free inside f: We add an edge connecting v to v' vertically (see v_2 in Fig. 4b) or horizontally, respectively. If v is incident to a horizontal and a vertical edge, we subdivide the horizontal edge of C_f incident to v' with a subdivision vertex s and we add a vertical edge from v to s. See v_3 in Fig. 4a. If $\alpha = \frac{5}{4}\pi$, e add an edge connecting v to v' that is vertical if v is incident to a horizontal edge or horizontal otherwise. See v_4 in Fig. 4b. If $\alpha < \pi$, then α is strictly convex and α' is reflex, i.e., we do a symmetric operation. More precisely, we subdivide the facial cycle of f if necessary and select the orientation of the edge connecting v and v' based on the edges incident to the v'.

Consider now the outer face f_o. We construct C_{f_o} similarly to how we constructed C_f for an internal face, inverting the construction as in this case C_{f_o} is external to f_o while C_f was internal with respect to f. Refer to Fig. 5b, that depicts an octilinear drawing realizing the shadow representation $\mathcal{R}'$ of the shadow graph G', where G and $\mathcal{R}$ are depicted in Fig. 5a. In the following, we denote by G' and $\mathcal{R}'$ the shadow graph and representation of G and $\mathcal{R}$.

Lemma 2. $\mathcal{R}' \neq \emptyset$ *if and only if* $\mathcal{R} \neq \emptyset$.

Proof. Given an octilinear drawing of G' one can obtain an octilinear drawing of G by removing the shadow vertices and their incident edges; given an octilinear drawing of G one can obtain a drawing of G' by adding the vertices of C_f and the corresponding edges following the construction of $\mathcal{R}'$, eventually scaling the

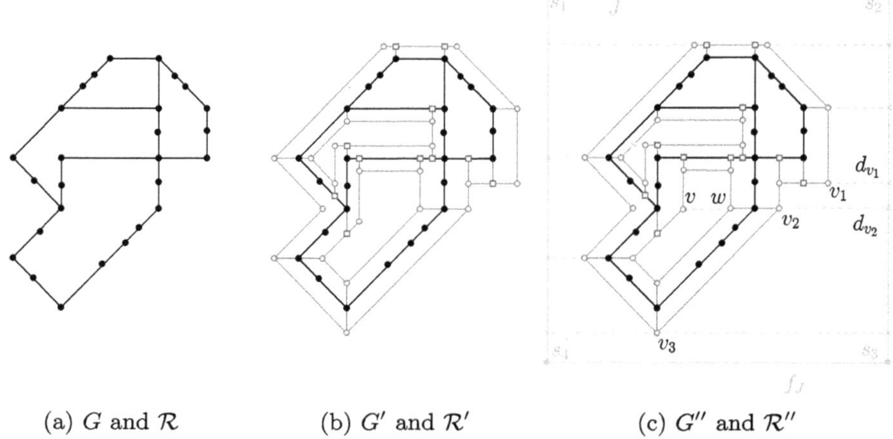

(a) G and $\mathcal{R}$ (b) G' and $\mathcal{R}'$ (c) G'' and $\mathcal{R}''$

Fig. 5. (a) Graph G and its representation $\mathcal{R}$. (b) Shadow graph G' and shadow representation $\mathcal{R}'$ obtained from Graph G and its representation $\mathcal{R}$. (c) Planar extension $\mathcal{R}''$ of $\mathcal{R}'$.

drawing to create space for such new vertices and edges when considering the internal faces. To this end, also note that the lengths of edges connecting vertices of f with vertices of C_f can be chosen arbitrarily small. $\qquad\square$

For every face of the shadow graph containing a reflex angle the number of vertices forming π angles inside is bounded by the number of reflex angles inside (as the angles forming π angles are exactly the angles created by subdivisions when defining the face of the shadow graph). Thus, we have the following:

Lemma 3. *The total length of the non-convex faces of G' is $O(\omega)$.*

Proof. Observe that an octilinear representation with $O(\omega)$ reflex angles has $O(\omega)$ inflex angles. Concerning a face f of G with ω_f reflex angles, we could still have $\Omega(n)$ vertices on its boundary, since many of them could form angles of π inside f. The same does not hold by construction for the shadow faces in G'. By construction, a shadow face f' of G' constructed for a face f with ω_f reflex corners has $O(\omega_f)$ vertices as only strictly convex, i.e., inflex, and reflex corners plus subdivision vertices that were introduced together with reflex corners (at most one subdivision vertex per reflex corner) define its boundary. All the vertices forming an angle of π inside f correspond to an edge in f'; see for instance the path p_x that is incident to f in Fig. 4a and that corresponds to the edge e_x incident to f' in Fig. 4b. By construction, all the faces of G' incident to vertices of G that were forming angles of π in the original representation inside f have no reflex angle inside. Hence, the total length of the shadow faces, summing over all faces F_r with reflex corners is $\sum_{f \in F_r} O(\omega_f) = O(\omega)$. $\qquad\square$

Planar Extensions and FPT Algorithm. Given the shadow graph G' and its shadow representation $\mathcal{R}'$ as introduced above, an *extension* of $\mathcal{R}'$ (and consequently of G', which is the graph that $\mathcal{R}'$ represents) is defined as follows. We first add a 4-cycle J consisting of four additional vertices s_1, s_2, s_3, and s_4 in G'. The angles around these vertices are $\frac{\pi}{2}$ inside the cycle and $\frac{3\pi}{2}$ outside. We define the outer face f_J of G' to be equal to the face of this cycle having the reflex angles outside. Hence, the rest of the graph has to be inside J. Let G'' be the plane graph obtained so far, which consists of G' and J. Let $\mathcal{R}''$ be the representation of G'' obtained as described above.

Consider G'' and $\mathcal{R}''$ and a shadow face f' of G'', whose border is C_f. Let v be a vertex of C_f forming a reflex angle in f'. Let w be another vertex of C_f. Let e be an edge of C_f or, if f is the outer face of G, it could be one of the four edges of J. Observe that in this case v is incident to the face between the two disconnected components J and G' of G''. We define an operation that we call *alignment*, that takes as input either v and w or v and e: In the first case, we add an edge connecting v to w; in the second case, we subdivide e and we add an edge connecting v to the subdivision vertex d_v added in e. In both cases, we change the angles in the representation accordingly so that the new edge is horizontal. Figure 5c shows the octilinear representation of $\mathcal{R}'$ depicted in Fig. 5b after performing alignments (added edges are dashed). Vertex v is aligned to w and both v_1 and v_2 are aligned to (s_2, s_3), the respective dummy vertices are d_{v_1} and d_{v_2}.

We now define the pair $\langle G_+, \mathcal{R}_+ \rangle$, constructed starting from G'' and $\mathcal{R}''$ and iteratively choosing one possible alignment for a vertex v forming a reflex angle in a face of $\mathcal{R}''$. When two or more vertices align at the same edge (a, b), we chose an ordering of the subdivision vertices associated to such vertices when traversing such subdivided edge from a to b. A single vertex can be considered at most twice by this procedure, as adding two horizontal edges to $\mathcal{R}''$ in any vertex makes sure that no angle around the vertex is larger than π. See for example vertex v_3 in Fig. 5c. Also, the considered ordering could imply that the obtained graph with the given rotation system at the vertices and the given outer face is not planar. For example, consider v_1 and v_2 that are both aligned to edge (s_2, s_3), in Fig. 5c. If the order of d_{v_1} and d_{v_2} was switched, the corresponding graph with the given rotation system and f_J as outer face would not be planar.

Given one of the possible extensions $\langle G_+, \mathcal{R}_+ \rangle$, of G' and $\mathcal{R}'$ we have a graph that might be not planar given that an outer face and that the rotation system of the vertices in $\mathcal{R}'$ is fixed. We say that an extension is *planar* if the graph G'_+ with the rotation system defined by $\mathcal{R}'_+$ is planar and J as the cycle bounding the outer face. For reason of space the proofs for the following two observations are in the full version of the paper.

Lemma 4. *There exists $2^{O(\omega^2)}$ possible extensions of G' and $\mathcal{R}'$.*

Lemma 5. *$\mathcal{R}' \neq \emptyset \Leftrightarrow$ There exists an extension $\mathcal{R}'_+ \neq \emptyset$ of $\mathcal{R}'$.*

We are now ready to combine our results into an FPT algorithm.

Theorem 3. OCTILINEAR REALIZABILITY *is FPT with respect to ω when the input graph is biconnected.*

Proof. Let $\mathcal{R}$ denote an octilinear representation of a graph G. If G has a vertex with degree at least 9 or if it is not planar, we reject the instance. Otherwise, we construct the shadow graph G' of G and the shadow representation $\mathcal{R}'$ of $\mathcal{R}$. Then we consider all the possible extensions of $\mathcal{R}'$ and we test the realizability of each extension using Lemma 1, which can be done as each extension has only four angles in the outer face, which are reflex angles, and no reflex angle inside by construction. We have that one of such extensions is realizable if and only if $\mathcal{R}$ is realizable by Lemmas 2 and 5. We now discuss the running time of this procedure. In order to construct G' and $\mathcal{R}'$ it suffices to traverse each face of G $O(1)$ times and, for each vertex and each edge, to perform $O(1)$ operations consisting in the case analysis depending on the angle at that vertex inside the face. Such operations are vertex addition, edge addition, and edge subdivision operations. Hence, constructing G' and $\mathcal{R}'$ can be done in polynomial time. To this end, note in particular, that the total number of vertices in the shadow graph is upper bounded by $17n$ as each vertex is incident to at most 8 faces and for each face we may create a shadow vertex plus a subdivision vertex in G'. This also limits the total number of edges of the shadow graph to $51n - 6$ (recall that maximal planr graphs have $3n - 6$ edges). Both of these values are independent of ω. Next, there are $2^{O(\omega^2)}$ possibilities to for the extension of $\mathcal{R}'$. Each of these extensions can be done in polynomial time and the consecutive realizability test using $\mathcal{R}'$ runs in polynomial time. Thus, the computational time of the entire procedure is of type $f(n)2^{O(\omega^2)}$ where $f(n)$ is a polynomial function on variable n. $\qquad\square$

4.3   An XP Algorithm for Octilinear Compaction

We assume there is no reflex angle in any internal face and no inflex angle in the outer face in $\mathcal{R}$. Observe that $\omega \in \{3, 4\}$ in this case since the outer face always contains at least three reflex corners. Moreover, by definition of OCTILINEAR COMPACTION, we may assume that $\mathcal{R}$ admits a realizing drawing. Note that this property can also be tested using Theorem 3 since $\omega \in \{3, 4\}$. We now describe an XP algorithm on parameter δ. Our goal is first proving that there exists a drawing of G realizing $\mathcal{R}$ such that the area of Γ depends exponentially only on the variable δ. In order to do that, given G and $\mathcal{R}$, we define the *skeleton graph* G^* and the *skeleton representation* $\mathcal{R}^*$. Refer to Fig. 6.

We define $\mathcal{R}^*$ given $\mathcal{R}$, the definition of G^* given G is then implicit. The vertices in $\mathcal{R}^*$ are the ones incident to at least one diagonal edge in $\mathcal{R}$. These vertices are represented by squares in Fig. 6 and diagonals are represented thicker. Let u and v be two vertices of $\mathcal{R}^*$: We add an edge with slope $+1$ or -1 if such edge is also present in $\mathcal{R}$; we add a horizontal (resp. vertical) edge if there is a horizontal (resp. vertical) path connecting u and v in $\mathcal{R}$.

Lemma 6. *There exists a drawing Γ of G realizing $\mathcal{R}$ with area $O(n^2 \cdot A(\delta))$, where $A(\delta)$ is a computable function.*

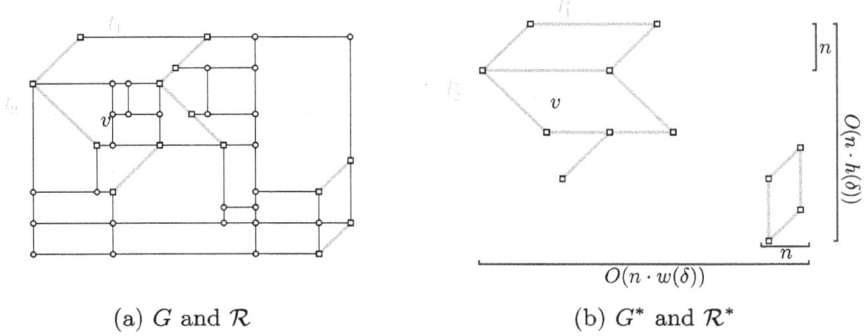

(a) G and $\mathcal{R}$ (b) G^* and $\mathcal{R}^*$

Fig. 6. (a) The graph G and its representation $\mathcal{R}$. (b) The compressed graph and representation G^* and $\mathcal{R}^*$. The figures illustrate the proof of Lemma 6.

Proof. We remove edges and vertices and perform a smoothing operation of vertices from Γ to obtain a drawing of a graph isomorphic to G^*. By construction of $\mathcal{R}^*$, this is a drawing of G^* realizing $\mathcal{R}^*$. We denote by Γ^* such drawing. Since G^* has $O(\delta)$ vertices, we can also assume that Γ^* has an area that only depends on δ. Let $h^*(\delta)$ and $w^*(\delta)$ the height and width of the smallest axis-parallel rectangle containing Γ^*. We scale Γ^* so that the minimum height or width of any of its edges is at least n. The smallest axis-parallel rectangle containing the drawing Γ^*, computed as described above, has now width $O(w^*(\delta) \cdot n)$ and height $O(h^*(\delta) \cdot n)$. We can now construct a realization $\tilde{\Gamma}$ of $\mathcal{R}$ different from Γ, using Γ to enrich Γ^* and then we prove that its area is $O(n^2 \cdot A(\delta))$.

We first consider every horizontal line of Γ containing at least a vertex and we map it to a horizontal line of Γ^* so that: The set of crossings of the diagonal edges the two lines have in the two drawings is the same (observe that every diagonal edge in Γ corresponds exactly to a diagonal edge of Γ^*); the order of the horizontal lines is the same in the two drawings. We do the same for the vertical lines. We now consider Γ^* and the horizontal and vertical lines described so far, and we place every vertex of G in the intersection of the corresponding vertical and horizontal line (either subdividing an edge or placing it incident to no edges). See, for example, vertex v in Fig. 6: The lines l_1 and l_2 are mapped to the lines l_1^* and l_2^*, respectively, and v is placed according to the vertical line l_1^* and the horizontal line l_2^*. This construction is always possible, as every edge of Γ^* has height or width (or both, in case of diagonal edges) equal or higher than n. Finally, after all the vertices are placed, we add the remaining edges. Let $\tilde{\Gamma}$ be the obtained drawing, which by construction is a realization of $\mathcal{R}$. The area of $\tilde{\Gamma}$ is $O(w^*(\delta) \cdot n) \times O(w^*(\delta) \cdot n) = O(n^2 \cdot A(\delta))$ with $A(\delta) = h^*(\delta \cdot w^*(\delta))$. $\square$

Lemma 6 implies in particular that we can assume, without loss of generality, that there exists a realization of $\mathcal{R}$ where every diagonal has height (and width) at most $A(\delta)$. We can consequently consider all the possible combinations of height (and width) of the diagonals and, for each combination use the flow algorithm described in Sect. 4.1 to minimize the area, once the flow along the diagonal is fixed. Among all the drawings computed with this procedure, we finally select the one having minimum area. As this procedure is exhaustive, this is the drawing realizing $\mathcal{R}$ having minimum area. Hence, we have the following:

Theorem 4. OCTILINEAR COMPACTION *is XP when there is no reflex angle in any internal face and no inflex angle in the outer face in* $\mathcal{R}$.

Remark 1. Lemma 1 implies that $A(\delta)$ can be encoded in $\mathcal{O}(\delta)$ bits, i.e., it is at most exponential in δ. Also observe that the conditions of Theorem 4 imply $\omega \in \{3, 4\}$, that is, our XP-algorithm solves the problem in general when $\omega = 3$.

5 Open Problems

For some $\kappa < 8$, it may be that OCTILINEAR REALIZABILITY becomes polynomial time solvable. In addition, our FPT algorithm may be parameterizable for a suitable definition of the kitty corners concept that is used for orthogonal graph drawings. We have strong evidence that the result of Theorem 3 can be easily extended in the general case, when the input graph is not biconnected. Finally, it remains open whether there is a constant factor approximation for OCTILINEAR COMPACTION or if the problem is inapproximable.

References

1. Beck, H.: Tube Map. Waterlow & Sons Ltd., London (1933)
2. Bekos, M.A., et al.: On turn-regular orthogonal representations. J. Graph Algorithms Appl. **26**(3), 285–306 (2022)
3. Bekos, M.A., Förster, H., Kaufmann, M.: On smooth orthogonal and octilinear drawings: relations, complexity and Kandinsky drawings. Algorithmica **81**(5), 2046–2071 (2019)
4. Bekos, M.A., Gronemann, M., Kaufmann, M., Krug, R.: Planar octilinear drawings with one bend per edge. J. Graph Algorithms Appl. **19**(2), 657–680 (2015)
5. Bekos, M.A., Kaufmann, M., Krug, R.: On the total number of bends for planar octilinear drawings. J. Graph Algorithms Appl. **21**(4), 709–730 (2017)
6. Bridgeman, S.S., Di Battista, G., Didimo, W., Liotta, G., Tamassia, R., Vismara, L.: Turn-regularity and optimal area drawings of orthogonal representations. Comput. Geom. **16**(1), 53–93 (2000)
7. Di Battista, G., Eades, P., Tamassia, R., Tollis, I.G.: Graph Drawing: Algorithms for the Visualization of Graphs. Prentice-Hall, Hoboken (1999)
8. Didimo, W., Gupta, S., Kindermann, P., Liotta, G., Wolff, A., Zehavi, M.: Parameterized approaches to orthogonal compaction. In: Gasieniec, L. (ed.) SOFSEM 2023. Lecture Notes in Computer Science, vol. 13878, pp. 111–125. Springer (2023)
9. Didimo, W., Kaufmann, M., Liotta, G., Ortali, G., Patrignani, M.: Rectilinear-upward planarity testing of digraphs. In: Iwata, S., Kakimura, N. (eds.) ISAAC 2023. LIPIcs, vol. 283, pp. 26:1–26:20. Schloss Dagstuhl (2023)
10. Esser, A.M.: Orthogonal compaction: Turn-regularity, complete extensions, and their common concept. In: Cláudio, A.P., Bouatouch, K., Chessa, M., Paljic, A., Kerren, A., Hurter, C., Trémeau, A., Farinella, G.M. (eds.) VISIGRAPP 2019. Communications in Computer and Information Science, vol. 1182, pp. 179–202. Springer (2019)

11. Evans, W.S., Fleszar, K., Kindermann, P., Saeedi, N., Shin, C., Wolff, A.: Minimum rectilinear polygons for given angle sequences. Comput. Geom. **100**, 101820 (2022)
12. Giacomo, E.D., Didimo, W., Liotta, G., Montecchiani, F., Ortali, G.: On the parameterized complexity of bend-minimum orthogonal planarity. Algorithmica **86**(10), 3231–3251 (2024)
13. Hong, S., Merrick, D., do Nascimento, H.A.D.: Automatic visualisation of metro maps. J. Vis. Lang. Comput. **17**(3), 203–224 (2006)
14. Karmarkar, N.: A new polynomial-time algorithm for linear programming. Comb. **4**(4), 373–396 (1984)
15. Klau, G.W., Mutzel, P.: Optimal compaction of orthogonal grid drawings. In: Cornuéjols, G., Burkard, R.E., Woeginger, G.J. (eds.) IPCO 1999. Lecture Notes in Computer Science, vol. 1610, pp. 304–319. Springer (1999)
16. Nöllenburg, M., Wolff, A.: Drawing and labeling high-quality metro maps by mixed-integer programming. IEEE Trans. Vis. Comput. Graph. **17**(5), 626–641 (2011)
17. Nöllenburg, M.: Automated drawings of metro maps, Institut fur Theoretische Informatik, Universität Karlsruhe (TH) (Master's thesis, 2005). https://i11www.iti.kit.edu/extra/publications/n-admm-05da.pdf
18. Object management group: business process model and notation (2014). https://www.omg.org/spec/BPMN/
19. Object management group: unified modeling language (2017). https://www.omg.org/spec/UML/
20. Patrignani, M.: On the complexity of orthogonal compaction. Comput. Geom. **19**(1), 47–67 (2001)
21. Stott, J.M., Rodgers, P., Martinez-Ovando, J.C., Walker, S.G.: Automatic metro map layout using multicriteria optimization. IEEE Trans. Vis. Comput. Graph. **17**(1), 101–114 (2011)
22. Tamassia, R.: On embedding a graph in the grid with the minimum number of bends. SIAM J. Comput. **16**(3), 421–444 (1987)
23. Transport for London: Harry Beck's Tube map. https://tfl.gov.uk/corporate/about-tfl/culture-and-heritage/art-and-design/harry-becks-tube-map. Accessed 26 July 2025

Hardness and Parameterized Tractability of the Weak Graph Distance

Maike Buchin[ID], Wolf Kißler[(✉)][ID], and Fabian Kubon

Faculty of Computer Science, Ruhr University Bochum, Bochum, Germany
{maike.buchin,wolf.kissler,fabian.kubon}@ruhr-uni-bochum.de

Abstract. The weak graph distance is a distance measure for embedded and immersed graphs which is motivated by its application to geographic networks. However, so far its computational complexity has only been partially understood, which is what we address in this paper.

First, we extend previous NP-hardness results for deciding the directed version of this distance, showing that it remains NP-hard in various restricted settings as well as hardness of approximation.

Then we present algorithmic results for deciding the weak graph distance in polynomial time under regularity conditions when G_1 is immersed and G_2 is planar embedded. Furthermore, we show two different FPT approaches for the case where both graphs are immersed in $\mathbb{R}^2$, parameterized in two ways of bounding the number of candidate solutions.

1 Introduction

Geometric graphs are widely used natural representations for various kinds of geometric networks, such as road or railway networks, river networks or anatomical networks. Here, we consider (straight line) embedded and immersed graphs, which are drawings of graphs in $\mathbb{R}^d$ in which the edges are drawn as line segments connecting their incident vertices. In embedded graphs, the edges may not intersect except in the vertices, whereas in immersed graphs, edge crossings are allowed. Given multiple such models of the same network or representations of related networks, one is typically interested in comparing the models.

In recent years, many different distance measures for embedded and immersed graphs have been proposed, cf. [6] for a survey of the metric properties of some notable approaches. These approaches include generalizations of distance measures originally defined on other structures such as the Fréchet distance and the contour tree distance, as well as distances based on the local persistent homology of the graphs [1], simultaneous traversals of both graphs [3] or (suitably constrained) edit distances [8,12]. Akitaya et al. [2] proposed the strong and weak graph distance, which are based on the strong and weak Fréchet distance for polygonal curves, respectively. Inherently, these distances are directed since they are based on mappings between the two graphs. A key advantage of these measures is that they capture both geometric and topological (dis)similarity. As discussed by Akitaya et al. [2], first experiments on reconstructions of real road networks showed promising results.

E. Di Giacomo and D. Mondal (Eds.): WALCOM 2026, LNCS 16444, pp. 64–78, 2026.
https://doi.org/10.1007/978-981-95-7127-7_5

However, the strong graph distance is NP-complete to approximate within a 1.10566 ratio even on plane (i.e., embedded in $\mathbb{R}^2$) graphs. The best known exact algorithm runs in time $O(Fm_2^{2F-1})$ where F is the number of faces of the first graph and m_2 is the number of edges of the second graph [2]. For the weak graph distance, they present a quadratic-time decision algorithm on spike-free (i.e., cycles are embedded in a nice way, cf. Def. 6) plane graphs. Akitaya et al. also showed that when both graphs are immersed in $\mathbb{R}^2$, the weak graph distance is NP-complete to decide.

Hence we are interested in whether (a variant of) the weak graph distance is tractable on realistic networks, in particular those with few edge crossings. Thus, in Sect. 2 we first extend the hardness result of [2] in showing that the directed weak graph distance is NP-complete to approximate in various more restricted scenarios. Cf. Table 1 for a summary of the complexity of deciding the directed weak graph distance $\vec{\delta}_{wG}(G_1, G_2)$ of two graphs G_1, G_2. Preliminary versions of Theorems 1 and 2 were presented at EuroCG 2024 [7].

In Sect. 3, we present new tractability results. For the weak graph distance between a strongly spike-free (i.e., spike-free and whenever an edge of a cycle intersects another edge, the ε-balls around the vertices do not intersect the ε-tube around the other edge, cf. Def. 7) immersed graph G_1 and a plane graph G_2, we show that the quadratic-time decision algorithm from [2] is still correct. Using that, we present two parameterized approaches for deciding whether the weak graph distance $\vec{\delta}_{wG}(G_1, G_2)$ is at most ε: First, an FPT algorithm parameterized in γk_2, where k_2 is the number of crossings of G_2 and γ is the maximum number of edges of G_1 that are ε-close to a crossing of G_2. Second, we introduce the notion of crossing-dependent placements, which are essentially parts of G_2 that a vertex of G_1 can be mapped onto iff a crossing of G_2 is used to map some incident edge. We explicitly assign to each vertex, which (or none) of its crossing-dependent placements it should be mapped onto. The running time of our algorithm depends on the number of such assignments and thus may still be exponential.

Table 1 summarizes (in)tractability results for the weak graph distance. Due to space constraints, most proofs are only sketched and full proofs will be contained in the full version.

1.1 Preliminaries

The weak graph distance is defined on embedded and immersed graphs. In our notation, an *embedding* is (essentially) a crossing-free drawing in $\mathbb{R}^d$ and an *immersion* is any drawing that may also contain crossings. A *plane graph* is a graph that is planar embedded in $\mathbb{R}^2$. Cf. [6] for more formal definitions.

In the following, let $G_1 = (V_1, E_1)$ and $G_2 = (V_2, E_2)$ be graphs immersed in $\mathbb{R}^2$ using straight-line immersions, i.e., edges are immersed as line segments. Also, let $|V_1| = n_1$, $|E_1| = m_1$, $|V_2| = n_2$ and $|E_2| = m_2$. We will slightly abuse notation and refer to the immersions of graphs, edges and vertices using the same notation as for the abstract graphs.

Table 1. Overview of complexity of deciding $\vec{\delta}_{wG}(G_1, G_2)$ in different scenarios

d	Planarity G_1	G_2	Additional Properties G_1	G_2	Complexity	Source
2	✓	✓	spike-free		P	[2]
any	–	–	tree		P	[2]
2	–	✓	strongly spike-free		P	Corollary 1
2	–	–	strongly spike-free		FPT in γk_2	Theorem 6
2	–	–	strongly spike-free		FPT in $\prod \#$c.d. placements	Theorem 7
≥ 2	–	–			NPC	[2]
≥ 2	✓	–	spike-free		α-approx. is NPC	Theorem 1
≥ 3	✓	✓	spike-free		α-approx. is NPC	Theorem 2
≥ 2	–	–	spike-free	size $O(1)$	α-approx. is NPC	Theorem 3
≥ 2	✓	✓			≈ 1.1-approx. is NPC	Theorem 4

A crossing is a point in $\mathbb{R}^d$ where the interiors of multiple edges overlap. Formally, we define a *crossing* c of an immersed graph G as a tuple (e, p), such that e is an edge of G and $p \in \mathbb{R}^d$ is an interior point of e's immersion that is also an interior point of the immersion of a different edge of G. Note that this definition influences how we count the numbers of crossings, since each point where a crossing occurs produces multiple such tuples. For example, if all edges cross in a single point, there are linearly many crossings in the above sense. For simplicity, we typically assume that no edges cross/overlap in a segment of positive length, and that no vertices are superimposed on other vertices/non-incident edges.

Next, to formally introduce the weak graph distance, we give some relevant notation from [2]. First, recall the weak Fréchet distance [4, defined as *nonmonotone Fréchet-metric*]:

Definition 1. *Let $s_1, s_2 : [0, 1] \to \mathbb{R}^2$ be curves. Define their* weak Fréchet distance *by*

$$\delta_{wF}(s_1, s_2) := \inf_{\alpha, \beta} \max_{t \in [0,1]} d(s_1(\alpha(t)), s_2(\beta(t))),$$

where α, β range over all continuous self-surjections of $[0, 1]$ that keep the endpoints fixed and d is the standard Euclidean metric.

That is, the weak Fréchet distance is the maximum distance attained between two points visited at the same time by an optimal pair of non-monotonous traversals of both curves. A well-known intuitive explanation of the (strong) Fréchet distance δ_F is that if s_1, s_2 are trajectories of a person walking their dog, then $\delta_F(s_1, s_2)$ is the minimum required leash length. The weak Fréchet distance fits the same intuition in the case where both person and dog are allowed to backtrack along their trajectory. Next, we define graph mappings, which are continuous mappings between the graphs:

Definition 2. *A graph mapping* $s\colon G_1 \to G_2$ *is a map that maps*

1. *each vertex* $v \in V_1$ *to a point* $s(v)$ *on an edge of* G_2 *and*
2. *each edge* $\{u, v\} \in E_1$ *to a simple path from* $s(u)$ *to* $s(v)$ *in* G_2.

The weak graph distance is now defined as the maximum weak Fréchet distance between an edge and its image under a (globally) optimal graph mapping:

Definition 3. *For immersed graphs* G_1, G_2, *define the directed weak graph distance as*

$$\vec{\delta}_{wG}(G_1, G_2) := \min_{s:G_1 \to G_2} \max_{e \in E_1} \delta_{wF}(e, s(e)),$$

where s ranges over all graph mappings and e and $s(e)$ refer to the corresponding immersions as curves in $\mathbb{R}^2$. *The* (undirected) *weak graph distance between* G_1 *and* G_2 *is defined as*

$$\delta_{wG}(G_1, G_2) := \max(\vec{\delta}_{wG}(G_1, G_2), \vec{\delta}_{wG}(G_2, G_1)).$$

Throughout this paper, we consider the decision problem for the directed distance, i.e., given immersed graphs G_1 and G_2 and a threshold ε, we want to decide whether $\vec{\delta}_{wG}(G_1, G_2) \le \varepsilon$. The undirected version can be decided by deciding both directed distances, whereas computing the distance can be done via a parametric search over the critical values as described in [2]. Lastly, we outline the general decision algorithm from [2]. For that, we introduce some useful notation and then define placements:

Definition 4. *Let* $G = (V, E)$ *be a graph immersed in* $\mathbb{R}^d$ *and* $\varepsilon > 0$. *For* $v \in V$, *we denote by* $B_\varepsilon(v) := \{x \in \mathbb{R}^d \mid d(v, x) \le \varepsilon\}$ *the closed ε-ball around (the immersion of)* v. *Moreover, for* $e \in E$, *let* $T_\varepsilon(e) := \{x \in \mathbb{R}^d \mid d(x, e) \le \varepsilon\}$ *denote the closed ε-tube around* e.

Definition 5. *An ε-placement of a vertex v is a connected component of* $G_2 \cap B_\varepsilon(v)$. *A weak ε-(edge-)placement of an edge* $e = \{u, v\} \in E_1$ *is a path P in* G_2 *that connects placements of u and v, respectively, such that* $\delta_{wF}(e, P) \le \varepsilon$. *A weak ε-placement of* G_1 *is a graph mapping* $s\colon G_1 \to G_2$ *that maps each edge to a weak ε-placement.*

Furthermore, we call a vertex placement C_v weakly valid if each adjacent vertex u has a placement C_u such that C_v and C_u are connected by a weak ε-placement of $\{u, v\}$. Otherwise, we call the placement weakly invalid.

By definition, a graph mapping $s\colon G_1 \to G_2$ is a weak ε-placement iff s realizes $\vec{\delta}_{wG}(G_1, G_2) \le \varepsilon$. Also, by definition of δ_{wF}, a weak placement of an edge e is simply a path within $G_2 \cap T_\varepsilon(e)$ that connects the corresponding vertex placements. The general decision algorithm for the weak graph distance now proceeds as summarized in Algorithm 1.

In Step 2, we identify for each edge $e = \{u, v\}$ of G_1 the pairs of respective placements C_u of u and C_v of v such that there exists a weak edge placement of e that connects C_u and C_v. Equivalently, these are the pairs of respective

---

Algorithm 1. General Decision Algorithm for the Weak Graph Distance [2]

1: Compute vertex placements.
2: Compute mutual reachability information for vertex placements.
3: Recursively prune weakly invalid placements.
4: Decide if there exists a placement for the whole graph G_1.

---

placements within the same connected component of $G_2 \cap T_\varepsilon(e)$. Thus, a vertex placement is weakly valid iff there exists a placement of each adjacent vertex such that the placements are mutually reachable. Note that Step 3 needs to be repeated recursively, since a deleted invalid placement might have been the last reachable placement of some placement of an adjacent vertex, making that placement invalid as well. Akitaya et al. [2] provide more details on the algorithm and show that steps 1–3 can be performed in $O(m_1 m_2)$ time for general immersed graphs. However, existence of a weakly valid placement for each vertex does not imply existence of a weak placement of the whole graph, cf. Figure 1b. Thus, Step 4 is non-trivial, in fact Akitaya et al. showed that it is NP-complete in general. However, Akitaya et al. also show that Step 4 is in fact trivial if both graphs are plane and the embedding of G_1 meets the following regularity condition:

Definition 6. *Let G be an immersed graph and C be a simple cycle of G. C is called* spike-free *if the shorter side of the minimum bounding box of C has length at least 2ε and, for each three consecutive vertices $u, v, w \in C$, the ε-ball around u does not intersect the ε-tube around the edge $\{v, w\}$. G is called* spike-free *if each simple cycle of G is spike-free.*

If G_1 is spike-free, its cycles may always be placed "outward", guaranteeing both the existence of consistent placements for each cycle and the compatibility of the placements of adjacent cycles. Figure 1 shows several examples where the proof of Akitaya et al. does not apply: In Fig. 1a, the whole spiky cycle G_1 can be placed on the orange edge, but cannot be placed "outward". Figure 1b shows a plane graph G_1 and an immersed graph G_2. Although every vertex of G_1 has exactly two placements, the cycle cannot be placed: After traversing the placement of the cycle starting in any placement of a vertex, we necessarily end up at the opposite placement of that vertex.

2   Hardness of Deciding the Weak Graph Distance

We extend the hardness results by Akitaya et al. [2] to various additional cases, including approximation and when the first graph is spike-free and the second has constant size.

Theorem 1. *Approximating $\vec{\delta}_{wG}(G_1, G_2)$ within any constant ratio $\alpha \geq 1$ remains NP-complete even if G_1 is a spike-free plane graph and G_2 is immersed.*

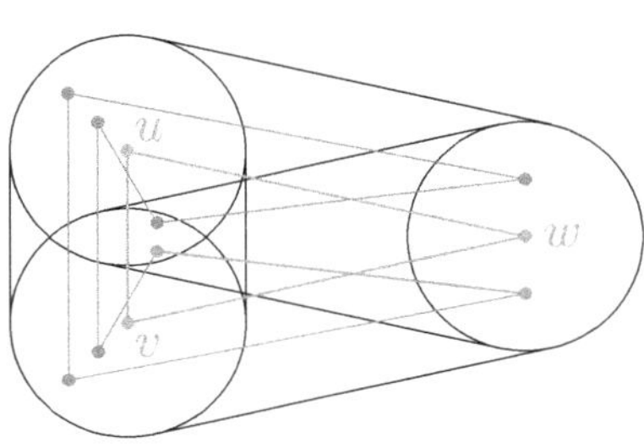

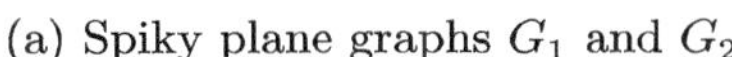

(a) Spiky plane graphs G_1 and G_2

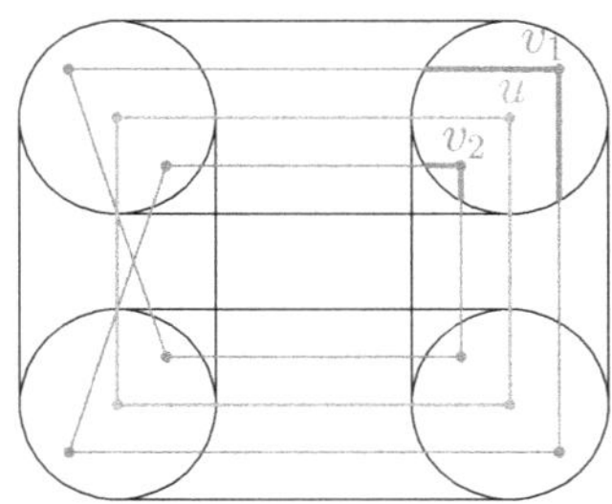

(b) Plane graph G_1 and immersed graph G_2

Fig. 1. Graphs G_1 (blue) and G_2 (red, orange) where no outer placement of G_1 onto G_2 exists. (Color figure online)

Theorem 2. *Approximating $\vec{\delta}_{wG}(G_1, G_2)$ within any constant ratio $\alpha \geq 1$ is NP-complete for graphs G_1, G_2 embedded in $\mathbb{R}^d$ for all $d \geq 3$ even if G_1 is spike-free.*

Theorem 3. *Approximating $\vec{\delta}_{wG}(G_1, G_2)$ within any constant ratio $\alpha \geq 1$ is NP-complete even if G_1, G_2 are immersed in $\mathbb{R}^2$, G_1 is spike-free and G_2 has constant size.*

For these results, the proofs (included in the full version due to space constraints) all utilize a reduction from 3-colorability of (4-regular) planar graphs, cf. [10,14] for the NP-completeness of these problems. The main idea is to construct a graph G_1 that roughly mimicks the input graph G, and a graph G_2 in which all vertices of G_1 have three placements representing the color choices, such that the edges of G_1 can be placed iff the vertices are mapped onto different "colors". In Theorem 1, we construct a grid embedding of G_1 [9], whereas in Theorem 2, we embed both graphs on the moment curve [13]. For Theorem 3, we map multiple edges onto the same paths in G_2 to keep the size of G_2 constant.

Figure 2 illustrates the above proof idea for the case of Theorem 1 on a single edge of the planar input graph G, where the blue edge is an edge of G_1 and each vertex has a "yellow", a "red" and a "purple" placement in G_2 such that placements of adjacent vertices are mutually reachable iff they belong to different "colors". The vertex $\widehat{uv}$ is inserted between u and v to prevent zig-zagging in the edge placement to change colors.

For the hardness of approximation results, we observe that placing the vertices of G_2 closer to their counterparts in G_1 results in a lower distance in the case where G is 3-colorable, while keeping the original lower bound in the case where G is not 3-colorable intact.

Next we confirm the conjecture by Akitaya et al. [2] that the weak graph distance is NP-complete to decide when both graphs are plane and we do not demand spike-freeness of G_1:

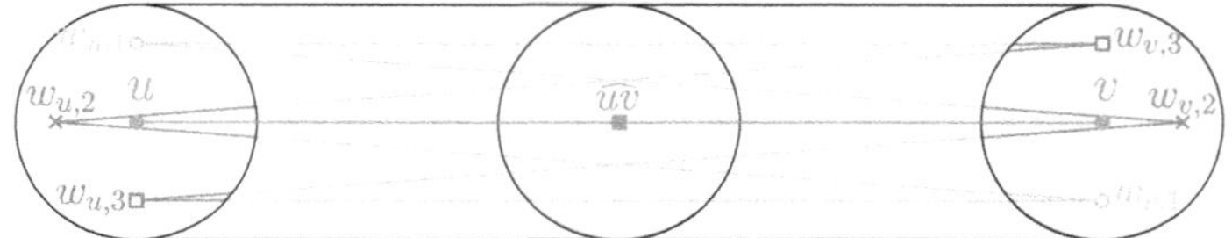

Fig. 2. Illustration of the reduction in Thm. 1 on a single edge $\{u, v\}$ of G.

Theorem 4. *For plane graphs G_1 and G_2, approximating $\vec{\delta}_{wG}(G_1, G_2)$ within a 1.10566 factor is NP-complete.*

Proof (sketch). We show that the edges of any plane graph G_1 can be subdivided by inserting degree 2 vertices such that for the subdivided (spiky) plane graph $\tilde{G}_1$, $\vec{\delta}_{wG}(\tilde{G}_1, G_2) = \vec{\delta}_G(G_1, G_2)$. Thus, hardness carries over from the strong graph distance, cf. [2, Theorem 6].

Note that for each such subdivision, $\vec{\delta}_{wG}(G_1, \tilde{G}_1) = \vec{\delta}_{wG}(\tilde{G}_1, G_1) = 0$ and that by the above, $\vec{\delta}_{wG}(G_1, G_2)$ can be arbitrarily smaller than $\vec{\delta}_{wG}(\tilde{G}_1, G_2)$. This implies that the weak graph distance does not fulfill the triangle inequality.

3   Decision Algorithms for the Weak Graph Distance

3.1   Polynomial Algorithm for the Immersed-Embedded Case

Here, we consider deciding the directed weak graph distance $\vec{\delta}_{wG}(G_1, G_2)$ in the case where G_1 is immersed in $\mathbb{R}^2$ and G_2 is plane. This extends the result of Akitaya et al. [2], who considered the case where both graphs are plane. There, the idea is that although existence of weakly valid placements does not guarantee that all placements are mutually compatible, a placement that is certain to be compatible with all other placements can always be constructed if G_1 is spike-free, cf. Definition 6. In that case, cycles are placed "outward" by choosing *outermost vertex placements*, that is, weakly valid placements that lie most towards the outside of the face bounded by the cycles, and by connecting outermost vertex placements by *outermost paths* defined similarly. An *outer placement* of a cycle consists of outermost paths for all its edges. When two adjacent (non-spiky) cycles are placed outwards, their placements are guaranteed to overlap because of the geometric construction and the definition of weak validity. Such overlaps are necessarily at vertices of G_2 because of planarity, so the placements can be merged by placing shared edges arbitrarily between the overlap vertices. Additionally, such a placement is also an outer placement of the face bounded by the merged cycle. Since we consider the case where G_1 is immersed, its cycles might now be self-intersecting as well as intersect other cycles and paths. To still be able to work with outermost placements, we extend spike-freeness to ensure that intersecting edges of cycles are well-behaved:

Definition 7. *Let $G = (V, E)$ be an immersed graph and C be a simple cycle of G. We say that C is* strongly spike-free *if C is spike-free (cf. Definition 6) and for any edge $e = \{u_1, u_2\}$ of C and all edges $e' = \{v_1, v_2\}$ in E that intersect e, $B_\varepsilon(u_i) \cap T_\varepsilon(e') = B_\varepsilon(v_j) \cap T_\varepsilon(e) = \varnothing$ for $i, j \in \{1, 2\}$. An immersed graph G is* strongly spike-free *if each simple cycle of G is strongly spike-free.*

For embedded graphs, spike-freeness and strong spike-freeness are equivalent. However, for immersed graphs, strong spike-freeness is a stronger assumption. As such, it would be interesting to investigate whether it suffices to assume a weaker regularity condition (e.g., spike-freeness) for immersed graphs. Note that it suffices to assume that each connected component of G_1 is strongly spike-free, because a family of weak placements for the connected components directly gives rise to a weak placement (i.e. graph mapping) of the whole graph.

We now state our tractability result:

Theorem 5. *Let G_1 be a strongly spike-free immersed graph and G_2 be an embedded graph. If each vertex of G_1 has at least one weakly valid ε-placement after running steps 1–3 of Algorithm 1, then G_1 has a weak ε-placement and therefore, $\vec{\delta}_{wG}(G_1, G_2) \leq \varepsilon$.*

Before we prove this theorem, we observe the following useful property.

Lemma 1. *In the situation of Theorem 5, let $e = \{u_1, u_2\}$ be an edge of some simple cycle C of G_1 and $e' = \{v_1, v_2\}$ be an edge of G_1 that intersects e. Then, all pairs of weakly valid placements of u_1 and u_2 are mutually reachable.*

Proof (sketch). The proof is by a geometric argument: Due to the well-separation of the ε-tubes and -balls, we guarantee overlaps of the placements of both edges, and by planarity of G_2, these overlaps are necessarily at vertices, which allows constructing a merged placement.

Proof (Theorem 5). Since our proof extends the proof for the plane case [2, Lemma 7], we first very briefly outline that proof, with some more details mentioned in the full version: In the plane case, we iteratively merge and place faces (simple cycles), bridge paths[1] connecting faces, and tree-substructures (i.e., trees connected to the remaining graph in at most one vertex), in that order. Faces are placed outward, which allows placing adjacent faces consistently since the outer placements must overlap, and we obtain an outer placement of the merged face. A path that connects faces must approach at least one of the faces from the outside, and thus, when we start placing the path from the other face, that placement must at some point intersect the outer placement of that face, which can again be used to obtain a consistent placement. Lastly, tree-substructures can always be placed from their root.

---

[1] Note: In standard terminology, both these edges and the edges of the tree-substructures would be called bridges, but for the graph distances, we need to handle paths that connect faces separately, which we call bridge paths.

Next, we discuss the modifications needed in the immersed case. We discuss these in some more detail since similar ideas will be used in the subsequent proofs. The main idea is to add a fourth category of edges, which are the intersecting edges of cycles, which we place last. Moreover, up until that final step, we essentially consider the graph $\overline{G_1}$ that arises from G_1 by removing all crossing edges from cycles. $\overline{G_1}$ is still not plane in general because bridges between cycles and tree-substructures might still intersect each other. Also, the topology may change: Connected components might decompose into multiple connected components (potentially degenerating to isolated vertices), and cyclic parts might become bridges or tree-substructures. The order in which we place edges is:

1. Non-intersecting edges of cycles in $\overline{G_1}$
2. Bridge edges between cycles of $\overline{G_1}$
3. Edges of tree-substructures of $\overline{G_1}$
4. Intersecting edges of cycles (i.e., edges of $E(G_1)\backslash E(\overline{G_1})$)

This is still a partition of the edges of G_1, but crucially, all cycles that are considered in the first step are spike-free and plane, and the bridges considered in the second step may intersect themselves or other parts of the graph, but may not intersect cycles (of $\overline{G_1}$), in particular not the cycles that they connect. As in the plane case, we consider each connected component in isolation, constructing a family of valid ε-placements for the connected components, which in turn gives rise to an ε-placement of the whole graph. However, for the first three steps, we consider the connected components of $\overline{G_1}$, so we still need to make sure that these placements are compatible in the fourth step whenever multiple connected components of $\overline{G_1}$ belong to a single connected component of G_1.

As to step 1: Let C be a connected component of $\overline{G_1}$. By construction, all faces of C are spike-free plane cycles, so they can be placed outward by the same arguments as in [2]. Moreover, the outer placements of any pair of adjacent faces can also be merged as in the plane case, obtaining an outer placement of the merged face. As described above, this process can be iterated to place all faces of C consistently.

For step 2, consider a path P that connects two cycles C_1 and C_2 that have been placed onto outermost placements. Crucially, P still approaches at least one of the cycles (without restriction: C_2) from the outside, since P does not intersect either cycle (P may approach C_1 from the inside if the cycles are nested into each other). The arguments from [2] carry over to the case where P has crossings, namely: Starting at the outer placement of C_1, we can place P onto some path Q because of weak validity. Because P approaches C_2 from the outside and C_2 is placed onto some outer placement O, Q is guaranteed to intersect O within the intersection of the ε-tubes around P and C_2. Because of the planarity of G_2, that intersection is at a vertex of G_2 and thus, the respective parts of the placements can be combined to place P, still using the previously assigned outer placements of both vertices.

In step 3, there are two cases to consider: Tree-substructures that belong to some connected component of $\overline{G_1}$ that contains a cycle, and tree-substructures that form a connected component (including isolated vertices). In the former case, we can choose the vertex v in which T is connected to the remaining connected component as the root. v will have already been placed onto some weakly valid placement in the previous steps. As in the plane case, we can now iteratively place the incident edges due to the definition of weak validity. Again, this will never create inconsistencies since T is a tree. In the latter case, no vertex of T has previously been placed, so we choose an arbitrary vertex as the root and place that vertex arbitrarily. Afterwards, the arguments from the former case hold true.

Lastly, for step 4, consider the intersecting edges of cycles of G_1 that we disregarded before. Both incident vertices have been placed previously to construct valid placements of their respective connected components in $\overline{G_1}$, and Lemma 1 guarantees that these two vertex placements are mutually reachable. Thus, the edge can be placed consistently.

Piecing everything together, in the scenario of the theorem, the existence of weakly valid placements of all vertices still implies $\vec{\delta}_{wG}(G_1, G_2) \leq \varepsilon$ as claimed.

Corollary 1. *For a strongly spike-free immersed graph G_1 and an embedded graph G_2, it can be decided whether $\vec{\delta}_{wG}(G_1, G_2) \leq \varepsilon$ in $O(m_1 n_2)$ time.*

Proof. As detailed in [2], steps 1–3 of Algorithm 1 can be performed in $O(m_1 m_2)$ time for general graphs. Since G_2 is plane, $m_2 = O(n_2)$. By Theorem 5, the only additional work needed is to confirm whether each vertex of G_1 retains a weakly valid placement, which can be done in linear time.

3.2   FPT Algorithm Under Geometric Assumptions

Here, we consider the case where G_1 and G_2 are immersed graphs such that G_1 is strongly spike-free, G_2 has k_2 many crossings and there is some small $\gamma \in \mathbb{N}$ such that for any crossing $c = (e_2, p)$ of G_2, at most γ many edges of G_1 are ε-close to p. This seems like a reasonable assumption for realistic networks, e.g., road maps, since we would typically expect the number of edges close to any given point to be significantly lower than the total size of the network. In this scenario, only these $\leq \gamma$ many edges of G_1 may be mapped through the crossing c of G_2 in an ε-placement. Since placing any other edge e onto a path P through c would yield the lower bound $\delta_{wF}(e, P) \geq d(e, p) > \varepsilon$, so the corresponding graph mapping would also realize a distance greater than ε.

Our idea is to explicitly assign for each edge $e_1 \in E_1$, which of the crossings that are ε-close to it its placement should use. Then, we mark the placements of the incident vertices of e_1 that are in the same connected component of $G_2 \cap T_\varepsilon(e_1)$ as all the assigned crossings as valid, and all other placements as invalid. Next, we perform step 3 of Algorithm 1, i.e. the recursive pruning, to mark all placements of other vertices invalid if they are incompatible with our assignment. Now, we want to check if this assignment of edges to crossings can be extended to a complete graph mapping. In that extension, we only need to consider mapping onto G_2 without its crossings. Our regular definition of a vertex placement (cf. Definition 5) does not specify the exact point within a placement onto which the vertex is mapped, which may influence which crossings need to be used. Thus, we define crossingless placements:

Definition 8. *Let G_1, G_2 be immersed graphs. We define the graph G_2' as follows: Whenever edges $e_1 = \{u_1, v_1\}, \ldots, e_\ell = \{u_\ell, v_\ell\}$ of G_2 intersect in some point p, replace these edges by vertices u_i', v_i', $i \in [\ell]$, superimposed at p as well as edges $\{u_i, u_i'\}$ and $\{v_i', v_i\}$.*

Then, for $\varepsilon > 0$, a crossingless placement of some vertex v of G_1 regarding G_2 is a connected component of $G_2' \cap B_\varepsilon(v)$. We define mutual reachability and (weak) validity analogous to Definition 5, using the graph G_2' for the graph traversal unless we currently consider an edge to which at least one crossing is assigned, in which case we force using exactly the assigned crossings and forbid using any other crossings.

That is, we disconnect the edges of G_2 at crossings and consider connected components of the resulting graph (restricted to the corresponding ε-balls). We do not typically allow superimposed vertices, but allowing this restricted usage does not influence our algorithms. In particular, since these vertices always have degree 1, whenever we guarantee an overlap of placements, that overlap will not lie at such a vertex (since in the overlaps we guarantee, placements necessarily change their relative position or become the same placement). Also, note that if an intersection occurs at distance precisely ε to some vertex, then there might be cases where it is not actually possible to place the vertex at distance ε without using the crossing, but crucially, the infimum of all distances for which the vertex is placeable without using the crossing remains ε, and thus so does the distance under that crossing assignment.

Recall that we define crossings as pairs (e, p) s.t. e intersects (at least) one other edge in p. Thus, G_2' has exactly $m_2 + k_2$ many edges since each crossing leads to exactly one subdivision, increasing the edge count by one. For the remainder of this section, we use the term (vertex) placement to refer to a crossingless placement. We now state our result.

Theorem 6. *Let G_1 and G_2 be graphs immersed in $\mathbb{R}^2$ such that G_1 is strongly spike-free, G_2 has at most k_2 crossings, and for each crossing (e_2, p) of an edge $e_2 \in E_2$, at most γ many edges of G_1 are ε-close to p. Then, it can be decided in $O(2^{\gamma k_2} m_1 (m_2 + k_2) + m_2 \log m_2)$ time whether $\vec{\delta}_{wG}(G_1, G_2) \le \varepsilon$.*

Proof (sketch). We compute the intersections of G_2 in time $O(m_2 \log m_2 + k_2)$ with Balaban's output-sensitive algorithm [5], and the set D of all pairs (e_1, c), where $c = (e_2, p)$ is a crossing of G_2 that is ε-close to $e_1 \in E_1$. Since $|D| \le \gamma k_2$, $|\mathcal{P}(D)| = O(2^{\gamma k_2})$.

We iterate over the subsets $S \subseteq D$, which we interpret as assignments of specific crossings to (exactly) use in the placements of each edge. Mutual reachability of crossingless placements under such an assignment can be distinguished into the following cases:

1. For edges e of G_1 that have no assigned crossings in S, pairs of respective crossingless placements are mutually reachable iff they lie in the same connected component of $T_\varepsilon(e) \cap G_2'$.
2. For edges e of G_1 that have at least one assigned crossing, we "reconnect" the edges corresponding to the assigned crossings in G_2', i.e., replace the subdivided edges by the original edge from G_2. Then, if all assigned crossings lie in the same connected component when restricted to the ε-tube $T_\varepsilon(e)$, all pairs of vertex placements within that connected component are mutually reachable. If not, no vertex placements are mutually reachable.

In total, these steps have time and space complexity $O(m_1(m_2 + k_2))$, as well as the recursive pruning step that we perform afterwards.

Now, if all vertices of G_1 retain at least one weakly valid placement after the pruning step, the idea is that the placements can be merged as in Theorem 5. However, we now have a fifth class of edges: the edges that were assigned at least one crossing, which we place last. For the first four classes, the proof proceeds along the same lines as the one in Theorem 5, i.e., we construct outermost crossingless placements of the edges, which are still guaranteed to overlap and thus be mergeable since they do not contain crossings. Finally, since we only retained placements within one connected component per edge of type 5, the incident vertices' placements are guaranteed to be mutually reachable.

Conversely, if $\vec{\delta}_{wG}(G_1, G_2) > \varepsilon$, then there can be no crossing assignment for which every vertex retains a weakly valid (crossingless) placement. Hence, the algorithm sketched above correctly decides the weak graph distance.

3.3 FPT Algorithm Based on Crossing-Dependent Placements

Here, we consider another parameterized approach for deciding the weak graph distance. Instead of parameterizing in the number of crossings of G_2, we observe that the main problems posed by crossings concern mutual reachability of vertex placements of vertices in G_1. This observation inspires the following definition to formalize the intuition of a placement being affected by crossings:

Definition 9. *Let $G_1 = (V_1, E_1)$ and $G_2 = (V_2, E_2)$ be immersed graphs. Further, let $u, v \in V_1$ be adjacent vertices and C_u and C_v be weakly valid placements of u and v, respectively. C_u and C_v are* crossing-dependently reachable *if they are mutually reachable (in the sense of step 2 of Algorithm 1) but any path connecting them must traverse a crossing of G_2. A vertex placement C_u is* crossing-dependent *if one of the following conditions holds:*

- *If all crossings are removed from C_u, C_u decomposes into multiple connected components.*
- *There exists a placement C_v of an adjacent vertex v such that C_u and C_v are crossing-dependently reachable.*

For a vertex $v \in V_1$, we define the set of all crossing-dependent placements of v:

$$\mathcal{X}_v := \{C_v \mid C_v \text{ is a crossing-dependent placement of } v\}.$$

Note that even if a vertex has crossing-dependent placements, it does not follow that all placements of that vertex are crossing-dependent. Our algorithm proceeds as follows: First, run steps 1–3 of Algorithm 1 to obtain the weakly valid placements in the sense of Definition 5. Then, check connectedness and mutual reachability of vertex placements again, considering the modified graph G_2' (cf. Definition 8) instead of G_2. By doing so, we see which placements are crossing-dependent. Next, we iterate over all assignments that assign each vertex either a specific crossing-dependent placement, or a placeholder ⊔ to indicate we want to assign a non crossing-dependent placement. For vertices that have been assigned a crossing-dependent placement, we mark all other placements invalid, including the placements that are not crossing-dependent. For all other vertices, we mark all crossing-dependent placements invalid. For each assignment, we repeat Step 3 of Algorithm 1 to recursively prune all placements that have now become invalid. We will show that this suffices to decide the weak graph distance.

The number of assignments of vertices to crossing/non-crossing placements does not depend on the total number of crossings and can be small even for graphs with many crossings. If any vertex in G_1 has at most δ many (crossing-dependent) placements in G_2 and there are κ vertices in G_1 that have crossing-dependent placements, there are up to $(\delta + 1)^\kappa$ such assignments. Although we expect both δ and κ to be small, in general we only know $\delta \leq m_2$ and $\kappa \leq n_1$.

Theorem 7. *Let G_1 be a strongly spike-free immersed graph and G_2 be an immersed graph. Further, let*

$$M := \prod_{v \in V_1} (\mathcal{X}_v \cup \{⊔ \mid v \text{ has some placement that is not crossing dependent}\}),$$

where $\prod$ denotes the cartesian product. One can decide in $O(|M|m_1 m_2 + m_1 m_2^2)$ time and $O(m_1 m_2^2)$ space whether $\vec{\delta}_{wG}(G_1, G_2) \leq \varepsilon$.

Proof (sketch). We show that if an assignment of vertices to crossing-dependent placements yields a weakly valid placement for all vertices, then $\vec{\delta}_{wG}(G_1, G_2) \leq \varepsilon$; the converse can be verified quickly. Similarly to Theorem 6, we consider the edges for which both incident vertices have been placed onto crossing-dependent placements in a fifth and final step of the merging procedure.

Crucially, in the first four steps of the merging procedure, i.e., the steps as described in Theorem 5, we restrict our attention to edges of the corresponding types for which it additionally holds that at most one of their incident vertices is placed onto a crossing-dependent placement. By definition of crossing-dependent placements, this means that any pair of placements of the incident vertices that is mutually reachable remains mutually reachable when forbidding the usage of crossings of G_2 (equivalently, when performing the graph traversal on the graph G_2' instead of G_2). As such, the (outer) placements we construct within the merging procedure can also be constructed with regard to G_2' instead of G_2, thus avoiding the usage of crossings and still retaining the property that whenever two such placements overlap for geometric reasons, that overlap will be located at a vertex, allowing for the placements to be merged.

Lastly, consider edges $e = \{u, v\}$ in G_1 for which both u and v were assigned a crossing-dependent placement. Since these placements C_u and C_v were the only remaining placements of their respective vertices before the pruning step and remained valid after the pruning step, we can conclude that C_u and C_v are mutually reachable (potentially requiring the traversal of crossings of G_2). As such, e can be placed consistently.

Regarding the running time, we first compute the crossings of G_2, which is possible in $O(m_2^2)$ time. Afterwards, we compute the placements of G_1 regarding both G_2 and G_2' and decide mutual reachability, which is possible in $O(m_1 m_2^2)$ total time and space (since G_2' has $O(m_2 + k_2) = O(m_2^2)$ many edges). All placements which become disconnected or for which the reachability information differs between G_2 and G_2' are crossing-dependent. The main loop then iterates over M, which can be implicitly achieved via nested traversals of the lists of crossing-dependent placements of the vertices and thus, the space consumption of the algorithm will only be $O(m_1 m_2^2)$. For each assignment, we need to repeat Step 3 of Algorithm 1 (recursive pruning), which takes $O(m_1 m_2)$ time.

The figure below illustrates the merits and limitations of Theorems 6 and 7: In the example on the left, each vertex of G_1 has exactly one placement (which is crossing-dependent), but each crossing of G_2 is ε-close to linearly many edges of G_1. In the example on the right, each vertex of G_1 has a near-linear number of crossing-dependent placements, but each crossing of G_2 is ε-close to only a single edge of G_1 (Fig. 3).

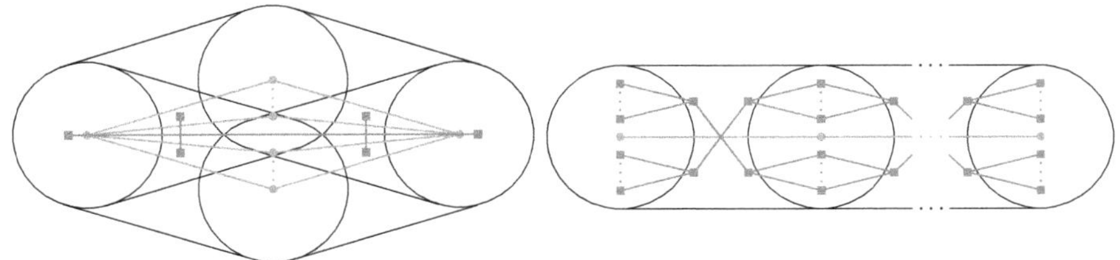

Fig. 3. Two pairs of graphs G_1 (blue) and G_2 (red) that illustrate the search spaces in Theorems 6 and 7. Left: Large search space in Theorem 6, small search space in Theorem 7. Right: Small search space in Theorem 6, large search space in Theorem 7. (Color figure online)

4   Conclusion

We have shown several new intractability and tractability results for the weak graph distance as summarized in Table 1. While these results improve our understanding of the complexity of deciding the weak graph distance, some questions remain unanswered, one of them being whether the weak graph distance admits an FPT algorithm when G_1 is (strongly) spike-free and when parameterized in the numbers of crossings of both graphs, even though it seems reasonable to suspect that this is not the case.

The crossing-rigid weak graph distances are variations in which only crossings may be mapped onto crossings, which seem reasonable for realistic graphs, in particular because they are fixed-parameter tractable [7,11]. Further examining these measures and their applicability in realistic graphs, e.g., by conducting experiments, are interesting avenues for future research.

References

1. Ahmed, M., Fasy, B.T., Wenk, C.: Local persistent homology based distance between maps. In: Proceedings of the 22nd ACM SIGSPATIAL International Conference on Advances in Geographic Information Systems, pp. 43–52 (2014). https://doi.org/10.1145/2666310.2666390
2. Akitaya, H.A., Buchin, M., Kilgus, B., Sijben, S., Wenk, C.: Distance measures for embedded graphs. Comput. Geom. **95**, 101743 (2021). https://doi.org/10.1016/j.comgeo.2020.101743
3. Alt, H., Efrat, A., Rote, G., Wenk, C.: Matching planar maps. J. Algor. **49**(2), 262–283 (2003). https://doi.org/10.1016/S0196-6774(03)00085-3
4. Alt, H., Godau, M.: Computing the Fréchet distance between two polygonal curves. Int. J. Comput. Geom. Appl. **5**, 75–91 (1995). https://doi.org/10.1142/S0218195995000064
5. Balaban, I.J.: An optimal algorithm for finding segments intersections. In: Proceedings of the 11th International Symposium on Computational Geometry, pp. 211–219. ACM (1995). https://doi.org/10.1145/220279.220302
6. Buchin, M., Chambers, E., Fang, P., Fasy, B.T., Gasparovic, E., Munch, E., Wenk, C.: Distances between immersed graphs: metric properties. La Matematica **2**, 197–222 (2023). https://doi.org/10.1007/s44007-022-00037-8
7. Buchin, M., Kißler, W.: Hardness and modifications of the weak graph distance. In: EuroCG 2024 (2024). https://eurocg2024.math.uoi.gr/data/uploads/paper_49.pdf
8. Cheong, O., Gudmundsson, J., Kim, H.S., Schymura, D., Stehn, F.: Measuring the similarity of geometric graphs. In: Proceedings of the 8th International Symposium on Experimental Algorithms, pp. 101–112 (2009). https://doi.org/10.1007/978-3-642-02011-7_11
9. Chrobak, M., Payne, T.H.: A linear-time algorithm for drawing a planar graph on a grid. Inf. Process. Lett. **54**(4), 241–246 (1995). https://doi.org/10.1016/0020-0190(95)00020-D
10. Dailey, D.P.: Uniqueness of colorability and colorability of planar 4-regular graphs are NP-complete. Disc. Math. **30**(3), 289–293 (1980). https://doi.org/10.1016/0012-365X(80)90236-8
11. Kubon, F.: On the Weak Graph Distance and its Three Crossing-Rigid Variants. Master's thesis, Ruhr University Bochum (2025)
12. Majhi, S., Wenk, C.: Distance measures for geometric graphs. Comput. Geom. **118**, 102056 (2024). https://doi.org/10.1016/j.comgeo.2023.102056
13. Matousek, J.: Lectures on Discrete Geometry, vol. 212. Springer, Heidelberg (2013). https://doi.org/10.1007/978-1-4613-0039-7
14. Stockmeyer, L.: Planar 3-colorability is polynomial complete. ACM SIGACT News **5**(3), 19–25 (1973). https://doi.org/10.1145/1008293.1008294

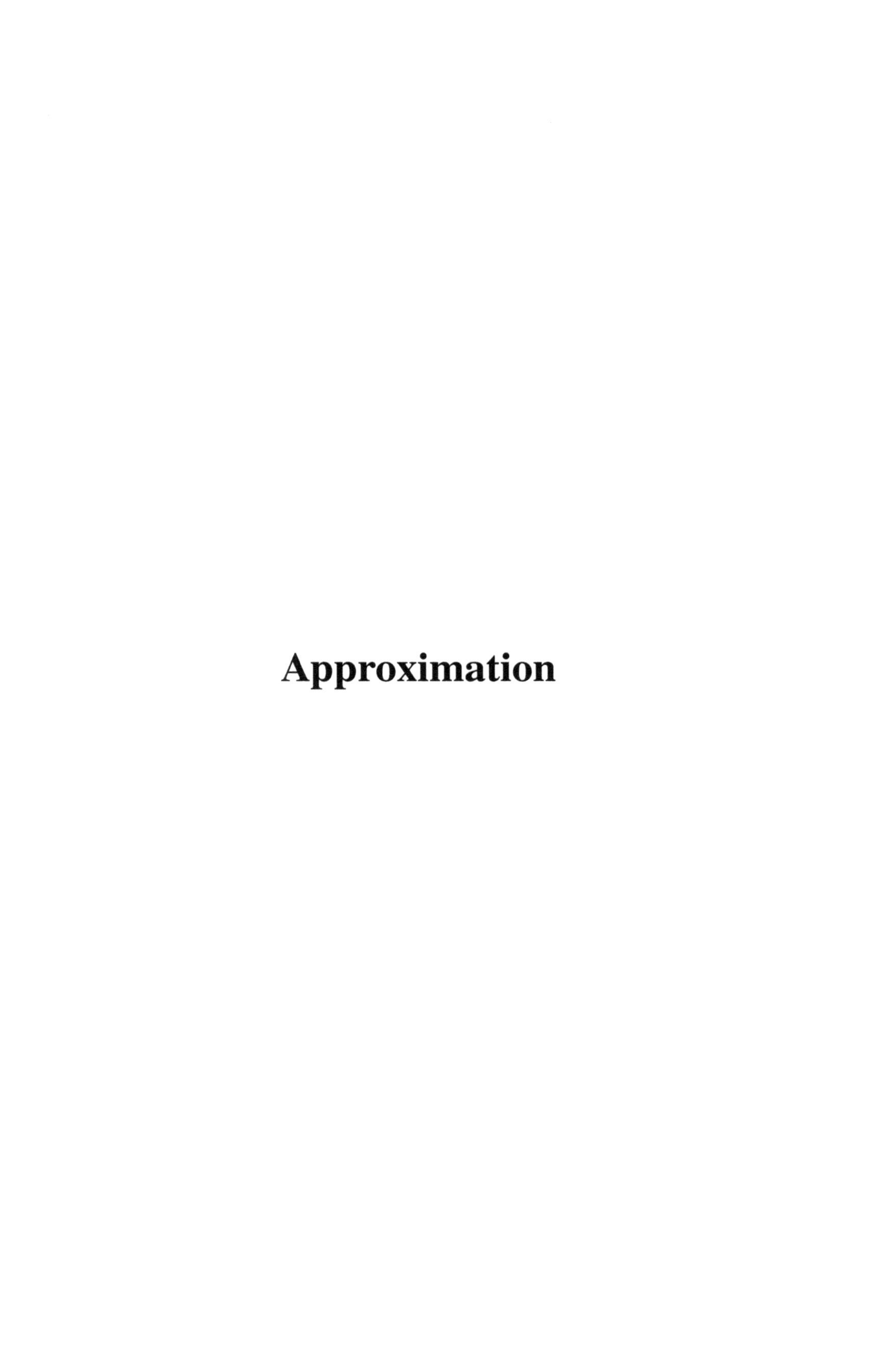

Approximation

Hardness and Approximation Results for Extending Unique Neighborhood Networks

Siam Habib[1($\boxtimes$)], Debajyoti Mondal[2], Sadia Sharmin[1], and Md. Saidur Rahman[1]

[1] Graph Drawing and Information Visualization Laboratory, Department of Computer Science and Engineering (CSE), Bangladesh University of Engineering and Technology (BUET), Dhaka, Bangladesh
`hsiam261@gmail.com`, {`sadiasharmin,saidurrahman`}`@cse.buet.ac.bd`
[2] Department of Computer Science, University of Saskatchewan, Saskatoon, Canada
`d.mondal@usask.ca`

Abstract. A graph G is called a unique neighborhood network (UNN) if it does not contain any vertex whose neighborhood is a subset of the neighborhood of another vertex in G. UNNs are used to model peer-authentication problems in communication network and quantum cryptography. A natural problem in this context is to extend an existing UNN by adding a new vertex. In this paper, we study the minimum-cost UNN extension problem, where the input is a UNN G and the goal is to add a new vertex u to G by making u adjacent to a minimum set of vertices such that the unique neighborhood property is preserved. Here the cost is the number of edges added to u. In practice, this can be seen as inserting a new user into a network that is already modeled as a UNN, and the goal is to restore the UNN property with minimum cost. In this paper, we give a polynomial-time $\mathcal{O}(\log n)$-approximation for the problem and a linear-time algorithm for bounded-treewidth graphs. In contrast, we show that the minimum-cost UNN extension problem cannot be approximated within a factor of $o(\log n)$ in polynomial time unless $\mathcal{P} = \mathcal{NP}$. As a byproduct, we obtain a similar hardness of approximation for the minimum-cost UNN completion problem that seeks to minimize the number of edge insertions to transform a graph into a UNN.

1 Introduction

Let $G = (V, E)$ be a simple graph with the vertex set V and the edge set E. The *neighborhood* $N_G(v)$ of a vertex $v \in V(G)$ consists of the set of vertices that are adjacent to v. We define the *closed neighborhood* $N_G[v]$ of a vertex v as $N_G(v) \cup \{v\}$. A *Unique Neighborhood Network* (UNN) is a graph G where every distinct pair $u, v \in V(G)$, we have $N_G(u) \nsubseteq N_G(v)$ and $N_G(v) \nsubseteq N_G(u)$. For example, a cycle of five vertices is a UNN because none of the neighborhoods is a subset of another, whereas a cycle of 4 vertices is not a UNN because the neighborhoods of the first and third (or second and fourth) vertices are the same.

E. Di Giacomo and D. Mondal (Eds.): WALCOM 2026, LNCS 16444, pp. 81–95, 2026.
https://doi.org/10.1007/978-981-95-7127-7_6

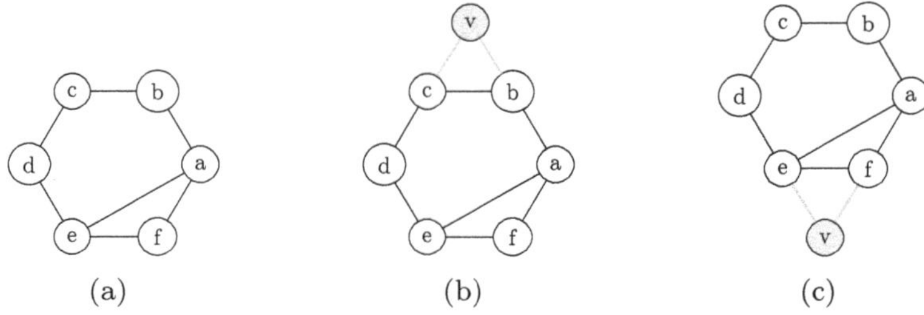

Fig. 1. (a) A UNN G. (b) A UNN extension of G with an extension vertex v. (c) An extension of G that is not a UNN extension.

Similarly a path a_1, a_2, a_3, a_4 of four vertices is not a UNN since the neighbor of a_1 is a subset of the neighbors of a_3. UNNs were first introduced by Rass [20] who showed that large random graphs produced by stochastic processes under different models such as the ErdősRényi model [7,10] and the random geometric graph model [19] have a tendency of being UNNs [21].

UNNs find applications in computer networking and quantum symmetric cryptography in the context of peer-authentication for secure end-to-end communication [21], where a UNN is useful to realize authentication protocols. For communication efficiency, it is desirable to minimize the size of secret keys and thus when constructing UNNs, a natural goal is to avoid creating vertices of large neighborhood. UNNs can be thought of as a stronger variant of point determining graphs [9] (also known as mating graphs [2]), where the inclusion of one neighborhood by another is allowed, but they must be different. UNNs are also closely related with open neighborhood locating dominating codes [23] (also known as strongly identifying codes [12]) where, the neighborhoods of the vertices have an unique intersection with a specific subset of vertices. Open neighborhood locating dominating code is special case of discriminating code in graphs [3,14]. There is a rich body of work on discriminating codes that are motivated by its application on sensor network coverage and localization problems [6,16,22] and the concept of UNN is also applicable to these settings. For example, consider a set of digital modules being maintained as a UNN. If a module becomes faulty, then its neighbors can sense the fault and send a distress signal to the control center without revealing the identity of the faulty sensor for security reasons. The control center can locate the faulty sensor by identifying the neighborhood.

In this paper, we study the UNN Extension Problem and the UNN Completion Problem. The UNN extension problem seeks to add a new vertex to an existing UNN. This is inspired by the scenario where a new user (i.e., a new vertex) wants to join an existing UNN-based protocol and this user should be added with a relatively small cost so that the resulting graph remains a UNN. Formally, we call a graph G' an *extension* of a UNN G if G' is obtained by adding a new vertex $v \notin V(G)$ to G and making v adjacent to zero or more vertices in $V(G)$. We refer to v as the *extension vertex* (Fig. 1). The *cost of an extension* is the number of vertices that become adjacent to v, i.e., $|N_{G'}(v)|$. If an extension G' of a UNN G is also a UNN, then G' is called *a UNN extension.* Note that a UNN extension must add one or more edges to v because otherwise, the neighborhood of v (which is empty) is a subset of the neighborhood of other vertices. The *minimum-cost UNN extension* problem seeks for a UNN extension with the minimum cost.

The UNN Completion Problem seeks to model the task of transforming a general graph into a UNN through the insertion of edges. Formally, given a graph G, we define a graph H to be a *UNN completion* of G if H is obtained from G by adding zero or more edges, and H satisfies the UNN property. The *cost of a UNN completion* is defined as the number of edge insertions required to obtain the UNN completion, i.e., $|E(H)| - |E(G)|$. Given any graph G, the *UNN Completion Problem* seeks to find a UNN completion of G with minimum cost. More generally, the problem of finding a minimum set of edge insertions to restore some property of a graph is known as a minimum edge fill problem or a Π-Completion problem. Such problems are central to the study of graph classes, with notable examples including chordal graphs [8,24], interval graphs [8,11,15], split graphs [18], co-chordal graphs [18], and perfect graphs [18].

Contributions. We give a polynomial-time algorithm to compute an $\mathcal{O}(\log n)$ approximation for the minimum-cost UNN Extension problem. In contrast, we prove that the minimum-cost UNN Extension problem cannot be approximated within a factor of $o(\log n)$ unless $\mathcal{P} = \mathcal{NP}$. The crux of our approach is the identification of a special type of dominating set that helps to show that the neighborhood of an extension vertex is related to such a dominating set in the complement of the graph. We then obtain the results using careful reduction techniques. Furthermore, we show that the problem can be expressed in monadic second order logic, and thus linear-time solvable for bounded-treewidth UNNs. Finally, we establish a reduction from the UNN extension problem to the UNN completion problem and show that the UNN completion problem also cannot be approximated within a factor of $o(\log n)$ in polynomial time unless $\mathcal{P} = \mathcal{NP}$.

The rest of the paper is organized as follows. Section 2 introduces notation and preliminary results. Section 3 proves some properties of a UNN extension. Section 4 and 5 give the approximation and hardness results for the UNN extension problem, respectively. Section 6 handles the UNN extension problem for bounded treewidth UNNs. Section 7 proves hardness results for the UNN completion problem. Finally, Sect. 8 concludes the paper.

2 Preliminaries

In the rest of the paper we consider only graphs that are finite, simple, undirected, without any multi-edges or self-loops.

Dominating Set and Its Variants. The *dominating set* of a graph $G = (V, E)$, is a set of vertices $D \subseteq V$ such that for every vertex $v \in V$, either v or a neighbor of v belongs to D. The size of a minimum dominating set of G is referred to as its *domination number* $\gamma(G)$. The *total dominating set* of G is a set of vertices

$D \subseteq V$ such that every vertex $v \in V$ is adjacent to some vertex in D. Note that a dominating set may contain a vertex that is not adjacent to any other vertices in the set, but a total dominating set cannot contain such a vertex. It is known that every graph without any isolated vertex has a total dominating set. The *minimum dominating set problem* and the *minimum total dominating set problem* focus on finding a dominating set and a total dominating set of minimum size, respectively. Both these problems are known to be $\mathcal{NP}$-hard [8,17]. The size of a minimum total dominating set of a graph G is referred to as its *total domination number* $\gamma_t(G)$. For graphs with isolated vertices, its total domination number is defined as infinity.

Concept of Suppressors and Suppressed Nodes. Let v, w be two vertices in a graph G. We say that v *suppresses* w (equivalently, w is suppressed by v) if $N_G(w) \subseteq N_G(v)$, and denote it by $w \preceq_G v$. We will use the notation $w \npreceq_G v$ to denote that w is not suppressed by v. For any two vertices w and v if $w \npreceq_G v$ and $v \npreceq_G w$, we say that w and v are *incomparable* in G and will be denoted by $w|_G v$. The proof of the following lemma follows from the definition.

Lemma 1. *The following statements are equivalent.*

- *G is a UNN.*
- *No two distinct vertices in G are comparable.*
- *No vertex v in $V(G)$ is a suppressor of any other vertex in G.*
- *No vertex v in $V(G)$ is suppressed by any other vertex in G.*

We now observe some properties of suppressors and suppressed vertices.

Remark 1. Let u, v be two adjacent vertices in a graph G. Then $u|_G v$.

Remark 2. Let G be an arbitrary graph and let H be an induced subgraph of G. If $u \npreceq_H v$ for some vertices $u, v \in V(H)$, then $u \npreceq_G v$. Similarly, if $u|_H v$ holds, then we must have $u|_G v$.

For a graph G, the *complement graph* $\overline{G}$ is a graph with the same vertex set as that of G and there exists an edge (a, b) in $\overline{G}$ if and only if a and b are not adjacent in G. By $G - \{v\}$ we denote a graph obtained from G by deleting the vertex v from G.

Lemma 2. *Let G be a graph and let v be a vertex in G. If $N_G(v)$ is a dominating set in $\overline{G - \{v\}}$ then v is not suppressed by any other vertex in G, and vice versa.*

Proof. Assume first that v is not suppressed by any other vertex in G. We now show that $N_G(v)$ must be a dominating set in $\overline{G - \{v\}}$. Consider an arbitrary vertex $u \in V(G - \{v\})$. If u and v are adjacent in G, then u belongs to $N_G(v)$. If u and v are not adjacent in G, then since v is not suppressed by u, there exists a vertex $v' \in N_G(v)$ that is not adjacent to u in G. Therefore, v' is adjacent to u in $\overline{G - \{v\}}$. Thus, $N_G(v)$ is a dominating set of $\overline{G - \{v\}}$.

We now show that if $N_G(v)$ is a dominating set in $\overline{G - \{v\}}$, then v cannot be suppressed by any other vertex in G. Suppose for a contradiction that a vertex

$u(\neq v)$ suppresses v in G. We know that either $u \in N_G(v)$ or a neighbor u' of u in $G - \{v\}$ is in $N_G(v)$. If $u \in N_G(v)$ then, u cannot suppress v in G (Remark 1). Similarly, if $u' \in N_G(v)$, then u cannot suppress v in G because u' is adjacent to v but not to u in G. $\square$

Lemma 3. *Let G be a graph and let v be a vertex in G. If every vertex $u \in (V(G) - N_G[v])$ has a neighbor $u' \notin N_G(v)$, then v is not a suppressor in G and vice versa.*

Proof. If every vertex $u \in (V(G) - N_G[v])$ has a neighbor that does not belong to $N_G(v)$, then by definition v cannot suppress any other vertex in G. Assume now that v is not a suppressor and consider an arbitrary vertex $u \in V(G) - N_G[v]$. Since v is not a suppressor, again by definition u has a neighbor $u' \in V(G)$ that is not a neighbor of v. $\square$

3   Relating Dominating Set to the UNN Extensions

In this section we show that an extension of a UNN G would be a UNN if and only if the neighborhood of the extension vertex forms a special type of dominating set (Lemma 6). This will later be used in subsequent sections to obtain the approximation and hardness results.

Lemma 4. *Let G' be an extension of a UNN G with the extension vertex v^*. Then G' is a UNN extension if and only if the following conditions hold:*

(a) v^* is not suppressed by any other vertex in G'.
(b) v^* does not suppress any other vertex in G'.

Proof. If G' is a UNN extension of G, then G' is a UNN and by the definition of UNN, G' cannot contain any suppressor or suppressed vertex. Therefore, v^* neither suppresses nor is suppressed by another vertex in G'.

Assume now that v^* neither suppresses nor is suppressed by another vertex in G', i.e., for every vertex $u(\neq v^*)$ in G', u and v^* are incomparable. Every pair of vertices $p(\neq v^*)$ and $q(\neq v^*)$ are incomparable in G and remains so in G' even when we add v^* (Remark 2). Therefore, by Lemma 1 G' is a UNN. $\square$

We can combine Lemmas 2, 3, and 4 to obtain the following lemma.

Lemma 5. *Let G' be an extension of a UNN G with the extension vertex v^*. Then G' will be a UNN extension if and only if the following conditions hold:*

(a) $N_{G'}(v^*)$ is a dominating set in $\overline{G}$.
(b) Every vertex $u \in V(G) - N_{G'}(v^*)$ has a neighbor $u' \notin N_{G'}(v^*)$ in G.

We refer to a dominating set as *a good dominating set* if it satisfies the Condition (b) of Lemma 5, which is formalized as follows.

Definition 1 (Good Dominating Set). *A dominating set D of a graph G is good if and only if for every vertex $u \in V(G) - D$, u has at least one neighbor outside of D in $\overline{G}$.*

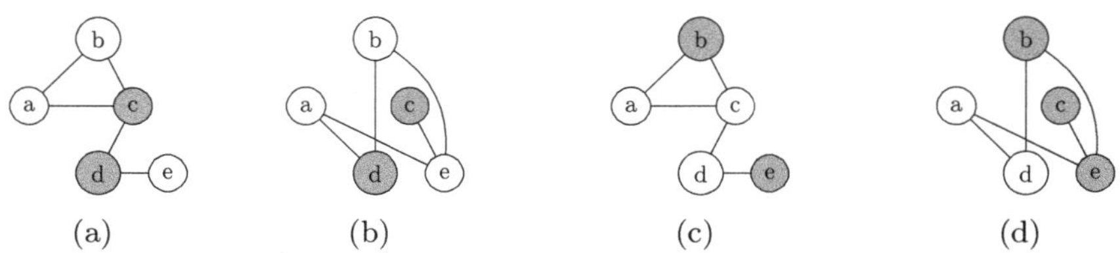

Fig. 2. (a) Example of a good dominating set D (in blue) in a graph G. (b) $\overline{G}$, where each vertex not in D (non-blue) has a non-blue neighbor. (c) A bad dominating set D' in the same graph G shown in blue. (d) D' shown in $\overline{G}$ with the bad vertices in red (all neighbors of a bad vertex in $\overline{G}$ are blue). (Color figure online)

By $\gamma_g(G)$ we refer to the *good domination number*, i.e., the size of a minimum cardinality good dominating set. Given a dominating set D in G, we call a vertex $v \in V(G) - D$ *bad* if it has no neighbor outside D in $\overline{G}$ (equivalently $V(G) - D \subseteq N_G[v]$). It is now straightforward to observe that a dominating set in G is good if and only if G does not contain any bad vertex.

In Fig. 2a and Fig. 2c we show examples of a good and a bad dominating set, respectively (highlighted in blue) for the same graph G. Using the definition of a good dominating set, we can reword Lemma 5 as follows.

Lemma 6. *Let G' be an extension of a UNN G with extension vertex v^*. Then G' is a UNN extension if and only if $N_{G'}(v^*)$ is a good dominating set of $\overline{G}$.*

4   An Approximation Algorithm

In this section we show a strict reduction from the minimum-cost UNN extension problem to the dominating set problem. Since there is a polynomial-time $\mathcal{O}(\log n)$-approximation algorithm for the minimum dominating set problem [13], such a reduction gives a polynomial-time $\mathcal{O}(\log n)$-approximation algorithm for the minimum-cost UNN extension. Specifically, given a UNN G and a dominating set D in $\overline{G}$, we first show how to find a good dominating set D' in $\overline{G}$ such that $|D'| \leq |D|$ (Lemma 7). Since a good dominating set in $\overline{G}$ corresponds to a UNN extension (Lemma 6), this will allow us to design an approximation for minimum-cost UNN Extension.

Lemma 7. *Let G be a UNN and let D be a dominating set in $\overline{G}$. Then there exists a good dominating set D' of $\overline{G}$ that is no larger than D and can be computed in polynomial time.*

Proof. We consider some cases depending on $\gamma(\overline{G})$.

Case 1: $\gamma(\overline{\mathbf{G}}) = 1$. We find a vertex v in $\overline{G}$ that is adjacent to every other vertex in $\overline{G}$. We set D' to be $\{v\}$. Thus $|D'| \leq |D|$. We now show that D' is good. Note that v must be isolated in G. But since G is a UNN, this can only be true if v is the only vertex in G. In that case D' is good.

Case 2: $\gamma(\overline{\mathbf{G}}) = \mathbf{2}$. Here $|D| \geq 2$. We find a good dominating set D' by searching over all $\binom{V}{2}$ pairs of vertices and checking if the choice of the pair is a good dominating set of $\overline{G}$. Note that such a good dominating set always exists. We prove this claim via contradiction.

Assume now that no such good dominating set exists. Since $\gamma(\overline{G}) = 2$, there exists a dominating set of size two. Let $S = \{x, y\}$ be such a dominating set. Then by our assumption there exists a vertex u that is bad with respect to S in $\overline{G}$. Since u is bad, every neighbor of u in G is in S. If u has no neighbors in G, then $\gamma(\overline{G}) = 1$, which is a contradiction. Thus, u must have either one or two neighbors in G. Note that u cannot have more than two neighbors in G since that would entail that u has neighbors outside of S.

If u has two neighbors in G, then both x and y are adjacent to u in G and thus S cannot dominate u in $\overline{G}$. Therefore u can have only one neighbor in G and that neighbor is also present in S. Without loss of generality assume that the neighbor is x. If x had any neighbor w in G other than u, then w would suppress u in G which cannot happen. Therefore, x only has one neighbor in G and that is u. Now we claim that $S' = \{x, u\}$ is a good dominating set of $\overline{G}$.

Since every vertex $v \in V(G) - \{u, x\}$ is adjacent to x in $\overline{G}$, S' is a dominating set of $\overline{G}$. Furthermore, no vertex in $V(G) - S$ has any neighbors in S' in G. Since $\gamma(\overline{G}) > 1$, G has no isolated vertices. Therefore, every vertex in $V(G) - S'$ has at least one neighbor. Hence S' is also good, which contradicts our assumption.

Case 3: $\gamma(\overline{\mathbf{G}}) > \mathbf{2}$. Without loss of generality, we may assume that D is minimal, i.e., $D - \{p\}$ is not a dominating set for $\overline{G}$ for any $p \in D$. If D is a good dominating set in $\overline{G}$, we may set $D' = D$. Otherwise, we consider two subcases.

Case 3.1: $\overline{\mathbf{G}}$ **has no isolated vertices.** Let u be a bad vertex with respect to D in $\overline{G}$. Since $\overline{G}$ has no isolated vertex and D is minimal, every vertex in D must have a neighbor outside D in $\overline{G}$. Using this observation we can construct a new set D' as follows.

- Initialize D' to an empty set and add u to D'. Since D is a dominating set in $\overline{G}$, we know that u is adjacent to at least one vertex $q \in D$ in $\overline{G}$.
- Iterate over each vertex $v(\neq q)$ in D. Note that v has a neighbor u' outside D in $\overline{G}$. If v does not have a neighbor in D' already, then add u' to D'.

We added the vertex u for the vertex $q \in D$ and for each remaining vertex in D, we added at most one vertex to D'. Therefore, $|D'|$ cannot be larger than $|D|$. In the following, we first show that D' is a dominating set of $\overline{G}$ and then show that D' is a good dominating set.

D' is a dominating set of $\overline{G}$: Let v be a vertex in $\overline{G}$. If $v \in D$, then a neighbor v' of v is in D'. If $v \notin D$, then by the definition of a bad vertex, v is dominated by the bad vertex u since $V(G) - D \subseteq N_{\overline{G}}[u]$.

D' is good in $\overline{G}$: Suppose for a contradiction that there exists a vertex v that is bad with respect to D' in $\overline{G}$. Then, v dominates every vertex in $V(G) - D'$ in $\overline{G}$. Since u is a bad vertex with respect to D in $\overline{G}$, u dominates all the vertices in $V(G) - D$ in $\overline{G}$. Since $D' \subseteq (V(G) - D)$, u dominates all the vertices in D'. Consequently, $\{u, v\}$ is a dominating set of $\overline{G}$, which contradicts that $\gamma(\overline{G}) > 2$.

Case 3.2: $\overline{G}$ has one or more isolated vertices. Let S be the set of vertices that are isolated in $\overline{G}$. Therefore, any dominating set of $\overline{G}$ must include all vertices in S.

If $\overline{G}$ only contains isolated vertices (i.e. if G is a complete graph), then $D(= S)$ is the minimum dominating set of $\overline{G}$, which is also good. We thus set D' to be D.

Otherwise, $G - S$ is a UNN such that $\overline{G} - S$ has no isolated vertices and $D - S$ is a dominating set in $\overline{G} - S$. Thus one of the cases among Cases 1, 2 and Case 3.1 must hold for $\overline{G} - S$ and thus we can find a good dominating set S' that is no larger than $|D - S|$. Therefore, $|S \cup S'| \le |D|$. It is now straightforward to verify that $S' \cup S$ is a good dominating set in $\overline{G}$ and hence we can choose D' to be $S' \cup S$. $\qquad\square$

We now use Lemma 7 to show the following.

Lemma 8. *If G is a UNN, then $\gamma(\overline{G}) = \gamma_g(\overline{G})$.*

Proof. Let D_m be a minimum dominating set of $\overline{G}$. Then using Lemma 7 we can find a good dominating set D' from D_m such that $|D'| \le |D_m| = \gamma(\overline{G})$. Since every good dominating set is also a dominating set, we have $\gamma(\overline{G}) \le \gamma_g(\overline{G}) \le |D'|$. Consequently, $\gamma_g(\overline{G}) = \gamma(\overline{G})$. $\qquad\square$

Given a dominating set D of $\overline{G}$, we can construct a UNN extension G' by adding a vertex $v^* \notin V(G)$ and connecting it to the vertices of D' where D' is the good dominating set constructed from D using the construction of Lemma 7. By Lemma 6, G' is a UNN. Since the cost of the extension is $|D'|$, the approximation ratio for the UNN extension problem is $\frac{|D'|}{\gamma_g(\overline{G})} = \frac{|D'|}{\gamma(\overline{G})} \le \frac{|D|}{\gamma(\overline{G})}$, which is the approximation ratio for the minimum dominating set in $\overline{G}$. Since the dominating set problem has a polynomial-time $\mathcal{O}(\log n)$-approximation algorithm [13], the strict reduction implies the following result.

Theorem 1. *Given a UNN G with n vertices, we can compute a UNN extension G' of G in polynomial time such that the cost is at most $\mathcal{O}(\log n)$ times the cost of a minimum UNN extension of G.*

5   Hardness of Approximation

In this section we show that the minimum-cost UNN extension problem cannot be approximated within a factor of $o(\log n)$ in polynomial time unless $\mathcal{P} = \mathcal{NP}$. We achieve this in two steps. In the first step we show a strict reduction from the minimum total dominating set problem to a minimum blue dominating set (a variant of dominating set) problem in the complement graphs of UNNs (Sect. 5.1). In the second step we show a strict reduction from the blue dominating set problem in complements of UNNs to the minimum-cost UNN extension problem (Sect. 5.2). We will use the concept of *UNNification* that given a graph G, constructs a UNN U_G as follows.

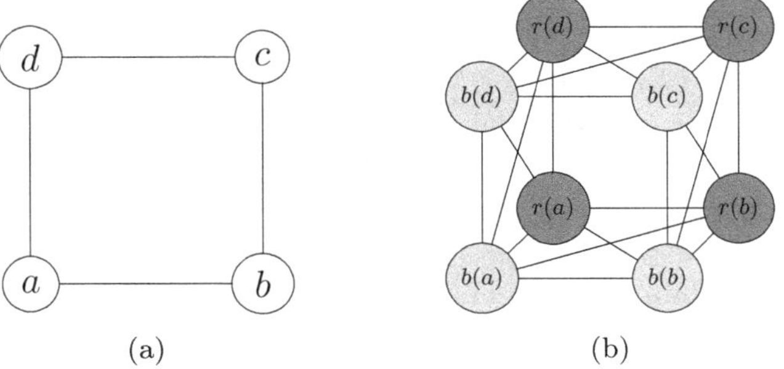

Fig. 3. (a) A graph G. (b) The UNNification of G.

Definition 2. *UNNification of a graph G is another graph U_G such that:*

- *For every vertex $v \in V(G)$, U_G has two copies of v, one red (denoted as $r(v)$) and one blue (denoted as $b(v)$), and $r(v)$ and $b(v)$ are adjacent in U_G.*
- *For distinct vertices $u, v \in V(G)$, their corresponding blue vertices $b(u)$ and $b(v)$ are adjacent in U_G if and only if u and v are adjacent in G.*
- *For distinct vertices $u, v \in V(G)$, $b(u)$ and $r(v)$ are adjacent in U_G if and only if u and v are adjacent in G.*
- *For distinct vertices $u, v \in V(G)$, the corresponding red vertices $r(u)$ and $r(v)$ are adjacent in U_G if and only if u and v are incomparable in G.*

Figure 3 shows an example of a graph G (on the left) and its UNNification U_G on the right. Let p and q be a red and blue vertex in U_G, respectively. Then $r^{-1}(p)$ and $b^{-1}(q)$ return their corresponding vertices in G. For notational convenience, we will extend the functions $b(v)$ and $r(v)$ to sets of vertices over G. For a set $S \subseteq V(G)$, $b(S)$ (similarly, $r(S)$) denotes the blue (red) vertices in U_G corresponding to the vertices in S. We say that two vertices $u, v \in V(U_G)$ are siblings if their corresponding vertices in G are the same. We now have the following theorem, whose proof is omitted due to space constraints.

Theorem 2. *Let G be a graph. Then its UNNification U_G is a UNN.*

The following two lemmas and corollary state some properties of $\overline{U_{\overline{G}}}$ which will later be used in our reduction.

Lemma 9. *Let G be an arbitrary graph. Let G' be the complement graph of the UNNification of $\overline{G}$. Let $u, v \in V(G')$ be adjacent in G'. Then the corresponding vertices in G are adjacent.*

Proof. Since u and v are adjacent in G', they are non-adjacent in $\overline{G'} = U_{\overline{G}}$. Note that in a UNNification, each vertex is colored either blue or red. We now consider two cases.

Case 1 (At least one of u and v are blue): Without loss of generality lets assume u is blue. Let u' and v' respectively be the vertices corresponding to u

and v in G. Since u is not adjacent to v in $U_{\overline{G}}$, u' and v' must be distinct. Now if u' is not adjacent to v' in G, then u' is adjacent to v' in $\overline{G}$ and thus $u = b(u')$ is adjacent to $b(v')$ and $r(v')$ in $U_{\overline{G}}$. Based on the color of v, v is either $b(v')$ or $r(v')$. In either case, u is adjacent to v in $U_{\overline{G}}$ which is a contradiction. Thus u' must be adjacent to v' in G.

Case 2 (Both u and v are red): By Definition of UNNification, the corresponding vertices $u' = r^{-1}(u)$ and $v' = r^{-1}(v)$ are comparable in $\overline{G}$. Note that since u and v are not the same, $u' \neq v'$. If u' and v' were adjacent in $\overline{G}$, they would be incomparable in $\overline{G}$ and thus u' and v' are not adjacent in $\overline{G}$. Thus, they are adjacent in G. $\square$

Lemma 10. *Let G be an arbitrary graph. Let G' be the complement graph of the UNNification of $\overline{G}$. Let $u, v \in V(G)$ be adjacent in G. Then $b(u)$ must be adjacent to both $b(v)$ and $r(v)$ in G'.*

Proof. Since u and v are adjacent in G, they are non-adjacent in $\overline{G}$. Thus by the definition of UNNification, $b(u)$ is not adjacent to $b(v)$ or $r(v)$ in the UNNification of $\overline{G}$. Therefore, $b(u)$ is adjacent to both $b(v)$ and $r(v)$ in G'. $\square$

Corollary 1. *Let G be an arbitrary graph. Let G' be the complement graph of the UNNification of $\overline{G}$. Let v be a red vertex in G' and u be its blue sibling. Then $N_{G'}(v) \subseteq N_{G'}(u)$.*

5.1 Reducing the Total Dominating Set Problem in G to the Blue Dominating Set Problem in $\overline{U_{\overline{G}}}$

In this section we give a reduction from the total dominating set problem in a graph G without any isolated vertices to finding a special type of dominating set called a blue dominating set in $G' = \overline{U_{\overline{G}}}$.

A *blue dominating set* in $\overline{U_{\overline{G}}}$ is a special type of dominating set that only contains blue vertices. The size of the minimum blue dominating set is the *blue domination number* of $\overline{U_{\overline{G}}}$, which we denote as $\gamma^b(\overline{U_{\overline{G}}})$. In this section we will show that for any total dominating set D in G, we can find a blue dominating set D' of G' such that $|D'| = |D|$ in polynomial time (Lemma 11). Furthermore, given a blue dominating set D' in G', we can find a total dominating set D in G such that $|D| = |D'|$ (Lemma 12).

Lemma 11. *Let D' be a dominating set in $G' = \overline{U_{\overline{G}}}$ where all the vertices are blue. Then the set $D = b^{-1}(D') = \{u \in V(G) \mid b(u) \in D'\}$ is a total dominating set in G.*

Proof. It suffices to show that every vertex $u \in V(G)$ is adjacent to a vertex in D. Consider the vertex $r(u)$ in G'. We know that $r(u)$ is not in D' since it is red. Since D' is a dominating set in G', D' must contain a blue vertex $b(v)$ that is adjacent to $r(u)$ in G'. By construction of D, we have $v \in D$. Finally, by Lemma 9, u is adjacent to v in G. $\square$

Lemma 12. *Let D be a total dominating set of G. Then $b(D) = \{b(u) \in V(G') \mid u \in D\}$ is a dominating set in $G' = \overline{U_{\overline{G}}}$.*

Proof. It suffices to show that $b(D)$ dominates every vertex $v' \in V(G')$. Let v be the vertex corresponding to v' in G and $u \in D$ be a vertex that is adjacent to v. Note that u must exist since D is a total dominating set. Since $u \in D$, $b(u) \in b(D)$. Again since u and v are adjacent in G, by Lemma 10, $b(u)$ is adjacent to both $b(v)$ and $r(v)$ in G'. Hence, $b(u)$ dominates v'. $\qquad\square$

Lemma 12 and 11 imply a strict approximation-preserving reduction

Lemma 13. *Let G be an arbitrary graph. Let $G' = \overline{U_{\overline{G}}}$ be the complement of the UNNification of $\overline{G}$. Then $\gamma_t(G) = \gamma^b(G')$.*

5.2   Reducing the Blue Dominating Set Problem in $\overline{U_{\overline{G}}}$ to the Minimum-Cost UNN Extension Problem

We now give a strict reduction from the problem of finding a blue dominating set in the complement of the UNNification, i.e., $G' = \overline{U_{\overline{G}}}$ to the problem of finding a UNN extension to the UNN $\overline{G'} = U_{\overline{G}}$. To this end, it suffices to show that the blue domination number of G', $\gamma^b(G')$ is equal to the min-cost UNN Extension size and given any UNN Extension of $\overline{G'}$, we can construct a blue dominating set of G' that is not larger.

Lemma 14. *Let G be a graph without any isolated vertices and let $G' = \overline{U_{\overline{G}}}$. Given any UNN extension of $U_{\overline{G}}$, a blue dominating set D' of G' can be computed in polynomial such that D' is not larger than the cost of the UNN extension.*

Proof. Let G'' be the UNN extension of $U_{\overline{G}}$ and let D be the neighborhood of the extension vertex of G''. Then, the cost of G'' is $|D|$. Therefore we will have to find a blue dominating set D' of G' such that $|D'|$ is not larger than $|D|$.

To construct the set D', we will use the concept of loneliness-free set. Let M be a subset of $V(G')$. We say that a vertex $v' \in M$ is *lonely* in M if v' is red and its blue sibling is not present in M. We call M *loneliness-free* if no vertex $v' \in M$ is lonely. We construct D' in two steps. If D is not loneliness-free, then we first create a loneliness-free dominating set L of G' such that $|L| = |D|$. We then construct the required dominating set D' from L.

Step 1 (Creating L from D): Let S be the set of lonely vertices in D and $T = b(r^{-1}(S))$ be the set of the blue siblings of all the lonely vertices in D. We will construct the set L by deleting all lonely vertices from D and adding their blue siblings to it, i.e. $L = (D \setminus S) \cup T$. We claim that L is the required loneliness free dominating set of G'. It is evident that $|L| = |D|$ and L can be computed in polynomial time. We thus need to show that L is indeed loneliness free and L is a dominating set in G'.

Suppose for a contradiction that v' is a lonely vertex in L. By definition of a lonely vertex, v' must be red. By the construction of L, all red vertices of L are also in D. Thus, v' must also belong to D. If v' was lonely in D, then v' cannot be in L. Thus, the blue sibling of v' must be present in D. Since all blue vertices of D are also present in L, v' cannot be lonely in L.

We now show that L is a dominating set of G'. Let u' be an arbitrary vertex in G'. Due to Lemma 6, D is a dominating set in G' and therefore there exists a vertex $v' \in D$ that dominates u' in G'. We now show that u' remains dominated by a vertex in L.

Case 1 (v' is not lonely in D): Since we only delete lonely vertices from D in our construction of L, $v' \in L$ and thus u' is dominated by L.

Case 2 (v' is lonely in D and $u' = v'$): Let v be the vertex corresponding to v' in G and let $w' = b(v)$ be the blue sibling of v'. Since v' is lonely in D, $w' \notin D$. Hence there exists a vertex x' in D that is adjacent to w'. Let x be the vertex in G corresponding to x'. Then regardless of the color of x', $b(x)$ is present in L. Since x' and v' are adjacent in G', by Lemma 9 x and v are adjacent in G and thus due to Lemma 10 $b(x)$ is adjacent to v'.

Case 3 (v' is lonely in D and $u' \neq v'$): Since u', v' are adjacent in G', by Lemma 9 the corresponding vertices u and v in G are adjacent. By Lemma 10 $b(v)$, which is the blue sibling of v', is adjacent to both $b(u)$ and $r(u)$. Here one of $b(u)$ and $r(u)$ is u'. Since v' is lonely in D, we have $b(v) \in L$, and hence u' remains dominated in L.

Step 2 (Creating D' from L): We create D' from L in polynomial time as follows. Add every blue vertex $u' \in L$ to D'. For every red vertex $v' \in L$, pick an arbitrary blue neighbour of v' in G' and add it to D'. Note that since G has no isolated vertex, such a blue neighbor of v' must exist in G'.

It is straightforward to observe that $|D'| \leq |L|$. It thus remains to show that D' is a dominating set of G'. Let v' be a vertex of G'. Since L is a dominating set of G', either v' belongs to L or there exists a vertex $u' \in L$ that is adjacent to v'. If v' belongs to L, then by construction of D' either v' or a neighbor of v' belongs to D'. We thus consider the scenario when $v' \notin L$.

If u' is blue, by our construction u' remains in D' and thus v' remains dominated by D'. If u' is red, let w' be the blue sibling of u'. By Corollary 1, we know that $N_{G'}(u') \subseteq N_{G'}(w')$. Therefore w' must be adjacent to v' as well. Notice that since L is loneliness-free, $w' \in L$. Since w' is blue, it remains in D' as well. Thus v' remains dominated by $w' \in D'$. $\qquad\square$

Lemma 14 implies that $\gamma^b(G') \leq \gamma_g(G') = \gamma(G')$. Since a blue dominating set is also a dominating set, $\gamma^b(G') = \gamma(G') = \gamma_g(G')$.

Lemma 15. *Let G be a graph without any isolated vertices. Let $G' = \overline{U_{\overline{G}}}$ be the complement of the UNNification of $\overline{G}$. Then $\gamma^b(G') = \gamma_g(G') = \gamma(G')$.*

By Lemma 6, $\gamma_g(G')$ corresponds a minimum-cost UNN extension of $U_{\overline{G}}$. Assume that there exists an algorithm $\mathcal{A}$ that produces a UNN extension H for $U_{\overline{G}}$ with cost $|N_H(v^*)|$, where v^* is the extension vertex. Then the approximation ratio for the minimum-cost UNN extension is $\frac{|N_H(v^*)|}{\gamma(G')}$. By Lemma 15, we have $\frac{|N_H(v^*)|}{\gamma(G')} = \frac{|N_H(v^*)|}{\gamma^b(G')} = \frac{|N_H(v^*)|}{\gamma_t(G)}$. By Lemma 14, we can construct a blue dominating set D' such that $|D'| \leq |N_H(v^*)|$. Therefore, $\frac{|D'|}{\gamma^b(G')} \leq \frac{|N_H(v^*)|}{\gamma(G')}$. By Lemma 11, we can construct a total dominating set D'' for G such that $|D''| = |D'|$. Hence $\frac{|D''|}{\gamma_t(G)} \leq \frac{|N_H(v^*)|}{\gamma(G')}$. Since the total dominating set problem is hard to approximate within a factor of $c \log n$, for some constant $c > 0$, we obtain the following.

Theorem 3. *The minimum-cost UNN extension problem cannot be approximated in polynomial time within a factor of $o(\log n)$ unless $\mathcal{P} = \mathcal{NP}$.*

6   Bounded-Treewidth UNNs

In this section we show that the min-cost UNN extension problem can be solved in linear time for bounded treewidth graphs. We refer to [5] for the definition of treewidth. Courcelle [4] showed that for bounded treewidth graphs, every decision problem definable in monadic second-order logic (MSOL) can be solved in time linear in the number of vertices. Furthermore, if an optimization problem can be modelled as searching for a set of vertices of minimum or maximum cardinality that satisfies some condition definable by MSOL, then it can also be solved in linear time [1].

By Lemma 6, finding the min-cost UNN extension is equivalent to finding a minimum good dominating set D of $\overline{G}$. We can model this problem in MSOL as minimizing the size of a set D that satisfies the formula $\text{DomSet}(\overline{G}, D) \wedge \text{Good}(\overline{G}, D)$. Here $\text{DomSet}(\overline{G}, D) = \forall_{v \in V} \exists_{u \in D} \neg adj(u, v)$ checks if D is a dominating set in $\overline{G}$ and $\text{Good}(\overline{G}, D) = \forall_{v \in V}(v \in D \vee (\exists_{u' \in V} adj(u', v) \wedge u' \notin D))$ verifies that D is good by checking if all vertices outside of D has a neighbor outside of D in G.

The following result is now a consequence of Courcelle's Theorem.

Theorem 4. *Let $G = (V, E)$ be a bounded treewidth UNN. One can find a UNN extension of minimum cost in linear time.*

7   Hardness of Approximation for the UNN Completion Problem

In this section we show that the UNN Completion Problem can not be approximated upto a factor of $o(\log n)$ in polynomial time unless $\mathcal{P} = \mathcal{NP}$ via strict reduction from the UNN extension problem.

Consider any UNN G. If we had a black box for solving the UNN completion problem, we could solve the UNN extension Problem for G by first constructing a graph $H = G \cup \{v^*\}$ for some $v^* \notin V(G)$ and then finding a UNN Completion H' of H using our black box.

Let $E' = E(H') \setminus E(H)$ be the set of edges inserted to H'. Let $E_1 \subseteq E'$ be the set of edges in E' that are incident on v^* and let D be the neighbors of v^* in

H'. Then $|D| = |E_1| \leq |E'|$. Since H' is a UNN, by Lemma 2 D is a dominating set of $\overline{H'} - \{v^*\}$. Since $\overline{H'} - \{v^*\}$ is a spanning subgraph of $\overline{G}$, therefore D is also a dominating set of $\overline{G}$. Then, by using the construction from Lemma 7, we can find a good dominating set D' of $\overline{G}$ such that $|D'| \leq |D|$ and use it to construct an UNN Extension G' of G with cost $|D'| \leq |D| \leq |E'|$. Then, the approximation ratio of the UNN extension G' is $\frac{|D'|}{\gamma_g(G)}$.

Let s^* denote the cost of a minimum UNN completion of H. Then, the approximation ratio of our UNN completion H' is $\frac{|E'|}{s^*}$. Note that any UNN extension of G with the extension vertex v^* is a valid UNN completion of H. So $\gamma_g(G) \geq s^*$. Thus, $\frac{|D'|}{\gamma_g(G)} \leq \frac{|E'|}{s^*}$. Thus our reduction is a strict reduction.

Theorem 5. *The minimum-cost UNN completion problem cannot be approximated in polynomial time within a factor of $o(\log n)$ unless $\mathcal{P} = \mathcal{NP}$.*

8 Conclusion

In this paper we have explored the minimum-cost UNN extension and UNN completion problems. We show that a minimum-cost UNN extension is $\mathcal{O}(\log n)$ approximable in polynomial time and cannot be approximated within a factor of $o(\log n)$ in polynomial time unless $\mathcal{P} = \mathcal{NP}$. Since total dominating set parameterized by the size of the solution is $W[2]$-hard [5] and our hardness reduction is also an FPT reduction, minimum-cost UNN extension is $W[2]$-hard when parameterized by solution size. However, we show that the problem is solvable in linear time for bounded treewidth graphs. We have also shown the minimum-cost UNN completion problem to be log-APX-hard. Furthermore, since minimum-cost UNN extension problem is $W[2]$-hard, our reduction establishes the same for the UNN completion problem. A natural direction for future research can be to find non-trivial approximation algorithm for the UNN completion problem.

Acknowledgements. The work of Debajyoti Mondal is supported in part by the Natural Sciences and Engineering Research Council of Canada (NSERC). The work of Md. Saidur Rahman and Sadia Sharmin is done under RISE Internal Research Grant 2021-01-06 and partly supported by "Basic Research Grant" of BUET. The work of Siam Habib is partially supported by a travel grant from "Dynamic Solution Innovators (DSi), House 177, Lane 2, New DOHS Mohakhali, Dhaka 1206, Bangladesh".

References

1. Arnborg, S., Lagergren, J., Seese, D.: Easy problems for tree-decomposable graphs. J. Algorithms **12**(2), 308–340 (1991)
2. Bull, J.J., Pease, C.M.: Combinatorics and variety of mating-type systems. Evolution 667–671 (1989)
3. Charbit, E., Charon, I., Cohen, G., Hudry, O.: Discriminating codes in bipartite graphs. Electron. Notes Discret. Math. **26**, 29–35 (2006)

4. Courcelle, B.: The monadic second-order logic of graphs. I. Recognizable sets of finite graphs. Inf. Comput. **85**(1), 12–75 (1990)

5. Cygan, M., et al.: Parameterized Algorithms, vol. 5. Springer (2015)

6. Dey, S., Foucaud, F., Nandy, S.C., Sen, A.: Discriminating codes in geometric setups. In: 31st International Symposium on Algorithms and Computation (ISAAC), pp. 24:1–24:16 (2020)

7. Erdős, P., Rényi, A.: On the evolution of random graphs. Publ. Math. Inst. Hung. Acad. Sci **5**(1), 17–60 (1960)

8. Garey, M.R., Johnson, D.S.: Computers and intractability. A Guide to the (1979)

9. Gessel, I.M., Li, J.: Enumeration of point-determining graphs. J. Comb. Theory Ser. A **118**(2), 591–612 (2011)

10. Gilbert, E.N.: Random graphs. Ann. Math. Stat. **30**(4), 1141–1144 (1959)

11. Goldberg, P.W., Golumbic, M.C., Kaplan, H., Shamir, R.: Four strikes against physical mapping of DNA. J. Comput. Biol. **2**(1), 139–152 (1995)

12. Honkala, I., Laihonen, T., Ranto, S.: On strongly identifying codes. Discret. Math. **254**(1–3), 191–205 (2002)

13. Johnson, D.S.: Approximation algorithms for combinatorial problems. In: Proceedings of the Fifth Annual ACM Symposium on Theory of Computing, pp. 38–49 (1973)

14. Karpovsky, M.G., Chakrabarty, K., Levitin, L.B.: On a new class of codes for identifying vertices in graphs. IEEE Trans. Inf. Theory **44**(2), 599–611 (1998)

15. Kashiwabara, T., Fujisawa, T.: An NP-complete problem on interval graphs. In: IEEE Symposium of Circuits and Systems, pp. 82–83 (1979)

16. Laifenfeld, M., Trachtenberg, A., Cohen, R., Starobinski, D.: Joint monitoring and routing in wireless sensor networks using robust identifying codes. Mob. Netw. Appl. **14**, 415–432 (2009)

17. Laskar, R., Pfaff, J., Hedetniemi, S.M., Hedetniemi, S.T.: On the algorithmic complexity of total domination. SIAM J. Algebraic Discrete Methods **5**(3), 420–425 (1984)

18. Natanzon, A., Shamir, R., Sharan, R.: Complexity classification of some edge modification problems. Discret. Appl. Math. **113**(1), 109–128 (2001)

19. Penrose, M.D.: Random Geometric Graphs. Oxford Studies in Probability, vol. 5. Oxford University Press, Oxford (2003)

20. Rass, S.: Perfectly secure communication, based on graph-topological addressing in unique-neighborhood networks (2023). https://arxiv.org/abs/1810.05602

21. Rass, S.: Tell me who you are friends with and I will tell you who you are: unique neighborhoods in random graphs. Theoret. Comput. Sci. **988**, 114386 (2024)

22. Ray, S., Starobinski, D., Trachtenberg, A., Ungrangsi, R.: Robust location detection with sensor networks. IEEE J. Sel. Areas Commun. **22**(6), 1016–1025 (2004)

23. Seo, S.J., Slater, P.J.: Open neighborhood locating dominating sets. Australas. J Comb. **46**, 109–120 (2010)

24. Yannakakis, M.: Computing the minimum fill-in is NP-complete. SIAM J. Algebraic Discrete Methods **2**(1), 77–79 (1981)

Approximating the Average-Case Graph Search Problem with Non-uniform Costs

Michał Szyfelbein$^{(\boxtimes)}$ (iD)

Gdańsk University of Technology, Gdańsk, Poland
`s193307@student.pg.edu.pl`

Abstract. Consider the following generalization of the classic binary search problem: A searcher is required to find a hidden target vertex x in a graph G. To do so, they iteratively perform queries to an oracle, each about a chosen vertex v. After each such call, the oracle responds whether the target was found and if not, the searcher receives as a reply the connected component in $G - v$ which contains x. Additionally, each vertex v may have a different query cost $c(v)$ and a different weight $w(v)$. The goal is to find the optimal querying strategy which minimizes the weighted average-case cost required to find x. The problem is NP-hard even for uniform weights and query costs. Inspired by the progress on the edge query variant of the problem [SODA '17], we establish a connection between searching and vertex separation. By doing so, we provide an $O(\sqrt{\log n})$-approximation algorithm for general graphs and a $(4 + \epsilon)$-approximation algorithm for the case when the input is a tree.

Keywords: Graph Searching · Binary Search · Decision Trees · Vertex Separators · Graph Theory · Approximation Algorithm

1 Introduction

Searching in graph structures is a fundamental problem in computer science, with applications ranging from machine learning to operations research. The input graph G is assumed to contain a hidden *target* element x which the searcher is required to locate. In the classical setting, while exploring the graph, the searcher usually obtains only a local information about the placement of x. This principle underlines the well-known BFS and DFS strategies, which allow the searcher to locate the target when the only information received, while visiting a node, is whether it is the target. We study a more global search model in which the searcher is allowed to perform arbitrary *queries*, each about a chosen vertex v. A *response* to such query consists of information whether $v = x$, and if not, which of the connected components of $G - v$ contains x. When the input graph is a path, this is equivalent to binary searching in a linearly ordered set. In case of trees, the problem was studied among multiple names including: Tree Search Problem

---

Full version available at https://arxiv.org/abs/2511.06564.

[7,8], Search Trees on Trees [3,4], Hub Labeling [1] and more. For general graphs the problem is known as Vertex Ranking [5,22,23,25], Elimination Trees [24] and Hierarchical Clustering [6,9,10].

In order to efficiently locate the target, the searcher needs a *strategy* of searching which allows them to locate the target efficiently. The strategy is an *adaptive* algorithm, which given previous responses in finite (and polynomial) time outputs the next query to be performed. We model this strategy as a rooted tree which we will call a *decision tree* D, whose nodes are vertices of G. We will require that each edge outgoing from the root r of D is associated with a unique response to a query to r in G and that the same holds for all decision subtrees of $D - r$ (in the respective components of $G - v$). When searching according to D, the root r of D is queried first. If $r \neq x$ the searcher moves down along the edge of r associated with the response, thereby entering the subtree D_u rooted at the next queried vertex u. This process recurses until the target is found.

We will allow each vertex $v \in V(G)$ to have an arbitrary *query cost* $c(v)$. Intuitively, $c(v)$ represents the number of time units required for a query to v to return an answer. We will also allow each vertex to have an arbitrary *weight* $w(v)$, which denotes how frequently the vertex is searched for. For a target vertex $x \in V(G)$ and a decision tree D, let $Q_G(D, x)$ denote the sequence of queries performed along the unique path in D from the root $r(D)$ to x. The *weighted average-case cost* of D on G is defined as:

$$c_G(D) = \sum_{x \in V(G)} w(x) \cdot \sum_{q \in Q_G(D,x)} c(q).$$

The *Graph Search Problem* (GSP) is to find the decision tree which minimizes the above cost.

1.1 Motivations and Applications

Fast retrieval of information in graph structures is a well-studied problem, beginning with the seminal work of Knuth [19]. When the underlying search space provides non-local information about the target, the search process can be accelerated, since each query rules out a large set of possible target locations. The goal of designing search strategies is to find ways of exploiting this property as effectively as possible.

Searching arises in various practical problems, but it can also be reformulated, to fit a wide range of real-life applications. Searching in graphs can be used to model a variety of problems, that may initially appear unrelated. These include: scheduling of parallel database join operations [14,20,21], automated bug detection in computer code [2,15,16,27], parallel Cholesky factorization of matrices [13], VLSI-layouts [26], hierarchical clustering of data [6,9,10] and parallel assembly of multi-part products from their components [11,18].

We focus on the average-case version of the problem rather than the worst-case, since it is natural to assume that the search strategies we design are intended to be used repeatedly. This motivates the introduction of a weight function, as some vertices may serve as targets more frequently than others. We also allow arbitrary query costs, since performing a query may require significant resources, such as time or money. To the best of our knowledge, this particular variant of the problem has not yet been investigated in the literature.

1.2   Organization of the Paper and Our Results

In Sect. 2, we introduce all necessary notions, preliminaries, and formal definitions required for the analysis, including decision trees and their cost (Sect. 2.1), as well as cuts and separators (Sect. 2.2).

In Sect. 3, we consider the case in which the input graph is a tree. Using a pseudoexact FPTAS for the Weighted α-Separator Problem, in Sect. 3.1 we present a $(4 + \epsilon)$-approximation algorithm running in $O(n^4/\epsilon^2)$ time.

In Sect. 4, we shift our focus to general graphs. Using the $O(\sqrt{\log n})$-approximation algorithm for the Min-Ratio Vertex Cut Problem from [17], we obtain a polynomial-time $O(\sqrt{\log n})$-approximation algorithm for the Graph Search Problem.

We also show that the problem is NP-hard even when restricted to trees with $\Delta(T) \leq 16$ and to trees with $\mathrm{diam}(T) \leq 8$. Due to the space limitations we omit the proof of this fact. For a complete version containing this hardness result, the aforementioned pseudoexact FPTAS for the Weighted α-separator problem and other excluded proofs, see the full version of the article.

2   Notions and Preliminaries

We assume that every graph G considered is simple and connected. By $uv \in E(G)$ we denote an edge connecting vertices u and v in G. Let $v \in V(G)$. By $G - v$ we denote the set of connected components resulting from deleting v from G. For a set $S \subseteq V(G)$, $G - S$ denotes the set of connected components resulting from deleting all vertices in S from G. For a family of subsets $\mathcal{F}$ of $V(G)$, we define $G - \mathcal{F} = G - \bigcup_{S \in \mathcal{F}} S$. The set of neighbors of a vertex $v \in V(G)$ is denoted by $N_G(v) = \{u \in V(G) : uv \in E(G)\}$, and the set of neighbors of a subgraph $\mathcal{G}$ of G is $N_G(\mathcal{G}) = \bigcup_{v \in V(\mathcal{G})} N_G(v) - V(\mathcal{G})$. For any function $f : V(G) \to \mathbb{N}$ and any set $S \subseteq V(G)$, we define $f(S) = \sum_{v \in S} f(v)$, and for a graph G, $f(G) = f(V(G))$.

Let T be a tree. If T is rooted, its root is denoted by $r(T)$. For any vertex $v \in V(T)$, let $\mathcal{C}_{T,v} = \left\{ c_1, c_2, \ldots, c_{\deg^+_{T,v}} \right\}$ be the set of children of v. By T_v we will denote the subtree of T rooted at v with all its descendants, and by $T_{v,i}$ we will denote the subtree of T consisting of v and $T_{c_1}, \ldots, T_{c_i}$. For a subset $S \subseteq V(T)$, $T\langle S \rangle$ denotes the minimal connected subtree of T containing all vertices in S.

A *Graph Search Instance* consists of a triple (G, c, w), where $c : V(G) \to \mathbb{N}$ is the cost of querying a vertex, and $w : V(G) \to \mathbb{N}$ is the weight of a vertex. During the *Search Process*, the searcher is allowed to iteratively perform *queries*, each asking about a chosen vertex v. After time $c(v)$, the query returns an answer. If the answer is affirmative, then v is the target, otherwise, a unique connected component of $G - v$ containing the target x is returned. Each answer allows the searcher to narrow the subset of vertices consistent with previous responses, and have not yet been eliminated as possible locations of x. We call such set a *candidate* set and the connected subgraph of G induced by it a *candidate* subgraph. This process continues until the position of the target is revealed (Fig. 1).

2.1 Decision Trees

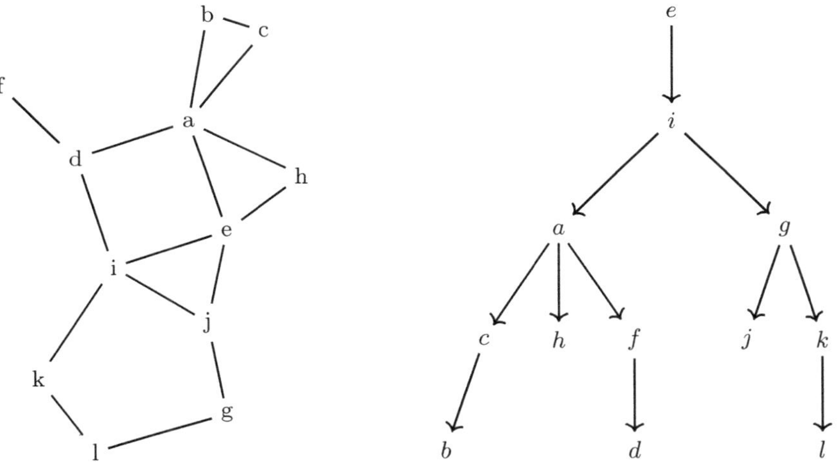

Fig. 1. Sample input graph (on the left) and a decision tree for it (on the right).

A decision tree is a pair $D = (V(D), E(D))$, where $V(D) = V(G)$ are vertices and $E(D)$ are edges of D. It is required that each child of $q \in V(D)$ corresponds to a distinct response to the query at q, with respect to the subtree of candidate solutions remaining after performing all previous queries.

Let $Q_G(D, x)$ denote the sequence of queries made to locate $x \in V(G)$ using decision tree D. We define the *Graph Search Problem* as follows:

Graph Search Problem (GSP)

Input: Graph G, a query cost function $c : V \to \mathbb{N}$ and a weight function $w : V \to \mathbb{N}$.

> **Output:** A decision tree D, minimizing the weighted average search cost:
> $$c_G(D) = \sum_{x \in V(G)} w(x) \cdot \sum_{q \in Q(D,x)} c(q).$$

If a given decision tree is optimal, we denote its cost by $\mathtt{OPT}(G)$. Let D' be a subtree of a valid decision tree D for T containing $r(D)$. We say that D' is a *partial* decision tree for D, and we define its cost analogously to that of D, although we only count the queries belonging to D'.

We also introduce the following reinterpretation of the cost function. For each node $v \in D$, let $G_{D,v}$ be the subgraph of G in which v is queried when using D. Then, the contribution of v to the total cost is $w(G_{D,v}) \cdot c(v)$, and therefore we have the following simple lemma:

Lemma 1.
$$c_G(D) = \sum_{v \in V(G)} w(G_{D,v}) \cdot c(v).$$

2.2   Cuts and Separators

To obtain a tight lower bound on the cost of our solution, we establish a connection between GSP and vertex cuts/separator problems. Our analysis is inspired by the edge query variant of the GSP [6]. However, since the query model is different, to obtain our results we use connection to the vertex separation rather then edge cutting. The *Weighted α-Separator Problem* is as follows:

> ### Weighted α-Separator Problem
>
> **Input:** Graph G, a cost function $c : V \to \mathbb{N}$, a weight function $w : V \to \mathbb{N}$ and a real number α.
> **Output:** A set $S \subseteq V(G)$ called α-*separator*, such that for every $H \in G - S$, $w(H) \leq w(G)/\alpha$ and $c(S)$ is minimized.

We also define the Min-Ratio Vertex Cut Problem:

> ### Min-Ratio Vertex Cut Problem
>
> **Input:** Graph $G = (V(G), E(G))$, the cost function $c : V \to \mathbb{N}$ and the weight function $w : V \to \mathbb{N}$.
> **Output:** A partition (A, S, B) of $V(G)$ called *vertex-cut*, such that there are no $u \in A$ and $v \in B$ for which $uv \in E(G)$, minimizing the ratio:
> $$\alpha_{c,w}(A, S, B) = \frac{c(S)}{w(A \cup S) \cdot w(B \cup S)}.$$

2.3 Levels of OPT and Basic Bounds

We begin with additional notation. For any graph G and decision tree D, denote by $\mathcal{R}_D(G) = \{V(G_{D,v}) : v \in V(G)\}$ the family of all candidate subsets of D in G.

Let D^* be an arbitrary decision tree for the Graph Search Problem such that $c_G(D^*) = \mathsf{OPT}(G)$. We denote by $\mathcal{L}_k^*$ the subfamily of $\mathcal{R}_{D^*}(G)$ consisting of all maximal elements H of $\mathcal{R}_{D^*}(G)$ with $w(H) \le k$, that is, if some superset H' of H belongs to $\mathcal{R}_{D^*}(G)$, then $w(H') > k$. We call such a set the k-th $level$ of $\mathsf{OPT}(G)$. Let $S_k^* = V(G) - \mathcal{L}_k^*$. These are the vertices belonging to the separator at the k-th level.

Notice that S_k^* forms a Weighted $w(G)/k$-separator of G. Furthermore, for any $H_1, H_2 \in \mathcal{R}_D(G)$, we have $H_1 \cup H_2 \neq \emptyset$ if and only if $H_1 \subseteq H_2$ or $H_2 \subseteq H_1$, so $\mathcal{R}_D(G)$ is laminar. Therefore, for any $k_1 \neq k_2$, we have $\mathcal{L}_{k_1}^* \cap \mathcal{L}_{k_2}^* = \emptyset$.

We begin with the following lemma, which allows us to decompose the value of $\mathsf{OPT}(G)$ into cost of the separators at each level:

Lemma 2.
$$\mathit{OPT}(G) = \sum_{k=0}^{w(G)-1} c(S_k^*).$$

Proof. Consider any vertex v. For every $0 \le k < w(G_{D^*,v})$, $v \notin \bigcup_{H \in \mathcal{L}_k^*} H$, so $v \in S_k^*$ and the contribution of v to the cost is $w(G_{D^*,v}) \cdot c(v)$:

$$\sum_{k=0}^{w(G)-1} c(S_k^*) = \sum_{v \in V(G)} \sum_{k=0}^{w(G_{D^*,v})-1} c(v) = \mathsf{OPT}(G)$$

where the second equality is by Lemma 1.

Using the above lemma one easily obtains the following lower bound on the cost of the optimal solution:

Lemma 3.
$$2 \cdot \mathit{OPT}(G) = 2 \cdot \sum_{k=0}^{w(T)-1} c(S_k^*) \ge \sum_{k=0}^{w(T)} c\left(S_{\lfloor k/2 \rfloor}^*\right).$$

Intuitively, the above reindexing allows us to lower bound the value of $\mathsf{OPT}(G)$ by the sum of minimal costs required to separate each candidate subgraph at some level k into candidate subgraphs on levels at most $\lfloor k/2 \rfloor$.

In order to upper bound the cost of decision trees produced by our algorithms we will use the following lemma:

Lemma 4. *Let $\mathcal{G}$ be any subgraph of G and $0 \le \beta \le 1$. Then:*

$$\beta \cdot w(\mathcal{G}) \cdot c\left(S_{\lfloor w(\mathcal{G})/2 \rfloor}^* \cap \mathcal{G}\right) \le \sum_{k=(1-\beta)w(\mathcal{G})+1}^{w(\mathcal{G})} c\left(S_{\lfloor k/2 \rfloor}^* \cap \mathcal{G}\right).$$

Proof. The inequality is due to the fact that as k decreases, more vertices belong to the separator.

We will also make use of the following lemma, which allows us to attach decision trees below already constructed partial decision trees without ambiguity:

Lemma 5. *Let $\mathcal{G}$ be a connected subgraph of a graph G and D be a partial decision tree for G having no queries to vertices of $\mathcal{G}$, but containing at least one query to $N_G(V(\mathcal{G}))$. Additionally, let Q be the set of all queries to vertices from $N_G(V(\mathcal{G}))$ in D. Then $D\langle Q \rangle$ forms a path in D.*

Proof. The proof of this lemma is available in the full version of the paper.

This will come useful in the following scenario: Let D be a partial decision tree for G satisfying the conditions of the lemma and $D_{\mathcal{G}}$ be any partial decision tree for a subgraph $\mathcal{G}$ of G. Since the queries in Q form a path, without ambiguity, we can attach $D_{\mathcal{G}}$ to D below the last query of Q, thereby obtaining a valid partial decision tree for G.

3   Searching in Trees

In this section, we present a $(4 + \epsilon)$-approximation algorithm for the case where the input graph is a tree. To achieve this, we establish a connection between searching in trees and the Weighted α-Separator Problem. This connection provides a lower-bounding scheme for our recursive algorithm, which at each level of recursion, constructs a decision tree using the α-separator obtained by the following procedure:

Theorem 1. *Let S be an optimal weighted α-separator for (T, c, w, α). For any $\delta > 0$ there exists an algorithm* `SeparatorFPTAS`*, which returns a separator S', such that:*

1. *$c(S') \leq c(S)$.*
2. *$w(H) \leq \frac{(1+\delta)\cdot w(T)}{\alpha}$ for every $H \in T - S'$.*
3. *The algorithm runs in $O(n^3/\delta^2)$ time.*

Proof. The algorithm and the proof are available in the full version of the paper.

3.1   How to Search in Trees

Below, we show how to use the `SeparatorFPTAS` procedure to construct a solution for the Tree Search Problem. At each level of the recursion, the algorithm greedily finds an (almost) optimal weighted α-separator of T, denoted S_T, and then builds an arbitrary decision tree D_T using the vertices in S_T (which can be done in $O(n^2)$ time).

Next, for each $H \in T - S_T$, the procedure is called recursively, and each resulting decision tree D_H is attached below the appropriate query in D_T. The resulting decision tree is then returned by the procedure (Fig. 2).

Theorem 2. *For any $\epsilon > 0$, there exists $(4+\epsilon)$-approximation algorithm for the Tree Search Problem running in time $O\left(n^4/\epsilon^2\right)$.*

Proof. The procedure is as follows:

Algorithm 1: The $(4+\epsilon)$-approximation algorithm for the Tree Search Problem

---

  Procedure DecisionTree(T, c, w, ϵ):

    $S_T \leftarrow$ SeparatorFPTAS$\left(T, c, w, \alpha = 2, \delta = \frac{\epsilon}{4+\epsilon}\right)$.

    $D_T \leftarrow$ arbitrary partial decision tree for T, built from vertices of S_T.

    foreach $H \in T - S_T$ **do**

      $D_H \leftarrow$ DecisionTree(H, c, w, ϵ).

      Hang D_H in D_T below the last query to $v \in N_T(H)$.

    return D_T.

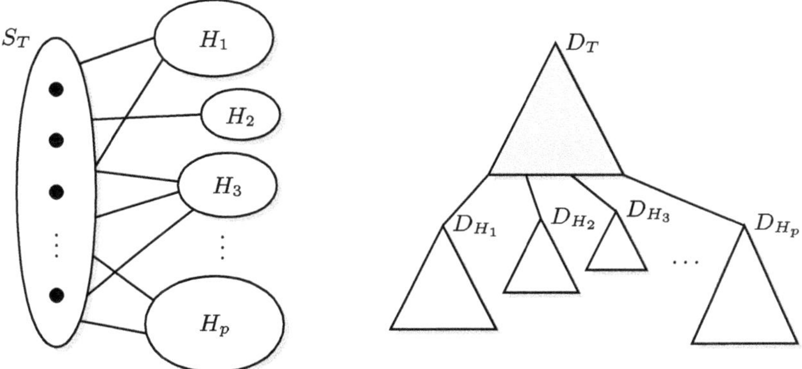

Fig. 2. The separator S_T produced by the algorithm and the structure of the decision tree built using S_T.

Let $\mathcal{T}$ be a subtree of T for which the procedure was called and let $S_{\mathcal{T}}^* = S_{\lfloor w(\mathcal{T})/2 \rfloor}^* \cap \mathcal{T}$. By Theorem 1, we have that $c(S_{\mathcal{T}}) \leq c(S_{\mathcal{T}}^*)$. Using $\beta = \frac{1-\delta}{2}$ and applying Lemma 4 we have that the contribution of the decision tree $D_{\mathcal{T}}$ is bounded by:

$$w(\mathcal{T}) \cdot c(S_{\mathcal{T}}) \leq w(\mathcal{T}) \cdot c(S_{\mathcal{T}}^*) \leq \frac{2}{1-\delta} \cdot \sum_{k=\frac{1+\delta}{2} \cdot w(\mathcal{T})+1}^{w(\mathcal{T})} c\left(S_{\lfloor k/2 \rfloor}^* \cap \mathcal{T}\right).$$

To bound the cost of the whole solution we will firstly show the following lemma which is necessary to proceed:

Lemma 6.

$$\sum_{\mathcal{T}} \sum_{k=\frac{1+\delta}{2}\cdot w(\mathcal{T})+1}^{w(\mathcal{T})} c\left(S^*_{\lfloor k/2 \rfloor} \cap \mathcal{T}\right) \le \sum_{k=0}^{w(\mathcal{T})} c\left(S^*_{\lfloor k/2 \rfloor}\right).$$

Proof. Fix a value of $\mathcal{T}$ and k. Their contribution to the cost is $c\left(S^*_{\lfloor k/2 \rfloor} \cap \mathcal{T}\right)$. Consider which candidate subtrees contribute such a term. As $S_{\mathcal{T}}$ is a weighted $\frac{2}{1+\delta}$-separator, we have that $\mathcal{T}$ is the minimal candidate subtree, such that $w(\mathcal{T}) \ge k \ge \frac{(1+\delta)\cdot w(\mathcal{T})}{2} + 1 > w(H)$, for every $H \in \mathcal{T} - S_{\mathcal{T}}$. This means that if for every $H \in \mathcal{T} - S_{\mathcal{T}}$, $w(H) < k$, then $\mathcal{T}$ contributes such a term. Since for all $H_1, H_2 \in \mathcal{T} - S_{\mathcal{T}}$ we have that $H_1 \cap H_2 = \emptyset$, $\left(S^*_{\lfloor k/2 \rfloor} \cap H_1\right) \cup \left(S^*_{\lfloor k/2 \rfloor} \cap H_2\right) = \emptyset$, the claim follows by summing over all values of k.

We are now ready to bound the cost of the solution. Let D be the decision tree returned by the procedure. Using the fact that by definition $\frac{4}{1-\delta} = 4 + \epsilon$, we have:

$$c_T(D) \le \sum_{\mathcal{T}} w(\mathcal{T}) \cdot c(S_{\mathcal{T}}) \le \frac{2}{1-\delta} \cdot \sum_{\mathcal{T}} \sum_{k=\frac{1+\delta}{2}\cdot w(\mathcal{T})+1}^{w(\mathcal{T})} c\left(S^*_{\lfloor k/2 \rfloor} \cap \mathcal{T}\right)$$

$$\le \frac{2}{1-\delta} \cdot \sum_{k=0}^{w(\mathcal{T})} c\left(S^*_{\lfloor k/2 \rfloor}\right) \le \frac{4}{1-\delta} \cdot \mathtt{OPT}(T) = (4+\epsilon) \cdot \mathtt{OPT}(T)$$

where the third inequality is due to Lemma 6 and the last inequality is by Lemma 3.

As $1/\delta = \frac{4+\epsilon}{\epsilon} = 1 + 4/\epsilon$ and each $v \in V(T)$ belongs to the set $S_{\mathcal{T}}$ exactly once, we have that the overall running time is at most $O(n^4/\epsilon^2)$ as required.

4   Searching in General Graphs

To construct decision trees for general graphs we exploit a connection to a different problem, namely the Min-Ratio Vertex Cut Problem. Let $\alpha_{c,w}(G)$ denote the optimal value of such a vertex cut. We invoke the following result:

Theorem 3 ([17])**.** *Given a graph $G = (V(G), E(G))$, the cost function $c : V \to \mathbb{N}$ and the weight function $w : V \to \mathbb{N}$, there exists a polynomial-time algorithm, which computes a partition (A, S, B), such that:*

$$\alpha_{c,w}(A, S, B) = O\left(\sqrt{\log n}\right) \cdot \alpha_{c,w}(G).$$

Combining the latter procedure, with the following result, yields an $O(\sqrt{\log n})$-approximation algorithm for the Graph Search Problem:

Theorem 4. *Let f_n be the approximation ratio of any polynomial time algorithm for the Min-Ratio Vertex Cut Problem. Then, there exists an $O(f_n)$-approximation algorithm for the GSP, running in polynomial time.*

Proof. Let `AlgorithmMinCut` be the procedure which achieves the f_n-approximation ratio for the Min-Ratio Vertex Cut Problem. The algorithm is as follows:

Algorithm 2: The f_n-approximation algorithm for the Graph Search Problem.

> **Procedure** `DecisionTree`(G, c, w)**:**
>> $A_G, S_G, B_G \leftarrow$ `AlgorithmMinCut`(G, c, w).
>> $D_G \leftarrow$ arbitrary partial decision tree for G, built from vertices of S_G.
>> **foreach** $H \in G - S_G$ **do**
>>> $D_H \leftarrow$ `DecisionTree`(H, c, w).
>>> Hang D_H in D_G below the last query to $v \in N_G(H)$.
>>
>> **return** D_G.

Let $\mathcal{G}$ be any subgraph of G, for which the procedure was called and let $S_{\mathcal{G}}^* = S_{\lfloor w(\mathcal{G})/2 \rfloor}^* \cap \mathcal{G}$. The next lemma, shows that we can use $S_{\mathcal{G}}^*$ and $\mathcal{G} - S_{\mathcal{G}}^*$ to build a vertex cut of $\mathcal{G}$, thus providing us with an upper bound on the cost of the solution built by the algorithm:

Lemma 7. *Let $\mathcal{H} = \mathcal{G} - S_{\mathcal{G}}^*$ and let $\lambda = 6 + 2\sqrt{5}$ be the unique, positive solution of the equation $\frac{1}{4} - \frac{1}{2\sqrt{\lambda}} = \frac{1}{\lambda}$. Then, we can partition $\mathcal{H}$ into two sets, $\mathcal{A}$ and $\mathcal{B}$ such that for $A = \bigcup_{H \in \mathcal{A}} V(H)$ and $B = \bigcup_{H \in \mathcal{B}} V(H)$, we have:*

$$w\left(A \cup S_{\mathcal{G}}^*\right) \cdot w\left(B \cup S_{\mathcal{G}}^*\right) \geq w(\mathcal{G})^2 / \lambda.$$

Proof. The proof of this lemma is available in the full version of the paper.

Using the fact, that the partition $\left(A, S_{\mathcal{G}}^*, B\right)$ in the above lemma is a vertex cut of $\mathcal{G}$, we have the following upper bound on the optimal value of the min-ratio-vertex cut of $\mathcal{G}$, $\alpha_{c,w}(\mathcal{G})$:

$$\alpha_{c,w}(\mathcal{G}) \leq \alpha_{c,w}\left(A, S_{\mathcal{G}}^*, B\right) = \frac{c\left(S_{\mathcal{G}}^*\right)}{w\left(A \cup S_{\mathcal{G}}^*\right) \cdot w\left(B \cup S_{\mathcal{G}}^*\right)} \leq \frac{\lambda \cdot c\left(S_{\mathcal{G}}^*\right)}{w(\mathcal{G})^2}.$$

Let $(A_{\mathcal{G}}, S_{\mathcal{G}}, B_{\mathcal{G}})$, be the partition returned by `AlgorithmMinCut`$(\mathcal{G}, c, w)$. Using Theorem 3, we get that:

$$\alpha_{c,w}(A_{\mathcal{G}}, S_{\mathcal{G}}, B_{\mathcal{G}}) = \frac{c(S_{\mathcal{G}})}{w(A_{\mathcal{G}} \cup S_{\mathcal{G}}) \cdot w(B_{\mathcal{G}} \cup S_{\mathcal{G}})} \leq f_n \cdot \frac{\lambda \cdot c\left(S_{\mathcal{G}}^*\right)}{w(\mathcal{G})^2}.$$

Without loss of generality we may assume that $w(A_{\mathcal{G}}) \geq w(B_{\mathcal{G}})$. Let $\beta = w(B_{\mathcal{G}} \cup S_{\mathcal{G}}) / w(\mathcal{G})$. We have that $(1 - \beta) \cdot w(\mathcal{G}) = w(A_{\mathcal{G}})$, so we conclude that the contribution of the decision tree $D_{\mathcal{G}}$ is bounded by:

$$w(\mathcal{G}) \cdot c(S_{\mathcal{G}}) \leq \lambda \cdot f_n \cdot \frac{w(A_{\mathcal{G}} \cup S_{\mathcal{G}}) \cdot w(B_{\mathcal{G}} \cup S_{\mathcal{G}})}{w(\mathcal{G})} \cdot c(S_{\mathcal{G}}^*)$$

$$\leq \lambda \cdot f_n \cdot w(B_{\mathcal{G}} \cup S_{\mathcal{G}}) \cdot c(S_{\mathcal{G}}^*) \leq \lambda \cdot f_n \cdot \sum_{k=w(A_{\mathcal{G}})+1}^{w(\mathcal{G})} c\left(S_{\lfloor k/2 \rfloor}^* \cap \mathcal{G}\right)$$

where the last inequality is by Lemma 4.

As before, to bound the cost of the whole solution we will firstly show the following lemma. The argument is mostly the same as for the Lemma 6, however, there are few differences and we include it for completeness:

Lemma 8.

$$\sum_{\mathcal{G}} \sum_{k=w(A_{\mathcal{G}})+1}^{w(\mathcal{G})} c\left(S_{\lfloor k/2 \rfloor}^* \cap \mathcal{G}\right) \leq \sum_{k=0}^{w(G)} c\left(S_{\lfloor k/2 \rfloor}^*\right).$$

Proof. The proof of this lemma is available in the full version of the paper.

We are now ready to bound the cost of the solution. Let D be the decision tree returned by the procedure. We have:

$$c_G(D) \leq \sum_{\mathcal{G}} w(\mathcal{G}) \cdot c(S_{\mathcal{G}})$$

$$\leq \lambda \cdot f_n \cdot \sum_{\mathcal{G}} \sum_{k=w(A_{\mathcal{G}})+1}^{w(\mathcal{G})} c\left(S_{\lfloor k/2 \rfloor}^* \cap \mathcal{G}\right) \leq \lambda \cdot f_n \cdot \sum_{k=0}^{w(G)} c\left(S_{\lfloor k/2 \rfloor}^*\right)$$

$$\leq 2 \cdot \lambda \cdot f_n \cdot \mathtt{OPT}(G) = \left(12 + 4\sqrt{5}\right) \cdot f_n \cdot \mathtt{OPT}(G)$$

where the third inequality is due to Lemma 8 and the last inequality is by Lemma 3.

5   Conclusions and Further Work

We have presented an $O\left(\sqrt{\log n}\right)$-approximation algorithm for the general Graph Search Problem, and a $(4 + \epsilon)$-approximation algorithm for the case when the input graph is a tree. To the best of our knowledge, these are the first approximation results for the vertex query and non-uniform costs variant of the problem.

Moreover, the $(4 + \epsilon)$-approximation algorithm for trees and its analysis can easily be adapted to obtain a $(4 + \epsilon)$-approximation for the edge-query variant, thus improving the 6.75-approximation being a direct corollary of [6,9,10].

Both presented algorithms behave greedily with respect to the cut/separator found by the respective subprocedure. The construction of a partial decision tree at each level of the recursion is currently arbitrary. A possible further improvement would be to replace this step with a more sophisticated procedure that reduces the approximation ratio, particularly in the case where the input graph is a tree.

The advantage of our solution lies in its simplicity, making it suitable for practical applications. This also places the average-case version of the problem in stark contrast to the worst-case variant, for which achieving the state-of-the-art $O(\sqrt{\log n})$-approximation for trees required a non-trivial dynamic programming procedure and an intricate analysis [12]. An open question is whether a constant-factor approximation exists for the latter as well.

The current state-of-the-art algorithm for the Min-Ratio Vertex Cut, which we use as an essential subroutine for our GSP algorithm, is based on an SDP relaxation of the problem. An interesting direction for future research is to formulate a similar relaxation for GSP itself and to prove that it has an integrality gap of at most $O(\sqrt{\log n})$. Another question is whether the approximation ratio for GSP is inherently limited by the approximation ratio of Min-Ratio Vertex Cut Problem, or if there exists a method to improve the approximation without relying on vertex cuts.

On the hardness side, we have shown that the problem is NP-hard even when restricted to trees with $\Delta(T) \leq 16$ and to trees with $\mathrm{diam}(T) \leq 8$. Note, that for general graphs, it can be argued that the problem cannot be approximated within any constant factor unless the Small Set Expansion Hypothesis fails, by reusing the argument for the edge-query variant provided in [6]. An open question is whether one can establish an inapproximability result for some constant $c \leq 4$ in the case when the input graph is a tree.

Acknowledgments. The author would like to thank Dariusz Dereniowski for guidance and preliminary discussions.

Disclosure of Interests. The authors have no competing interests to declare that are relevant to the content of this article.

References

1. Angelidakis, H.: Shortest path queries, graph partitioning and covering problems in worst and beyond worst case settings. arXiv abs/1807.09389 (2018). https://api.semanticscholar.org/CorpusID:51718679
2. Ben-Asher, Y., Farchi, E., Newman, I.: Optimal search in trees. SIAM J. Comput. **28**(6), 2090–2102 (1999). https://doi.org/10.1137/S009753979731858X
3. Berendsohn, B., Golinsky, I., Kaplan, H., Kozma, L.: Fast approximation of search trees on trees with centroid trees (2022). https://doi.org/10.48550/arXiv.2209.08024
4. Berendsohn, B., Kozma, L.: Splay trees on trees, pp. 1875–1900 (2022). https://doi.org/10.1137/1.9781611977073.75
5. Bodlaender, H.L., et al.: Rankings of graphs. SIAM J. Discret. Math. **11**(1), 168–181 (1998). https://doi.org/10.1137/S0895480195282550
6. Charikar, M., Chatziafratis, V.: Approximate hierarchical clustering via sparsest cut and spreading metrics. In: Proceedings of the Twenty-Eighth Annual ACM-SIAM Symposium on Discrete Algorithms, SODA 2017, pp. 841–854. Society for Industrial and Applied Mathematics, USA (2017)

7. Cicalese, F., Jacobs, T., Laber, E., Molinaro, M.: Improved approximation algorithms for the average-case tree searching problem. Algorithmica **68**(4), 1045–1074 (2012). https://doi.org/10.1007/s00453-012-9715-6

8. Cicalese, F., Keszegh, B., Lidický, B., Pálvölgyi, D., Valla, T.: On the tree search problem with non-uniform costs. Theoret. Comput. Sci. **647**, 22–32 (2016). https://doi.org/10.1016/j.tcs.2016.07.019

9. Cohen-addad, V., Kanade, V., Mallmann-trenn, F., Mathieu, C.: Hierarchical clustering: objective functions and algorithms. J. ACM **66**(4) (2019). https://doi.org/10.1145/3321386

10. Dasgupta, S.: A cost function for similarity-based hierarchical clustering. In: Proceedings of the Forty-Eighth Annual ACM Symposium on Theory of Computing, STOC 2016, pp. 118–127. Association for Computing Machinery, New York (2016). https://doi.org/10.1145/2897518.2897527

11. Dereniowski, D.: Edge ranking of weighted trees. Discret. Appl. Math. **154**(8), 1198–1209 (2006). https://doi.org/10.1016/j.dam.2005.11.005

12. Dereniowski, D., Kosowski, A., Uznanski, P., Zou, M.: Approximation strategies for generalized binary search in weighted trees. In: Chatzigiannakis, I., Indyk, P., Kuhn, F., Muscholl, A. (eds.) 44th International Colloquium on Automata, Languages, and Programming (ICALP 2017). Leibniz International Proceedings in Informatics (LIPIcs), vol. 80, pp. 84:1–84:14. Schloss Dagstuhl – Leibniz-Zentrum für Informatik, Dagstuhl, Germany (2017). https://doi.org/10.4230/LIPIcs.ICALP.2017.84

13. Dereniowski, D., Kubale, M.: Cholesky factorization of matrices in parallel and ranking of graphs. In: Wyrzykowski, R., Dongarra, J., Paprzycki, M., Waśniewski, J. (eds.) Parallel Processing and Applied Mathematics, pp. 985–992. Springer, Heidelberg (2004)

14. Dereniowski, D., Kubale, M.: Efficient parallel query processing by graph ranking. Fundam. Inform. **69**, 273–285 (2006). https://doi.org/10.3233/FUN-2006-69302

15. Dereniowski, D., Wrosz, I.: Constant-factor approximation algorithm for binary search in trees with monotonic query times. In: Szeider, S., Ganian, R., Silva, A. (eds.) 47th International Symposium on Mathematical Foundations of Computer Science (MFCS 2022). Leibniz International Proceedings in Informatics (LIPIcs), vol. 241, pp. 42:1–42:15. Schloss Dagstuhl – Leibniz-Zentrum für Informatik, Dagstuhl, Germany (2022). https://doi.org/10.4230/LIPIcs.MFCS.2022.42

16. Dereniowski, D., Wrosz, I.: Searching in trees with monotonic query times (2024). https://arxiv.org/abs/2401.13747

17. Feige, U., Hajiaghayi, M., Lee, J.R.: Improved approximation algorithms for minimum-weight vertex separators. In: Proceedings of the Thirty-Seventh Annual ACM Symposium on Theory of Computing, STOC 2005, pp. 563–572. Association for Computing Machinery, New York (2005). https://doi.org/10.1145/1060590.1060674

18. Iyer, A.V., Ratliff, H.D., Vijayan, G.: Parallel assembly of modular products—an analysis. Technical Report 88-06, Georgia Institute of Technology, Atlanta, Georgia (1988)

19. Knuth, D.: The Art of Computer Programming, vol. 3: Sorting and Searching. Addison-Wesley (1973)

20. Makino, K., Uno, Y., Ibaraki, T.: On minimum edge ranking spanning trees. J. Algorithms **38**(2), 411–437 (2001). https://doi.org/10.1006/jagm.2000.1143. https://www.sciencedirect.com/science/article/pii/S019667740091143X

21. Makino, K., Uno, Y., Ibaraki, T.: Minimum edge ranking spanning trees of threshold graphs. In: Bose, P., Morin, P. (eds.) Algorithms and Computation, pp. 428–440. Springer, Heidelberg (2002)
22. Mozes, S., Onak, K., Weimann, O.: Finding an optimal tree searching strategy in linear time. In: Teng, S. (ed.) Proceedings of the Nineteenth Annual ACM-SIAM Symposium on Discrete Algorithms, SODA 2008, San Francisco, California, USA, 20–22 January 2008, pp. 1096–1105. SIAM (2008). http://dl.acm.org/citation.cfm?id=1347082.1347202
23. Onak, K., Parys, P.: Generalization of binary search: searching in trees and forest-like partial orders. In: 2006 47th Annual IEEE Symposium on Foundations of Computer Science (FOCS 2006), pp. 379–388 (2006). https://doi.org/10.1109/FOCS.2006.32
24. Pothen, A.: The complexity of optimal elimination trees. Technical Report CS-88-16, Pennsylvania State University, Department of Computer Science, Penn State University CS-88-16 (1988)
25. Schäffer, A.A.: Optimal node ranking of trees in linear time. Inf. Process. Lett. **33**(2), 91–96 (1989). https://doi.org/10.1016/0020-0190(89)90161-0
26. Sen, A., Deng, H., Guha, S.: On a graph partition problem with application to VLSI layout. Inf. Process. Lett. **43**(2), 87–94 (1992). https://doi.org/10.1016/0020-0190(92)90017-P. https://www.sciencedirect.com/science/article/pii/002001909290017P
27. Szyfelbein, M.: Searching in trees with k-up-modular cost functions (2025). https://arxiv.org/abs/2504.17887

Linear Time Small Coresets for k-Mean Clustering of Segments with Applications

David Denisov[1]($\boxtimes$), Shlomi Dolev[2], Dan Feldman[1], and Michael Segal[2]

[1] University of Haifa, Haifa, Israel
daviddenisovphd@gmail.com
[2] Ben-Gurion University of the Negev, Beer-Sheva, Israel

Abstract. We study the k-means problem for a set $\mathcal{S} \subseteq \mathbb{R}^d$ of n segments, aiming to find k centers $X \subseteq \mathbb{R}^d$ that minimize $D(\mathcal{S}, X) := \sum_{S \in \mathcal{S}} \min_{x \in X} D(S, x)$, where $D(S, x) := \int_{p \in S} |p - x| dp$ measures the total distance from each point along a segment to a center. Variants of this problem include handling outliers, employing alternative distance functions such as M-estimators, weighting distances to achieve balanced clustering, or enforcing unique cluster assignments. For any $\varepsilon > 0$, an ε-coreset is a weighted subset $C \subseteq \mathbb{R}^d$ that approximates $D(\mathcal{S}, X)$ within a factor of $1 \pm \varepsilon$ for any set of k centers, enabling efficient streaming, distributed, or parallel computation. We propose the first coreset construction that provably handles arbitrary input segments. For constant k and ε, it produces a coreset of size $O(\log^2 n)$ computable in $O(nd)$ time. Experiments, including a real-time video tracking application, demonstrate substantial speedups with minimal loss in clustering accuracy, confirming both the practical efficiency and theoretical guarantees of our method.

Keywords: Clustering · k-means · Segment clustering · Non-convex optimization · Coresets · Approximation algorithms · Video tracking

Due to space limit see the full version of this paper, including additional experiments and all the omitted proofs, in https://arxiv.org/abs/2511.12564.

1 Introduction

Segment clustering generalizes point and line clustering and arises in applications such as video tracking, facility location, and spatial data analysis. While point clustering is well-studied [1] and line clustering has a few provable approximations [18], no provable algorithms exist for segments. Segments are more challenging because they represent continuous sets of points, making the loss function an integral rather than a sum. Nevertheless, segments naturally model real-world structures such as maps, road networks, and image features. Efficient segment clustering enables the scalable analysis of such structured data, supporting tasks such as object tracking, pattern detection, and optimization over

E. Di Giacomo and D. Mondal (Eds.): WALCOM 2026, LNCS 16444, pp. 110–124, 2026.
https://doi.org/10.1007/978-981-95-7127-7_8

vectorized inputs. In this work, we introduce novel compression and approximation algorithms for segment clustering that provide both theoretical guarantees and practical performance, extending classical point-based methods to structured, high-dimensional inputs. Applications of our methods are demonstrated in Sect. 4 through a real-time video tracking system leveraging segment coresets for fast and accurate processing.

1.1 Segment Clustering

To formalize the problem, we introduce the following definitions and notations.

Definition 1. *For two given vectors u, v in $\mathbb{R}^d$, a segment is a function $\ell :$ $[0,1] \to \mathbb{R}^d$ that is defined by $\ell(x) = u + vx$, for every $x \in [0,1]$. For every pair $(p, p') \in \mathbb{R}^d \times \mathbb{R}^d$ of points, let $D(p, p') := \|p - p'\|_2$ denote the Euclidean distance between p and p'.*

Formally, we generalize the classical k-means problem to segments, defining the segment clustering problem as follows.

Problem 1 (Segment clustering). Let S be a set of n segments in $\mathbb{R}^d$, and let $k \geq 1$ be an integer. Let $\mathcal{Q}$ be the set of all pairs (C, w), where C is a set of $|C| = k$ points in $\mathbb{R}^d$ called *centers*, and $w : C \to [0, \infty)$ is a weight function representing the influence of each center. The *k-segment mean* of S, which we aim to find, is a set $(C, w) \in \mathcal{Q}$ that minimizes

$$\text{loss}(C, w) := \sum_{s \in S} \int_0^1 \min_{c \in C} w(c) D\big(c, s(x)\big) \, dx. \tag{1}$$

The integral exists due to the continuity of the Euclidean distance. The corresponding loss loss is illustrated in Fig. 1.

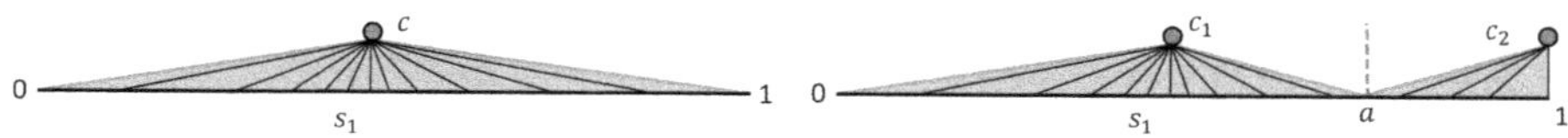

Fig. 1. Visual illustration of Problem 1. Left: the loss of a single center $c \in \mathbb{R}^2$ to a segment s_1, with the green area representing the integrated distance. Right: the loss for two centers $C = \{c_1, c_2\}$, where the green area is split at $a \in (0, 1)$, assigning points closer to c_1 or c_2 accordingly. (Color figure online)

In this paper, we aim to derive a provable approximation to Problem 1, i.e., the k-segment mean, which has broad applicability as demonstrated in our experiments. Our results naturally extend to the setting where each segment is assigned to a single cluster. The compression remains the same, but solving the problem over the compressed set corresponds to set clustering [16], which is less straightforward than k-means; in contrast, k-means (e.g., via k-means++ [1]) is efficiently implemented in many standard libraries.

1.2 Coresets

Informally, for an input set L of segments, a coreset C is a (weighted) set of points that approximates the loss $\mathrm{loss}(L, (Q, w))$ for any weighted set (Q, w) of size k up to a factor of $1 \pm \varepsilon$, with computation time depending only on $|C|$. Ideally, C is much smaller than L, making it efficient. For more on the motivation behind coresets, see "Why coresets?" in [8,17].

Coreset for Segments. The segment clustering problems from Sects. 1.1 and 2.1 admit a small coreset, a weighted subset of points sampled from the segments. While many coresets are subsets of the input, there are exceptions [15,24], including ours, which is a weighted set of points rather than the original segments. There is extensive work on fitting weighted centers to weighted points [1,10], but little on fitting points to segments or handling non-discrete integrals. Thus, a coreset represented as weighted points provides a practical and usable solution for segment clustering.

Coreset for Convex Sets and Hyperplane Fitting. Convex sets can be seen as unions of infinitely many segments, which allows us to construct small coresets for $n \geq 1$ convex sets (not given due to lack of space). Moreover, our focus on fitting weighted centers naturally extends the approach to hyperplanes [10].

Relation to Our Goal. We aim to derive a provable approximation to the k-segment mean. Given the extensive work on fitting weighted centers to weighted points, we compute a coreset as a weighted set of points. Any provable or optimal solution over this coreset directly yields an approximation to the original problem. A formal statement appears in Corollary 1.

1.3 Novelty

Our main contribution is a provable reduction of the segment clustering problem to point clustering, yielding a coreset construction via prior work [2,10]. For a function $f : \mathbb{R} \to \mathbb{R}$ and integer $n \geq 1$, we call $\sum_{i=1}^{n} f(i/n)$ an n-*discrete integral* of f over $[0, 1]$.

Discrete Integrals. Previous work [12] shows that a weighted subset of size $O(\log n/\varepsilon^2)$ approximates the n-discrete integral for certain functions. We generalize this to the continuous integral $\int_0^1 f(x)dx$, yielding the first provably small coreset for this setting. Applying [12] directly to a segment would yield an infinite coreset, whereas we provide a finite one (with proven guarantees).

Riemann Sums. Our approach is related to Riemann sums [22], but unlike prior work, we provide a hard bound for finite samples (Theorem 3).

Relation to Prior Coreset Results. Theorem 3 generalizes previous coreset constructions for points [10] to handle infinitely many points along segments. Similarly, while [18] constructs coresets for lines. However, their approach does not trivially extend to segments.

Paper Structure

Section 2 presents our main theoretical results and algorithms. Section 3 compares our approach to previous line-clustering work [18], and Sect. 4 demonstrates a video tracking and detection method based on our results.

2   Theoretical Results

In the following section, we present our main theoretical results, which essentially outline an efficient data reduction scheme for Problem 1 defined in the previous section. Note that the loss function we define supports cluster weighting and a distance function as in [10], which allows immediate generalization of the support for M-estimators, outlier support, and more from [10].

2.1   Notation and Definitions

Notations. Throughout this paper, we assume that $k, d \geq 1$ are integers and denote by $\mathbb{R}^d$ the set of all d-dimensional real vectors. A *weighted set* is a pair (P, w) where P is a finite set of points in $\mathbb{R}^d$ and $w : P \to [0, \infty)$ is a *weight function*. For simplicity, we denote $\log(x) := \log_2(x)$.

Definition 2. *Let $r \geq 0$, and let $h : \mathbb{R} \to [0, \infty)$ be a non-decreasing function. We say that h is r-log-Lipschitz iff, for every $c \geq 1$ and $x \geq 0$, we have $h(cx) \leq c^r h(x)$.*

Definition 3. *A symmetric-r function is a function $f : \mathbb{R} \to \mathbb{R}$, such that there is $a \in \mathbb{R}$ and an r-log-Lipschitz function $\tilde{f} : [0, \infty) \to [0, \infty)$, that for every $x \in \mathbb{R}$ satisfies $f(x) = \tilde{f}(|x - a|)$.*

In the following definitions we define lip, a global r-log-Lipschitz function, with $r, t, d^* \in [0, \infty)$, and D, a distance function; those definitions are inspired by [10].

Definition 4 (global parameters). *Let* $\mathrm{lip} : \mathbb{R} \to \mathbb{R}$ *be a symmetric-r function for some $r \geq 0$, and suppose that for every $x \in [0, \infty)$ we can compute $\mathrm{lip}(x)$ in t time, for some integer $t \geq 1$; see Definition 2. For every weighted set (C, w) of size $|C| = k$ and $p \in \mathbb{R}^d$ let $D((C, w), p) := \min_{c \in C} \mathrm{lip}(w(c)D(c, p))$.*

Definition 5 (VC-dimension). *For every $P \subset \mathbb{R}^d, r \geq 0$ and any weighted set (C, w) of size k let*

$$\mathrm{ball}(P, (C, w), r) := \left\{ p \in P \mid D((C, w), p) \leq r \right\}.$$

Let $B := \left\{ \mathrm{ball}(P, (C, w), r) \mid (P, r, (C, w)) \in \tilde{Q} \right\}$, where $\tilde{Q}$ is the union over every $P \subset \mathbb{R}^d, r \geq 0$ and any weighted set (C, w) of size k. Let d^* denote the VC-dimension associated with lip, which is the smallest positive integer such that for every finite $S \subset \mathbb{R}^d$ we have

$$\left| \left\{ (S \cap \beta) \mid \beta \in B \right\} \right| \leq |S|^{d^*}.$$

We emphasize that the choice of the distance function D dictates the appropriate values of r, t, d^*.

Utilizing this definition, we define an algorithm, as stated in the following theorem, which follows from Theorem 5.1 of [10].

Theorem 1. *Let $k' \in (k+1)^{O(k)}$. There is an algorithm* CORE-SET$(P, k, \varepsilon, \delta)$ *that gets: (i) A set P of points in $\mathbb{R}^d$. (ii) An integer $k \geq 1$. (iii) Input parameters $\varepsilon, \delta \in (0, 1/10)$, such that its output $(S, w) :=$ CORE-SET$(P, k, \varepsilon, \delta)$ is a weighted set satisfying Claims (i)–(iii) as follows:*

(i) With probability at least $1 - \delta$, for every weighted set (C, w) of size $|C| = k$, we have

$$\left| \sum_{p \in P} D((C, w'), p) - \sum_{p \in S} \left(w(p) D((C, w'), p) \right) \right| \leq \varepsilon \cdot \sum_{p \in P} D((C, w'), p).$$

(ii) $|S| \in \dfrac{k' \cdot \log^2 n}{\varepsilon^2} \cdot O\left(d^* + \log\left(\dfrac{1}{\delta} \right) \right).$

(iii) The computation time of a call to CORE-SET$(P, k, \varepsilon, \delta)$ *is in*

$$ntk' + tk' \log(n) \cdot \log\left(\log(n)/\delta \right)^2 + \frac{k' \log^3 n}{\varepsilon^2} \cdot \left(d^* + \log\left(\frac{1}{\delta} \right) \right).$$

Definition 6 (Loss function). *Let $\ell : [0, 1] \to \mathbb{R}^d$ be a segment; see Definition 1. We define the* fitting loss *of a weighted set (C, w) of size k, as*

$$\text{loss}((C, w), \ell) = \int_0^1 D((C, w), \ell(x)) dx.$$

Given a finite set L of segments and a weighted set (C, w) of size $|C| = k$, we define the loss of fitting (C, w) to L as

$$\text{loss}((C, w), L) = \sum_{\ell \in L} \text{loss}((C, w), \ell). \tag{2}$$

Given such a set L, the goal is to recover a weighted set (C, w) of size k that minimizes $\text{loss}((C, w), \ell)$.

Definition 7 ((ε, k)-coreset). *Let ℓ be a segment, and let $\varepsilon > 0$ be an error parameter; see Definition 1. A weighted set (S, w) is an (ε, k)-coreset for ℓ if for every weighted set $Q := (C, w)$ of size $|C| = k$ we have*

$$\left| \text{loss}(Q, \ell) - \sum_{p \in S} w(p) D(Q, p) \right| \leq \varepsilon \cdot \text{loss}(Q, \ell).$$ *More generally a weighted set (S, w) is an (ε, k)-coreset for a set L of segments if for every weighted set $Q := (C, w)$ of size $|C| = k$ we have*

$$\left| \text{loss}(Q, L) - \sum_{p \in S} w(p) \cdot D(Q, p) \right| \leq \varepsilon \cdot \text{loss}(Q, L).$$

2.2   Algorithms

Algorithm 1. Algorithm 1 and its corresponding theorem (Theorem 2) are our main theoretical results. The algorithm provides a simple deterministic coreset for the single-segment case ($n = 1$) in Definition 6 (Eq. (2)). The parameter r depends on the distance function D (e.g., $r = 1$ for absolute error, $r = 2$ for MSE). Its novelty lies in bounding the contribution of each point on the segment to the loss, which is constant. While a uniform sample could approximate this at the cost of failure probability and larger size, we use a deterministic construction generalizing [24]. We emphasize that the algorithm is not a heuristic but comes with provable guarantees. Although simple in form, it relies on non-trivial analysis, which enables this very simplicity.

Algorithm 1: SEG-CORESET(ℓ, k, ε); see Theorem 2.

    Input   : A segment $\ell : [0,1] \to \mathbb{R}^d$, an integer $k \geq 1$, and error $\varepsilon \in (0, 1/10]$; see Definition 1.
    Output: A weighted set (S, w), which is an (ε, k)-coreset of ℓ; see Definition 7.

1 $\varepsilon' := \left\lceil \dfrac{4k \cdot (20k)^{r+1}}{\varepsilon} \right\rceil$ `//` r `is as defined in Definition 4.`

2 $S := \{\ell(i/\varepsilon') \mid i \in \{0, \cdots, \varepsilon'\}\}$ `// note that` $\varepsilon' \in \dfrac{O(k)^{r+2}}{\varepsilon}$.

3 Let $w(p) := 1/\varepsilon'$ for all $p \in S$.
4 **return** (S, w).

The following theorem states the desired properties of Algorithm 1; the proof is not given due to lack of space.

Theorem 2. *Let $\ell : [0,1] \to \mathbb{R}^d$ be a segment, and let $\varepsilon \in (0, 1/10]$; see Definition 1. Let (S, w) be the output of a call to* SEG-CORESET(ℓ, k, ε); *see Algorithm 1. Then (S, w) is an (ε, k)-coreset for ℓ; see Definition 7.*

Setting $\tilde{f}(x) = x^2$ and using previous results, we obtain a provable k-means approximation for segments by constructing coresets per segment (Theorem 2) and applying $O(\log n)$ repetitions of [1] on their union.

Corollary 1. *Let L be a set of n segments, and let $\varepsilon, \delta \in (0, 1/10]$. We can compute in $O(nk^3 \varepsilon \log(1/\delta))$ time, a set $P \subset \mathbb{R}^d, |P| = k$ such that with probability at least $1 - \delta$ we have*

$$\mathrm{loss}(P, L) \in O(\log k) \min_{P' \in \mathbb{R}^d, |P'|=k} \mathrm{loss}(P', L).$$

Algorithm 2. We use Algorithm 1 to convert each segment into points, then apply the compression scheme from [10] on their union.

Overview of Algorithm 2. Given a set L of n segments, an integer $k \geq 1$, and parameters $(\varepsilon, \delta) \in (0, 1/10]$, the algorithm outputs a weighted set (P', w')

that, with probability at least $1 - \delta$, is an (ε, k)-coreset for L. Each segment is first reduced to points using Algorithm 1, producing a weighted set. The union of these sets is then processed via CORE-SET (Theorem 1) to further reduce the coreset size.

Algorithm 2: CORESET$(L, k, \varepsilon, \delta)$; see Theorem 3.

    Input   : A finite set L of segments, an integer $k \geq 1$, and input parameters $\varepsilon, \delta \in (0, 1/10]$.

    Output: A weighted set (P, w), which, with probability at least $1 - \delta$, is an (ε, k)-coreset of L; see Definition 7.

1 For every $\ell \in L$ let $(P'_\ell, w_\ell) :=$ SEG-CORESET$(\ell, k, \varepsilon/2)$ // see Algorithm 1, the division by 2 accounts for the output being further reduced.

2 $P' := \bigcup_{\ell \in L} P'_\ell$

3 $(P, w') :=$ CORE-SET$(P', k, \varepsilon/4, \delta)$// see Theorem 1.

4 $\varepsilon' := \left\lceil \dfrac{8k \cdot (20k)^{r+1}}{\varepsilon} \right\rceil$ // r is as defined in Definition 4.

5 Set $w(p) := w'(p)/\varepsilon'$ for every $p \in P'$.

6 **return** (P, w).

The structure of Algorithm 2 is illustrated in Fig. 2; note that we have substituted CORE-SET in Definition 4 by [2]. For an explanation on motion vectors, see Sect. 4.2.

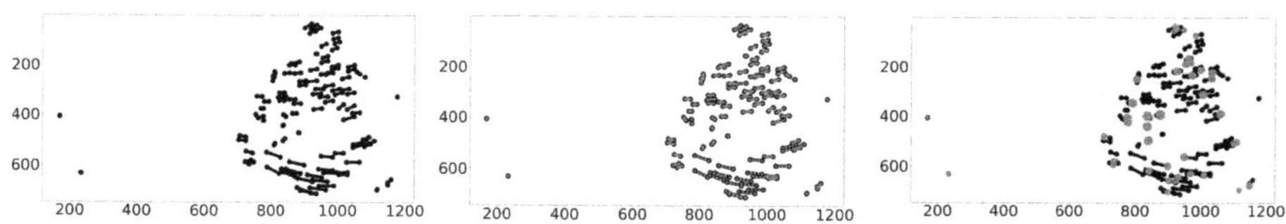

Fig. 2. Illustration of Algorithm 2. *Left:* The input $n = 100$ segments, obtained from the motion vectors of the video [23]. *Middle:* The union of the outputs of the calls to Algorithm 1 computed at Line 1 in the call to Algorithm 2 is the red points (all with equal weight) and black lines representing the input segments. *Right:* The final coreset returned by Algorithm 2, which is obtained by further sampling from the outputs of Algorithm 1 via [2]. Here, the size of each point is proportional to its weight.

The following theorem states the desired properties of Algorithm 2; the proof is not given due to lack of space.

Theorem 3. *Let L be a set of n segments, and let $\varepsilon, \delta \in (0, 1/10]$. Let $m := \dfrac{8kn \cdot (20k)^{r+1}}{\varepsilon}$ and $k' \in (k + 1)^{O(k)}$. Let (P, w) be the output of a call to* CORESET$(L, k, \varepsilon, \delta)$; *see Algorithm 2. Then Claims (i)–(iii) hold as follows:*

(i) With probability at least $1 - \delta$, we have that (P, w) is an (ε, k)-coreset of L; see Definition 7.

(ii) $|P| \in \dfrac{k' \cdot \log^2 m}{\varepsilon^2} \cdot O\left(d^* + \log\left(\dfrac{1}{\delta}\right)\right).$

(iii) The (ε, k)-coreset (P, w) can be computed in time

$$mtk' + tk' \log(m) \cdot \log^2\left(\log(m)/\delta\right) + \frac{k' \log^3 m}{\varepsilon^2} \cdot \left(d^* + \log\left(\frac{1}{\delta}\right)\right).$$

3 Empirical Evaluation

We evaluate the quality of our approximation against the line-clustering method from [18]. For consistency, we measure error using the sum of squared distances (corresponding to the Gaussian noise model [11]), which allows reduction of the task to the standard k-means, solved with the k-means++ approximation [1]. Accordingly, we apply only Algorithm 1 to reduce each segment to a small set of representative points (coreset). Our implementation (Python 3.8, NumPy) is tested on both synthetic and real datasets, with open-source code available in [7].

Methods. We compare:

Our-Method. For each segment, compute a coreset of 10 points using SEG-CORESET, then apply k-means++ [1] to cluster the resulting points.

Line-Clustering. Line-clustering from [18], applied to segments extended to lines.

Note that Line-Clustering is Designed for Line-Clustering Rather than Segment-Clustering. However, since [18] evaluated their approach on road segments extended to infinite lines in the OpenStreetMap dataset [19], we include it here as the closest available baseline. To our knowledge, no provable or commonly used methods exist for segment-clustering, making this the only relevant comparison.

Loss. Since the optimization problem is not elementary, we approximate it using the same coreset construction as our method, but with a larger size. For each segment $\ell \in L$, we compute $(P_\ell, w_\ell) := \text{SEG-CORESET}(\ell, k, 0.1)$ with $|P_\ell| = 10{,}000$, a larger coreset used to obtain a more robust approximation, and define $P = \bigcup_{\ell \in L} P_\ell$. The loss is then $\sum_{p \in P} \min_{c \in C} D(p, c)^2 / |P|$, i.e., the mean squared error (MSE).

Data. We use three datasets for comparison.
(i). Synthetic segments in $\mathbb{R}^{10}$, where each endpoint is sampled uniformly from $[-1, 1]^{10}$; the sample size varies across experiments.
(ii). Motion vectors (see [30]) from a short clip of [23], sampled independently, with probability proportional to segment length. This setting demonstrates applications in tracking and object detection (see Sect. 4.4).

(iii). Road segments from OpenStreetMap [19] (via Geofabrik), using longitude/latitude coordinates and sampling proportional to segment length. We focus on the one million longest segments in Malaysia, Singapore, and Brunei, where fitted centers highlight key points in the road network relevant to facility location and infrastructure planning.

Hardware. We used a PC with an Intel Core i5-12400F, NVIDIA GTX 1660 SUPER (GPU), and 32GB of RAM.

Segment Count. 200 to 1000 in steps of 100.

Repetitions. Each experiment was repeated 40 times.

Results. Our results are presented in Fig. 3. In all the figures, the values presented are the medians across the tests, along with error bars that present the 25% and 75% percentiles.

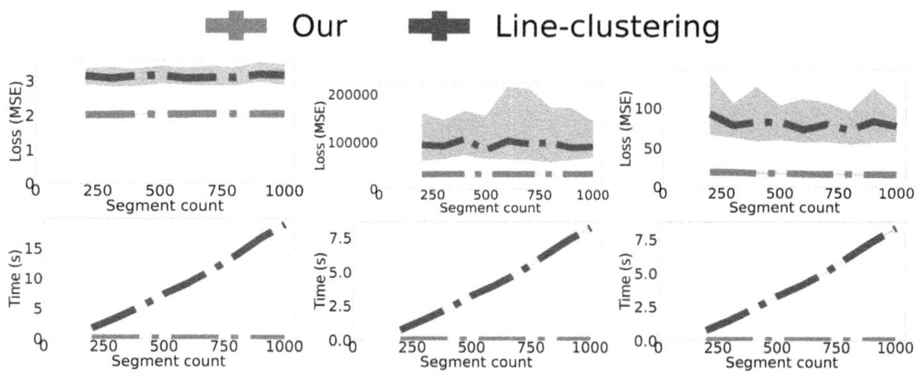

Fig. 3. The results of the experiment from Sect. 3. The top row is the legend for all the plots, the middle row corresponds to the loss, and the bottom row to the times. The columns correspond (left to right) to the synthetic segments (dataset **(i)**), the motion vectors (dataset **(ii)**), and the Malaysia, Singapore, and Brunei roads (dataset **(iii)**).

As shown in Fig. 3, OUR-METHOD consistently achieves lower loss and significantly faster runtimes compared to LINE-CLUSTERING. Our runtime is dominated by the k-means++ call, reflecting the simplicity of our reduction.

4   Application: Video Tracking and Detection

Goal. These tests demonstrate that using the coreset from Algorithm 1 enables real-time tracking and detection of moving objects in video. The next paragraph provides a summary of the video tracking and detection setup.

4.1 Preliminaries on Video Tracking and Detection

Tracking objects in RGB videos is a well-studied problem, with thousands of papers in recent years [31]. Neural networks achieve state-of-the-art results, but training and data labeling are costly, and real-time performance ($\approx$ 30 fps) often requires mid-level GPUs. Moreover, small perturbations can "fool" networks [25], and it is unclear whether more sophisticated attacks could bypass recent robustness improvements [5]. Although our method does not use object detection, real-time methods that do, e.g., [6], are available.

Our proposed method leverages motion vectors for video tracking and detection, as detailed below.

4.2 Motion Vectors

Motion vectors, computed in real time by standard video encoders such as H.264 [30] and H.265 [20], map blocks from one frame to another, typically minimizing a loss function (e.g., MSE of RGB values) so that only the vectors and residual differences need to be stored. We focus on the simple case of mapping each frame to its immediate predecessor (illustrated in Fig. 4). Because our method utilizes only motion vectors and not RGB data, it enables real-time object tracking while preserving some degree of privacy.

Fig. 4. Illustration of motion vectors. Left: a snapshot from [23] with a sub-sample of motion vectors to the next frame (blue: left, red: right). Right: the corresponding vectors. While most vectors follow the movement, some noise is visible, e.g., on the rock and grass in the left half. (Color figure online)

4.3 The Proposed Method

For simplicity, the experiments use the sum of squared distances between objects, corresponding to Gaussian noise [11].

Tracking Method. In general, we set k equal to the number of objects tracked. For our test, tracking a single object, we use $k = 2$ (for $k = 1$ no clustering needs to be done and the tracking can be done as in [21]) and track the largest cluster by the number of points assigned.

For each 10 consecutive frames:
(i). Append each motion vector's angles to $(0, 1)$ and $(1, 0)$, scaled so that $180°$ corresponds to the largest endpoint, yielding 4-dimensional vectors. **(ii).** If there are more than 1000 vectors, sample 1000 uniformly without replacement. **(iii).** Treat each 4-D vector as a segment and apply Algorithm 1 with coreset size 10. **(iv).** Compute k-means for the resulting points via [1]. **(v).** Track the mean start and end positions of motion vectors in the largest cluster.

Note that although each motion vector is represented as a segment, we include additional values to form 4-D segments rather than simple 2-D endpoints, which helps preserve the vector's orientation.

Software. We implemented our algorithms in Python 3.8 [29] utilizing [3, 4, 27], and [13]. The source code is provided in [7].

4.4   Big Buck Bunny Test

For this test, we used a 400-frame clip of the Big Buck Bunny video [23] at 720×1280p, streamed in real time from a file. This video is widely used in the video tracking community and is licensed under Creative Commons Attribution 3.0, allowing free reuse with proper attribution [23] (https://peach.blender.org/about). We selected this segment because it shows a bunny walking across a stationary background, allowing visual validation of tracking results.

Hardware. In this section, we used a standard Laptop with an Intel Core i3-1115G4 processor and 16GB of RAM.

Figure 5 shows examples from the clip, with the complete tracking results available in [7]. In the left column, the motion vectors point opposite to the bunny's actual movement, occurring when the bunny first enters the frame, and the closest color matches are from already visible areas. Nonetheless, the cluster center still aligns with the bunny's center, so the tracked position remains visually accurate. Once the bunny is fully in view and walking (bottom images), the predicted movement aligns with the actual motion, and the cluster center follows the object correctly.

Fig. 5. A subset of results from the experiment in Sect. 4.4. The top row displays the motion vector clusters, and the bottom row shows the cluster centers and their corresponding mean directions. The left column captures the bunny rising from the cave, while the middle and right columns show the bunny walking right at slightly different times.

Running Time. Our tracking algorithm processed the 400-frame clip in 0.28 s, achieving over 1,400 fps. For comparison, YOLOv8 [26] (an improvement over YOLOv5 [14]) takes 33.4 s on the same clip, or 12 fps [7].

4.5   Re-running for Single-Board Computer

We reran the video tracking test (Sect. 4.4) on the Libre Computer AML-S905X-CC (Le Potato, https://libre.computer/products/aml-s905x-cc/), a small single-board computer similar to Raspberry Pi [28], using the official Raspberry Pi OS. Due to limited ARM support, motion vectors were pre-extracted and transferred as a Numpy array. This test demonstrates that our method can run in real time on low-end hardware. Results were essentially unchanged, with only minor differences resulting from clustering tie-breaking and sample noise. The tracking algorithm ran in 4.23 s for 400 frames, achieving a frame rate of over 94 fps.

Conclusion and Future Work

This paper provides a coreset construction that gets a set of segments in $\mathbb{R}^d$ and returns a small weighted set of points (coreset) that approximates its sum of (fitting) distances to any weighted k-centers, up to a factor of $1 \pm \varepsilon$. This was achieved by reducing the problem of fitting k-centers to segments to the problem of fitting k-centers to points. This reduction enabled us to leverage the extensive literature on k-means to obtain an efficient approximation and achieve low running times in the experimental sections.

While we have not analyzed the variants of the problem where an additional constraint requires each segment to be assigned to a single center, we believe that this would follow from incorporating our result with Sect. 15.2 of [9]. We have not considered this direction since our method is based on reducing the

122   D. Denisov et al.

segments to a set of points and applying the classic k-means clustering, which can be computed efficiently via [1]. Even after compression, this variation on Problem 1 yields a complex set clustering problem.

Our results support loss functions where the integral in (1) at Problem 1 is not necessarily elementary, and as such, direct minimization seems to us unfeasible.

Acknowledgements. David Denisov was (partially) funded by the Israeli Science Foundation (Grant No. 465/22). Shlomi Dolev was (partially) funded by the Israeli Science Foundation (Grant No. 465/22) and by Rita Altura trust chair in Computer Science. Michael Segal was (partially) funded by Israeli Science Foundation (Grant No. 465/22), US Army Research Office under Grant Number W911NF-22-1-0225, grant from Israel Ministry of Science No. 0005355) and grant from the Israel Innovation Authority.

References

1. Arthur, D., Vassilvitskii, S.: k-means++: the advantages of careful seeding. In: Proceedings of the Eighteenth Annual ACM-SIAM Symposium on Discrete Algorithms, SODA 2007, pp. 1027–1035 (2007)
2. Bachem, O., Lucic, M., Krause, A.: Scalable k -means clustering via lightweight coresets. In: Proceedings of the 24th ACM SIGKDD International Conference on Knowledge Discovery & Data Mining, KDD 2018, pp. 1119–1127. Association for Computing Machinery, New York (2018). https://doi.org/10.1145/3219819.3219973
3. Bommes, L., Lin, X., Zhou, J.: Mvmed: fast multi-object tracking in the compressed domain. In: 2020 15th IEEE Conference on Industrial Electronics and Applications (ICIEA), pp. 1419–1424 (2020). https://doi.org/10.1109/ICIEA48937.2020.9248145. https://github.com/LukasBommes/mv-extractor
4. Bradski, G.: The OpenCV library. Dr. Dobb's J. Softw. Tools (2000)
5. Chen, B., Chin, T.J., Klimavicius, M.: Occlusion-robust object pose estimation with holistic representation. In: Proceedings of the IEEE/CVF Winter Conference on Applications of Computer Vision, pp. 2929–2939 (2022)
6. Comaniciu, D., Ramesh, V., Meer, P.: Real-time tracking of non-rigid objects using mean shift. In: Proceedings IEEE Conference on Computer Vision and Pattern Recognition. CVPR 2000 (Cat. No. PR00662), vol. 2, pp. 142–149 (2000). https://doi.org/10.1109/CVPR.2000.854761
7. Denisov, D.: Official code for all the algorithms presented in this paper. https://github.com/DavidDenisov/Segment-Clustering
8. Feldman, D.: Core-sets: updated survey. In: Ros, F., Guillaume, S. (eds.) Sampling Techniques for Supervised or Unsupervised Tasks. USL, pp. 23–44. Springer, Cham (2020). https://doi.org/10.1007/978-3-030-29349-9_2
9. Feldman, D., Langberg, M.: A unified framework for approximating and clustering data. In: Proceedings of the Forty-Third Annual ACM Symposium on Theory of Computing, STOC 2011, pp. 569–578. Association for Computing Machinery, New York (2011). https://doi.org/10.1145/1993636.1993712
10. Feldman, D., Schulman, L.J.: Data reduction for weighted and outlier-resistant clustering. In: Proceedings of the Twenty-Third Annual ACM-SIAM Symposium on Discrete Algorithms, SODA 2012, pp. 1343–1354. Society for Industrial and Applied Mathematics, USA (2012)

11. Guo, D., Wu, Y., Shitz, S.S., Verdú, S.: Estimation in gaussian noise: properties of the minimum mean-square error. IEEE Trans. Inf. Theory **57**(4), 2371–2385 (2011)

12. Har-Peled, S.: Coresets for discrete integration and clustering. In: Arun-Kumar, S., Garg, N. (eds.) FSTTCS 2006. LNCS, vol. 4337, pp. 33–44. Springer, Heidelberg (2006). https://doi.org/10.1007/11944836_6

13. Harris, C.R., et al.: Array programming with NumPy. Nature **585**(7825), 357–362 (2020). https://doi.org/10.1038/s41586-020-2649-2

14. Jocher, G., et al.: ultralytics/yolov5: v5. 0-yolov5-p6 1280 models, aws, supervise. ly and youtube integrations. Zenodo (2021)

15. Jubran, I., Sanches Shayda, E.E., Newman, I.I., Feldman, D.: Coresets for decision trees of signals. In: Ranzato, M., Beygelzimer, A., Dauphin, Y., Liang, P., Vaughan, J.W. (eds.) Advances in Neural Information Processing Systems, vol. 34, pp. 30352–30364. Curran Associates, Inc. (2021). https://proceedings.neurips.cc/paper_files/paper/2021/file/fea9c11c4ad9a395a636ed944a28b51a-Paper.pdf

16. Jubran, I., Tukan, M., Maalouf, A., Feldman, D.: Sets clustering. In: III, H.D., Singh, A. (eds.) Proceedings of the 37th International Conference on Machine Learning, vol. 119, pp. 4994–5005. PMLR (2020). https://proceedings.mlr.press/v119/jubran20a.html

17. Maalouf, A., Statman, A., Feldman, D.: Tight sensitivity bounds for smaller coresets, KDD 2020, pp. 2051–2061. Association for Computing Machinery, New York (2020). https://doi.org/10.1145/3394486.3403256

18. Marom, Y., Feldman, D.: k-means clustering of lines for big data. In: Wallach, H., Larochelle, H., Beygelzimer, A., d'Alché-Buc, F., Fox, E., Garnett, R. (eds.) Advances in Neural Information Processing Systems, vol. 32. Curran Associates, Inc. (2019). https://proceedings.neurips.cc/paper_files/paper/2019/file/6084e82a08cb979cf75ae28aed37ecd4-Paper.pdf

19. OpenStreetMap contributors: Planet dump (2017). https://planet.osm.org. https://www.openstreetmap.org

20. Pereira, F.C., Ebrahimi, T.: The MPEG-4 Book. Prentice Hall PTR, USA (2002)

21. Riechmann, M., Gardiner, R., Waddington, K., Rueger, R., Leymarie, F.F., Rueger, S.: Motion vectors and deep neural networks for video camera traps. Ecol. Inform. **69**, 101657 (2022). https://doi.org/10.1016/j.ecoinf.2022.101657. https://www.sciencedirect.com/science/article/pii/S1574954122001066

22. Riemann, B.: Uber die fläche vom kleinsten inhalt bei gegebener begrenzung. (bearbeitet von k. hattendorff.). In: Abhandlungen der Königlichen Gesellschaft der Wissenschaften in Göttingen (Proceedings of the Royal Philosophical Society at Göttingen), vol. 13 (1868). http://resolver.sub.uni-goettingen.de/purl?PPN250442582_0013

23. Roosendaal, T.: Big buck bunny. In: SIGGRAPH Asia 2008, p. 62. Association for Computing Machinery, New York (2008). https://doi.org/10.1145/1504271.1504321

24. Rosman, G., Volkov, M., Feldman, D., Fisher III, J.W., Rus, D.: Coresets for k-segmentation of streaming data supplementary material. In: Ghahramani, Z., Welling, M., Cortes, C., Lawrence, N., Weinberger, K. (eds.) Advances in Neural Information Processing Systems, vol. 27. Curran Associates, Inc. (2014). https://proceedings.neurips.cc/paper_files/paper/2014/file/bca82e41ee7b0833588399b1fcd177c7-Paper.pdf

25. Su, J., Vargas, D., Sakurai, K.: One pixel attack for fooling deep neural networks. IEEE Trans. Evol. Comput. (2017). https://doi.org/10.1109/TEVC.2019.2890858

26. Terven, J., Cordova-Esparza, D.: A comprehensive review of yolo: from yolov1 to yolov8 and beyond. arXiv preprint arXiv:2304.00501 (2023)
27. Thakur, A., et al.: abhitronix/vidgear: Vidgear v0.2.5 (2022). https://doi.org/10.5281/zenodo.6046843
28. Upton, E., Halfacree, G.: Raspberry Pi User Guide. Wiley (2016)
29. Van Rossum, G., Drake, F.L.: Python 3 Reference Manual. CreateSpace, Scotts Valley, CA (2009)
30. Wiegand, T., Sullivan, G., Bjontegaard, G., Luthra, A.: Overview of the h.264/avc video coding standard. IEEE Trans. Circuits Syst. Video Technol. **13**(7), 560–576 (2003). https://doi.org/10.1109/TCSVT.2003.815165
31. Zou, Z., Chen, K., Shi, Z., Guo, Y., Ye, J.: Object detection in 20 years: a survey. Proc. IEEE **111**(3), 257–276 (2023). https://doi.org/10.1109/JPROC.2023.3238524

Cartesian Forest Matching

Bastien Auvray[1,2(✉)], Julien David[1,3], Richard Groult[1,2],
and Thierry Lecroq[1,2]

[1] CNRS NormaSTIC FR 3638, Caen, France
{bastien.auvray,richard.groult,thierry.lecroq}@univ-rouen.fr,
julien.david@unicaen.fr
[2] Univ Rouen Normandie, LITIS UR 4108, 76000 Rouen, France
[3] Normandie University, UNICAEN, ENSICAEN, CNRS, GREYC, Caen, France

Abstract. In this paper, we introduce the notion of Cartesian Forest, which generalizes Cartesian Trees, in order to deal with duplicates in ordered sequences. We show that algorithms that solve both exact and approximate Cartesian Tree Matching can be adapted to solve Cartesian Forest Matching in linear time in the worst-case for exact matching and linear time on average for approximate matching. We adapt the notion of Cartesian Tree Signature to Cartesian Forests. We also show a one-to-one correspondence between Cartesian Forests and Schröder Trees.

Keywords: Cartesian Tree · Cartesian Forest · Pattern Matching · Approximate Pattern Matching · Schröder Tree

1 Introduction

Pattern matching consists of searching for one or all the occurrences of a pattern in a text. It is an essential task in many computer science applications. It can take different forms. For instance it can be done online when the pattern can be preprocessed or offline when the text can be preprocessed. Occurrences can be exact or approximate. When the pattern and the text are sequences of characters, it is known as string matching. When searching patterns in time series data, the notion of pattern matching is a bit more involved. Solutions can use Cartesian Trees that were introduced by Vuillemin in 1980 [23].

Cartesian Tree Matching has been introduced by Park *et al.* [19,20]. Given a pattern, it consists of finding the factors of a text that share the same Cartesian Tree as the Cartesian Tree of the pattern.

Since then it has gained a lot of interest. Efficient solutions for practical cases for online search were given in [21]. Expected linear time algorithms are given in [3] for approximate Cartesian Tree Matching with one difference.

Indexing structures in the Cartesian Tree pattern matching framework are presented in [13,16,18]. Methods for computing regularities are given in [12] and methods for computing palindromic structures are presented in [9]. An algorithm for episode matching (given two sequences p and t, finding all minimal

E. Di Giacomo and D. Mondal (Eds.): WALCOM 2026, LNCS 16444, pp. 125–139, 2026.
https://doi.org/10.1007/978-981-95-7127-7_9

length factors of t that contains p as a subsequence) in Cartesian Tree framework is presented in [17]. Practical methods for finding longest common Cartesian substrings of two strings appeared in [7]. Very recently, dynamic programming approaches for approximate Cartesian Tree pattern matching with edit distance have been considered in [14] and longest common Cartesian Tree subsequences are computed in [22]. Efficient algorithms for determining if two equal-length indeterminate strings match in the Cartesian Tree framework are given in [11].

Cartesian Trees are defined (Sect. 2) on arbitrary sequences where there can be values with multiple occurrences. In that case, ties should be broken in some way. In this article, in order to better deal with equal values we introduce *Cartesian Forests* (Sect. 3). The main idea is to deal with ties in the sequence by adding siblings in the tree-like structure, hence resulting in a forest.

We show that algorithms that solve both exact and approximate Cartesian Tree Matching (Sect. 4 and Sect. 5) can be adapted to solve Cartesian Forest Matching in linear time in the worst-case for exact matching and linear time on average for approximate matching. We then show a one to one correspondence between Cartesian Forests and Schröder Trees (Sect. 6). More specifically we give algorithms for computing a Schröder Tree and a special type of words with parentheses given a Cartesian Forest. We also adapt the notion of Cartesian Tree Signature introduced in [6] to Cartesian Forests (Sect. 7), and show how this notion can be used to experimentally improve the computation (Sect. 8).

2 Preliminaries

In this paper, a sequence is always defined on an ordered alphabet. For a given sequence x, $|x|$ denotes the length of x. A sequence v is *factor* of a sequence x if $x = uvw$ for any sequences u and w. A sequence u is a *prefix* (resp. *suffix*) of a sequence x if $x = uv$ (resp. $x = vu$). A prefix of a sequence that is also a suffix is called a *border*. For a sequence x of length m, $x[i]$ is the i-th element of x and $x[i \dots j]$ represents the factor of x starting at the i-th element and ending at the j-th element, for $1 \leq i \leq j \leq m$.

2.1 Cartesian Trees

Definition 1 (Cartesian Tree $C(x)$). *Given a totally ordered sequence x of length m, the* Cartesian Tree *of x, denoted by $C(x)$, is the binary tree recursively defined as follows:*

- *if x is empty, then $C(x)$ is the empty tree;*
- *if $x[1 \dots m]$ is not empty and $x[i]$ is the smallest value of x, $C(x)$ is the binary tree with i as its root, the Cartesian Tree of $x[1 \dots i-1]$ as the left subtree and the Cartesian Tree of $x[i+1 \dots m]$ as the right subtree.*

We denote by $\mathrm{rb}(T)$ the list of nodes on the *right branch* of a Cartesian Tree T. Also, $C_i(x) = C(x[1 \dots i])$ to simplify the notations. The Cartesian Tree of a sequence can be built online in linear time and space [10]. Informally, let x be a sequence such that $C_{i-1}(x)$ is already known. In order to build $C_i(x)$, one only needs to find the nodes $j_1 < \cdots < j_k$ in $\mathrm{rb}(C_{i-1}(x))$ such that $x[j_1] > x[i]$. Then, j_1 will be the root of the left subtree of node i. If j_1 is the root of $C_{i-1}(x)$ then i becomes the root of $C_i(x)$, otherwise let j_0 be the parent node of j_1, then node i will be the root of the right subtree of j_0. Then $\mathrm{rb}(C_i(x)) = \mathrm{rb}(C_{i-1}(x)) \setminus (j_1, \ldots, j_k) \cup (i)$. All these operations can be easily done by implementing rb with a stack. The amortized cost of such an operation can be shown to be constant.

Cartesian Trees are defined on arbitrary sequences where there can be multiple occurrences of the smallest value. In that case, ties should be broken in some way. Usually the first occurrence of the smallest value is chosen to be the root of the tree.

Given a Cartesian Tree $C(x)$ of a sequence x of length m, a linear representation of $C(x)$, denoted LN_x, is a table of integers of length m, and the set of linear representations of a given length is in bijection with the set of Cartesian Trees of the same size. A linear representation of Cartesian Trees was introduced by Park et $al.$ [19]: the parent-distance representation (see example Fig. 3). Using this representation and classical string algorithms (Knuth-Morris-Pratt (KMP) [15] and Aho-Corasick [1] algorithms), they obtained linear-time solutions for single and multiple pattern Cartesian Tree matching (see Fig. 1 for an example).

Let $\mathrm{small}_x(i) = \max(\{k \mid x[k] < x[i] \text{ with } 1 \le k < i\} \cup \{0\})$.

Definition 2 (Parent-distance representation PD_x). *Given a sequence x of length m, the* parent-distance representation *of x is an integer sequence $\mathrm{PD}_x[1 \ldots m]$, which is defined as follows:*

$$\mathrm{PD}_x[i] = \begin{cases} i - \mathrm{small}_x(i) & \textit{if } \mathrm{small}_x(i) > 0 \\ 0 & \textit{otherwise.} \end{cases}$$

Since the parent-distance representation has a one-to-one mapping with Cartesian Trees, it can replace them without loss of information.

The parent-distance table is not the only linear representation of Cartesian Trees. Another linear representation is based on the number of nodes skipped by a new node when building online the Cartesian Tree of a sequence. Informally, for each node i, it stores the number of nodes ($j_1 < \cdots < j_k$) deleted from the right path when computing $C_i(x)$ from $C_{i-1}(x)$. We say that node i skipped nodes ($j_1 < \cdots < j_k$) and that these nodes are skipped by node i. The left subtree of a Cartesian Tree $C(x)$ is denoted $left(C(x))$ and the length of its right branch $\mathrm{RB}(C(x)) = |\mathrm{rb}(C(x))|$.

See Fig. 1 for an example.

Definition 3 (Skipped-number representation SN_x). *Given a sequence $x[1 \ldots m]$, the* Skipped-number representation *of x is a sequence $SN_x[1 \ldots m]$ of integers such that $SN_x[i]$ is the length of the right-branch of the left-subtree of the subtree rooted at node i of the Cartesian Tree of x, that is $SN_x[i] = \mathrm{RB}(left(C_i(x)))$.*

Both linear representations can be computed in linear time [3,6,19].

i	1	2	3	4	5	6	7	8	9
$x[i]$	3	1	6	4	8	6	7	5	9
$\mathrm{PD}_x[i]$	0	0	1	2	1	2	1	4	1
$SN_x[i]$	0	1	0	1	0	1	0	2	0
$CBord[i]$	0	1	1	2	3	4	5	2	3

Fig. 1. The parent-distance representation, the skipped-number representation and the Cartesian border table of a sequence x.

2.2 Exact Cartesian Tree Matching

Given two sequences p and t of length m and n respectively, the *Exact Cartesian Tree Matching* problem (CTM) consists of finding all the factors of t that have the same Cartesian Tree of the one of p. This can be done in time $O(m+n)$ by using a KMP-like algorithm. To achieve this complexity, a *Cartesian border table* is used. Given the Parent-Distance representation of a sequence x, the Parent-Distance of a factor $x[i \ldots j]$ of x satisfies:

$$\mathrm{PD}_{x[i \ldots j]}[k] = \begin{cases} 0 & \text{if } \mathrm{PD}_x[i+k-1] \geq k, \\ \mathrm{PD}_x[i+k-1] & \text{otherwise.} \end{cases}$$

Definition 4 (Cartesian Border Table $CBord$). *The Cartesian border table of x is defined in the following way: $CBord[0] = 0$ and, for $1 \leq k < i$, $CBord[i] = \max\{k \mid C_k(x) = C(x[i-k+1 \ldots i])\}$.*

Figure 1 shows the Cartesian border table of $x = (3, 1, 6, 4, 8, 6, 7, 5, 9)$.

2.3 Approximate Cartesian Tree Matching

Given two sequences p and t of length m and n respectively, the *Approximate Cartesian Tree Matching* problem (ACTM) consists of finding all the factors of t that have the same Cartesian Tree of the one of p up to some differences.

Unfortunately, the linear-time solutions obtained in [19] for the exact case do not work anymore in the general approximate case. Also we conjecture that there is no algorithm that solves approximate pattern matching with a linear worst-case complexity.

In [3], the authors provide the following MetaAlgorithm that can solve ACTM with up to one difference (substitution, insertion, deletion, swap) in quadratic worst-case time complexity and in linear time in average. The test at line 5, using the $\sim$ symbol, signifies that depending on the considered notion of error, we will not test equality between linear representation but rather test an equivalence.

3   Cartesian Forests

A limitation of Cartesian Trees is that it does not take into account equal values in the original sequence. For instance, the sequence $(1, 1, 1, 1, 1)$ shares the same

---

Algorithm 1: METAALGORITHM(p, t)

Input : Two sequences p and t of length m and n
Output: The number of positions j such that p is equivalent to $t[j \ldots j + m - 1]$

1 $occ \leftarrow 0$
2 $x \leftarrow t[1 \ldots m]$
3 $\mathrm{LN}_p, \mathrm{LN}_x \leftarrow$ linear representations of $C(p)$ and $C(x)$
4 **for** $j \in \{1, \ldots, n - m + 1\}$ **do**
5   **if** $\mathrm{LN}_p \sim \mathrm{LN}_x$ **then**
6         $occ \leftarrow occ + 1$
7      $x \leftarrow t[j + 1 \ldots j + m]$
8   Update LN_x
9 **return** occ

Cartesian Tree as $(1, 2, 3, 4, 5)$ when ties are broken by ordering equal values according to their positions. In order to better deal with equal values we now introduce Cartesian Forests.

Recall that a planar tree, or general tree, is a tree in which each node possesses a sequence of children, or subtrees. In the following, *trees* designates planar trees.

Informally, during the construction of $C_i(x)$ from $C_{i-1}(x)$, assume that node i should be inserted in the right branch between nodes j_0 and j_1 such that $x[j_0] \leq x[i] < x[j_1]$. In a Cartesian Tree, node i becomes the root of the right subtree of node j_0 and node j_1 becomes the root of the left subtree of node i. In a Cartesian Forest, node i becomes the root of the right subtree of node j_0 only if $x[j_0] < x[i]$ while it becomes the right sibling of node j_0 if $x[j_0] = x[i]$ and node j_1 remains in the right subtree of node j_0.

A *forest* is a sequence of trees. An empty forest is an empty sequence that does not contain any tree. The children of a node can be seen as a *sub-forest*. In the following, a node of a tree in a *Cartesian Forest*, instead of having a sequence of children, can possess a *left* sequence of children and a *right* sequence of children, which we call the *left sub-forest* and the *right sub-forest*. The leftmost tree of a Cartesian Forest is actually the only one that can possess a left sub-forest and a right sub-forest, whereas all the other trees can only have a right sub-forest. Figure 2 illustrates this with an example of a Cartesian Forest.

Definition 5 (Cartesian Forest $F(x)$ of a sequence x). *Given a sequence x of length m, the* Cartesian Forest *of x, denoted by $F(x)$, is recursively defined as follows:*

- *if x is empty, then $F(x)$ is the empty forest;*
- *if $x[1 \ldots m]$ is not empty, let $(r_1, \ldots, r_k)$ be the ordered sequence of all the k positions of the smallest value of x and let $r_{k+1} = m + 1$, then $F(x)$ is a forest composed of k trees whose roots are $r_1, \ldots, r_k$, such that:*
  - *the* left *sub-forest of r_1 is the Cartesian Forest $F(x[1 \ldots r_1 - 1])$, and*
  - *for $1 \leq i \leq k$, the* right *sub-forest of r_i is the Cartesian Forest $F(x[r_i + 1 \ldots r_{i+1} - 1])$.*

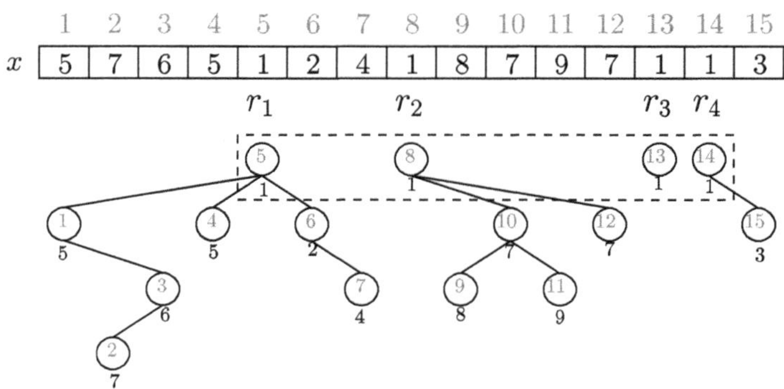

Fig. 2. The Cartesian Forest associated to an ordered sequence x. Roots r_2 to r_4 are created since the sequence contains a value equal to the one at position 5. The values between two roots r_i and r_{i+1} belong to the tree enrooted in r_i. Therefore, except for the leftmost tree of a Cartesian Forest, a tree cannot have a left sub-forest. This idea is true at every level and can be seen in this example at positions 4 and 12.

4   Exact Cartesian Forest Matching

We adapt the linear representations of Cartesian Trees, such as the Parent-Distance [19] and the Skipped-Number representation [6] to Cartesian Forests. Basically, for a sequence x, when computing $C_i(x)$ from $C_{i-1}(x)$, the Parent-Distance of i is equal to the distance between i and its parent in $C_i(x)$ and its Skipped-Number is the number of nodes removed from the right branch of $C_{i-1}(x)$ compared to the right branch of $C_i(x)$ (see [3]).

The main idea of the adaptation to Cartesian Forests is simple: both representations involve the comparisons of two values $x[j]$ and $x[k]$ in the original sequence x, for any $j < k$. Suppose we have either $x[j] > x[k]$ or $x[j] < x[k]$, then we are in the case of Cartesian Trees and the linear representation is the same. The idea for the Parent-Distance of the Cartesian Forest of a sequence x is that its absolute value at a position i gives the distance between a node i and its parent or left sibling in the Cartesian Forest associated to the sequence $x[1 \dots i]$. Let $\mathrm{equal}_x(i) = \max(\{k \mid x[k] = x[i] \text{ with } 1 \leq k < i\} \cup \{0\})$.

Definition 6 (Forest Parent-Distance representation F-PD$_x$). *Given a sequence x of length m, the* Forest Parent-Distance representation *of x is an integer sequence* F-PD$_x[1 \dots m]$ *defined as follows:*

$$\text{F-PD}_x[i] = \begin{cases} i - \mathrm{small}_x(i) & \text{if } \mathrm{small}_x(i) > \mathrm{equal}_x(i) \\ -(i - \mathrm{equal}_x(i)) & \text{if } \mathrm{small}_x(i) < \mathrm{equal}_x(i) \\ 0 & \text{otherwise.} \end{cases}$$

The referent table that we define next contains the position of the next smaller (or equal) value in the sequence.

Definition 7 (The referent table F-ref$_x$). *Given a sequence $x[1 \ldots m]$, the referent table of x is a sequence of sets F-ref$_x[1 \ldots m]$ such that*

$$\text{F-ref}_x[i] = \begin{cases} \min_{i < j \le m}\{j \mid x[j] \le x[i]\} & \text{if such } j \text{ exists} \\ -1 & \text{otherwise.} \end{cases}$$

Definition 8 (Forest Skipped-Number representation F-SN$_x$). *Given a sequence $x[1 \ldots m]$, the* Forest Skipped-Number representation *of x is an integer sequence F-SN$_x[1 \ldots m]$ such that*

$$\text{F-SN}_x[i] = \begin{cases} |\{j < i \mid \text{F-ref}_x[j] = i\}| & \text{if } \text{small}_x(i) > \text{equal}_x(i) \\ -|\{j < i \mid \text{F-ref}_x[j] = i\}| & \text{if } \text{small}_x(i) < \text{equal}_x(i) \end{cases}$$

Figure 3 shows examples of Forest Parent-Distance representations, Forest Skipped-Number representations and referent tables of two Cartesian Forests.

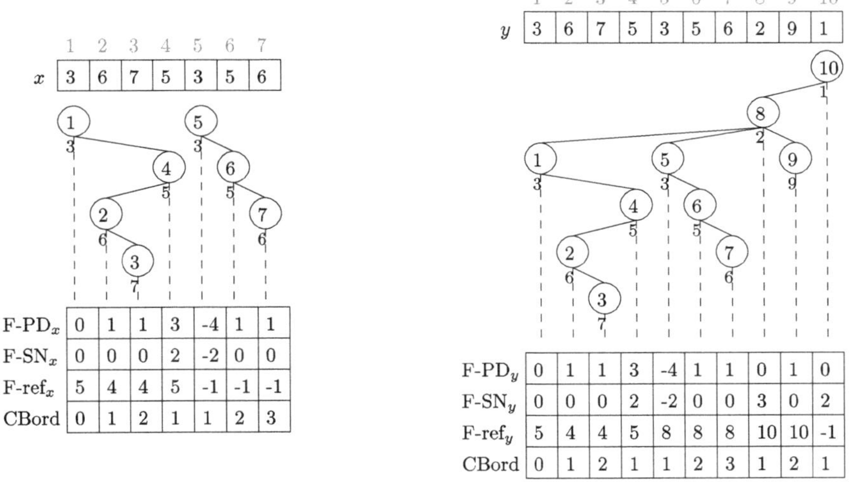

Fig. 3. Two sequences x and y, their associated Cartesian Forests $F(x)$ and $F(y)$, and their corresponding Forest Parent-Distance representations, Forest Skipped-Number representations and referent tables. As one can see, x is a prefix of y and the forest $F(x)$ is transformed into a sequence of left subtrees in $F(y)$.

Given two sequences x and y, we will denote $x \approx_{CF} y$ when the two sequences share the same Cartesian Forest.

Definition 9 (Cartesian Forest Matching (CFM)). *Given two sequences $p[1 \ldots m]$ and $t[1 \ldots n]$, find every position j, with $1 \le j \le n - m + 1$, such that $t[j \ldots j + m - 1] \approx_{CF} p[1 \ldots m]$.*

Example 1. Let $t = (5, 7, 3, 6, 3, 7, 2, 8, 2, 4, 3, 3)$ and $p = (2, 3, 1, 4, 1, 5)$ respectively be the text and the pattern. We have two occurrences of p in t as $t[1 \ldots 6] = (5, 7, 3, 6, 3, 7) \approx_{CF} p$ and $t[5 \ldots 10] = (3, 7, 2, 8, 2, 4) \approx_{CF} p$.

Note that on totally ordered sequences, CFM and CTM are equivalent. When the sequences contains duplicate, Cartesian Forests do not require to preprocess the sequences and are able to distinguish some sequences that would have been considered equivalent using Cartesian Trees. Also, Definition 4 still applies for Cartesian Forests.

Proposition 1. *Algorithm 1 solves the CFM problem in $\mathcal{O}(n)$ space and $\mathcal{O}(mn)$ time in the worst-case and $\mathcal{O}(n)$ time on average.*

Proposition 2. *Using a Cartesian Forest border table, the CFM problem can be solved in $\mathcal{O}(n)$ time and space by adapting the KMP approach used in [19].*

Note that, in order to prove both propositions, it is sufficient to prove that two sequences share the same Cartesian Forest if and only if they share the same linear representation. It can done by adapting the proof from [19].

5   Approximate Cartesian Forest Matching

The above approach for exact CFM can be extended to *approximate CFM* with one difference in a similar way as approximate CTM with one difference is done in [3]. Given a sequence x of length m and i a position $1 \leq i < m$, let y be the sequence defined by swapping the elements in positions i and $i + 1$ in x. In [3], it is shown that there are at most 3 mismatches between the *Skipped-number* representation of the Cartesian Tree of x and the *Skipped-number* representation of the Cartesian Tree of y. The same result holds for the Forest Skipped-Number representation of the Cartesian Forest of x and the Forest Skipped-Number representation of the Cartesian Forest of y considering that two elements match if their absolute values are equal. Thus, based on the results in [3], the *CFM with one swap* of a pattern of length m in a text of length n can be done in time $O(mn)$ in the worst case and in linear time in average.

Approximate CTM with one mismatch, one insertion or one deletion is solved by comparing the Parent-Distance representations from left to right and the Parent-Distance representations from right to left of the Cartesian Trees of x and y. The Forest Parent-Distance representation from right to left of a Cartesian Forest can be defined in a similar way as for a Cartesian Tree. And then, *approximate CFM with one mismatch, one insertion or one deletion* can be solved in a similar way than approximate CTM with one mismatch, one insertion or one deletion. Thus, based on the results in [3], the CFM with one mismatch, one insertion or one deletion of a pattern of length m in a text of length n can be done in time $O(mn)$ in the worst case and in linear time in average.

6   Combinatorics of the Cartesian Forests

6.1   Recursive Definition

Definition 10 (Cartesian Forest F). *A Cartesian Forest F can be:*

- *empty,*
- *a sequence of k trees rooted in $(r_1, \ldots, r_k)$, with $k \geq 1$, such that r_1 has a left Cartesian sub-Forest and a right Cartesian sub-Forest, and r_i has only a right Cartesian sub-Forest for all $i \in \{2, \ldots, k\}$ (the left sub-Forest is necessarily empty).*

Figure 4 shows examples of one Cartesian Forest and two forests that are not Cartesian.

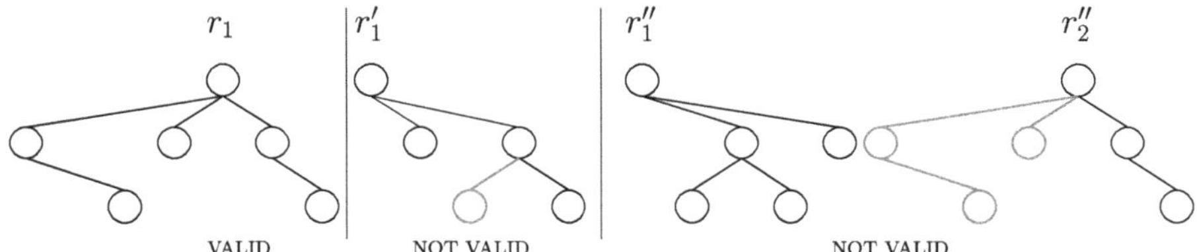

Fig. 4. On the left is a valid Cartesian Forest. In the middle, it is not a valid Cartesian Forest because the second tree in the right sub-forest of r_1' has a left sub-forest. On the right, the forest is not Cartesian: the second tree has a left sub-forest.

According to Definition 10, a Cartesian Forest F can be either empty (denoted by $\emptyset$) or contains at least one node. This node - the leftmost root - has a left sub-forest and a right sub-forest, but it can also have siblings. A sibling S is a particular Cartesian Forest that cannot have a left sub-forest. Therefore, we obtain the following recursive decomposition:
$$
\begin{cases}
F = \overset{\bullet}{\underset{F \quad F}{\wedge}} \times S + \emptyset \\
S = \overset{\bullet}{\underset{F}{\searrow}} \times S + \emptyset
\end{cases}
$$

6.2   Generating Function

Let $F(z) = \sum_{n \geq 0} f_n z^n$ be the generating function of Cartesian Forests, where f_n counts the number of Cartesian Forests with n nodes, and $S(z)$ be the generating function of siblings. The translation between the recursive decomposition is made using the Symbolic Method [8]:

$$
\begin{cases}
F(z) = z \cdot F^2(z) \cdot S(z) + 1 \\
S(z) = z \cdot F(z) \cdot S(z) + 1
\end{cases}
$$

from which we obtain that $S(z) = \frac{1}{1 - z \cdot F(z)}$ and $F(z) = 1 + \frac{z \cdot F^2(z)}{1 - z \cdot F(z)}$.

It can be shown from here that $F(z) = \frac{1+z+\sqrt{1-6z+z^2}}{4z}$, whose associated coefficients are known to be Schröder–Hipparchus numbers, also called super-Catalan numbers (Sequence $A001003$ on OEIS) enumerated by the following formula: $f_n = \sum_{i=1}^{n} \frac{1}{n}\binom{n}{i}\binom{n}{i-1}2^{i-1}$, which, amongst other things, counts the number of ways of inserting parentheses into a sequence of $n+1$ symbols, where each pair of parentheses surrounds at least two symbols or parenthesized groups, and without any parentheses surrounding the entire sequence. In [4], the authors show that:

$$f_n \sim \frac{\sqrt{3\sqrt{2}-4}}{4\sqrt{n^3\pi}}(3+2\sqrt{2})^n \sim 0.07(5.828)^n n^{-\frac{3}{2}}.$$

In the following subsections, we describe bijections between Cartesian Forests and classical combinatorial objects counted by the Schröder-Hipparchus numbers. We recall that since those objects all share the same generating function, there exists a bijection between those sets. Any injection from one set to the other is a bijection.

6.3   A Bijection with Schröder Trees

Definition 11 (Schröder Tree ST). *A Schröder Tree is a tree whose internal nodes have two or more subtrees.*

Schröder Trees with $n+1$ leaves are counted by f_n. Given a Cartesian Forest F with k roots $(r_1, \ldots, r_k)$, we denote left(r_1) the left sub-forest of the root r_1. For $i \in \{1, \ldots, k\}$, we denote right(r_i) the right sub-forest of r_i. From a Cartesian Forest with $n \geq 1$ nodes, Algorithm 2 gives a recursive approach to building a Schröder Tree with $n+1$ leaves.

Algorithm 2 is an injective function: at each level, the number of subtrees added to the Schröder Tree is exactly $k+1$, where k is the number of trees in the Cartesian Forest. Since both objects share the same generating function, we obtain the result announced in Lemma 1.

6.4   A Bijection with Parentheses Words

Definition 12 (Parentheses Word w). *A Parentheses Word w is a word over the alphabet $\{(\,,\,\square\,,\,)\}$ such that either $w = \square$ or $w = (w_1 \cdots w_k)$ where $k \geq 2$ and each w_i is a Parentheses Word.*

Do note that, in this definition, unlike the more commonly found definition of these words, we allow parentheses to surround the entire sequence. We notably do so in order to simplify Algorithm 3 and to make the bijection more apparent to the reader. But since every parentheses word that is not $\square$ is contained between a pair of parentheses, it is not necessary to represent it. Informally, one may consider the symbols of the word as separators between the nodes and each group of parentheses as a (sub-)forest. Using the same arguments as for Schröder Tree, Algorithm 3 is an injective function.

Figure 5 shows the correspondence between Cartesian Forests, Schröder Trees and Parentheses Words.

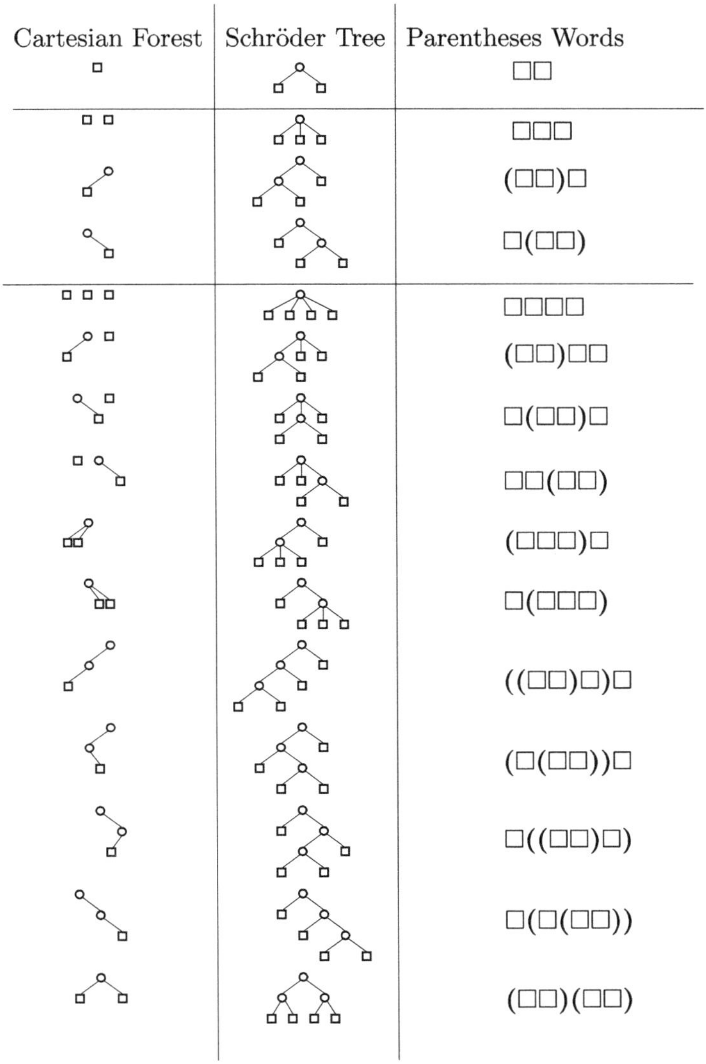

Fig. 5. Correspondence between Cartesian Forests with n nodes, Schröder Trees with $n+1$ leaves and sequences of length $n+1$ with parentheses when $n \in \{1, 2, 3\}$ (respect. top, middle, bottom). □ symbols represent leaves and ○ symbols represent internal nodes. Do note that we omit the parentheses surrounding the entire sequence this time around, to match the classical representation of Parentheses Words.

Lemma 1. *Algorithms 2 and 3 are bijective functions that map a Cartesian Forest with n nodes to a Schröder Tree with $n + 1$ leaves (Algorithm 2), and to a Parentheses Word with $n + 1$ symbols (Algorithm 3). They both have a $\Theta(n)$ time-complexity and a $\Theta(n)$ worst-case space complexity.*

Algorithm 2: CFTOST(F)	**Algorithm 3:** CFTOW(F)
Input : A Cartesian Forest F w. n nodes and k roots, s.t. $F = (r_1, \ldots, r_k)$ **Output:** A Schröder Tree with $n + 1$ leaves	**Input** : A Cartesian Forest F with n nodes and k roots, s.t. $F = (r_1, \ldots, r_k)$ **Output:** A sequence with parentheses and $n + 1$ □

```
Algorithm 2: CFToST(F)
1  ST ← a root;
2  if F is not empty then
3      (c₁, ..., c_{k+1}) ← Create a
         tuple of subtrees;
4      c₁ ← CFToST(left(r₁));
5      for i ∈ {2, ..., k + 1} do
6          cᵢ ←
             CFToST(right(rᵢ₋₁));
7      ST ← Add subtrees
         (c₁, ..., c_{k+1}) to ST;
8  return ST;
```

```
Algorithm 3: CFToW(F)
1  w ← □;
2  if F is not empty then
3      (w₁ ··· w_{k+1}) ← Create a tuple
         of words;
4      w₁ ← CFToW(left(r₁));
5      for i ∈ {2, ..., k + 1} do
6          wᵢ ←
             CFToW(right(rᵢ₋₁));
7      w ← (w₁ · w₂ ··· w_{k+1});
8  return w;
```

Fig. 6. Algorithms that take a Cartesian Forest as input and return the corresponding Schröder Tree (algorithm on the left) or Parentheses Word (algorithm on the right). As one can see, both methods are very similar. In both cases, two different inputs imply two different outputs. Hence both functions are injections. Since there is the same number of objects of the same size in all three cases, the functions are bijections.

7   Cartesian Forest Signature and Cartesian Forest Matching Using a Filter

In [6], the authors propose a *signature* (or perfect hash) of a Cartesian Tree, based on the *Skipped-number* representation. Given a Cartesian Tree with n nodes, its signature is an integer with at most $2n$ bits. In this section, we extend this notion of signature to Cartesian Forests, obtaining a signature with at most $3n$ bits.

Definition 13 (Cartesian Forest Signature). *Given the Forest Skipped-Number representation* F-SN$_x$ *of a sequence* x *(see Definition 8), its signature is defined in the following way: for positions* $i \in \{1, \ldots, n\}$ *in the Forest Skipped-Number representation, each value* F-SN$_x[i]$ *is encoded by a sequence of bits, that are concatenated to obtain an integer.*

- *The first bits concern the sign of* F-SN$_x[i]$*: If* F-SN$_x[i] = 0$*, it is equal to* 0*; The case* F-SN$_x[i] < 0$ *is encoded by* 10*, and the case* F-SN$_x[i] > 0$ *is encoded by* 11*.*
- *If* F-SN$_x[i] \neq 0$*, the following bits are a unary encoding of* $|$F-SN$_x[i]|$*:* $|$F-SN$_x[i]| - 1$ *bits equal to* 1 *followed by a bit equals to* 0*.*

Since the total number of skipped nodes cannot exceed n, a signature contains at most $3n$ bits. This representation could be used efficiently as a perfect hash for small patterns, to obtain an efficient algorithm to solve the Cartesian Forest Matching problem: one only needs to update the signature as one would update the Forest Skipped-Number representation of the chunk of text compared to the pattern, using bitwise operations. But it becomes inefficient when $3n$ exceeds the size of a register, as one cannot use those operations efficiently anymore. Therefore, in order to accelerate the Cartesian Forest Matching in a Rabin-Karp fashion we now introduce a filtering method.

Definition 14 (Cartesian Forest τ-Filter). *Given a sequence x of length n and its linear representation LN_x, its τ-Filter, denoted by $\mathcal{F}il_x$, is a sequence of τ bits such that $\mathcal{F}il_x[i] = 0$ if $\mathrm{LN}_x[n - \tau + i] = 0$ and $\mathcal{F}il_x[i] = 1$ otherwise.*

Algorithm 1 can be adapted: line 5 is only tested if $\mathcal{F}il_x$ is equal to $\mathcal{F}il_p$. τ can be adapted to match the size of a classical register, that is 32, 64 or 128 bits. Therefore, comparing two filters can be made in constant time. The update of the filter can also be made in constant time by tracking the positions in LN_x that have been updated (see [3] for more details on the update function of the linear representations). In the following Section, we only implemented the filter for the Forest Skipped-Number representation.

8   Experiments

In this Section, we implemented a Cartesian Forest τ-Filter using $\tau = 64$. The first set of experiments considers the uniform distribution over text and patterns of respective length n and m, over a k letter alphabet. Figure 7 sums up the details and observations. As one can see, the Forest Parent-Distance version is the slowest (`PD meta`), then comes the one using the Forest Skipped-Number representation (`SN meta`). The method using a filter (`SN hash`) is even faster, but only with a slight margin. Finally, the KMP version [19] is the fastest one (`PD KMP`). Note that we tried to implement the Skip-Search algorithm [21] without the SIMD instructions, but the experimental results were disappointing. The average cost slightly increases with k, which is probably due to the cost of the update function, since the average Forest Parent-Distance increases with the size of the alphabet. Note that, whatever the linear representation is, the number of comparisons at Line 5 in Algorithm 1 is the same, since they encode the same prefix. Therefore, it is likely that the extra cost of the Forest Parent-Distance representation is due to the fact that both requires to use a stack in their implementations but more values are pushed in the stack, in the Forest Parent-Distance case.

The experiment in Fig. 8 uses the random generator from [5]. It shows the algorithm remains efficient even with (not too) low entropy. It generates sequences uniformly amongst those of a fixed length over a fixed alphabet and a fixed Collision Entropy (that is Rényi Entropy with $\alpha = 2$). The Collision Entropy is a function of the probability for two random variables (a letter in the pattern p and the text t) to be equal. The lower the entropy is, the higher average complexity of the algorithms. As a matter of fact, if the entropy reaches

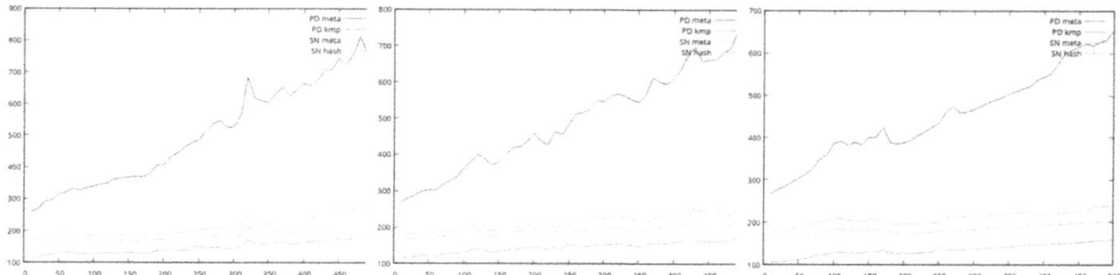

Fig. 7. We randomly generated 10 000 patterns of size m and texts of size 1 000. The x-axis represents m and the y-axis is the average time taken by each algorithm in microseconds. From left to right, the alphabet size is respectively equal to 2, 4 and m.

0, then both the text and the pattern contain the repetition of a unique symbol, which corresponds to the worst-case complexity of Algorithm 1. Though, as one can see in Fig. 8, the average cost of the algorithms quickly drops and stabilizes. The KMP version is the most efficient (this is in line with the results of [2] showing that the KMP algorithm is efficient for order preserving matching) and is (unsurprisingly) very stable, even when the entropy is low.

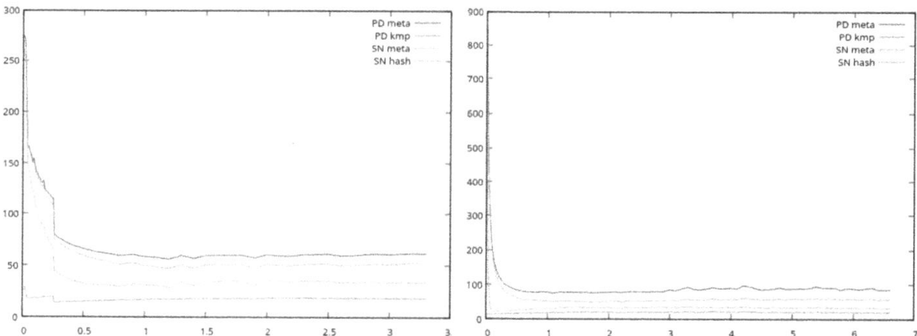

Fig. 8. For each value of the collision entropy (x-axis), 10 000 random texts of length 1 000 and patterns of length 100 were generated, and the four algorithms were applied to them. On the left $m = 10$, and on the right $m = 100$. The y-axis is the average time in microseconds. The added value of using the filter method, compared to the simple Forest Skipped-Number version, is more important when the entropy is low, whereas the difference in efficiency between the Forest Parent-Distance version and the Forest Skipped-Number version drops.

References

1. Aho, A.V., Corasick, M.J.: Efficient string matching: an aid to bibliographic search. Commun. ACM **18**(6), 333–340 (1975)
2. Amir, O., Amir, A., Fraenkel, A., Sarne, D.: On the practical power of automata in pattern matching. SN Comput. Sci. **5**(4), 400 (2024)

3. Auvray, B., David, J., Ghazawi, S., Groult, R., Landau, G.M., Lecroq, T.: Approximate Cartesian Tree matching with one difference (2025). https://arxiv.org/abs/2505.09236. Submitted at TCS, Under revision
4. Bouvel, M., Chauve, C., Mishna, M., Rossin, D.: Average-case analysis of perfect sorting by reversals. Discret. Math. Algorithms Appl. **3**(03), 369–392 (2011)
5. David, J.: Random Generation of Source Vectors with a fixed determining property (2023). https://hal.science/hal-04616899. Submitted at TCS, Under revision
6. Demaine, E.D., Landau, G.M., Weimann, O.: On Cartesian Trees and range minimum queries. Algorithmica **68**(3), 610–625 (2014)
7. Faro, S., Lecroq, T., Park, K., Scafiti, S.: On the longest common Cartesian substring problem. Comput. J. **66**(4), 907–923 (2023)
8. Flajolet, P., Sedgewick, R.: Analytic Combinatorics. Cambridge University Press (2009)
9. Funakoshi, M., Mieno, T., Nakashima, Y., Inenaga, S., Bannai, H., Takeda, M.: Computing maximal palindromes in non-standard matching models. In: IWOCA, pp. 165–179 (2024)
10. Gabow, H.N., Bentley, J.L., Tarjan, R.E.: Scaling and related techniques for geometry problems. In: STOC, pp. 135–143 (1984)
11. Gawrychowski, P., Ghazawi, S., Landau, G.M.: On indeterminate strings matching. In: CPM, pp. 14:1–14:14 (2020)
12. Kikuchi, N., Hendrian, D., Yoshinaka, R., Shinohara, A.: Computing covers under substring consistent equivalence relations. In: SPIRE, pp. 131–146 (2020)
13. Kim, S., Cho, H.: A compact index for Cartesian Tree matching. In: CPM, pp. 18:1–18:19 (2021)
14. Kim, S., Han, Y.: Approximate Cartesian Tree pattern matching. In: DLT, pp. 189–202 (2024)
15. Knuth, D.E., Morris, J.H., Jr., Pratt, V.R.: Fast pattern matching in strings. SIAM J. Comput. **6**(1), 323–350 (1977)
16. Nishimoto, A., Fujisato, N., Nakashima, Y., Inenaga, S.: Position heaps for Cartesian-Tree matching on strings and tries. In: SPIRE, pp. 241–254 (2021)
17. Oizumi, T., Kai, T., Mieno, T., Inenaga, S., Arimura, H.: Cartesian Tree subsequence matching. In: CPM, pp. 14:1–14:18 (2022)
18. Osterkamp, E.M., Köppl, D.: Extending the Burrows-Wheeler transform for Cartesian Tree matching and constructing it. In: CPM, pp. 26:1–26:17 (2025)
19. Park, S., Amir, A., Landau, G., Park, K.: Cartesian Tree matching and indexing. In: CPM, pp. 1–14 (2019)
20. Park, S.G., Bataa, M., Amir, A., Landau, G.M., Park, K.: Finding patterns and periods in Cartesian Tree matching. Theoret. Comput. Sci. **845**, 181–197 (2020)
21. Song, S., Gu, G., Ryu, C., Faro, S., Lecroq, T., Park, K.: Fast algorithms for single and multiple pattern Cartesian Tree matching. Theoret. Comput. Sci. **849**, 47–63 (2021)
22. Tsujimoto, T., Shibata, H., Mieno, T., Nakashima, Y., Inenaga, S.: Computing longest common subsequence under Cartesian-Tree matching model. In: IWOCA, pp. 369–381 (2024)
23. Vuillemin, J.: A unifying look at data structures. Commun. ACM **23**(4), 229–239 (1980)

Streaming Algorithms for Products of Probabilities

Markus Lohrey$^{(\boxtimes)}$ [iD], Leon Rische, Louisa Seelbach [iD], and Julio Xochitemol [iD]

Department ETI, Universität Siegen, 57076 Siegen, Germany
`lohrey@eti.uni-siegen.de`, `leon.rische@me.com`,
`Julio.JXochitemol@uni-siegen.de`

Abstract. We consider streaming algorithms for approximating a product of input probabilities up to multiplicative error of $1 - \epsilon$. It is shown that every randomized streaming algorithm for this problem needs space $\Omega(\log n + \log b - \log \epsilon - \Theta(1))$, where n is length of the input stream and b is the bit length of the input numbers. This matches an upper bound from Alur et al. up to a constant multiplicative factor. Moreover, we consider the threshold problem, where it is asked whether the product of the input probabilities is below a given threshold. It is shown that every randomized streaming algorithm for this problem needs space $\Omega(n \cdot b)$. Finally, also the sliding window variant of the approximation problem for a product of input probabilities is considered.

Keywords: Streaming algorithms · Lower bounds · Computing products

1 Introduction

The streaming model is one of the central concepts in data science to deal with huge data sets, where random access is no longer feasible. It is based on the processing of an input stream of data values. Every data value has to be processed immediately when it arrives. For a good introduction to streaming algorithms see [3,8]. In this paper we consider the problem of computing probabilities in the streaming model. More precisely, consider a stream of probabilities $q_1, q_2, \ldots, q_n$ where every q_i is given as a fraction r_i/s_i of integers r_i, s_i with $r_i \geq 0$ and $1 \leq r_i \leq s_i$. The goal is to compute a good approximation of the product probability $\prod_{1 \leq i \leq n} q_i$. This problem has been first studied by Alur et al. in an automata theoretic setting [2]. It was shown that a $(1 - \epsilon)$-approximation[1] of the product $\prod_{1 \leq i \leq n} q_i$ can be computed by a deterministic streaming algorithm in space $\log_2(n^2 b/\epsilon) = 2 \log n + \log b - \log \epsilon$. Here, n is the length of the stream and b is the maximal bit length of the numerators r_i and denominators s_i in the stream of probabilities $q_i = r_i/s_i$.[2]

---

[1] A $(1-\epsilon)$-approximation of a number x is any number x' with $(1-\epsilon)x < x' < x/(1-\epsilon)$.

[2] The space bound stated in [2] is actually larger than $2 \log n + \log b - \log \epsilon$ since the authors in [2] also count the space for internal computations; see Sect. 2.1 for more details on this.

E. Di Giacomo and D. Mondal (Eds.): WALCOM 2026, LNCS 16444, pp. 140–153, 2026.
https://doi.org/10.1007/978-981-95-7127-7_10

Our first result is a lower bound that matches the upper bound $2\log n + \log b - \log \epsilon$ up to a constant factor. This is also true for randomized streaming algorithms that produce a $(1-\epsilon)$-approximation of the product $\prod_{1 \leq i \leq n} q_i$ with high probability. Our lower bound only holds if $\log(1/\epsilon) \leq b < \log(1/(1-\epsilon)) \cdot n/2$. This is a reasonable setting. If we consider the approximation parameter ϵ as a small constant then the requirement becomes $\Omega(1) \leq b \leq \mathcal{O}(n)$. Since the length n of the stream is usually very large, it is reasonable to assume $b \leq \mathcal{O}(n)$. In a realistic scenario one would even require $b \leq \mathcal{O}(\log n)$.

Our second result is a new lower bound for the threshold problem: the input consists again of a stream of probabilities $q_1, q_2, \ldots, q_n$, where $q_i = r_i/s_i$ and the integers r_i and s_i need at most b bits. In addition, there is a threshold probability $t = r/s$, where r and s also need at most b bits. The question is whether $\prod_{1 \leq i \leq n} q_i < t$ holds. It was shown in [2, Theorem A.1] that every deterministic streaming algorithm for the general threshold problem, where the q_i and t are not restricted to the interval $[0,1]$, has to store $\Omega(n)$ many bits when $b = \log n$. It was asked whether this lower bound also holds when all q_i and t are probabilities. We prove that this is indeed the case and moreover improve the lower bound from $\Omega(n)$ to $\Omega(n \cdot b)$ if $b \geq \Omega(\log n)$ for randomized streaming algorithms. Note that the threshold problem can always be solved in space $\Omega(n \cdot b)$ by naively computing the product $\prod_{1 \leq i \leq n} q_i$. On the other hand, if n is very large compared to b, then there is a better (deterministic) algorithm that only needs $\mathcal{O}(2^b)$ bits; see Theorem 5.

In the final section we consider the product approximation problem for rational probabilities in the sliding window setting. Here, the goal is to maintain a $(1-\epsilon)$-approximation of the product of the last m probabilities from the input stream. The parameter m is called the window size. We provide a space lower bound of $\Omega(m \cdot (\log b - \log \epsilon - \Theta(1)))$ for randomized sliding window algorithms, where as before b is the bit length of the input probabilities. Currently, we are not able to match this lower bound by an upper bound. In Sect. 5 we provide two incomparable upper bounds: $2mb$ (which is trivial) and $m \cdot (\log m + \log b - \log \epsilon)$.

1.1 Related Work

The first paper that studied the sliding window model formally was [5]. It is shown there that for a stream of positive integers of bit length b, a $(1-\epsilon)$-approximation of the sum of the numbers in a sliding window of length m can be maintained in space $\mathcal{O}((1/\epsilon)(\log m + b)\log m)$. A matching lower bound is shown as well. A survey on sliding window algorithms can be found in [1].

2 Preliminaries

For a possibly infinite set A we denote with A^* the set of all finite words (or sequences) over the set A. We write $A^{\leq n}$ for the subset of words of length at most n. With log we always denote the logarithm to the base two. For a positive rational number $q = r/s$ with $r, s \in \mathbb{N}$ and $\gcd(r, s) = 1$ we define the *bit size* of

q as $\|q\| = \max\{\lceil \log r \rceil, \lceil \log s \rceil\}$. In addition, we define $\|0\| = 0$. Let $\mathbb{P} = \{q \in \mathbb{Q} : 0 \le q \le 1\}$ be the set of rational probabilities and $\mathbb{P}_b = \{q \in \mathbb{P} : \|q\| \le b\}$ be the set of rational probabilities of bit size at most b.

2.1 Streaming Algorithms

We only consider streaming problems where the input values are rational probabilities and there is a set of allowed rational output probabilties (this also covers the case, where the output is either 0 or 1). In our context, the set of allowed output probabilities consists of all approximations of some rational probability that is uniquely determined by the input stream.

Formally, a streaming problem is specified by a function $F : \mathbb{P}^* \to 2^{\mathbb{P}}$ mapping a sequence $q_1 q_2 \cdots q_n$ of rational probabilities to a set $F(q_1 q_2 \cdots q_n)$ of rational probabilities. Below, we slice the set of all input sequences $\mathbb{P}^*$ into the sets $\mathbb{P}_b^{\le n}$ consisting of input sequences of length at most n, where in addition all input numbers have bit size at most b.

A deterministic streaming algorithm can be formalized by a family of deterministic finite automata $\mathcal{D} = (\mathcal{A}_{n,b})_{n,b \in \mathbb{N}}$, where the automaton $\mathcal{A}_{n,b}$ is responsible for inputs from the slice $\mathbb{P}_b^{\le n}$. Hence, the input alphabet of $\mathcal{A}_{n,b}$ is $\mathbb{P}_b$. Let $Q_{n,b}$ be the set of states of $\mathcal{A}_{n,b}$. Its elements can be seen as the memory states of the streaming algorithm. There is a distinguished initial memory state $s_{0,n,b} \in Q_{n,b}$. The dynamics of the streaming algorithm is described by the state transition function $\delta_{n,b} : Q_{n,b} \times \mathbb{P}_b \to Q_{n,b}$ such that for a memory state $s \in Q_{n,b}$ and an incoming data value $q \in \mathbb{P}_b$, $\delta_{n,b}(s, q) \in Q_{n,b}$ is the new memory state. In addition, $\mathcal{A}_{n,b}$ has an output function $o_{n,b} : Q_{n,b} \to \mathbb{P}$ instead of a set of final states. The automaton $\mathcal{A}_{n,b}$ computes a function $f_{\mathcal{A}_{n,b}} : \mathbb{P}_b^{\le n} \to \mathbb{P}$ as follows: Starting from the initial state, the automaton reads the input stream $w \in \mathbb{P}_b^*$. After reading the last input number from w, the automaton is in a certain state $s \in Q_{n,b}$. The output value $f_{\mathcal{A}_{n,b}}(w)$ is then $o_{n,b}(s)$. A deterministic streaming algorithm $\mathcal{D} = (\mathcal{A}_{n,b})_{n,b \in \mathbb{N}}$ for the streaming problem F has to satisfy $f_{\mathcal{A}_{n,b}}(w) \in F(w)$ for all $n, b \in \mathbb{N}$ and $w \in \mathbb{P}_b^{\le n}$. Finally, the space complexity of the streaming algorithm $\mathcal{D}$ is the mapping $\mathsf{space}_{\mathcal{D}}(n, b) = \lceil \log_2 |Q_{n,b}| \rceil$, which is the number of bits required to encode the states in binary notation. It is therefore the number of bits that $\mathcal{A}_{n,b}$ has to store.

Some comments regarding our definition are appropriate. First, our definition of a deterministic streaming algorithm is non-uniform in the sense that for every input slice $\mathbb{P}_b^{\le n}$ we have a separate algorithm (the automaton $\mathcal{A}_{n,b}$). In principle the function $(n, b) \mapsto \mathcal{A}_{n,b}$ may be not computable. In a real streaming algorithm, this function will be certainly computable. On the other hand, the main focus in this paper is on lower bounds, and lower bounds for non-uniform algorithms also hold for uniform algorithms. The few streaming algorithms that we present in this paper are moreover clearly uniform.

Related to the previous point is the fact that we ignore in our definition of the space complexity $\mathsf{space}_{\mathcal{D}}(n, b) = \lceil \log_2 |Q_{n,b}| \rceil$ the space needed for computing the mapping $\delta_{n,b}$, i.e., the space needed to compute the next memory state from the previous memory state and the input data value. This space is what we called the space for internal computations in footnote 2. In general, this internal space will also grow with n and b. In a non-uniform streaming algorithm, the function $(n, b, q, s) \mapsto \delta_{n,b}(q, s)$ might be even uncomputable. The justification for ignoring the space for internal computation is the same as for non-uniformity. A lower bound for our measure of space complexity also holds for the more realistic measure that incorporates the internal space needed for computing the state transformation mapping. Another point is that lower bound techniques for streaming algorithms (which are typically based on information theoretic arguments or communication complexity) can only yield lower bounds on our measure of space complexity: the classical application of communication lower bounds to streaming lower bounds constructs from a streaming algorithm (for a certain streaming problem) a communication protocol (for a certain communication problem). Thereby the messages exchanged between Alice to Bob are memory states of the streaming algorithm. Hence, a lower bound for the communication problem yields a lower bound for the bit length of the memory states of the streaming algorithm. This line of argument does not take the memory for internal computations in the streaming algorithm into account.

Randomized streaming algorithms can be defined analogously to deterministic streaming algorithms by taking for every $\mathcal{A}_{n,b}$ a probabilistic finite automaton (see [9] for details on probabilistic automata). For an input stream $w \in \mathbb{P}_b^*$, we then obtain a random variable $\mathsf{R}_{\mathcal{A}_{n,b}}^w$, whose support is the set of all output values $o_{n,b}(s)$ for $s \in Q_{n,b}$. For an output value $q \in \mathbb{P}$, $\mathsf{R}_{\mathcal{A}_{n,b}}^w(q)$ is the probability that on input w the probabilistic automaton $\mathcal{A}_{n,b}$ returns q. A randomized streaming algorithm $\mathcal{R} = (\mathcal{A}_{n,b})_{n,b \in \mathbb{N}}$ for the streaming problem F has to satisfy

$$\mathsf{Prob}[\mathsf{R}_{\mathcal{A}_{n,b}}^w \in F(w)] \geq \frac{2}{3} \tag{1}$$

for all $n, b \in \mathbb{N}$ and $w \in \mathbb{P}_b^{\leq n}$.

We consider the following streaming problems in this paper, where $0 \leq \epsilon < 1$ is an approximation ratio and $q_1, q_2, \ldots, q_n \in \mathbb{P}$:

- The *approximation of products of probabilities*, APP_ϵ for short, is

$$\mathsf{APP}_\epsilon(q_1 q_2 \cdots q_n) = \left\{ q \in \mathbb{P} : (1 - \epsilon) \prod_{i=1}^{n} q_i < q < \frac{1}{1 - \epsilon} \prod_{i=1}^{n} q_i \right\}.$$

- The *threshold problem for products of probabilities*, TPP for short, is

$$\mathsf{TPP}(q_1 q_2 \cdots q_n) = \begin{cases} 1 & \text{if } \prod_{i=2}^{n} q_i < q_1 \\ 0 & \text{otherwise.} \end{cases}$$

It is easy to see that for these streaming problems the error probability $\leq \frac{1}{3}$ from (1) can be reduced to any $\lambda < 1/3$ by running $\mathcal{O}(\log(1/\lambda))$ many independent copies of $\mathcal{A}_{n,b}$. In the case of APP_ϵ, the output value will be the median of the output values of the independent copies of $\mathcal{A}_{n,b}$, whereas for TPP (where the output is boolean) the output bit is obtained by a majority vote.

2.2 Communication Complexity

Lower bounds for randomized streaming algorithms are typically derived from lower bounds on randomized communication complexity [6,11]. We only use the one-way setting, where Alice sends a single message to Bob.

Consider a function $f \colon X \times Y \to \{0,1\}$ for some finite sets X and Y, which we call a communication problem. A *randomized one-way (communication) protocol* P for f consists of two parties called Alice and Bob. The input for Alice (resp., Bob) is an element $x \in X$ and a random choice $r \in R$ (resp., $y \in Y$ and a random choice $s \in S$). Here, R and S are finite sets and the random choices of Alice and Bob are independent. The goal of Alice and Bob is to compute $f(x,y)$ with high probability. For this, Alice computes from here input $x \in X$ and random choice $r \in R$ a message $a(x,r) \in \{0,1\}^*$ and sends it to Bob. Bob then computes from $a(x,r)$, his input $y \in Y$ and random choice $s \in S$ the final output $b(a(x,r),y,s) \in \{0,1\}$ of the protocol. We also assume a probability distribution on the set R (resp., S) of Alice's (resp., Bob's) random choices. The protocol P *computes* f if for all $(x,y) \in X \times Y$ we have

$$\mathsf{Prob}_{r \in R, s \in S}[b(a(x,r),y,s) = f(x,y)] \geq \frac{2}{3}. \tag{2}$$

The cost of the protocol is the maximum length of $a(x,r)$ taken over all $(x,r) \in X \times R$. The *randomized one-way communication complexity* of f is the minimal cost among all (one-way) randomized protocols that compute f. Here, the size of the finite sets R and S is not restricted. The choice of the constant $2/3$ in (2) is arbitrary in the sense that changing the constant to any $1 - \lambda$ only changes the communication complexity by a constant (depending on λ), see [6, p. 30]. Also note that we only use the private version of randomized communication protocols, where Alice and Bob make private random choices from the sets R and S, respectively, and their choices are not known to the other party (in contrast to the public version of randomized communication protocols).

We will use the following communication problems in this paper: The *greater-than problem* GT_m is defined as follows:

- Alice's input is a number $a \in \{1,\ldots,m\}$.
- Bob's input is a number $a' \in \{1,\ldots,m\}$.
- Bob's goal is to determine whether $a > a'$.

The *index-greater-than problem* $\mathsf{IGT}_{m,n}$ is defined as follows:

- Alice's input is a sequence $a_1 a_2 \cdots a_n$ with $a_i \in \{1,\ldots,m\}$ for all i.
- Bob's input is a number $a' \in \{1,\ldots,m\}$ and a number $i \in \{1,\ldots,n\}$.
- Bob's goal is to determine whether $a_i > a'$.

Theorem 1. *The following hold:*

- GT_m *has randomized one-way communication complexity* $\Theta(\log m)$ [7].
- $\mathsf{IGT}_{m,n}$ *has randomized one-way communication complexity* $\Theta(n \cdot \log m)$ [4].

3   Lower Bound for the Product Approximation Problem

In this section we consider the problem APP_ϵ. The following upper bound was shown in [2].

Theorem 2 ([2]). *For every $0 < \epsilon < 1/2$ there is a deterministic streaming algorithm for APP_ϵ with space complexity $\log(n^2 b/\epsilon) = 2\log n + \log b - \log \epsilon$.*

Let us briefly explain the idea since it will be needed in Sect. 5. Assume that the input stream is $q_1 q_2 \cdots q_n$. Define $\epsilon' = \epsilon/n$. For an $a \in \mathbb{N}$ we define the interval $B_{a,\epsilon'} = ((1 - \epsilon')^{a+1}, (1 - \epsilon')^a]$, called a *bucket* in [2]. The streaming algorithm computes for every input number q_i the unique $a_i \in \mathbb{N}$ such that $q_i \in B_{a_i,\epsilon'}$ and computes the sum $a = \sum_{i=1}^n a_i$. At the end it returns $(1 - \epsilon')^a$. It can be shown that this is a $(1 - \epsilon)$-approximation of the product of the q_i. Moreover, every bucket index a_i can be bounded by nb/ϵ (in [2] the bound $4nb/\epsilon$ is stated, but it easy to see that nb/ϵ suffices.) Hence, every partial sum of the a_i is bounded by $n^2 b/\epsilon$ for which $2\log n + \log b - \log \epsilon$ bits suffice.

The space bound stated in [2] also contains the space needed for computing the memory state transition function (called $\delta_{n,b}$ in Sect. 2.1). This is mainly the space for computing the bucket indices a_i. Recall from Sect. 2.1 that we ignore this internal space in our model. We match the upper bound $2\log n + \log b - \log \epsilon$ by a lower bound up to a factor of two for a reasonable range of the parameters n, b, and ϵ.

Theorem 3. *Let $0 < \epsilon < 1/2$. Every deterministic streaming algorithm for APP_ϵ has space complexity at least $\log n + \log b - \log \epsilon - \Theta(1)$ when n, b, and ϵ satisfy $-\log \epsilon \leq b < -\log(1 - \epsilon) \cdot n/2$.*

Proof. Fix an ϵ with $0 < \epsilon < 1/2$ and let $\delta = -\log(1 - \epsilon)$. Note that $0 < \delta < 1$. Let $\mathcal{D}$ be a deterministic streaming algorithm for APP_ϵ. It suffices to show that $\mathsf{space}_\mathcal{D}(2n, b) \geq \log n + \log b - \log \delta - \Theta(1)$ under the assumption

$$-\log \epsilon \leq b < \delta \cdot n. \tag{3}$$

We can then subsitute n by $n/2$. Moreover, it can be easily shown that the difference $\log \epsilon - \log \delta = \log \epsilon - \log \log(1/(1-\epsilon))$ belongs to the interval $(-1, \log \ln 2) \subseteq (-1, -0.52)$ when $0 < \epsilon < 1/2$. Thus, we have $\log \epsilon - \Theta(1) = \log \delta - \Theta(1)$.

By (3) (which yields $1/2^b \leq \epsilon < 1/2$) there is a $k \in \mathbb{N}$ with $3 \leq k \leq 2^b$ and

$$\frac{1}{k} \leq \epsilon \leq \frac{1}{k-1} \tag{4}$$

and hence

$$\frac{k-2}{k-1} \leq 1 - \epsilon \leq \frac{k-1}{k}. \tag{5}$$

We consider streams of length at most $2n$ of rational numbers from the set

$$S = \left\{ \frac{k-1}{k}, \frac{k-2}{k-1}, \frac{1}{2^b} \right\} \subseteq \mathbb{P}. \tag{6}$$

Note that $\|q\| \leq b$ for all $q \in S$. Let us define

$$y := \left\lfloor \frac{nb}{\delta} \right\rfloor. \tag{7}$$

For $j \geq 0$ we define the j-th bucket as the interval

$$B_j = \left((1 - \epsilon)^{j+1}, (1 - \epsilon)^j\right] \subseteq (0, 1].$$

Note that every real number $r \in (0, 1]$ belongs to a unique bucket.

Claim. For every j with $0 \leq j \leq y$ there are numbers $q_1, q_2, \ldots, q_\ell \in S$ with $\ell \leq 2n$ and $\prod_{i=1}^{\ell} q_i \in B_j$.

Proof. Let $0 \leq j \leq y$. Choose the number $m \geq 0$ maximal such that $2^{-bm} \geq (1 - \epsilon)^j$, which is equivalent to $2^{bm} \leq 2^{\delta j}$, i.e., $bm \leq \delta j$. We therefore have

$$m = \left\lfloor \frac{j\delta}{b} \right\rfloor \leq \frac{j\delta}{b} \leq \frac{y\delta}{b} \overset{(7)}{\leq} n.$$

We also have

$$(1 - \epsilon)^j \leq 2^{-bm} \leq 2^{-b\left(\frac{j\cdot\delta}{b} - 1\right)} = 2^b \cdot 2^{-j\cdot\delta} = 2^b(1 - \epsilon)^j \overset{(3)}{<} 2^{\delta n}(1 - \epsilon)^j$$
$$= (1 - \epsilon)^{-n}(1 - \epsilon)^j = (1 - \epsilon)^{j-n}.$$

It follows that if B_s is the unique bucket containing 2^{-bm} then $\max\{0, j - n\} \leq s \leq j$. Hence, we have $0 \leq j - s \leq n$. We now define the first $m \leq n$ numbers of the stream as $q_1 = q_2 = \cdots = q_m = 2^{-b}$. Hence we have

$$q_1 q_2 \cdots q_m = 2^{-bm} \in B_s, \tag{8}$$

where s satisfies $0 \leq j - s \leq n$.

In the second step we choose the numbers $q_{m+1}, \ldots, q_{m+j-s}$ inductively such that $q_1 q_2 \cdots q_{m+j-s} \in B_j$. Note that $m + j - s \leq 2n$. Assume that $q_{m+1}, \ldots, q_{m+d} \in S$ have been chosen for some $0 \leq d < j - s$ such that $q_1 q_2 \cdots \cdots q_{m+d} \in B_{s+d}$. Note that for $d = 0$ this holds by (8). We choose q_{m+d+1} by a case distinction depending on $q_1 q_2 \cdots \cdots q_{m+d}$. It might happen that $q_1 q_2 \cdots \cdots q_{m+d}$ satisfies both cases 1 and 2 below. Then we can choose q_{m+d+1} according to one of them. Recall also that k is such that $3 \leq k \leq 2^b$ and (5) holds.

Case 1. We have

$$q_1 q_2 \cdots q_{m+d} \in \left((1 - \epsilon)^{s+d+1}, \frac{k}{k-1}(1 - \epsilon)^{s+d+1}\right] \subseteq B_{s+d}.$$

Inclusion in B_{s+d} holds since $k(1 - \epsilon)/(k - 1) \leq 1$ by (5). Then we get

$$q_1 q_2 \cdots q_{m+d} \cdot \frac{k-1}{k} \in \left(\frac{k-1}{k}(1 - \epsilon)^{s+d+1}, (1 - \epsilon)^{s+d+1}\right]$$
$$\subseteq \left((1 - \epsilon)^{s+d+2}, (1 - \epsilon)^{s+d+1}\right] = B_{s+d+1},$$

where the inclusion in the second line follows from $(k-1)/k \geq (1-\epsilon)$; see (5). We can therefore choose $q_{m+d+1} = (k-1)/k \in S$.

Case 2. We have

$$q_1 q_2 \cdots q_{m+d} \in \left(\frac{k-1}{k-2}(1-\epsilon)^{s+d+2}, (1-\epsilon)^{s+d} \right] \subseteq B_{s+d},$$

where the inclusion in B_{s+d} follows from $(k-1)(1-\epsilon)/(k-2) \geq 1$; see again (5). Also note that $k \geq 3$ so that $k-2$ is not zero. Then we get

$$q_1 q_2 \cdots q_{m+d} \cdot \frac{k-2}{k-1} \in \left((1-\epsilon)^{s+d+2}, \frac{k-2}{k-1}(1-\epsilon)^{s+d} \right]$$

$$\subseteq \left((1-\epsilon)^{s+d+2}, (1-\epsilon)^{s+d+1} \right] = B_{s+d+1},$$

where the inclusion in the second line follows again from (5). We can therefore choose $q_{m+d+1} = (k-2)/(k-1) \in S$.

It remains to show that the cases 1 and 2 cover all possibilities, i.e., that

$$\frac{k}{k-1}(1-\epsilon)^{s+d+1} \geq \frac{k-1}{k-2}(1-\epsilon)^{s+d+2}.$$

After cancelling $(1-\epsilon)^{s+d+1}$ we obtain

$$\frac{k}{k-1} \geq \frac{k-1}{k-2}(1-\epsilon) = \frac{k-1}{k-2} - \epsilon \cdot \frac{k-1}{k-2},$$

or, equivalently,

$$\epsilon \geq \frac{k-2}{k-1} \cdot \left(\frac{k-1}{k-2} - \frac{k}{k-1} \right) = \frac{k-2}{k-1} \cdot \frac{1}{(k-1)(k-2)} = \frac{1}{(k-1)^2}.$$

But this is true, because by (4) and $k \geq 3$ we have $\epsilon \geq k^{-1} \geq (k-1)^{-2}$. This concludes the proof of the claim. $\qquad\square$

Consider the buckets B_{3j} for $0 \leq j \leq \lfloor y/3 \rfloor$. By our claim, for every $0 \leq j \leq \lfloor y/3 \rfloor$ there is a number $a_j \in B_{3j}$ such that a_j can be written as a product of length at most $2n$ over the set S. Let s_j be the corresponding input stream whose product is a_j.

Consider now i and j with $0 \leq i < j \leq \lfloor y/3 \rfloor$. We have $a_j < a_i$. Moreover,

$$\frac{a_j}{1-\epsilon} \leq \frac{(1-\epsilon)^{3j}}{1-\epsilon} = (1-\epsilon)^{3j-1} \leq (1-\epsilon)^{3i+2} = (1-\epsilon)(1-\epsilon)^{3i+1} < (1-\epsilon)a_i.$$

This means that the deterministic streaming algorithm $\mathcal{D}$ for APP_ϵ must yield different output numbers for the input streams s_i and s_j. In particular, the streaming algorithm must arrive in different memory states for the input streams s_i and s_j. Since this holds for all $0 \leq i < j \leq \lfloor y/3 \rfloor$, the algorithm must have at least $\lfloor y/3 \rfloor + 1 \geq y/3$ memory states and therefore must store at least $\log(y/3) = \log n + \log b - \log \delta - \Theta(1)$ bits. $\qquad\square$

We now extend Theorem 3 to randomized streaming algorithms.

Theorem 4. *Let $0 < \epsilon < 1/2$. Every randomized streaming algorithm for* APP_ϵ *has space complexity* $\Omega(\log n + \log b - \log \epsilon - \Theta(1))$ *when n, b, and ϵ satisfy* $-\log \epsilon \leq b < -\log(1 - \epsilon) \cdot n/2$.

Proof. We can show this result by a standard reduction to the one-way communication complexity of GT_m; see Theorem 1. Let $\mathcal{R}$ be a randomized streaming algorithm for APP_ϵ. As in the proof of Theorem 3, we define $\delta = -\log(1 - \epsilon)$. It suffices to show that $\mathsf{space}_\mathcal{R}(2n, b) \geq \Omega(\log n + \log b - \log \delta - \Theta(1))$ under the assumption (3). We can assume that the error probability of $\mathcal{R}$ is bounded by $1/6$. Consider the number y and the buckets B_{5i} for $0 \leq i \leq \lfloor y/5 \rfloor$ from the proof of Theorem 3. Let s_i be an input stream of length at most $2n$ whose product yields a number $a_i \in B_{5i}$.

We show that $\mathcal{R}$ yields a randomized one-way protocol for $\mathsf{GT}_{\lfloor y/5 \rfloor + 1}$. Assume that i is the input for Alice and j is the input for Bob, where $0 \leq i, j \leq \lfloor y/5 \rfloor$. Alice starts by running the streaming algorithm $\mathcal{R}$ on the input stream s_i. Thereby she uses here random bits to simulate the probabilistic choices of $\mathcal{R}$. Let α_i by the final memory state of $\mathcal{R}$ obtained by Alice. She sends α_i to Bob. Bob then runs $\mathcal{R}$ on input s_j; let α_j be the final memory state of $\mathcal{R}$ obtained by Bob. Bob then applies the output function of $\mathcal{R}$ to the memory states α_i and α_j and obtains rational numbers r_i and r_j respectively. With probability at least $(5/6)^2 \geq 2/3$ the output numbers r_i and r_j satisfy $a_i(1 - \epsilon) \leq r_i \leq a_i/(1 - \epsilon)$ and $a_j(1 - \epsilon) \leq r_j \leq a_j/(1 - \epsilon)$, and thus $r_i \in B_{5i+1} \cup B_{5i} \cup B_{5i-1}$ and $r_j \in B_{5j+1} \cup B_{5j} \cup B_{5j-1}$. Bob can then distinguish the cases $i < j$, $i \geq j$, and $i > j$:

- If $i < j$ then $r_i > r_j$ and between the unique buckets containing r_j and r_i there are at least two buckets, namely $B_{5i+2}, \ldots, B_{5j-2}$.
- If $i > j$ then $r_i < r_j$ and between the unique buckets containing r_i and r_j there are at least two buckets, namely $B_{5j+2}, \ldots, B_{5i-2}$.
- If $i = j$ then $r_i, r_j \in B_{5i+1} \cup B_{5i} \cup B_{5i-1}$ and there is at most one bucket between the unique buckets containing r_i and r_j.

With Theorem 1, this shows that the communicated memory state α_i must have bit length $\Omega(\log(y/5)) = \Omega(\log n + \log b - \log \delta - \Theta(1))$. $\qquad\square$

4   The Threshold Problem

In this section we consider the threshold problem TPP. We start with two simple upper bounds:

Theorem 5. *There are deterministic streaming algorithms for* TPP *with space complexity $2nb$ and $1.443 \cdot 2^b$ (the latter for b large enough), respectively.*

Proof. Let $B = 2^b$. For notational convenience we consider streams of length $n+1$. Consider such an input stream $q_0 q_1 \cdots q_n \in \mathbb{P}^*$ of $n+1$ rational probabilities with $\|q_i\| \leq b$ for all $0 \leq i \leq n$. Recall that q_0 is the threshold value. A streaming algorithm for TPP can simply store all the q_i with at most $2(n+1)b$ bits in total.

Another way to store the product $\prod_{i=1}^{k} q_i$ for $1 \leq k \leq n$ is to store for every prime $p \leq B$ how often it appears in the product, where an occurrence in the denominator is counted negative. By the prime number theorem, the number of primes $p \leq B$ is asymptotically $B/\ln(B)$ and therefore bounded by $1.4427 \cdot B/b$ for b large enough (note that $1.4427 > \log_2(e)$). The total number of occurrences of a prime p in the product is between $-nb$ and nb. This leads to the space bound $1.4427 \cdot B \cdot (\log n + \log b + 1)/b$.

This can be improved to $1.443 \cdot B$ as follows: We can assume that $q_0 > 0$, otherwise the algorithm can output 0. Since $\|q_0\| \leq b$ we must have $q_0 \geq 1/B$. Moreover, we can ignore all 1's in the stream $q_1 q_2 \cdots q_n$. If $q_i < 1$ then $q_i \leq (B-1)/B$. Assume now that the number of i with $1 \leq i \leq n$ and $q_i \leq (B-1)/B$ is at least $B \cdot \ln B$. Then, the product $\prod_{i=1}^{n} q_i$ is upper bounded by

$$\left(\frac{B-1}{B} \right)^{B \cdot \ln B} < \left(\frac{1}{e} \right)^{\ln B} = \frac{1}{B} \leq q_0.$$

Hence, the algorithm can safely output 1. This means that we can replace in the above space bound $1.4427 \cdot B \cdot (\log n + \log b + 1)/b$ the value $\log n$ by $\log(B \cdot \ln B) \leq b + \log(b) + \mathcal{O}(1)$ which gives the space bound $1.443 \cdot B$ for b large enough (this also includes additional space $\mathcal{O}(b)$ for storing q_0 and a binary counter up to $B \cdot \ln B$ for the number of $q_i \leq (B-1)/B$). $\qquad\square$

Let us now come to lower bounds for TPP.

Theorem 6. *Fix a constant $\gamma > 0$. Every deterministic streaming algorithm $\mathcal{D}$ for* TPP *satisfies* $\mathsf{space}_{\mathcal{D}}(2n, (2+\gamma)(\log_2 n + b)) \geq nb$ *for n and b large enough depending on γ.*

Proof. Let $B = 2^b$. We take the first $nB+n+1$ prime numbers $p_1, p_2, ..., p_{nB+n+1}$. It is known that $p_k \leq k \cdot (\ln k + \ln \ln k)$ for $k \geq 6$ [10, 3.13]. In all numbers appearing in the two input streams that we will construct, the numerator and denominator are bounded by p_{nB+n+1}^2. Therefore, the bit size of all the numbers involved is bounded by

$$\lceil \log_2 p_{nB+n+1}^2 \rceil = \lceil 2 \log_2 p_{nB+n+1} \rceil \leq (2+\gamma)(\log_2 n + b) \tag{9}$$

if n and b are large enough (depending on the constant γ). Assume for the following that (9) holds. For every $1 \leq i \leq n$ we define the set

$$Q_i = \left\{ \frac{p_j}{p_{j+1}} : (i-1)(B+1) + 1 \leq j \leq i(B+1) - 1 \right\}.$$

Thus, Q_i consists of B fractions of two consecutive primes. Note that the sets Q_i are pairwise disjoint. We then consider the set of all words

$$S_{n,b} = \{ q_1 q_2 \cdots q_n \in \mathbb{P}^n : q_i \in Q_i \text{ for all } 1 \leq i \leq n \}. \tag{10}$$

Clearly, $|S_{n,b}| = B^n$. Moreover, for two different words $q_1 q_2 \cdots q_n, r_1 r_2 \cdots r_n \in S_{n,b}$ the products $\prod_{i=1}^{n} q_i$ and $\prod_{i=1}^{n} r_i$ are also different due to the uniqueness of prime factorizations. We set the threshold (the first number in the stream) to

$$t = \frac{p_1}{p_{nB+n+1}}. \tag{11}$$

Assume now that there is a deterministic streaming algorithm $\mathcal{D}$ for TPP such that $\mathsf{space}_{\mathcal{D}}(2n, (2+\gamma)(\log_2 n + b)) < n \cdot b = \log_2 |S_{n,b}|$. Then there must exist two different words $u = q_1 q_2 \cdots q_n$ and $v = r_1 r_2 \cdots r_n$ in $S_{n,b}$ such that after reading tu and tv, $\mathcal{D}$ is in the same memory state. W.l.o.g. we assume that $\prod_{i=1}^{n} q_i < \prod_{i=1}^{n} r_i$. We continue the words tu and tv, respectively, with the length-n word $w = (s_1/r_1)(s_2/r_2) \cdots (s_n/r_n)$, where

$$s_i = \frac{p_{(i-1)(B+1)+1}}{p_{i(B+1)+1}}. \tag{12}$$

Note that $s_i/r_i \leq 1$ and that

$$\prod_{i=1}^{n} s_i = \prod_{i=1}^{n} \frac{p_{(i-1)(B+1)+1}}{p_{i(B+1)+1}} = \frac{p_1}{p_{nB+n+1}} = t.$$

Finally, we have

$$\prod_{i=1}^{n} q_i \prod_{i=1}^{n} r_i^{-1} \prod_{i=1}^{n} s_i < \prod_{i=1}^{n} s_i = t \text{ and} \tag{13}$$

$$\prod_{i=1}^{n} r_i \prod_{i=1}^{n} r_i^{-1} \prod_{i} s_i = \prod_{i=1}^{n} s_i = t. \tag{14}$$

But this is a contradiction, since $\mathcal{D}$ arrives in the same memory state after reading the streams tuw and tvw. $\qquad\square$

From the construction in the proof of Theorem 6 we can easily obtain a lower bound for randomized streaming algorithms for TPP:

Theorem 7. *Fix a constant $\gamma > 0$. Every randomized streaming algorithm $\mathcal{R}$ for TPP satisfies* $\mathsf{space}_{\mathcal{R}}(2n, (2+\gamma)(\log_2 n + b)) \geq \Omega(n \cdot b)$.

Proof. We transform a randomized streaming algorithm $\mathcal{R}$ for TPP into a randomized one-way protocol for GT_{B^n}.

Take the set of words $S_{n,b}$ from (10), the numbers s_i from (12) and the threshold t from (11). Let $f : \{1, 2, \ldots, B^n\} \to S_{n,b}$ be the unique bijection such that for all $1 \leq i, j \leq B^n$ with $f(i) = q_1 q_2 \cdots q_n$ and $f(j) = r_1 r_2 \cdots r_n$ we have: $i > j$ if and only if $\prod_{k=1}^{n} q_k < \prod_{k=1}^{n} r_k$.

On input i, Alice computes the word $f(i) = q_1 q_2 \cdots q_n$, reads $t\, q_1 \cdots q_n$ into $\mathcal{R}$ and sends the resulting memory state α of $\mathcal{R}$ to Bob. Bob then computes from his input j the word $f(j) = r_1 r_2 \cdots r_n$ and reads, starting from memory state α, the word $(s_1/r_1)(s_2/r_2) \cdots (s_n/r_n)$ into the algorithm. From the final output of $\mathcal{R}$, Bob can decide with probability at least $2/3$ whether $i > j$:

- If $\prod_{k=1}^{n} q_k < \prod_{k=1}^{n} r_k$ (i.e., $i > j$) then $\mathcal{R}$ outputs 1 with probability at least $2/3$ because

$$\prod_{k=1}^{n} q_k \prod_{k=1}^{n} \frac{s_k}{r_k} < \prod_{k=1}^{n} s_k = t.$$

– If $\prod_{k=1}^{n} q_k \geq \prod_{k=1}^{n} r_k$ (i.e., $i \leq j$) then $\mathcal{R}$ outputs 0 with probability at least $2/3$ because

$$\prod_{k=1}^{n} q_k \prod_{k=1}^{n} \frac{s_k}{r_k} \geq \prod_{k=1}^{n} s_k = t.$$

It follows from Theorem 1 that the communicated memory state α of $\mathcal{R}$ must have bit length $\Omega(\log B^n) = \Omega(n \cdot b)$. $\qquad\square$

If we require $b \geq \Omega(\log_2 n)$ then the term $(2 + \gamma)(\log_2 n + b)$ in Theorem 7 becomes $\Theta(b)$ and we obtain:

Corollary 1. *Every randomized streaming algorithm $\mathcal{R}$ for the threshold problem* TPP *satisfies* $\mathsf{space}_{\mathcal{R}}(n, b) \geq \Omega(n \cdot b)$ *whenever* $b \geq \Omega(\log_2 n)$.

The condition $b \geq \Omega(\log_2 n)$ cannot be completely avoided in Corollary 1, since this would contradict the upper bound $1.443 \cdot 2^b$ in Theorem 5.

Also note that the lower bound for APP_ϵ in Theorem 3 holds for input streams over a set of only 3 numbers (namely the set S in (6)). The lower bound from Corollary 1 does not hold for a constant number of probabilities, since we can then store the product with $\mathcal{O}(\log n)$ bits.

5   Sliding Window Products

In this section we consider the following sliding window variant $\mathsf{SWAPP}_{m,\epsilon}$ of APP_ϵ, where m is the size of the sliding window.

$$\mathsf{SWAPP}_{m,\epsilon}(q_1 q_2 \cdots q_n) = \left\{ q \in \mathbb{P} : (1 - \epsilon) \prod_{i=n-m+1}^{n} q_i < q < \frac{1}{1 - \epsilon} \prod_{i=n-m+1}^{n} q_i \right\}$$

(here, we set $q_i = 1$ for $i \leq 0$). A randomized sliding window algorithm for $\mathsf{SWAPP}_{m,\epsilon}$ is a collection of probabilistic finite automata $\mathcal{R} = (\mathcal{A}_b)_{b \in \mathbb{N}}$ such that for every $b \in \mathbb{N}$ and every input stream $w \in \mathbb{P}_b^*$ we have:

$$\mathsf{Prob}[\mathsf{R}_{\mathcal{A}_b}^w \in \mathsf{SWAPP}_{m,\epsilon}(w)] \geq \frac{2}{3}.$$

As usual for sliding window algorithms, we measure the space complexity of a sliding window algorithm for $\mathsf{SWAPP}_{m,\epsilon}$ in the window size m (instead of the total stream length n), the bit size b and the approximation ratio ϵ.

Theorem 8. *For* $0 < \epsilon < 1/2$, *there are deterministic sliding window algorithms for* $\mathsf{SWAPP}_{m,\epsilon}$ *with space complexity* $m \cdot (\log m + \log b - \log \epsilon)$ *and* $2mb$, *respectively.*

Proof. The upper bound $2mb$ is clear, since the whole window content can be stored with $2mb$ bits. For the upper bound $m \cdot (\log m + \log b - \log \epsilon)$ we use the algorithm for APP_ϵ from [2]. Recall the sketch after Theorem 2. Our sliding window algorithm stores for a window content $q_1 q_2 \cdots q_m$ the sequence of bucket indices $a_1 a_2 \cdots a_m$, where $q_i \in B_{a_i, \epsilon'}$ for $\epsilon' = \epsilon/m$. By the analysis in [2], $(1 - \epsilon')^a$ (with $a = \sum_{i=1}^{m} a_i$) is a $(1 - \epsilon)$-approximation of $\prod_{i=1}^{m} q_i$. Moreover, every a_i is bounded by mb/ϵ and therefore can be stored with $\log m + \log b - \log \epsilon$ bits. $\quad\square$

Theorem 9. *Let $0 < \epsilon < 1/2$. Every randomized sliding window algorithm for* $\mathsf{SWAPP}_{m,\epsilon}$ *requires space* $\Omega(m \cdot (\log b - \log \epsilon - \Theta(1)))$.

Proof. Let $\delta := -\log(1 - \epsilon) \in (0,1)$. Since $\log \epsilon - \Theta(1) = \log \delta - \Theta(1)$ (see the beginning of the proof of Theorem 3), it suffices to show the space lower bound $\Omega(m \cdot (\log b - \log \delta - \Theta(1)))$.

Define the following integers:

$$\alpha = \lceil 4\delta \rceil, \qquad c = \left\lfloor \frac{b}{\alpha} \right\rfloor. \tag{15}$$

Consider the set $A = \{2^{-i \cdot \alpha} : 0 \leq i \leq c - 1\} \subseteq \mathbb{P}$ of size c. Also note that $\|q\| \leq c \cdot \alpha \leq b$ for all $q \in A$. We prove the theorem by a reduction from the communication problem $\mathsf{IGT}_{c,m}$.

Let $\mathcal{R}$ be a randomized sliding window algorithm for $\mathsf{SWAPP}_{m,\epsilon}$. We can assume that the error probability of $\mathcal{R}$ is bounded by $1/6$. We obtain a randomized one-way communication protocol for $\mathsf{IGT}_{c,m}$ as follows: The input of Alice is a sequence $a_1, a_2, \ldots, a_m$ with $a_i \in A$ and the input of Bob is an index $1 \leq i \leq m$ and a number $a \in A$. Bob wants to find out whether $a_i > a$. Alice sends the memory state of the randomized sliding window algorithm after reading $a_1 a_2 \cdots a_m$ to Bob. This allows Bob to compute with high probability $(1 - \epsilon)$-approximations P_1' and P_2' of the products

$$P_1 = a_i a_{i+1} \cdots a_m \quad \text{and} \quad P_2 = a_{i+1} \cdots a_m a,$$

respectively. For this, Bob continues the sliding window algorithm from the memory state obtained from Alice with the input numbers $1, 1, \ldots, 1$ and $1, 1, \ldots, 1, a$, respectively, where the number of 1's is $i - 1$. With probability at least $1 - 1/3$ the computed approximations P_1' and P_2' satisfy

$$(1 - \epsilon)P_1 < P_1' < P_1/(1 - \epsilon) \quad \text{and} \quad (1 - \epsilon)P_2 < P_2' < P_2/(1 - \epsilon). \tag{16}$$

Assume for the following that (16) holds. We claim that from the quotient P_1'/P_2', Bob can determine whether $a_i > a$ holds. Note that $P_1/P_2 = a_i/a$. Moreover, (16) implies

$$(1 - \epsilon)^2 \cdot \frac{a_i}{a} < \frac{P_1'}{P_2'} < \frac{1}{(1 - \epsilon)^2} \cdot \frac{a_i}{a}.$$

If $a_i \leq a$ then $P_1'/P_2' < 1/(1 - \epsilon)^2$. On the other hand, if $a_i > a$ then by the choice of the set A we must have $a_i/a \geq 2^\alpha \geq 2^{4\delta} = 1/(1 - \epsilon)^4$, which yields $P_1'/P_2' > 1/(1 - \epsilon)^2$. Hence, from the quotient P_1'/P_2' Bob can indeed distinguish the cases $a_i > a$ and $a_i \leq a$. By Theorem 1, $\mathcal{R}$ must use space $\Omega(m \cdot \log c) = \Omega(m \cdot (\log b - \log \delta - \Theta(1)))$. $\qquad\square$

6   Open Problems

For the space complexity of $\mathsf{SWAPP}_{m,\epsilon}$, there is a gap between the upper bounds in Theorem 8 and the lower bound from Theorem 9 that we would like to close.

For APP_ϵ, a small gap (by a multiplicative factor of 2) arises from the upper bound in Theorem 2 and the lower bound in Theorem 3. One may also ask, whether the condition $-\log \epsilon \le b < -\log(1 - \epsilon) \cdot n/2$ in Theorem 3 can be relaxed.

Acknowledgements. This work has been supported by the DFG research project LO748/13-2 (Streaming Automata Theory).

References

1. Aggarwal, C.C. (ed.): Data Streams - Models and Algorithms, Advances in Database Systems, vol. 31. Springer (2007). https://doi.org/10.1007/978-0-387-47534-9
2. Alur, R., Chen, Y., Jothimurugan, K., Khanna, S.: Space-efficient query evaluation over probabilistic event streams. In: Proceedings of the 35th Annual ACM/IEEE Symposium on Logic in Computer Science, LICS 2020, pp. 74–87. ACM (2020). https://doi.org/10.1145/3373718.3394747
3. Blum, A., Hopcroft, J., Kannan, R.: Foundations of Data Science. Cambridge University Press (2020). https://doi.org/10.1017/9781108755528
4. Braverman, V., Grigorescu, E., Lang, H., Woodruff, D.P., Zhou, S.: Nearly optimal distinct elements and heavy hitters on sliding windows. In: Proceedings of the 21st International Conference on Approximation Algorithms for Combinatorial Optimization Problems, and the 22nd International Conference on Randomization and Computation, APPROX/RANDOM 2018. LIPIcs, vol. 116, pp. 7:1–7:22. Schloss Dagstuhl - Leibniz-Zentrum für Informatik (2018). https://doi.org/10.4230/LIPIcs.APPROX-RANDOM.2018.7
5. Datar, M., Muthukrishnan, S.: Estimating rarity and similarity over data stream windows. In: Möhring, R., Raman, R. (eds.) ESA 2002. LNCS, vol. 2461, pp. 323–335. Springer, Heidelberg (2002). https://doi.org/10.1007/3-540-45749-6_31
6. Kushilevitz, E., Nisan, N.: Communication complexity. Cambridge University Press (1997). https://doi.org/10.1017/CBO9780511574948
7. Miltersen, P.B., Nisan, N., Safra, S., Wigderson, A.: On data structures and asymmetric communication complexity. J. Comput. Syst. Sci. **57**(1), 37–49 (1998). https://doi.org/10.1006/JCSS.1998.1577
8. Muthukrishnan, S.: Data streams: algorithms and applications. Found. Trends Theor. Comput. Sci. **1**(2) (2005). https://doi.org/10.1561/0400000002
9. Paz, A.: Introduction to Probabilistic Automata (Computer Science and Applied Mathematics). Academic Press, Inc. (1971). https://doi.org/10.1016/C2013-0-11297-4
10. Rosser, J.B., Schoenfeld, L.: Approximate formulas for some functions of prime numbers. Ill. J. Math. **6**(1), 64–94 (1962). https://doi.org/10.1215/ijm/1255631807
11. Roughgarden, T.: Communication complexity (for algorithm designers). Found. Trends Theor. Comput. Sci. **11**(3–4), 217–404 (2016). https://doi.org/10.1561/0400000076

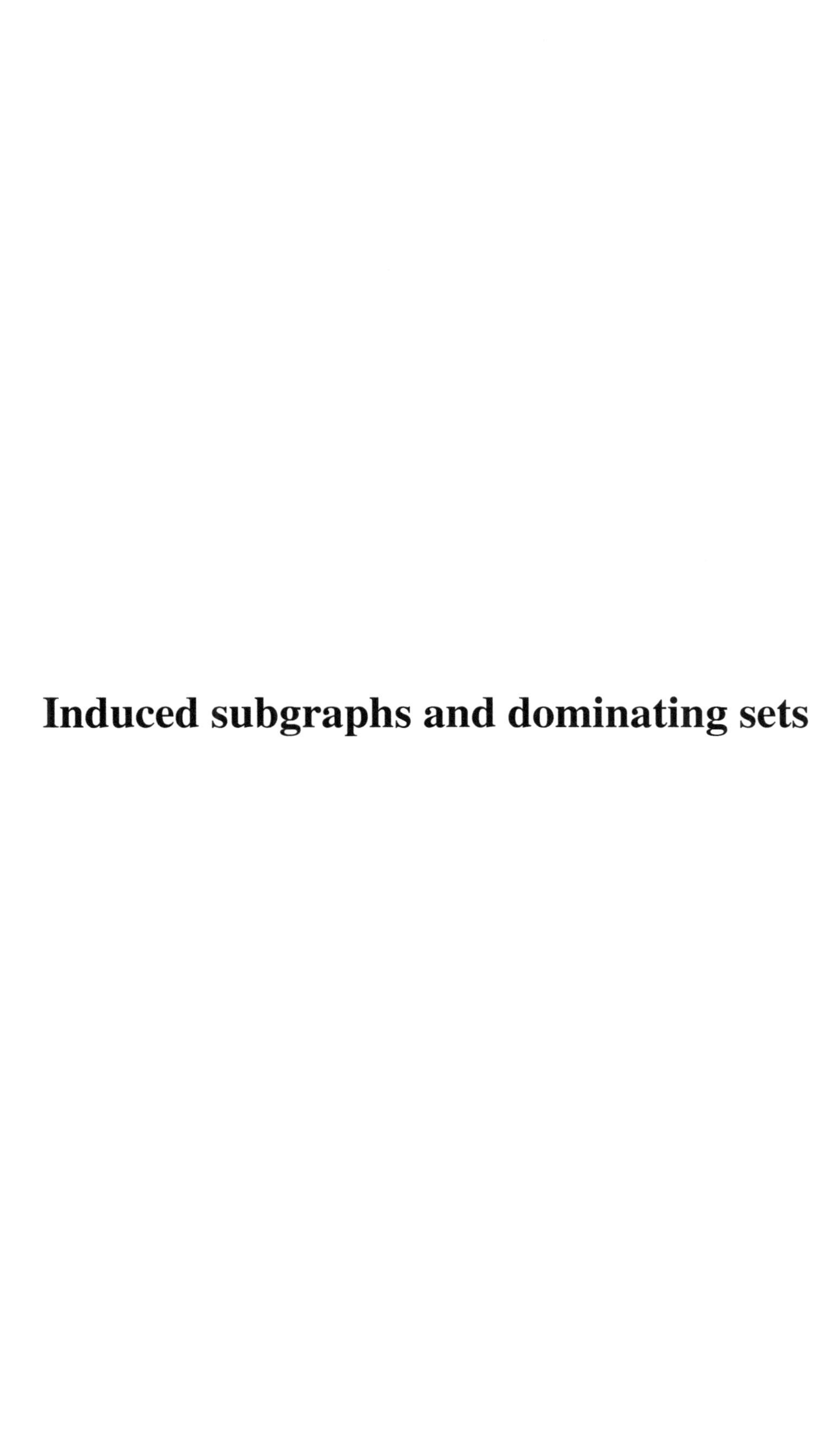

Induced subgraphs and dominating sets

Finding Order-Preserving Subgraphs

Haruya Imamura[1]([✉]) [iD], Yasuaki Kobayashi[2] [iD], Yota Otachi[3] [iD],
Toshiki Saitoh[1] [iD], Keita Sato[4], Asahi Takaoka[4] [iD], Ryo Yoshinaka[5] [iD],
and Tom C. van der Zanden[6] [iD]

[1] Kyushu Institute of Technology, Iizuka, Fukuoka, Japan
imamura.haruya389@mail.kyutech.jp, toshikis@ai.kyutech.ac.jp
[2] Hokkaido University, Sapporo, Hokkaido, Japan
koba@ist.hokudai.ac.jp
[3] Nagoya University, Nagoya, Aichi, Japan
otachi@nagoya-u.jp
[4] Muroran Institute of Technology, Muroran, Hokkaido, Japan
takaoka@muroran-it.ac.jp
[5] Tohoku University, Sendai, Miyagi, Japan
ryoshinaka@tohoku.ac.jp
[6] Maastricht University, Maastricht, The Netherlands
t.vanderzanden@maastrichtuniversity.nl

Abstract. (INDUCED) SUBGRAPH ISOMORPHISM and MAXIMUM COMMON (INDUCED) SUBGRAPH are fundamental problems in graph pattern matching and similarity computation. In graphs derived from time-series data or protein structures, a natural total ordering of vertices often arises from their underlying structure, such as temporal sequences or amino acid sequences. This motivates the study of problem variants that respect this inherent ordering. This paper addresses ORDERED (INDUCED) SUBGRAPH ISOMORPHISM (O(I)SI) and its generalization, MAXIMUM COMMON ORDERED (INDUCED) SUBGRAPH (MCO(I)S), which seek to find subgraph isomorphisms that preserve the vertex orderings of two given ordered graphs. Our main contributions are threefold: (1) We prove that these problems remain NP-complete even when restricted to small graph classes, such as trees of depth 2 and threshold graphs. (2) We establish a gap in computational complexity between OSI and OISI on certain graph classes. For instance, OSI is polynomial-time solvable for interval graphs with their interval orderings, whereas OISI remains NP-complete under the same setting. (3) We demonstrate that the tractability of these problems can depend on the vertex ordering. For example, while OISI is NP-complete on threshold graphs, its generalization, MCOIS, can be solved in polynomial time if the specific vertex orderings that characterize the threshold graphs are provided.

Keywords: Ordered (induced) subgraph isomorphism · graph classes · (in)tractability

1   Introduction

Given two graphs G and H, (INDUCED) SUBGRAPH ISOMORPHISM asks whether G contains a graph isomorphic to H as an (induced) subgraph. MAXIMUM COMMON (INDUCED) SUBGRAPH, which is a generalization of (INDUCED) SUBGRAPH ISOMORPHISM, is a problem to find, given two graphs G and H, a graph Z with the maximum number of edges (vertices) such that Z is an (induced) subgraph of both G and H. It is known that, if the smaller graph H is disconnected, the problems are NP-complete even for disjoint unions of paths [6,8]. Hence, when considering graph classes that include disjoint unions of paths, we particularly focus on the case where H is connected. Kijima et al. [15] showed the NP-completeness of SUBGRAPH ISOMORPHISM on proper interval graphs, bipartite permutation graphs, and trivially perfect graphs. They also gave polynomial-time algorithms for chain graphs, cochain graphs, and threshold graphs. Konagaya et al. [17] studied the complexity of the problem when H belongs to a small class of graphs. For INDUCED SUBGRAPH ISOMORPHISM, the problem is NP-complete on interval graphs [18] and cographs [6], while it can be solved in polynomial time for proper interval graphs, bipartite permutation graphs [11], and threshold graphs [1].

Many graph classes discussed above are intersection graphs of some geometric objects. For example, an interval graph is the intersection graph of a family of intervals on the real line. We can naturally define a vertex ordering (i.e., a total order on the vertex set) from the set of intervals. In this setting, isomorphism mappings are expected to be order-preserving; that is, a vertex u precedes a vertex v in H if and only if $f(u)$ precedes $f(v)$ in G for a subgraph isomorphism f from H to G. Such a mapping is naturally relevant in pattern-matching problems on graphs where each vertex corresponds to an event in a time series. As another application in bioinformatics, a protein contact network is a graph constructed based on the distances between the carbon atoms of amino acids in a protein, and the resulting graph is an intersection graph of unit balls with an inherent vertex ordering [7]. In these applications, we aim to perform pattern matching or compute similarity on graphs with vertex orderings. Therefore, we address the problems ORDERED (INDUCED) SUBGRAPH ISOMORPHISM (O(I)SI) and MAXIMUM COMMON ORDERED (INDUCED) SUBGRAPH (MCO(I)S): Given two graphs along with their vertex orderings, we aim to find an order-preserving subgraph isomorphism or an order-preserving maximum common subgraph (see Sect. 2 for the formal definition).

Here, we discuss the computational complexity of the problems. The problems O(I)SI are generally NP-complete because they include CLIQUE due to the vertex symmetry in complete graphs. On the other hand, when the two input graphs have the same number of vertices, both OISI and OSI can be trivially solved by providing vertex orderings. Note that in this case of non-ordered problems, INDUCED SUBGRAPH ISOMORPHISM is equivalent to the graph isomorphism problem, whereas SUBGRAPH ISOMORPHISM includes HAMILTONIAN CYCLE and is therefore NP-complete. Bose et al. [2] showed that the ORDER PRESERVING SUBSEQUENCE (OPS) problem for two integer sequences is NP-

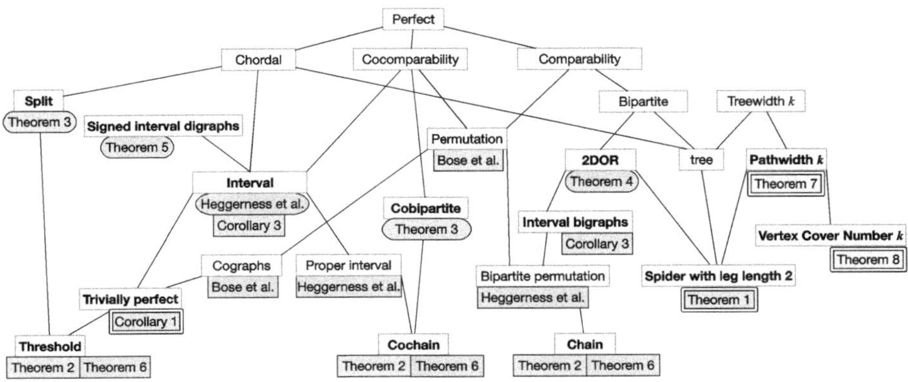

Fig. 1. Illustration of the inclusion relations among graph classes along with the (in)tractability of OSI and OISI. Red, blue, and green denote NP-completeness, polynomial-time solvability, and fixed-parameter tractability, respectively. OSI, OISI, and both problems are represented by squashed rectangles, single-stroked rectangles, and double-stroked rectangles, respectively. (Color figure online)

complete. This problem is equivalent to OISI on permutation graphs with naturally defined vertex orderings of permutation graphs. In contrast, they also proposed a polynomial-time algorithm for cographs with the same vertex orderings. Guillemot and Marx [10] presented a linear-time algorithm for OISI on permutation graphs with the orderings when the number of vertices in H is constant. Heggernes et al. [11] implicitly provided a polynomial-time algorithm for OSI on interval graphs with interval orderings that characterize interval graphs. In the preliminary version of [11], they [12] also provided a polynomial-time algorithm for OISI on interval graphs with interval orderings, but they have been pointed out that it contains a bug in [11]. These studies have considered only the natural orderings that characterize the graph classes, without addressing arbitrary orderings.

Our Contributions: In this paper, we discuss the O(I)SI problems and their generalization, MCO(I)S, with a focus not only on graph classes but also on the properties of orderings. The contributions of this paper are as follows, and they are illustrated in Fig. 1.

We prove in Sect. 3.1 that both OSI and OISI are NP-complete, even on trees. The trees obtained in our reductions are spiders with leg length 2, which are two-directional orthogonal ray graphs (or 2DOR graphs) with pathwidth 2. By a similar reduction, we can also show that both problems are NP-complete for trivially perfect graphs (Corollary 1), a subclass of interval graphs, and for graphs consisting of disjoint edges (Corollary 2). Furthermore, for OISI, we show NP-completeness for three smaller graph classes: threshold graphs, chain graphs, and cochain graphs in Sect. 3.2. The result on threshold graphs (Corollary 3) implies that OISI remains NP-complete on interval graphs even if the vertex orderings are restricted to interval orderings, which indicates that a bug in the algorithm of [12] is unlikely to be fixable. We also prove that OSI remains NP-complete even on split graphs with perfect elimination orderings and cobipartite graphs with cocomparability orderings in Sect. 3.3.

For OSI, we propose polynomial-time algorithms for two graph classes related to interval graphs and interval bigraphs in Sect. 4.1, provided that the vertex orderings are "good": One is the class of 2DOR graphs with comparability weak orderings, and the other is the class of signed-interval digraphs with Min orderings. In Sect. 3.1, we show that for these graph classes, OSI is NP-complete (without restrictions on vertex orderings). This indicates that there is a difference in the complexity between OSI and OISI: For these graph classes, OSI can be solved in polynomial time using a "good" ordering, whereas OISI remains NP-complete even with such orderings (Corollary 3).

We propose algorithms for MCO(I)S, a generalization of O(I)SI. For MCOIS, we propose polynomial-time algorithms for threshold graphs, chain graphs, and cochain graphs with inclusion orderings in Sect. 4.2. These graph classes can be characterized by inclusion orderings, and our algorithms are based on dynamic programming on these orderings. For MCO(I)S, we propose fixed-parameter algorithms (FPT algorithms), parameterized by the pathwidth of the input vertex ordering in Sect. 4.3. Here, the pathwidth of a vertex ordering is defined based on an alternative characterization of pathwidth, known as the vertex separation number [16]. For these graphs, the problems are NP-complete on arbitrary orderings in Sects. 3.1 and 3.2. These results suggest that the vertex orderings have a significant impact on the complexity of our problems. Finally, we present FPT algorithms for MCO(I)S, parameterized by the vertex cover number of the input graph in Sect. 4.4. We note that the algorithm works on arbitrary vertex orderings and that the pathwidth of the input vertex ordering can be arbitrarily large even when the vertex cover number is bounded.

2   Definitions

For a natural number n, we denote the set of integers $\{1, 2, \ldots, n\}$ by $[n]$.

Unless stated otherwise, graphs are assumed to be simple and undirected. A graph G is an ordered pair (V_G, E_G), where V_G is the vertex set of G and E_G is the edge set of G. Let n_G be the number of vertices of G. The neighborhood of a vertex v is the set $N(v) = \{u \in V_G \mid \{u, v\} \in E_G\}$, and the degree of v is $|N(v)|$ denoted by $d(v)$. If $d(v) = 1$, v is a *leaf*. A sequence of pairwise distinct vertices $P = (v_1, v_2, \ldots, v_k)$ is a *path* from v_1 to v_k of length $k - 1$ in G provided that $\{v_i, v_{i+1}\} \in E_G$ for $i \in [k - 1]$. A graph G is connected if for any two vertices u, v, there exists a path from u to v. For a subset $V' \subseteq V_G$, the subgraph of G induced by V' is denoted as $G[V']$. If every pair of vertices in V' is adjacent (resp. non-adjacent) in G, V' is called a *clique* (resp. an *independent set*). A graph G is *bipartite* if the vertex set of G can be partitioned into two independent sets. A graph $\overline{G} = (V_G, \overline{E_G})$ is called the *complement* of G, where $\overline{E_G} = \{\{u, v\} \mid \{u, v\} \notin E_G\}$. A graph G is *cobipartite* if $\overline{G}$ is bipartite.

A vertex ordering of G is a linear ordering $\sigma = (v_1, \ldots, v_n)$ of the vertices of G. We write $v_i \prec_\sigma v_j$ if $i < j$. A graph $H = (V_H, E_H)$ is *subgraph-isomorphic* to a graph $G = (V_G, E_G)$ if there exists an injective map f from V_H to V_G such that $\{u, v\} \in E_H$ implies $\{f(u), f(v)\} \in E_G$. The map f is called a *subgraph isomorphism*. If the map f satisfies $\{u, v\} \in E_H \Leftrightarrow \{f(u), f(v)\} \in E_G$, then f is an *induced subgraph isomorphism* and H is *induced subgraph-isomorphic* to G. Let σ and τ be vertex orderings of G and H, respectively. An injective map $f\colon V_H \to V_G$ is (σ, τ)-*injective*, if $f(u) \prec_\sigma f(v)$, for every pair u, v of vertices in H with $u \prec_\tau v$. For two graphs G and H, if a (σ, τ)-injective map f is a (induced) subgraph isomorphism from H to G, f is an *ordered (induced) subgraph isomorphism*. If there is an ordered (induced) subgraph isomorphism from H to G, then H is *ordered (induced) subgraph-isomorphic* to G. The problems addressed in this paper are defined as follows.

ORDERED (INDUCED) SUBGRAPH ISOMORPHISM (O(I)SI)
Input: Graphs $G = (V_G, E_G)$ and $H = (V_H, E_H)$ with vertex orderings σ and τ, respectively, where $|V_G| \geq |V_H|, |E_G| \geq |E_H|$.
Question: Is H ordered (induced) subgraph-isomorphic to G?

MAXIMUM COMMON ORDERED (INDUCED) SUBGRAPH (MCO(I)S)
Input: Graphs $G = (V_G, E_G)$ and $H = (V_H, E_H)$ with vertex orderings σ and τ, respectively, and a positive integer k.
Question: Is there a graph Z with $|E_Z| \geq k$ ($|V_Z| \geq k$) and its vertex ordering such that Z is ordered (induced) subgraph-isomorphic to both G and H.

Let a triple (Z, f_G, f_H) be a solution of G and H for MCO(I)S, where f_G and f_H are ordered subgraph isomorphisms from Z to G and H, respectively. A vertex $v \in V_G$ is *matched* with $u \in V_H$ if $f_G^{-1}(v) = f_H^{-1}(u)$.

3   NP-Completeness

3.1   O(I)SI on Spiders with Leg Length 2

We show that both OSI and OISI are NP-complete even for spiders with leg length 2. A tree is called a *spider* if it has just one vertex of degree 3 or more, which is called the *central vertex*. Each path from the central vertex to a leaf is called a *leg*. Our reduction is from ORDER PRESERVING SUBSEQUENCE (OPS), which asks, given two permutations π over $[n]$ and ρ over $[k]$ with $n \geq k$, whether there exists an index subset $\{i_1, i_2, \ldots, i_k\} \subseteq [n]$ with $i_1 < i_2 < \cdots < i_k$ such that for any $j_1, j_2 \in [k]$, $\pi(i_{j_1}) < \pi(i_{j_2})$ if and only if $\rho(j_1) < \rho(j_2)$. This problem is known to be NP-complete [2].

Theorem 1. *OSI and OISI are NP-complete for spiders with leg length 2.*

Sketch of proof. We construct two spiders T_G and T_H and their vertex orderings from the instance π over $[n]$ and ρ over $[k]$ of OPS. The vertex set of the spider T_G is $V_G = \{v_0, v_1 \ldots, v_{2n}\}$ with ordering $\sigma = (v_0, v_1, \ldots, v_{2n})$. The central vertex is v_0 and the legs are $(v_0, v_i, v_{n+\pi(i)})$ for $i \in [n]$. Similarly, the spider T_H has vertices $V_H = \{u_0, u_1 \ldots, u_{2k}\}$ with ordering $\tau = (u_0, u_1, \ldots, u_{2k})$ and legs $(u_0, u_i, u_{k+\rho(i)})$ for $i \in [k]$.

Assume indices $i_1, \ldots, i_k \in [n]$ with $i_1 < i_2 < \cdots < i_k$ witness that (π, ρ) is a "yes" instance of OPS. We define a mapping $f \colon V_H \to V_G$ by $f(u_0) = v_0$ and $f(u_j) = v_{i_j}$ and $f(u_{k+\rho(j)}) = v_{n+\pi(i_j)}$ for $j \in [k]$. We can see that f is an ordered subgraph isomorphism from T_H to T_G. Conversely, suppose that there exists an ordered subgraph isomorphism f from T_H to T_G. Let $i_j \in [n]$ be such that $f(u_j) = v_{i_j}$ for $j \in [k]$. We find that those indices $i_1, \ldots, i_k$ provide a witness for (π, ρ) to be a "yes" instance of OPS. $\qquad\square$

The graphs used in the proof of Theorem 1 are actually 2DOR graphs, defined in Sect. 4.1, with pathwidth 2. This contrasts with the tractability results of Theorems 4 and 7, which demonstrate that if the given orderings are "inherent" to the graph structures, the problem may become easier to solve.

A graph is *trivially perfect* if it contains neither an induced path of length four nor an induced cycle of length four [3]. A trivially perfect graph can be obtained from the graph used in the proof of Theorem 1 by adding edges between the center vertex and the leaves.

Corollary 1. *OSI and OISI are NP-complete even on trivially perfect graphs.*

Although this paper is focused on connected graphs, we present here, as an exception, a result concerning disconnected graphs. We used the central vertices of the spiders in the proof of Theorem 1 to make the graphs connected. The reduction still works if we remove the central vertices.

Corollary 2. *OSI and OISI are NP-complete even for graphs consisting of disjoint edges.*

3.2   OISI on Threshold/Chain/Cochain Graphs

Let $G = (V_G, E_G)$ be a graph with a bipartition (X, Y) of V_G. For a vertex $x \in X$, $N_Y(x)$ denotes the neighbor set of x in Y, that is, $N_Y(x) = N(x) \cap Y$. For a vertex set $X' \subseteq X$, $N_Y(X') = \bigcup_{x \in X'} N_Y(x)$. The neighbor sets $N_X(y)$ for $y \in Y$ and $N_X(Y')$ for $Y' \subseteq Y$ are defined similarly. An ordering $(x_1, x_2, \ldots, x_{|X|})$ on X is an *inclusion ordering* if $N_Y(x_i) \subseteq N_Y(x_j)$ for every i, j with $i < j$. Note that if there exists an inclusion ordering on X, then there also exists an inclusion ordering $(y_1, y_2, \ldots, y_{|Y|})$ on Y such that if $N_X(y_i) \subseteq N_X(y_j)$ for every i, j with $i < j$. Suppose that X (and hence Y) has an inclusion ordering in G. We say that G is *threshold* if X is a clique and Y is an independent set, G is *chain* if G is bipartite, and G is *cochain* if G is cobipartite with respect to X and Y. An *inclusion ordering* of G is defined as the inclusion ordering on Y followed by the inclusion ordering on X.

Theorem 2. *OISI is NP-complete even for threshold graphs, chain graphs, and cochain graphs.*

Interval Orderings and Interval Bigraph Orderings: Let $G = (V_G, E_G)$ be a graph with V_G. A vertex ordering σ of G is an *interval ordering* if for every triple u, v, w in $V(G)$ with $u \prec_\sigma v$ and $v \prec_\sigma w$, $\{u, w\} \in E_G$ implies

$\{v, w\} \in E_G$. A graph G is an interval graph if and only if it admits an interval ordering [12,20]. Note that the class of interval graphs is a superclass of threshold graphs [9]. A vertex ordering σ of G is an *interval bigraph ordering* if for every triple $u, v \in X_G$ and $w \in Y_G$ or $u, v \in Y_G$ and $w \in X_G$ with $u \prec_\sigma v$ and $v \prec_\sigma w$, and $\{u, w\} \in E_G$ implies $\{v, w\} \in E_G$. A graph G is an interval bigraph if and only if it admits an interval bigraph ordering [13]. Note that the class of interval bigraphs is a superclass of chain graphs.

Corollary 3. *OISI is NP-complete even on threshold graphs with interval orderings, and chain graphs with interval bigraph orderings.*

3.3 OSI on Split Graphs and Cobipartite Graphs

An edge that joins two non-consecutive vertices in a cycle is called a *chord*. A graph G is *chordal* if every cycle with length at least 4 has a chord in G. A vertex ordering $\sigma = (v_1, \ldots, v_n)$ is a *perfect elimination ordering* of G if $N(v_i) \cap \{v_{i+1}, \ldots, v_n\}$ is a clique for all $i \in [n]$. A graph G is a chordal graph if and only if there is a perfect elimination ordering of G [3]. A graph G is *split* if the vertex set of G can be partitioned into an independent set and a clique. A graph G is a split graph if and only if G and the complement of G are chordal [3].

A graph G is a *cocomparability graph* if the complement $\overline{G}$ of G has a comparability ordering (see Sect. 4.1 for the definition). A vertex ordering $\sigma = (v_1, \ldots, v_n)$ is a *cocomparability ordering* if for every triple $i, j, k \in [n]$ with $i < j < k$, $\{v_i, v_k\} \in E_G$ implies $\{v_i, v_j\} \in E_G$ or $\{v_j, v_k\} \in E_G$. A graph G is cocomparability if and only if it has a cocomparability ordering [20].

Theorem 3. *OSI is NP-complete even on split graphs with perfect elimination orderings and cobipartite graphs with cocomparability orderings.*

4 Tractable Cases

4.1 OSI on 2DOR Graphs and Signed-Interval Digraphs

As stated in the introduction, Heggernes et al. [11] implicitly presented a polynomial-time algorithm for OSI on interval graphs with interval orderings. Building on this algorithm, we present polynomial-time algorithms for two-directional orthogonal ray graphs (or 2DOR graphs) and signed-interval digraphs in this section. These three graph classes share a common ordering characterization, known as min orderings [14], and 2DOR graphs and signed-interval digraphs can be seen as bigraph and digraph analogs of interval graphs, respectively.

2DOR Graphs with Comparability Weak Orderings. A bipartite graph $G = (V_G, E_G)$ with bipartition (X_G, Y_G) is called a *two-directional orthogonal ray graph* (or *2DOR graph*) [19] if there exist a family of rightward rays (i.e., half-lines) R_x for $x \in X_G$ and a family of downward rays R_y for $y \in Y_G$ in the plane such that for any $x \in X_G$ and $y \in Y_G$, $\{x, y\} \in E_G$ if and only if R_x intersects R_y. The class of 2DOR graphs admits a characterization known as a weak ordering [19]. A pair of linear orders, σ_X on X_G and σ_Y on Y_G, is called a *weak ordering* if for every $x_1, x_2 \in X_G$ and $y_1, y_2 \in Y_G$ with $x_1 \prec_{\sigma_X} x_2$ and $y_1 \prec_{\sigma_Y} y_2$, $\{x_1, y_2\} \in E_G$ and $\{x_2, y_1\} \in E_G$ imply $\{x_1, y_1\} \in E_G$. A bipartite graph is a 2DOR graph if and only if it admits a weak ordering. In this paper, we also refer to a vertex ordering σ of G as a weak ordering if the pair of suborderings induced by X_G and Y_G forms a weak ordering of G.

Now, we introduce the comparability weak ordering. A vertex ordering σ of a graph $G = (V_G, E_G)$ is called a *comparability ordering* if for every triple $u, v, w \in V_G$ with $u \prec_\sigma v \prec_\sigma w$, $\{u, v\} \in E_G$ and $\{v, w\} \in E_G$ imply $\{u, w\} \in E_G$; see [5,20] and [3, Section 7.4]. For a bipartite graph G, we refer to its vertex ordering as a *comparability weak ordering* if it is both a comparability ordering and a weak ordering of G. Note that every 2DOR graph admits a comparability weak ordering, which is given by the concatenation of σ_X and σ_Y. Moreover, a bipartite graph is a 2DOR graph if and only if it admits a comparability weak ordering.

The input to OSI consists of two bipartite graphs G and H with comparability weak orderings $\sigma = (v_1, \ldots, v_{n_G})$ and $\tau = (u_1, \ldots, u_{n_H})$, respectively. Our algorithm determines whether H is ordered subgraph-isomorphic to G and, if so, outputs an ordered subgraph isomorphism f. It proceeds in the following three steps.

1. Initialize an order-preserving mapping f by setting $f(u_i) := v_i$ for each i with $1 \le i \le n_H$.
2. If f is a subgraph isomorphism from H to $G[\{f(u_i) \mid 1 \le i \le n_H\}]$, output f and stop. Otherwise, proceed to the next step.
3. Let u_p and u_q be an arbitrary pair of vertices of H with $p < q$ such that $\{u_p, u_q\} \in E_H$ but $\{f(u_p), f(u_q)\} \notin E_G$; such indices exist as f is not a subgraph isomorphism. Let $v_{p'} = f(u_p)$ and $v_{q'} = f(u_q)$. We consider the following two cases:
   Case 1: If $\{v_k, v_{q'}\} \notin E_G$ for any k with $p' < k < q'$, update f as follows:
     (a) If $q' + 1 \le n_G$, set $f(u_q) := v_{q'+1}$; otherwise, output No and stop.
     (b) If $f(u_{q+1}) = v_{q'+1}$, update $q := q + 1$ and $q' := q' + 1$, then return to Step 3(a) in Case 1. Otherwise, return to Step 2.
   Case 2: If there exists some k with $p' < k < q'$ such that $\{v_k, v_{q'}\} \in E_G$, update f as follows:
     (a) If $p' + 1 \le n_G$, set $f(u_p) := v_{p'+1}$; otherwise, output No and stop.
     (b) If $f(u_{p+1}) = v_{p'+1}$, update $p := p + 1$ and $p' := p' + 1$, then return to Step 3(a) in Case 2. Otherwise, return to Step 2.

Theorem 4. OSI *on 2DOR graphs with comparability weak orderings can be solved in* $O(n_G n_H^3 + n_G^2 n_H)$ *time, where* $n_G = |V_G|$ *and* $n_H = |V_H|$.

We note that a close analysis of the proof shows that the algorithm works correctly even if the vertex ordering τ of graph H is arbitrary.

Signed-Interval Digraphs with min Orderings. A directed graph (or *digraph*) $G = (V_G, E_G)$, which may contain loops and opposite arcs, is called a *signed-interval digraph* [14] if it admits a vertex ordering $\sigma = (v_1, \ldots, v_{n_G})$ such that for any two arcs $(v_i, v_j), (v_{i'}, v_{j'}) \in E_G$, there exists an arc $(v_{\min\{i,i'\}}, v_{\min\{j,j'\}}) \in E_G$. Such an ordering is called a *min ordering* of G.

Theorem 5. OSI *on signed-interval digraphs with min orderings can be solved in* $O(n_G n_H^3 + n_G^2 n_H)$ *time, where* $n_G = |V_G|$ *and* $n_H = |V_H|$.

4.2   MCOIS on Threshold/Chain/Cochain Graphs

In Sect. 3.2, we proved the NP-hardness for OISI on threshold graphs, chain graphs, and cochain graphs. In this section, we propose polynomial-time algorithms for a more general problem, MCOIS, on these graph classes, assuming that the input vertex orderings are restricted to inclusion orderings. Note that the restriction on vertex orderings for these algorithms is natural in the sense that inclusion orderings characterize these graph classes. Let $G = (X \cup Y, E)$ with a bipartition (X, Y), where X (and Y) admits an inclusion ordering. Recall that an *inclusion ordering* of G is defined as the inclusion ordering on Y followed by the inclusion ordering on X. We here present a polynomial-time algorithm for MCOIS on threshold graphs with inclusion orderings.

Let $G = (X_G \cup Y_G, E_G)$ and $H = (X_H \cup Y_H, E_H)$ be threshold graphs and $(y_1^G, \ldots, y_{|Y_G|}^G, x_1^G, \ldots, x_{|X_G|}^G)$ and $(y_1^H, \ldots, y_{|Y_H|}^H, x_1^H, \ldots, x_{|X_H|}^H)$ be inclusion orderings of G and H, respectively. We assume that X_G and X_H are both cliques (or independent sets) in G and H, respectively; otherwise, the answer is trivial. The vertices in X_G are matched with at most one vertex in Y_H, and those in Y_G are matched with at most one vertex in X_H. By guessing the match, we only consider the case that the vertices in X_G (resp. Y_G) are matched with the vertices in X_H (resp. Y_H) in an optimal solution (Z, f_G, f_H), where f_G and f_H are ordered induced subgraph isomorphisms from Z to G and H, respectively. Such an optimal solution satisfies one of the followings: (1) no vertex in X_G is matched with a vertex in X_H, or (2) there exists a pair of the largest indices (i, j) such that $f_G^{-1}(x_i^G) = f_H^{-1}(x_j^H)$. In case (1), we match as many vertices of Y_G and Y_H as possible; that is, $\min\{|Y_G|, |Y_H|\}$. In case (2), the vertices $Y_G \setminus N_{Y_G}(x_i^G)$ are matched with the vertices $Y_H \setminus N_{Y_H}(x_j^H)$ because they are isolated in $G[Y_G \cup \{x_i^G\}]$ and $H[Y_H \cup \{x_j^H\}]$. Thus, the number of vertices of the optimal solution is the sum of (i) the number of vertices of the connected ordered induced subgraph of $G_i := G[X_i^G \cup N_{Y_G}(X_i^G)]$ and $H_j := H[X_j^H \cup N_{Y_H}(X_j^H)]$, and (ii) $\min\{|Y_G \setminus N_{Y_G}(x_i^G)|, |Y_H \setminus N_{Y_H}(x_j^H)|\}$, where $X_i^G = \{x_1^G, \ldots, x_i^G\}$ and $X_j^H = \{x_1^H, \ldots, x_j^H\}$.

It remains for us to compute (i) the number of vertices of a connected optimal ordered induced subgraph Z_{ij} of G_i and H_j using dynamic programming. For $i \in [|X_G|] \cup \{0\}, j \in [|X_H|] \cup \{0\}$, the entry of $\mathsf{dp}[i, j]$ represents $|V_{Z_{ij}}|$ for G_i and H_j when x_i^G is matched with x_j^H. If $i = 0$ or $j = 0$, then $\mathsf{dp}[i, j] = 0$ since there is no vertex in G_i or H_j. For $i \in [|X_G|]$ and $j \in [|X_H|]$, let a triple $T = (Z_{ij}, f_{G_i}, f_{H_j})$ be an optimal solution for G_i and H_j. Let $i' < i$ and $j' < j$ be the largest indices such that $f_{G_i}^{-1}(x_{i'}^G) = f_{H_j}^{-1}(x_{j'}^H)$. The triple T can be computed from $(Z_{i'j'}, f_{G_{i'}}, f_{H_{j'}})$ of an optimal solution for $G_{i'}$ and $H_{j'}$ as follows; x_i^G is matched with x_j^H, and the vertices in $N_{Y_G}(x_i^G) \setminus N_{Y_G}(x_{i'}^G)$ are matched with the vertices in $N_{Y_H}(x_j^H) \setminus N_{Y_H}(x_{j'}^H)$, as they have not been matched yet and can be matched. To consider the case that there is no such pair (i', j'), we assume that $N_{Y_G}(x_0^G) = N_{Y_H}(x_0^H) = \emptyset$ for $i' = j' = 0$. Thus, we compute $\mathsf{dp}[i, j]$ as follows:

$$\mathsf{dp}[i, j] = \max_{i' \leq i, j' \leq j} (\mathsf{dp}[i', j'] + d_{ij} + 1),$$

where $d_{ij} = \min\{|N_{Y_G}(x_i^G) \setminus N_{Y_G}(x_{i'}^G)|, |N_{Y_H}(x_j^H) \setminus N_{Y_H}(x_{j'}^H)|\}$. Finally, we obtain the size of the maximum common ordered induced subgraph by computing

$$\max_{i \in [|X_G|], j \in [|X_H|]} (\mathsf{dp}[i, j] + \min\{|Y_G \setminus N_{Y_G}(x_i^G)|, |Y_H \setminus N_{Y_H}(x_j^H)|\}).$$

We analyze the time complexity of the algorithm. The algorithm first guesses the $n_G \cdot n_H + 1$ cases: $x_G \in X_G$ matches with $y_H \in Y_H$, $y_G \in Y_G$ matches with $x_H \in X_H$, and no such match exists. There are $O(n_G n_H)$ table entries, each of which can be computed in $O(n_G n_H)$ time. Thus, the running time to fill the DP table is bounded by $O(n_G^2 n_H^2)$. After computing the DP table, it takes $O(n_G n_H)$ time to obtain the size of an optimal solution. Therefore, the overall time complexity of our algorithm is $O(n_G^3 n_H^3)$.

Our algorithm does not use the neighborhood relation within X or Y except for trivial solutions, and the vertex partition can be determined uniquely from the inclusion ordering. Thus, the algorithm for threshold graphs can be extended to chain graphs and cochain graphs.

Theorem 6. *MCOIS can be solved in polynomial time on threshold graphs, chain graphs, and cochain graphs with inclusion orderings.*

4.3 MCO(I)S Parameterized by Pathwidth

Let $G = (V_G, E_G)$ be a graph with n_G vertices. A *path decomposition* of G is a sequence of vertex subsets, called *bags*, $\mathcal{P} = (X_1, X_2, \ldots, X_r)$ satisfying: (1) $\bigcup_{i \in [r]} X_i = V_G$, (2) $\forall \{u, v\} \in E_G, \exists i \in [r], \{u, v\} \subseteq X_i$, and (3) for $v \in V_G$, the bags containing v are consecutive. The *width* of a path decomposition is the size of the largest bag minus 1. The *pathwidth* of a graph G is the minimum width over all possible path decompositions of G.

A path decomposition $(X_0, X_1, X_2, \ldots, X_{2n_G})$ of a graph $G = (V_G, E_G)$ is *nice* if it satisfies the following additional constraints: (4) $X_0 = X_{2n_G} = \emptyset$, and (5) for every $i \in [2n_G]$, there is either a vertex $x^i \notin X_{i-1}$ such that $X_i = X_{i-1} \cup \{x^i\}$, or a vertex $x^f \in X_i$ such that $X_i = X_{i-1} \setminus \{x^f\}$. For a nice path decomposition, if $X_i = X_{i-1} \cup \{x^i\}$, X_i is called an *introduce node* and X_i introduces x^i, and if $X_i = X_{i-1} \setminus \{x^f\}$, X_i is called a *forget node* and X_i forgets x^f. Given a path decomposition of G with width w, we can compute a nice path decomposition of G with width w in polynomial time [4]. For a nice path decomposition $\mathcal{P}$ of G, a vertex ordering of $\sigma = (x_1, \ldots, x_{n_G})$ is called an *introduce ordering* if for any two vertices $x_i, x_j \in V_G$, we have $x_i \prec_\sigma x_j$ if and only if the index of the bag where x_i is introduced is smaller than that of x_j. This ordering is equivalent to the ordering defined in terms of the vertex separation number [16]. Based on this equivalence, we also say that w is the pathwidth of the vertex ordering σ.

In this section, we present FPT algorithms for MCO(I)S parameterized by the pathwidth w of the vertex orderings of G and H, assuming that their nice path decompositions $\mathcal{P}_G = (X_0, X_1, \ldots, X_{2n_G})$ and $\mathcal{P}_H = (Y_0, Y_1, \ldots, Y_{2n_H})$ of width at most w and their introduce orderings $\sigma = (x_1, \ldots, x_{n_G})$ and $\tau = (y_1, \ldots, y_{n_H})$ associated with $\mathcal{P}_G$ and $\mathcal{P}_H$, respectively, are given as input. The algorithms are based on dynamic programming on the nice path decompositions. Let G_i (resp. H_i) be an induced subgraph $G[\bigcup_{j \in [i]} X_j]$ (resp. $H[\bigcup_{j \in [i]} Y_j]$) for an integer $i \in [2n_G]$ (resp. $i \in [2n_H]$). Note that $G_{2n_G} = G$ and $H_{2n_H} = H$.

Our algorithms compute an optimal solution $(Z_{ij}, f_{G_i}, f_{H_j})$ of G_i and H_j, where f_{G_i} and f_{H_j} are ordered (induced) subgraph isomorphisms from Z_{ij} to G_i and H_j, respectively. The key idea of the algorithm is to match the introduce vertices x^i and y^i only when both X_i and Y_j are introduce nodes. It is easy to see that the ordering constraint at the introduced vertices is always satisfied since x^i and y^i are the largest in the suborderings induced by V_{G_i} and V_{H_j}, respectively. We need to determine whether or not we can match x^i and y^i for MCOIS or to count the number of additional edges for MCOS when x^i matches y^i. To this end, we maintain four subsets X^f, X^c of X_i, and Y^f, Y^c of Y_j as a state in our dynamic programming. The first two subsets indicate that any vertex in X^f is matched with a *forgotten* vertex, and any vertex in X^c is matched with a *current* vertex in Y_j. Analogously, any vertex in Y^f and any vertex in Y^c are respectively matched with a forgotten vertex and a current vertex. Note that $|X^c| = |Y^c|$ has to hold. Since f_{G_i} and f_{H_j} are ordered (induced) subgraph isomorphisms, the matching between X^c and Y^c is uniquely determined.

More specifically, we maintain the following table. For integers $i \in [2n_G] \cup \{0\}$ and $j \in [2n_H] \cup \{0\}$, the value $\mathsf{dp}[(i, X^f, X^c), (j, Y^f, Y^c)]$ is the number of edges (vertices) in an optimal solution $(Z_{ij}, f_{G_i}, f_{H_j})$ of G_i and H_j satisfying X^f, X^c, Y^f, and Y^c conditions. If there is no (f_{G_i}, f_{H_j}), then the entry is $-\infty$. Since $X_{2n_G} = Y_{2n_H} = \emptyset$, $G_{2n_G} = G$ and $H_{2n_H} = H$, the entry $\mathsf{dp}[(2n_G, \emptyset, \emptyset), (2n_H, \emptyset, \emptyset)]$ is stored the number of edges (vertices) in a maximum common ordered (induced) subgraph of G and H.

To analyze the time complexity of the algorithms, we estimate the size of the table $\mathsf{dp}[(i, X^f, X^c)(j, Y^f, Y^c)]$. For each $i \in [2n_G]$, there are at most 3^w possible ways to form the first triple $(i, \cdot, \cdot)$ because X_i is partitioned into X^f, X^c, and the remaining vertices. Similarly, for each $j \in [2n_H]$, there are at most 3^w possible ways to form the second triple $(j, \cdot, \cdot)$. Thus, the number of possible entries in the DP table is $O(9^w)$. Each entry can be computed in time polynomial in w.

Here, we carefully observe the algorithm for the non-induced problem, MCOS. A crucial difference from the MCOIS algorithm is that we do not take into account the edges between x^i and X^f, nor y^i and y^f, respectively. Thus, we do not need to maintain X^{f} and Y^{f}. This improves the size of the DP table to $O(4^w)$.

Theorem 7. *Let $\mathcal{P}_G$ and $\mathcal{P}_H$ be path decompositions of graphs G and H with width at most w, and σ and τ be introduce orderings of G and H, respectively. We can solve MCOIS and MCOS in time $O^*(9^w)$ and $O^*(4^w)$, respectively.*[1]

4.4   MCO(I)S Parameterized by Vertex Cover Number

We now show that both MCOS and MCOIS are fixed-parameter tractable parameterized by vertex cover number of both input graphs, even if there is no restriction on vertex orderings.

Theorem 8. MCOS *and* MCOIS *admit* $O^*(2^{p^2})$*-time algorithms, where p is the maximum of vertex cover numbers of input graphs.*

The algorithms for both problems are based on almost the same dynamic programming approach.

Let G and H be the input graphs of MCOS with vertex orderings $(u_1, u_2, \ldots, u_{|V(G)|})$ and $(v_1, v_2, \ldots, v_{|V(H)|})$, respectively. Let $U_{\leq i} = \{u_1, \ldots, u_i\}$ and $U_{>i} = \{u_{i+1}, \ldots, u_{|V(G)|}\}$. We define $V_{\leq j}$ and $V_{>j}$ analogously. Our goal is to find a subgraph Z of G with maximum number of edges that is ordered subgraph-isomorphic to H.

We assume that a vertex cover S of G with size at most p is given and Z contains all vertices in S. We can justify this assumption by first finding a minimum vertex cover S' of G by an FPT algorithm parameterized by p [4], trying all $2^{|S'|}$ subsets S of S', and then updating G as $G := G - (S' \setminus S)$. For simplicity, we assume that $|S| = p$ in the following.

Let $T_1, \ldots, T_q$ be the twin classes of H, that is, each T_i is a maximal vertex such that $N_H(u) \setminus \{v\} = N_H(v) \setminus \{u\}$ holds for any two vertices $u, v \in T_i$. These classes can be computed in polynomial time. Note that $q \leq 2^p + p$ as H has vertex cover number at most p.

Definition of the DP Table. The key idea of the algorithm is to guess for each $s \in S$ the type of its image in H. That is, we guess a mapping $\phi \colon S \to [q]$ such that $f(s) \in T_{\phi(s)}$ for each $s \in S$, where f is an ordered subgraph isomorphism from a subgraph of G to H.

Now we define the DP table dp_ϕ. For $i \in [n_G] \cup \{0\}$ and $j \in [n_H] \cup \{0\}$, the entry $\mathsf{dp}_\phi[i, j]$ stores the maximum number of edges in an ordered subgraph Z of G with $S \subseteq V(Z) \subseteq S \cup U_{\leq i}$ such that there is an ordered subgraph isomorphism f from Z to H that satisfies

- $f((S \cap U_{\leq i}) \cup (V(Z) \setminus S)) \subseteq V_{\leq j}$,
- $f(S \cap U_{>i}) \subseteq V_{>j}$, and
- $f(s) \in T_{\phi(s)}$ for each $s \in S$,

---

[1] The O^* notation supresses polynomial factors in the input size.

where we set $\mathsf{dp}_\phi[i,j] = -\infty$ if no subgraph of G satisfies these conditions. When $\mathsf{dp}_\phi[i,j] \neq -\infty$, we call a pair (Z,f) satisfying the conditions above and having the maximum number of edges $|E(Z)| = \mathsf{dp}_\phi[i,j]$ a *certificate* for $\mathsf{dp}_\phi[i,j]$.

Observe that, when $(i,j) = (|V(G)|, |V(H)|)$, the conditions above boil down to the last one, which asks that the ordered subgraph isomorphism f respects ϕ. Thus, for finding the maximum size of an ordered common subgraph of G and H under ϕ, it suffices to compute $\mathsf{dp}_\phi[|V(G)|, |V(H)|]$.

Computing the DP Table Entries. *Preprocessing.* We first test for each pair (i,j) whether $\mathsf{dp}_\phi[i,j] \neq -\infty$. Let $s_1, \ldots, s_{|S|}$ be the vertices in S ordered in this way in G. We check the existence of $|S|$ vertices $w_1, \ldots, w_{|S|} \in V(H)$ ordered in this way in H such that

- $w_h \in T_{\phi(s_h)}$ for each $s_h \in S$,
- $\{w_h \mid 1 \leq h \leq |S \cap U_{\leq i}|\} \subseteq V_{\leq j}$, and
- $\{w_h \mid |S \cap U_{\leq i}| + 1 \leq h \leq |S|\} \subseteq V_{>j}$.

If such vertices exist, we can greedily find them by taking the first vertex satisfying the conditions for each h. Such vertices guarantee that $\mathsf{dp}_\phi[i,j] \geq 0$ since we can set $Z = (S, \emptyset)$ and set $f(s_h) = w_h$ for each h. If such vertices do not exist, then we can conclude that $\mathsf{dp}_\phi[i,j] = -\infty$.

Case 1: $\min(i,j) = 0$ and $\mathsf{dp}_\phi[i,j] \neq -\infty$. In this case, the definition of dp_ϕ implies that $V(Z) = S$ as follows. If $i = 0$, then $S \subseteq V(Z) \subseteq S \cup U_{\leq 0}$ implies $S = V(Z)$ as $U_{\leq 0} = \emptyset$. If $j = 0$, then $f((S \cap U_{\leq i}) \cup (V(Z) \setminus S)) \subseteq V_{\leq j}$ implies $V(Z) \setminus S = \emptyset$ as $V_{\leq j} = \emptyset$, and thus $V(Z) = S$ as $S \subseteq V(Z)$. Since $\mathsf{dp}_\phi[i,j] \neq -\infty$, there exists an ordered subgraph isomorphism from a subgraph of $G[S]$ to H respecting ϕ. Thus, $\mathsf{dp}_\phi[i,j]$ can be computed by counting the number of edges $\{s, s'\}$ in $G[S]$ such that the type $\phi(s)$ vertices and the type $\phi(s')$ vertices are adjacent.

Case 2: $\min(i,j) \neq 0$, $u_i \in S$, and $\mathsf{dp}_\phi[i,j] \neq -\infty$. Let (Z,f) be a certificate for $\mathsf{dp}_\phi[i,j]$. Recall that Z is an ordered subgraph of G and f is an ordered subgraph isomorphism from Z to H. Since $u_i \in S$, its image $v_h := f(u_i)$ belongs to $V_{\leq j}$. Furthermore, $v_h \in T_{\phi(u_i)}$ holds since f obeys ϕ. We can see that the pair (Z,f) satisfies all conditions for $\mathsf{dp}_\phi[i-1, h-1]$ as well. On the other hand, for a pair (Z', f') that satisfies the conditions for $\mathsf{dp}_\phi[i-1, h-1]$, it is always possible to change $f'(u_i)$ to v_h without changing Z' since $f'(u_i) \in V_{>h-1}$. Hence,

$$\mathsf{dp}_\phi[i,j] = \max\{\mathsf{dp}_\phi[i-1, h-1] \mid h \leq j, v_h \in T_{\phi(u_i)}\}.$$

Case 3: $\min(i,j) \neq 0$, $u_i \notin S$, and $\mathsf{dp}_\phi[i,j] \neq -\infty$. Let (Z,f) be a certificate for $\mathsf{dp}_\phi[i,j]$. Since $u_i \notin S$, we need to consider two subcases of $u_i \notin V(Z)$ and $u_i \in V(Z)$, and then take the maximum of them. If $u_i \notin V(Z)$, we can simply set $\mathsf{dp}_\phi[i,j] = \mathsf{dp}_\phi[i-1, j]$. Now assume that $u_i \in V(Z)$ and $f(u_i) = v_h$ for some $h \leq j$. Let f' be the restriction of f to $V(Z) \setminus \{u_i\}$. Observe that the pair $(Z - u_i, f')$ satisfies all conditions for $\mathsf{dp}_\phi[i-1, h-1]$, and thus it achieves the maximum edge number $\mathsf{dp}_\phi[i-1, h-1]$. Since this pair $(Z - u_i, f')$ for

$\mathsf{dp}_\phi[i-1, h-1]$ is extended to the pair (Z, f) for $\mathsf{dp}_\phi[i, j]$ by setting $f(u_i) = v_h$, the increase $\mathsf{dp}_\phi[i,j] - \mathsf{dp}_\phi[i-1, h-1]$ can be computed by counting the number of edges $\{u_i, s\}$ with $s \in S$ such that v_h is adjacent to the type $\phi(s)$ vertices. Thus, by setting $\mu(i, h)$ to this number, we can write $\mathsf{dp}_\phi[i, j] = \mathsf{dp}_\phi[i-1, h-1] + \mu(i, h)$. The discussion so far implies that

$$\mathsf{dp}_\phi[i, j] = \max\{\mathsf{dp}_\phi[i-1, j], \max\{\mathsf{dp}_\phi[i-1, h-1] + \mu(i, h) \mid h \le j\}\}.$$

Running Time. There are q^p $(\le (2^p + p)^p \in \mathrm{O}(2^{p^2}))$ candidates for the type assignment ϕ for S. For each ϕ, we can compute $\mathsf{dp}_\phi[i, j]$ in time polynomial in $|V(G)|$ and $|V(H)|$ for all (i, j). We output the maximum of $\mathsf{dp}_\phi[|V(G)|, |V(H)|]$ among all possible ϕ. In total, the running time is bounded by $\mathrm{O}^*(2^{p^2})$.

References

1. Belmonte, R., Heggernes, P., van 't Hof, P.: Edge contractions in subclasses of chordal graphs. Discret. Appl. Math. **160**(7-8), 999–1010 (2012)
2. Bose, P., Buss, J.F., Lubiw, A.: Pattern matching for permutations. Inf. Process. Lett. **65**(5), 277–283 (1998)
3. Brandstädt, A., Le, V.B., Spinrad, J.P.: Graph Classes: A Survey. Society for Industrial and Applied Mathematics (1999)
4. Cygan, M., et al.: Parameterized Algorithms. Springer (2015)
5. Damaschke, P.: Forbidden ordered subgraphs. In: Topics in Combinatorics and Graph Theory, pp. 219–229. Physica-Verlag HD (1990)
6. Damaschke, P.: Induced subgraph isomorphism for cographs in NP-complete. In: WG 1990. LNCS, vol. 484, pp. 72–78. Springer (1990)
7. Di Paola, L., De Ruvo, M., Paci, P., Santoni, D., Giuliani, A.: Protein contact networks: an emerging paradigm in chemistry. Chem. Rev. **113**(3), 1598–1613 (2013)
8. Garey, M.R., Johnson, D.S.: Computers and Intractability: A Guide to the Theory of NP-Completeness. W. H. Freeman (1979)
9. Golumbic, M.C.: Algorithmic Graph Theory and Perfect Graphs. Elsevier Science (2004)
10. Guillemot, S., Marx, D.: Finding small patterns in permutations in linear time. In: SODA 2014, pp. 82–101. SIAM (2014)
11. Heggernes, P., van 't Hof, P., Meister, D., Villanger, Y.: Induced subgraph isomorphism on proper interval and bipartite permutation graphs. Theor. Comput. Sci. **562**, 252–269 (2015)
12. Heggernes, P., Meister, D., Villanger, Y.: Induced subgraph isomorphism on interval and proper interval graphs. In: Cheong, O., Chwa, K.-Y., Park, K. (eds.) ISAAC 2010. LNCS, vol. 6507, pp. 399–409. Springer, Heidelberg (2010). https://doi.org/10.1007/978-3-642-17514-5_34
13. Hell, P., Huang, J.: Certifying LexBFS recognition algorithms for proper interval graphs and proper interval bigraphs. SIAM J. Discret. Math. **18**(3), 554–570 (2004)
14. Hell, P., Huang, J., McConnell, R.M., Rafiey, A.: Min-orderable digraphs. SIAM J. Discret. Math. **34**(3), 1710–1724 (2020)
15. Kijima, S., Otachi, Y., Saitoh, T., Uno, T.: Subgraph isomorphism in graph classes. Discret. Math. **312**(21), 3164–3173 (2012)

16. Kinnersley, N.G.: The vertex separation number of a graph equals its path-width. Inf. Process. Lett. **42**(6), 345–350 (1992)
17. Konagaya, M., Otachi, Y., Uehara, R.: Polynomial-time algorithms for subgraph isomorphism in small graph classes of perfect graphs. Discret. Appl. Math. **199**, 37–45 (2016)
18. Marx, D., Schlotter, I.: Cleaning interval graphs. Algorithmica **65**(2), 275–316 (2013)
19. Shrestha, A.M.S., Tayu, S., Ueno, S.: On orthogonal ray graphs. Discrete Appl. Math. **158**(15), 1650–1659 (2010)
20. Wood, D.R.: Characterisations of intersection graphs by vertex orderings. Australas. J Comb. **34**, 261–268 (2006)

Finding a Maximum Common (Induced) Subgraph: Structural Parameters Revisited

Tesshu Hanaka[1] , Yuto Okada[2] , Yota Otachi[2(✉)] , and Lena Volk[3]

[1] Kyushu University, Fukuoka, Japan
hanaka@inf.kyushu-u.ac.jp
[2] Nagoya University, Nagoya, Japan
pv.20h.3324@s.thers.ac.jp, otachi@nagoya-u.jp
[3] Technische Universität Darmstadt, Darmstadt, Germany
volk@mathematik.tu-darmstadt.de

Abstract. We study the parameterized complexity of the problems of finding a maximum common (induced) subgraph of two given graphs. Since these problems generalize several NP-complete problems, they are intractable even when parameterized by strongly restricted structural parameters. Our contribution in this paper is to sharply complement the hardness of the problems by showing fixed-parameter tractable cases: both induced and non-induced problems parameterized by max-leaf number and by neighborhood diversity, and the induced problem parameterized by twin cover number. These results almost completely determine the complexity of the problems with respect to well-studied structural parameters. Also, the result on the twin cover number presents a rather rare example where the induced and non-induced cases have different complexity.

Keywords: Maximum common (induced) subgraph · Structural parameter · Fixed-parameter tractability · Twin cover · Max-leaf number

1 Introduction

In this paper, we study structural parameterizations of MAXIMUM COMMON SUBGRAPH and MAXIMUM COMMON INDUCED SUBGRAPH and resolve a few new cases: twin cover number, max-leaf number, and neighborhood diversity. Our results almost completely determine the complexity of the problems in the hierarchy of well-studied graph parameters [23], while one important case of the induced version parameterized by cluster vertex deletion number remains unsettled.

Partially supported by JSPS KAKENHI Grant Numbers JP22H00513, JP24H00697, JP25K03076, JP25K03077, by JST, CRONOS, Japan Grant Number JPMJCS24K2, and by JST SPRING, Grant Number JPMJSP2125.

Given graphs G_1, G_2 and an integer h, MAXIMUM COMMON SUBGRAPH (MCS) asks whether there exists a graph H with at least h *edges* such that both G_1 and G_2 contain a subgraph isomorphic to H. MAXIMUM COMMON INDUCED SUBGRAPH (MCIS) is a variant of MCS that asks for a common *induced* subgraph with at least h *vertices*. These problems are known to be intractable even in highly restricted settings. Indeed, most of the known hardness results hold already for their special cases SUBGRAPH ISOMORPHISM and INDUCED SUBGRAPH ISOMORPHISM as described later.

Given graphs G and H, SUBGRAPH ISOMORPHISM (SI) asks whether G contains a subgraph isomorphic to H. INDUCED SUBGRAPH ISOMORPHISM (ISI) is a variant of SI that asks the existence of an *induced* subgraph of G isomorphic to H. By setting $G_1 = G$, $G_2 = H$, and $h = |E(H)|$ ($h = |V(H)|$), we can see that SI (ISI) is a special case of MCS (MCIS, respectively). The problems SI and ISI (and thus MCS and MCIS as well) are NP-complete since they generalize many other NP-complete problems. For example, if H is a complete graph, then SI and ISI coincide with CLIQUE [12, GT19].

1.1 Background of the Target Setting

The problems MCS, MCIS, SI, and ISI have been studied extensively in many different settings. In this paper, we focus on the setting where *both* input graphs have the same restriction on their structures.

We can see that SI and ISI are NP-complete even if both G and H are path forests (i.e., disjoint unions of paths). The hardness for SI appeared in an implicit way already in the book of Garey and Johnson [12, p. 105]. The hardness for ISI can be shown basically in the same way by adding a small number of vertices to each connected component of G as separators [6]. Note that these hardness results imply NP-completeness of the case where both graphs are of bandwidth 1, feedback edge set number 0, and distance to path forest 0. In this direction, we may ask the following question about their common special case, the max-leaf number.

Q: *Are MCS and MCIS tractable when parameterized by max-leaf numbers of both input graphs?*

By answering this question in the affirmative, we complement the hardness results in a sharp way.

For the case where both G and H are disjoint unions of complete graphs, the NP-completeness of SI follows directly from the path forest case by replacing each connected component with a complete graph with the same number of vertices. This implies that SI is NP-complete when both graphs are of twin cover number 0. One of the main motivating questions in this work is whether such a reduction is possible for the induced case ISI. At least the same reduction does not work as we cannot embed two or more disjoint cliques into one clique as an induced subgraph. Actually, it is not difficult to see that if both G and H are disjoint unions of complete graphs, ISI is polynomial-time solvable. On the other hand, ISI is known to be NP-complete even on cographs [6] (i.e., graphs of modular-width 2). Now the question here is as follows.

> Q: *Can we solve MCIS efficiently when input graphs are close to a disjoint union of complete graphs in some sense?*

We partially answer this question by presenting a fixed-parameter algorithm parameterized by twin cover number.

In the context of structural parameters, Abu-Khzam [1] showed that MCIS is fixed-parameter tractable parameterized by $\mathsf{vc}(G_1)+\mathsf{vc}(G_2)$, where vc denotes the vertex cover number of a graph (see also [2]). This result was later generalized by Gima et al. [13], who showed that both MCS and MCIS are fixed-parameter tractable parameterized by $\mathsf{vi}(G_1) + \mathsf{vi}(G_2)$, where vi denotes the vertex integrity of a graph. On the other hand, Bodlaender et al. [4] showed that SI is NP-complete on forests of treedepth 3. Their proof can be easily modified to show that ISI is NP-complete on the same class [13]. Note that $\mathsf{td}(G) \leq \mathsf{vi}(G) \leq \mathsf{vc}(G)+1$ for every graph G, where td denotes the treedepth of a graph. Bodlaender et al. [4] also showed that SI is fixed-parameter tractable parameterized by $\mathsf{nd}(G) + \mathsf{nd}(H)$, where nd denotes the neighborhood diversity of a graph, which is another generalization of vertex cover number in the sense that $\mathsf{nd}(G) \leq \mathsf{vc}(G) + 2^{\mathsf{vc}(G)}$ for every graph G [19]. A natural question would be whether we can get the same results for MCS and MCIS.

> Q: *Are MCS and MCIS tractable when parameterized by neighborhood diversity of both input graphs?*

We show that the idea of the previous algorithm [4] can be applied to MCS and MCIS almost directly.

1.2   Our Results

As mentioned above, our results can be summarized as follows.

1. MCS and MCIS are fixed-parameter tractable parameterized by $\mathsf{ml}(G_1) + \mathsf{ml}(G_2)$, where ml denotes the max-leaf number of a graph.
2. MCIS is fixed-parameter tractable parameterized by $\mathsf{tc}(G_1) + \mathsf{tc}(G_2)$, where tc denotes the twin cover number of a graph.
3. MCS and MCIS are fixed-parameter tractable parameterized by $\mathsf{nd}(G_1) + \mathsf{nd}(G_2)$, where nd denotes the neighborhood diversity of a graph.

Note that although the problems MCS and MCIS are defined as decision problems, our positive results can be easily modified to output optimal solutions.

See Fig. 1 for the summary of the results.[1] Formal definitions of the graph parameters will be given as needed.

Due to the space limitation, the proofs of some results (marked with an asterisk $*$) are omitted and can be found in the full version.

---

[1] In Fig. 1, the abbreviations mean clique-width (cw), treewidth (tw), pathwidth (pw), feedback vertex set number (fvs), feedback edge set number (fes), distance to path forest (dpf), bandwidth (bw), modular-width (mw), shrub-depth (sd), and cluster vertex deletion number (cvd).

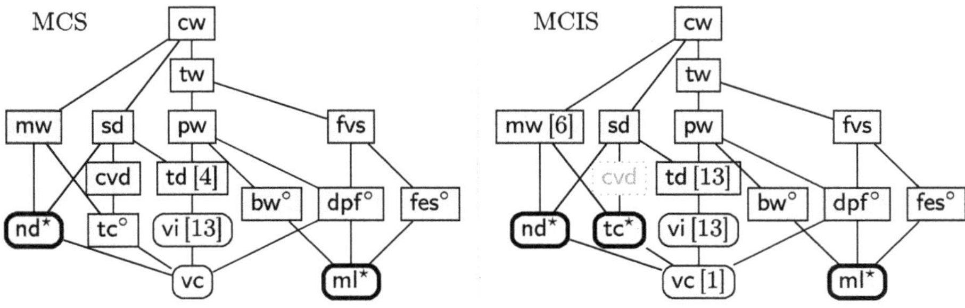

Fig. 1. The complexity of MCS (left) and MCIS (right) when a structural parameter of both input graphs is bounded. The normal rectangles and the rounded rectangles represent paraNP-complete cases and fixed-parameter tractable cases, respectively. The results marked with $\star$ are shown in this paper and the ones with $\circ$ are corollaries of the observations in Sect. 1.1. A connection between two parameters means that the one above is upper-bounded by a function of the one below (e.g., $\mathsf{tw}(G) \le \mathsf{pw}(G)$).

1.3 Related Results

Marx and Pilipczuk [22] studied the parameterized complexity of SI and presented comprehensive results for many combinations of (possibly different) structural parameters of G and H. For the setting where both input graphs satisfy the same condition, restricting the graph class that they belong to is another natural direction. In this setting, Kijima et al. [17] studied SI and Heggernes et al. [14] studied ISI both on interval graphs and related graph classes.

Jansen and Marx [16] considered SI in the setting where only H belongs to a restricted hereditary graph class and presented a dichotomy between randomized polynomial-time solvable cases and NP-complete cases.

If we restrict a structural parameter of H only, almost all studied cases are known to be hard. Both SI and ISI are already W[1]-hard parameterized by $|V(H)|$ as Clique parameterized by the solution size [7] reduces to this case. We can see that ISI is already NP-complete when $\mathsf{vc}(H) = 0$ (Independent Set [12, GT20]) and when $\mathsf{ml}(H) = 2$ (Induced Path [12, GT23]). For SI, we can see that it is NP-complete when $\mathsf{nd}(H) = 1$ and $\mathsf{tc}(H) = 0$ (Clique [12, GT19]), when $\mathsf{ml}(H) = 2$ (Hamiltonian Path [12, GT39]), and when $\mathsf{vi}(H) = 3$ and $\mathsf{tc}(H) = 0$ (Partition Into Triangles [12, GT11]). The only positive result known in this setting is that SI belongs to XP parameterized by $\mathsf{vc}(H)$; indeed, a result of Bodlaender et al. [3, Theorem 14] implies that MCS belongs to XP parameterized by $\min\{\mathsf{vc}(G_1), \mathsf{vc}(G_2)\}$.

2 Preliminaries

We assume that the reader is familiar with the concept of fixed-parameter tractability. See a standard textbook (e.g., [5]) for the terms not defined in this paper.

Let G be a graph. We denote by $\Delta(G)$ the maximum degree of G. The set of connected component of G is denoted by $\mathrm{cc}(G)$. For a vertex $v \in V(G)$, the *(open) neighborhood* of v is $N_G(v) = \{u \mid \{u, v\} \in E(G)\}$. For a set $S \subseteq V(G)$, let $N_G(S) = \bigcup_{v \in S} N_G(v) \setminus S$. We omit the subscript G when the graph G is clear from the context. For $S \subseteq V(G)$, we denote by $G[S]$ the subgraph of G induced by S, and we denote the subgraph $G[V(G) \setminus S]$ by $G - S$.

Let G be a graph. Two vertices $u, v \in V(G)$ are *twins* if $N(u) \setminus \{v\} = N(v) \setminus \{u\}$. That is, two vertices are twins if they have the same neighborhood when ignoring the adjacency between them. Clearly, being twins is an equivalence relation. A *twin class* of G is a maximal set of twins in G. Observe that a twin class in a graph is either a clique or an independent set of the graph.

Let G and H be graphs. An injective mapping $\eta \colon V(H) \to V(G)$ is a *subgraph isomorphism* (resp. an *induced subgraph isomorphism*) from H to G if for $u, v \in V(H)$, $\{u, v\} \in E(H)$ only if (resp. if and only if) $\{\eta(u), \eta(v)\} \in E(G)$.

For a positive integer n, we denote by $[n]$ the set of positive integers not greater than n, i.e., $[n] = \{1, \ldots, n\}$. Also, let $[n]_0 = [n] \cup \{0\}$.

3   MCIS Parameterized by Twin Cover Number

Let G be a graph. An edge between twin vertices of G is a *twin edge* in G. If an edge is not a twin edge, we call it a *non-twin edge*. A set $S \subseteq V(G)$ is a *twin cover* of G if every non-twin edge of G has at least one of its endpoints in S. In other words, S is a twin cover of G if and only if S is a vertex cover of $G - F$, where F is the set of twin edges of G. Note that for a twin cover S, each connected component K of $G - S$ is a complete graph and each vertex in K has the same neighborhood in S.

The *twin cover number* of G, denoted $\mathsf{tc}(G)$, is the minimum size of a twin cover of G. Since finding a minimum twin cover is fixed-parameter tractable parameterized by $\mathsf{tc}(G)$ [11], we assume that a minimum twin cover is given as part of the input when designing a fixed-parameter algorithm parameterized by $\mathsf{tc}(G)$.

Theorem 3.1. MAXIMUM COMMON INDUCED SUBGRAPH *is fixed-parameter tractable parameterized by* $\mathsf{tc}(G_1) + \mathsf{tc}(G_2)$.

Proof. Let $\langle G_1, G_2, h \rangle$ be an instance of MCIS and S_1 and S_2 be minimum twin covers of G_1 and G_2, respectively. Let $p = \max\{\mathsf{tc}(G_1), \mathsf{tc}(G_2)\}$. We first guess the sets of vertices $S_1' \subseteq S_1$ and $S_2' \subseteq S_2$ that are included in a maximum common induced subgraph of G_1 and G_2. There are $2^{|S_1 \cup S_2|}$ ($\leq 2^{2p}$) candidates for this guess. We update G_i and S_i as $G_i := G_i - (S_i \setminus S_i')$ and $S_i = S_i'$ for each $i \in \{1, 2\}$. Now our goal is to find a maximum common induced subgraph H of G_1 and G_2 with an induced subgraph isomorphism η_i from H to G_i such that $S_i \subseteq \eta_i(V(H))$ for each $i \in \{1, 2\}$. Since $\mathsf{tc}(H) \leq p$, we can guess from $2^{O(p^2)}$ candidates the subgraph H' of H induced by a minimum twin-cover of H. Let $T = V(H')$, i.e., $H' = H[T]$.

Guessing η_i on T. Fix $i \in \{1,2\}$. We show that we can guess η_i restricted to T from a small number of candidates. We first guess the partition of T into $T_{\to S_i} := T \cap \eta_i^{-1}(S_i)$ and $T_{\not\to S_i} := T \setminus \eta_i^{-1}(S_i)$ from $2^{|T|}$ ($\leq 2^p$) candidates. Next we guess η_i restricted to $T_{\to S_i}$ from $|S_i|^{|T_{\to S_i}|}$ ($\leq p^p$) candidates. To guess η_i on $T_{\not\to S_i}$, the following claim is crucial for bounding the number of candidates.

Claim 3.2. If $u \in T$ with $\eta_i(u) \notin S_i$ and K is the connected component of $G_i - S_i$ containing $\eta_i(u)$, then $K \cap \eta_i(V(H)) \subseteq \eta_i(T)$.

Proof (Claim 3.2). Suppose to the contrary that there is a vertex $v \notin T$ such that $\eta_i(v) \in K$. Observe that $\eta_i(u)$ and $\eta_i(v)$ are twins in G_i as they both belong to K. This implies that u and v are twins in H since η_i is an induced subgraph isomorphism from H to G_i. By the minimality of T as a twin cover, there is a non-twin edge e of H that $T \setminus \{u\}$ does not hit. Observe that e has u as an endpoint, and so, let $e = \{u, w\}$. Since u and v are twins, w is a neighbor of v as well. On the other hand, since u and w are not twins, v and w are not twins. This contradicts the assumption that T is a twin cover of H as T does not hit the non-twin edge $\{v, w\}$. $\diamond$

Intuitively, Claim 3.2 means that once we decide to map a vertex in T to a connected component K of $G_i - S_i$, then all vertices mapped to $V(K)$ have to belong to T. This allows us to guess η_i on $T_{\not\to S_i}$ from at most $(p+1)^{p \cdot 2^p + p}$ candidates as follows.

1. For $X \subseteq S_i$, let $\mathcal{K}_X = \{K \in cc(G_i - S_i) \mid N_{G_i}(V(K)) = X\}$.
2. For each $X \subseteq S_i$, we guess a vector in $[|T_{\not\to S_i}|]_0^{|T_{\not\to S_i}|}$ that represents how $\mathcal{K}_X$ contains vertices of $\eta_i(T_{\not\to S_i})$. For example, $(2, 1, 0, 0, \dots)$ means that "the class $\mathcal{K}_X$ contains three vertices of $\eta_i(T_{\not\to S_i})$ in total, two vertices in one clique and the other one in another clique." There are at most $(|T_{\not\to S_i}| + 1)^{|T_{\not\to S_i}|}$ options for each $X \subseteq S_i$, and thus, at most $(|T_{\not\to S_i}| + 1)^{|T_{\not\to S_i}| \cdot 2^{|S_i|}}$ ($\leq (p+1)^{p \cdot 2^p}$) options in total.
3. By Claim 3.2, we can greedily use the smallest cliques in $\mathcal{K}_X$ that together satisfy the guessed vector. Furthermore, since each component of $G_i - S_i$ consists of twin vertices, we can pick arbitrary the guessed number of vertices from the selected components. This gives us a complete guess of $\eta_i(T_{\not\to S_i})$.
4. Now we can guess η_i on $T_{\not\to S_i}$ from $|T_{\not\to S_i}|^{|T_{\not\to S_i}|}$ ($\leq p^p$) candidates as we already know the set $\eta_i(T_{\not\to S_i})$.

So far, we guessed η_i on $T_{\to S_i}$ and on $T_{\not\to S_i}$. By combining them, we get η_i on T as $T_{\to S_i} \cup T_{\not\to S_i} = T$. At this point, we reject the current guess if η_i is not an isomorphism from H' ($= H[T]$) to $G_i[\eta_i(T)]$ for each $i \in \{1, 2\}$. Also, we reject the current guess if $\eta_i(T)$ is not a twin cover of $G_i[S_i \cup \eta_i(T)]$, which is an induced subgraph of $G_i[\eta_i(V(H))]$. Note that $\eta_i(T)$ is not necessarily a twin cover of G_i itself.

Cleaning up G_i. Now we show that when $\eta_i(T)$ is not a twin cover of G_i, we can find a vertex in G_i not belonging to $\eta_i(V(H))$.

Claim 3.3 If $\eta_i(T)$ is not a twin cover of G_i, then G_i contains a non-twin edge between a vertex in $S_i \setminus \eta_i(T)$ and a vertex in $V(G_i) \setminus (S_i \cup \eta_i(T))$.

Proof (Claim 3.3). Let $\{u, v\}$ be a non-twin edge in G_i with $\{u, v\} \cap \eta_i(T) = \emptyset$. Since S_i is a twin cover of G_i, at least one of u and v belongs to $S_i \setminus \eta_i(T)$. If exactly one of them belongs to $S_i \setminus \eta_i(T)$, then the other belongs to $V(G_i) \setminus (S_i \cup \eta_i(T))$, and thus the claim holds. In the following, we assume that both u and v belongs to $S_i \setminus \eta_i(T)$.

Since $\{u, v\} \in E(G_i)$ and $\eta_i(T)$ is a twin cover of $G_i[S_i \cup \eta_i(T)]$, the vertices u and v belong to the same connected component of $G_i[S_i \cup \eta_i(T)] - \eta_i(T)$. Since u and v are not twins in G_i, there exists a vertex w in G_i adjacent to exactly one of them, say u. Since $u, v \in S_i \setminus \eta_i(T)$ and $G[S_i \setminus \eta_i(T)]$ $(= G_i[S_i \cup \eta_i(T)] - \eta_i(T))$ is a disjoint union of complete graphs, $w \notin S_i \setminus \eta_i(T)$. Since $\eta_i(T)$ is a twin cover of $G_i[S_i \cup \eta_i(T)]$, u and v have the same neighbors in $\eta_i(T)$, and thus $w \notin \eta_i(T)$. Hence, we can conclude that $w \notin S_i \cup \eta_i(T)$. This implies the claim since $\{u, w\}$ is a non-twin edge (because of v) with $u \in S_i \setminus \eta_i(T)$ and $w \notin S_i \cup \eta_i(T)$.   ◇

By Claim 3.3, if $\eta_i(T)$ is not a twin cover of G_i, then we can find (in polynomial time) a non-twin edge $\{x, y\}$ in G_i such that $x \in S_i \setminus \eta_i(T)$ and $y \in V(G_i) \setminus (S_i \cup \eta_i(T))$. Since $\eta_i(T)$ is a twin cover of $G_i[\eta_i(V(H))]$ and $S_i \subseteq \eta_i(V(H))$, it holds that $y \notin \eta_i(V(H))$. Hence, we can safely remove such y from G_i. By exhaustively applying this reduction, we eventually obtain an induced subgraph G_i' of G_i such that $\eta_i(V(H)) \subseteq V(G_i')$ and $\eta_i(T)$ is a twin cover for G_i'. After the reduction we update G_i as $G_i := G_i'$ for each $i \in \{1, 2\}$.

Finding a Maximum Solution Under the Guesses. The remaining task is to extend the already guessed parts of H and η_i to an entire solution. Since we already know the correspondence between the twin covers $\eta_1(T)$ of G_1 and $\eta_2(T)$ of G_2, we only need to *match* the connected components of $G_1 - \eta_1(T)$ and $G_2 - \eta_2(T)$ in an optimal way. Since each connected component of $G_i - \eta_i(T)$ is a complete graph, it can contain the image of at most one connected component of $H - T$. This allows us to reduce the task to the maximum weighted bipartite matching problem as follows.

We construct a bipartite graph $B = (V_1, V_2; F)$, where $V_1 = \mathrm{cc}(G_1 - \eta_1(T))$ and $V_2 = \mathrm{cc}(G_2 - \eta_2(T))$. Two vertices $K_1 \in V_1$ and $K_2 \in V_2$ are adjacent in B if they have the *same* adjacency to the corresponding twin covers; i.e., $\{K_1, K_2\} \in F$ if $\eta_1^{-1}(N_{G_1}(K_1)) = \eta_2^{-1}(N_{G_2}(K_2))$. (Note that $N_{G_i}(K_i) \subseteq \eta_i(T)$.) For $\{K_1, K_2\} \in F$, we set $w(\{K_1, K_2\}) = \min\{|V(K_1)|, |V(K_2)|\}$. We call a vertex of B corresponding to a connected component of $G_i[S_i \setminus \eta_i(T)]$ *special*. (Note that a connected component of $G_i[S_i \setminus \eta_i(T)]$ is a connected component of $G_i - \eta_i(T)$ as well.) We compute a matching M of B with the maximum weight under that condition that M contains all special vertices. Such a matching can be computed in polynomial time by using an algorithm for finding a maximum-weight degree-constrained subgraph [10]. As we describe in the next paragraph, we can construct H with $|T| + w(M)$ vertices, where $w(M)$ is the total weight of the edges in M. Thus, we set $|T| + w(M)$ to the optimal value under the current guess.

Given M in the previous paragraph, we extend the already guessed parts of H and η_i as follows. Let $\{K_1, K_2\} \in M$ with $K_1 \in V_1$ and $K_2 \in V_2$. We add a complete graph K_H of $\min\{|V(K_1)|, |V(K_2)|\}$ vertices into H and make it adjacent to the subset $\eta_1^{-1}(N_{G_1}(K_1))$ $(= \eta_2^{-1}(N_{G_2}(K_2)))$ of T. For $i \in \{1, 2\}$, we extend η_i by mapping $V(K_H)$ to arbitrarily $\min\{|V(K_1)|, |V(K_2)|\}$ vertices in K_i. After all, we have $|V(H)| = |T| + w(M)$, where $w(M)$ is the total weight of the edges in M. Note that this solution satisfy the condition that $S_i \setminus \eta_i(T) \subseteq \eta_i(V(H))$ as M contains all special vertices. $\qquad\square$

A set $S \subseteq V(G)$ is a *cluster vertex deletion set* of G if every connected component of $G - S$ is a complete graph. The *cluster vertex deletion number* of G, denoted $\mathsf{cvd}(G)$, is the minimum size of a cluster vertex deletion set of G. It is known that finding a minimum cluster vertex deletion set is fixed-parameter tractable parameterized by $\mathsf{cvd}(G)$ [15]. From their definitions, we have $\mathsf{cvd}(G) \leq \mathsf{tc}(G)$ for every graph G.

Although the complexity of MCIS parameterized by $\mathsf{cvd}(G_1) + \mathsf{cvd}(G_2)$ remains unsettled, we have a couple of partial results: an XP-algorithm and an FPT approximation scheme. These results are omitted in this version and will be contained in the full version.

4   MCS and MCIS Parameterized by Max-Leaf Number

The *max-leaf number* of a connected graph G, denoted $\mathsf{ml}(G)$, is the maximum number of leaves in a spanning tree of G. For a disconnected graph G, we define its max-leaf number as the sum of the max-leaf number of its connected components; that is, $\mathsf{ml}(G) = \sum_{C \in \mathsf{cc}(G)} \mathsf{ml}(C)$. From the definition of $\mathsf{ml}(G)$, we can see that $\mathsf{ml}(G) \geq |\mathsf{cc}(G)|$ and $\mathsf{ml}(G) \geq \Delta(G)$. (To see the latter, consider a BFS tree rooted at a vertex of the maximum degree.)

In this section, we show that both MCS and MCIS fixed-parameter tractable parameterized by max-leaf number of both graphs.

Theorem 4.1. MAXIMUM COMMON SUBGRAPH *is fixed-parameter tractable parameterized by* $\mathsf{ml}(G_1) + \mathsf{ml}(G_2)$.

Theorem 4.2. MAXIMUM COMMON INDUCED SUBGRAPH *is fixed-parameter tractable parameterized by* $\mathsf{ml}(G_1) + \mathsf{ml}(G_2)$.

In our algorithms, we do not need to compute a spanning tree with the maximum number of leaves (although it is actually fixed-parameter tractable parameterized by the number of leaves [24]). Instead, we use a polynomial-time computable structure described below.

By $V_{\neq 2}(G)$, we denote the non-degree-2 vertices of G. It is known that, using the result of Kleitman and West [18], the number of non-degree-2 vertices can be bounded from above by a linear function of max-leaf number (see e.g., [8]). We include a proof of the following statement in the full version. (Indeed, we prove a slightly stronger bound of $4\mathsf{ml}(G) - 6$, which is actually tight.)

Lemma 4.3 (Folklore). *For every graph G, $|V_{\neq 2}(G)| \leq 4\mathsf{ml}(G)$.*

We call a trail (i.e., a walk in which no edges are repeated) in a graph G a *maximal degree-2 trail* if all internal vertices are of degree-2 in G and both end vertices are non-degree-2 vertices in G. Further, if G has simple cycles as connected components (which we call *isolated cycles* in the following), we also consider them as maximal degree-2 trails by selecting an arbitrary vertex in the cycle as its endpoint (in the following called *designated* vertex). Let $\mathcal{T}_2(G)$ be the set of maximal degree-2 trails. Note that an element of $\mathcal{T}_2(G)$ is either a path or a cycle in G. Note also that we can compute $\mathcal{T}_2(G)$ in polynomial time.

Using Lemma 4.3, we can bound the number of maximal degree-2 trails as follows.

Lemma 4.4. $(*)$ *For every graph G, $|\mathcal{T}_2(G)| \leq 2\mathsf{ml}(G)^2$.*

Although a subgraph (or an induced subgraph) of a graph of bounded max-leaf number may have unbounded max-leaf number in general, the next lemma shows that a maximum common (induced) subgraph of graphs with bounded max-leaf number always has bounded max-leaf number.

Lemma 4.5. $(*)$ *For graphs G_1 and G_2 with max-leaf number at most ℓ, every maximum common (induced) subgraph H has max-leaf number at most $24\ell^5$.*

The *smoothing* of a graph G is the graph obtained by repeatedly deleting a vertex of degree 2 and its incident edges and then adding an edge between its neighbors, which may be the same vertex, until no degree-2 vertices remain except for isolated loops. Note that the smoothing of a simple graph may have loops and multi-edges. We begin by proving Theorem 4.1 and then adapt the proof to show Theorem 4.2.

Proof (Theorem 4.1). Let $\langle G_1, G_2, h \rangle$ be an instance of MAXIMUM COMMON SUBGRAPH and let $\ell = \max\{\mathsf{ml}(G_1), \mathsf{ml}(G_2)\}$.

First of all, we guess the smoothing S of a maximum common subgraph of G_1 and G_2. By Lemma 4.5, every maximum common subgraph of G_1 and G_2 has max-leaf number bounded by $24\ell^5$. Then, Lemma 4.3 yields, that a maximum common subgraph of G_1 and G_2 has at most $96\ell^5$ non-degree-2 vertices. Further, it has maximum degree at most $24\ell^5$ and contains at most $24\ell^5$ many isolated loops. Thus, as S, we guess a graph on at most $96\ell^5$ vertices with maximum degree $24\ell^5$, possibly with multi-edges and loops, but without any degree-2 vertices, and additionally with up to $24\ell^5$ isolated loops. Note that we can assume S to have no isolated vertices, as it is the smoothing of a maximum common subgraph. Clearly, the number of possible options for S is bounded by a function depending only on ℓ.

Before we proceed to the next step, we need to introduce some notation. We call an alternating sequence of elements in $V_{\neq 2}(G_i)$ and $\mathcal{T}_2(G_i)$ *valid*, if

- it contains at least one element of $\mathcal{T}_2(G_i)$,
- it does not visit an element of $V_{\neq 2}(G_i)$ multiple times with the only exception that it may start and end at the same vertex,

- every element T of $\mathcal{T}_2(G_i)$ in the sequence is in between its two end vertices, or at the beginning or end of the sequence and next to one of its end vertices, or the only element in the sequence.

Note that the second and third condition together imply that all trails, except for the first and last ones, in such a valid sequence are pairwise disjoint as well. In particular, a valid sequence corresponds to a path or a cycle in G_i. In such a sequence, we call all elements besides the first and last ones *inner*. Inner trails in a sequence always correspond to paths in G_i, with the only exception of a sequence with three elements starting and ending at the same vertex and a degree-2 cycle between them. Further, in the following, we do not distinguish between a sequence and the same sequence in reversed order.

Now, we guess a mapping η_i from S to G_i with the following properties:

- For each vertex $v \in V(S)$ with $\deg(v) > 2$, guess a vertex $\eta_i(v) \in V(G_i)$ with $\deg(\eta_i(v)) \geq \deg(v)$.
- For each vertex $v \in V(S)$ with $\deg(v) = 1$, guess either a vertex $\eta_i(v) \in V(G_i)$ with $\deg(\eta_i(v)) \neq 2$, or guess a maximal degree-2 trail $\eta_i(v) \in \mathcal{T}_2(G_i)$.
- For each isolated loop $l = \{u, u\} \in E(S)$, guess either a vertex $\eta_i(u) \in V(G_i)$ with $\deg(\eta_i(u)) > 2$, or guess an isolated cycle $\eta_i(u)$ in G_i.
- Ensure that until now each vertex in $V(G_i)$ is guessed at most once, i.e., $|\eta_i^{-1}(u)| \leq 1$ for all $u \in V(G_i)$.
- For each edge $e \in E(S)$ with end vertices u and v (note that there could be multiple edges with this property, and it might hold $u = v$), guess a valid alternating sequence $\eta_i(e)$ of elements in $V_{\neq 2}(G_i)$ and $\mathcal{T}_2(G_i)$ starting and ending with $\eta_i(u)$ and $\eta_i(v)$.
- Ensure that for all edges $e, e' \in E(S)$ the inner elements of $\eta_i(e)$ and $\eta_i(e')$ are pairwise disjoint.
- Ensure that for every edge $e \in E(S)$ all inner elements of $\eta_i(e)$ are pairwise disjoint to all elements in $\eta_i(V(S))$.

Since the numbers of vertices and edges in S, and the numbers of non-degree-2 vertices, maximal degree-2 trails, and connected components in G_i is bounded by a function depending only on ℓ, the number of guesses is bounded by a function of ℓ as well.

Further, for the smoothing of every maximum common subgraph of G_1 and G_2, there exists a mapping with the properties above, as it is induced by the corresponding embeddings. Vica versa, this mapping later gives rise to an embedding of the resulting graph to G_1 and G_2.

Before we proceed, note that if for an edge e with endpoints u and v, we have $\eta_i(u) = \eta_i(v) \in V(G_i)$, then $\eta_i(e)$ has an inner element. Further, if $\eta_i(e)$ has no inner elements, then at least one of $\eta_i(u)$ and $\eta_i(v)$ belongs to $\mathcal{T}_2(G_i)$. If both of them belong to $\mathcal{T}_2(G_i)$, then they coincide and $\eta_i(e)$ only contains this one element.

Based on the graph S and the mappings η_i, we can now reduce the problem of finding a maximum common subgraph of G_1 and G_2 to INTEGER LINEAR PROGRAMMING (ILP) with bounded number of variables. It is known that ILP parameterized by the number of variables is fixed-parameter tractable [20] even with a linear objective function to maximize (see e.g., [9]).

For each edge $e \in E(S)$ introduce a variable l_e. This variable describes the length of the path in a maximal common subgraph corresponding to the edge e in its smoothed graph S. Further, for every $v \in V(S)$ with $\deg(v) = 1$ such that $\eta_i(v) \in \mathcal{T}_2(G_i)$ is a path with distinct end vertices x_i and y_i, introduce two variables v_{x_i} and v_{y_i}. These variables describe where on the path $\eta_i(v)$ the vertex v is mapped in an embedding, giving the distances from v to the end vertices x_i and y_i. Analogously, for every $v \in V(S)$ with $\deg(v) = 1$ such that $\eta_i(v) \in \mathcal{T}_2(G_i)$ is a cycle with designated vertex z_i, introduce a variable d_{v,z_i}. This variable describes where on the cycle $\eta_i(v)$ the vertex v is mapped in an embedding, by giving the distance from v to z_i.

Now the objective is to maximize $\sum_{e \in E(S)} l_e$ under the following constraints. Note that some of the following constraints are not linear (like taking one of two values or having an equality to an absolute value). We can handle this issue by guessing the correct (linear) constraints corresponding to an optimal solution since the numbers of variables and constraints depend only on ℓ.

1. For all $l_e, v_{x_i}, v_{y_i}, d_{v,z_i}$: $l_e, v_{x_i}, v_{y_i}, d_{v,z_i} \geq 1$.
2. For all v_{x_i}, v_{y_i}: $v_{x_i} + v_{y_i} = |E(\eta_i(v))|$.
3. For all d_{v,z_i}: $d_{v,z_i} \leq \lfloor |E(\eta_i(v))|/2 \rfloor$.
4. For all $e \in E(S)$ with end vertices $u \neq v$ such that $\eta_i(e)$ contains no inner elements (note that at most one of $\eta_i(u)$ and $\eta_i(v)$ belongs to $V(G_i)$):
   - If $\eta_i(v) \in V(G_i)$ (the case of $\eta_i(u) \in V(G_i)$ is symmetric), then $\eta_i(u) \in \mathcal{T}_2(G_i)$ is a path or a cycle.
     - If $\eta_i(u)$ is a path with end vertices $\eta_i(v)$ and y_i: $l_e = u_{\eta_i(v)}$.
     - If $\eta_i(u)$ is a cycle with the designated vertex $\eta_i(v)$:
       $l_e \in \{d_{u,\eta_i(v)}, |E(\eta_i(u))| - d_{u,\eta_i(v)}\}$.
   - If $\eta_i(u), \eta_i(v) \notin V(G_i)$, then $\eta_i(v) = \eta_i(u) \in \mathcal{T}_2(G_i)$ is a path or a cycle.
     - If $\eta_i(u)$ is a path with end vertices x_i and y_i: $l_e = |u_{x_i} - v_{x_i}|$.
     - If $\eta_i(u)$ is a cycle with the designated vertex z_i: $l_e = |d_{u,z_i} - d_{v,z_i}|$.
5. For all loops $e \in E(S)$ with end vertex u such that $\eta_i(e)$ contains no inner elements (note that this can only happen if e is an isolated loop for which $\eta_i(u)$ is an isolated cycle c in G_i): $l_e = |E(c)|$.
6. For all $e \in E(S)$ with end vertices u, v such that $\eta_i(e)$ contains inner elements:

$$l_e = a + b + \sum_{p \text{ inner path in } \eta_i(e)} |E(p)|, \quad \text{where}$$

$$a = \begin{cases} 0 & \text{if } \eta_i(e) \text{ starts with a vertex in } G_i, \\ v_{y_i} & \text{if } \eta_i(e) \text{ starts with } (\eta_i(v), y_i, \dots) \text{ and } \eta_i(v) \text{ is a path}, \\ d & \text{if } \eta_i(e) \text{ starts with } (\eta_i(v), z_i, \dots) \text{ and } \eta_i(v) \text{ is a cycle}, \\ & \quad \text{and } d \in \{d_{v,z_i}, |E(\eta_i(v))| - d_{v,z_i}\}, \end{cases}$$

and b is defined analogously for the end of the sequence.

7. For every maximal degree-2 trail $T_i \in \mathcal{T}_2(G_i)$, which is not an inner element of some $\eta_i(e)$: Let x_i and y_i be the end vertices of T_i. Then, there is at most one edge $e \in E(S)$ such that $x_i \in \eta_i(e)$ and analogously there is at most one edge $e' \in E(S)$ such that $y_i \in \eta_i(e')$.

$$|E(T_i)| \geq (a+1) + b + \sum_{f \in E(S) \text{ with } \eta_i(f)=(T_i)} (l_f + 1), \qquad \text{where}$$

$$a = \begin{cases} 0 & \text{if no such edge } e \text{ exists,} \\ v_{x_i} & \text{if } T_i \text{ is a path and } v \text{ is the end vertex of } e \text{ with } \eta_i(v) = T_i \\ & \text{and } \eta_i(e) \text{ starts or ends with } (T_i, x_i, \dots) \text{ or } (\dots, x_i, T_i), \\ d & \text{if } T_i \text{ is a cycle and } v \text{ is in the case as above and} \\ & d \in \{d_{v,x_i}, |E(\eta_i(v))| - d_{v,x_i}\}, \end{cases}$$

and b is defined analogously for e'. Note that in case T_i is a cycle, b is 0.

In the first condition, we require $l_e \geq 1$, as the subdivision of any edge is at least of length 1 (namely in case the edge was not subdivided at all). Further, we demand $v_{x_i}, v_{y_i}, d_{v,z_i} \geq 1$, as in case of lengths 0 we can instead consider the initial guess of η_i where we guessed the corresponding vertex for $\eta_i(v)$ instead of the maximal degree-2 trail. The second and third conditions ensure, that the position of the vertex v in the embedding is well-defined. The fourth and fifth conditions ensure that the length l_e of the edges which get completely embedded into one maximal degree-2 trail in G_i is well-defined. Analogously, the sixth condition ensures that the length l_e of the remaining edges (i.e., the ones which get embedded into more than one maximal degree-2 path) is well-defined. The seventh condition then ensures, that every maximal degree-2 trail in G_i is long enough for all parts which get mapped there.

To see that the seventh condition suffices to ensure that all parts which are mapped to a maximal degree-2 trail T_i can get properly embedded, note that T_i is in the image of at most two edges $e, e' \in E(S)$ such that $\eta_i(e), \eta_i(e')$ does not only contain T. Then, the requirement on the pairwise disjointness of the valid sequences already ensures, that for both of these edges different end vertices of the trail are used (or, in case the trail is a cycle only one such edge exists). Further, all edges which get completely mapped to T_i correspond to paths in G_i and hence, it only remains to ensure that the total length of all edges that get mapped to T_i fit. □

Proof (Theorem 4.2). We proceed analogously to the proof of Theorem 4.1 and only describe the differences here. Again, we begin by guessing a graph S with the same properties. Note that this time we cannot assume S to have no isolated vertices, as we are in the induced setting. Thus, when we guess the mapping η_i from S to G_i, we additionally guess for each vertex $v \in V(S)$ with $\deg(v) = 0$, either a vertex $\eta_i(v) \in V(G_i)$ with $\deg(\eta_i(v)) \neq 2$, or guess a maximal degree-2 trail $\eta_i(v) \in \mathcal{T}_2(G_i)$. It remains the same that we afterwards ensure that all vertices in $V(G_i)$ are guessed at most once, and further all inner elements of $\eta_i(E(S))$ are disjoint to all elements in $\eta_i(V(S))$.

When guessing S we additionally demand the following: If $\eta_i(u), \eta_i(v) \in V(G_i)$ are adjacent, then there is the edge $\{u, v\} \in E(S)$. Otherwise we reject this guess. This ensures, that all vertices of the resulting graph with degree greater than 2 can be embedded in an induced way. Thus, it remains to ensure that all vertices with degree not greater than 2 are embedded correctly. We do this by adapting the ILP. Note that we can consider the same objective function as in the non-induced case, because the number of non-degree-2 vertices (including isolated vertices) is already fixed by the choice of S and hence, the number of vertices in the resulting graph is maximized if and only if the sum over the l_e is maximized.

First of all, we also need to introduce variables v_{x_i}, v_{y_i} and d_{u,z_i} for the isolated vertices of S which get mapped by η_i to a maximal degree-2 trail. They also need to satisfy constraints (1) to (3). Then, we further include the following two additional constraints to the ILP:

N1. For all v_{x_i} such that there exists some $u \in V(S)$ with $\eta_i(u) = x_i$ and $\{u, v\} \notin E(S)$: $v_{x_i} \geq 2$.
N2. For all d_{v,z_i} such that there exists some $u \in V(S)$ with $\eta_i(u) = z_i$ and $\{u, v\} \notin E(S)$: $d_{v,z_i} \geq 2$.

These two additional constraints ensure that only positions for the degree-0 and degree-1 vertices of the resulting graph are considered, which can correspond to an induced embedding.

Finally, we replace the equation in constraint (7) in the ILP with

$$|E(T_i)| \geq (a + 2) + b + 2k + \sum_{f \in E(S) \text{ with } \eta_i(f)=(T_i)} (l_f + 2),$$

where k is the number of isolated vertices $v \in V(S)$ with $\eta_i(v) = (T_i)$.

This change in constraint (7) ensures, that within each trail T_i of G_i, there is enough space to embed the different components which get mapped to T_i in a way such that they are not adjacent. Since all components which get mapped to T_i are isolated vertices or paths (in the non-induced case they were all paths), it suffices again to check if the length of the trails fit.

Hence, it remains to ensure that all isolated loops can be embedded in an induced way. The isolated loops $e = \{u, u\}$ for which we guessed $\eta_i(u) \in V(G_i)$ are covered by the additional assumption on the initial guessing of S and the new ILP. All other isolated loops get embedded into an isolated cycle in G_i of correct length (ensured by constraint 5) and hence, are embedded as an induced subgraph. $\square$

5   MCS and MCIS Parameterized by Neighborhood Diversity

The *neighborhood diversity* of a graph G, denoted $\mathsf{nd}(G)$, is the number of twin classes in G. Clearly, the twin classes of a graph (and thus its neighborhood diversity as well) can be computed in polynomial time. Recall that a twin class is a clique or an independent set. By the definition of twins, the connection between two twin classes is either *full* or *empty*; that is, there are either no or all possible edges between them.

We solve MCIS parameterized by neighborhood diversity by solving instances of INTEGER LINEAR PROGRAMMING (ILP) parameterized by the number of variables [20]. For MCS, we use INTEGER QUADRATIC PROGRAMMING (IQP) parameterized by the number of variables [21,25].

Theorem 5.1 (∗)   MAXIMUM COMMON INDUCED SUBGRAPH *is fixed-parameter tractable parameterized by* $\mathsf{nd}(G_1) + \mathsf{nd}(G_2)$.

Theorem 5.2 (∗) MAXIMUM COMMON SUBGRAPH *is fixed-parameter tractable parameterized by* $\mathsf{nd}(G_1) + \mathsf{nd}(G_2)$.

6   Conclusion

In this paper, we showed fixed-parameter tractable cases for MCS and MCIS. Given our results, the parameterized complexity of these problems with respect to well-studied structural parameters is almost completely understood (see Fig. 1). Filling the missing part (i.e., MCIS parameterized by $\mathsf{cvd}(G_1) + \mathsf{cvd}(G_2)$) would be the natural next step.

References

1. Abu-Khzam, F.N.: Maximum common induced subgraph parameterized by vertex cover. Inf. Process. Lett. **114**(3), 99–103 (2014). https://doi.org/10.1016/J.IPL. 2013.11.007
2. Abu-Khzam, F.N., Bonnet, É., Sikora, F.: On the complexity of various parameterizations of common induced subgraph isomorphism. Theor. Comput. Sci. **697**, 69–78 (2017). https://doi.org/10.1016/J.TCS.2017.07.010
3. Bodlaender, H.L., Hanaka, T., Jaffke, L., Ono, H., Otachi, Y., van der Zanden, T.C.: Hedonic seat arrangement problems. Auton. Agents Multi Agent Syst. **39**(2), 33 (2025). https://doi.org/10.1007/S10458-025-09711-X
4. Bodlaender, H.L., et al.: Subgraph isomorphism on graph classes that exclude a substructure. Algorithmica **82**(12), 3566–3587 (2020). https://doi.org/10.1007/ S00453-020-00737-Z
5. Cygan, M., et al.: Parameterized Algorithms. Springer (2015). https://doi.org/10. 1007/978-3-319-21275-3
6. Damaschke, P.: Induced subgraph isomorphism for cographs in NP-complete. In: WG 1990. Lecture Notes in Computer Science, vol. 484, pp. 72–78 (1990). https:// doi.org/10.1007/3-540-53832-1_32
7. Downey, R.G., Fellows, M.R.: Parameterized Complexity. Springer (1999). https:// doi.org/10.1007/978-1-4612-0515-9
8. Fellows, M.R., Lokshtanov, D., Misra, N., Mnich, M., Rosamond, F.A., Saurabh, S.: The complexity ecology of parameters: an illustration using bounded max leaf number. Theory Comput. Syst. **45**(4), 822–848 (2009). https://doi.org/10.1007/ S00224-009-9167-9

9. Fellows, M.R., Lokshtanov, D., Misra, N., Rosamond, F.A., Saurabh, S.: Graph layout problems parameterized by vertex cover. In: Hong, S.-H., Nagamochi, H., Fukunaga, T. (eds.) ISAAC 2008. LNCS, vol. 5369, pp. 294–305. Springer, Heidelberg (2008). https://doi.org/10.1007/978-3-540-92182-0_28

10. Gabow, H.N.: An efficient reduction technique for degree-constrained subgraph and bidirected network flow problems. In: STOC 1983, pp. 448–456 (1983). https://doi.org/10.1145/800061.808776

11. Ganian, R.: Improving vertex cover as a graph parameter. Discret. Math. Theor. Comput. Sci. **17**(2), 77–100 (2015). https://doi.org/10.46298/DMTCS.2136

12. Garey, M.R., Johnson, D.S.: Computers and Intractability: A Guide to the Theory of NP-Completeness. W. H. Freeman (1979)

13. Gima, T., Hanaka, T., Kiyomi, M., Kobayashi, Y., Otachi, Y.: Exploring the gap between treedepth and vertex cover through vertex integrity. Theor. Comput. Sci. **918**, 60–76 (2022). https://doi.org/10.1016/J.TCS.2022.03.021

14. Heggernes, P., van 't Hof, P., Meister, D., Villanger, Y.: Induced subgraph isomorphism on proper interval and bipartite permutation graphs. Theor. Comput. Sci. **562**, 252–269 (2015). https://doi.org/10.1016/J.TCS.2014.10.002

15. Hüffner, F., Komusiewicz, C., Moser, H., Niedermeier, R.: Fixed-parameter algorithms for cluster vertex deletion. Theory Comput. Syst. **47**(1), 196–217 (2010). https://doi.org/10.1007/s00224-008-9150-x

16. Jansen, B.M.P., Marx, D.: Characterizing the easy-to-find subgraphs from the viewpoint of polynomial-time algorithms, kernels, and turing kernels. In: SODA 2015, pp. 616–629 (2015). https://doi.org/10.1137/1.9781611973730.42

17. Kijima, S., Otachi, Y., Saitoh, T., Uno, T.: Subgraph isomorphism in graph classes. Discret. Math. **312**(21), 3164–3173 (2012). https://doi.org/10.1016/J.DISC.2012.07.010

18. Kleitman, D.J., West, D.B.: Spanning trees with many leaves. SIAM J. Discret. Math. **4**(1), 99–106 (1991). https://doi.org/10.1137/0404010

19. Lampis, M.: Algorithmic meta-theorems for restrictions of treewidth. Algorithmica **64**(1), 19–37 (2012). https://doi.org/10.1007/s00453-011-9554-x

20. Lenstra, H.W., Jr.: Integer programming with a fixed number of variables. Math. Oper. Res. **8**(4), 538–548 (1983). https://doi.org/10.1287/MOOR.8.4.538

21. Lokshtanov, D.: Parameterized integer quadratic programming: variables and coefficients. CoRR abs/1511.00310 (2015)

22. Marx, D., Pilipczuk, M.: Everything you always wanted to know about the parameterized complexity of subgraph isomorphism (but were afraid to ask). In: STACS 2014. LIPIcs, vol. 25, pp. 542–553. Schloss Dagstuhl - Leibniz-Zentrum für Informatik (2014). https://doi.org/10.4230/LIPICS.STACS.2014.542

23. Sorge, M., Weller, M.: The graph parameter hierarchy (2019). https://manyu.pro/assets/parameter-hierarchy.pdf

24. Zehavi, M.: The k-leaf spanning tree problem admits a klam value of 39. Eur. J. Comb. **68**, 175–203 (2018). https://doi.org/10.1016/J.EJC.2017.07.018

25. Zemmer, K.: Integer polynomial optimization in fixed dimension. Ph.D. thesis, ETH Zurich (2017). https://doi.org/10.3929/ethz-b-000241796

Large Induced Subgraphs of Bounded Degree in Outerplanar and Planar Graphs

Marco D'Elia[iD] and Fabrizio Frati[✉][iD]

Roma Tre University, Rome, Italy
{marco.delia,fabrizio.frati}@uniroma3.it

Abstract. In this paper, we study the following question. Let $\mathcal{G}$ be a family of planar graphs and let $k \geq 3$ be an integer. What is the largest value $f_k(n)$ such that every n-vertex graph in $\mathcal{G}$ has an induced subgraph with degree at most k and with $f_k(n)$ vertices? Similar questions, in which one seeks a large induced forest, or linear forest, or d-degenerate graph, rather than a large induced graph of bounded degree, have been studied for decades and have given rise to some of the most fascinating and elusive conjectures in Graph Theory. We tackle our problem when $\mathcal{G}$ is the class of the outerplanar graphs or the class of the planar graphs. In both cases, we provide upper and lower bounds on the value of $f_k(n)$. For example, we prove that every n-vertex planar graph has an induced subgraph with degree at most 3 and with $\frac{5n}{13} > 0.384n$ vertices, and that there exist n-vertex planar graphs whose largest induced subgraph with degree at most 3 has $\frac{4n}{7} + O(1) < 0.572n + O(1)$ vertices.

1 Introduction

The study of induced subgraphs has a central role in Graph Theory. As a prominent example, establishing the truth of the Albertson and Berman conjecture [1,3], stating that every n-vertex planar graph contains an induced forest with at least $\frac{n}{2}$ vertices, is one of the most famous and frequently mentioned (see, e.g., [5,11,18,19,28,29,31,32,35,36,39]) open problems in Graph Theory. The conjectured bound $\frac{n}{2}$ is the best possible [1]. The best known lower bound is $\frac{2n}{5}$, which comes from the acyclic 5-colorability of planar graphs [10]. Other popular related conjectures state that every n-vertex bipartite planar graph has an induced forest with at least $\frac{5n}{8}$ vertices [1] and that every n-vertex planar graph has an induced linear forest with at least $\frac{4n}{9}$ vertices [32]. In the former case, the best known lower bound is $\lceil \frac{4n+3}{7} \rceil$ [39], while in the latter case it is $\lceil \frac{n}{3} \rceil$ [33].

Combinatorial problems on induced graphs usually have the following form: Given a class $\mathcal{G}$ of n-vertex graphs, what is the largest function $f(n)$ such that every graph in $\mathcal{G}$ has an induced subgraph with $f(n)$ vertices and satisfying

Work partially supported by the European Union, Next Generation EU, Mission 4, Component 1, CUP J53D23007130006 PRIN proj. 2022ME9Z78 "NextGRAAL: Next-generation algorithms for constrained GRAph visuALization".

Table 1. Summary of the best known coefficients of the linear term for $k = 3$.

	Forests	Outerplanar graphs	Planar graphs
LB	0.75 [14]	$0.\overline{6}$ [Theorem 3]	> 0.384 [Theorem 6]
UB	0.75 [14]	$0.\overline{6}$ [Theorem 1]	< 0.572 [Theorem 5]

certain properties? Since every induced subgraph of a clique is a clique, the class $\mathcal{G}$ usually contains sparse graphs – often $\mathcal{G}$ is the class of the n-vertex planar graphs or a subclass of it. Research on this topic has addressed the problem of determining the maximum size of: (i) an induced forest that can be found in every planar graph [1,3,36], outerplanar graph [24], 2-outerplanar graph [11], planar graph with given girth [4,18,21,28,29,35], graph with bounded treewidth [14], or cubic graph [9,25,38]; (ii) an induced linear forest that can be found in a planar graph [19,32,33] or outerplanar graph [32]; and (iii) a k-degenerate induced graph that can be found in a planar graph [20,23,31].

In this paper, we study the maximum size of an induced graph of bounded degree that one is guaranteed to find in an outerplanar graph or in a planar graph. It was surprising to us, given that the degree is one of the most important, natural, and used graph parameters, that this problem has not been studied systematically before. A set of vertices which induces a graph of degree at most k in a graph is sometimes called a k-*stable set* [22]. Our results are as follows.

- In Sect. 3, we show that every n-vertex outerplanar graph has an induced subgraph with degree at most 3 and with at least $\frac{2}{3}n$ vertices and an induced subgraph with degree at most k and with at least $\frac{2k+1}{2k+5}n$ vertices (for even $k \geq 4$) or at least $\frac{2k+2/3}{2k+14/3}n$ vertices (for odd $k \geq 5$). On the other hand, for any integer $k \geq 3$, we show that there exists an n-vertex outerplanar graph (in fact, outerpath) such that any induced subgraph with degree at most k has at most $\frac{k-1}{k}n + O(1)$ vertices. We also show an algorithm that matches the last bound for the class of the n-vertex outerpaths.
- In Sect. 4, we show that every n-vertex planar graph has an induced subgraph with degree at most 3 and with at least $\frac{5}{13}n$ vertices, an induced subgraph with degree at most 4 and at least $\frac{27}{59}n$ vertices, and, for any integer $k \geq 5$, an induced subgraph with degree at most k and at least $\frac{k-2}{k+1}n$ vertices. These results actually follow from bounds with respect to the number of vertices *and edges* for (not necessarily planar) graphs. We also show that, for any integer $k \geq 3$, there exists an n-vertex planar graph such that any induced subgraph with degree at most k has at most $\frac{2k-2}{2k+1}n + O(1)$ vertices and an n-vertex planar graph such that any induced subgraph with degree at most k has at most $\frac{k+2}{k+4}n + O(1)$ vertices.

Table 1 shows the coefficients of the linear terms in the size of the induced subgraphs of degree at most 3 that we can find in the considered graph classes.

Related Work. Although the question of determining the maximum size of an induced subgraph of bounded degree that one is guaranteed to find in an outerplanar graph or in a planar graph was not addressed explicitly before, some results from the literature have an implication or a relationship to our problem. Interestingly, linear lower bounds on the size of such subgraphs can be derived from known results on different graph parameters. However, our lower bounds have larger multiplicative coefficient of the linear term; making this coefficient as large as possible is the typical objective in this research field, think again about the Albertson and Berman conjecture [1,3] and the related results mentioned above.

- For $k = 0$, a k-stable set is just called *independent set*; the literature about finding large independent sets in planar graphs is one among the richest in graph theory, see, e.g. [2,6–8,13,34]. A consequence of the four-color theorem is that every planar graph admits an independent set with $\frac{n}{4}$ vertices. This defines an induced graph whose degree is bounded by k, for any $k \geq 0$.
- Fountoulakis, Kang, and McDiarmid [22] studied the size of an induced subgraph of bounded degree that one is expected to find in a random graph.
- Results of Cowen, Cowen, and Woodall [15] on (c, k)-*defective colorings* (see also [40]) imply the existence of sets with $\frac{n}{2}$ and $\frac{n}{3}$ of vertices in an n-vertex outerplanar or planar graph, resp., that induce graphs with degree at most 2. Our lower bounds are larger than these values, already for $k = 3$.
- The size of an induced linear forest that one is guaranteed to find in an n-vertex outerplanar graph (the optimal size is $\lceil \frac{4n+2}{7} \rceil$ [32]) or in an n-vertex planar graph (the best known lower bound is $\lceil \frac{n}{3} \rceil$ [33]) is a lower bound for the size of an induced graph with degree at most k, for any integer $k \geq 2$. Our lower bounds are larger than these values, already for $k = 3$.
- Chappell and Pelsmajer [14] proved that n-vertex graphs with treewidth t contain induced *forests* with degree at most k and at least $\lceil \frac{2kn+2}{tk+k+1} \rceil$ vertices. This provides a lower bound of $\lceil \frac{2kn+2}{3k+1} \rceil$ on the size of an induced subgraph with degree at most k in an n-vertex outerplanar graph. Our lower bounds are larger for any $k \geq 3$ (and sufficiently large n).
- Chappell and Pelsmajer [14] proved (and attributed the result to Chappell, Gimbel, and Hartman) that n-vertex forests contain induced forests with degree at most k and at least $\lceil \frac{k+1}{k+2} n \rceil$ vertices. This bound is the best possible.
- Edge-counting arguments show that sparse graphs, like planar graphs, have many vertices (and hence large induced subgraphs) with bounded degree. The linear lower bounds one gets in this way are worse than ours, see, e.g., [26,37]. In the same spirit, Bose, Dujmović, and Wood [12] find, in any graph G with bounded treewidth w, a large subgraph of bounded treewidth $t \leq w$ whose vertices have bounded degree in G. Their bound for this problem implies that every n-vertex outerplanar graph G (in fact, every graph with treewidth 2) has a subset of at least $\frac{k-3}{k-1} n$ vertices whose degree in G is at most k. This lower bound is weaker than ours for every integer $k \geq 3$.
- Finally, Knauer and Ueckerdt [27] study induced graphs with components of small size. Their bounds imply a $\frac{k+1}{w+k+2} n$ lower bound for the size of an induced graph of degree at most k that one is guaranteed to find in a graph of treewidth at most w. For outerplanar graphs, this yields a $\frac{k+1}{k+4} n$ lower bound, which is weaker than our lower bounds for every integer $k \geq 3$.

Table 2. Results on outerplanar graphs with different values of k.

	$k = 3$	$k = 4$	$k = 5$	$k = 6$	$k = 7$
LB	$0.\overline{6}$	> 0.69	$0.\overline{72}$	> 0.76	> 0.78
UB	$0.\overline{6}$	0.75	0.8	$0.8\overline{3}$	< 0.858

Complete proofs can be found in the full version of the paper [16].

2 Preliminaries

We use standard terminology in Graph Theory [17]. For a graph G, we let $V(G)$ and $E(G)$ be its vertex and edge sets, respectively. For a set $\mathcal{I} \subseteq V(G)$, we denote by $G[\mathcal{I}]$ the *induced subgraph* of G, that is, the graph whose vertex set is $\mathcal{I}$ and whose edge set consists of every edge $uv \in E(G)$ such that $u, v \in \mathcal{I}$. The *degree of a vertex v* is the number of its incident edges, and the *degree of a graph* is the maximum degree of any of its vertices. A set $\mathcal{D} \subseteq V(G)$ is a *dominating set* if every vertex of G is in $\mathcal{D}$ or is adjacent to a vertex in $\mathcal{D}$.

A planar drawing of a graph is *outerplanar* if every vertex is incident to the outer face. An *outerplanar graph* is a graph that admits an outerplanar drawing; such a graph is *maximal* if adding any edge to it results in a graph that contains a multiple edge or is not outerplanar. A maximal outerplanar graph G has a unique outerplanar drawing, up to a reflection and a homeomorphism of the plane. We call *outerplane embedding*, and denote it by $\mathcal{O}_G$, the topological information associated with such a drawing. The outer face of $\mathcal{O}_G$ is delimited by a Hamiltonian cycle, while the internal faces by cycles with 3 vertices. An edge uv of G splits G into two outerplanar graphs, called *uv-split subgraphs* of G, induced by the vertices encountered when walking in clockwise and counterclockwise direction along the outer face of $\mathcal{O}_G$ from u to v. The vertices u and v, and the edge uv, belong to both such subgraphs. The *weak dual* of $\mathcal{O}_G$ is the graph with a vertex for each internal face of $\mathcal{O}_G$ and with an edge between two vertices if the corresponding faces of $\mathcal{O}_G$ share an edge. The weak dual of $\mathcal{O}_G$ is a tree with degree at most 3. We say that G is a *maximal outerpath* if its weak dual is a path. An *outerpath* is a subgraph of a maximal outerpath.

3 Induced Subgraphs of Outerplanar Graphs

In this section, we consider the size of the largest induced subgraph of degree at most k that one is guaranteed to find in an n-vertex outerplanar graph. Table 2 shows the coefficients of the linear terms of our bounds, for various values of k.

We start with an upper bound, which comes from outerpaths, see Fig. 1.

Theorem 1. *For every $n \geq 1$ and for every integer $k \geq 3$, there exists an n-vertex outerpath G with the following property. Consider any set $\mathcal{I} \subseteq V(G)$ such that $G[\mathcal{I}]$ has degree at most k. Then $|\mathcal{I}| \leq \frac{k-1}{k}n + 2$.*

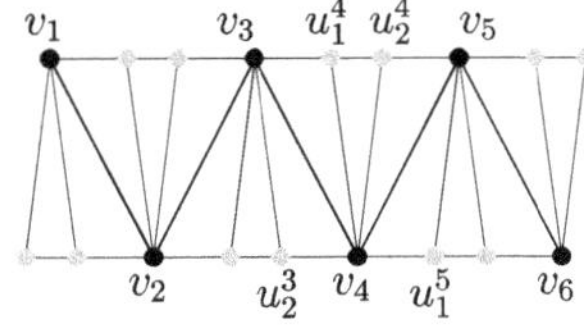

Fig. 1. Illustration for Theorem 1, with $n = 18$, $k = 3$, and $h = 6$.

Proof. Let G be the n-vertex outerpath in Fig. 1, defined as follows. Let $h = \lfloor \frac{n}{k} \rfloor$. The graph contains a path $v_1, \ldots, v_h$, vertices $u_1^i, \ldots, u_{k-1}^i$, for $i = 1, \ldots, h$, and edges $u_j^i u_{j+1}^i$, $v_i u_j^i$, $v_{i-1} u_1^i$, and $u_{k-1}^i v_{i+1}$, whenever their end-vertices are in G. Additional $n - h \cdot k$ isolated vertices are added to G, so that it has n vertices.

Consider any set $\mathcal{I} \subseteq V(G)$ such that $G[\mathcal{I}]$ has degree at most k. We prove that there exists a set $\mathcal{I}' \subseteq V(G)$ such that $G[\mathcal{I}']$ has degree at most k, such that $|\mathcal{I}'| = |\mathcal{I}|$, and such that $\mathcal{I}'$ does not contain any of $v_2, \ldots, v_{h-1}$. Suppose that $\mathcal{I}$ contains a vertex v_i, for some $i \in \{2, \ldots, h-1\}$. Then there is a vertex w among $u_{k-1}^{i-1}, u_1^i, u_2^i, \ldots, u_{k-1}^i, u_1^{i+1}$ not in $\mathcal{I}$, as otherwise v_i would have degree $k+1$ in $G[\mathcal{I}]$. Replace v_i in $\mathcal{I}$ with w. Clearly, the size of $\mathcal{I}$ remains the same. Also, $G[\mathcal{I}]$ still has degree at most k. Indeed, the only vertices that have degree larger than k in G are $v_1, \ldots, v_h$. Also, w is a neighbor of v_i, which however is not in $\mathcal{I}$ after the replacement, and might be a neighbor of v_{i-1} or v_{i+1}. However, v_{i-1} or v_{i+1} are also neighbors of v_i, hence if they are in $\mathcal{I}$, their degree in $G[\mathcal{I}]$ does not increase. The repetition of this replacement results in the desired set $\mathcal{I}'$. It remains to observe that $|\mathcal{I}'| \leq n - (h - 2) = n - \lfloor \frac{n}{k} \rfloor + 2 \leq \frac{k-1}{k} n + 2$.

The following theorem proves that the upper bound of Theorem 1 is tight, up to additive constants, for n-vertex outerpaths.

Theorem 2. *For every $n \geq 1$, for every n-vertex outerpath G, and for every integer $k \geq 3$, there exists a set $\mathcal{I} \subseteq V(G)$ such that $G[\mathcal{I}]$ has degree at most k and $|\mathcal{I}| \geq \frac{k-1}{k} n$.*

Proof Sketch. W.l.o.g., assume that G is a maximal outerpath. Let e be an edge, called *special edge*, incident to the outer face of the outerplane embedding $\mathcal{O}_G$ of G and to an internal face that is an extreme of the weak dual of $\mathcal{O}_G$. Let u and v be the end-vertices of e, where u is called *special vertex*. We prove, by induction on n, that there exists a set $\mathcal{I} \subseteq V(G)$ such that: (i) $|\mathcal{I}| \geq \frac{k-1}{k} n$; (ii) $G[\mathcal{I}]$ has degree at most k; and (iii) if the special vertex u is in $\mathcal{I}$, then u has at most $k - 1$ neighbors in $\mathcal{I}$. This implies the statement of the theorem.

In the base case, which happens if $n \leq k + 1$, one can just set $\mathcal{I} := V(G)$ (if $n \leq k$) or $\mathcal{I} := V(G) \setminus \{x\}$ (if $n = k + 1$), where x is any vertex of G.

If $n \geq k + 2$, we identify an edge wz with the following properties. Let G_R be the wz-split subgraph of G not containing uv and let $G_L = G[V(G) \setminus V(G_R)]$. We require that G_L has at least k vertices and that it contains a vertex h such that, by defining $\mathcal{I} = \mathcal{I}_L \cup \mathcal{I}_R$, where $\mathcal{I}_L := V(G_L) \setminus \{h\}$ and $\mathcal{I}_R \subseteq V(G_R)$ is the set

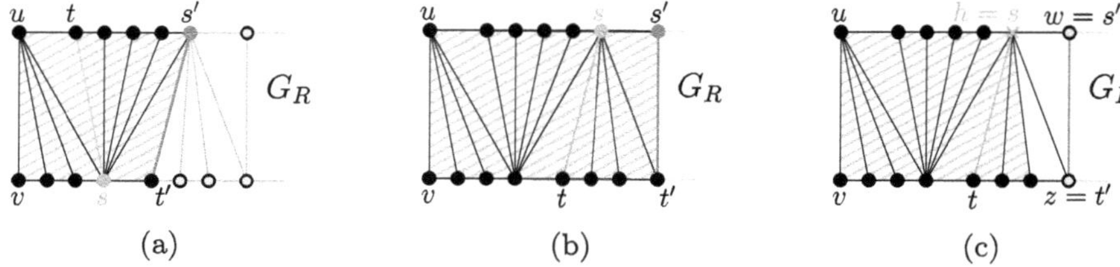

Fig. 2. Illustration for the proof of Theorem 2. Graph G_L is striped, graph G_R is gray or not shown, the old active vertex and edge are orange, while the new ones are red.

obtained by applying induction on G_R, the conditions of the claim are satisfied. In order to identify wz and h, we work iteratively. We initialize $G_L = G[\{u, v\}]$ and $G_R = G[V(G) \setminus \{u, v\}]$. Only one of u and v has at least two neighbors in G_R; such a vertex is called *active vertex* and uv is the *active edge*. While G_L has at most $k + 1$ vertices, let s and st be the current active vertex and edge. We move to G_L all the neighbors of s that are in G_R. Let s' and t' be the neighbors of s such that the edge $s't'$ exists and its distance from uv in $\mathcal{O}_G$ is maximum. If G_L still has at most $k + 1$ vertices, as in Fig. 2(a), we continue the iteration with $s't'$ as the new active edge and the one of s' and t' with at least two neighbors in G_R as the new active vertex. If G_L has at least $k + 2$ vertices, as in Fig. 2(b), we stop the iteration, move s' and t' back to G_R, as in Fig. 2(c), and define $h = s$ and $wz = s't'$. We define $\mathcal{I}_L = V(G_L) \setminus \{h\}$. We apply induction on G_R with wz as a special edge and with the one between w and z with more than two neighbors in G_R as special vertex. The induction returns us a set $\mathcal{I}_R \subseteq V(G_R)$ and we define $\mathcal{I} := \mathcal{I}_L \cup \mathcal{I}_R$. In the full version of the paper [16] we show that $\mathcal{I}$ satisfies the required properties. We mention here that the one between w and z selected as special vertex has at most $k - 1$ neighbors from G_R in $\mathcal{I}$ (this is where we use the condition on the degree of the special vertex), and at most two neighbors, one of which is h, in G_L, hence that vertex has degree at most k in $G[\mathcal{I}]$.   □

We now deal with general outerplanar graphs. We exhibit three lower bounds, one for $k = 3$, one for $k \geq 4$ with k even, and one for $k \geq 5$ with k odd. The first one is tight, by Theorem 1. We can assume that the input is an n-vertex maximal outerplanar graph G. We first present some tools.

Lemma 1. *Let $\ell \geq 1$ be an integer and let xy be an edge incident to the outer face of $\mathcal{O}_G$. If $n \geq \ell + 2$, there exists an edge $uv \in E(G)$ such that a uv-split subgraph H of G satisfies the following properties:*

- *H has a number h of vertices such that $\ell + 2 \leq h \leq 2\ell + 1$; and*
- *if uv is not the edge xy, then H does not contain the edge xy.*

For a subgraph H of G, denote by $\overline{H}$ the subgraph of G induced by $V(G) \setminus V(H)$. The strategy we use in order to find a large set $\mathcal{I} \subseteq V(G)$ of vertices such that $G[\mathcal{I}]$ has degree at most k is the following. By Lemma 1, we find an edge uv such that, in a uv-split subgraph H of G, we can select a set $\mathcal{I}_H \subset V(H)$ of vertices that contains the fraction of the vertices in $V(H)$ we are aiming for ($\frac{2}{3}$ for $k = 3$, $\frac{2k+1}{2k+5}$ for $k \geq 4$ with k even, and $\frac{2k+2/3}{2k+14/3}$ for $k \geq 5$ with k odd), that does not contain u and v, and such that $G[\mathcal{I}_H]$ has degree at most k. We also apply induction to find a large set of vertices $\mathcal{I}_{\overline{H}}$ in $\overline{H}$ such that $G[\mathcal{I}_{\overline{H}}]$ has degree at most k. Then the set $\mathcal{I} = \mathcal{I}_H \cup \mathcal{I}_{\overline{H}}$ has the required number of vertices and $G[\mathcal{I}]$ has degree at most k. This is formalized in the following lemma.

Lemma 2. *Let $k \geq 3$ be an integer and let $0 < c < 1$ be a real number. Suppose that G contains an edge uv such that one of the two uv-split subgraphs of G, say H, has a set $\mathcal{I}_H$ of vertices such that $\mathcal{I}_H$ contains neither u nor v, such that $|\mathcal{I}_H| \geq c \cdot |V(H)|$, and such that $G[\mathcal{I}_H]$ has degree at most k. Suppose also that $\overline{H}$ has a set $\mathcal{I}_{\overline{H}}$ of vertices such that $|\mathcal{I}_{\overline{H}}| \geq c \cdot |V(\overline{H})|$ and such that $G[\mathcal{I}_{\overline{H}}]$ has degree at most k. Then the set $\mathcal{I} := \mathcal{I}_H \cup \mathcal{I}_{\overline{H}}$ contains at least $c \cdot n$ vertices and $G[\mathcal{I}]$ has degree at most k.*

We now present our theorem for induced subgraphs of degree at most 3.

Theorem 3. *For every $n \geq 1$ and for every n-vertex outerplanar graph G, there exists a set $\mathcal{I} \subseteq V(G)$ such that $G[\mathcal{I}]$ has degree at most 3 and $|\mathcal{I}| \geq \frac{2}{3}n$.*

Proof Sketch. Suppose that G is maximal. The proof is by induction on n. In the base case we have $n \leq 11$. Then an edge-counting argument proves that we can find a set $\mathcal{I}$ with $\frac{|\mathcal{I}|}{n} \geq \frac{2}{3}$ such that $G[\mathcal{I}]$ has degree at most 3.

Suppose that $n \geq 12$. We define a uv-split subgraph H of G where we can select a large set $\mathcal{I}_H \subseteq V(H)$ of vertices, with $u, v \notin \mathcal{I}_H$, such that $G[\mathcal{I}_H]$ has degree at most 3. Applying induction on $\overline{H}$ to get a set $\mathcal{I}_{\overline{H}} \subseteq V(\overline{H})$ where $G[\mathcal{I}_{\overline{H}}]$ has degree at most 3 and $|\mathcal{I}_{\overline{H}}| \geq \frac{2}{3}|V(\overline{H})|$, and then applying Lemma 2 allows us to find the desired set $\mathcal{I} = \mathcal{I}_H \cup \mathcal{I}_{\overline{H}}$. We distinguish 16 cases, labeled from $h = 6$ to $h = 21$, where $h = |V(H)|$. Eventually, we use Lemma 1 with $\ell = 10$ to find a uv-split subgraph with a number of vertices between 12 and 21, however our arguments require to consider smaller uv-split subgraphs, as well. Specifically, if we can find a uv-split subgraph H with $h \in \{6, 9, 10, 12, 13, \ldots, 21\}$, we can select a large set $\mathcal{I}_H \subseteq V(H)$ of vertices, with $u, v \notin \mathcal{I}_H$, such that $G[\mathcal{I}_H]$ has degree at most 3, which allows us to apply Lemma 2 to find the desired set $\mathcal{I}$ of vertices in G. If H has 7, 8, or 11 vertices, we can select a large set $\mathcal{I}_H \subseteq V(H)$ of vertices such that $G[\mathcal{I}_H]$ has degree at most 3, however this requires u and/or v to be part of $\mathcal{I}_H$. The fact that u and/or v belong to $\mathcal{I}_H$ prevents us from employing Lemma 2 to find the desired set of vertices in G, however the existence of such a set $\mathcal{I}_H$ is used when uv-split subgraphs with more vertices are considered. Whenever we consider a uv-split subgraph H, we denote by w the vertex such that the cycle (u, v, w) bounds an internal face of $\mathcal{O}_H$. Also, we denote by H_1 the uw-split subgraph of G not containing v and by H_2 the vw-split subgraph of G not containing u, with $h_1 = |V(H_1)|$ and $h_2 = |V(H_2)|$, where $h_1 \geq 2$, $h_2 \geq 2$, and $h_1 + h_2 = h + 1$, given that w is in H_1 and H_2.

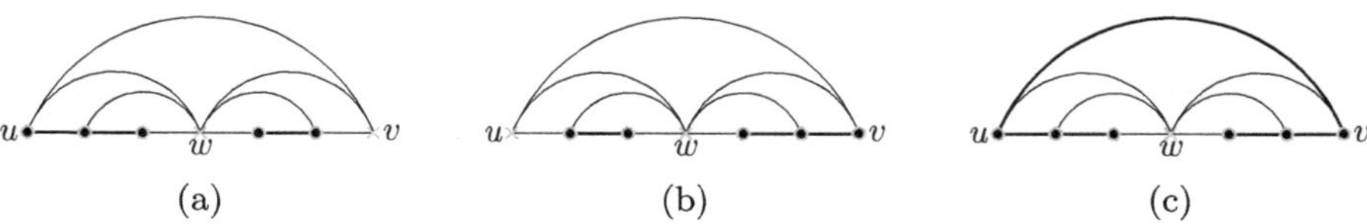

Fig. 3. A uv-split subgraph H with 7 vertices in which w has degree 6. (a) A 5-over-7 set with u. (b) A 5-over-7 set with v. (c) A 6-over-7 set with u and v. Vertices in the sets are represented by disks, other vertices by crosses.

Case h = 6: Suppose that G has a uv-split subgraph H with 6 vertices. We define $\mathcal{I}_H := V(H) \setminus \{u, v\}$. Note that $|\mathcal{I}_H|/|V(H)| = \frac{4}{6} = \frac{2}{3}$. Also, $\mathcal{I}_H$ contains 4 vertices, hence $G[\mathcal{I}_H]$ has degree at most 3. By Lemma 2, the set $\mathcal{I} := \mathcal{I}_H \cup \mathcal{I}_{\overline{H}}$ contains at least $\frac{2}{3}n$ vertices and $G[\mathcal{I}]$ has degree at most 3, as desired. We thus assume that G has no uv-split subgraph with 6 vertices.

Case h = 7: Suppose that G has a uv-split subgraph H with 7 vertices. Then $h_1 \geq 3$, as otherwise H_2 would have 6 vertices. Symmetrically, $h_2 \geq 3$. If there is a vertex in $V(H) \setminus \{w\}$ not adjacent to w, then we define $\mathcal{I}_H := V(H) \setminus \{u, v\}$. Then $|\mathcal{I}_H|/|V(H)| = \frac{5}{7} > \frac{2}{3}$. Also, $G[\mathcal{I}_H]$ has degree at most 3, since w has at most 3 neighbors in $\mathcal{I}_H$, by assumption. By Lemma 2, the set $\mathcal{I} := \mathcal{I}_H \cup \mathcal{I}_{\overline{H}}$ contains at least $\frac{2}{3}n$ vertices and $G[\mathcal{I}]$ has degree at most 3, as desired.

We thus assume that, if G has a uv-split subgraph H with 7 vertices, the degree of w in H is 6. This implies that H admits a *5-over-7 set with u*, a *5-over-7 set with v*, and a *6-over-7 set with u and v*, which are defined as follows. The former is a set $\mathcal{I}_H$ of vertices such that $|\mathcal{I}_H| = 5$, such that $u \in \mathcal{I}_H$ and $v \notin \mathcal{I}_H$, such that $G[\mathcal{I}_H]$ has degree at most 3, and such that the degree of u in $G[\mathcal{I}_H]$ is at most 1; the set $\mathcal{I}_H := V(H) \setminus \{v, w\}$ satisfies these properties, see Fig. 3(a). The definition of a 5-over-7 set with v is analogous, with v in place of u and vice versa, and is realized by the set $\mathcal{I}_H := V(H) \setminus \{u, w\}$, see Fig. 3(b). A 6-over-7 set with u and v is a set $\mathcal{I}_H$ of vertices such that $|\mathcal{I}_H| = 6$, such that $u, v \in \mathcal{I}_H$, such that $G[\mathcal{I}_H]$ has degree at most 3, and such that the degree of u and v in $G[\mathcal{I}_H]$ is at most 2; the set $\mathcal{I}_H := V(H) \setminus \{w\}$ satisfies these properties, see Fig. 3(c). Although such sets cannot exclude the existence of a uv-split subgraph with 7 vertices, since they contain u or v, we will use them in later cases.

Case h = 8: If G contains a uv-split subgraph H with 8 vertices, then, since G has no uv-split subgraph with 6 vertices, either one of h_1 and h_2 is 2 and the other one is 7, or one of h_1 and h_2 is 4 and the other one is 5.

Suppose that $h_1 = 2$ and $h_2 = 7$, the case $h_1 = 7$ and $h_2 = 2$ is symmetric. By assumption, H_2 admits a 5-over-7 set with w, say $\mathcal{I}_{H_2}^w$, and a 6-over-7 set with w and v, say $\mathcal{I}_{H_2}^{wv}$. This implies that H admits a *6-over-8 set with u* and a *6-over-8 set with v*. The definition of these sets naturally extends the one of a 5-over-7 set with u/v; these sets are realized by $\mathcal{I}_{H_2}^w \cup \{u\}$, see Fig. 4(a), and $\mathcal{I}_{H_2}^{wv}$, see Fig. 4(b). Also, H admits a *6-over-8 set with u-deg-1* or *with v-deg-1*. The former (latter) is a 6-over-8 set with u (v) where the degree of u (v) in $G[\mathcal{I}_H]$ is

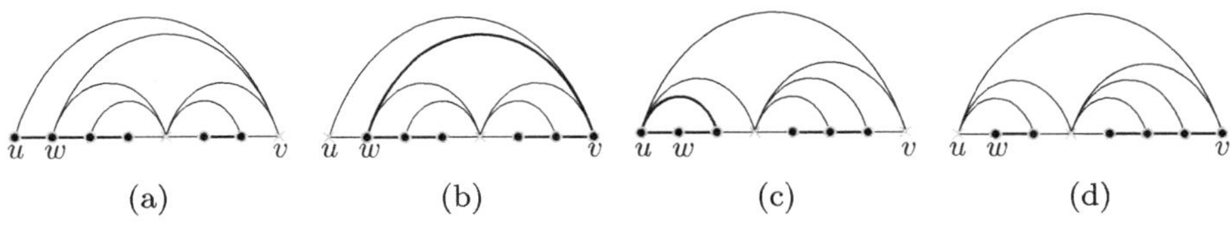

Fig. 4. Some uv-split subgraphs H with 8 vertices; in (a)–(b) we have $h_1 = 2$ and $h_2 = 7$, while in (c)–(d) $h_1 = 4$ and $h_2 = 5$. (a) A 6-over-8 set with u. (b) A 6-over-8 set with v. (c) A 6-over-8 set with u. (d) A 6-over-8 set with v.

at most 1. Indeed, if $h_1 = 2$ and $h_2 = 7$, then H admits a 6-over-8 set with u-deg-1, see again Fig. 4(a), while if $h_1 = 7$ and $h_2 = 2$, then H admits a 6-over-8 set with v-deg-1. Although such sets cannot exclude the existence of a uv-split subgraph with 8 vertices, we will use them in later cases.

Suppose that $h_1 = 4$ and $h_2 = 5$, the case $h_1 = 5$ and $h_2 = 4$ is symmetric. If w has at most 3 neighbors among the vertices in $V(H) \setminus \{u, v\}$, then we can define $\mathcal{I}_H := V(H) \setminus \{u, v\}$. Then $|\mathcal{I}_H|/|V(H)| = \frac{6}{8} > \frac{2}{3}$. Also, $G[\mathcal{I}_H]$ has degree at most 3. This is true for w by assumption. If $z \in V(H_2)$ and $z \neq w$ then z is not adjacent to the 2 vertices of H_1 different from u and w and is not adjacent to z itself. Since u and v are not in $\mathcal{I}_H$, we have that z has at most 3 neighbors in $G[\mathcal{I}_H]$. The argument for the case in which $z \in V(H_1)$ and $z \neq w$ is analogous. We thus assume that, if G has an edge uv such that a uv-split subgraph H of G has 8 vertices and if the uw-split subgraph H_1 has 4 or 5 vertices, then w has degree at least 6 in H. This again implies that H admits a 6-over-8 set with u and a 6-over-8 set with v, see $V(H) \setminus \{v, w\}$ in Fig. 4(c) and $V(H) \setminus \{u, w\}$ in Fig. 4(d). Also, H admits a 6-over-8 set with u-deg-1 (if $h_1 = 5$ and $h_2 = 4$) or a 6-over-8 set with v-deg-1 (if $h_1 = 4$ and $h_2 = 5$, see again Fig. 4(d)).

Case h $=$ 9: Suppose that G contains a uv-split subgraph H with 9 vertices. Since G has no xy-split subgraph with 6 vertices, either one of h_1 and h_2 is 2 and the other one is 8, or one is 3 and the other one is 7, or both h_1 and h_2 are 5. If $h_1 = 2$ and $h_2 = 8$ (the case $h_1 = 8$ and $h_2 = 2$ is symmetric), we define $\mathcal{I}_H := \mathcal{I}_{H_2}^w$, where $\mathcal{I}_{H_2}^w$ is a 6-over-8 set with w. If $h_1 = 3$ and $h_2 = 7$ (the case $h_1 = 7$ and $h_2 = 3$ is symmetric), we define $\mathcal{I}_H := \mathcal{I}_{H_2}^w \cup \{t\}$, where $\mathcal{I}_{H_2}^w$ is a 5-over-7 set with w and t is the vertex of H_1 different from u and w. Finally, if $h_1 = 5$ and $h_2 = 5$, we define $\mathcal{I}_H := V(H) \setminus \{u, v, w\}$. In all three cases, $G[\mathcal{I}_H]$ has degree at most 3 and $|\mathcal{I}_H|/|V(H)| = \frac{6}{9} = \frac{2}{3}$. Since u and v are not in $\mathcal{I}_H$, by Lemma 2, the set $\mathcal{I} := \mathcal{I}_H \cup \mathcal{I}_{\overline{H}}$ contains at least $\frac{2}{3}n$ vertices and $G[\mathcal{I}]$ has degree at most 3, as desired.

In **Case h $=$ 10**, a set $\mathcal{I}_H$ with the required properties can be found similarly to Case $h = 9$. In **Case h$=$11**, a set $\mathcal{I}_H$ with the required properties might not exist. Similarly to Cases $h = 7$ and $h = 8$, we define 8-*over*-11 *sets with u* and *with v*, where u and v have degree at most 1 in the graphs induced by these sets. We prove that H always contains such sets; the proof exploits the existence of a 6-over-8 set with w-deg-1 or with v-deg-1 in H_2 if $h_2 = 8$. Although such sets cannot exclude the existence of a uv-split subgraph with 11 vertices, they

Table 3. Results on planar graphs with different values of k.

	$k = 3$	$k = 4$	$k = 5$	$k = 6$	$k = 7$
LB	> 0.384	> 0.457	0.5	> 0.571	0.625
UB	< 0.572	$0.\overline{6}$	$0.\overline{72}$	< 0.769	0.8

can be used in later cases. In fact, in **Cases h=12–20**, we prove that a set $\mathcal{I}_H$ such that $|\mathcal{I}_H|/|V(H)| \geq \frac{2}{3}$, such that $G[\mathcal{I}_H]$ has degree at most 3, and such that u and v are not in $\mathcal{I}_H$ can always be found. This proof exploits arguments similar to Case $h = 9$. By Lemma 2, the set $\mathcal{I} := \mathcal{I}_H \cup \mathcal{I}_{\overline{H}}$ contains at least $\frac{2}{3}n$ vertices and $G[\mathcal{I}]$ has degree at most 3, as desired. We thus assume that G has no uv-split subgraph with a number of vertices between 12 and 20.

Case h = 21: By Lemma 1 with $\ell = 10$, we have that G has a uv-split subgraph with $h = 21$ vertices. Since G has no xy-split subgraph with a number of vertices between 12 and 20, we have that h_1 and h_2 are both 11. By assumption, H_1 admits a 8-over-11 set with w, say $\mathcal{I}_{H_1}^w$, and H_2 admits a 8-over-11 set with w, say $\mathcal{I}_{H_2}^w$. Then we define $\mathcal{I}_H := \mathcal{I}_{H_1}^w \cup \mathcal{I}_{H_2}^w$. Note that $|\mathcal{I}_H|/|V(H)| = \frac{15}{21} > \frac{2}{3}$. The degree of w in each of $G[\mathcal{I}_{H_1}^w]$ and $G[\mathcal{I}_{H_2}^w]$ is at most 1, hence its degree in $G[\mathcal{I}_H]$ is at most 2. Since the degree in $G[\mathcal{I}_H]$ of every vertex in $\mathcal{I}_{H_i}^w \setminus \{w\}$ is the same as its degree in $G[\mathcal{I}_{H_i}^w]$, for $i = 1, 2$, the degree of $G[\mathcal{I}_H]$ is at most 3. Since u and v do not belong to $\mathcal{I}_H$, by Lemma 2, the set $\mathcal{I} := \mathcal{I}_H \cup \mathcal{I}_{\overline{H}}$ contains at least $\frac{2}{3}n$ vertices and $G[\mathcal{I}]$ has degree at most 3. $\square$

The next theorem deals with induced subgraphs with degree at most $k \geq 4$.

Theorem 4. *For every $n \geq 1$, for every n-vertex outerplanar graph G, and for every integer $k \geq 4$, there exists a set $\mathcal{I} \subseteq V(G)$ such that $G[\mathcal{I}]$ has degree at most k and $|\mathcal{I}| \geq \frac{2k+1}{2k+5}n$ (if k is even) or $|\mathcal{I}| \geq \frac{2k+2/3}{2k+14/3}n$ (if k is odd).*

4   Induced Subgraphs of Planar Graphs

We consider the size of the largest induced subgraph of degree at most k that one is guaranteed to find in an n-vertex planar graph. Table 3 shows the coefficients of the linear terms of our bounds, for various values of k. We start with two upper bounds $\frac{2k-2}{2k+1}n + O(1)$ and $\frac{k+2}{k+4}n + O(1)$. The first one is tighter for $k < 10$, while the second one for $k > 10$. The proof of the first upper bound employs the graph in Fig. 5(a) and is similar to the one of Theorem 1, while the proof of the second upper bound employs the graph in Fig. 5(b) and is elementary.

Theorem 5. *For every $n \geq 1$ and for every integer $k \geq 3$, let $\mu = \min\{\frac{2k-2}{2k+1}n + 5, \frac{k+2}{k+4}n + 2\}$. There exists an n-vertex planar graph G such that any set $\mathcal{I} \subseteq V(G)$ where $G[\mathcal{I}]$ has degree at most k is such that $|\mathcal{I}| \leq \mu$.*

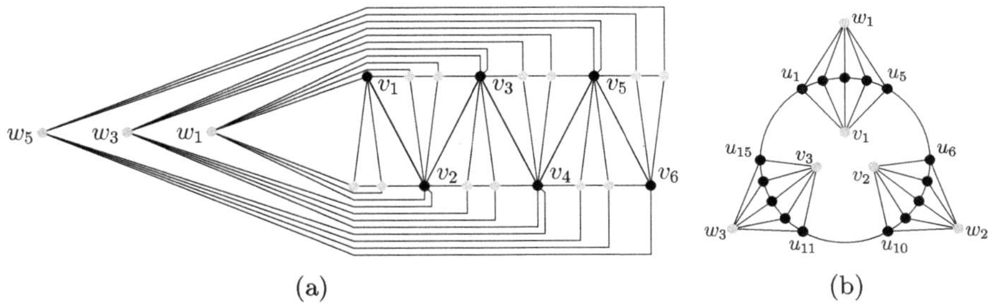

Fig. 5. Illustration for the upper bound in Theorem 5, with $k = 3$. (a) The graph for the case $\mu = \frac{2k-2}{2k+1}n + 5$. (b) The graph for the case $\mu = \frac{k+2}{k+4}n + 2$.

We now focus on the lower bound. We present an algorithm, called GREEDYREMOVAL, that given a (not necessarily planar) graph G and an integer $k \geq 3$, defines a large set of vertices $\mathcal{I} \subseteq V(G)$ such that $G[\mathcal{I}]$ has degree at most k. As the name suggests, the algorithm follows a greedy strategy: It repeatedly removes from the current graph $\mathcal{G}$ (initially $\mathcal{G} = G$) a vertex with maximum degree, until $\mathcal{G}$ has degree k. When this happens, $V(\mathcal{G})$ is the desired set $\mathcal{I}$. To get better bounds, we tweak this strategy if $k = 3$ or $k = 4$. In these cases, once $\mathcal{G}$ has degree $k + 1$, we either repeatedly remove vertices of degree $k + 1$, or we act as follows. We augment $\mathcal{G}$ to a $(k + 1)$-regular graph, in which we find a small dominating set $\mathcal{D}$. We remove from $\mathcal{G}$ the vertices in $\mathcal{D}$. Now $\mathcal{G}$ has degree k and $V(\mathcal{G})$ is the desired set $\mathcal{I}$. We choose the strategy that removes the smallest number of vertices. This algorithm is the basis for the following theorem.

Theorem 6. *For every $n \geq 1$, for every graph G with n vertices, m edges, and degree $d \geq 4$, and for every integer $3 \leq k \leq d - 1$, there exists a set $\mathcal{I} \subseteq V(G)$ such that $G[\mathcal{I}]$ has degree at most k and such that:*

1. *if $d = 4$ and $k = 3$, then $|\mathcal{I}| \geq \frac{7}{11}n - \frac{24}{11}$;*

2. *if $d = 5$ and $k = 3$, then $|\mathcal{I}| \geq \frac{35}{78}n - \frac{8}{13}$;*

3. *if $d \geq 6$ and $k = 3$, then $|\mathcal{I}| \geq \frac{10}{13}n - \frac{5}{39}m - \frac{8}{13}$;*

4. *if $d = 5$ and $k = 4$, then $|\mathcal{I}| \geq \frac{9}{14}n - \frac{1}{2}$;*

5. *if $d \geq 6$ and $k = 4$, then $|\mathcal{I}| \geq \frac{54}{59}n - \frac{9}{59}m - \frac{35}{59}$; and*

6. *if $d \geq 6$ and $k \geq 5$, then $|\mathcal{I}| \geq n - \frac{m}{k+1}$.*

Proof Sketch. First, note that the degree of the current graph $\mathcal{G}$ does not increase throughout the algorithm's execution. For any $i = d, d - 1, \ldots, k$, let $\mathcal{G}_i$ be the graph $\mathcal{G}$ the first time it has degree at most i; thus, $\mathcal{G}_d = G$ and $V(\mathcal{G}_k) = \mathcal{I}$. Let r_i be the number of vertices of degree i removed by GREEDYREMOVAL; these vertices are in $\mathcal{G}_i$ and not in $\mathcal{G}_{i-1}$. Also, let $r_6^+ = \sum_{j=6}^{d} r_j$. We now analyze GREEDYREMOVAL. We distinguish the six cases in the statement.

Case 1: $d = 4$ and $k = 3$. We show that $G = \mathcal{G}_4$ can be augmented to a 4-regular graph by adding at most 6 vertices to it. While there exist two non-adjacent vertices with degree smaller than 4, we add an edge between them. Eventually, the vertices of degree smaller than 4 induce a clique K_x in $\mathcal{G}_4$, with $x \leq 4$. Then it suffices to add at most 6 vertices (and some further edges) to $\mathcal{G}_4$ in order to obtain a 4-regular graph $\mathcal{H}_4$, see [16]. Every h-vertex 4-regular graph has a dominating set of size at most $\frac{4}{11}h$ [30], hence $\mathcal{H}_4$ has a dominating set $\mathcal{D}$ of size at most $\frac{4}{11}(n+6)$. Removing the r_4 vertices of $\mathcal{G}_4$ that are in $\mathcal{D}$ results in a graph $\mathcal{G}_3$ with degree at most 3. We thus get **(Eq. 1)** $r_4 \leq \frac{4}{11}(n+6)$. The vertex set of $\mathcal{G}_3$ is our desired set $\mathcal{I}$, where $|\mathcal{I}| = n - r_4 \geq \frac{7}{11}n - \frac{24}{11}$.

Case 2: $d = 5$ and $k = 3$. The algorithm first removes r_5 vertices from G obtaining $\mathcal{G}_4$, and then removes r_4 vertices from $\mathcal{G}_4$ resulting in $\mathcal{G}_3$, whose vertex set is $\mathcal{I}$. We upper bound $r_4 + r_5$ in two different ways. First, by (Eq. 1) and since the number of vertices of $\mathcal{G}_4$ is $n - r_5$, we get **(Eq. 2)** $r_4 + r_5 \leq \frac{4}{11}(n - r_5 + 6) + r_5 = \frac{4n + 7r_5 + 24}{11}$. Second, the number of vertices of degree 4 that are removed from $\mathcal{G}_4$ is at most $\frac{1}{4}$ times the number of edges of $\mathcal{G}_4$, since the removal of a degree-4 vertex removes 4 edges from $\mathcal{G}_4$. Also, the number of edges of $\mathcal{G}_4$ is equal to the number of edges of $\mathcal{G}_5$, which is at most $\frac{5n}{2}$ given that $\mathcal{G}_5$ has degree 5, minus 5 times the number of vertices of degree 5 that are removed from $\mathcal{G}_5$. Hence, we get **(Eq. 3)** $r_4 + r_5 \leq \frac{\frac{5n}{2} - 5r_5}{4} + r_5 = \frac{5n - 2r_5}{8}$. We thus get **(Eq. 4)** $r_4 + r_5 \leq \frac{43}{78}n + \frac{8}{13}$, which comes from (Eq. 2) when $r_5 \leq \frac{23n}{78} - \frac{32}{13}$ and from (Eq. 3) when $r_5 \geq \frac{23n}{78} - \frac{32}{13}$. Therefore, it follows that $|\mathcal{I}| = n - r_4 - r_5 \geq \frac{35}{78}n - \frac{8}{13}$.

Case 3: $d \geq 6$ and $k = 3$. The algorithm first removes r_{6+} vertices from G to obtain $\mathcal{G}_5$, then removes r_5 vertices from $\mathcal{G}_5$ to obtain $\mathcal{G}_4$, and finally removes r_4 vertices from $\mathcal{G}_4$, resulting in $\mathcal{G}_3$, whose vertex set is $\mathcal{I}$. We upper bound $r_4 + r_5 + r_{6+}$ in three different ways. First, by (Eq. 4) and since the number of vertices of $\mathcal{G}_5$ is $n - r_{6+}$ we get: **(Eq. 5)** $r_4 + r_5 + r_{6+} \leq \frac{43}{78}(n - r_{6+}) + \frac{8}{13} + r_{6+} = \frac{43n + 35r_{6+} + 48}{78}$. Second, by (Eq. 1) and since the number of vertices of $\mathcal{G}_4$ is $n - r_5 - r_{6+}$, we get **(Eq. 6)** $r_4 + r_5 + r_{6+} \leq \frac{4}{11}(n - r_5 - r_{6+} + 6) + r_5 + r_{6+} = \frac{4n + 7r_5 + 7r_{6+} + 24}{11}$. Third, as in Case 2, the number of vertices of degree 4 that are removed from $\mathcal{G}_4$ is at most $\frac{1}{4}$ times the number of edges of $\mathcal{G}_4$. Also, the number of edges of $\mathcal{G}_4$ is at most the number m of edges of G, minus 5 times the number of vertices of degree 5 that are removed from $\mathcal{G}_5$, minus 6 times the number of vertices of degree at least 6 that are removed from G. Hence, we get: **(Eq. 7)** $r_4 + r_5 + r_{6+} \leq \frac{m - 5r_5 - 6r_{6+}}{4} + r_5 + r_{6+} = \frac{m - r_5 - 2r_{6+}}{4}$. We thus get **(Eq. 8)** $r_4 + r_5 + r_{6+} \leq \frac{3}{13}n + \frac{5}{39}m + \frac{8}{13}$. (Eq. 8) comes from (Eq. 5) when $r_{6+} \leq -\frac{5}{7}n + \frac{2}{7}m$, from (Eq. 6) when $r_5 + r_{6+} \leq -\frac{19}{91}n + \frac{55}{273}m - \frac{32}{13}$, and from (Eq. 7) when $r_{6+} \geq -\frac{5}{7}n + \frac{2}{7}m$ and $r_5 + r_{6+} \geq -\frac{19}{91}n + \frac{55}{273}m - \frac{32}{13}$. It follows that $|\mathcal{I}| = n - r_4 - r_5 - r_{6+} \geq \frac{10}{13}n - \frac{5}{39}m - \frac{8}{13}$.

Cases 4 and 5 are similar to Cases 1 and 2, resp., see [16].

Case 6: $d \geq 6$ and $k \geq 5$. The number of vertices of degree at least $k+1$ removed from G is at most $\frac{1}{k+1}$ times the number of edges of G, since each vertex removal removes at least $k + 1$ edges from G. Hence, we get **(Eq. 9)** $\sum_{j=k+1}^{d} r_j \leq \frac{m}{k+1}$. Therefore, it follows that $|\mathcal{I}| = n - \sum_{j=k+1}^{d} r_j \geq n - \frac{m}{k+1}$. $\square$

Since a planar graph has at most $3n - 6$ edges, we get the following corollary.

Corollary 1. *For every $n \geq 1$, for every planar graph G with n vertices, and for every integer $k \geq 3$, there exists a set $\mathcal{I} \subseteq V(G)$ such that $G[\mathcal{I}]$ has degree at most k and such that (i) if $k = 3$, then $|\mathcal{I}| \geq \frac{5}{13}n + \frac{2}{13}$; (ii) if $k = 4$, then $|\mathcal{I}| \geq \frac{27}{59}n + \frac{19}{59}$; and (iii) if $k \geq 5$, then $|\mathcal{I}| \geq \frac{k-2}{k+1}n + \frac{6}{k+1}$.*

Similarly, we can get better lower bounds for sparse families of planar graphs. For example, n-vertex planar graphs of girth at least g have at most $\frac{g(n-2)}{g-2}$ edges, and hence have a set $\mathcal{I} \subseteq V(G)$ such that $G[\mathcal{I}]$ has degree at most k and such that $|\mathcal{I}| \geq \frac{(25g-60)n-14g+48}{39(g-2)}$, $|\mathcal{I}| \geq \frac{(45g-108)n-17g+70}{59(g-2)}$, and $|\mathcal{I}| \geq \frac{(gk-2k-2)n+2g}{(g-2)(k+1)}$ if $k = 3$, $k = 4$, or $k \geq 5$, respectively. Thus, n-vertex bipartite planar graphs have a set $\mathcal{I} \subseteq V(G)$ such that $G[\mathcal{I}]$ has degree at most 3 and such that $|\mathcal{I}| \geq \frac{20}{39}n - \frac{4}{39}$, which is larger than the maximum size of an induced linear forest [19].

5   Conclusions and Open Problems

In this paper, we proved bounds for the value of the function $f_k(n)$ such that every n-vertex outerplanar/planar graph has an induced subgraph with $f_k(n)$ vertices and degree at most k. Our lower bounds are based on polynomial-time algorithms that compute the desired subgraphs. An obvious direction for further research is to close the gaps between our upper and lower bounds. See again Table 1 in Sect. 1 for the case $k = 3$. We believe that our techniques might yield further improvements on the considered problems, as explained below.

Let c_k be the maximum value such that every n-vertex outerplanar graph G has an induced subgraph with degree at most k and $c_k \cdot n + o(n)$ vertices. The use of Lemmas 1 and 2 that led to prove Theorems 3 and 4 could allow us to prove a $d_k \cdot n + O(1)$ lower bound, where d_k is arbitrarily close to c_k. Indeed, by employing "large" values of ℓ in Lemma 1, the effect of the constraint imposed by Lemma 2 of excluding two vertices u and v from those inducing a large subgraph with degree at most k in the uv-split subgraph of G becomes negligible. This intuition could be used to experimentally determine c_k, for small values of k, up to some decimal digits. It would be interesting to understand whether $c_k = \frac{k-1}{k}$, which would match our upper bound from Theorem 1 and which is indeed the right bound for $k = 3$, by Theorem 3. Also, our techniques could be extended to find large bounded-degree induced subgraphs in graphs of bounded treewidth.

For planar graphs, we employed a greedy algorithm which repeatedly removes a vertex with maximum degree, until the graph has at most the desired degree k. For $k = 3$ and $k = 4$, our algorithm stops removing vertices when the graph G has degree at most $k + 1$. Then G is augmented to $(k + 1)$-regular and finally it achieves degree at most k by removing a small dominating set [30, 41]. The greedy algorithm without the dominating set strategy still produces a lower

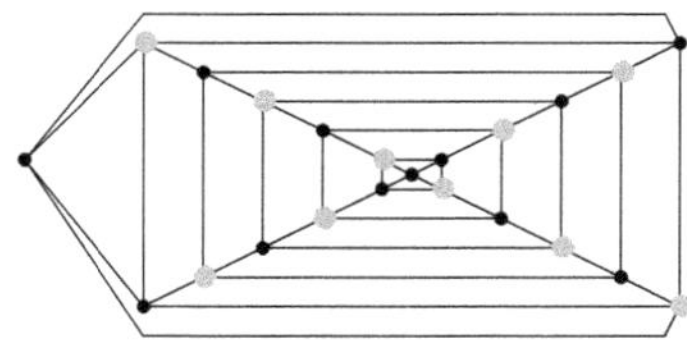

Fig. 6. A 4-regular planar graph in which repeatedly removing degree-4 vertices (shown as large disks) might result in the removal of $\frac{n}{2} - O(1)$ vertices.

bound better than $\lceil \frac{n}{3} \rceil$. However, if $d = 4$ and $k = 3$, it only gives us an $\frac{n}{2}$ lower bound, see Fig. 6, which is worse than the $\frac{7n}{11}$ lower bound we get by using dominating sets; consequently, not employing the dominating set strategy leads to worse bounds also for $d \geq 5$. We remark that our greedy algorithm does not exploit the planarity of the graph, other than for bounding its number of edges. Hence, we find interesting to understand whether planarity can be exploited more deeply to get better bounds. In particular, lower bounds better than the ones presented in this paper could be obtained if one could prove that every planar graph with maximum degree 4 or with maximum degree 5 has a subset of vertices whose size is smaller than $\frac{4n}{11}$ or $\frac{5n}{14}$, respectively, such that every degree-4 vertex is in the set or adjacent to a vertex in the set.

Acknowledgments. Thanks to Ross J. Kang and Michael J. Pelsmajer for pointing out to us relevant literature for the problem we studied.

References

1. Akiyama, J., Watanabe, M.: Maximum induced forests of planar graphs. Graphs Comb. **3**(1), 201–202 (1987)
2. Albertson, M.O.: A lower bound for the independence number of a planar graph. J. Comb. Th. Ser. B **20**(1), 84–93 (1976)
3. Albertson, M.O., Berman, D.M.: A conjecture on planar graphs. Graph theory and related topics, p. 357 (1979)
4. Alon, N., Mubayi, D., Thomas, R.: Large induced forests in sparse graphs. J. Graph Th. **38**(3), 113–123 (2001)
5. Angelini, P., Evans, W.S., Frati, F., Gudmundsson, J.: SEFE without mapping via large induced outerplane graphs in plane graphs. J. Graph Th. **82**(1), 45–64 (2016)
6. Appel, K., Haken, W.: Every planar map is four colorable. I. Discharging. Illinois J. Math. **21**(3), 429–490 (1977)
7. Appel, K., Haken, W., Koch, J.: Every planar map is four colorable. II. Reducibility. Illinois J. Math. **21**(3), 491–567 (1977)
8. Biedl, T., Wilkinson, D.F.: Bounded-degree independent sets in planar graphs. Theory Comput. Syst. **38**(3), 253–278 (2005)
9. Bondy, J.A., Hopkins, G., Staton, W.: Lower bounds for induced forests in cubic graphs. Canad. Math. Bull. **30**(2), 175–178 (1987)
10. Borodin, O.V.: On acyclic colorings of planar graphs. Discret. Math. **25**(3), 211–236 (1979)

11. Borradaile, G., Le, H., Sherman-Bennett, M.: Large induced acyclic and outerplanar subgraphs of 2-outerplanar graph. Graphs Comb. **33**(6), 1621–1634 (2017)
12. Bose, P., Dujmovic, V., Wood, D.R.: Induced subgraphs of bounded degree and bounded treewidth. Contributions Discret. Math. **1**(1) (2006)
13. Caro, Y., Roditty, Y.: On the vertex-independence number and star decomposition of graphs. Ars Combin. **20**, 167–180 (1985)
14. Chappell, G.G., Pelsmajer, M.J.: Maximum induced forests in graphs of bounded treewidth. Electron. J. Comb. **20**(4), 8 (2013)
15. Cowen, L.J., Cowen, R., Woodall, D.R.: Defective colorings of graphs in surfaces: Partitions into subgraphs of bounded valency. J. Graph Th. **10**(2), 187–195 (1986)
16. D'Elia, M., Frati, F.: Large induced subgraphs of bounded degree in outerplanar and planar graphs. CoRR abs/2412.14784 (2024)
17. Diestel, R.: Graph Theory. Graduate Texts in Mathematics, 4th edn, vol. 173. Springer (2012)
18. Dross, F., Montassier, M., Pinlou, A.: Large induced forests in planar graphs with girth 4. Discret. Appl. Math. **254**, 96–106 (2019)
19. Dross, F., Montassier, M., Pinlou, A.: A lower bound on the order of the largest induced linear forest in triangle-free planar graphs. Disc. Math. **342**, 943–950 (2019)
20. Dvorák, Z., Kelly, T.: Induced 2-degenerate subgraphs of triangle-free planar graphs. Electron. J. Comb. **25**(1), 1 (2018)
21. Fertin, G., Godard, E., Raspaud, A.: Minimum feedback vertex set and acyclic coloring. Inf. Process. Lett. **84**(3), 131–139 (2002)
22. Fountoulakis, N., Kang, R.J., McDiarmid, C.: The t-stability number of a random graph. Electron. J. Comb. **17**(1) (2010)
23. Gu, Y., Kierstead, H.A., Oum, S., Qi, H., Zhu, X.: 3-degenerate induced subgraph of a planar graph. J. Graph Th. **99**(2), 251–277 (2022)
24. Hosono, K.: Induced forests in trees and outerplanar graphs. Proc. Faculty Sci. Tokai Univ. **25**, 27–29 (1990)
25. Kelly, T., Liu, C.: Size of the largest induced forest in subcubic graphs of girth at least four and five. J. Graph Th. **89**(4), 457–478 (2018)
26. Kirkpatrick, D.G.: Optimal search in planar subdivisions. SIAM J. Comput. **12**(1), 28–35 (1983)
27. Knauer, K., Ueckerdt, T.: Clustered independence and bounded treewidth. CoRR abs/2303.13655 (2023)
28. Kowalik, L., Luzar, B., Skrekovski, R.: An improved bound on the largest induced forests for triangle-free planar graphs. Disc. Math. Th. Comp. Sc. **12**, 87–100 (2010)
29. Le, H.: A better bound on the largest induced forests in triangle-free planar graph. Graphs Comb. **34**(6), 1217–1246 (2018)
30. Liu, H., Sun, L.: On domination number of 4-regular graphs. Czechoslov. Math. J. **54**, 889–898 (2004)
31. Lukot'ka, R., Mazák, J., Zhu, X.: Maximum 4-degenerate subgraph of a planar graph. Electron. J. Comb. **22**(1), 1 (2015)
32. Pelsmajer, M.J.: Maximum induced linear forests in outerplanar graphs. Graphs Comb. **20**(1), 121–129 (2004)
33. Poh, K.S.: On the linear vertex-arboricity of a planar graph. J. Graph Th. **14**(1), 73–75 (1990)
34. Robertson, N., Sanders, D.P., Seymour, P.D., Thomas, R.: The four-colour theorem. J. Comb. Theory B **70**(1), 2–44 (1997)
35. Salavatipour, M.R.: Large induced forests in triangle-free planar graphs. Graphs Comb. **22**(1), 113–126 (2006)

36. Shi, L., Xu, H.: Large induced forests in graphs. J. Graph Th. **85**(4), 759–779 (2017)
37. Snoeyink, J., van Kreveld, M.J.: Linear-time reconstruction of Delaunay triangulations with applications. In: Burkard, R., Woeginger, G. (eds.) 5th European Symposium on Algorithms (ESA 1997). LNCS, vol. 1284, pp. 459–471. Springer (1997)
38. Staton, W.: Induced forests in cubic graphs. Discret. Math. **49**(2), 175–178 (1984)
39. Wang, Y., Xie, Q., Yu, X.: Induced forests in bipartite planar graphs. J. Comb. **8**(1), 93–166 (2017)
40. Wood, D.R.: Defective and clustered graph colouring. Electron. J. Comb. #**DS23**, 1–71 (2018)
41. Xing, H.M., Sun, L., Chen, X.G.: An upper bound for domination number of 5-regular graphs. Czechoslov. Math. J. **56**, 1049–1061 (2006)

Complexity of Perfect $(1, 2)$-Dominating Sets in Low-Degree Graphs

Urszula Bednarz[1], Jan Kratochvíl[2]([✉]), and Adrian Michalski[1]

[1] Faculty of Mathematics and Applied Physics, Rzeszow University of Technology, al. Powstańców Warszawy 8, 35-959 Rzeszów, Poland
{ubednarz,a.michalski}@prz.edu.pl
[2] Faculty of Mathematics and Physics, Charles University, Malostranské nám. 25, 11800 Prague, Czech Republic
honza@kam.mff.cuni.cz

Abstract. We study the problem of the existence of perfect (1,2)-dominating sets in graphs. A perfect (1,2)-dominating set is a $(1, 2)$-dominating set that is not double dominating and has the property that every vertex outside the set has exactly one neighbor in it. While previous work focused mainly on graphs with large maximum degree, we investigate perfect (1,2)-dominating sets in graphs of small degree, namely subcubic graphs and 4-regular graphs. We prove that every connected subcubic graph which is not isomorphic to one of the graphs in the family $\{K_1, K_2, K_3, K_4, K_4^-\}$ admits a perfect (1,2)-dominating set, and such a set can be constructed in polynomial time. For 4-regular graphs, we provide a characterization in terms of induced 3-regular subgraphs, and show that the problem of deciding the existence of a perfect (1,2)-dominating set is NP-complete. Finally, we extend the study to independent perfect (1,2)-dominating sets in cubic planar graphs, establishing a direct relation with 1-perfect codes. In particular, we show that deciding whether a cubic planar graph contains such a set is NP-complete.

Keywords: dominating set · NP-completeness · independent set · perfect code · planar graph

1 Introduction

In general, we use the standard terminology and notation of graph theory; see, for example, [5]. Let $G = (V(G), E(G))$ be an undirected, simple graph with vertex set $V(G)$ and edge set $E(G)$. In this paper, we will consider only connected graphs. By $\deg_G(x)$ we denote the degree of a vertex x in the graph G. If all vertices of a graph have the same degree equal to r, we say that the graph is *r-regular*. Graphs that are 3-regular are called *cubic*. If the maximum degree of a graph does not exceed 3, then the graph is called *subcubic*. By $d_G(u, v)$ we denote the distance of u and v in G, i.e., the length of the shortest path connecting u to v (measured by the number of edges).

© The Author(s), under exclusive license to Springer Nature Singapore Pte Ltd. 2026
E. Di Giacomo and D. Mondal (Eds.): WALCOM 2026, LNCS 16444, pp. 203–214, 2026.
https://doi.org/10.1007/978-981-95-7127-7_14

A subset $S \subseteq V(G)$ is an *independent set* in G if no two vertices of S are adjacent.

A subset $D \subset V(G)$ is *dominating* if every vertex of G not belonging to D has a neighbour in D. Dominating sets were introduced by Ore [11] and Berge [3] and have been extensively studied over the past decades. Besides classical dominating sets, many other variants have been considered, obtained either by restricting or generalizing the original definition. More information can be found, for example, in [8].

In 2008, *secondary dominating sets* were introduced in [9]. For a positive integer k, a set D is called $(1, k)$-*dominating* if for every vertex $v \in V(G) \setminus D$, there exist two distinct vertices $u, w \in D$ such that $vu \in E(G)$ and $d_G(v, w) \leq k$. The authors proved that every dominating set with at least two vertices is a $(1, 4)$-dominating set. Hence, the only interesting cases remaining are $k \in \{1, 2, 3\}$.

The case $k = 1$ is equivalent to the well-known concept of *double dominating sets*, also called 2-*dominating*, which appeared in the literature as early as 1985, when the definition of multiple dominating sets was introduced in [7]. A subset D is double dominating if every vertex outside D has at least two neighbours in D.

The case $k = 2$ has emerged as a natural extension with interesting properties. The concept of $(1, 2)$-dominating sets, their structural properties, applications, and computational complexity have been studied, among others, in [12, 13, 16, 17].

From the definition of a $(1, k)$-dominating set it follows that every double dominating set is also a $(1, 2)$-dominating set. To distinguish between these concepts, Michalski et al. [14] introduced the notion of *proper* $(1, 2)$-*dominating sets*. A proper $(1, 2)$-dominating set is a $(1, 2)$-dominating set which is not double dominating. This means that if D is a $(1, 2)$-dominating set, then there must exist at least one vertex outside D having exactly one neighbour in D. Since the set $V(G)$ is considered to be double dominating, it is not a proper $(1, 2)$-dominating set. Therefore, it was crucial to study the problem of existence of such sets in graphs. In [14], the authors proved that among connected graphs, only complete graphs do not have a proper $(1, 2)$-dominating set.

A natural extension of the concept of proper $(1, 2)$-dominating sets was to study $(1, 2)$-dominating sets such that *every* vertex not belonging to the $(1, 2)$-dominating set has exactly one neighbour in this set. Therefore, Bednarz and Pirga [2] introduced *perfect* $(1, 2)$-*dominating sets*.

Definition 1. *A set D of vertices of a graph G is a* perfect $(1,2)$-dominating set *if it is* $(1, 2)$-*dominating, not double dominating, and every vertex in* $V(G) \setminus D$ *has exactly one neighbour in D.*

Since $V(G)$ itself is double dominating, it cannot be a perfect $(1, 2)$-dominating set, making the problem of existence nontrivial. Moreover, it turns out to be much more difficult than in the case of proper $(1, 2)$-dominating sets.

So far, the existence of perfect $(1, 2)$-dominating sets has been studied mainly in graphs with large maximum degree. In [2], the authors investigated graphs with maximum degree equal to $|V(G)| - 1$ or $|V(G)| - 2$. They gave a complete characterization in the case that the graph under consideration contains at most two vertices of the maximum degree. In [1], a characterization of perfect $(1, 2)$-dominating sets was obtained for graphs that have exactly three vertices of maximum degree equal to $|V(G)| - 2$. These characterizations imply polynomial time algorithms for deciding the existence of a perfect $(1, 2)$-dominating set in such a graph.

Our Results. In this paper, we turn to graphs with relatively small maximum degree, namely subcubic graphs and 4-regular graphs. In Theorem 1, we prove that apart from a finite number of graphs, every graph of maximum degree at most three contains a perfect $(1, 2)$-dominating set. In particular, every connected subcubic graph with at least 5 vertices contains a perfect $(1, 2)$-dominating set. Moreover, a perfect $(1, 2)$-dominating set can be constructed in polynomial time (if it exists). It follows without saying that the existence of perfect $(1, 2)$-dominating sets in subcubic graphs can be decided in polynomial time. The situation changes dramatically when the maximum degree of the input graph increases by just one. In Corollary 1, we prove that deciding the existence of a perfect $(1, 2)$-dominating set in a 4-regular graph is NP-complete. We deduce this NP-hardness result from showing, in Theorem 2, that deciding the existence of a non-empty induced 3-regular subgraph in a 4-regular graph is NP-complete. We believe that this result is interesting on its own. To the best of our knowledge, so far it has only been known that deciding the existence of a non-empty 3-regular subgraph (not necessarily induced) in a graph of maximum degree 4 is NP-complete. Thus, we are strengthening this in two directions: proving the NP-hardness for *induced* subgraphs, and having the input graph 4-regular (and not just maximum degree 4).

In Sect. 4, we consider the problem of the existence of independent perfect $(1, 2)$-dominating sets and we show that it is NP-complete even in 3-regular planar graphs.

2 Perfect (1,2)-Domination in Subcubic Graphs

We denote by $\mathcal{B} = \{K_1, K_2, K_3, K_4, K_4^-\}$ the family consisting of all complete graphs on at most four vertices together with the diamond graph K_4^- (obtained from K_4 by deleting one edge). It is easy to see that for each of these graphs, every subset of vertices of cardinality greater than 1 is double dominating, hence these graphs do not admit a perfect $(1, 2)$-dominating set. We will prove that they are the only subcubic graphs of this property.

Theorem 1. *Let G be a connected nonempty subcubic graph which does not belong to $\mathcal{B}$. Then G contains a perfect (1,2)-dominating set. Such a set can be constructed in polynomial time.*

Proof. We will proceed by proving several claims. In all of them, we assume that G is a nonempty connected subcubic graph that does not belong to the set $\mathcal{B}$.

Claim 1. If G contains a vertex of degree 1, then G admits a perfect $(1,2)$-dominating set.

Proof of Claim 1. Let x be a vertex of degree 1. Then $D = V(G) \setminus \{x\}$ is a perfect (1,2)-dominating set. The only vertex outside D is x, it is dominated by its only neighbor, and since $G \neq K_2$, this neighbor is adjacent to another vertex of D, which is the required vertex of D at distance 2 from x (there may be more such vertices). $\qquad\Box$

Claim 2. If G contains a triangle with all vertices of degree 3, then G admits a perfect (1,2)-dominating set.

Proof of Claim 2. Let x, y, z be pairwise adjacent vertices of G such that $\deg_G(x) = \deg_G(y) = \deg_G(z) = 3$. Let $\bar{u}$ denote the other neighbor of u, for $u = x, y, z$. It may happen that, e.g., $\bar{x} = \bar{y}$, but it cannot happen that $\bar{x} = \bar{y} = \bar{z}$ because $G \neq K_4$. Hence, the set $\{\bar{x}, \bar{y}, \bar{z}\}$ consists of at least two different vertices. Set $D = V(G) \setminus \{x, y, z\}$. Every $u \in \{x, y, z\}$ has exactly one neighbor in D, its buddy $\bar{u}$. And $\{\bar{x}, \bar{y}, \bar{z}\} \setminus \{\bar{u}\}$ contains at least one vertex, which is the required vertex of D at distance 2 from u. Hence D is a perfect (1,2)-dominating set. $\qquad\Box$

Claim 3. If G contains a cycle with all vertices of degree 3, then G admits a perfect (1,2)-dominating set.

Proof of Claim 3. Consider a shortest cycle C in G whose all vertices are of degree 3 (if there exists at least one such cycle, there exists a shortest one as well). If C is a triangle, we are done by Claim 2.

Hence, suppose that C has at least 4 vertices. Note that since C is a shortest cycle, it must be induced. Again, let $\bar{u}$ be the neighbour of u outside of C, for $u \in V(C)$. Since C is induced and $\deg_G(u) = 3$, $\bar{u}$ is uniquely defined. Set $D = V(G) \setminus V(C)$. Every vertex $u \in V(C)$ is dominated by exactly one vertex of D, namely by $\bar{u}$. Let v and w be the two neighbours of u on C. We observe that if $\bar{u} = \bar{v} = \bar{w}$, then $\deg_G(\bar{u}) = 3$ and G contains a triangle $u, v, \bar{u}$ with all vertices of degree 3. Claim 2 would imply that G has a perfect (1,2)-dominating set in such a case. Hence, suppose that $\bar{v} \neq \bar{u}$ or $\bar{w} \neq \bar{u}$, say w.l.o.g. that $\bar{v} \neq \bar{u}$ (see Fig. 1). Then $\bar{v}$ is the requested vertex of D at distance 2 from u. Since these arguments hold for all vertices u of C, D is a perfect (1,2)-dominating set. $\qquad\Box$

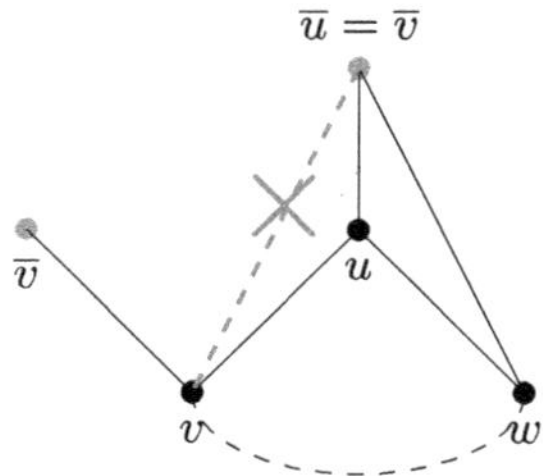

Fig. 1. Illustration of the proof of *Claim 3*

Claim 4. If G contains two adjacent vertices of degree 2, then G admits a perfect (1,2)-dominating set.

Proof of Claim 4. Let x, y be adjacent vertices of degree 2. In view of the already proven Claim 1, we may assume that all vertices of G have degrees greater than 1. Let $\overline{x}$ be the neighbour of x different from y, and similarly, let $\overline{y}$ be the neighbour of y different from x.

If $\overline{x} = \overline{y}$ (see the left side of Fig. 2), this vertex must have another neighbour z (otherwise G would be isomorphic to K_3). If $\overline{x} \neq \overline{y}$ (see the right side of Fig. 2), $\overline{x}$ has another neighbour z_1 and $\overline{y}$ has another neighbour z_2, because $\deg_G(\overline{x}) > 1$ and $\deg_G(\overline{y}) > 1$.

Set $D = V(G) \setminus \{x, y\}$. Then each $u \in \{x, y\}$ is dominated by exactly one vertex of D, namely by $\overline{u}$, and it has a vertex at distance 2 in D, namely z in the former case and z_1 or z_2 in the latter one. Hence D is a perfect (1,2)-dominating set. $\qquad\square$

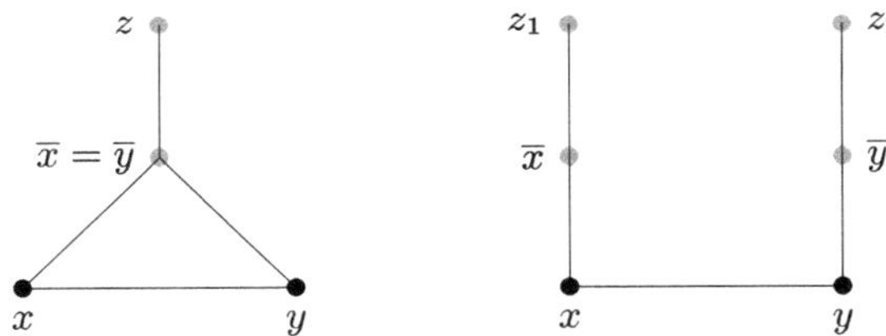

Fig. 2. Illustration of the proof of *Claim* 4

Claim 5. If G contains a cycle with at least two vertices of degree 2, then G admits a perfect (1,2)-dominating set.

Proof of Claim 5. Suppose G contains a cycle passing through two different vertices of degree 2, say x and y, and without loss of generality suppose that all vertices on one of the arcs of this cycle between x and y are of degree 3. Let C be a shortest cycle that passes through x and y with this property. The cycle consists of two paths P_1 and P_2, which are the components of $C - \{x, y\}$. In other words, $C = x P_1 y P_2$, and we assume that all the vertices of P_1 have degree 3. The cycle C is not necessarily induced, but since it is a shortest one, there is no chord with both end-vertices in $\{x, y\} \cup V(P_1)$, as well as no chord with both end-vertices in $\{x, y\} \cup V(P_2)$ (there may be, however, edges of type ab with $a \in P_1$ and $b \in P_2$). In view of Claim 4, we may assume that both P_1 and P_2 contain at least one vertex each (otherwise G has a perfect (1,2)-dominating set according to the claim).

Now set $D = V(G) \setminus (\{x, y\} \cup V(P_1))$. The vertices of D are green in Fig. 3. We claim that D is a perfect (1,2)-dominating set in G.

For every vertex $u \in V(P_1)$, let $\bar{u}$ be the neighbour of u outside $\{x, y\} \cup V(P_1)$ (there is such a neighbour, since $\deg_G(u) = 3$ and there is no edge uv with $v \in \{x, y\} \cup V(P_1)$ other than the two edges of the cycle C incident with u). This $\bar{u}$ is in D and it is the only vertex of D that dominates u. Vertices x and y are dominated by their neighbours in P_2, and these are their only neighbours in D. Let x' be the vertex of P_1 adjacent to x. Then $\overline{x'}$ is a D vertex at distance 2 from x, and the neighbour of x on P_2 is a D vertex at distance 2 from x'. Similarly for y', the neighbour of y on P_1. For an inner vertex u of P_1, let u' be one of its neighbours on P_1. Then u is dominated by $\bar{u}$ and $\overline{u'}$ is the requested D vertex at distance 2 from u.

Obviously, D is nontrivial, as it does not contain x nor y. $\qquad\square$

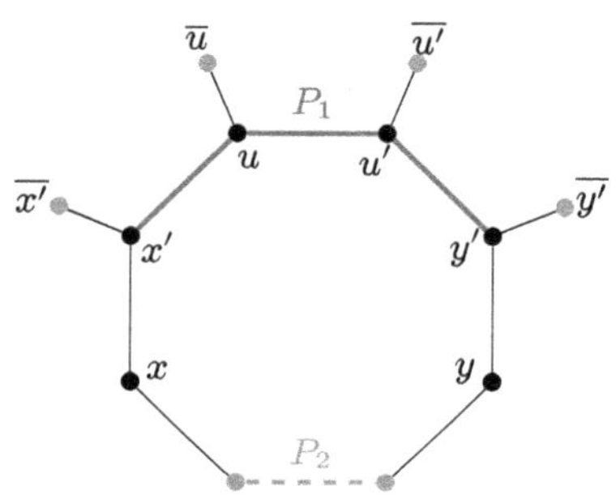

Fig. 3. Illustration of the proof of *Claim* 5

It remains to discuss the case where every cycle in G contains exactly one vertex of degree 2. We will show that no subcubic graph has this property unless it satisfies the assumptions of one of the claims above.

Claim 6. Suppose G is such that every cycle contains exactly one vertex of degree 2. Then G contains a vertex of degree 1.

Proof of Claim 6. Construct an auxiliary graph G' by suppressing the vertices of degree 2, which means that the paths consisting of vertices of degree 2 will be replaced by edges. These new edges will be called 'red' to retain the information on how they originated. Of course, G' may have parallel edges, but no loops (a loop would be created by a cycle containing at least two vertices of degree 2). If G has no vertices of degree 1, G' is 3-regular. The assumption on G translates into the property that every cycle of G' contains exactly one red edge.

Now observe that parallel edges in G' must form edge-cuts. This is because two edges with the same end-vertices, say x, y, must have originated from a triangle x, y, z in G, with z being a vertex of degree 2. If the two edges did not form a cut, there would be a path P in G' connecting x and y. Now both $P + xy$ and $P + \{xz, yz\}$ are cycles in G, and if in the latter one, z is the unique vertex of degree 2, then the former one has no vertex of degree 2, contradicting the assumptions. This case also covers the fact that G' cannot be a dipole with 3 parallel edges.

Now consider the block structure of G'. The blocks are either bridges, or dipoles of two parallel edges (out of which exactly one is red), or 2-connected components which are simple graphs (as proven in the previous paragraph) with, possibly, some edges red. The block structure is known to be acyclic. Consider a leaf block, say B. A leaf block contains only one articulation (of G'), and thus B is a simple 3-regular graph with exactly one vertex of degree 2. As such, it is not outerplanar (every 2-connected outerplanar graph contains at least two vertices of degree 2), and therefore it contains a subdivision of K_4 or a subdivision of $K_{2,3}$. In either case, B contains a subdivision of a Theta graph – it contains two vertices, say x, y, connected by three internally disjoint paths P_1, P_2, P_3. Now, each two of these paths form a cycle in G', and thus the union of any two of them contains exactly one red edge, which is impossible. $\square$

It follows from Claim 6 that the assumptions of one of Claims 1–5 must be satisfied, and hence G admits a perfect (1,2)-dominating set.

For the claim about the algorithmic aspect, we argue as follows. Let n be the number of vertices of the input graph G. We first check in linear time if G contains a vertex of degree 1. We construct a perfect (1,2)-dominating set along the lines of the proof of Claim 1 in the affirmative case and proceed in the negative one. Then we construct the auxiliary graph G' and its block structure, clearly in polynomial time, and consider a leaf block B.

In linear time, we check all red edges of B. If one of them originates from a path with more than one vertex of degree 2, we proceed via the proof of Claim 4. If not, we know that we must find a 'good' cycle in B. We can do this by trying all pairs of vertices of B as candidates for the two vertices of degree 3 of a subdivided Theta graph, and for each choice we check if there are three internally disjoint paths connecting them by a network flow algorithm. Once we identify a subdivision of a Theta graph, we check which two of the paths give together a cycle with no or more than one red edges, and construct a perfect (1,2)-dominating set along the lines of the proof of Claim 3 or Claim 5. $\square$

3   Perfect (1,2)-Domination in 4-Regular Graphs

In this section, we will present a certain characterization of 4-regular graphs with perfect $(1,2)$-dominating sets with the help of induced 3-regular subgraphs. However, we will show that, despite this characterization, the problem of the existence of perfect $(1,2)$-dominating sets is NP-complete for 4-regular graphs.

Proposition 1. *If a 4-regular graph has the property that no edge belongs to more than 2 triangles, then it contains a perfect (1,2)-dominating set if and only if it contains a non-empty induced 3-regular subgraph.*

Proof. Let G be a 4-regular graph. Suppose $D \subset V(G)$ is a perfect (1,2)-dominating set. Set $T = V(G) \setminus D$. Since every vertex $u \in T$ has exactly one neighbour in D and $\deg_G(u) = 4$, u has exactly 3 neighbours in T. Hence $G[T]$ is a 3-regular subgraph. It is induced, and non-empty (since $D \neq V(G)$).

On the other hand, if $T \subseteq V(G)$ is a non-empty set of vertices such that $G[T]$ is 3-regular, then every vertex of T has exactly one neighbour in $D = V(G) \setminus T$. We need to show that every vertex $u \in T$ has at least one vertex of D at distance 2. Let u' be the neighbour of u in D. If u' has a neighbour u'' in D, this is the requested D-vertex at distance 2 from u. If not, u' has all its 4 neighbors in T (one of them being u). Let the other neighbours be u_1, u_2, u_3. The vertex u cannot be adjacent to all three of them, since then the edge uu' would belong to three triangles $uu_iu', i = 1, 2, 3$. Hence u has another neighbour v in T (since $G[T]$ is 3-regular). This v has a neighbour $v' \in D$ and this v' is the requested vertex of D at distance 2 from u. $\qquad\square$

Theorem 2. *Deciding the existence of a non-empty induced 3-regular subgraph is NP-complete even if the input graph is 4-regular and has the property that no edge belongs to more than 2 triangles.*

Proof. Plesnik [15] has proved that deciding the existence of a non-empty induced 3-regular subgraph in a bipartite graph of maximum degree 4 is NP-complete. Although in Theorem 1 of [15] the result is only claimed for subgraphs (i.e., not necessarily induced ones), the proof actually provides this stronger result. So suppose G is a bipartite graph of maximum degree 4 (and minimum degree 3, as vertices of degree less than 3 cannot belong to any 3-regular subgraph, and can thus be discarded). We will augment G with several copies of the graph H depicted in Fig. 4, thus creating a 4-regular graph G'.

The graph G has an even number, say $2k$, of vertices of degree 3, so pair them into pairs $a_i, b_i, i = 1, 2, \ldots, k$. For each i, take a copy of H and unify its vertex a with a_i, and its vertex b with b_i. Thus, G' is 4-regular and none of its edges belong to more than 2 triangles (the edges of G belong to no triangles, since G is bipartite, and the edges of H belong to at most two triangles each). Note also that G is an induced subgraph of G'.

We claim that G' contains a non-empty induced 3-regular subgraph if and only if G does. The "if" part is obvious – every induced subgraph of G is also an induced subgraph of G'.

The opposite implication follows from the following claim.

Claim. Let $T \subseteq V(H)$ be such that all vertices of $H[T]$, except possibly a, b, have degree 3. Then $T \setminus \{a, b\} = \emptyset$.

Proof of Claim. By case analysis, see Fig. 4. The graph H is in the upper left corner; then we illustrate the cases described in the following. A vertex in T is blue (or green), and a vertex not in T is white. Each case that ends with a contradiction ends with a vertex in T that has 4 neighbours (i.e., too many) in T. Such a vertex is marked with green. The membership of a and b is irrelevant, and so these vertices are marked in red.

Case 1. $1 \in T, 2 \notin T$: Since 1 is in T, it needs exactly 3 neighbours in T, so $3, 4, 5 \in T$. Since $4 \in T$, it needs exactly 3 neighbours in T, and since $2 \notin T$,

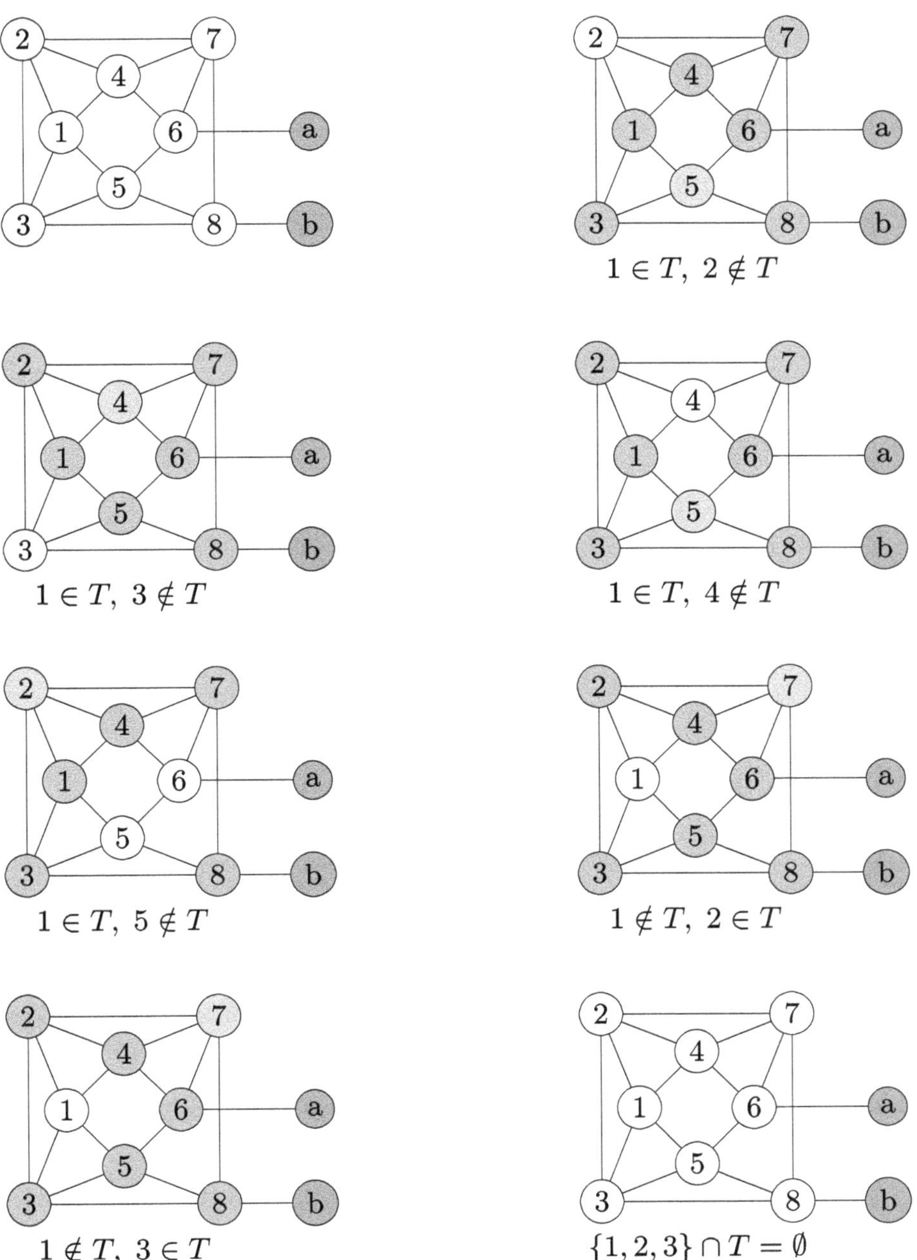

Fig. 4. The augmenting graph H and the case analysis of its (im)possible 3-regular subgraphs (Color figure online)

it must be $6, 7 \in T$. Similarly, since $3 \in T$ needs 3 neighbours in T, it must be $5, 8 \in T$. But then vertex 5 is in T and has all 4 neighbours in T, a contradiction.

Case 2. $\overline{1 \in T, 3 \notin T}$: $1 \in T$ implies $2, 4, 5 \in T$. $2 \in T$ implies $7 \in T$, $5 \in T$ implies $6, 8 \in T$. Then $4 \in T$ has 4 neighbours in T, a contradiction.

Case 3. $\overline{1 \in T, 4 \notin T}$: $1 \in T$ implies $2, 3, 5 \in T$. $2 \in T$ implies $7 \in T$, $7 \in T$ implies $6, 8 \in T$. Then $5 \in T$ has 4 neighbours in T, a contradiction.

Case 4. $\overline{1 \in T, 5 \notin T}$: $1 \in T$ implies $2, 3, 4 \in T$. $3 \in T$ implies $8 \in T$, $8 \in T$ implies $7 \in T$. Then $2 \in T$ has 4 neighbours in T, a contradiction.

We have analyzed all cases where $1 \in T$. Next, we look at the cases where $1 \notin T$.

Case 5. $\overline{1 \notin T, 2 \in T}$: $2 \in T$ implies $3, 4, 7 \in T$. $3 \in T$ implies $5, 8 \in T$, $4 \in T$ implies $6, 7 \in T$. Then $7 \in T$ has 4 neighbours in T, a contradiction.

Case 6. $\overline{1 \notin T, 3 \in T}$: $3 \in T$ implies $2, 5, 8 \in T$ and we are in Case 5.

Case 7. $\overline{\{1, 2, 3\} \cap T = \emptyset}$: The vertex 4 has 2 neighbours outside of T, so it cannot be in T. Similarly, $7 \notin T$, and hence $6 \notin T$. The vertex 5 has 2 neighbours outside of T, and so it cannot be in T. Similarly $8 \notin T$. This completes the proof of the Claim.

Suppose G' has a non-empty induced 3-regular subgraph induced by a vertex set $T \subseteq V(G')$. The Claim implies that $T \subseteq V(G)$ and thus $G[T]$ is a non-empty 3-regular induced subgraph of G. $\square$

Corollary 1. *Deciding whether a 4-regular graph contains a perfect (1,2)-dominating set is NP-complete.*

4 Independent Perfect $(1, 2)$-Domination in 3-Regular Planar Graphs

In the literature, independent dominating sets have been widely studied, where to the domination requirement one adds the additional condition of independence. In particular, properties of independent $(1, 2)$-dominating sets were investigated in [9]. It is natural to consider perfect $(1, 2)$-dominating sets with the additional condition that these sets are independent. We can find direct relations between such sets and other concepts, well known in graph theory, namely perfect codes.

In [4] Biggs defined a *t-perfect code* in a graph as a subset S of vertices which is t-independent (the distance between any pair of vertices from S is greater than t) and every vertex not in S is at distance at most t from exactly one vertex in S. Perfect codes have been extensively studied in literature, in particular in regular graphs [6, 10] and distance-regular graphs [4].

It is easy to see that every independent perfect $(1, 2)$-dominating set is a 1-perfect code. However, the opposite statement is not true since a 1-perfect code does not need to be $(1, 2)$-dominating. To illustrate this, in Fig. 5 the red vertices form a perfect $(1, 2)$-dominating set (and thus also a 1-perfect code), while the green vertices form a 1-perfect code, which is not $(1, 2)$-dominating, as the blue vertex has no green vertex at distance 2 apart from its neighbour. However, under some assumptions, the concepts become equivalent.

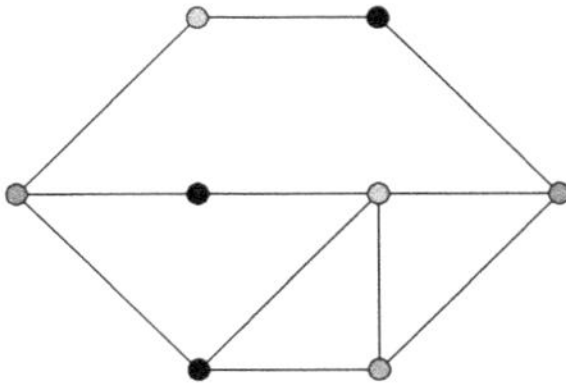

Fig. 5. Difference between an independent perfect $(1, 2)$-dominating set and a 1-perfect code

Proposition 2. *Let G be a cubic graph such that no edge belongs to two or more triangles. Then every 1-perfect code in G is an independent perfect $(1, 2)$-dominating set, and vice versa.*

Proof. We have already noted that every independent perfect $(1, 2)$-dominating set in any graph is a 1-perfect code. For the opposite implication, suppose that $S \subset V(G)$ is a 1-perfect code. Then S is independent and every vertex outside of S is dominated by exactly one vertex of S. It remains to be shown that every vertex not in S has at least one S-vertex at distance 2.

So consider a vertex $x \notin S$, and let $u \in S$ be its only neighbour in S. Since G is 3-regular, u has two more neighbours, say y and z. The vertex x also has 2 more neighbors, and since the edge ux does not belong to two or more triangles, x cannot be adjacent to both y and z. Hence, x is adjacent to a vertex $t \notin \{y, z\}$. This t cannot be in S (x would be dominated by two vertices u and t in such a case), and thus t has a neighbour $v \in S$. This v is a code vertex at distance 2 from x. $\square$

In [10] Kratochvíl proved that deciding whether a cubic planar graph contains a 1-perfect code is NP-complete. The reduction is such that the constructed graph does not contain K_4^- as a subgraph. Hence, no edge belongs to more than one triangle, and so the graph constructed in the reduction contains a 1-perfect code if and only if it contains a perfect $(1, 2)$-dominating set. This proves the following corollary.

Corollary 2. *It is NP-complete to decide whether a given planar 3-regular graph contains an independent perfect $(1, 2)$-dominating set.*

5   Conclusion and Open Problems

We have studied the computational complexity of the existence of perfect $(1, 2)$-dominating sets in graphs of small maximum degree. We showed that this problem is polynomial time solvable in graphs of maximum degree 3 and NP-complete in 4-regular graphs. It is natural to ask the following question.

Problem 1. What is the computational complexity of deciding the existence of perfect $(1, 2)$-dominating sets in k-regular graphs for fixed $k > 4$?

Our NP-hardness reduction for perfect $(1,2)$-dominating sets in 4-regular graphs is based on the NP-hardness of existence of non-empty induced 3-regular subgraphs in 4-regular graphs. The extension of this result to regular input graphs of higher valency is unclear to us. Therefore, we ask the following open question.

Problem 2. For fixed integers $3 \leq d < k$, what is the computational complexity of deciding the existence of a non-empty induced d-regular subgraph of an input k-regular graph?

Acknowledgments. The second author acknowledges support of Czech Science Foundation by Research grant No. GAČR 23-04949X.

Disclosure of Interests. The authors have no competing interests to declare that are relevant to the content of this article.

References

1. Bednarz, U.: On the existence of perfect $(1,2)$-dominating sets in graphs. Symmetry **17**(3), 405 (2025)
2. Bednarz, U., Pirga, M.: $(1,2)$-PDS in graphs with the small number of vertices of large degrees. Opuscula Math. **45**(1), 53–62 (2025)
3. Berge, C.: The Theory of Graphs and Its Applications. Wiley, New York (1962)
4. Biggs, N.: Perfect codes in graphs. J. Comb. Theory (B) **15**, 289–296 (1973)
5. Diestel, R.: Graph Theory. Springer, New York (2005)
6. Dvořáková-Rulićová, I.: Perfect codes in regular graphs. Comment. Math. Univ. Carol. **29**(1), 793–83 (1988)
7. Fink, J.F., Jacobson, M.S.: n-domination in graphs. In: Graph Theory with Applications to Algorithms and Computer Science, pp. 283–300, Wiley, New York (1985)
8. Haynes, T.W., Hedetniemi, S.T., Slater, P.J.: Fundamentals of Domination in Graphs. Marcel Dekker Inc., New York (1998)
9. Hedetniemi, S.M., Hedetniemi, S.T., Knisely, J., Rall, D.F.: Secondary domination in graphs. AKCE J. Graphs. Combin. **5**(2), 103–115 (2008)
10. Kratochvíl, J.: Regular codes in regular graphs are difficult. Discret. Math. **133**(1–3), 191–205 (1994)
11. Ore, O.: Theory of Graphs, vol. 38. American Mathematical Society, Providence (1962)
12. Michalski, A., Bednarz, P.: On independent secondary dominating sets in generalized graph products. Symmetry **13**, 2399 (2021)
13. Michalski, A., Włoch, I.: On the existence and the number of independent $(1,2)$-dominating sets in the G-join of graphs. Appl. Math. Comput. **377**, 125155 (2020)
14. Michalski, A., Włoch, I., Dettlaff, M., Lemańska, M.: On proper $(1,2)$-dominating sets in graphs. Math. Models Appl. Sci. **45**, 7050–7057 (2022)
15. Plesník, J.: A note on the complexity of finding regular subgraphs. Discret. Math. **49**(2), 161–167 (1984)
16. Raczek, J.: Polynomial algorithm for minimal $(1,2)$-dominating set in networks. Electronics **11**(3), 300 (2022)
17. Raczek, J.: Complexity issues on of secondary domination number. Algorithmica **86**, 1163–1172 (2024)

Complexity

A Complexity Analysis of the c-Closed Vertex Deletion Problem

Lisa Lehner, Christian Komusiewicz, and Luca Pascal Staus[✉]

Friedrich Schiller University Jena, Jena, Germany
{c.komusiewicz,luca.staus}@uni-jena.de

Abstract. A graph is c-closed when every pair of nonadjacent vertices has at most $c - 1$ common neighbors. In c-CLOSED VERTEX DELETION, the input is a graph G and an integer k and we ask whether G can be transformed into a c-closed graph by deleting at most k vertices. We study the classic and parameterized complexity of c-CLOSED VERTEX DELETION. We obtain, for example, NP-hardness for the case that G is bipartite with bounded maximum degree. We also show upper and lower bounds on the size of problem kernels for the parameter k and introduce a new parameter, the number x of vertices in *bad pairs*, for which we show a problem kernel of size $\mathcal{O}(x^3 + x^2 \cdot c)$. Here, a pair of nonadjacent vertices is bad if they have at least c common neighbors. Finally, we show that c-CLOSED VERTEX DELETION can be solved in polynomial time on unit interval graphs with depth at most $c + 1$ and that it is fixed-parameter tractable with respect to the neighborhood diversity of G.

1 Introduction

One of the most striking features of social networks is that they tend to show a high degree of *triadic closure*. This is the phenomenon that two agents which both interact with the same third agent are likely to interact directly, too. A formal way of quantifying this observation is graph *closure*, introduced by Fox et al. [4]. Here, a network is called *c-closed* if all pairs of vertices with at least c common neighbors are adjacent and the closure number of a network is the smallest number c such that the network is c-closed. The hypothesis driven by the triadic closure phenomenon is now that networks should have a relatively small closure number and this is indeed the case for real-world data [4,12]. Apart from giving further evidence for theories about the mechanisms that underly the creation of real-world networks, this observation has consequences for the development of algorithms for real-world networks: Many hard computational problems are fixed-parameter tractable when parameterized by c or by combinations of c and other parameters. Informally, this means that these problems can be solved (more) efficiently when the input networks have a small closure number.

For example, the enumeration of maximal cliques is fixed-parameter tractable with respect to the closure number c [4]. More precisely, all maximal cliques can be enumerated in $3^{c/3} \cdot n^{\mathcal{O}(1)}$ time, where n denotes the number of vertices in the graph. This result has been extended to the enumeration of different types of near-cliques [1,8]. Further algorithmic applications of c-closure were given for variants of DOMINATING SET [5,7,8], INDEPENDENT SET [7,8], and VERTEX COVER [6,9] and for problems related to finding certain small subgraphs [10].

A drawback of c-closure is that it is defined as the *maximum* of the sizes of common neighborhoods of nonadjacent vertices (plus one) which makes it sensitive to local changes: all it takes to drastically increase the c-closure of a graph is the absence of one edge between two vertices with many common neighbors. Ideally, we would like to obtain more robust positive algorithmic results that hold not only for c-closed graphs but also for those graphs which are *almost* c-closed. We follow the natural way of quantifying this by counting the number k of vertex deletions that are necessary to obtain a c-closed graph. Using simple arguments, one can show that positive results for INDEPENDENT SET and CLIQUE can be extended from c-closed graphs to almost c-closed graphs.

Proposition 1. *Let $G = (V, E)$ be a graph and S a size-k vertex set in G such that $G - S$ is c-closed. Then, we can (1) enumerate all maximal cliques of G in $2^k \cdot (3^{c/3}) \cdot n^{\mathcal{O}(1)}$ time, and (2) determine in $2^k \cdot f(\ell+c) \cdot n^{\mathcal{O}(1)}$ time whether G contains an independent set of size at least ℓ.*

Proof. To determine whether G contains an independent set I on ℓ vertices, we use the following known algorithm which can be formulated for vertex deletion sets S to any hereditary graph class for which independent sets can be found efficiently. Branch into all cases for $S' := S \cap I$. In each branch, first check whether S' is an independent set. If this is the case, then delete all vertices from $G - S$ that are adjacent to S' (as they cannot be contained in an independent set that includes S') and determine whether the resulting graph has an independent set of size $\ell - |S'|$. Since the resulting graph is c-closed, this can be done in $f(\ell + c) \cdot n^{\mathcal{O}(1)}$ time [7]. Since S has size k, we consider altogether 2^k cases for S' which gives the claimed running time.

To enumerate all maximal cliques of G proceed as follows. For each subset S' of S, enumerate all maximal cliques K such that $K \cap S = S'$. To do this, delete from $G - S$ all vertices that are not adjacent to all vertices of S'. Then enumerate all maximal cliques in the resulting graph. Since this graph is c-closed, this enumeration can be done in $3^{c/3} \cdot n^{\mathcal{O}(1)}$ time. For each enumerated clique K' we check in polynomial time whether $S' \cup K'$ is a maximal clique in G and output $S' \cup K'$ if this is the case. The running time bound now follows from the fact that 2^k cases for S' are considered. The correctness follows from the fact that only maximal cliques are output and that each maximal clique K is output when the branch $S' = K \cap S$ is considered. $\qquad\square$

Hence, it may be worthwhile to find out whether a graph G is almost c-closed and to identify the respective vertex deletion set S in that case. Motivated by this, we investigate the complexity of the corresponding decision problem.

c-CLOSED VERTEX DELETION (c-CVD)
Input: An undirected graph $G = (V, E)$ and a positive integer k.
Question: Is there a set $S \subseteq V$ with $|S| \le k$ such that $G - S$ is c-closed?

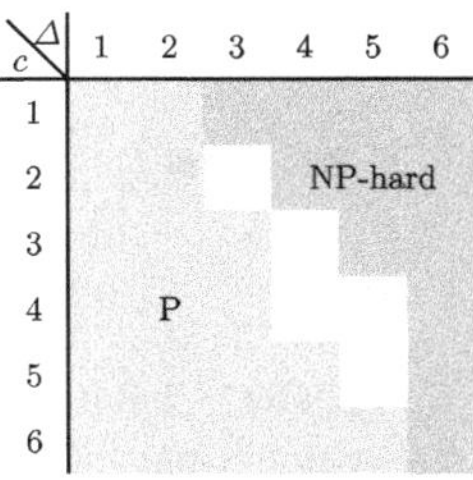

Fig. 1. Hardness for different values of c and Δ. Red means NP-hard, green means polynomial-time solvable, and gray means unknown. (Color figure online)

Unless mentioned otherwise, throughout the paper we assume that c is constant.

Known Results and Further Related Work. Since being c-closed is hereditary, c-CVD is NP-hard for every fixed value of c [14]. Assuming the exponential time hypothesis (ETH), one cannot even solve c-CVD in $2^{o(n+m)}$ time, where m is the number of edges in the graph [11]. The case $c = 1$ is equivalent to CLUSTER VERTEX DELETION since a graph is 1-closed if and only if it does not contain a P_3, a path on three vertices, as induced subgraph. CLUSTER VERTEX DELETION is NP-hard even when restricted to graphs with maximum degree 3 [16]. This gives the following complexity dichotomy: For maximum degree $\Delta \leq 2$, 1-CVD is easily solvable in linear time, all other cases are NP-hard.

A further related parameter is the *weak* closure number γ. This is the smallest number γ such that every induced subgraph of G contains *some* vertex v that has at most γ common neighbors with each of its nonneighbors. The value of γ is upper-bounded by the closure number c and by the degeneracy of the input graph. As a consequence, it assumes very small values in real-world data [4,12]. There are several problems where a small weak closure number γ can be exploited, for example CLIQUE ENUMERATION [4], DOMINATING SET [15], INDEPENDENT SET [8], INDUCED MATCHING, CAPACITATED VERTEX COVER and CONNECTED VERTEX COVER [9]. As we show in Sect. 2, the weak closure number γ of a graph is unrelated to $k + c$ where k is defined as the vertex deletion distance to c-closed graphs. More precisely, we show that there are graphs with unbounded γ-value where $k + c$ is constant and that there are graphs where γ is constant but $k + c$ is unbounded.

Our Results. In Sect. 3, we make progress towards a maximum degree-based complexity dichotomy for c-CVD; the results are shown in Fig. 1. In particular, our results show that for all $c \geq 6$, c-CVD is NP-hard when restricted to graphs with maximum degree $\Delta \geq c$ and polynomial-time solvable when restricted to graphs with maximum degree $\Delta \leq c - 1$. We also show that for $c = \Delta = 3$, the problem is polynomial-time solvable so the complexity landscape seems more complicated for smaller values of c. Our results also show that the problem is NP-hard on bipartite graphs.

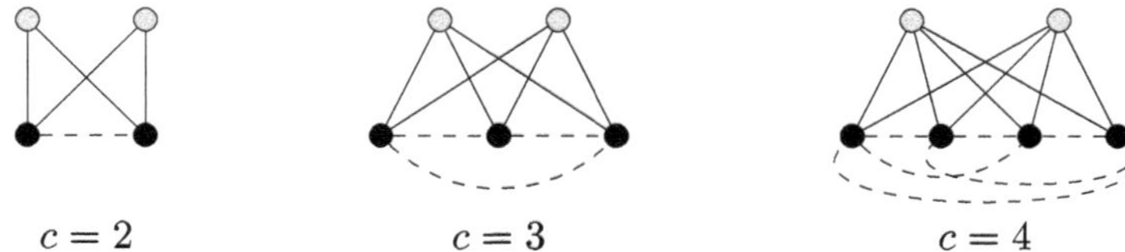

$$c = 2 \qquad\qquad c = 3 \qquad\qquad c = 4$$

Fig. 2. Minimal forbidden subgraphs (FSG) for $c = 2, 3, 4$. The grey vertices are the bad pair, the black vertices are the connecting vertices, the full edges are the critical edges, and the dashed edges are optional edges.

In Sect. 4, we then consider the parameterized complexity of the problem with respect to the natural parameter k. We show that a kernel with $\mathcal{O}(k^{c+2})$ vertices can be computed in time $\mathcal{O}(n^2 m)$ and also provide a lower bound of $\mathcal{O}(k^{c-\epsilon})$ for the bit size of the kernel. To obtain kernels whose size depends only polynomially on c, we introduce a further natural parameter, the number x of *vertices in bad pairs* in G, that is, the total number of vertices that have at least c common neighbors with some nonneighbor. This parameter can be thought of as being larger than k since deleting all vertices in a bad pair makes the graph c-closed. We show that c-CVD admits a kernel with $\mathcal{O}(x^3 \cdot c)$ vertices.

Finally, in Sect. 5 we show that c-CVD is fixed-parameter tractable with respect to the neighborhood diversity and that it can be solved in polynomial time on unit interval graphs with maximum clique size $c + 1$.

Missing proofs are marked with (★) and can be found in the full version [13]. For the relevant definitions of parameterized complexity, refer to [3].

2   Preliminaries

An *undirected graph* $G = (V, E)$ consists of a *vertex set* V and an *edge set* $E \subseteq \{\{u, v\} \mid u, v \in V \wedge u \neq v\}$. We let $V(G)$ and $E(G)$ denote the vertices and edges of G, respectively, and define $n :- |V(G)|$ and $m :- |E(G)|$. Two vertices u and v are *neighbors* if $\{u, v\} \in E(G)$. The *neighborhood* of a vertex v in G is $N_G(v) :- \{u \in V(G) \mid \{u, v\} \in E(G)\}$. The *degree* of v is $\deg_G(v) :- |N_G(v)|$. The *maximum degree* of G is $\Delta_G :- \max_{v \in V(G)} \deg(v)$. We drop the subscript G if G is clear from context. A graph G' is a *subgraph* of G if $V(G') \subseteq V(G)$ and $E(G') \subseteq E(G)$. Let $U \subseteq V(G)$. The graph $G[U] :- (U, \{e \in E(G) \mid e \subseteq U\})$ is the *subgraph of G induced by* U. We write $G - U :- G[V(G) \setminus U]$ to denote the subgraph of G obtained by deleting the vertices in U.

Let G be a graph. We call G a *clique* if it contains every possible edge. We call G an *independent set* if it contains no edges. We call G *bipartite* if $V(G)$ can be partitioned into two sets V_1 and V_2 such that $G[V_1]$ and $G[V_2]$ are independent sets. We call G a *split graph* if $V(G)$ can be partitioned into two sets V_1 and V_2 such that $G[V_1]$ is a clique and $G[V_2]$ is an independent set.

A graph property Π is *hereditary* if it is *closed under vertex deletion*. In other words, any induced subgraph of a graph with property Π also has property Π. All graph properties defined above, including c-closed graphs, are hereditary. [4]

Problem-Specific Definitions. A graph $G = (V, E)$ is *c-closed* for some positive integer c if every pair of nonadjacent vertices has at most $c-1$ common neighbors. The *closure number* of G is the smallest number c such that G is c-closed. We call a pair of nonadjacent vertices with at least c common neighbors a *bad pair*. The common neighbors of a bad pair are its *connecting vertices*. Clearly, a graph is c-closed if and only if it contains no bad pair. In other words, the c-closed property can be defined based on the following forbidden subgraphs: A *minimal forbidden subgraph* (FSG) for being c-closed is a graph with $c + 2$ vertices such that two of the vertices form a bad pair and the remaining c vertices are their connecting vertices. The adjacency between the connecting vertices can be arbitrary. Given an FSG G', we refer to the edges between the bad pair and the connecting vertices as *critical edges* of G'. We call an edge e *critical* in G if G contains an FSG where e is a critical edge. Figure 2 shows the FSGs for $c = 2, 3, 4$. Note that a bad pair can have more than c connecting vertices if it is the bad pair in multiple FSGs. Moreover, a minimal non-c-closed subgraph can have several bad pairs; for convenience, we let FSGs often refer to a combination of a subgraph and a bad pair.

Relation to Weak Closure. The parameter combination $c + k$ is unrelated to the closure number and the weak closure number γ of a graph for any fixed c. Note that c is not the closure number of the graph but instead the desired closure number after deleting up to k vertices. We first show that $c + k$ can be arbitrarily smaller than γ. Let s be any arbitrarily large integer and let G be a graph containing a clique with $2s$ vertices and two additional vertices v_1 and v_2. The vertex v_1 is connected to s of the clique vertices and v_2 is connected to the other s clique vertices. Clearly, we can make this graph 1-closed by removing v_1 and v_2. This means we have $c + k \leq c + 2$ for any fixed $c \geq 1$. In contrast, we have $\gamma \geq s + 1$ since each vertex has at least one nonneighbor with s common neighbors.

Conversely, $c + k$ can also be arbitrarily larger than the closure number of a graph: Let G be a graph with an independent set of size c and two vertices v_1 and v_2 that are adjacent to each vertex in the independent set. A graph consisting of $s \gg c + 1$ copies of this component is $c + 1$-closed. To make it c-closed we need to remove one vertex from each of the s components.

3 Bounded Maximum Degree Δ

We now study the complexity of c-CVD for different values of c and Δ. Figure 1 shows an overview of these results.

3.1 NP-Hard Cases

We reduce from VERTEX COVER which is NP-hard when the maximum degree is 3 and all degree-3 vertices have distance at least 3 to each other [11].

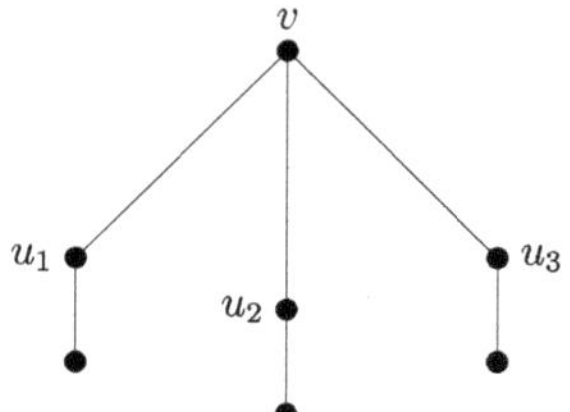

(a) VERTEX COVER instance with $\Delta = 3$.

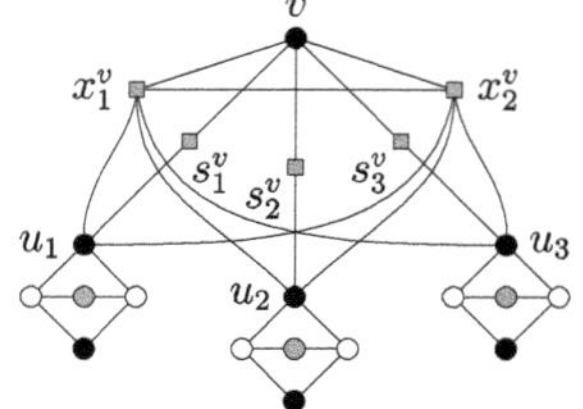

(b) 3-CVD INSTANCE WITH $\Delta = 5$.

Fig. 3. Construction of a 3-CVD instance with maximum degree 5 from a VERTEX COVER instance with maximum degree 3. The vertices from the VERTEX COVER instance are black. The gray square vertices are the connecting vertices for the FSGs corresponding to the edges incident with the degree-3 vertex. The white circle vertices are the bad pair vertices for the FSGs corresponding to other edges with the gray circle vertices as additional connecting vertices.

Theorem 1. ($\bigstar$) c-CLOSED VERTEX DELETION *is NP-hard for $c \geq 2$ on bipartite graphs with maximum degree $\Delta = \max(c, 6)$.*

The following theorem proves further hard cases for $c = 2$ and $c = 3$.

Theorem 2. c-CLOSED VERTEX DELETION *is NP-hard for $c = 2$ and $\Delta = 4$ and for $c = 3$ and $\Delta = 5$.*

Proof. Let (G, k) be a VERTEX COVER instance with maximum degree 3 where all degree-3 vertices have distance at least 3 to each other. We want to replace each edge with an FSG but we will use a special construction for edges that have a degree-3 vertex as an endpoint. We construct a new graph G' as follows: Each vertex $v \in V(G)$ becomes a vertex in G'. For each edge $e \in E(G)$ where both endpoints have at most degree 2 we add two bad pair vertices b_1^e, b_2^e and $c - 2$ connecting vertices $w_1^e, \ldots, w_{c-2}^e$. We add edges from the bad pair vertices to all connecting vertices and the two endpoints of e. With this, G' currently has a maximum degree of c while the vertices in $V(G') \cap V(G)$ all have at most degree 2. For the remaining edges we look at each degree 3 vertex individually. Let $v \in V(G)$ be a degree 3 vertex in G and let u_1, u_2, u_3 be the three neighbors of v in G. We now add the vertices s_1^v, s_2^v, s_3^v to G' and connect them to v. Additionally, we connect u_i to s_i^v for each $i \in \{1, 2, 3\}$. If $c = 2$ we add one additional vertex x_1^v and connect it to v, u_1, u_2, u_3. If instead $c = 3$ we add two vertices x_1^v, x_2^v and connect them to v, u_1, u_2, u_3 and eachother. The construction is visualized in Fig. 3. The s_i^v vertices have degree 2 while v, the x_i^v vertices, and the u_i vertices have degree 4 and 5 for $c = 2$ and $c = 3$, respectively. That means G' has a maximum degree of 4 for $c = 2$ and 5 for $c = 3$. This finishes the construction of the c-CVD instance (G', k).

We will first show that for every FSG H in G' we have $V(H) \cap V(G) = e$ for some edge $e \in E(G)$. Let H be some FSG in G'. First, observe that H cannot contain three vertices from $V(G)$: The vertices in $V(G)$ are never adjacent in G'.

The FSG H can therefore only contain three of them if they are all connecting vertices and $c = 3$. The only vertices that have at least three neighbors from $V(G)$ are the x_1^v, x_2^v vertices. These two must therefore form the bad pair which is not possible because they are adjacent. Next, we show that H cannot contain two vertices from $V(G)$ that are not adjacent in G: The vertices in H must have at most distance 2 to each other. The only vertices in $V(G)$ that are not connected in G for which this is the case are the neighbors u_1, u_2, u_3 of some degree-3 vertex v in G. However, for $c = 3$ they only have the vertices x_1^v and x_2^v as common neighbors and those two vertices cannot form a bad pair since they are adjacent. For $c = 2$ they only have x_1^v as a common neighbor which also cannot form a bad pair on its own. Finally, we show that H has to contain at least two vertices from $V(G)$: Any FSG must contain a cycle of length 4 as a subgraph. That means if H contains at most one vertex from $V(G)$, then H must contain a path on three vertices from $V(G') \setminus V(G)$ as a subgraph. In G' the only instances of such subgraphs are the vertices b_1^e, b_2^e, w_1^e for some edge $e \in E(G)$ when $c = 3$. The only way these vertices can form an FSG is by including the endpoints of e as connecting vertices which are both in $V(G)$. Summarizing, every FSG H in G' contains the two endpoints of some edge $e \in E(G)$ and no other vertices from $V(G)$, as claimed. By construction, for each edge $e \in E(G)$ we also have *at least* one FSG H with $H \cap V(G) = e$. With the help of these two facts, we can now show that

$$(G, k) \text{ is a yes-instance} \Leftrightarrow (G', k) \text{ is a yes-instance.}$$

($\Rightarrow$) Since each FSG in G' contains the two endpoints of some edge in G we know that any solution S for (G, k) is also a solution for (G', k).

($\Leftarrow$) Let S' be a solution for (G', k). We now construct a set S that is a solution for (G, k). The FSGs in G' share a vertex if and only if the corresponding edges in G share an endpoint. This can easily be seen from the construction of G' by observing that each FSG must have diameter 2. That means if a vertex in S' appears in multiple FSGs we just add the vertex in which the corresponding edges overlap to S. And if a vertex in S' only appears in one FSG, then we add an arbitrary endpoint of the corresponding edge to S. The solution S now clearly contains at most k vertices from $V(G)$ and covers all edges in $E(G)$. $\qquad\square$

3.2   Polynomial-Time Solvable Cases

If $\Delta < c$, then c-CVD is obviously polynomial-time solvable since any input graph would already be c-closed. To show less trivial cases, we use the following reduction rule that removes edges that are not critical edges in FSGs and do not create new FSGs when removed. Recall that an edge is critical in G if it is between a bad pair vertex and one of its connecting vertices of some FSG.

Reduction Rule 1. *Remove an edge $\{u, v\} \in E(G)$ if it is not critical in G and $|N(u) \cap N(v)| < c$.*

Lemma 1. *Rule 1 is correct.*

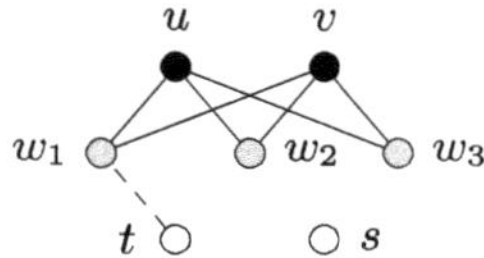

(a) FSG with additional dashed critical edge.

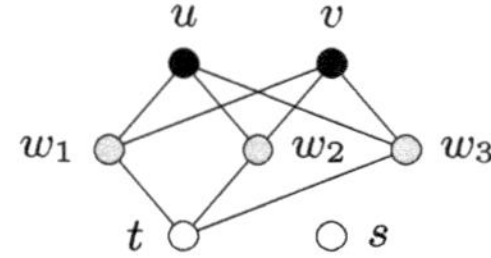

(b) Case 2: t is in a bad pair. All vertices must have maximum degree.

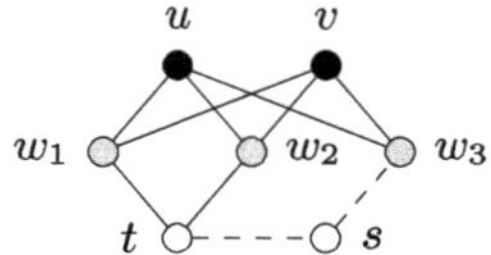

(c) Case 3: w_1 is in a bad pair. We prove that the dashed edges cannot exist.

Fig. 4. Connected components for max degree 3 graphs with $c = 3$ after exhaustively applying Rule 1. The black vertices are the bad pair in the initial FSG. The gray vertices are the connecting vertices of the bad pair and the white vertices are additional vertices that may or may not be part of the component.

Proof. Let (G, k) be a c-CVD instance and let G' be the graph that is created after Rule 1 removed edge e from G. We show that

$$(G, k) \text{ is a yes-instance} \Leftrightarrow (G', k) \text{ is a yes-instance.}$$

($\Rightarrow$) Let S be a solution for (G, k). Since S covers all FSGs in G, the only way that S is not a solution for (G', k) is if removing e creates a new FSG H in G'. Removing an edge cannot add a new common neighbor to a pair of vertices. That means e is an edge between the bad pair of H. But then the bad pair in H has at least c common neighbors in G which contradicts the precondition of Rule 1. Hence, S is also a solution for (G', k).

($\Leftarrow$) Let S' be a solution for (G', k). Since S' covers all FSGs in G', the only way that S' is not also a solution for (G, k) is if removing e destroys an FSG H in G. This is only possible if removing e reduces the number of common neighbors of the bad pair in H. However, that means that e is an edge between a bad pair vertex and a connecting vertex in H which contradicts that e is not critical due to the precondition of Rule 1. Hence, S' is also a solution for (G, k).   □

We can now use this rule to show that c-CVD is polynomial-time solvable for $c = \Delta = 2$ and $c = \Delta = 3$. In both proofs we use the following observation: if Rule 1 cannot be applied to a graph G with $c = \Delta$, then every edge in G must be a critical edge. This is because there cannot be an edge where the two endpoints have at least c common neighbors as they would then have degree at least $c + 1$.

Theorem 3. (★) *c-CVD can be solved in $\mathcal{O}(n)$ time on graphs with $c = \Delta = 2$ and on graphs with $c = \Delta = 3$.*

Proof (for $c = \Delta = 3$). Let (G, k) be a c-CVD instance with $c = \Delta = 3$ and let G' be the graph obtained by exhaustively applying Rule 1 to G. We show that each connected component in G' has at most six vertices and can therefore be solved in constant time. The main steps of the proof are illustrated in Fig. 4. Clearly, each connected component with at least one edge must contain an FSG. Let H be such an FSG with the bad pair u and v and the connecting vertices w_1, w_2, and w_3. The bad pair vertices already have degree 3 and can therefore not be adjacent to any other vertex. The connecting vertices have degree 2 and are therefore adjacent to at most one further vertex. If no additional edge exists, then the component only has five vertices and we are done. Without loss of generality, assume that w_1 is adjacent to a vertex t (see Fig. 4a). The edge $\{w_1, t\}$ must be critical edge which leads to three possible cases.

In the first case t is equal to w_2 or w_3. Without loss of generality assume $t = w_2$. To further increase the size of the component w_3 must now connect to a new vertex via a bridge edge. However, any edge in an FSG is part of at least one cycle which means a bridge edge cannot be critical.

In the second case t is part of a bad pair. Since $c = \Delta$ we know that all neighbors of a bad pair vertex must be connecting vertices. Here this means that w_1 is a connecting vertex. Consequently, the other bad pair vertex is either u or v since it must be adjacent to w_1. But then w_2 and w_3 are also connecting vertices and t is adjacent to them. This leads to a graph with six vertices all of which have degree 3 (see Fig. 4b).

In the third case w_1 is part of a bad pair and t is not. Since the other vertex of the bad pair must be connected to all three neighbors of w_1 that vertex can only be w_2 or w_3. Without loss of generality assume that it is w_2. This leaves w_3 and t as the only vertices with degree less than 3 (see Fig. 4c). If we connect them with an edge we again have a graph with 6 vertices all of which have degree 3. To increase the size of the component to more than 6 at least one of these two vertices must be connected to a seventh vertex s with an edge. This new edge again implies that one of the two endpoints must be in a bad pair. We already showed that t being part of a bad pair leads to a component of size 6. And if w_3 was part of a bad pair, then we would need to connect s to w_1 or w_2 as these are the only vertices that could be the second vertex of the bad pair. But that is not possible because these vertices already have degree 3. This only leaves the case where s is part of a bad pair. If s is adjacent to w_3, then the other vertex of the bad pair must be u or v which is not possible since s cannot be adjacent to w_1 or w_2. If s is adjacent to t, then w_1 or w_2 must be the other vertex of the bad pair. But that is also not possible because s cannot be adjacent to u or v. Hence, we cannot connect to a seventh vertex and each connected component contains at most six vertices.

The running time can be seen as follows: We can apply Rule 1 in $\mathcal{O}(n)$ time on graphs with constant maximum degree. Then we can go through the $\mathcal{O}(n)$ connected components and solve each in constant time. $\square$

4   Problem Kernel

In this section, we first show that c-CVD does not admit a kernel of bit size $\mathcal{O}(k^{c-\epsilon})$ unless $\mathrm{coNP} \subseteq \mathrm{NP/poly}$ using a reduction from c-HITTING SET. We then derive a problem kernel for c-CVD with $\mathcal{O}(k^{c+2})$ vertices using a reduction to $c+2$-HITTING SET. Finally, we show a problem kernel of size $\mathcal{O}(x^2 \cdot (k+c))$ where x is the number of vertices that are in at least one bad pair.

4.1   Hitting Set Reductions

We start by defining d-HITTING SET:

> d-HITTING SET
> *Input:*   A family $\mathcal{A}$ of subsets of a universe U such that each subset has size at most d, and a positive integer k.
> *Question:*   Is there a set $H \subseteq U$ with $|H| \leq k$ such that H contains at least one element of each set in $\mathcal{A}$?

We now define a reduction from d-HITTING SET to c-CVD with $c = d$ by turning the elements in each set into the connecting vertices of an FSG.

Let $(U, \mathcal{A}, k)$ be an instance of d-HITTING SET. We first add additional elements to the sets in $\mathcal{A}$ such that each set has exactly size d. We construct the graph G as follows: Each element of U becomes a vertex in G. Additionally, for each set $A \in \mathcal{A}$ we add two new vertices v_A and u_A. For each vertex $v \in A$ we add the edges $\{v, v_A\}$ and $\{v, u_A\}$ such that v_A and u_A form a bad pair with the vertices in A as their connecting vertices. Finally, we add all edges between vertices in U such that U becomes a clique. With this we obtain the c-CVD instance (G, k) with $c = d$.

Lemma 2 (★) *The reduction from d-HITTING SET to c-CVD is correct.*

From this reduction we can obtain two results. First, the constructed graph is a split graph which means that c-CVD is NP-hard on split graphs.

Corollary 1. *For every $c \geq 2$, c-CVD is NP-hard on split graphs.*

Let us remark that the case $c = 1$ can be solved in polynomial time on split graphs [2]. The second result is a lower bound for the bit size of a problem kernel for c-CVD. For this we can use the result that d-HITTING SET does not have a compression with bit size $\mathcal{O}(k^{d-\epsilon})$ for any $\epsilon > 0$ unless CONP $\subseteq$ NP/poly [3]. Since k does not change in the reduction and we have $c = d$ we immediately obtain the following result for c-CVD.

Theorem 4. c-CLOSED VERTEX DELETION *with $c \geq 2$ does not have a kernel with bit size $\mathcal{O}(k^{c-\epsilon})$ for any $\epsilon > 0$ unless CONP $\subseteq$ NP/poly.*

Next, we want to show that c-CVD admits a kernel with $\mathcal{O}(k^{c+2})$ vertices. For this we first prove the following lemma.

Lemma 3. *Let (G, k) be an instance of c-CVD and let u, v be a bad pair in G with at least $k + c$ connecting vertices. Any solution for (G, k) must contain at least one of u and v.*

Proof. Assume that (G, k) has a solution S that does not contain u or v. Then, u and v still have at least c connecting vertices in $G - S$ as S contains at most k connecting vertices of u and v. Hence, u and v form an FSG with these connecting vertices in $G - S$. This contradicts S being a solution for (G, k). □

We now define a reduction from c-CVD to d-HITTING SET with $d = c + 2$ by turning (most of) the FSGs into sets of the d-HITTING SET instance. Let (G, k) be an instance of c-CVD. The universe U consists of all vertices $V(G)$. For each bad pair u, v in G we do the following: If u and v have less than $k + c$ common neighbors, then, for each FSG where u, v is the bad pair, then we add a new subset to $\mathcal{A}$ that contains all vertices in the FSG. Otherwise, then we choose an arbitrary subset C containing exactly $k + c$ of these common neighbors. Then, for each FSG where u, v is the bad pair and the connecting vertices are all contained in C, we add a new subset to $\mathcal{A}$ that contains all vertices in the FSG. This gives the d-HITTING SET instance $(U, \mathcal{A}, k)$ with $d = c + 2$.

Lemma 4. ($\bigstar$) *The reduction from c-CVD to d-HITTING SET is correct.*

René van Bevern showed that d-HITTING SET has an *expressive* kernel with $\mathcal{O}(k^d)$ elements and sets [17]. Expressive kernels are defined as follows:

Definition 1 ([17]). *A kernelization algorithm for d-HITTING SET is expressive if, given an instance $(U, \mathcal{A}, k)$, it outputs an instance $(U', \mathcal{A}', k')$ such that*

1. *$U' \subseteq U$ and $\mathcal{A}' \subseteq \mathcal{A}$,*
2. *any vertex set of size at most k is a minimal hitting set for $(U, \mathcal{A})$ if and only if it is a minimal hitting set for $(U', \mathcal{A}')$, and*
3. *it outputs a certificate for $(U', \mathcal{A}', k')$ being yes if and only if $(U, \mathcal{A}, k)$ is.*

We can now use this result to obtain the following kernel for c-CVD.

Theorem 5. *For $c \geq 2$, c-CLOSED VERTEX DELETION admits a kernel with $\mathcal{O}(k^{c+2})$ vertices that can be computed in $\mathcal{O}(n^3 + n^2 m)$ time.*

Proof. Let (G, k) be a c-CVD instance, let $(U, \mathcal{A}, k)$ be the $c + 2$-HITTING SET instance constructed with the reduction described above, and let $(U', \mathcal{A}', k')$ be the instance obtained by using the kernelization algorithm from René van Bevern [17]. Since k does not change in the kernelization, we can replace k' by k. Our kernel is now the c-CVD instance (G', k) that is constructed by removing any vertex in $V(G) \setminus U'$ and their incident edges from G.

We first prove equivalence of the instances, that is, we show that

$$(G, k) \text{ is a yes-instance} \Leftrightarrow (G', k) \text{ is a yes-instance.}$$

($\Rightarrow$) Let S be a solution for (G, k). Since the property of being c-closed is hereditary and we only deleted vertices to obtain G' from G, it is clear that S is also a solution for (G', k).

($\Leftarrow$) Let S be a minimal solution for (G', k). Each set in $H \in \mathcal{A}'$ also exists in $\mathcal{A}$ by the first property in Definition 1. Hence, H is the vertex set of some FSG in G. Since we did not remove any vertices in H when creating G', we know that H is also the vertex set of some FSG in G'. This means that S is a solution for $(U', \mathcal{A}', k)$. According to the second property in Definition 1, S is also a solution for $(U, \mathcal{A}, k)$. Finally, the proof of Lemma 4 shows that S is also a solution for (G, k).

We have now shown that (G', k) is equivalent and since the vertex set $V(G')$ is equal to U' we know that G' has $\mathcal{O}(k^{c+2})$ vertices. It remains to bound the running time of the kernelization. The kernelization consists of three main steps. First, the reduction from (G, k) to $(U, \mathcal{A}, k)$. Second, the expressive kernelization of $(U, \mathcal{A}, k)$. Finally, creating the induced subgraph $G' = G[U']$. Observe that if $n \leq k^{c+2}$, then the kernelization can simply output (G, k). Hence, we assume $n > k^{c+2}$ in the following.

For the first step, we start by looking at each pair of vertices u and v and enumerating their common neighbors in $\mathcal{O}(n \cdot m)$ time. If u and v have less than c common neighbors we do not need to do anything. Otherwise, we construct the set C of at most $k + c$ common neighbors, and then add $\mathcal{O}((k + c)^c)$ sets of size $c + 2$ to $\mathcal{A}$. Since c is constant and we can assume that $n > k^{c+2}$ we can say that we only add $\mathcal{O}(n)$ sets of constant size to $\mathcal{A}$ for each pair of vertices in G. This bounds the size of $\mathcal{A}$ to $\mathcal{O}(n^3)$ and the running time of the first step to $\mathcal{O}(n^2 m)$. The second step can be done in $\mathcal{O}(n + |\mathcal{A}|)$ time for constant d [17]. Combined with the bound on $|\mathcal{A}|$ from above we get a running time of $\mathcal{O}(n^3)$ for the second step. The final step can be done in $\mathcal{O}(n + m)$ time but since both terms are dominated by the running times of the first two steps we get the total running time of $\mathcal{O}(n^3 + n^2 m)$. $\qquad\square$

4.2   Kernel with Parameter x

To obtain a kernel with polynomial dependence on c, we use a new parameter $x := |X|$ where X is the set of vertices in $V(G)$ that are part of at least one bad pair in G. With this parameter we can obtain a kernel with $\mathcal{O}(x^2 \cdot (k + c))$ vertices. To obtain this kernel we first introduce a new reduction rule and then show that exhaustively applying this rule to a graph G will reduce its size to the desired bound. We may assume $x > k$ since any instance with $x \leq k$ can be trivially solved by removing all vertices of X. We say that a bad pair is *heavy* if it has more than $k + c$ connecting vertices in $V(G) \setminus X$. Otherwise, the bad pair is *weak*.

Reduction Rule 2 *Mark all vertices in X and all vertices that are connecting vertices of at least one weak bad pair. Then, remove an arbitrary unmarked vertex.*

Lemma 5. *Rule 2 is correct.*

Proof. Let (G, k) be an instance of c-CVD and let (G', k) be the instance that is obtained by Rule 2 removing vertex v from G. We need to show that

$$(G, k) \text{ is a yes-instance} \Leftrightarrow (G', k) \text{ is a yes-instance.}$$

($\Rightarrow$) Let S be a solution for (G, k) which means that $G - S$ is c-closed. Since the property of being c-closed is hereditary, we know that $(G - S) - \{v\} = (G' - S)$ is also c-closed. Hence, $S \setminus \{v\}$ is a solution for (G', k).

($\Leftarrow$) Let S' be a solution for (G', k). Assume towards a contradiction that $G - S'$ is not c-closed and therefore contains some FSG H. If $v \notin V(H)$, then H must also be a subgraph of $G' - S'$ which contradicts S' being a solution for (G', k).

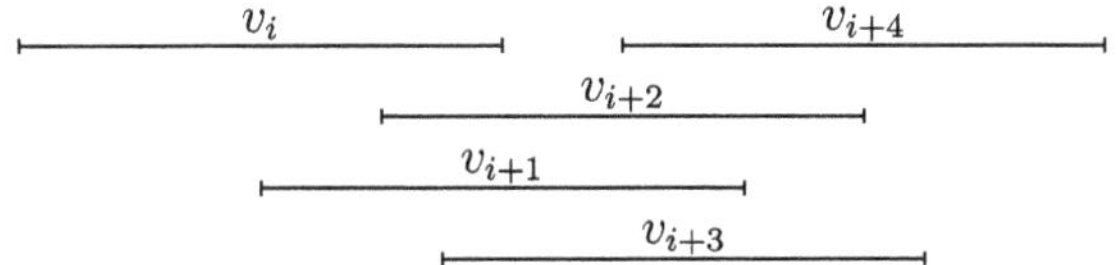

Fig. 5. Example FSG for $c = 3$ in a unit interval graph with depth $c + 1 = 4$. The vertices v_i and v_{i+4} form the bad pair while the other three vertices are the connecting vertices.

If $v \in V(H)$, then v must be a connecting vertex in H since it is not in X. Moreover, the bad pair in H is a heavy bad pair in G since v was not marked by Rule 2. This means that the bad pair in H must be a bad pair in G' with at least $k + c$ connecting vertices. Lemma 3 now tells us that S' contains a vertex from this bad pair. This contradicts H being an FSG in $G - S'$. $\qquad\square$

Exhaustively applying Rule 2 leads to a kernel of the desired size.

Lemma 6. *c-*Closed Vertex Deletion *has a problem kernel with $\mathcal{O}(x^2 \cdot (k + c))$ vertices where x is the number of vertices that are part of at least one bad pair.*

Proof. Let (G, k) be an instance of c-CVD such that Rule 2 cannot be applied anymore. This means that all vertices in G are marked and are therefore part of a bad pair or connecting vertices of at least one weak bad pair. Graph G can contain at most $\binom{x}{2} = \frac{x \cdot (x-1)}{2} \in \mathcal{O}(x^2)$ weak bad pairs. Each of those can have at most $k + c$ connecting vertices. Since each vertex is a connecting vertex of at least one of those pairs and $|X| = x$, we obtain $\mathcal{O}(x^2 \cdot (k + c))$ as an upper bound for the number of vertices in G. $\qquad\square$

Since we can assume $x > k$, we can also write the bound as $\mathcal{O}(x^3 + x^2 \cdot c)$.

5   Easy Special Cases

Unit Interval Graphs. A graph G is an *interval graph* if each vertex v corresponds to a closed interval $I_v = [l(v), r(v)]$ on the real line and two vertices are connected with an edge if and only if their intervals overlap. A *unit interval graph* is an interval graph where all intervals have the same length. The *depth* of an interval graph is the largest number of intervals that overlap in the same point. The depth is the same as the size of a largest clique in the graph. We assume that $(v_1, v_2, \ldots, v_n)$ is an ordering of $V(G)$ with nondecreasing starting points.

We now describe an algorithm that solves c-CVD on unit interval graphs with depth at most $c + 1$ in linear time $\mathcal{O}(c \cdot n)$. For this we first analyze how FSGs look in such graphs. Figure 5 shows an example FSG for $c = 3$.

Lemma 7. ($\bigstar$) *Let H be an FSG in a unit interval graph G with depth $\leq c+1$. We then have $V(H) = \{v_i, v_{i+1}, \ldots, v_{i+c}, v_{i+c+1}\}$ for some $1 \leq i \leq n$ where v_i and v_{i+c+1} form the bad pair and $v_{i+1}, \ldots, v_{i+c}$ are the connecting vertices.*

We can now solve c-CVD on unit interval graphs with depth at most $c+1$ with a simple greedy algorithm. For each vertex v_i it checks if v_i is the first vertex of the bad pair in an FSG. If that is the case, then the algorithm adds the last vertex of the FSG to the solution set.

Theorem 6. (★) c-CLOSED VERTEX DELETION *can be solved in* $\mathcal{O}(n \cdot c)$ *time when the input is a unit interval graph with maximal clique size* $c+1$.

Neighborhood Diversity. The *neighborhood diversity* of a graph, nd(G), is the number of *neighborhood classes* of G which are the equivalence classes of the relation $\sim$ with $u \sim v :\Leftrightarrow N(u) \setminus \{v\} = N(v) \setminus \{u\}$. We first show that for each neighborhood class there are essentially only $c+1$ possibilities to consider.

Lemma 8. *Let* S *be a minimum-size solution for a* c-CVD *instance* (G, k) *with* $c \geq 2$. *Then for each neighborhood class* D *we either have* $D \setminus S = D$ *or* $|D \setminus S| < c$.

Proof. Assume towards a contradiction, that $D \setminus S$ is a proper subset of D that contains at least c vertices and let v be a vertex of $D \cap S$. Since S is a minimal solution, $S \setminus \{v\}$ is not a solution, that is, $G - (S \setminus \{v\})$ contains an FSG H one of whose vertices is v. This FSG can contain at most c vertices of D, since the two bad pair vertices have different neighborhoods than the connecting vertices. Hence, v can be replaced in H by some vertex of $D \setminus S$, giving an FSG H' which is contained in $G - S$. This contradicts that S is a solution. □

With this we can now obtain the following via branching.

Theorem 7. (★) c-CVD *with* $c \geq 2$ *can be solved in* $(c+1)^{\text{nd}} \cdot n^{\mathcal{O}(1)}$ *time*.

Next we formulate the optimization version of c-CVD where we want to find the smallest set S such that $G - S$ is c-closed as an ILP with $\mathcal{O}(\text{nd})$ variables which implies fixed-parameter tractability with respect to nd alone. The ILP relies mainly on the observation that we only care about the number of vertices that are removed from each neighborhood class and not about the specific vertices.

Theorem 8. (★) c-CVD *can be solved in* $f(\text{nd}) \cdot n^{\mathcal{O}(1)}$ *time*.

6   Conclusion

We have introduced c-CLOSED VERTEX DELETION and provided a first overview of its complexity. Several interesting questions are left open: First, it remains to settle the remaining five open cases for the complexity on bounded-degree graphs. Second, one could close the gap between the upper and lower bounds on the kernel size for the solution size parameter k. Third, a complexity classification for interval graphs and unit interval graphs remains open. In addition, it is of interest to identify further problems which can be solved efficiently on almost c-closed graphs. Finally, we introduced the bad pairs-related parameter which could be useful for other vertex deletion problems where the forbidden subgraphs are large but contain distinguished pairs of vertices. Identifying such problems and providing a generic definition of a bad pair-related parameter could be fruitful.

References

1. Behera, B., Husić, E., Jain, S., Roughgarden, T., Seshadhri, C.: FPT algorithms for finding near-cliques in c-closed graphs. In: Braverman, M. (ed.) Proceedings of 13th ITCS. LIPIcs, vol. 215, pp. 17:1–17:24, Dagstuhl, Germany. Schloss Dagstuhl – Leibniz-Zentrum für Informatik (2022)
2. Cao, Y., Ke, Y., Otachi, Y., You, J.: Vertex deletion problems on chordal graphs. Theor. Comput. Sci. **745**, 75–86 (2018)
3. Cygan, M., et al.: Parameterized Algorithms. Springer (2016)
4. Fox, J., Roughgarden, T., Seshadhri, C., Wei, F., Wein, N.: Finding cliques in social networks: a new distribution-free model. SIAM J. Comput. **49**(2), 448–464 (2020)
5. Kanesh, L., Madathil, J., Roy, S., Sahu, A., Saurabh, S.: Further exploiting c-closure for FPT algorithms and kernels for domination problems. SIAM J. Disc. Math. **37**(4), 2626–2669 (2023)
6. Koana, T., Komusiewicz, C., Nichterlein, A., Sommer, F.: Covering many (or few) edges with k vertices in sparse graphs. In: Berenbrink, P., Monmege, B. (eds.) Proceedings of 39th STACS. LIPIcs, vol. 219, pp. 42:1–42:18, Dagstuhl, Germany. Schloss Dagstuhl – Leibniz-Zentrum für Informatik (2022)
7. Koana, T., Komusiewicz, C., Sommer, F.: Exploiting c-closure in kernelization algorithms for graph problems. SIAM J. Discret. Math. **36**(4), 2798–2821 (2022)
8. Koana, T., Komusiewicz, C., Sommer, F.: Computing dense and sparse subgraphs of weakly closed graphs. Algorithmica **85**(7), 2156–2187 (2023)
9. Koana, T., Komusiewicz, C., Sommer, F.: Essentially tight kernels for (weakly) closed graphs. Algorithmica **85**(6), 1706–1735 (2023)
10. Koana, T., Nichterlein, A.: Detecting and enumerating small induced subgraphs in c-closed graphs. Disc. Appl. Math. **302**, 198–207 (2021)
11. Komusiewicz, C.: Tight running time lower bounds for vertex deletion problems. ACM Trans. Comput. Theory **10**(2), 1–18 (2018)
12. Komusiewicz, C., Morawietz, N., Sommer, F., Staus, L.P.: The parameter report: an orientation guide for data-driven parameterization (2025)
13. Lehner, L., Komusiewicz, C., Staus, L.P.: A complexity analysis of the c-closed vertex deletion problem (2025)
14. Lewis, J.M., Yannakakis, M.: The node-deletion problem for hereditary properties is NP-complete. J. Comput. Syst. Sci. **20**(2), 219–230 (1980)
15. Lokshtanov, D., Surianarayanan, V.: Dominating set in weakly closed graphs is fixed parameter tractable. In: Bojańczyk, M., Chekuri, C. (eds.) Proceedings of 41st FSTTCS, vol. 213, pages 29:1–29:17, Dagstuhl, Germany. Schloss Dagstuhl – Leibniz-Zentrum für Informatik (2021)
16. Rusu, I.: Cluster vertex deletion problems on cubic graphs. arXiv preprint arXiv:2505.07443 (2025)
17. van Bevern, R.: Towards optimal and expressive kernelization for d-hitting set. Algorithmica **70**(1), 129–147 (2014)

On the Computational Complexity of Covering Multi-Interface Networks

Cristina Bazgan[1], Morgan Chopin[2], André Nichterlein[3],
and Camille Richer[1,2]

[1] Université Paris-Dauphine, PSL Research University, CNRS, UMR 7243,
LAMSADE, Paris, France
`cristina.bazgan@lamsade.dauphine.fr`, `camille.richer@dauphine.eu`
[2] Orange Research, Châtillon, France
`morgan.chopin@orange.com`
[3] Algorithmics and Computational Complexity, Technische Universität Berlin, ,
Germany
`andre.nichterlein@tu-berlin.de`

Abstract. We investigate the computational complexity of MINIMUM COVERAGE IN MULTI-INTERFACE NETWORKS (MIN CMI), which has applications in the field of Internet of Things (IoT). In this problem, we are given an undirected graph where each vertex represents a device (*e.g.*, smartphones, environmental sensors, smart speakers) capable of activating multiple interfaces (*e.g.*, Bluetooth, Wi-Fi, 4G/5G) to establish connections with one another. The objective is to ensure the connectivity requirement of the network while minimizing its total energy consumption by activating only the necessary interfaces.

Contributing to the computational complexity landscape of MIN CMI, we unify and extend previously known algorithmic results and demonstrate their limits. In particular, we analyze two scenarios based on the number of different interfaces across the network. When this number is unbounded, we show that the problem remains fixed-parameter intractable with respect to the number p of active interfaces per node within highly restricted graph classes such as stars and cliques. Additionally, we show that even in the case that p is a small constant, strong structural parameters like vertex cover number, feedback edge number, or distance to clique do not help to obtain fixed-parameter tractability. On the positive side, we provide a polynomial-time algorithm for cographs when $p = 2$. When the number of different interfaces is bounded by a constant, the problem, while still NP-hard, allows for more tractable special cases, including sparse graphs with small separators (*i.e.*, low treewidth) and dense graphs.

Keywords: Multi-interface Network · Wireless Network ·
Parameterized Complexity

E. Di Giacomo and D. Mondal (Eds.): WALCOM 2026, LNCS 16444, pp. 232–246, 2026.
https://doi.org/10.1007/978-981-95-7127-7_16

1   Introduction

The rapid proliferation of wireless devices within the Internet of Things (IoT) is significantly impacting various critical domains such as healthcare monitoring, where wearable sensors and connected devices allow for continuous tracking of patient health and early detection of medical conditions; autonomous driving, where IoT devices facilitate real-time communication between vehicles to enhance safety and efficiency; and agriculture, where IoT technologies optimize resource management and improve crop growth through the use of smart sensors and automated systems. This trend is expected to intensify with the advent of disruptive networking technologies, particularly those introduced by 5G+ networks, which will play a crucial role in the future evolution of IoT networks [18].

The oversight of such complex networks, which may include millions of heterogeneous devices, has led to numerous challenging combinatorial optimization problems that must be addressed to develop efficient decision-making algorithms for their effective management. In this paper, we are particularly interested in the question of efficiently connecting devices with one another through their interfaces — such as Bluetooth, Wi-Fi, and 4G/5G — while minimizing their energy consumption and thus extending their battery-life. This question has attracted a significant amount of attention due to its applications to IoT networks [2,12,16].

We investigate the computational complexity of the associated combinatorial optimization problem, namely the MIN COVERAGE IN MULTI-INTERFACE NETWORKS (MIN CMI) problem. In this problem, a network is represented by an undirected graph whose vertices are a set of heterogeneous devices and whose edges are the required connections, see Figure 1 for an illustrative example. Each device has a set of available interfaces (denoted by Greek letters in Figure 1). We assume that if two devices must be connected, they share at least one common available interface. To establish that connection, the two devices must activate a common interface, spending energy which is modeled by costs. We study the problem of finding which interfaces to activate in each device so as to establish every connection while minimizing the total energy consumption of the network. To control the energy spent by individual devices, there is a further constraint that each of them can activate a limited number p of interfaces. The associated feasibility problem, called CMI, asks for a solution establishing all required connections using at most p active interfaces per device, regardless of the overall energy consumption.

Related work. Given its practical relevance, many variants of MIN CMI have been studied [1–5,8,12,16]. The first description of the problem, called MINIMUM ENERGY COMMUNICATION INTERFACES [8], used more general cost functions and mutually exclusive interfaces.

MIN CMI has been studied without the limit on the number p of active interfaces in a device: Some results on the problem COST MINIMIZATION IN MULTI-INTERFACE NETWORKS [16] can be transferred to our problem with the additional constraint, *e.g.*, NP-hardness on planar graphs when a device can activate three or more interfaces, or a polynomial-time algorithm when there are only

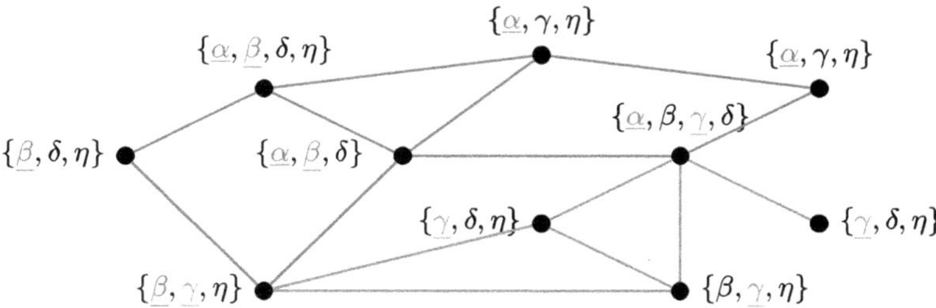

Fig. 1. Example of an instance of MIN CMI with $k = 5$ different interfaces across the network, unit costs, and where each device can activate at most $p = 2$ interfaces. In the solution represented, active interfaces are colored and underlined, and every edge is colored according to an interface covering it. That solution has cost 14 and activates 3 interfaces overall.

two different interfaces in the network. In another close problem called MINI-MUM MAXIMUM-COST COVERAGE IN MULTI-INTERFACE NETWORKS [12]—still without the limit p—the objective is changed to minimize the maximum energy spent in a single device. With unit interface cost, this is the problem of minimizing the maximum number of active interfaces needed in a vertex to establish all connections in the network. Thus some hardness proofs and polynomial-time algorithms for this problem can be directly transferred to ours.

The closest variant of MIN CMI that has been studied in the literature, denoted CMI(2), restricts to $p = 2$ active interfaces in each device [1–3,5]. This case is NP-hard already on a bipartite graph with maximum degree $\Delta = 4$ [2] and the proof can be adapted to also have a constant number of different interfaces $k = 18$ in the network [12]. Polynomial-time algorithms are given for special graph classes. We can group them into two main categories. For very dense graphs (complete graphs, complete bipartite graphs [2]), the large number of constraints limits the number of solutions to consider, allowing to enumerate them. For sparse graphs (cycles, trees, paths [2], series-parallel graphs [5]), the algorithms rely on the structural properties of the graph. Specifically, all these classes have treewidth at most two and the proposed dynamic programming algorithms rely on that property. Also the fixed-parameter tractability of the problem has been studied; the corresponding algorithms rely on tree-likeness [1,3]: treewidth, carvingwidth and branchwidth are three graph parameters that measure how tree-like a graph is and they are within constant factors of each other, and treewidth is upper-bounded by pathwidth [3].

See Tables 1 and 2 for an overview of results for MIN CMI and CMI.

Our contributions. We provide a unified and extended view of previously known algorithmic results for CMI and MIN CMI by identifying new tractable and intractable subcases, many of which are listed in Tables 1 and 2. In particular, we study the problems when the number k of interfaces is unbounded and when it is bounded–either by a constant or as part of the parameter. In the former case, CMI remains W[2]-hard with respect to p even in stars and cliques (Prop.

Table 1. Overview on complexity of CMI and MIN CMI with respect to the number p of interfaces allowed to be activated per vertex, the number k of different interfaces in the network, and the maximum degree Δ. All polynomial-time results also hold for $k \in \omega(1)$ and all NP-hardness results hold for unit costs. Observe that when $p \geq \Delta$, CMI is trivial as a vertex will not activate more than one interface per neighbor. For $\Delta \leq 2$, both CMI and MIN CMI are polynomial-time solvable [2].

Δ		$k=2$ any p	$k=3$ $p \leq 2$	$k=3$ $p=3$	$k \in O(1)$ $p \leq 2$	$k \in O(1)$ $p \geq 3$
$=3$	CMI			P [12]	?	P $(p \geq \Delta)$
	MIN CMI			?	**NP-h** (Thm. 4)	**NP-h** (Thm. 4)
$=4$	CMI	P [16]		P [12]		?
	MIN CMI	P [16]		?	NP-h [2,12]	**NP-h** (Thm. 4)
≥ 5	CMI			P [12]		NP-h [12]
	MIN CMI		?	NP-h [16]		NP-h [12]

Table 2. Overview on complexity w.r.t. special graph classes. All hardness results hold for unit costs. MIN CMI and CMI are polynomial on paths and cycles for any p.

graph class		$p=2$	$p \in O(1) (\geq 3)$	$p \in \omega(1)$
trees	(MIN) CMI	P [1]	P (Prop. 3)	
constant treewidth	(MIN) CMI	P [1]	P (Prop. 3)	
complete graphs	(MIN) CMI	P [2]	?	**W[2]-h** (Prop. 1 & [12])
complete bipartite	(MIN) CMI	P [2]	P (Thm. 2)	**W[2]-h** (Prop. 1 & [12])
cographs	(MIN) CMI	P (Thm. 3)	?	**W[2]-h** (Prop. 1 & [12])
planar	CMI		?	
	MIN CMI	**NP-h** (Thm. 4)	**NP-h** (Thm. 4 & [16])	

1). Additionally, as one of our main results we show that even for $p = 2$ CMI is W[1]-hard even with respect to parameters posing strong structural restrictions like vertex cover number, feedback edge number, or distance to clique (Prop. 2, Thm. 1 and Corollary 1). We complement these negative results by exhibiting a polynomial-time algorithm for MIN CMI in cographs provided that every device activates at most two interfaces (Thm. 3) and for complete bipartite graphs for any constant number of active interfaces (Thm. 2). For a bounded number k of interfaces, which is a more realistic setting, we identify more tractable cases. In particular, we show that MIN CMI is fixed-parameter tractable with respect to the two combined parameters k plus treewidth (Prop. 3) and k plus cluster vertex deletion number (Prop. 5).

Organization. The rest of the paper is organized as follows. Section 2 provides necessary definitions and a formal statement of the problem. Section 3 focuses on the case of an unbounded number k of interfaces and Section 4 on the case of a limited number of interfaces ($k \in O(1)$ or k part of the parameter). Finally, the conclusion outlines future research directions. Due to space restrictions, some proofs (marked by $\star$) are deferred to the full version.

2 Definitions and Problem Statement

Parameterized Algorithmics. A parameterized problem $\Pi \subseteq \Sigma^* \times \mathbb{N}$ is a set of pairs (I, k), where I denotes the problem instance and k is the parameter. Problem Π is *fixed-parameter tractable* (FPT) if there exists an algorithm (called FPT algorithm for short) solving any instance of Π in $f(k) \cdot |I|^c$ time, where f is some computable function and c is some constant. A *parameterized reduction* from a parameterized problem $\Pi \subseteq \Sigma^* \times \mathbb{N}$ to a parameterized problem $\Pi' \subseteq \Sigma^* \times \mathbb{N}$ is a function which maps any instance $(I, k) \in \Sigma^* \times \mathbb{N}$ to another instance $(I', k') \in \Sigma^* \times \mathbb{N}$ such that (1) (I', k') can be computed from (I, k) in FPT time, (2) $k' \leq g(k)$ for some computable function g, and (3) $(I, k) \in \Pi \iff (I', k') \in \Pi'$. If Π is W$[i]$-hard, $i \geq 1$, then such a parameterized reduction shows that also Π' is W$[i]$-hard, that is, presumably not fixed-parameter tractable.

The *vertex cover number* is the size of a minimum set of vertices whose removal results in an edgeless graph. The *distance to clique* is the minimum number of vertices to remove to obtain a clique. The *feedback edge number* is the number of edges to remove to make the graph acyclic. A *tree decomposition* for a graph $G = (V, E)$ is a pair (T, β), where T is a tree and β is a mapping that assigns to every node in T a subset of V called *a bag* in such a way that (i) every vertex and every edge of G is in some bag and (ii) for every v, the set $\{x \mid v \in \beta(x)\}$ induces a subtree of T. The width of a tree decomposition (T, β) is the maximum number of vertices assigned to a node minus one, that is, $\max_x |\beta(x)| - 1$. The *treewidth* of G is the minimum width of any tree decomposition for G, see, *e.g.*, [11] for more details. A graph is a *cograph* if every connected induced subgraph of G has diameter at most 2. Equivalently, a cograph is a graph without induced paths of length three (P_4-free) [7].

Problem Definition. A network is represented by a simple undirected graph $G = (V, E)$ where V is the set of $|V| = n$ devices and E is the set of $|E| = m$ required connections. Without loss of generality, we assume that G is connected. The set of interfaces is $[k] = \{1, \dots, k\}$. Every vertex has a set of *available interfaces* given by $\lambda \colon V \to 2^{[k]}$. We assume that for each edge $uv \in E$, $\lambda(u) \cap \lambda(v) \neq \emptyset$, otherwise this connection cannot be established. Each vertex can activate at most $p \leq k$ interfaces. An allocation of *active interfaces* is a function $\lambda_A \colon V \to 2^{[k]}$ such that $\lambda_A(v) \subseteq \lambda(v)$. An edge $uv \in E$ is *covered* if $\lambda_A(u) \cap \lambda_A(v) \neq \emptyset$. We say that λ_A *covers* G if it covers every edge of G. A solution is an allocation of active interfaces λ_A covering G such that $|\lambda_A(v)| \leq p$ for every vertex v. A solution λ_A activates d interfaces *overall* if it activates d different interfaces over the network, that is, $|\bigcup_{v \in V} \lambda_A(v)| = d$. Every interface $\alpha \in [k]$ is associated with a cost $c(\alpha)$ representing energy consumption. The cost of an allocation of active interfaces λ_A is $c(\lambda_A) = \sum_{v \in V} \sum_{\alpha \in \lambda_A(v)} c(\alpha)$.

The formal statement of the problem is:

MIN COVERAGE IN MULTI-INTERFACE NETWORKS (MIN CMI)
Input: An undirected graph $G = (V, E)$, an allocation of available interfaces $\lambda \colon V \to 2^{[k]}$, an integer $p \leq k$, an interface cost function $c \colon [k] \to \mathbb{N}$.
Task: Find $\lambda_A \colon V \to 2^{[k]}$ covering G such that $\forall v \in V$, $\lambda_A(v) \subseteq \lambda(v) \wedge |\lambda_A(v)| \leq p$ minimizing $c(\lambda_A) = \sum_{v \in V} \sum_{\alpha \in \lambda_A(v)} c(\alpha)$.

Note that with the constraint on the maximum number of active interfaces per vertex, it is not guaranteed that an instance admits a solution. Thus we also study the feasibility problem CMI, where we ask whether there is a solution, regardless of its cost. In the optimization problem MIN CMI, we always assume that the instance is feasible, that is, the answer to the associated feasibility problem is yes. To study the complexity of MIN CMI, we also use its associated decision problem, where there is an additional integer b in the input and we ask whether there is a solution of cost at most b.

We can transfer all NP-hardness and W-hardness results for CMI to MIN CMI and all polynomial or FPT results for MIN CMI to CMI. A reduction from CMI to MIN CMI works as follows: we add an interface δ to every vertex so that the assumption on feasibility holds, and we give cost 1 to interface δ and cost zero to interfaces $1, \ldots, k$. Then an optimal solution to MIN CMI with cost > 0 implies that the instance is not feasible for CMI, whereas an optimal solution with cost 0 does not use δ and gives a feasible solution for CMI.

3   Unbounded Number of Interfaces

In this section, we present our results for the case where the number k of interfaces is unbounded. This case turns out to be algorithmically challenging with only few tractable subcases (*e.g.*, for $p = 2$ on cographs as shown at the end of the section). The computational difficulty of this setting stems from the fact that the graph structure is essentially irrelevant: D'Angelo et al. [12] showed that hard problems, like SET COVER, can be encoded in the interfaces of stars. It is straightforward to adapt this hardness for cliques.

Proposition 1 ($\star$). *CMI is W[2]-hard with respect to p on stars and on cliques.*

Smaller number of allowed interfaces. Having a rather large number p of allowed interfaces gives a lot of flexibility in reductions, as witnessed by the above result. Contrasting Prop. 1, CMI with $p = 2$ is polynomial-time solvable on cliques or trees [2]. Thus, smaller values of p seem, in general, easier than larger values of p. We next show that larger values of p are at least as difficult as smaller values:

Proposition 2 ($\star$). *Let $q \in \mathbb{N}$ be a constant with $q \geq 3$. There is a linear-time reduction from CMI with $p = 2$ to CMI with $p = q$ that preserves the feedback edge number.*

We are not aware of a simple, linear-time self-reduction for CMI that shrinks the number p of allowed interfaces. While $p = 2$ seems indeed to be easier than the general case, it remains computationally difficult as we demonstrate below. Indeed, we show that CMI with $p = 2$ is W[1]-hard with respect to the vertex cover number, the feedback edge number, and with respect to the (vertex-deletion) distance to a single clique. All these parameters impose quite strong restrictions on the graph structure and usually allow for fixed-parameter tractability for many problems [15].

We remark that the W[1]-hardness with respect to vertex cover number implies W[1]-hardness with respect to treewidth [13,14]. This contradicts the claimed fixed-parameter tractability with respect to treewidth [3, Thm 3], which is based on expressing CMI with $p = 2$ in MSO_2 (a variant of monadic second-order logic) and employing Courcelle's theorem [9][1]. However, they ignore the dependency on the formula size and provide a formula that is too large (size depending on n or k) for their claimed fixed-parameter tractability (see the proof of Prop. 4 for a detailed description of a possible formula with length depending on k).

Theorem 1. *CMI with $p = 2$ is W[1]-hard with respect to the vertex cover number.*

Proof. We provide a parameterized reduction from the problem MULTICOLORED CLIQUE (MCC) which is W[1]-hard with respect to the number of colors [19].

MULTICOLORED CLIQUE
Input: An undirected graph $G = (V, E)$, an integer ℓ and a partition of V into ℓ sets $V_1, \ldots, V_\ell$.
Task: Is there a multicolored clique in G, that is, a set of pairwise adjacent vertices containing exactly one vertex of each V_i?

Let $\mathcal{I} = (G = (V, E), \ell)$ be an instance of MCC with $V = V_1 \cup \ldots \cup V_\ell$. Assume $|V_i| = n$ for each color $i \in \{1, \ldots, \ell\}$. Set $E_{ij} = \{uv \in E : u \in V_i, v \in V_j\}$ and given some $v \in V$, set $E_{ij}^{\bar{v}} = \{e \in E_{ij} : v \notin e\}$. We construct in polynomial time an instance $\mathcal{I}' = (G' = (V', E'), \lambda)$ of CMI with $p = 2$ as follows (see Figure 2).

We encode the vertices and edges of the original graph G in the interfaces λ: we define the set of interfaces as $V \cup E \cup \{\delta\}$ where δ is a dummy interface. For each color $i \in \{1, \ldots, \ell\}$, create a *color vertex* x_i and set $\lambda(x_i) = V_i \cup \{\delta\}$. For all $i \neq j$ do:

---

1 Courcelle's theorem states that the problem of verifying that an MSO_2-formula ϕ holds in a graph G is fixed-parameter tractable with respect to the size of ϕ and the treewidth of the graph [9], cf. [11].

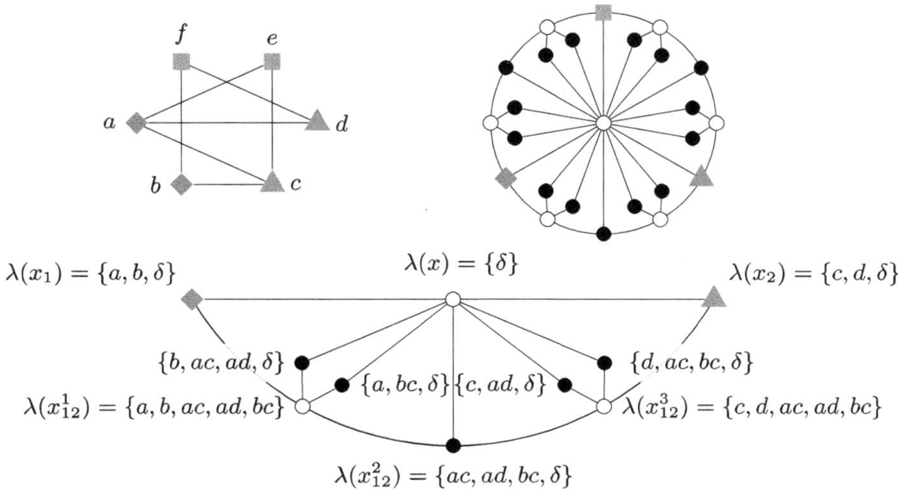

Fig. 2. At the top left is an instance of MCC (red diamond is color 1, blue triangle is color 2 and green square is color 3). At the top right is the graph of the reduced instance of CMI, where each colored vertex has the shape and color of the vertices of the MCC instance that it represents. The circle white vertices form a vertex cover. The figure at the bottom details the interface sets between two colors.

- Connect the pair $\{x_i, x_j\}$ with a path on three new vertices x_{ij}^1 (closest to x_i), x_{ij}^2 (in the middle) and x_{ij}^3 (closest to x_j).
- Set $\lambda(x_{ij}^1) = V_i \cup E_{ij}$, $\lambda(x_{ij}^2) = E_{ij} \cup \{\delta\}$ and $\lambda(x_{ij}^3) = V_j \cup E_{ij}$.
- For each $v \in V_i$, add a neighbor to x_{ij}^1 with interfaces $\{v, \delta\} \cup E_{ij}^{\bar{v}}$. That is, x_{ij}^1 has n new neighbors. Do the same for x_{ij}^3 for each $v \in V_j$.

Create a *dummy vertex* x with a single dummy interface $\lambda(x) = \{\delta\}$ and connect x to all vertices that have δ in their interface set. (This forces the selection of interface δ in all vertices where it appears.)

The interface (aside from δ) selected in color vertex x_i corresponds to the vertex of V_i selected in the multicolored clique. Covering the path between two color vertices x_i and x_j ensures that the two interfaces (aside from δ) selected in x_i and x_j correspond to two adjacent vertices in the MCC instance.

Clearly the above transformation is polynomial-time computable. The number of vertices in the reduced instance is $|V'| = \ell + \binom{\ell}{2}(2n+3) + 1$ and the total number of interfaces is $|V| + |E| + 1$. The vertex cover number is at most $2\binom{\ell}{2} + 1$ (take x and all x_{ij}^1, x_{ij}^3). We now show that $\mathcal{I}$ is a yes-instance if and only if $\mathcal{I}'$ is a yes-instance.

"$\Rightarrow$": Let $u_1, \ldots, u_\ell$ be a multicolored clique in G, where $u_i \in V_i$. Set the active interfaces as follows: $\lambda_A(x) = \{\delta\}$, $\lambda_A(x_i) = \{\delta, u_i\}$ for all $i = 1, \ldots, \ell$, $\lambda_A(x_{ij}^1) = \{u_i, u_i u_j\}$, $\lambda_A(x_{ij}^2) = \{\delta, u_i u_j\}$, and $\lambda_A(x_{ij}^3) = \{u_j, u_i u_j\}$ for all pairs of colors. Finally, among the n neighbors of x_{ij}^1, the one with interface set

$\{\delta, u_i\} \cup E_{ij}^{\bar{u}_i}$ activates $\{\delta, u_i\}$ and the $n-1$ others activate $\{\delta, u_i u_j\}$. Similarly the n neighbors of x_{ij}^3 activate respectively $\{\delta, u_j\}$ and $\{\delta, u_i u_j\}$. Clearly, this covers all edges.

"$\Leftarrow$": Suppose there is an assignment λ_A of at most two active interfaces to each vertex that covers all edges. Because of the dummy vertex x with $\lambda(x) = \{\delta\}$, all of its neighbors must activate interface δ and can only activate one additional interface.

Consider some x_i: it has available interfaces $\lambda(x_i) = V_i \cup \{\delta\}$ and its neighbor x_{ij}^1 has $\lambda(x_{ij}^1) = V_i \cup E_{ij}$, so $\exists \ u_i \in V_i : \lambda_A(x_i) = \{u_i, \delta\}$. Let $C = \{u_i \in \lambda_A(x_i) \cap V_i, \ i \in \{1, \ldots, \ell\}\}$. We show that C is a clique in G. Let $i \neq j$ be two colors. Since $\lambda(x_{ij}^2) = E_{ij} \cup \{\delta\}$ and $\lambda(x_{ij}^1) = V_i \cup E_{ij}$ and $\lambda(x_{ij}^3) = V_j \cup E_{ij}$, x_{ij}^2 must activate a single $e \in E_{ij}$ to cover the two edges incident to x_{ij}^1 and x_{ij}^3. Thus $\lambda_A(x_{ij}^1) = \{u_i, e\}$ and $\lambda_A(x_{ij}^3) = \{u_j, e\}$ where $u_i, u_j \in C$. We show that $u_i \in e$. Assume $u_i \notin e$, that is, $e = u'v$ where $u' \in V_i \setminus \{u_i\}$ and $v \in V_j$. Then the neighbor of x_{ij}^1 with interface set $\{\delta, u'\} \cup E_{ij}^{\bar{u}'}$ cannot have a common interface with x_{ij}^1. Thus $u_i \in e$ and symmetrically, $u_j \in e$ so $e = u_i u_j$. For all $i \neq j$, $u_i u_j \in E$ so C is a multicolored clique in G. $\qquad\square$

The reduction behind the above statement can be easily modified to show hardness with respect to different parameterizations:

Corollary 1. *CMI with $p = 2$ is W[1]-hard with respect to the feedback edge number and with respect to distance to clique.*

Proof. Both statements are obtained by adapting the reduction in the proof of Thm. 1.

Distance to Clique. Consider the same reduction as in Thm. 1. The reduced instance can be partitioned into a vertex cover (x and all x_{ij}^1, x_{ij}^3) and an independent set (the rest of the vertices). Make the independent set a clique. All vertices of the (former) independent set must activate δ because of x, so all edges of the clique are covered and the correctness proof is identical. The distance to clique of this reduced instance is the size of the vertex cover, that is $2\binom{\ell}{2} + 1$.

Feedback edge number. Consider the same reduction as in Thm. 1 except that instead of having a single dummy vertex x, there is one dummy vertex for every non-dummy vertex with interface δ. The number of vertices in G' roughly doubles (the number of edges does not change), the number of interfaces does not change and the transformation is still polynomial. The proof of correctness is identical. The only cycles in G' are in the clique with subdivided edges between the color vertices. Taking one edge from each subdivided edge gives a feedback edge set of size $\binom{\ell}{2}$. $\qquad\square$

Note that Prop. 2, and Corollary 1 imply that CMI is W[1]-hard with respect to the feedback edge number (and thus also with respect to treewidth [13,14]) for any constant $p \in O(1)$.

Smaller number of allowed interfaces & dense graphs. We complement the computational lower bounds listed above with a polynomial-time algorithm for CMI on complete bipartite graphs with $p \in O(1)$. The difference to the above hardness results for $p = 2$ is that a *minimal* solution covering all edges in a complete bipartite graph can only use a constant number of interfaces overall [2]. We say that λ_A is a *minimal* solution if removing any interface from any $\lambda_A(v)$ makes the solution infeasible (some edge is no longer covered). In other words, a solution λ_A is minimal, if for each vertex $v \in V$, each interface $\alpha \in \lambda_A(v)$, there is a neighbor $u \in N(v)$ with $\lambda_A(u) \cap \lambda_A(v) = \{\alpha\}$.

Lemma 1. *Any minimal feasible assignment of at most p interfaces for each vertex in a complete bipartite graph uses at most $(2p)^{2p}$ interfaces overall.*

Proof. Let $(G = (V, E), \lambda, p)$ be an instance of MIN CMI on a complete bipartite graph $G = (V = U \cup W, E)$. For $a, b \in \mathbb{N} \setminus \{0\}$, let $f(a, b)$ denote the maximum number of interfaces used overall in a minimal solution to G that assigns at most a and b many interfaces to each vertex of the two sides respectively; if no such assignment exists we set $f(a, b) = 0$. Let $\lambda_A \colon V \to 2^{[k]}$ be a minimal solution for G that uses a maximum number of interfaces overall (a solution can use at most k interfaces, so λ_A is well-defined).

For $a = 1$ or $b = 1$, we have $f(a, b) \leq \max\{a, b\}$: As the vertices on one side, say U can get at most one interface assigned, it follows that all vertices on the other side W need to have the same interfaces $\bigcup_{u \in U} \lambda_A(u)$ assigned. Since $a = 1$ or $b = 1$, $\lambda_A(u)$ can use at most $\max\{a, b\}$ interfaces overall. For $a \geq 2$ and $b \geq 2$, we will show the following inequality:

$$f(a, b) \leq 1 + f(a - 1, b) + f(a, b - 1) + \min\{(a - 1)f(a, b - 1), (b - 1)f(a - 1, b)\}$$

We will prove the inequality by appropriately splitting G in subgraphs. To this end, let $\alpha \in [k]$ be an interface used by λ_A and $uw \in E$ be an edge that is only covered by α, that is, $\lambda_A(u) \cap \lambda_A(w) = \{\alpha\}$. Note that such an edge exists since λ_A is a minimal solution. As α is used, we have the first term (the 1) in the inequality.

We partition U and W into two subsets depending on whether α is active or not: For $X \in \{U, W\}$, set $X_\alpha = \{x \in X \mid \alpha \in \lambda_A(x)\}$ and $X_{\overline{\alpha}} = X \setminus X_\alpha$. Note that all edges in $U_\alpha \times W_\alpha$ are covered by the interface α. It remains to cover the complete bipartite subgraphs $G[U_\alpha \cup W_{\overline{\alpha}}]$, $G[U_{\overline{\alpha}} \cup W_\alpha]$, and $G[U_{\overline{\alpha}} \cup W_{\overline{\alpha}}]$. The first two are completely independent of each other and provide at most $f(a, b-1)$ and $f(a-1, b)$ additional interfaces overall as α cannot be used to cover any edge in these subgraphs. This provides the second and third terms in the inequality.

It remains to justify the last term in the inequality. Let $\lambda_A(u) = \{\alpha, \beta_1, \ldots, \beta_q\}$ for $q \leq a - 1$. Denote $W_{\overline{\alpha}}^i = \{v \in W_{\overline{\alpha}} \mid \beta_i \in \lambda_A(v)\}$. As all edges between u and $W_{\overline{\alpha}}$ are covered, we have a cover of $W_{\overline{\alpha}}$ with these q parts: $W_{\overline{\alpha}} = \bigcup_{i \in [q]} W_{\overline{\alpha}}^i$. Similarly, we split $U_{\overline{\alpha}}$ as follows: $U_{\overline{\alpha}}^i = \{v \in U_{\overline{\alpha}} \mid \beta_i \notin \lambda_A(v)\}$ (note that we require β_i *not* being assigned to any vertex in $U_{\overline{\alpha}}^i$). Clearly, β_i covers all edges in $(U_{\overline{\alpha}} \setminus U_{\overline{\alpha}}^i) \times W_{\overline{\alpha}}^i$ and none of the edges in $U_{\overline{\alpha}}^i \times W_{\overline{\alpha}}^i$. Thus,

242   C. Bazgan et al.

to cover the edges in $G[U_{\overline{\alpha}}^i \cup W_{\overline{\alpha}}^i]$, we get at most $f(a, b-1)$ many additional interfaces overall as β_i is used in $W_{\overline{\alpha}}^i$. To cover all edges in $U_{\overline{\alpha}} \times W_{\overline{\alpha}}$, at most $q \cdot f(a, b-1) \leq (a-1) \cdot f(a, b-1)$ many additional interfaces overall can be introduced. When starting the analogous argument with vertex w, one obtains the upper bound $(b-1) \cdot f(a-1, b)$. This completes the argument for the last term.

The following rough upper bound $f(a, b) \leq (a+b)^{a+b}$ can be seen inductively: Each recursion reduces the sum of the arguments $a + b$ by one; the recursion depth for $f(a, b)$ is thus bounded by $a + b$. If we substitute in the highlighted inequality $f(a-1, b)$ and $f(a, b-1)$ by the bound $(a+b-1)^{a+b-1}$ from the induction hypothesis, then we obtain $f(a, b) \leq 1 + (a+b-1)^{a+b-1} \min\{a+1, b+1\}$. As $a \geq 2$ and $b \geq 2$, we obtain $f(a, b) \leq (a+b)^{a+b}$. The stated bound follows from setting $a = b = p$. $\qquad\square$

We remark that the bound given in the above lemma is only a very rough upper bound. For $p = 2$, a minimal feasible assignment uses at most four interfaces overall [2], which is less than the value $f(2, 2) = 7$ in the above proof computed by hand, and less than the upper bound $(2p)^{(2p)} = 256$. While this bound is far from being tight, it allows to brute-force the set of overall used interfaces in polynomial time and, hence, reduce to the case that k is constant. This yields a simple polynomial algorithm for complete bipartite graphs.

Theorem 2 ($\star$). *MIN CMI with constant p is polynomial-time solvable on complete bipartite graphs.*

For cographs however, even if $p = 2$, there might be instances where an optimal solution requires an unbounded number of interfaces (see Figure 3 for an example).

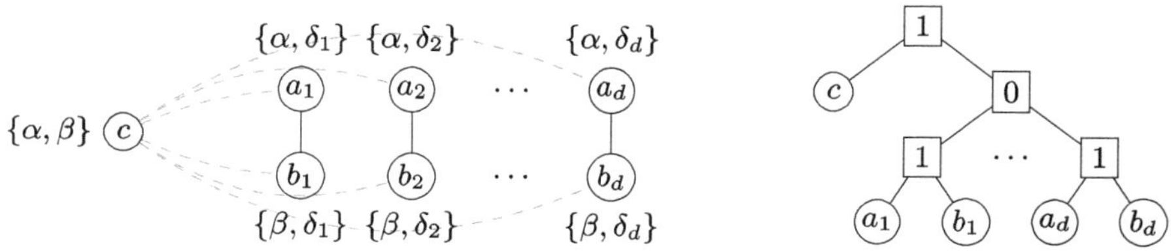

Fig. 3. On the left is an instance of MIN CMI on a cograph with $p = 2$ where the only feasible solution activates $d + 2$ interfaces: dashed edges are covered by α or β and solid edges by $\delta_1, \ldots, \delta_d$. On the right is the cotree, where join nodes are labeled 1 and disjoint union nodes are labeled 0.

Still, we show that MIN CMI is polynomial-time solvable on cographs for $p = 2$. We achieve this with dynamic programming on cotrees based on the observation that only a constant number of interfaces at each step are "relevant".

Theorem 3 ($\star$). *MIN CMI with $p = 2$ is polynomial-time solvable on cographs.*

4   Small Number of Interfaces

In this section, we focus on the case with a small number k of interfaces ($k \in O(1)$ or k part of the parameter), which is more relevant in practice as usually only few interfaces (such as Bluetooth, Wi-Fi, and 4G/5G) are employed. While still being NP-hard in general [2, 12], the restriction on the number of interfaces allows for more tractability. Our algorithmic results follow two rough settings: Either the graph always has small separators allowing for dynamic programming (that is, we exploit small treewidth) or the graph is so dense that a small number of allowed interfaces restricts the search space to a polynomial number of solutions. While we prove the dense result only for $p = 2$, we expect that the same approach with extensive case distinctions yields tractability for $p \in O(1)$.

Sparse Graphs. We first settle a particular special case of MIN CMI by showing NP-hardness on 3-regular graphs with $p \geq 2$ and $k = 4$. This fits into the general complexity landscape of suggesting that small maximum degree does not make the problem tractable, even for small values of p and k.

Theorem 4 ($\star$). *MIN CMI with $p \geq 2$, $k = 4$, and unit costs is NP-hard even on 3-regular planar graphs.*

Contrasting the W[1]-hardness of CMI with respect to treewidth, we provide an algorithm showing fixed-parameter tractability with respect to the combined parameter treewidth and k. The algorithm is based on standard dynamic programming already employed by Aloisio and Navarra [3] for $p = 2$ and combined parameter k and pathwidth; we extend this algorithm to any value of p and to the parameter treewidth. Note that computing a nice tree-decomposition of optimal width is fixed-parameter tractable with respect to the treewidth itself [11, 17].

Proposition 3 ($\star$). *MIN CMI can be solved in $O((k^p)^{(tw+2)}(tw + 1)n)$ time, given a nice tree-decomposition of width tw.*

Adapting the MSO_2-approach of Aloisio and Navarra [3] to work with MSO_1 and observing that the formula size depends only on k, we obtain fixed-parameter tractability with respect to the stronger parameterization k combined with cliquewidth. However, note that this is purely a classification result.

Proposition 4. *MIN CMI is fpt parameterized by $k + cw$ where cw is the clique-width of the graph.*

Proof. We consider a slight modification of the input, where the interfaces available to each vertex are given not by a function λ but by k subsets $L_1, \ldots, L_k \subseteq V$ where L_α is the set of vertices that have interface α available. Similarly, a solution to the problem is represented not by a function λ_A but by k subsets $A_1, \ldots, A_k \subseteq V$ where A_α is the set of vertices activating interface α. Then the constraints over a feasible solution can be expressed with a MSO_1 formula $\varphi(A_1, \ldots, A_k)$ over free vertex subset variables $A_1, \ldots, A_k$.

The constraint $\forall x \in V: \lambda_A(x) \subseteq \lambda(x)$ becomes $\forall \alpha \in [k]: A_\alpha \subseteq L_\alpha$. The constraint over the maximum number of active interfaces in a vertex $\forall x \in V: |\lambda(x)| \leq p$ can we written as a disjunction over all possible combinations $\Lambda \subseteq [k]$ of at most p interfaces. Denote $\binom{[k]}{p}$ the set of subsets of $[k]$ with size at most p. The constraint becomes $\forall x \in V: \bigvee_{\Lambda \in \binom{[k]}{p}} \left(\bigwedge_{\alpha \in \Lambda} x \in A_\alpha\right) \wedge \left(\bigwedge_{\alpha \notin \Lambda} x \notin A_\alpha\right)$. The constraint on edge coverage $\forall xy \in E: \lambda_A(x) \cap \lambda_A(y) \neq \emptyset$ becomes $\forall xy \in E: \exists \alpha \in [k]: x \in A_\alpha \wedge y \in A_\alpha$.

The length of $\varphi(A_1, \ldots, A_k)$ depends only on k and is at most $O(k2^k)$. The cost of a solution represented by $A_1, \ldots, A_k$ is $\sum_{\alpha=1}^{k} c(\alpha)|A_\alpha|$. Thus MIN CMI can be seen as a LINEMSOL problem and is fpt parameterized by the clique-width of G and the size of φ[10].

Dense Graphs. Similar to the sparse setting, we obtain fixed-parameter tractability with respect to k and the cluster vertex deletion number (the number of vertices to remove to obtain disjoint cliques). This contrasts the W[1]-hardness with respect to the distance to a single clique (see Corollary 1).

Proposition 5. *MIN CMI with $p = 2$ can be solved in $O\left(\binom{k}{2}^x (x^2 k + k^3 xn)\right)$ time where x is the cluster vertex deletion number.*

Proof. We provide an FPT-algorithm that relies on the polynomial-time algorithm for MIN CMI with $p = 2$ on complete graphs [2]. This last algorithm is based on two observations: (a) There is no optimal solution that activates exactly two interfaces overall: If there is a vertex activating a single interface, this interface would be enough for each vertex. If all vertices activate exactly two interfaces, then they all activate the same two and one interface would be enough too. (b) There is no optimal solution that activates four or more interfaces: Every vertex can activate at most two and every pair of vertices must share at least one interface.

Based on these observations, it is sufficient to test the following possibilities and keep the cheapest solution: (1) Test in $O(kn)$ time the feasible solutions that activate exactly one interface overall and keep the cheapest one. (2) Test in $O\left(\binom{k}{3}n\right)$ time the feasible solutions that activate exactly three interfaces overall: for each combination of three interfaces, a vertex activates the two cheapest available interfaces among them.

We adapt this approach for our FPT-algorithm. Let $\mathcal{I} = (G = (V, E), \lambda)$ be an instance of MIN CMI with $p = 2$ and let $X \subseteq V$ such that $G - X$ is a cluster graph. We can compute an optimal cluster vertex deletion set X, $|X| = x$, in time $O(c^x(n + m))$ for some constant $c < 2$ [6,20].

First, brute-force all assignments of active interfaces λ_A over X. For each assignment over X, check that all edges in X are covered then test if that assignment is extendable to G. Among all feasible solutions, we keep the cheapest one. For each cluster K of $G - X$, adapt the above algorithm for cliques:

1. Test the solutions that activate exactly one interface α to cover the edges within K. More precisely, for each interface $\alpha \in \bigcap_{u \in K} \lambda(u)$ covering all edges within K, test whether the edges between X and K can also be covered: For

each $v \in K$ with neighbors in X, compute $\bigcap_{u \in N(v) \cap X} \lambda_A(u)$ and if it does not contain α, v must also activate the cheapest interface in it (if any).

2. Test the solutions that activate three interfaces overall in K. For each choice of three interfaces, and each $v \in K$, if v has no neighbor in X, then it activates the two cheapest available interfaces among the three. Otherwise, if v has neighbors in X, then it activates the two cheapest available interfaces among the three such that $\lambda_A(v) \cap \lambda_A(u) \neq \emptyset$ for all $u \in N(v) \cap X$.

There are $\binom{k}{2}^x$ possibilities for the partial solution in X. Testing the feasibility of λ_A over X takes time $O(x^2 k)$. For each cluster K, testing the solutions with one interface takes time $O(k^2 x |K|)$ and testing the solutions with three interfaces takes time $O(k^3 x |K|)$. Thus with reading the graph and computing X, in total the algorithm takes time $O(\binom{k}{2}^x (m + x^2 k + k^3 xn))$. $\qquad\square$

5   Conclusion

In this paper, we explored the complexity of the MIN COVERAGE IN MULTI-INTERFACE NETWORKS problem. We showed that, when the number k of interfaces is unbounded, the problem remains W[2]-hard with respect to the number p of active interfaces per vertex even in very restricted graph classes such as stars and cliques and that strong structural parameters (vertex cover number, feedback edge number, distance to clique) do not make the problem fixed-parameter tractable even when p is a small constant. Despite these negative results, we were able to provide a polynomial-time algorithm for cographs when $p = 2$. Furthermore, in the case when k is constant, even though the problem remains NP-hard, we provided FPT-algorithms with respect to structural parameters related to the density of the graph namely treewidth and cluster vertex deletion.

We conjecture that our polynomial-time algorithm for cographs for $p = 2$ can be extended to constant $p \in O(1)$. This would also cover the open case for complete graphs and establishing an equivalence of Lemma 1 for these graphs could give valuable insights. Another promising research direction is to look for so-called above-guaranteed parameterizations. Indeed, assuming the cost of every interface is at least one then the decision version of MIN CMI parameterized by the solution cost ℓ admits a trivial kernelization: if the number n of vertices is greater than ℓ then we have a no-instance, otherwise $n < \ell$ and we have a kernel. However, the parameterized tractability of the problem with respect to the parameter $n - \ell$ is yet to be determined.

Acknowledgments. We thank an anonymous reviewer for pointing out Prop. 4.

Disclosure of Interests. The authors have no competing interests to declare that are relevant to the content of this article.

References

1. Aloisio, A.: Algorithmic aspects of distributing energy consumption in multi-interface networks. In: International Conference on Advanced Information Networking and Applications. pp. 114–123. Springer (2024)

2. Aloisio, A., Navarra, A.: Balancing energy consumption for the establishment of multi-interface networks. In: International Conference on Current Trends in Theory and Practice of Informatics. pp. 102–114. Springer (2015)
3. Aloisio, A., Navarra, A.: Constrained connectivity in bounded x-width multi-interface networks. Algorithms **13**(2) (2020)
4. Aloisio, A., Navarra, A.: Parameterized complexity of coverage in multi-interface iot networks: pathwidth. Internet of Things **28**, 101353 (2024)
5. Aloisio, A., Navarra, A., Mostarda, L.: Distributing energy consumption in multi-interface series-parallel networks. In: Workshops of the International Conference on Advanced Information Networking and Applications. pp. 734–744. Springer (2019)
6. Boral, A., Cygan, M., Kociumaka, T., Pilipczuk, M.: A fast branching algorithm for cluster vertex deletion. Theory of Computing Systems **58**(2), 357–376 (2016)
7. Brandstädt, A., Le, V.B., Spinrad, J.P.: Graph Classes: a Survey, SIAM Monographs on Discrete Mathematics and Applications, vol. 3. SIAM (1999)
8. Caporuscio, M., Charlet, D., Issarny, V., Navarra, A.: Energetic performance of service-oriented multi-radio networks: issues and perspectives. In: Proceedings of the 6th International Workshop on Software and Performance. pp. 42–45 (2007)
9. Courcelle, B.: The monadic second-order logic of graphs. i. recognizable sets of finite graphs. Information and Computation **85**(1), 12–75 (1990)
10. Courcelle, B., Makowsky, J.A., Rotics, U.: Linear time solvable optimization problems on graphs of bounded clique-width. Theory of Computing Systems **33**(2), 125–150 (2000)
11. Cygan, M., Fomin, F.V., Kowalik, L., Lokshtanov, D., Marx, D., Pilipczuk, M., Pilipczuk, M., Saurabh, S.: Parameterized Algorithms. Springer (2015)
12. d'Angelo, G., Di Stefano, G., Navarra, A.: Minimize the maximum duty in multi-interface networks. Algorithmica **63**, 274–295 (2012)
13. Fellows, M.R., Jansen, B.M.P., Rosamond, F.A.: Towards fully multivariate algorithmics: Parameter ecology and the deconstruction of computational complexity. Eur. J. Comb. **34**(3), 541–566 (2013)
14. Fellows, M.R., Lokshtanov, D., Misra, N., Mnich, M., Rosamond, F.A., Saurabh, S.: The complexity ecology of parameters: An illustration using bounded max leaf number. Theory of Computing Systems **45**(4), 822–848 (2009)
15. Ganian, R.: Improving vertex cover as a graph parameter. Discrete Mathematics & Theoretical Computer Science **17**(2), 77–100 (2015)
16. Klasing, R., Kosowski, A., Navarra, A.: Cost minimization in wireless networks with a bounded and unbounded number of interfaces. Networks **53**(3), 266–275 (2009)
17. Korhonen, T., Lokshtanov, D.: An improved parameterized algorithm for treewidth. In: Saha, B., Servedio, R.A. (eds.) Proceedings of the 55th Annual ACM Symposium on Theory of Computing (STOC 2023). pp. 528–541. ACM (2023)
18. Nguyen, D.C., Ding, M., Pathirana, P.N., Seneviratne, A., Li, J., Niyato, D., Dobre, O., Poor, H.V.: 6G Internet of Things: A comprehensive survey. IEEE Internet Things J. **9**(1), 359–383 (2021)
19. Pietrzak, K.: On the parameterized complexity of the fixed alphabet shortest common supersequence and longest common subsequence problems. J. Comput. Syst. Sci. **67**(4), 757–771 (2003)
20. Tsur, D.: Faster parameterized algorithm for cluster vertex deletion. Theory of Computing Systems **65**(2), 323–343 (2021)

Generalizing Brooks' Theorem via Partial Coloring Is Hard Classically and Locally

Jan Bok[1], Avinandan Das[2(✉)], Anna Gujgiczer[3],
and Nikola Jedličková[1]

[1] Department of Algebra, Faculty of Mathematics and Physics, Charles University,
Prague, Czech Republic
{jan.bok,nikola.jedlickova}@matfyz.cuni.cz
[2] Department of Computer Science, Aalto University, Espoo, Finland
avinandan.das@aalto.fi
[3] MTA–HUN-REN RI Lendület "Momentum" Arithmetic Combinatorics Research
Group, HUN-REN Alfréd Rényi Institute of Mathematics, Budapest, Hungary
gujgicza@renyi.hu

Abstract. We investigate the classical and distributed complexity of
k-partial c-coloring where $c = k$, a natural generalization of Brooks' theorem where each vertex should be colored from the palette $\{1, \ldots, c\} =
\{1, \ldots, k\}$ such that it must have at least $\min\{k, \deg(v)\}$ neighbors col-
ored differently. Das, Fraigniaud, and Rosén [OPODIS 2023] showed that
the problem of k-partial $(k+1)$-coloring admits efficient centralized and
distributed algorithms and posed an open problem about the status of the
distributed complexity of k-partial k-coloring. We show that the problem
becomes significantly harder when the number of colors is reduced from
$k + 1$ to k for every constant $k \geq 3$.

In the classical setting, we prove that deciding whether a graph admits
a k-partial k-coloring is NP-complete for every constant $k \geq 3$, revealing
a sharp contrast with the linear-time solvable $(k+1)$-color case. For the
distributed LOCAL model, we establish an $\Omega(n)$-round lower bound for
computing k-partial k-colorings, even when the graph is guaranteed to
be k-partial k-colorable. This demonstrates an exponential separation
from the $O(\log^2 k \cdot \log n)$-round algorithms known for $(k+1)$-colorings.

Our results leverage novel structural characterizations of "hard
instances" where partial coloring reduces to proper coloring, and we
construct intricate graph gadgets to prove lower bounds via indistinguishability arguments.

1 Introduction

Every graph of maximum degree Δ can be properly colored with at most $\Delta + 1$
colors (with a simple greedy algorithm). The famous *Brooks' theorem* [7] states
that Δ colors are enough for almost all graphs, except for two cases, complete
graphs and odd cycles, where $\Delta + 1$ colors are needed. This means that – except
for those two cases – every vertex v is colored from the palette $\{1, \ldots, \Delta\}$ in
such a way that each of its $\deg(v)$ neighbors receives different colors than v.

1.1 Partial Coloring

A natural generalization of proper coloring is *partial coloring* [2,6,17,21]. Given two integers $k \geq 0$ and $c \geq 1$, a *k-partial c-coloring* of a graph assigns to each vertex a color from $\{1, \ldots, c\}$ such that every vertex v has at least $\min\{k, \deg(v)\}$ neighbors with a color different from its own.

This notion generalizes several well-known coloring paradigms:

- For $k = \Delta$, where Δ is the maximum degree of the graph, k-partial $(k + 1)$-coloring corresponds to proper $(\Delta + 1)$-coloring and k-partial k-coloring corresponds to proper Δ-coloring.
- For $k = 1$, the condition captures the classical notion of *weak coloring* [21].

The problems of proper $(\Delta + 1)$-coloring [1,3,10,12,13,15,20,23,25], proper Δ-coloring [4,9,14,15,22] and *weak* coloring [2,6,17,21] have been studied extensively in the distributed computing paradigm. The notion of partial coloring brings all these problems in the same framework.

For every integer k and every graph G, a k-partial $(k + 1)$-coloring always exists. As outlined in [8] and implicitly in [2], a simple greedy algorithm achieves this: initialize all vertices with color 1, and process the vertices sequentially. For each vertex v, let $C(v)$ denote the set of colors in its neighborhood. If $|C(v)| = k + 1$, the vertex retains its color. Otherwise, we color it with any color from $\{1, \ldots, k + 1\} \setminus C(v)$. This construction guarantees that each vertex has enough differently colored neighbors at the end of the process. Moreover, once two adjacent vertices are assigned different colors during any stage of the algorithm, they permanently maintain this property.

From a distributed perspective, Das, Fraigniaud, and Rosén [8] studied the complexity of computing such a coloring in the **LOCAL** model. They designed an $O(\log n \cdot \log^2 k)$-round algorithm for k-partial $(k+1)$-coloring using the rounding framework from [15] which matches their round complexity of $O(\log n \cdot \log^2 \Delta)$ for proper $(\Delta + 1)$-coloring. The result suggested that "relaxing" the constraints of proper coloring to partial coloring made the problem "easier", as demonstrated by the improvement of the round complexity. They also posed an open problem; what is the distributed complexity of k-partial k-coloring? The problem of proper Δ-coloring admits polylogarithmic round complexity ([4,15]) in the deterministic **LOCAL** model. This question can also be viewed as whether a similar scaling of the complexity w.r.t. k extends to Δ-coloring.

While proper Δ-coloring is well understood in the classical (centralized) setting – with known characterizations and complexity classifications—the same is not true for k-partially k-colorable graphs. To the best of our knowledge, no characterization or complexity classification is known for the decision problem: given a graph G and integer k, is G k-partially k-colorable? The problem is far from being trivial. For example, our quest for characterization of k-partial k-colorability quickly yielded non-trivial obstructions (which, for instance, do not even have $(k + 1)$-clique as a subgraph, see Fig. 1 for such an example).

In this work, we initiate a systematic study of this problem. First, we investigate the classical complexity of recognizing k-partially k-colorable graphs. Then, leveraging structural insights, we develop distributed lower bounds for computing k-partial k-colorings in graphs that are guaranteed to admit such a coloring.

For distributed computing, we work in the standard LOCAL model of distributed computing [19]. In this model, the input graph represents a communication network where each node is a computational entity. The nodes communicate in synchronous rounds, exchanging unbounded-sized messages with their neighbors. Each node has a unique identifier (typically from a polynomial-size range) and knows only its own id and the size of the universe of the ids at the beginning of the computation. The complexity of a distributed algorithm is measured in the number of rounds until all nodes produce their output. In our context, each node must eventually decide its color based on the information available in its local neighborhood. A more detailed discussion of the LOCAL model is provided in Sect. 4.

2 Related Work

Proper $(\Delta + 1)$-Coloring. For many years, the fastest known deterministic algorithms for proper $(\Delta + 1)$-coloring in graphs with maximum degree Δ required essentially $2^{O(\sqrt{\log n})}$ rounds in n-node networks [1,23]. This long-standing upper bound was only recently surpassed when a polylogarithmic-round algorithm for $(\Delta + 1)$-coloring was introduced [25]. All prior algorithms relied on a particular type of graph decomposition, and the breakthrough in [25] demonstrated how to efficiently compute this decomposition within a polylogarithmic number of rounds.

A major advance followed in 2020, when it was shown that $(\Delta + 1)$-coloring can be achieved in $O(\log n \cdot \log^2 \Delta)$ rounds [15], this time without graph decomposition, by using rounding of a fractional solution. For large Δ, this algorithm still runs in $O(\log^3 n)$ rounds, which was subsequently improved to $\tilde{O}(\log^2 n)$ [12] and more recently to $\tilde{O}(\log^{5/3} n)$ [13].

The problem of proper $(\Delta + 1)$-coloring has also been studied exclusively with respect to round complexity as a function of the parameter Δ. This line of work has yielded algorithms for $(\Delta + 1)$-coloring that run in $\tilde{O}(\sqrt{\Delta}) + O(\log^* n)$ rounds [3,10,20]. However, these are efficient only when Δ is small—specifically, their running times are polylogarithmic in n only if Δ itself grows at most polylogarithmically with n.

Proper Δ-Coloring. The problem of proper Δ-coloring has been extensively studied in the distributed setting. The work of [22] established the *distributed Brooks' theorem*, showing that if all but one vertex of the graph are already Δ-colored, the coloring can be extended by recoloring a path of length $O(\log_\Delta n)$ from that vertex, leading to a distributed Δ-coloring algorithm with complexity $O(\Delta \cdot \text{poly} \log n)$ and a randomized version with complexity $O(\log^3 n / \log \Delta)$ rounds. Subsequent work has significantly improved these bounds. The work of [14] presented a randomized algorithm with round complexity $O(\log \Delta) + 2^{O(\sqrt{\log \log n})}$ for $\Delta \geq 4$. Later, [15] gave the first deterministic polylogarithmic-time algorithm for all $\Delta \geq 3$, with round complexity $O(\log^2 \Delta \log^2 n)$. The work of [9] designed the first randomized poly log log n-round LOCAL algorithm for the problem. Most recently, Bourreau et al. [4] obtained an $O(\log^4 \Delta + \log^2 \Delta \log n \log^* n)$ round deterministic LOCAL algorithm for $\Delta \geq 3$, which remains the state of the art for the problem.

Partial Coloring. Under certain restrictions, efficient algorithms and also lower bounds for partial coloring are known. For example, in (constant) Δ-regular graphs, there exists a deterministic algorithm that computes a k-partial 3-coloring in $O(\log^* n)$ rounds whenever $\Delta \geq 3k - 4$ and $k \geq 3$, and a k-partial k-coloring under the condition $\Delta \geq k+2$ and $k \geq 4$ (see [2]). The same work also establishes lower bounds: computing a 2-partial 2-coloring in Δ-regular graphs requires $\Omega(\log n)$ rounds deterministically and $\Omega(\log \log n)$ rounds randomized, for any $\Delta \geq 2$. More generally, any k-partial c-coloring in Δ-regular graphs with $k \geq \frac{\Delta(c-1)}{c} + 1$ also requires $\Omega(\log n)$ deterministic rounds and $\Omega(\log \log n)$ randomized rounds [2]. Additionally, Brandt [6] recently proved that 1-partial 2-coloring in odd-degree graphs cannot be solved in $o(\log^* \Delta)$ rounds, matching the 30-year-old upper bound from [21]. Finally, both k-partial $O(k^2)$-coloring and d-defective $O(\Delta^2/d^2)$-coloring can be computed in $O(\log^* n)$ rounds [17].

3   Our Results

The problem of k-partial k-coloring has been studied in the restrictive settings by [2] where they showed that for constant degree d-regular graphs, a k-partial k-coloring can be computed in $O(\log^* n)$ rounds provided that $d \geq k + 2$ and $k \geq 4$. Contrary to these results and the existence of efficient round complexity for k-partial $(k+1)$-coloring in general graphs, our results show that the problem of k-partial k-coloring becomes "hard" in the classical (centralized) setting and "global" in the local distributed setting for general (constant degree) graphs for every fixed integer $k \geq 3$. In Sect. 5, we characterize the hard instances of partial coloring which have to be properly colored even when relaxed to partial coloring restrictions. We leverage such structure to prove the next two theorems which form the crux of our contribution (see also Table 1).

Table 1. Complexity theoretic landscape of k-partial coloring with k and $k+1$ colors in the classical and distributed LOCAL models.

Model	Problem	Complexity	Source
Classical	k-partial $(k+1)$-coloring	$O(n)$ time	[8]
	k-partial k-coloring	NP-complete	This paper
LOCAL	k-partial $(k+1)$-coloring	$O(\log^2 k \cdot \log n)$ rounds	[8]
	k-partial k-coloring	$\Omega(n)$ rounds (const. k)	This paper

Classical Complexity. We proved that determining whether a given graph is k-partially k-colorable is NP-complete for every constant $k \geq 3$, which is in stark contrast to the problem of k-partial $(k+1)$-coloring which admits a simple $O(n)$ step greedy algorithm.

Theorem 1 *For every fixed integer $k \geq 3$, the problem of deciding whether an input graph G is k-partially k-colorable is NP-complete.*

Distributed Complexity. The problem of 2-partial 2-coloring on general graphs trivially admits an $\Omega(n)$ lower bound as 2-partial 2-coloring implies a proper 2-coloring on paths which readily admits an $\Omega(n)$ round lower bound (refer to Section 7.5 of [24]). We show that the problem of k-partial k-coloring of graphs which are k-partially k-colorable, admits an $\Omega(n)$ round lower bound for every constant $k \geq 3$, demonstrating an exponential separation as we reduce the number of colors from $k + 1$ to k.

Theorem 2 *For every fixed integer $k \geq 3$ and for any k-partial k-coloring LOCAL algorithm $\mathcal{A}$ for k-partially k-colorable graphs, there exists an n-node k-partially k-colorable graph G such that $\mathcal{A}$ requires $\Omega(n/k^3)$ rounds to color G.*

4 Distributed **LOCAL** Model

The standard LOCAL model of distributed computing is a *synchronous* model of distributed computing which was introduced in [19]. It comprises of an n-vertex communication network $G = (V, E)$. Each edge $\{u, v\} \in E$ is a bi-directional communication channel between u and v. Each vertex gets assigned a unique identifier through the function id $: V \to \{1, 2, \ldots, n^d\}$ for some constant $d \geq 1$ which is unique to each communication network. Initially, each node only knows its identifier, the size of the id, and potentially some input. The nodes communicate by exchanging messages along the edges of the graph. All nodes execute the same algorithm, which proceeds in synchronous rounds. In each round, every node sends one message to each of its neighbors, receives the messages from its neighbors, and performs some individual computation. After a certain number of rounds, every node outputs, and terminates. For example, let us consider the problem of proper $(\Delta + 1)$-coloring which has been extensively studied in this model. Here, each node of the graph G will initially have Δ as input and at the end of a LOCAL coloring algorithm execution, would output the color of the node, which is a number from the palette $[\Delta + 1]$.

The LOCAL model assumes a perfect world scenario where the topology of the graph G doesn't change over time and the communication channels (i.e. the edges) are perfect (i.e. the messages aren't dropped and they are delivered instantaneously). Each node $v \in V$ is a computer which is computationally unlimited. Also, there is no limit on the size of the message transmitted by the nodes.

The complexity of a LOCAL algorithm is measured by the number of synchronous rounds. There is an equivalent definition of LOCAL algorithm based on locality. Before describing the equivalent definition, we define the *t-neighborhood* of a vertex v in the graph G as follows. For $v \in V(G)$ and $t \geq 1$,

$$N_G^t[v] := G[\{u \in V(G) \ : \ \operatorname{dist}_G(u,v) \leq t\}].$$

In other words, the t-neighborhood of v in G is the graph induced by the vertices of G which are at distance at most t from v (in particular, $v \in N_G^t[v]$). A *t-radius view* of v is the t-neighborhood of v along with the id, input, and the initial state of each vertex in the neighborhood.

Any $(t+1)$-round LOCAL algorithm $\mathcal{A}$ is equivalent to each vertex computing its output independently provided given access to its t-radius view. Each node v implements the *gather algorithm* where at round 1, node v sends its id, input, and state to each node $u \in N_G(v)$ and receives $\operatorname{id}(u)$, the input, and state of u. At the second round, it sends the collection of ids, states, and inputs to its neighbors, collects the same from them, and computes the 1-radius view of v. At each round $i \geq 3$, node v sends its $(i-2)$-radius view to each node $u \in N_G(v)$, receives the $(i-2)$-radius view of u, and trivially computes its $(i-1)$-radius view. Once v computes its t-radius view, it "simulates" $\mathcal{A}$ offline and computes its output. A comprehensive exposition of this equivalent definition can be found in the lecture notes [26].

4.1   Proving Lower Bounds by Constructing Indistinguishable Graphs

The lower bound proofs based on indistinguishability graphs rely on the fact, that given two graphs G_1 and G_2, and assignment of ids of their respective vertices such that there exist two vertices v_1 in G_1 and v_2 in G_2 sharing the same id and t-radius view implies that any t-round LOCAL algorithm executing on the vertices v_1 and v_2 cannot *distinguish* between G_1 and G_2 and will produce the same output for for v_1 and v_2. This follows from the radius view based definition of the LOCAL algorithm. The technique of using indistinguishability to prove LOCAL lower bounds have been used extensively in the literature (for example, refer to Sect. 7.5 of [24], and [18,19]).

For designing lower bounds for our problem, we create two n-vertex graphs $G_1^{\{\ell,k\}}$ and $G_2^{\{\ell,k\}}$ such that there exist two pairs of vertices $u_1, v_1 \in G_1^{\{\ell,k\}}$ and $u_2, v_2 \in G_2^{\{\ell,k\}}$ such that the following properties hold.

1. There exists an assignment of ids for the vertices of $G_1^{\{\ell,k\}}$ and $G_2^{\{\ell,k\}}$ such that u_1 and u_2 receive the same id and have the same t-radius view for every $t = o(n)$. The same holds for v_1 and v_2 as well.
2. Any k-partial k-colorings χ_1 and χ_2 of $G_1^{\{\ell,k\}}$ and $G_2^{\{\ell,k\}}$, respectively, satisfy that $\chi_1(u_1) = \chi_1(v_1)$ while $\chi_2(u_2) \neq \chi_2(v_2)$.

For $t = o(n)$, any t-round algorithm $\mathcal{A}$ executing on u_1, u_2, v_1 and v_2 cannot distinguish $G_1^{\{\ell,k\}}$ and $G_2^{\{\ell,k\}}$ and therefore, assigns the same color to u_1, u_2 and v_1, v_2 thus incorrectly coloring at least one of the two graphs.

5   Hard Instances of Partial Coloring

In this section, we identify the structural properties that make a graph instance hard for k-partial k-coloring. The goal is to understand which parts of a graph force any partial coloring algorithm to behave like a proper coloring algorithm.

Given a graph G and integers k and c, any algorithm for computing a k-partial c-coloring must implicitly address two subtasks:

1. Identifying a subgraph $G' \subseteq G$ which must be properly colored to satisfy the k-partial coloring condition—this identification is solely dependent on k and not on the number of colors c.
2. Assigning colors to G' in a way that yields a proper coloring using c colors.

The complexity of k-partial coloring is therefore inherently tied to the structure of this critical subgraph G'. We formally characterize such subgraphs in the lemma below, which highlights the role of a specific graph parameter in determining where proper coloring becomes unavoidable.

To this end, we introduce the parameter Δ_E for a graph $G = (V, E)$ as follows:

$$\Delta_E := \max_{e=\{u,v\}\in E} \min\{\deg_G(u), \deg_G(v)\}.$$

Lemma 1 *A Δ_E-partial coloring of G is a proper coloring of G.*

This lemma tells us that when $k = \Delta_E$, any valid k-partial coloring is in fact a proper coloring. Hence, the subgraph induced by edges realizing the maximum in Δ_E forms the core of difficulty for partial coloring—it forces proper coloring behaviour.

These graphs thus form the *hard instances* for k-partial k-coloring. As we will observe in the following sections (Observations 1 and 2), all our lower bound constructions for centralized and distributed settings are captured by this characterization.

Key Observations. For a given graph G, it is possible that $\Delta_{E(G)} < \Delta(G)$. See Fig. 1 for such an example.

In fact, the following holds: For every graph G, $\mathcal{K}(G) \leq \Delta_{E(G)} \leq \Delta(G)$ where $\mathcal{K}(G)$ is the degeneracy of the graph G. While the inequality $\Delta_{E(G)} \leq \Delta(G)$ follows immediately from the definitions, here is a simple argument that $\mathcal{K}(G) \leq \Delta_{E(G)}$. Partition $V(G)$ into two sets: $H = \{v \in V(G) \mid \deg(v) > \Delta_{E(G)})\}$ and $L = V(G) \smallsetminus H$. The set H is an independent set: if $u, v \in H$ were adjacent, the vertices u and v would be incident to at least $\Delta_{E(G)} + 1$ edges, contradicting the definition of Δ_E. Now, order the vertices so that all vertices of H appear first (in an arbitrary order), followed by all vertices of L (also in an arbitrary order). In this ordering, each vertex has at most $\Delta_{E(G)}$ neighbors appearing later, and thus $\mathcal{K}(G) \leq \Delta_{E(G)}$.

Therefore, the problems of k-partial $(k+1)$-coloring and k-partial k-coloring are generalizations of proper $(\Delta_E + 1)$-coloring and proper Δ_E-coloring, respectively. These in turn are non-trivial generalizations[1] of proper $(\Delta + 1)$-coloring and proper Δ-coloring.

The k-partial $(k+1)$-coloring algorithm of [8] implicitly provides an algorithm for (Δ_E+1)-coloring in $O(\log^2 \Delta_E \cdot \log n)$ rounds by adapting the rounding-based coloring algorithm by Ghaffari and Kuhn [15]. Our results show that reducing even one color from the palette makes the problem "hard" in the classical model as well as in the LOCAL model.

6   Classical Complexity

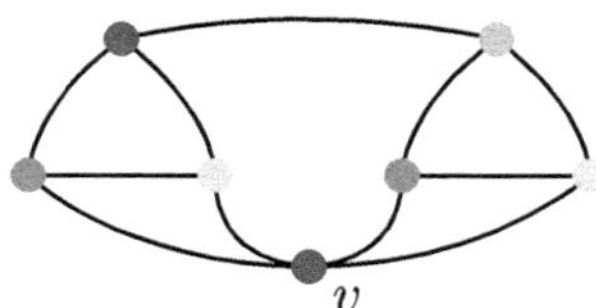

Fig. 1. A nontrivial obstruction for 3-partial 3-coloring, showing an example of a graph with $\Delta_E = 3$ which does not admit a proper 3-coloring, and therefore, a 3-partial 3-coloring. Also, observe that for this instance, $\Delta = 4$ and therefore, $\Delta_E < \Delta$.

This section is dedicated to proving the following theorem.

Theorem 1 *For every fixed integer $k \geq 3$, the problem of deciding whether an input graph G is k-partially k-colorable is NP-complete.*

We prove Theorem 1 by reduction from the decision problem of proper k-colorability of an input graph. It was shown to be NP-complete for every constant $k \geq 3$ [11]. Given an input graph G of proper k-colorability, we transform G into G' by replacing each edge $\{u, v\} \in E(G)$ with the *edge gadget* $e_{\{u,v\}}$ which we define as follows.

Edge Gadget. The edge gadget $e_{\{u,v\}}$ is constructed by adding additional vertices $\{(u, v, 1), (u, v, 2), \ldots, (u, v, k)\}$ and edges between them such that these vertices form a k-clique. Moreover, the vertex u is made adjacent to $(u, v, 1), (u, v, 2)$, $\ldots$, and $(u, v, k-1)$ and the vertex v is adjacent to (u, v, k). See Fig. 2 for an example. The transformation trivially runs in polynomial time. We now prove the following property of the graph G'.

---

[1]   A problem Π is a *generalization* of a problem Π' if every instance of Π' can be expressed as an instance of Π such that any algorithm for Π also solves Π' with the same complexity.

Fig. 2. The edge gadget $e_{\{u,v\}}$ with $k = 5$ (in red circle, connected with grey edges to the vertices u and v). Vertices u and v are forced to take different colors. (Color figure online)

Lemma 2. *Any k-partial k-coloring γ of G' satisfies that $\gamma(u) \neq \gamma(v)$ for each edge $\{u, v\} \in E(G)$.*

Proof. By the construction of the edge gadget $e_{\{u,v\}}$, for each $i \in [k]$, $\deg_{G'}((u, v, i)) = k$, therefore by the definition of k-partial coloring, the vertex (u, v, i) must be properly colored with respect to all its neighbors. By the fact that the set $\{(u, v, 1), \ldots, (u, v, k)\}$ forms a k-clique and that the vertex u is adjacent to all vertices of the k-clique except (u, v, k), the only possibility is that $\gamma(u) = \gamma((u, v, k))$. Since (u, v, k) is adjacent to v and has to be properly colored with respect to it, the claim immediately follows. $\qquad\square$

The next claim proves the correctness of the reduction.

Lemma 3. *The graph G is properly k-colorable if and only if the graph G' is k-partially k-colorable.*

Proof. Let us assume that G is properly k-colorable and let γ be such a coloring of G. We define a coloring γ' on G' by extending the coloring of γ on the vertices of $e_{\{u,v\}}$ for each edge $\{u, v\}$ of G as follows. Assign $\gamma'((u, v, k)) := \gamma(u)$. For the rest of the vertices, assign each vertex with a unique color from the palette $[k] \smallsetminus \{\gamma(u)\}$. The resultant coloring γ' is a proper coloring and therefore, a k-partial k-coloring of G'.

Now, let us assume that G' is k-partially k-colorable. Let γ' be such a coloring of G'. We can deduce a proper k-coloring γ of G from γ' by simply assigning $\gamma(v) := \gamma'(v)$ for all $v \in V$. The fact that γ is a proper coloring of G follows from Lemma 2. $\qquad\square$

Observation 1. *The graph G' satisfies that $\Delta_{E(G')} = k$ and therefore, any k-partial k-coloring is also a proper k-coloring of G'.*

Proof. Consider an arbitrary Δ_E-partial coloring γ of G. For an edge $\{u, v\} \in E$, $\deg_G(u) \leq \Delta_E$ or $\deg_G(v) \leq \Delta_E$. Without loss of generality, let us assume that this holds true for vertex u. Therefore, by the definition of partial-coloring, each neighbor of u must be colored properly with respect to u by γ, and hence, the edge $\{u, v\}$ is not monochromatic and γ is a proper coloring of G. $\qquad\square$

Now we complete the proof of Theorem 1. If there exists a polynomial time algorithm for deciding k-partial k-colorability we can construct a polynomial time algorithm for k-colorability, which is a contradiction. Given an instance G of proper k-colorability, we use the transformation described above to create an instance G' in polynomial time. Then we can use the k-partial k-colorability decision algorithm on G'. By Lemma 3, this gives the answer whether G is properly k-colorable or not.

7   Distributed Complexity

In this section, we prove that the problem of k-partial k-coloring is "global". More precisely, we prove the following theorem.

Theorem 2. *For every fixed integer $k \geq 3$ and for any k-partial k-coloring LOCAL algorithm $\mathcal{A}$ for k-partially k-colorable graphs, there exists an n-node k-partially k-colorable graph G such that $\mathcal{A}$ requires $\Omega(n/k^3)$ rounds to color G.*

To prove Theorem 2, we first present the following graph gadget which has also been used in [16] and [5] in the context of distributed certification.

Path of Cliques. For integers $k \geq 3$ and $\ell \geq 2$, the graph ℓ-PATH OF k-CLIQUES is defined on the vertex set $V = [k] \times [\ell]$ with respect to $\ell - 1$ permutations τ_1, τ_2, ..., $\tau_{\ell-1} : [k] \to [k]$. Let the set of vertices $\{(1, i), \dots, (k, i)\}$ be denoted by V_i for each $i \in \ell$. The graph induced on V_i is a k-clique, denoted by K_k^i. For each $i \in [\ell - 1]$, each vertex of V_i is adjacent to every vertex of V_{i+1} except the perfect matching specified by the permutation τ_i, i.e. for vertices a and b from V_i and V_{i+1} respectively, $\{(a, i), (b, i+1)\} \in E$ if and only if $b \neq \tau_i(a)$. We call the set of such edges between V_i and V_{i+1} as *anti-matching* specified by τ_i. See Fig. 3 for an example. Also observe that ℓ-PATH OF k-CLIQUES is uniquely determined by the permutations $\tau_1, \dots, \tau_{\ell-1}$.

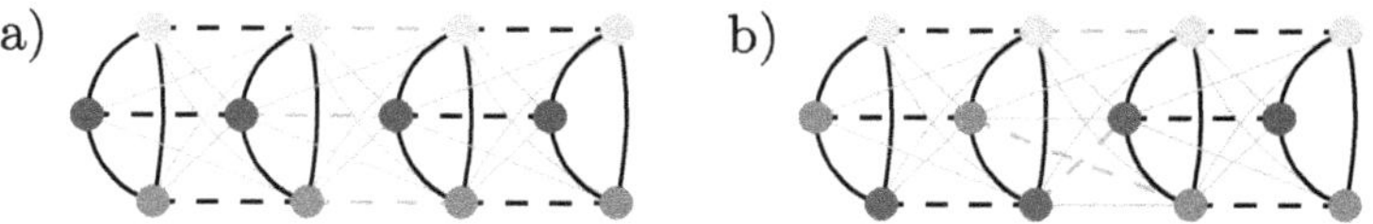

Fig. 3. Two 4-path of 3-cliques which differ only on the permutation τ_2 and the effect of this difference on the coloring of the two graphs. Anti-matchings induced by τ_1, τ_2, τ_3 are depicted with dashed lines, and the one induced by τ_2 being in violet color. (Color figure online)

Lemma 4. *For every integer $k \geq 3$ and $\ell \geq 2$, every ℓ-PATH OF k-CLIQUES is properly k-colorable.*

Proof. We prove this by induction on ℓ. For $\ell = 2$, consider a graph G which is a 2-PATH OF k-CLIQUES specified by an arbitrary injective mapping τ_1 on $[k]$. A proper k-coloring χ of G can be specified as follows. The vertices of K_k^1 can be colored properly with k colors where each vertex is assigned a unique color. The incomplete coloring χ can be extended to K_k^2 by assigning $\chi((i, 2))$ the same color as $\chi((\tau_2^{-1}(i), 1))$ for each $i \in [k]$. This is the only possible color that the vertex $(i, 2)$ can take as it is adjacent to every other vertex in K_k^1 except $(\tau_2^{-1}(i), 1)$.

Let the statement be true for $\ell = j$ for some $j > 1$. Consider a graph G which is an arbitrary $(j+1)$-PATH OF k-CLIQUES specified by arbitrary injective mappings $\tau_1, \ldots, \tau_j$ on $[k]$. By induction hypothesis, $G[K_k^1 \cup \ldots \cup K_k^j]$ can be k-colored properly. Let the coloring be χ. It can be extended to K_k^{j+1} by essentially the same arguments as in the base case. The coloring $\chi((i, j+1))$ is assigned the same color as $\chi((\tau_j^{-1}(i), j))$ for each $i \in [k]$. This is the only possible color that the vertex $(i, j+1)$ can take as it is adjacent to every other vertex in K_k^j except $(\tau_j^{-1}(i), j)$. $\qquad\square$

Lemma 5. *For an ℓ-PATH OF k-CLIQUES G specified by injective mappings $\tau_1, \ldots, \tau_{\ell-1}$ and a proper k-coloring χ of G, we have that $\chi((i, 1)) = \chi((\tau_1 \circ \ldots \circ \tau_{\ell-1}(i), \ell))$ for all $i \in [k]$ and $\ell \geq 2$.*

Proof. We prove this by induction on ℓ. For $\ell = 2$, consider a vertex $(i, 2) \in K_k^2$. As there is an anti-matching between K_k^1 and K_k^2, the vertex $(i, 2)$ is only not adjacent to $(\tau_1^{-1}(i), 1)$ in K_k^1. As K_k^1 is a k-clique, the only possible color that $(i, 2)$ can take, is $\chi((\tau_1^{-1}(i), 1))$ and hence the statement follows. Let the statement be true for $\ell = j$ for some $j > 2$. Consider an arbitrary $(j+1)$-PATH OF k-CLIQUES specified by arbitrary injective mappings $\tau_1, \ldots, \tau_j$ on $[k]$. By the same arguments as in the base case, we have that $\chi((i, j+1)) = \chi((\tau_j^{-1}(i), j))$ for each $i \in [k]$. By the induction hypothesis, the claim follows. $\qquad\square$

Construction of Indistinguishable Graphs. For parameters k and ℓ such that ℓ is even, we define the indistinguishability graphs $G_1^{\{\ell,k\}}$ and $G_2^{\{\ell,k\}}$ as follows. Consider two ℓ-PATH OF k-CLIQUES G_1 and G_2 where G_1 is specified by injective mappings $\tau_i = \mathbb{K}$ (where $\mathbb{K}$ represents the identity mapping) for each $i \in [\ell - 1]$ and G_2 is specified by $\tau_i = \mathbb{K}$ for each $i \in [\ell - 1] \setminus \{\ell/2\}$ and $\tau_{\ell/2}$ which is an arbitrary injective mapping on $[k]$ except $\mathbb{K}$. The graph G_1 is transformed to $G_1^{\{\ell,k\}}$ (and G_2 is transformed to $G_2^{\{\ell,k\}}$) by performing the graph transformation as described in the NP-completeness reduction in Sect. 6, i.e. replace each edge with the k-edge gadget. More specifically, the following transformations are performed.

1. For $(a, b) \in [k]^2$ such that $a \neq b$ and $i \in [\ell]$, replace $\{(a, i), (b, i)\}$ with $e_{\{(a,i),(b,i)\}}$.
2. For $(a, b) \in [k]^2$ and $i \in [\ell - 1]$ such that $b \neq \tau_i(a)$, replace $\{(a, i), (b, i+1)\}$ with $e_{\{(a,i),(b,i+1)\}}$.

Both the graphs $G_1^{\{\ell,k\}}$ and $G_2^{\{\ell,k\}}$ have $O(\ell k^3)$ vertices (as each edge of G_1 (and G_2) accounts for k vertices (of the k-clique) and for each $i \in [\ell]$, K_i has $O(k^2)$ edges and for each $j \in [\ell-1]$ the anti-matching between V_j and V_{j+1} has $O(k^2)$ edges).

Observation 2. *The graphs $G_1^{\{\ell,k\}}$ and $G_2^{\{\ell,k\}}$ are properly k-colorable. Moreover, $\Delta_{E(G_1^{\{\ell,k\}})} = \Delta_{E(G_2^{\{\ell,k\}})} = k$. Therefore, any arbitrary k-partial k-coloring is also a proper k-coloring of $G_1^{\{\ell,k\}}$. The same holds true for $G_2^{\{\ell,k\}}$.*

Proof. The graphs G_1 and G_2 are properly k-colorable by Lemma 4. As $G_1^{\{\ell,k\}}$ and $G_2^{\{\ell,k\}}$ are created by applying the same transformation on G_1 and G_2 respectively as in Sect. 6, the graphs $G_1^{\{\ell,k\}}$ and $G_2^{\{\ell,k\}}$ are k-partially k-colorable by Lemma 3, properly k-colorable, and $\Delta_{E(G_1^{\{\ell,k\}})} = \Delta_{E(G_2^{\{\ell,k\}})} = k$ by Observation 1. $\qquad\square$

Fact 1. *For arbitrary k-partial k-colorings χ_1 and χ_2 of $G_1^{\{\ell,k\}}$ and $G_2^{\{\ell,k\}}$ respectively, the following hold:*

1. *$\forall i \in [k]$, $\chi_1((i,1)) = \chi_1((i,\ell))$ in $G_1^{\{\ell,k\}}$,*
2. *$\exists j \in [k]$, $\chi_2((j,1)) \neq \chi_2((j,\ell))$ in $G_2^{\{\ell,k\}}$.*

Proof. Let χ_1' and χ_2' be k-colorings of G_1 and G_2 respectively which are defined as follows:

$$\forall i \in [k], j \in [\ell] : \chi_1'((i,j)) = \chi_1((i,j)),$$
$$\forall i \in [k], j \in [\ell] : \chi_2'((i,j)) = \chi_2((i,j)).$$

By the same arguments as in the proof of Lemma 2, every pair of adjacent vertices is colored properly in G_1 and G_2 and therefore, χ_1' and χ_2' are proper colorings of G_1 and G_2, respectively. By Lemma 5, for each $i \in [k]$, $\chi_1'((i,1)) = \chi_1'((i,\ell))$ in G_1 and therefore, $\chi_1((i,1)) = \chi_1((i,\ell))$ in $G_1^{\{\ell,k\}}$ as well and hence item 1 holds. By the same argument and the fact that $\tau_{\ell/2} \neq \not\!\!\!K$, there exists $j \in [k]$ such that $\chi_2'((j,1)) \neq \chi_2'((j,\ell))$ in G_2 and therefore, $\chi_2((j,1)) \neq \chi_2((j,\ell))$ in $G_2^{\{\ell,k\}}$ and item 2 holds. $\qquad\square$

Fact 2. *For each $i \in [k]$, the vertex $(i,1)$ has the same $3\left(\frac{\ell}{2}-1\right)$-neighborhood in $G_1^{\{\ell,k\}}$ and $G_2^{\{\ell,k\}}$. The same holds true for the vertex (i,ℓ). More precisely, the following holds:*

$$N_{G_1^{\{\ell,k\}}}^{3(\ell/2-1)}[(i,1)] = N_{G_2^{\{\ell,k\}}}^{3(\ell/2-1)}[(i,1)],$$
$$N_{G_1^{\{\ell,k\}}}^{3(\ell/2-1)}[(i,\ell)] = N_{G_2^{\{\ell,k\}}}^{3(\ell/2-1)}[(i,\ell)].$$

Proof. By the construction of the graphs G_1 and G_2, we have that the graphs induced on $(V_1 \cup \ldots \cup V_{\ell/2})$ by G_1 and G_2 are same. This is also true for the set $(V_{\ell/2+1} \cup \ldots V_\ell)$. More precisely, the following holds:

$$G_1[V_1 \cup \ldots \cup V_{\ell/2}] = G_2[V_1 \cup \ldots \cup V_{\ell/2}],$$
$$G_1[V_{\ell/2+1} \cup \ldots \cup V_\ell] = G_2[V_{\ell/2+1} \cup \ldots \cup V_\ell].$$

The graphs G_1 and G_2 only differ in the anti-matching between $V_{\ell/2}$ and $V_{\ell/2+1}$. Therefore, for each $i \in [k]$, the vertex $(i, 1)$ has the same $(\ell/2-1)$-neighborhood in G_1 and G_2. The same holds true for the vertex (i, ℓ), i.e.

$$N_{G_1}^{\ell/2-1}[(i, 1)] = N_{G_2}^{\ell/2-1}[(i, 1)],$$
$$N_{G_1}^{\ell/2-1}[(i, \ell)] = N_{G_2}^{\ell/2-1}[(i, \ell)].$$

By the construction of $G_1^{\{\ell,k\}}$ and $G_2^{\{\ell,k\}}$, as each edge $\{u, v\}$ is replaced consistently with the k-edge gadget $e_{\{u,v\}}$, and the distance between u and v in $e_{\{u,v\}}$ is 3, the vertex $(i, 1)$ (as well as (i, ℓ)) has the same $3(\ell/2 - 1)$-neighborhood in $G_1^{\{\ell,k\}}$ and $G_2^{\{\ell,k\}}$. $\square$

Now, we have all the tools to prove Theorem 2.

Proof (of Theorem 2). We prove this by contradiction. Let us assume that there exists an integer $k_0 \geq 3$ and a **LOCAL** algorithm $\mathcal{A}$ which k_0-partially colors every n-vertex k_0-partially k_0-colorable graph in $o(n/k_0^3)$ rounds. Then, there exists an integer $\ell_0 > 0$ such that for each even $\ell \geq \ell_0$, on every $O(\ell k_0^3)$-vertex k_0-partially k_0-colorable graph, the algorithm $\mathcal{A}$ terminates in at most $3(\ell/2-1)$ rounds.

Now let us create indistinguishability graphs $G_1^{\{\ell,k\}}$ and $G_2^{\{\ell,k\}}$ for k_0 and some even arbitrary $\ell \geq \ell_0$ as described in the section. For both $G_1^{\{\ell,k\}}$ and $G_2^{\{\ell,k\}}$, the ids of the vertices are set as follows.

1. For $a \in [k_0]$, $i \in [\ell]$, $\mathrm{id}((a, i)) := (a, i)$.
2. For $(a, b, j) \in [k_0]^3$ and $i \in [\ell-1]$ such that $b \neq \tau_i(a)$, $\mathrm{id}(((a, i), (b, i+1), j)) := ((a, i), (b, i+1), j)$.

The ids of the vertices of $G_1^{\{\ell,k\}}$ (and $G_2^{\{\ell,k\}}$) are unique and can be encoded in $O(\log n)$ bits and therefore can be mapped uniquely to the set $\{1, 2, \ldots, n^d\}$ for some constant d. Hence the identifiers are valid.

Let the colorings output by the algorithm $\mathcal{A}$ for $G_1^{\{\ell,k\}}$ and $G_2^{\{\ell,k\}}$ be χ_1 and χ_2, respectively. For each $i \in [k]$, the vertex with id $(i, 1)$ has the same $(3\ell/2 - 1)$-radius view in $G_1^{\{\ell,k\}}$ and $G_2^{\{\ell,k\}}$ (thanks to Fact 2 and the assignment of ids). The same is also true to the vertex with id (i, ℓ). Therefore, algorithm $\mathcal{A}$ at vertex with id $(i, 1)$ cannot distinguish between $G_1^{\{\ell,k\}}$ or $G_2^{\{\ell,k\}}$ after $(3\ell/2 - 1)$ rounds and outputs the same color for $(i, 1)$ (and (i, ℓ)) in $G_1^{\{\ell,k\}}$ and $G_2^{\{\ell,k\}}$.

More specifically, the following holds for vertices with ids $(i,1)$ and (i,ℓ) for each $i \in [k_0]$:

$$\chi_1((i,1)) = \chi_2((i,1)) \text{ and } \chi_1((i,\ell)) = \chi_2((i,\ell)).$$

Item 1 of Fact 1 implies that $\chi_2((i,1)) = \chi_2((i,\ell))$ for each $i \in [k_0]$ in $G_2^{\{\ell,k\}}$. However, this leads to contradiction, as due to item 2 of Fact 1, there must exist $j \in [k_0]$ such that $\chi_2((j,1) \neq \chi_2((j,\ell))$. $\qquad\square$

8 Conclusions

We systematically investigated the complexity of k-partial k-coloring in both classical and distributed settings, revealing fundamental limitations of this generalized coloring paradigm. Our results demonstrate a sharp threshold between the tractable $(k+1)$-color case and the computationally hard k-color variant. We proved that deciding k-partial k-colorability is NP-complete for all constants $k \geq 3$, even for graphs where partial coloring reduces to proper coloring (as characterized by Δ_E). This contrasts with the linear-time greedy algorithm for $(k+1)$-colorings.

In the LOCAL model, we established an $\Omega(n)$-round lower bound for computing k-partial k-colorings, showing an exponential separation from the $O(\log^2 k \cdot \log n)$-round algorithms for $(k+1)$-colorings, using graph constructions based on *paths of cliques* and *indistinguishability arguments*. To this end, we extend the connection of this weaker version of coloring to proper coloring for proving locality as well as classical lower bounds.

Acknowledgements. J.B. was supported by the European Union (ERC, POCOCOP, 101071674). The views and opinions expressed are those of the authors only and do not necessarily reflect those of the European Union or the European Research Council Executive Agency. Neither the European Union nor the granting authority can be held responsible for them. A.D. was supported in part by the Research Council of Finland, Grant 363558. A.G. was supported by the Lendület "Momentum" program of the Hungarian Academy of Sciences (MTA). N.J. was supported by the research grants PRIMUS/24/SCI/008 and UNCE/24/SCI/022 of Charles University, and SVV–2025–260822.

References

1. Awerbuch, B., Goldberg, A.V., Luby, M., Plotkin, S.A.: Network decomposition and locality in distributed computation. In: 30th IEEE Symposium on Foundations of Computer Science (FOCS), pp. 364–369 (1989)
2. Balliu, A., Hirvonen, J., Lenzen, C., Olivetti, D., Suomela, J.: Locality of not-so-weak coloring. In: Censor-Hillel, K., Flammini, M. (eds.) SIROCCO 2019. LNCS, vol. 11639, pp. 37–51. Springer, Cham (2019). https://doi.org/10.1007/978-3-030-24922-9_3

3. Barenboim, L., Elkin, M., Goldenberg, U.: Locally-iterative distributed $(\Delta + 1)$-coloring and applications. J. ACM **69**(1), 5:1–5:26 (2022)
4. Bourreau, Y., Brandt, S., Nolin, A.: Faster distributed Δ-coloring via ruling subgraphs. CoRR arxiv:2503.04320 (2025)
5. Bousquet, N., Feuilloley, L., Zeitoun,: Local certification of local properties: tight bounds, trade-offs and new parameters. In: 41st International Symposium on Theoretical Aspects of Computer Science (STACS), vol. 289 of LIPIcs, pp. 21:1–21:18. Schloss Dagstuhl - Leibniz-Zentrum für Informatik (2024)
6. Brandt, S.: An automatic speedup theorem for distributed problems. In: 38th ACM Symposium on Principles of Distributed Computing (PODC), pp. 379–388 (2019)
7. Brooks, R.L.: On colouring the nodes of a network. In: Mathematical Proceedings of the Cambridge Philosophical Society, vol. 37, pp. 194–197. Cambridge University Press (1941)
8. Das, A., Fraigniaud, P., Rosén, A.: Distributed partial coloring via gradual rounding. In: 27th International Conference on Principles of Distributed Systems (OPODIS), vol. 286 of LIPIcs, pp. 30:1–30:22. Schloss Dagstuhl - Leibniz-Zentrum für Informatik (2023)
9. Fischer, M., Halldórsson, M.M., Maus, Y.: Fast distributed Brooks' theorem. In: Bansal, N., Nagarajan, V. (eds.) Proceedings of the 2023 ACM-SIAM Symposium on Discrete Algorithms, SODA 2023, Florence, Italy, 22–25 January 2023, pp. 2567–2588. SIAM (2023)
10. Fraigniaud, P., Heinrich, M., Kosowski, A.: Local conflict coloring. In: Dinur, I. (ed.) 57th IEEE Symposium on Foundations of Computer Science (FOCS), pp. 625–634 (2016)
11. Garey, M.R., Johnson, D.S.: Computers and Intractability: A Guide to the Theory of NP-Completeness. W. H. Freeman, San Francisco (1979)
12. Ghaffari, M., Grunau, C.: Faster deterministic distributed MIS and approximate matching. In: 55th ACM Symposium on Theory of Computing (STOC), pp. 1777–1790 (2023)
13. Ghaffari, M., Grunau,: Near-optimal deterministic network decomposition and ruling set, and improved MIS. In: 2024 IEEE 65th Annual Symposium on Foundations of Computer Science (FOCS), pp. 2148–2179 (2024)
14. Ghaffari, M., Hirvonen, J., Kuhn, F., Maus, Y.: Improved distributed Delta-coloring. In: Newport, C., Keidar, I. (eds.) Proceedings of the 2018 ACM Symposium on Principles of Distributed Computing, PODC 2018, Egham, United Kingdom, 23–27 July 2018, pp. 427–436. ACM (2018)
15. Ghaffari, M., Kuhn, F.: Deterministic distributed vertex coloring: simpler, faster, and without network decomposition. In: 62nd IEEE Symposium on Foundations of Computer Science (FOCS), pp. 1009–1020 (2021)
16. Göös, M., Suomela, J.: Locally checkable proofs in distributed computing. Theory Comput. **12**(1), 1–33 (2016)
17. Kuhn, F.: Weak graph colorings: distributed algorithms and applications. In: 21st ACM Symposium on Parallelism in Algorithms and Architectures (SPAA), pp. 138–144 (2009)
18. Kuhn, F., Moscibroda, T., Wattenhofer, R.: What cannot be computed locally! In: Proceedings of the 23rd Annual ACM Symposium on Principles of Distributed Computing (PODC), pp. 300–309. ACM (2004)
19. Linial, N.: Locality in distributed graph algorithms. SIAM J. Comput. **21**(1), 193–201 (1992)

20. Maus, Y., Tonoyan, T.: Local conflict coloring revisited: linial for lists. In: 34th International Symposium on Distributed Computing (DISC), vol. 179 of LIPIcs, pp. 16:1–16:18. Schloss Dagstuhl - Leibniz-Zentrum für Informatik (2020)
21. Naor, M., Stockmeyer, L.J.: What can be computed locally? SIAM J. Comput. **24**(6), 1259–1277 (1995)
22. Panconesi, A., Srinivasan, A.: The local nature of Δ-coloring and its algorithmic applications. Combinatorica **15**(2), 255–280 (1995)
23. Panconesi, A., Srinivasan, A.: On the complexity of distributed network decomposition. J. Algor. **20**(2), 356–374 (1996)
24. Peleg, D.: Distributed Computing: A Locality-Sensitive Approach. Discrete Mathematics and Applications. Society for Industrial and Applied Mathematics (2000)
25. Rozhoň, V., Ghaffari, M.: Polylogarithmic-time deterministic network decomposition and distributed derandomization. In: 52nd ACM Symposium on Theory of Computing (STOC), pp. 350–363 (2020)
26. Wattenhofer, R., Uitto, J.: Locality lower bounds (Chapter 8). In: Lecture notes. From the course "Principles of Distributed Computing (PODC) All-Stars;; at ETH Zürich (2025). Accessed 17 Sept 2025

Space Efficient Algorithms for Parameterised Problems

Sheikh Shakil Akhtar[(✉)][ID], Pranabendu Misra[ID], and Geevarghese Philip[ID]

Chennai Mathematical Institute, Chennai, India
{shakil,pranabendu,gphilip}@cmi.ac.in

Abstract. We study "space efficient" FPT algorithms for graph problems with limited memory. Let n be the size of the input graph and k be the parameter. We present algorithms that run in time $f(k) \cdot n^{\mathcal{O}(1)}$ and use $g(k) \cdot (\log n)^{\mathcal{O}(1)}$ working space, where f and g are functions of k alone, for k-Path, MaxLeaf SubTree and Multicut in Trees. These algorithms are motivated by big-data settings where very large problem instances must be solved, and using $n^{\mathcal{O}(1)}$ memory is prohibitively expensive. They are also theoretically interesting, since most of the standard methods tools, such as deleting a large set of vertices or edges, are unavailable, and we must develop different ways to tackle them.

1 Introduction

With the increasing use of big data in practical applications, the field of *space-efficient algorithms* has increased in importance. Traditionally, the time required by an algorithm has been the primary focus of analysis. However, when dealing with large volumes of data we must also pay attention to the working space required to run the algorithm, otherwise it will be impossible to run the algorithm at all.

A few models of computation for space-efficient algorithms have been proposed, the most prominent among them being *streaming algorithms* [4]. Recently streaming algorithms have been studied in the parameterised setting [16], to solve parameterized versions of NP-hard problems. This model places strong restrictions on how many times one can access the input data. Consequently, we get strong lower bounds on the amount of space required, and this restricts the set of problems that can be efficiently solved [23] in this model. In particular, it is well-known that for most problems on graphs with n vertices, the space required by a streaming algorithm is at least $n \cdot (\log n)^{\Omega(1)}$; this is true even for something as basic as detecting if the input graph contains a cycle [28].

A more relaxed model allows for the input to be read as many times as needed, but restricts the amount of working space that the algorithm can use. This is the model that we consider in this work, where we study various problems on graphs. Let n be the number of vertices of the input graph and let k be the *parameter* which is—typically—the size of the solution that we are looking for. We study algorithms that run in time $f(k) \cdot n^{\mathcal{O}(1)}$ and use $g(k) \cdot (\log n)^{\mathcal{O}(1)}$ working

E. Di Giacomo and D. Mondal (Eds.): WALCOM 2026, LNCS 16444, pp. 263–275, 2026.
https://doi.org/10.1007/978-981-95-7127-7_18

space, where f and g are functions of k alone. We say that these algorithms are *Fixed-parameter tractable (FPT)* [18] and space-efficient. Note that with so little working space, many standard algorithmic tools and techniques become unavailable. Indeed, even something as simple as deleting some edges from the graph could become non-trivial to implement. This is because keeping track of a large arbitrary set of edges that was deleted, will require more memory than is allowed by the model. Therefore, developing algorithms under this model seems to require new ideas and methods.

The class of problems solvable in LOGSPACE, and related complexity classes are very well studied in computer science [6]. There has been a lot of work on streaming algorithms for problems on graphs [22, 28], and more recently on parameterised streaming algorithms [27]. Space efficient FPT algorithms have been studied earlier [8–12, 15, 20, 21, 25] for VERTEX COVER, d-HITTING SET, EDGE-DOMINATING SET, MAXIMAL MATCHING, FEEDBACK VERTEX SET, PATH CONTRACTION, LIST COLORING and other problems. Furthermore, a theory of hardness (primarily based on strictly using $g(k) \cdot O(\log n)$-space) is being developed [9–12, 20].

In this paper, we present algorithms that use $g(k) \cdot \log^{O(1)} n$-space and take $f(k) \cdot n^{O(1)}$-time for the following problems.

k-PATH
Input: An undirected graph $G = (V, E)$ and an integer k.
Parameter: k
Question: Does G have a path on k vertices?

MAXLEAF SUBTREE
Input: An undirected graph $G = (V, E)$ and an integer k.
Parameter: k
Question: Does G have a subtree with at least k leaves?

MULTICUT IN TREES
Input: An undirected tree $T = (V, E), n = |V|$, a collection H of m pairs of nodes in T and an integer k.
Parameter: k
Question: Does T have an edge subset of size at most k whose removal separates each pair of nodes in H?

We obtain the following results:

Theorem 1. *There is a deterministic algorithm, that solves the k-PATH problem, runs in time $n^{O(1)} \cdot 2^{k^2} \cdot k!$ and uses $O(2^{k^2} \cdot k! \cdot k \cdot \log n)$ working space.*

Theorem 2. *There is a deterministic algorithm which solves the MAXLEAF SUBTREE problem, in time $n^{O(1)} \cdot 4^k$ and uses $O(4^k \cdot k \cdot \log n)$ working space.*

Theorem 3. *There is a deterministic algorithm which solves the MULTICUT IN TREES problem, in time $n^{O(1)} \cdot 2^k$ and uses $O(2^k \cdot k \cdot \log n)$ working space.*

To the best of our knowledge, these are the first results on the above mentioned problems in the bounded space setting. While the above problems have theoretical interests, they are also used in practical applications as well. For example, the k-PATH problem can be used in studying protein-protein interaction [2], whereas the MAXLEAF SUBTREE problem can be used in the study of phylogenetic networks [19]. While it may be difficult to find an application of the MULTICUT IN TREES problem, as real world scenario often don't occurs as tree, one can get application of the more general MULTICUT IN GRAPHS problem. For instance, in [7], the authors use the MULTICUT IN GRAPHS problem to study reliability in communication networks.

2   Preliminaries

For any positive integer, i, we denote the set $\{1, \ldots, i\}$ by $[i]$. We will use standard notations from graph theory. For a graph $G = (V, E)$, with V as the vertex set and E as the edge set, we will assume an order on V and hence on the neighbourhood of every vertex in G. This will help us in avoiding many of the issues that can arise due to restriction in working space. Also, we use n to denote $|V|$. We consider V to be the set $\{v_1, v_2, \ldots, v_n\}$. For any $S \subseteq V$, we use $G[S]$ to denote the induced subgraph of G on S. And we use $G - S$ to denote $G[V \setminus S]$. Also, if $S = \{v\}$, for some $v \in V$, then we simply write $G - v$, instead of $G - \{v\}$. For any vertex $v \in V$, $N(v) := \{w \in V | vw \in E\}$ and $N[v] := N(v) \cup \{v\}$. A path on k vertices will have length $k - 1$, i.e., the number of edges in it.

A celebrated result of Reingold [29], gives a $\mathcal{O}(\log n)$ space, polynomial time deterministic algorithm for undirected st-connectivity, where s and t are two vertices in an input-graph G on n vertices. Let's call it $\mathcal{A}_{con}$.

Colour Coding: Colour Coding is an algorithmic technique to detect if a given input graph has a smaller subgraph, isomorphic to another graph ([3,5]). In other words, given a pattern k-vertex "pattern" graph H and an n-vertex input graph G, the goal is to find a subgraph of G isomorphic to H ([18]). While this techniqe can be used to efficiently detect small subgraphs like paths and cycles, it may not result in efficient algorithms for arbitrary H; for example, if H is a complete graph (the k-CLIQUE problem).

We implement the technique by the standard method of using a family of *universal hash functions*. In particular, if $G = (V, E)$ is an input graph with n vertices and the vertices are labeled as $\{v_1, v_2, \cdots, v_n\}$, then when we need to find a vertex subset of size k, we use the following family of hash functions. Let p be a prime number greater than n. We define the following family of hash functions.

$$\mathcal{H}_u := \{h_{a,b} : \{1, \ldots, n\} \to \{1, \ldots, k^2\} \mid a, b \in \{0, \ldots, p-1\}, a \geq 1, \forall i \in [n],$$
$$h_{a,b}(i) = ((ai + b) \bmod p) \bmod k^2\} \quad (1)$$

Given n and p, each of the above functions can be evaluated in $\mathcal{O}(n)$ time and $\mathcal{O}(\log n)$ space. The set $[k^2]$ is called the set of colours. Each member of $\mathcal{H}_u$ is a called a *colouring function* and given a colouring function, $h_{a,b}$, for each $i \in [n]$, $h_{a,b}(i)$ is called the colour of the vertex v_i.

A subset S of V will be called *properly coloured* or *colourful* under a given colouring function, if every member of S has a distinct colour, i.e., the function is injective when restricted to S. The family $\mathcal{H}_u$ has the property of being a universal hash family [13]. This means that the probability that a given subset of vertices, say S, where $|S| \leq k$, is *not* colourful under a colouring function chosen uniformly at random from the family $\mathcal{H}_u$, is at most $1/2$ (see, e.g., [14,17]). Thus, there is some function from the family $\mathcal{H}_u$, which when restricted to S will be injective. Given the prime number p, n and k, we can enumerate the functions in $\mathcal{H}_u$, lexicographically with respect to the pair (a, b), in time $\mathcal{O}(n^3)$ and space $\mathcal{O}(\log n)$, deterministically.

How Do We Get a Sufficiently Large p, Deterministically, Given the Space Restrictions? By Bertrand's Postulate (later Theorem), for any $n \geq 2$ there exists a prime p, such that $n < p < 2n$. Therefore, by simply testing the primality of each integer between $n+1$ and $2n-1$, we can obtain a prime $p > n$. As we will be dealing with an integer of absolute-value at most $2n$, thus the number of bits required to represent these integers will be at most $\log 2n$ which is $\mathcal{O}(\log n)$. To test the primality of a number q, we simply check if any integer $r < q$ divides it, which can be done in $\mathcal{O}(q \log n)$ time and $\mathcal{O}(\log q)$ space. Thus, by simply testing each integer between $n+1$ and $2n-1$, we have a deterministic algorithm that runs in time $\mathcal{O}(n^2 \log n)$ and space $\mathcal{O}(\log n)$, and outputs a prime $p > n$.

The celebrated AKS primality test by Agrawal et al. [1] gives a deterministic algorithm to check if an input integer q is prime or not, where the running time (and hence the space complexity) is $\log^{\mathcal{O}(1)} q$. Then we also have a deterministic algorithm to obtain a prime number $p > n$ in time $n \cdot \log^{\mathcal{O}(1)} n$ and space $\mathcal{O}(\log n)$.

Deleting Vertices and Edges: Typically when deleting vertices or edges from a graph, we make a copy of the given graph with those vertices or edges deleted. However, in our setting we cannot simply make a copy of a subgraph of the input graph, unless we can guarantee its size to be bounded by some function of k only. If on the other hand, as we shall encounter going forward, we can guarantee that the size of the set of deleted vertices (or edges) can be bounded by a function of k only, then we can *simulate* the deletion, i.e., we can make the algorithm act as if the set to be deleted has actually been deleted.

Let $S \subseteq V$ be a set of vertices that need to be deleted and $|S| \leq \alpha(k)$, where α is an increasing function from $\mathbb{N}$ to $\mathbb{N}$. As its size is bounded by a function of k only, so we can explicitly keep a copy of S in our working space and mark it as deleted. Thus, if we need to select a vertex from $G - S$, we go through elements in $V(G)$ and check if they are from S or not. Similarly, if $F \subseteq E$ be a set of edges that needs to be deleted, such that $|F| \leq \beta(k)$, where β is an increasing function from $\mathbb{N}$ to $\mathbb{N}$, then we can simulate the deletion by storing a copy of

F and marking it as deleted. If G' is the subgraph of G which is obtained after deletion of F from E, then we can simply select edges of G' by accessing the adjacency matrix of G via an oracle which checks if and edge is in F or not. This can be clearly implemented in $\mathcal{O}(n^2)$ time and $\mathcal{O}(\beta(k) \cdot \log n)$ space.

We will now proceed to describe the problems along with the algorithms.

3 k-PATH

Input: An undirected graph $G = (V, E)$ and an integer k.
Parameter: k
Question: Does G have a path on at least k vertices?

Theorem 1. *There is a deterministic algorithm, that solves the k-PATH problem, runs in time $n^{\mathcal{O}(1)} \cdot 2^{k^2} \cdot k!$ and uses $\mathcal{O}(2^{k^2} \cdot k! \cdot k \cdot \log n)$ working space.*

We will use the colour coding technique to design our algorithm. If G is a YES-instance, then there exists some subset S of V such that $|S| = k$ and there is a path of length $k-1$ on vertices of S. As mentioned earlier, there exists some colouring function $h_{a,b}$ from the family $\mathcal{H}_u$, such that $h_{a,b}$ will be injective when restricted to S. Thus, we can use the functions from $\mathcal{H}_u$ one by one to identify S. And we have already seen that it can be done in $\mathcal{O}(n^3)$ time and $\mathcal{O}(\log n)$ space. We fix one such colouring function, and describe the algorithm with respect to it.

Description of the Main Algorithm: We enumerate all possible permutations of all possible subsets of k elements chosen from the set $\{1, \ldots, k^2\}$. Let us consider one such permutation $\{c_1, \ldots, c_k\}$. For each $i \in [k]$, we define $n_i = |\{$Vertices of G which have the colour $c_i\}|$. If for some $i \in [k]$, $n_i = 0$, then that is not a valid permutation and we move on to the next permutation of k colours.

Construct an auxiliary path $\mathcal{P}$, such that $V(\mathcal{P}) = \{c_i \mid 1 \leq i \leq k\}$ and $E(\mathcal{P}) = \{c_i c_{i+1} \mid 1 \leq i \leq k-1\}$ (a path on the k colours). We will use Algorithm 1 to find a colourful path in G of size at least k, if one such path exists in G. If Algorithm 1 returns YES for any input, then we return YES. Otherwise, if Algorithm 1 return NO for all permutations of k-subsets of $[k^2]$, then we return NO.

In Algorithm 1, we (implicitly) construct an auxiliary graph $G^\star$ using the path $\mathcal{P}$ as follows: $V(G^\star) \leftarrow \{s, t\} \cup V(G)$ where s, t are two new vertices, $N(s) \leftarrow \{v \in V(G) \mid$ colour of v is $c_1\}$, $N(t) \leftarrow \{v \in V(G) \mid$ colour of v is $c_k\}$, and $E(G^\star) \leftarrow \{sv \mid v \in N(s)\} \cup \{tv \mid v \in N(t)\} \cup_{i=1}^{k-1} \{vw \mid vw \in E(G) \wedge$ colour of v is $c_i \wedge$ colour of w is $c_{i+1}\}$.

Observe that, any path between s and t in $G^\star$, if one exists, has at least k internal vertices, by construction. Otherwise, if there is a path with fewer vertices, then as s is adjacent to vertices of colour c_1 only and t is adjacent to vertices of colour c_k only, hence there must be an edge in this path between two vertices of colours c_i and c_j where $|i - j| \geq 2$, which contradicts the construction.

Lemma 1. *If Algorithm 1 is correct and G is a* No-*instance, then the above algorithm will return* No.

Proof. Suppose, the graph G is a No-instance, i.e. we don't have a path of length at least $k-1$ in G; which implies that we cannot have a colourful path of length at least $k-1$. Thus, for any permutation of any k-subset of the k^2 colours, one cannot get any properly coloured path of length at least $k-1$, starting from a vertex of the the first colour and ending at a vertex of the last colour. Thus, there is no path from s to t in $G^\star$. Hence, the algorithm $\mathcal{A}_{con}$ will return No. So our algorithm will correctly return No as the answer.

Algorithm 1. Finding a colourful path

1: **FindAPath($\mathcal{P}$)**
2: Add two vertices s and t which are not in the input graph G.
3: $N(s) \leftarrow \{v \in V(G) \mid \text{colour of } v \text{ is } c_1\}$
4: $N(t) \leftarrow \{v \in V(G) \mid \text{colour of } v \text{ is } c_k\}$
5: Construct an undirected graph $G^\star$ as follows.
6: $V(G^\star) \leftarrow \{s, t\} \cup V(G)$
7: $E(G^\star) \leftarrow \{sv \mid v \in N(s)\} \cup \{tv \mid v \in N(t)\} \cup_{i=1}^{k-1} \{vw \mid vw \in E(G) \wedge$ colour of v is $c_i \wedge$ colour of w is $c_{i+1}\}$
8: Pass the information of $G^\star$ to $\mathcal{A}_{con}$ to check for the connectivity of s and t
9: If s and t are connected in $G^\star$, then return Yes
10: If s and t are not connected in $G^\star$, then return No

---

Note that in Algorithm 1, the contruction of $G^\star$ is not done explicitly, as we do not have enough space for it. Instead, we provide access to their adjacency matrix via an oracle that can be implemented in $O(\log n)$-space and polynomial time. We first introduce vertices s and t which are not already in $V(G)$. Their neighbourhoods can be determined in $\mathcal{O}(n)$ time and $\mathcal{O}(\log n)$ space by scanning through $V(G)$ and determining the colour of each vertex. As for determining the other edges, then note that those vertices which do not have the colours in the input auxiliary path $\mathcal{P}$ are considered isolated. And the vertices which have a colour, say $c^\star$, from the set $\{c_1, c_2, \ldots, c_{k-1}\}$ then we only consider their neighbours which have the colour $c^\star + 1$. It is with this oracle access to $G^\star$ that we call $\mathcal{A}_{con}$ on $G^\star, s$ and t.

We prove the correctness of Algorithm 1 in the following lemma.

Lemma 2. *There exists a path on k-vertices in G, with the same colour config-uration as $\mathcal{P}$ if and only if, s and t are connected in $G^\star$.*

Proof. Suppose, G has path P with the same colour configuration as $\mathcal{P}$, i.e., $V(P) = \{w_i \mid 1 \le i \le k\}$, $E(P) = \{w_i w_{i+1} \mid 1 \le i \le k-1\}$ and for each $i \in [k]$, the colour of w_i is c_i. Then, by construction there exists a path from s to t in $G^\star$, as $w_1 \in N(s)$ and $w_k \in N(t)$. Therefore, the algorithm $\mathcal{A}_{con}$ on $G^\star, s$ and t will return Yes.

Conversely, suppose that s and t are connected in $G^\star$. Then, there exists a path between s and t in $G^\star$, say P'. Recall that, by the construction of $G^\star$, any path from s to t has at least k internal vertices. Observe that, the internal vertices and edges of P' are also present in G, and thus we obtain a path of length at least k in G.

Lemma 3. *A single run of Algorithm 1 takes $n^{\mathcal{O}(1)}$ time and $\mathcal{O}(k \cdot \log n)$ space.*

Proof. We know from [29], that a call to $\mathcal{A}_{con}$ will take up polynomial time and $\mathcal{O}(\log n)$ space. Apart from that, the rest of Algorithm 1 clearly takes up polynomial time. As for space then we need $\mathcal{O}(k \cdot \log n)$ for the description of $\mathcal{P}$, the vertices s and t and their neighbours.

Lemma 4. *The above main algorithm takes $n^{\mathcal{O}(1)} \cdot 2^{k^2} \cdot k!$ time and $\mathcal{O}(2^{k^2} \cdot k! \cdot k \cdot \log n)$ space.*

Proof. There are most $2^{k^2} \cdot k!$ possible permutations of k-elements chosen from the set $[k^2]$ and hence that many choices for the auxiliary path $\mathcal{P}$. We list all of them. The space needed for that is $2^{k^2} \cdot k! \cdot \log n$ and time is $2^{k^2} \cdot k!$.

For every choice of $\mathcal{P}$, we need to call Algorithm 1 at most once. By Lemma 3, the total time taken is $n^{\mathcal{O}(1)} \cdot 2^{k^2} \cdot k!$ and the space used it $\mathcal{O}(2^{k^2} \cdot k! \cdot k \cdot \log n)$.

We thus have a proof of Theorem 1.

Theorem 1. *There is a deterministic algorithm, that solves the k-PATH problem, runs in time $n^{\mathcal{O}(1)} \cdot 2^{k^2} \cdot k!$ and uses $\mathcal{O}(2^{k^2} \cdot k! \cdot k \cdot \log n)$ working space.*

4   Maxleaf Subtree

Input: An undirected graph $G = (V, E)$ and an integer k.
Parameter: k
Question: Does G have a subtree with at least k leaves?

Theorem 2. *There is a deterministic algorithm which solves the MAXLEAF SUBTREE problem, in time $n^{\mathcal{O}(1)} \cdot 4^k$ and uses $\mathcal{O}(4^k \cdot k \cdot \log n)$ working space.*

We will adapt the algorithm used in [26] to our bounded space setting and give a sketch of our algorithm along with the key ideas involved. The pseudocode along with other relevant details will be provided in the full version of the paper.

Before we describe the algorithm, we will introduce certain terms from [26], which will be needed.

Given a rooted tree T, we denote its root by $root(T)$. The set of leaves of T will be denoted by $leaves(T)$. A tree with k leaves will be called k-leaf tree. A non-leaf vertex of a tree is called an *inner vertex*. For a graph G and a rooted subtree T, we call T to be *inner-maximal* rooted tree if for every inner vertex v of T, $N_G(v) \subseteq V(T)$.

- For rooted trees T and T', we say that T' extends T, denoted by $T' \succeq T$, iff $root(T') = root(T)$ and T is an induced subgraph of T'. We write $T' \succ T$, when $T' \succeq T$ and $T' \neq T$.
- Given a rooted tree T, the algorithm distinguishes its leaves into two kinds. The *red* leaves R of T, are those which will remain as leaves in any other tree which extends T. The *blue* leaves B of T are those which may be inner vertices for some other tree which extends T.
- A *leaf-labelled tree* is 3-tuple (T, R, B), such that T is a rooted tree, $R \cup B = leaves(T)$ and $R \cap B = \emptyset$. A leaf-labelled rooted tree (T, R, B) is called *inner-maximal leaf-labelled rooted tree*, if T is inner-maximal. If (T, R, B) is a leaf-labelled tree and T' is a rooted tree such that $T' \succeq T$ and $R \subseteq leaves(T')$, we say that T' is a *(leaf-preserving) extension* of (T, R, B) denoted by $T' \succeq (T, R, B)$. A leaf-labelled rooted tree (T', R', B') *extends* a leaf-labelled rooted tree (T, R, B) , denoted by $(T', R', B') \succeq (T, R, B)$, iff $T' \succeq (T, R, B)$ and $R \subseteq R'$.

For each $v \in V(G)$, we define T_v be the tree rooted at v, such that $V(T_v) = N[v]$ and $E(T_v) = \{vw | w \in N(w)\}$. Notice that in order to describe T_v, we need not store a copy of the entire tree but suffice with just knowing the root.

Description of the Main Algorithm: We first check if there exists a vertex of degree at least k. If there is one such vertex, we return YES. Otherwise, we are in the case where each vertex in G has degree at most k and for every $v \in V(G)$, we call Algorithm 2, with the input $(v, \emptyset, N(v))$. Algorithm 2 will take $(root(T), R, B)$ as input for a leaf-labelled inner-maximal rooted tree (T, R, B) and check if it can be extended to a k-leaf rooted tree. Thus with $(v, \emptyset, N(v))$ as input, Algorithm 2 will check if there exists a k-leaf rooted subtree in G with v as its root.

Sketch of Algorithm 2: At first we check if the total number of leaves in the input tree is at least k. If so, then we return YES. If not and if there are no blue leaves left, then we return NO. Otherwise, we choose a blue vertex, say u. Notice that there are exactly two ways to treat u. In one case, u will remain a leaf in any further extension of the input tree and as such will become a read leaf. We make a recursive call corresponding to this. The other case is when u is no longer a leaf and it is possible to extend the input tree by adding neighbours of u not already in the tree. We make another recursive call corresponding to this. We make sure that Algorithm 2 always passes inner maximal trees as inputs to the recursive calls, when given an inner maximal tree as input. The key idea here is that, if the input is a YES instance, then it is enough to check if there is a inner maximal tree with at least k leaves.

Upon careful analysis, we observe that, if we were to use the algorithm in [26] without our modifications then we would have to counstruct the tree T explicitly, which would require $\Omega(n)$ space.

This is how we prove Theorem 2.

Algorithm 2. Finding a rooted tree with many leaves

```
 1: MaxLeaf(root(T), R, B)
 2: if |R| + |B| ≥ k then
 3:    return Yes
 4: end if
 5: if B = ∅ then
 6:    return No
 7: end if
 8: Choose u ∈ B
 9: if MaxLeaf(root(T), R ∪ {u}, B \ {u}) then
10:    return Yes//The branch where u remains a leaf
11: end if
12: B ← B \ {u}
13: N ← ExtendTree(u, root(T), R ∪ B ∪ {u}) //Let N be set of neighbours of u
    outside of T
14: i ← 0
15: w ← The first member of N
16: while (|leaves(T)| ≠ k) ∨ (i ≠ |N|) do
17:    Add w ∈ T as a neighbour of u
18:    w ← Next member of N in their order
19:    i ← i + 1
20: end while
21: if |leaves(T)| = k then
22:    return Yes
23: end if
24: while |N| = 1 do
25:    //Follow paths
26:    Let u be the unique element of N
27:    N' ← ExtendTree(u, root(T), R ∪ B ∪ {u})
28:    i ← 0
29:    w ← The first member of N'
30:    while (|leaves(T)| ≠ k) ∧ (i ≠ |N'|) do
31:       Add w ∈ T as a neighbour of u
32:       w ← Next member of N' in their order
33:       i ← i + 1
34:    end while
35:    if |leaves(T)| = k then
36:       return Yes
37:    end if
38:    N ← N'
39: end while
40: if N = ∅ then
41:    return No
42: end if
43: return MaxLeaf(root(T), R, B ∪ N)
```

Algorithm 3. Finding neighbours of a leaf to extend

1: ExtendTree$(u, root(T), leaves(T))$
2: **if** $u \notin leaves(T)$ **then**
3:     return "Wrong Input"
4: **end if**
5: $r \leftarrow root(T)$
6: $N \leftarrow \emptyset$
7: Let G^* be the induced subgraph of G on $V(G) \setminus leaves(T)$
8: (We construct G^* implicitly)
9: **for** $w \in N(u) \setminus leaves(T)$ **do**
10:     Use $\mathcal{A}_{con}$ to check for connectivity of w and r in G^*
11:     **if** w and r are not connected in G^* **then**
12:         //By inner-maximality of T
13:             $N \leftarrow N \cup \{w\}$
14:     **end if**
15: **end for**
16: return N

---

5 Multicut in Trees

Input: An undirected tree $T = (V, E), n = |V|$, a collection H of m pairs of nodes in T and an integer k.

Parameter: k

Question: Does T have an edge subset of size at most k whose removal separates each pair of nodes in H?

Theorem 3. *There is a deterministic algorithm which solves the* Multicut In Trees *problem, in time* $n^{\mathcal{O}(1)} \cdot 2^k$ *and uses* $\mathcal{O}(2^k \cdot k \cdot \log n)$ *working space.*

We can assume that the elements of H are in order. Let $H = \{(a_i, b_i) | 1 \le i \le m\}$.

We will adapt the algorithm used in [24] to our bounded space setting. Due to lack of space, the pseudocode of the relevant algorithms and the proofs of correctness, running time and space usage have not been provided here and readers are requested to refer to the full version of the paper. We give a sketch of the algorithm.

Before running the algorithm, we root T at an arbitrary vertex, say r. For two vertices u and v, their *least common ancestor* is a vertex w, such that it is an ancestor to both u and v and has the greatest depth in T measured by the distance from the root. Let E^* be the solution we will be constructing and initialise it to $\emptyset$. At first we check if deleting the edges already in E^*, will separate all the pairs in H. If so, then we return Yes or else we identify the pair with the deepest least common ancestor. Once we have identified the pair with the deepest least common ancestor with respect to the root, say $\{u, v\}$, we check if the lowest common ancestor, say w, is one of the pair $\{u, v\}$. If it is so, then we delete the edge incident on w and lying in the uniquely determined path between u and v and recursively call the function with the appropriate values. Otherwise, there are two edges incident on w from the path between u and v. We identify them and branch off to two recursive calls, with each call representing the deletion of one such edge.

We have to make sure that the space bound is respected. We do this by making calls to $\mathcal{A}_{con}$ whenever needed and making sure that the size of $E^\star$ never exceeds k. If we used the algorithm in [24], without our modifications, then we would have to delete edges in $E^\star$ explicitly to get a new tree. Explicitly storing this new tree would require $\Omega(n)$ space, which we are not allowed to use.

This is how we prove Theorem 3.

6   Conclusion

We discussed three graph theoretic problems in the space bounded settings and provided FPT-time algorithms for them, namely k-PATH, MAXLEAF SUBTREE and MULTICUT in Trees. It would be interesting to see whether other standard graph theoretic problems can also admit FPT algorithms when the working space is bounded, like the STEINER TREE problem. Also, improving the running time of the problems already solved here can be an interesting area of study.

References

1. Agrawal, M., Kayal, N., Saxena, N.: Primes is in p. Ann. Math. **160**(2), 781–793 (2004). http://www.jstor.org/stable/3597229
2. Alon, N., Dao, P., Hajirasouliha, I., Hormozdiari, F., Sahinalp, S.C.: Biomolecular network motif counting and discovery by color coding. Bioinformatics **24**(13), i241–i249 (2008). https://doi.org/10.1093/bioinformatics/btn163
3. Alon, N., Gutner, S.: Balanced hashing, color coding and approximate counting. In: Chen, J., Fomin, F.V. (eds.) IWPEC 2009. LNCS, vol. 5917, pp. 1–16. Springer, Heidelberg (2009). https://doi.org/10.1007/978-3-642-11269-0_1
4. Alon, N., Matias, Y., Szegedy, M.: The space complexity of approximating the frequency moments. In: Proceedings of the Twenty-Eighth Annual ACM Symposium on Theory of Computing, STOC '96, pp. 20–29. Association for Computing Machinery, New York (1996). https://doi.org/10.1145/237814.237823
5. Alon, N., Yuster, R., Zwick, U.: Color-coding: a new method for finding simple paths, cycles and other small subgraphs within large graphs. In: Proceedings of the Twenty-Sixth Annual ACM Symposium on Theory of Computing, STOC '94, pp. 326–335. Association for Computing Machinery, New York (1994). https://doi.org/10.1145/195058.195179
6. Arora, S., Barak, B.: Computational Complexity: A Modern Approach. Cambridge University Press, Cambridge (2009)
7. Barman, S., Chawla, S.: Region growing for multi-route cuts. In: Proceedings of the Twenty-First Annual ACM-SIAM Symposium on Discrete Algorithms, SODA '10, pp. 404–418. Society for Industrial and Applied Mathematics, USA (2010)

8. Bergougnoux, B., et al.: Space-efficient parameterized algorithms on graphs of low shrubdepth. In: Gørtz, I.L., Farach-Colton, M., Puglisi, S.J., Herman, G. (eds.) 31st Annual European Symposium on Algorithms (ESA 2023). Leibniz International Proceedings in Informatics (LIPIcs), vol. 274, pp. 18:1–18:18. Schloss Dagstuhl – Leibniz-Zentrum für Informatik, Dagstuhl (2023). https://doi.org/10.4230/LIPIcs.ESA.2023.18. https://drops.dagstuhl.de/entities/document/10.4230/LIPIcs.ESA.2023.18

9. Bodlaender, H.L., Groenland, C., Jacob, H.: List colouring trees in logarithmic space. In: Chechik, S., Navarro, G., Rotenberg, E., Herman, G. (eds.) 30th Annual European Symposium on Algorithms, ESA 2022, Berlin/Potsdam, Germany, 5–9 September 2022. LIPIcs, vol. 244, pp. 24:1–24:15. Schloss Dagstuhl - Leibniz-Zentrum für Informatik (2022). https://doi.org/10.4230/LIPICS.ESA.2022.24, https://doi.org/10.4230/LIPIcs.ESA.2022.24

10. Bodlaender, H.L., Groenland, C., Jacob, H., Jaffke, L., Lima, P.T.: Xnlp-completeness for parameterized problems on graphs with a linear structure. In: Dell, H., Nederlof, J. (eds.) 17th International Symposium on Parameterized and Exact Computation, IPEC 2022, Potsdam, Germany, 7–9 September 2022. LIPIcs, vol. 249, pp. 8:1–8:18. Schloss Dagstuhl - Leibniz-Zentrum für Informatik (2022). https://doi.org/10.4230/LIPICS.IPEC.2022.8

11. Bodlaender, H.L., Groenland, C., Nederlof, J., Swennenhuis, C.M.F.: Parameterized problems complete for nondeterministic FPT time and logarithmic space. In: 62nd IEEE Annual Symposium on Foundations of Computer Science, FOCS 2021, Denver, CO, USA, 7–10 February 2022, pp. 193–204. IEEE (2021). https://doi.org/10.1109/FOCS52979.2021.00027

12. Bodlaender, H.L., Szilágyi, K.: Xnlp-hardness of parameterized problems on planar graphs. In: Král, D., Milanic, M. (eds.) Graph-Theoretic Concepts in Computer Science - 50th International Workshop, WG 2024, Gozd Martuljek, Slovenia, 19–21 June 2024, Revised Selected Papers. LNCS, vol. 14760, pp. 107–120. Springer, Heidelberg (2024). https://doi.org/10.1007/978-3-031-75409-8_8

13. Carter, J., Wegman, M.N.: Universal classes of hash functions. J. Comput. Syst. Sci. **18**(2), 143–154 (1979). https://doi.org/10.1016/0022-0000(79)90044-8. https://www.sciencedirect.com/science/article/pii/0022000079900448

14. Chen, J., Chu, Z., Guo, Y., Yang, W.: Space limited graph algorithms on big data. In: Computing and Combinatorics: 28th International Conference, COCOON 2022, Shenzhen, China, 22–24 October 2022, Proceedings, pp. 255–267. Springer, Heidelberg (2023). https://doi.org/10.1007/978-3-031-22105-7_23

15. Chen, J., Guo, Y., Huang, Q.: Linear-time parameterized algorithms with limited local resources. Inf. Comput. **289**, 104951 (2022). https://doi.org/10.1016/j.ic.2022.104951. https://www.sciencedirect.com/science/article/pii/S0890540122001067

16. Chitnis, R., Cormode, G.: Towards a theory of parameterized streaming algorithms. In: Jansen, B.M.P., Telle, J.A. (eds.) 14th International Symposium on Parameterized and Exact Computation (IPEC 2019). Leibniz International Proceedings in Informatics (LIPIcs), vol. 148, pp. 7:1–7:15. Schloss Dagstuhl – Leibniz-Zentrum für Informatik, Dagstuhl (2019). https://doi.org/10.4230/LIPIcs.IPEC.2019.7. https://drops.dagstuhl.de/entities/document/10.4230/LIPIcs.IPEC.2019.7

17. Cormen, T.H., Leiserson, C.E., Rivest, R.L., Stein, C.: Introduction to Algorithms, 3rd edn. The MIT Press, Cambridge (2009)

18. Cygan, M., ET AL.: Parameterized Algorithms. Springer, Heidelberg (2015). https://doi.org/10.1007/978-3-319-21275-3
19. Davidov, N., et al.: Maximum covering subtrees for phylogenetic networks. IEEE/ACM Trans. Comput. Biol. Bioinf. **18**(6), 2823–2827 (2020). https://doi.org/10.1109/TCBB.2020.3040910
20. Elberfeld, M., Stockhusen, C., Tantau, T.: On the space and circuit complexity of parameterized problems: classes and completeness. Algorithmica **71**(3), 661–701 (2015). https://doi.org/10.1007/S00453-014-9944-Y
21. Fafianie, S., Kratsch, S.: A shortcut to (sun)flowers: kernels in logarithmic space or linear time. In: Italiano, G.F., Pighizzini, G., Sannella, D.T. (eds.) MFCS 2015. LNCS, vol. 9235, pp. 299–310. Springer, Heidelberg (2015). https://doi.org/10.1007/978-3-662-48054-0_25
22. Feigenbaum, J., Kannan, S., McGregor, A., Suri, S., Zhang, J.: On graph problems in a semi-streaming model. Theor. Comput. Sci. **348**(2), 207–216 (2005). https://doi.org/10.1016/j.tcs.2005.09.013. https://www.sciencedirect.com/science/article/pii/S0304397505005323
23. Ghosh, P., Kuchlous, S.: New algorithms and lower bounds for streaming tournaments. In: Chan, T., Fischer, J., Iacono, J., Herman, G. (eds.) 32nd Annual European Symposium on Algorithms (ESA 2024). Leibniz International Proceedings in Informatics (LIPIcs), vol. 308, pp. 60:1–60:19. Schloss Dagstuhl – Leibniz-Zentrum für Informatik, Dagstuhl (2024). https://doi.org/10.4230/LIPIcs.ESA.2024.60. https://drops.dagstuhl.de/entities/document/10.4230/LIPIcs.ESA.2024.60
24. Guo, J., Niedermeier, R.: Fixed-parameter tractability and data reduction for multicut in trees. Netw. Int. J. **46**(3), 124–135 (2005)
25. Kammer, F., Sajenko, A.: Space-efficient graph kernelizations. In: Chen, X., Li, B. (eds.) Theory and Applications of Models of Computation, pp. 260–271. Springer, Singapore (2024). https://doi.org/10.1007/978-981-97-2340-9_22
26. Kneis, J., Langer, A., Rossmanith, P.: A new algorithm for finding trees with many leaves. Algorithmica **61**, 882–897 (2011)
27. Lokshtanov, D., Misra, P., Panolan, F., Ramanujan, M.S., Saurabh, S., Zehavi, M.: Meta-theorems for Parameterized Streaming Algorithms‡, pp. 712–739 (2024). https://doi.org/10.1137/1.9781611977912.28. https://epubs.siam.org/doi/abs/10.1137/1.9781611977912.28
28. McGregor, A.: Graph stream algorithms: a survey. SIGMOD Rec. **43**(1), 9–20 (2014). https://doi.org/10.1145/2627692.2627694
29. Reingold, O.: Undirected st-connectivity in log-space. In: Proceedings of the Thirty-Seventh Annual ACM Symposium on Theory of Computing, STOC '05, pp. 376–385. Association for Computing Machinery, New York (2005). https://doi.org/10.1145/1060590.1060647

Subexponential and Parameterized Mixing Times of Glauber Dynamics on Independent Sets

Malory Marin[✉]

ENS de Lyon, CNRS, Université Claude Bernard Lyon 1, LIP, UMR 5668,
69342 Lyon Cedex 07, France
`malory.marin@ens-lyon.fr`

Abstract. Given a graph G, the hard-core model defines a probability distribution over its independent sets, assigning to each set of size k a probability of $\frac{\lambda^k}{Z}$, where $\lambda > 0$ is a parameter known as the *fugacity* and Z is a normalization constant. The Glauber dynamics is a simple Markov chain that converges to this distribution and enables efficient sampling. Its *mixing time*—the number of steps needed to approach the stationary distribution—has been widely studied across various graph classes, with most previous work emphasizing the dichotomy between polynomial and exponential mixing times, with a particular focus on sparse classes of graphs.

Inspired by the modern fine-grained approach to computational complexity, we investigate subexponential mixing times of the Glauber dynamics on geometric intersection graphs, such as disk graphs. We further study parameterized mixing times with respect to two structural parameters that can remain small even in dense graphs: the tree-independence number and the path-independence number. Building on a result of Dyer, Greenhill, and Müller, we show that Glauber dynamics mixes in polynomial time on graphs of bounded path-independence number, and in quasi-polynomial time when the tree-independence number is bounded. Moreover, we prove that both bounds are tight via a conductance argument, thereby resolving a question raised in their work.

This work provides a simple and efficient algorithm for sampling from the hard-core model. Unlike classical approaches that rely explicitly on geometric representations or on constructing decompositions such as tree decompositions or separator trees, our analysis only requires their existence to establish mixing time bounds-these structures are not used directly by the algorithm itself.

1 Introduction

Glauber Dynamics. An *independent set* in a graph G is a set of pairwise non-adjacent vertices. Determining the size of a maximum independent set in G, known as the *independence number* $\alpha(G)$, is one of the most fundamental problems in algorithmic graph theory. The hard-core model is a probability distribution over the set of independent sets of a graph $G = (V, E)$, parameterized

© The Author(s), under exclusive license to Springer Nature Singapore Pte Ltd. 2026
E. Di Giacomo and D. Mondal (Eds.): WALCOM 2026, LNCS 16444, pp. 276–290, 2026.
https://doi.org/10.1007/978-981-95-7127-7_19

by the *fugacity* $\lambda > 0$, where the probability of an independent set $I \subseteq V$ is given by $\pi_{G,\lambda}(I) = \lambda^{|I|}/Z_G(\lambda)$, where $Z_G(\lambda) = \sum_{J \in \mathcal{I}(G)} \lambda^{|J|}$ is the *partition function*, summing over $\mathcal{I}(G)$, the set of all independent sets of G. The *Glauber dynamics* on independent sets of a graph is a Markov chain whose stationary distribution corresponds exactly to the hard-core model. In this process, one starts with an arbitrary independent set and then repeatedly selects a vertex $v \in V$ uniformly at random. If v is in the current independent set, it is removed with probability $\frac{1}{\lambda+1}$; otherwise, it is added with probability $\frac{\lambda}{\lambda+1}$, provided that adding it maintains independence.

The *mixing time* of the Glauber dynamics refers to the number of steps required for the Markov chain to converge close to its stationary distribution $\pi_{G,\lambda}$. More formally, given two distributions μ and ν on $\mathcal{I}(G)$, the *total variation distance* $\Delta(\mu, \nu)$ is defined as $\Delta(\mu, \nu) = \frac{1}{2} \sum_{I \in \mathcal{I}(G)} |\mu(I) - \nu(I)|$. Given two independent sets $I, J \in \mathcal{I}(G)$ and an integer $t \geq 0$, let $\mu_I^t(J)$ denote the probability that the current independent set is J after t transitions, starting from the independent set I. The mixing time $\tau_{G,\lambda}$ is the smallest t such that $\max_{I \in \mathcal{I}(G)} \Delta(\mu_I^t, \pi_{G,\lambda}) \leq \frac{1}{4}$. The constant $\frac{1}{4}$ is chosen arbitrarily in $(0, \frac{1}{2})$, and running the Glauber dynamics during $\ell\tau_{G,\lambda}$ steps would guarantee a total variation distance of $2^{-\ell}$. A key question in the study of this chain is whether it exhibits *fast mixing*, meaning the mixing time is at most polynomial in the size of the graph. Recent results have established fast mixing under various conditions, such as when the graph has maximum degree d and the activity parameter λ is below a critical threshold $\lambda_c(d)$ [8], while it can be exponential when λ is above $\lambda_c(d)$ [16]. More generally, Sly [19] proved that unless NP = RP, no polynomial-time approximation scheme exists for the partition function when $\lambda_c(d) < \lambda < \lambda_c(d) + \varepsilon(d)$. Beyond bounded-degree graphs, Eppstein and Frishberg [11] demonstrated that Glauber dynamics mixes in time $n^{O(\mathrm{tw})}$ for graphs of treewidth tw. Similarly, Bezakova and Sun [4] showed that in chordal graphs with separator size b, the mixing time is $n^{O(\ln b)}$ for all $\lambda > 0$. Recently, Jerrum [12] established a dichotomy in H-free graphs, identifying structural conditions under which Glauber dynamics mixes efficiently for all $\lambda > 0$.

The three results mentioned above share a key feature: they hold for arbitrarily large fugacity $\lambda > 0$. However, they all focus on sparse graphs since in some dense graphs—such as complete bipartite graphs—Glauber dynamics typically mixes in exponential time (see Sect. 5). Nevertheless, there are exceptions: for instance, in complete graphs, Glauber dynamics exhibits rapid mixing. This suggests that fast mixing may still be achievable in certain dense graph classes, raising the following question, which we aim to explore:

> Does Glauber dynamics mix "reasonably" fast on some classes of dense graphs, for arbitrarily large fugacity?

In 2020, Dyer, Greenhill, and Müller [10] initiated the study of this question by identifying a class of dense graphs on which Glauber dynamics mixes rapidly:

the graphs of bounded bipartite pathwidth. This class, defined as the maximum pathwidth of a bipartite subgraph of a graph, notably includes all claw-free graphs and hence all line graphs. Their work suggests that the modern fine-grained perspective on complexity and algorithms, built upon graph parameters, provides a fruitful framework for characterizing graph classes in which Glauber dynamics exhibits fast mixing. In the following paragraph, we review results on subexponential and parameterized algorithms, which serve as the necessary background and motivation for our contribution.

Subexponential and Parameterized Algorithms. In general, most NP-hard problems require at least $2^{\Omega(n)}$ time on n-vertex graphs. However, for certain graph classes, such as planar graphs, many of these problems—including finding a maximum independent set—can be solved in time $2^{O(\sqrt{n})}$. Such a complexity of $2^{o(n)}$ is called *subexponential*. Other graph classes, such as H-minor-free graphs and unit disk graphs, also admit similar algorithms. These classes share a key property: they have *separator theorems*.

For instance, the celebrated planar separator theorem states that any n-vertex planar graph contains a balanced separator of size $O(\sqrt{n})$ [14]. This line of research has been extended to more general geometric intersection graphs, which, unlike planar graphs, can be dense. For example, while planar graphs have bounded clique size, the intersection graph of unit disks can have arbitrarily large cliques. More formally, let $\mathcal{X}$ be a collection of n subsets of $\mathbb{R}^d$ ($d \geqslant 2$). The intersection graph $G[\mathcal{X}]$ of $\mathcal{X}$ is the n-vertex graph with one vertex per subset of $\mathcal{X}$, and an edge between two vertices if and only if both corresponding subsets have a nonempty intersection. When all subsets are balls in $\mathbb{R}^d$, $G[\mathcal{X}]$ is called a *d-dimensional ball graph*. For the special case of $d = 2$, we say that $G[\mathcal{X}]$ is a *disk graph*. Early work established subexponential algorithms with running times of $O(n^{n^{1-\frac{1}{d}}})$, and even $2^{O(n^{1-\frac{1}{d}})}$ for d-dimensional geometric graphs [15, 20]. More recently, a robust framework was developed to obtain tight running times for geometric graphs [1]. This framework is based on a new type of separator, called *clique-based separators*, which has received further attention in [3]. The notion of a separator is closely linked to the concept of tree decompositions. A *tree decomposition* of a graph $G = (V, E)$ is a pair $\mathcal{T} = (T, \{X_t\}_{t \in V(T)})$, where T is a tree and $\{X_t\}_{t \in V(T)}$ is a collection of subsets of V (called *bags*) satisfying the following properties:

(a) $\bigcup_{t \in V(T)} X_t = V$, meaning every vertex of G appears in at least one bag.
(b) For every edge $(u, v) \in E$, there exists a node $t \in V(T)$ such that $\{u, v\} \subseteq X_t$.
(c) For each vertex $v \in V$, the set of nodes $\{t \in V(T) \mid v \in X_t\}$ induces a connected subtree of T.

The *width* of a tree decomposition is $\max_{t \in V(T)} |X_t| - 1$, and the *treewidth* of G, denoted $\mathrm{tw}(G)$, is the minimum width over all possible tree decompositions of G. A well-known algorithmic result states that a maximum independent set can be found in time $2^{O(\mathrm{tw}(G))} \cdot n$.

Such complexities of the form $f(k) \cdot n^{O(1)}$, where k is some parameter and f is a computable function, are called *fixed-parameter tractable* (FPT). A direct consequence of the planar separator theorem is that any planar graph has treewidth $O(\sqrt{n})$, leading to a $2^{O(\sqrt{n})}$ algorithm for finding a maximum independent set in planar graphs.

However, in dense geometric graphs, the presence of large cliques implies high treewidth, making this parameter unsuitable. To address this, a new parameter that allows for some density was introduced in [9]. The *independence number* of a tree decomposition is defined as $\max_{t \in V(T)} \alpha(G[X_t])$, where $\alpha(G[X_t])$ is the independence number of the subgraph induced by a bag. The *tree independence number*, denoted tree-$\alpha(G)$, is the minimum independence number over all possible tree decompositions of G. This parameter leads to alternative algorithmic results: a maximum independent set can be found in time $n^{\text{tree-}\alpha(G)}$. While this is not FPT, it remains polynomial, providing a useful trade-off for dense graphs where treewidth-based approaches are inefficient.

A *path decomposition* is a special case where T is a path. For simplicity, we define a path decomposition only as the ordered set of bags $\{X_i\}_{1 \leqslant i \leqslant p}$. The *pathwidth* of G, denoted pw(G), is the minimum width of any path decomposition of G. Similarly, the *path independence number* of G, denoted path-$\alpha(G)$, is the minimum independence number over all possible path decompositions of G.

The *bipartite pathwidth* of a graph G, introduced by Dyer, Greenhill, and Müller [10], denoted bpw(G), is defined as the maximum pathwidth of a bipartite induced subgraph of G. This parameter was specifically introduced to study the mixing time of Glauber dynamics, where the following result was established (up to some reformulation):

Theorem 1 [10]. *Given a n-vertex graph G, the mixing time of Glauber dynamics with fugacity $\lambda > e/n$ satisfies $\tau_{G,\lambda} = (2\tilde{\lambda}n)^{O(bpw(G))}$, where $\tilde{\lambda} = e^{|\ln \lambda|}$.*

From now on, we will write $\tilde{\lambda}$ to denote $e^{|\ln(\lambda)|}$. In the preliminaries, we will observe that bpw$(G) \leqslant 2$path-$\alpha(G)$, which immediately yields a polynomial bound on the mixing time for graphs of bounded path-independence number. Likewise, since every n-vertex graph satisfies tw$(G) = O(\ln(n))$pw(G) [13], it follows that the mixing time is quasi-polynomial whenever the *bipartite treewidth* of G, defined analogously as the maximum treewidth of a bipartite induced subgraph, is bounded. Moreover, as btw$(G) \leqslant 2$tree-$\alpha(G)$, the same quasi-polynomial bound applies when the tree-independence number is bounded.

The polynomial bound for bounded bipartite pathwidth is tight by a simple conductance argument (see Sect. 5). However, the tightness of the quasi-polynomial bound was left open by Dyer, Greenhill, and Müller.

Contributions and Organization. In the next section, we introduce the main definitions and tools required for this work.

In Sect. 3, we establish our first main result: a subexponential mixing time for intersection graphs of balls in d-dimensional space.

Theorem 2. *Let G be an intersection graph of n balls in $\mathbb{R}^d$ with $d \geqslant 2$. The mixing time of the Glauber dynamics with fugacity $\lambda > 0$ satisfies*

$$\tau_{G,\lambda} = (2\tilde{\lambda})^{O(n^{1-\frac{1}{d}})}.$$

Note that for such graphs, computing the partition function $Z_G(\lambda)$ exactly, or determining the activation probabilities p_v for all $v \in V(G)$, cannot be done in time $2^{o(n^{1-1/d})}$ under the ETH, by a simple reduction to the problem of computing the independence number $\alpha(G)$ within this class [1]. However, by using Glauber dynamics, we obtain a randomized algorithm to estimate the p_vs that is not only simple to implement but also matches the complexity of the best possible exact deterministic algorithm.

In fact, we prove a slightly more general result for hereditary graph classes that admit clique-based separators, which include intersection graphs of balls. Notably, this result also extends to intersection graphs of similarly-sized fat objects in d-dimensional space, and many other classes of graphs.

In Sect. 4, we investigate mixing times in graphs of bounded path independence number and tree independence number. We prove the following

Theorem 3. *Let G be a graph with n vertices. The mixing time of the Glauber dynamics with fugacity $\lambda > 0$ satisfies*

- *$\tau_{G,\lambda} = (b\tilde{\lambda})^{O(path\text{-}\alpha(G))} \cdot n^{O(1)}$ where b is the minimum width (plus one) of a path decomposition of path independence number $path\text{-}\alpha(G)$;*
- *$\tau_{G,\lambda} = n^{O(tree\text{-}\alpha(G))\cdot \ln(b\tilde{\lambda})}$ where b is the minimum width (plus one) of a tree decomposition of tree independence number $tree\text{-}\alpha(G)$.*

This theorem can be seen as a refinement of Theorem 1 in the context of path and tree-independence number since it takes into account the size of the bags in the bounds, while such a parameter do not exist in the context of bipartite pathwidth and treewidth. In particular, Theorem 3 generalizes the polynomial mixing time result for chordal graphs with bounded separators [4]. In addition, with minor modifications to the proof, we also obtain a result for pathwidth, establishing the first FPT mixing time bound for Glauber dynamics in the hard-core model. Bordewich et al. [6] previously proved an FPT mixing time bound in terms of pathwidth for a different Glauber dynamics, which they used to establish polynomial mixing time when treewidth is bounded. Our result on bounded pathwidth similarly implies polynomial mixing time for bounded treewidth, offering an alternative proof of the result by Eppstein and Frishberg [11]. Note that, in full generality (that is, when the bag size b may be as large as the number of vertices), Theorem 3 follows from Theorem 1 by simple inequalities between the parameters.

Finally, in Sect. 5, we establish lower bounds for all our results, relying on conductance arguments. Previous works proved that Theorem 2 is tight in dimension two [17], and not improvable by far in dimension at least 3 [7]. The matching lower bounds for pathwidth and path independence number of Theorem 3 are straightforward to obtain, as they involve well-known families of graphs, specifically complete bipartite graphs and slight variations thereof. Then, we prove a quasi-polynomial matching lower bound for the case of tree-independence number, which is more intricate and reveals connections to *TAR-reconfiguration*, a well-studied variant of independent set reconfiguration problems. More precisely, we prove the following.

Theorem 4. *For any $t \geqslant 1$, there exist infinitely many graphs G such that tree-$\alpha(G) = 2t$ and for any $\lambda \geqslant 1$,*

$$\tau_{G,\lambda} \geqslant (\lambda n)^{ct\ln(n)}$$

where $n = |V(G)|$ and c is a constant.

Since graphs of bounded tree-independence number have bounded bipartite treewidth, it answers a question of Dyer, Greenhill and Muller.

Due to space constraints, many proofs (those marked with a *) have been omitted. They are all available in the extended version of this paper, along with a more detailed introduction to Markov chains and a discussion on an application for this work in the context of Wi-Fi networks.

2 Preliminaries

Notations. Let G be a simple graph. We denote by $V(G)$ and $E(G)$ the set of vertices and the set of edges of G, respectively. When there is no ambiguity, we denote by n the number of vertices of G, and by m the number of edges of G.

Given a set $R \subseteq V(G)$, we use $G[R]$ to denote the subgraph induced by R, and $G - R$ to denote the graph induced by $V(G) \setminus R$. For a vertex $v \in V(G)$, we denote by $N(v)$ the *open neighborhood* of v, that is, $N(v) = \{u \in V(G) \mid uv \in E(G)\}$, and by $N[v]$ its *closed neighborhood*, defined as $N[v] = N(v) \cup \{v\}$. Given two sets A and B, we denote by $A \Delta B$ the *symmetric difference* of A and B, that is $(A \setminus B) \cup (B \setminus A)$.

For a subset $X \subseteq V(G)$, we write $Z_X(\lambda)$ instead of $Z_{G[X]}(\lambda)$ when there is no ambiguity. Given a graph G, we define $\mathcal{G}(G)$ as the *reconfiguration graph* of independent sets in G. Its vertex set is $\mathcal{I}(G)$, and two independent sets $I, J \in \mathcal{I}(G)$ are adjacent if and only if $|I \Delta J| \leqslant 1$. This graph serves as the support graph for the Glauber dynamics.

Relation Between Parameters. Notice that bipartite pathwidth and bipartite treewidth can be related to path- and tree-independence numbers in a straightforward way.

Proposition 1 (*). *For any graph G, the inequalities $bpw(G) \leqslant 2path\text{-}\alpha(G)+1$ and $btw(G) \leqslant 2tree\text{-}\alpha(G) + 1$ hold.*

Since any bipartite induced subgraph of a claw-free graph is a disjoint union of paths and cycles [10], the class of claw-free graphs has bipartite path-width at most 2. However, this class has unbounded tree-independence number. Indeed, line graphs of grids are claw-free, and their tree-independence number is unbounded, since they have bounded maximum degree but unbounded treewidth.

Conductance. The conductance of Markov chain $\mathcal{M}$ is defined as the edge expansion of the underlying transition graph. It is the main tool to prove lower bound on the mixing time of Glauber dynamics. One of the simplest methods to bound the conductance, introduced by Randall [17], involves partitioning the state space $\mathcal{I}(G)$ into three subsets $(\Omega_S, \Omega_1, \Omega_2)$ such that Ω_S *separates* Ω_1 and Ω_2. This means that for any pair of states $(I, J) \in \Omega_1 \times \Omega_2$ and any path $\gamma_{I,J} = (I = W_0, W_1, \ldots, W_\ell = J)$ in $\mathcal{G}(G)$, there exists at least one intermediate independent set W_i such that $W_i \in \Omega_S$. For the sake of completeness, we now formally state the following theorem, which encapsulates this argument and will be used in Sect. 5.

Theorem 5 [17,18]. *Let G be a graph and let $(\Omega_S, \Omega_1, \Omega_2)$ be a partition of $\mathcal{I}(G)$ such that Ω_S separates Ω_1 and Ω_2. For any $\lambda > 0$,*

$$\tau_{G,\lambda} \geqslant \frac{\ln 2}{4} \left(\frac{\pi_{G,\lambda}(\Omega_1)}{\pi_{G,\lambda}(\Omega_S)} - 2 \right)$$

Canonical Paths. The *canonical paths* technique provides a powerful method for upper bounding the mixing time. Let G be a graph, and for each pair of independent sets $I, J \in \mathcal{I}(G)$, fix a specific path $\gamma_{I,J} = (I = W_0, W_1, \ldots, W_\ell = J)$ in $\mathcal{G}(G)$. The path $\gamma_{I,J}$ is called the canonical path from I to J. Let $\Gamma := \{\gamma_{I,J} \mid I, J \in \mathcal{I}(G)\}$ be the set of all fixed canonical paths. The *congestion* through a transition $e = (W, W')$ where $|W \Delta W'| \leqslant 1$, is defined as

$$\rho(\Gamma, e) := \frac{1}{\pi_{G,\lambda}(W)P(W, W')} \sum_{\substack{K, L \in \mathcal{I}(G) \\ \gamma_{K,L} \text{ uses } e}} \pi_{G,\lambda}(K)\pi_{G,\lambda}(L)|\gamma_{K,L}|$$

where $|\gamma_{K,L}|$ is the length of the path $\gamma_{K,L}$ and $P(W, W')$ is the probability associated to the transition (W, W'). The overall congestion of the paths Γ is then

$$\rho(\Gamma) := \max_{e=(W,W'):P(W,W')>0} \rho(\Gamma, e).$$

We will use the following:

Theorem 6 [18]. *Let G be a graph and let Γ be a set of canonical paths between independent sets of G. For any $\lambda > 0$,*

$$\tau_{G,\lambda} \leqslant \rho(\Gamma) \ln \left(\frac{4}{\min_{I \in \mathcal{I}(G)} \pi_{G,\lambda}(I)} \right)$$

3   Subexponential Mixing Time on Geometric Graphs

This section is dedicated to proving a slightly more general result that will imply Theorem 2. First, we introduce the notion of clique-based separator trees, which serve as the main tool for our proof. Then, we establish that in any graph belonging to a class that admits clique-based separators, the Glauber dynamics mixes in subexponential time.

Clique-Based Separator. Let $a \in (0,1)$ and $w > 0$. Following from [1], we say that a triple (A, B, S) is a (a, w)-*clique based separation* of G if it verifies the following properties:

- (A, B, S) is a partition of G such that $(A \times B) \cap E(G) = \emptyset$;
- $|A| \leqslant a|V(G)|$ and $|B| \leqslant a|V(G)|$;
- there exists a set of cliques $C_1, ..., C_k$ $(k \geqslant 1)$ such that $S = \bigcup_{1 \leqslant i \leqslant k} C_i$ and such that

$$\sum_{1 \leqslant i \leqslant k} \log_2(|C_i| + 1) \leqslant w$$

The set S is then called a (a, w)-*clique based separator* of G. A triple (A, B, S) that satisfies only the first condition is simply called a *separation* of G, and S is a *separator* of G.

Observe that a clique-based separator generalizes the classical notion of a separator, where only the number of vertices in the separator matters. Suppose that (A, B, S) satisfies the first two conditions of the definition of a clique-based separator. Then, it also forms a $(a, |S|)$-clique-based separator by considering only cliques of size 1. This shows that clique-based separators are particularly interesting when the number of vertices in S is large compared to the total number of vertices in the graph.

Clique-Based Separator Tree. Let $a \in (0,1)$ be a constant and let g be a real-valued non-decreasing function. We say that $\mathcal{T} = (T, \{X_t\}_{t \in V(T)})$ is a (a, g)-*clique based separator tree* of a graph G if T is a rooted binary full tree with the following properties:

(a) Each node $t \in V(T)$ is associated with a set $X_t \subseteq V(G)$.
(b) The sets X_t, for $t \in T$, partition $V(G)$, i.e.,

$$\bigcup_{t \in V(T)} X_t = V(G), \quad \text{and} \quad X_t \cap X_{t'} = \emptyset \text{ for distinct } t, t' \in V(T).$$

(c) For each node $t \in V(T)$, let $V_t = \bigcup_s X_s$, where s ranges over the descendants of t (including t). Note that if t is an internal node with children u and v, then V_t is the disjoint union of X_t, V_u, and V_v. If t is a leaf, then $V_t = X_t$. For each internal node $t \in T$ with children u and v, the triplet (V_u, V_v, X_t) forms a $(a, g(|V_t|))$-clique based separation for the subgraph $G[V_t]$.
(d) For each leaf $t \in T$, we have $|V_t| \leqslant 1$.

A $(1, id_{\mathbb{R}})$-clique-based separator tree of G is simply called a *separator tree* of G, as the third condition reduces to a separation condition, while the fourth condition becomes trivial.

Bounding Mixing Time. Using the notion of a separator tree, we first define a set of canonical paths that will exhibit subexponential congestion. This will then lead to an upper bound on the mixing time.

Let G be a graph and $\mathcal{T} = (T, \{X_t\}_{t \in V(T)})$ be a separator tree of G. For the moment, $\mathcal{T}$ has not to be specifically clique-based separator tree. We define $\Gamma_{\mathcal{T},G}$ a set of canonical paths between the independent sets of G as follows. We define by induction on T, for any $t \in V(T)$ and I, J independent sets of $G[V_t]$, the canonical path $\gamma_{I,J}$ from I to J.

- If t is a leaf, then the path $\gamma_{I,J}$ consists in removing all the vertices of I and then add all vertices of J;
- otherwise, t has a left child $u \in V(T)$ and a right child $v \in V(T)$. $\gamma_{I,J}$ is obtained as follows:
  - go from I to $I \setminus X_t$ by deleting all the vertices of $I \cap X_t$;
  - go from $I \setminus X_t$ to $(J \cap V_u) \cup (I \cap V_v)$ by following the canonical path from $(I \cap V_u)$ to $(J \cap V_u)$ obtained by induction on u;
  - go from $(J \cap V_u) \cup (I \cap V_v)$ to $(J \cap V_u) \cup (J \cap V_v) = J \setminus X_t$ by following the canonical path from $(I \cap V_v)$ to $(J \cap V_v)$ obtained by induction on v;
  - go from $J \setminus X_t$ to J by adding all the vertices of $J \cap X_t$.

Notice that for any pair of independent sets (I, J) of G, the path $\gamma_{I,J}$ is defined. We set $\Gamma(\mathcal{T}, G)$ to be the set of such paths. For a vertex $v \in V(G)$, let $t_v \in V(T)$ be the unique node of T such that $v \in X_{t_v}$. In addition, let A_v be the set of vertices of G which appear in the bag of an ancestor node of t_v in T. Note that, in particular, $v \in A_v$.

The key property of this set of canonical paths is that, for any transition $(I, I \cup v)$ or $(I, I \setminus v)$, where $I \in \mathcal{I}(G)$ and $v \in V(G)$, the congestion of this transition is bounded by a function of $Z_{A_v}(\lambda)^2$, up to a polynomial factor. The next lemmas formalizes this argument. In the following, we denote by π the distribution $\pi_{G,\lambda}$ when there is no ambiguity.

Lemma 1 (*). *Let G be a graph, $\mathcal{T} = (T, \{X_t\}_{t \in V(T)})$ be a separator tree of G. Then, for any $\lambda > 0$,*

$$\rho(\Gamma_{\mathcal{T},G}) \leqslant 4n^2 \tilde{\lambda}^{2\alpha+1} \max_{v \in V(G)} |\mathcal{I}(G[A_v])|^2$$

where $\alpha = \max_{v \in V(G)} \alpha(G[A_v])$.

Lemma 1 gives an upper bound on the congestion of $\Gamma_{\mathcal{T},G}$ depending on $\max_{v \in V(G)} \alpha(G[A_v])$ and $\max_{v \in V(G)} |\mathcal{I}(G[A_v])|$. However, this result does not rely on any specific properties beyond $\mathcal{T}$ being a separator tree. Theorem 7 leverages the fact that $\mathcal{T}$ is a clique-based separator tree of G to derive a subexponential bound on the congestion, and consequently, on the mixing time.

Theorem 7 (*). *Let $\mathcal{C}$ be a hereditary class of graphs, such that any n-vertex graph $G \in \mathcal{C}$ has a $(a, cn^{1-b} \ln^{\delta}(n))$-clique based separator for some constants a, b, c and δ depending on $\mathcal{C}$. Let $G \in \mathcal{C}$ be a n-vertex graph. The mixing time of the Glauber dynamics on G with fugacity $\lambda > 0$ satisfies $\tau_{G,\lambda} = (2\tilde{\lambda})^{O(n^{1-b} \ln^{\delta}(n))}$.*

Theorem 2 is an immediate consequence of Theorem 7, since ball graphs in $\mathbb{R}^d$ admit clique-based separators [1,3]. Moreover, the same argument extends to broader classes of graphs, with intersection graphs of convex fat objects, similarly-sized fat objects, pseudo-disks, and map graphs being the most notable examples.

4 Parameterized Mixing Time

This section is dedicated to Theorem 3. All the proofs are available in the extended version, but all the upper bounds on the mixing time follow the same outline as in the previous section: we construct a set of canonical paths, which in this case depend on path decompositions rather than separator trees. We then show that the congestion of these paths is bounded by the desired function of the parameter. Finally, we extend these results to the tree independence number by establishing structural connections between path and tree decompositions.

Theorem 8 (*). *Let G be a graph with n vertices. The mixing time of the Glauber dynamics with fugacity $\lambda > 0$ satisfies*

- $\tau_{G,\lambda} = (2\tilde{\lambda})^{O(pw(G))} \cdot n^{O(1)}$;
- $\tau_{G,\lambda} = (b\tilde{\lambda})^{O(path\text{-}\alpha(G))} \cdot n^{O(1)}$ *where b is the minimum width (plus one) of a path decomposition of path independence number $path\text{-}\alpha(G)$.*

To establish bounds on the mixing time parameterized by the tree independence number, there are two possible approaches. One option is to use a separator-tree-based method, as in Sect. 3. The other is to relate the width of path decompositions to the width of tree decompositions. A result of this type is already known for sparse parameters: for any n-vertex graph, it is known that $\mathrm{pw}(G) = O(\ln(n))\mathrm{tw}(G)$ [13]. Proving a strengthening of this result for the dense case, we obtain the following result.

Theorem 9 (*). *Let G be a graph with n vertices. The mixing time of the Glauber dynamics on G with fugacity $\lambda > 0$ satisfies:*

- $\tau_{G,\lambda} = n^{O(tw(G)) \ln(\tilde{\lambda})}$;
- $\tau_{G,\lambda} = n^{O(tree\text{-}\alpha(G)) \cdot \ln(b\tilde{\lambda})}$, *where b is the minimum width (plus one) of a tree decomposition of tree independence number $tree\text{-}\alpha(G)$.*

5 Lower Bound on the Mixing Time of Glauber Dynamics

This section is dedicated to reviewing and establishing lower bounds on the mixing time of the Glauber dynamics, based on upper bounds on the conductance of the Markov chain. We demonstrate that our parameterized mixing time bounds for dense graphs are tight by constructing, in each case, a family of graphs that matches the upper bounds stated in Theorems 3.

Geometric Graphs. Lattices serve as key examples of graphs exhibiting exponential mixing times. Borgs et al. [7] proved that Glauber dynamics on the d-dimensional torus has at least subexponential mixing time, implying that the exponent in Theorem 7 is tight. Additionally, while the torus is not a disk graph, for sufficiently large sizes, it is a $(d+1)$-dimensional ball graph, setting a fundamental limitation on potential improvements to Theorem 2. In [17], Randall proved that Theorem 2 is tight in dimension two.

Pathwidth and Path Independence Number. Using a conductance argument, the simplest graph on which Glauber dynamics mixes in exponential time is the complete bipartite graph $K_{t,t}$. Indeed, to transition from an independent set contained entirely in one part to an independent set contained in the other, the dynamics must pass through the empty independent set, which has significantly lower energy compared to any other state. Observe that this graph has pathwidth exactly t. We show that this graph establishes the tightness of the FPT mixing time bound in terms of pathwidth.

Theorem 10 (*). *For any $c > 0$, there exists a n-vertex graph G such that for any $\lambda \geqslant 1$,*

$$\tau_{G,\lambda} \geqslant \left(\frac{\lambda+1}{2}\right)^{pw(G)} \cdot n^c$$

By replacing each vertex in $K_{t,t}$ with a clique, we establish the tightness of the mixing time in terms of the path independence number.

Theorem 11 (*). *For any $t \geqslant 1$, there exists infinitely many graphs G such that $\text{path-}\alpha(G) = t$ and for any $\lambda \geqslant 1$,*

$$\tau_{G,\lambda} \geqslant c_t \, (n\lambda)^t$$

where $n = |V(G)|$ and c_t is a constant which only depends on t.

Tree-Independence Number. Having proven Theorems 10 and 11, it is natural to ask whether these results extend to establish the optimality of the mixing time for treewidth and tree-independence number. For treewidth, an equivalent question was already raised by Eppstein and Frishberg [11], who asked whether an FPT mixing time in terms of treewidth is possible. For the tree-independence number, Dyer, Greenhill, and Müller posed the weaker question of whether the quasi-polynomial mixing time bound for bounded bipartite treewidth is tight. We answer this in the affirmative: by showing that the quasi-polynomial mixing time is optimal for graphs with bounded tree-independence number, we resolve their question. This follows from Proposition 1, since graphs of bounded tree-independence number also have bounded bipartite treewidth.

The proof strategy for Theorems 10 and 11 can be summarized as follows: identify a graph that has exactly two distinct "high-energy states" (such as two distinct maximum independent sets) and demonstrate that any transformation

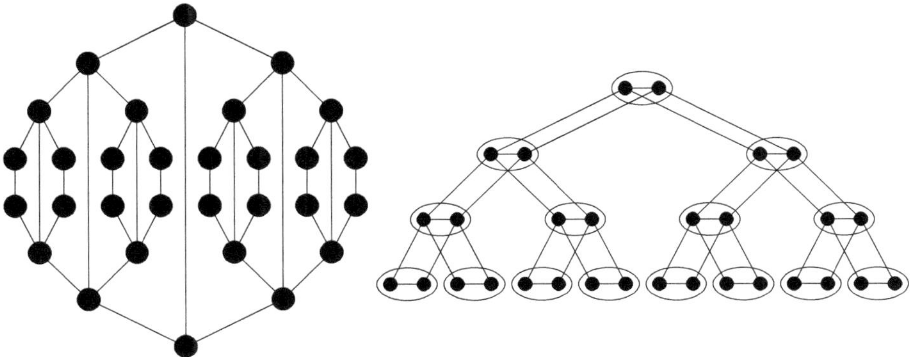

Fig. 1. Two representations of the graph G_3. In the right representation, two vertices are in the same bag if and only if they belong to the same set X_t for some $t \in V(T_3)$.

from one state to the other must pass through a low-energy state. A natural first step is to find such a graph that also has bounded treewidth.

We construct a family of graphs G_k for $k \geqslant 0$. Let T_k be the complete binary tree of depth k. The graph G_k is obtained by replacing each node $t \in V(T_k)$ with a set $X_t = \{u_t, v_t\}$ of two adjacent vertices. For each edge $tt' \in E(T_k)$, we add a matching of two edges $u_t u_{t'}$ and $v_t v_{t'}$. The graph G_3 is illustrated in Fig. 1. Building on the work from [2], the graph G_k can be seen as the simplest graph of treewidth 2 with unbounded TAR reconfiguration threshold. We next prove some basic results about those graphs. In particular, the second point of the following lemma is based on the isoperimetric value of a binary tree [5].

Lemma 2 (*). *Let $k \geqslant 0$. The following properties hold.*

(a) *The graph G_k has $2^{k+2} - 2$ vertices and exactly two distinct maximum independent sets (MIS) I_k and J_k of size $\alpha(G_k) = 2^{k+1} - 1$, and they form a partition of $V(G_k)$.*
(b) *There is a constant c_1 such that for every path $(W_i)_{0 \leqslant i \leqslant p}$ from I_k to J_k in $\mathcal{G}(G_k)$, there exists some $0 \leqslant i \leqslant p$ such that $|W_i| \leqslant \alpha(G_k) - \lfloor c_1 \cdot k \rfloor$.*
(c) *For any $d \geqslant 0$, G_k contains at most $2^{2d+1}\alpha(G_k)^d$ independent sets of size exactly $\alpha(G_k) - d$.*

Proof Theorem 4. Consider the graph $H_{k,t}$ obtained by replacing each vertex $v \in V(G_k)$ with a set of vertices Y_v inducing a disjoint union of t cliques of size 2^{2k}. For each edge $uv \in E(G_k)$, we add all possible edges between Y_u and Y_v. For any subset $V' \subseteq V(H_{k,t})$, define $p(V') = \{u \in V(G_k) \mid Y_u \cap S \neq \emptyset\}$, and observe that if V' is an independent set of $H_{k,t}$, $p(V')$ is an independent set of G_k. Let $I_{k,t}$ be a maximum independent set of $H_{k,t}$ such that $p(I_{k,t}) = I_k$. Notice that $|I_{k,t}| = t\alpha(G_k)$. We partition $\mathcal{I}(H_{k,t})$ as follows:

- Ω_S: independent sets S such that $p(S)$ is an independent set of G_k of size exactly $s = \alpha(G_k) - \lfloor c_1 k \rfloor$, where c_1 is the constant of Lemma 2(b).

- Ω_I: independent sets I for which there exists a path of independent sets in $\mathcal{G}(H_{k,t})$ from $I_{k,t}$ to I avoiding Ω_S.
- Ω_J: the remaining independent sets.

By definition, Ω_S separates Ω_I and Ω_J in the reconfiguration graph $\mathcal{G}(H_{k,t})$. Indeed, suppose there exists a path from an independent set $I \in \Omega_I$ to an independent set $J \in \Omega_J$ that avoids Ω_S. Then, by the definition of Ω_I, there must also be a path from $I_{k,t}$ to J that avoids Ω_S. This implies that $J \in \Omega_I$, contradicting the assumption that $J \in \Omega_J$.

We denote by π the distribution $\pi_{H_{k,t},\lambda}$. To bound the conductance, we first establish an upper bound on $\pi(\Omega_S)$ and a lower bound on $\min(\pi(\Omega_I), \pi(\Omega_J))$.

Claim (*). $\pi(\Omega_S) \leqslant 2^{\lfloor c_1 k \rfloor (k+3)+1}((2^{2k}\lambda + 1)^t - 1)^s.$

Claim (*). $\min(\pi(\Omega_I), \pi(\Omega_J)) \geqslant ((2^{2k}\lambda + 1)^t - 1)^{\alpha(G_k)}.$

Using Theorem 5, we obtain:

$$\tau_{H_{k,t},\lambda} \geqslant \frac{\ln(2)}{4}\left(\frac{((2^{2k}\lambda + 1)^t - 1)^{\alpha(G_k)}}{2^{\lfloor c_1 k \rfloor (k+3)+1}((2^{2k}\lambda + 1)^t - 1)^s} - 2\right)$$

$$= \frac{\ln(2)}{4}\left(\frac{((2^{2k}\lambda + 1)^t - 1)^{\lfloor c_1 k \rfloor}}{2^{\lfloor c_1 k \rfloor (k+3)+1}} - 2\right)$$

Let $n = |V(H_{k,t})|$. We have $n = t2^{2k}(2^{k+2}-2) \leqslant t2^{3k+2}$, which implies $\log_2(n) \leqslant \log_2(t) + 3k + 2$. Rearranging, we obtain $k \geqslant \frac{1}{3}(\log_2(n) - \log_2(t) - 2)$.

It is straightforward to observe that there exists a threshold k_t such that for any $k \geqslant k_t$, the following inequalities hold:

$$\begin{cases} k \geqslant \frac{\ln(2)}{4}\log_2(n), \\[2mm] ((2^{2k}\lambda + 1)^t - 1)^{\lfloor c_1 k \rfloor} \geqslant 2^{\frac{4}{3}\lfloor c_1 k \rfloor kt + 3c_1 k + 3}\lambda^{\frac{2}{3}c_1 kt} \end{cases}$$

Thus, for any $k \geqslant k_t$, we obtain

$$\tau_{H_{k,t},\lambda} \geqslant \lambda^{\frac{2}{3}c_1 kt} 2^{\lfloor c_1 k \rfloor k\left(\frac{4}{3}t-1\right)}$$

and finally since $k \geqslant \frac{\log_2(n)}{4}$ and $\frac{4}{3}t - 1 \geqslant \frac{t}{3}$ whenever $t \geqslant 1$, there exists a small enough constant $c > 0$ such that

$$\tau_{H_{k,t},\lambda} \geqslant (\lambda n)^{ct \ln n}$$

$\square$

6   Conclusion

We showed that the mixing time of the Glauber dynamics for the hard-core model is subexponential on geometric intersection graphs. Additionally, we established tight parameterized mixing time bounds for this Markov chain, enabling a more fine-grained understanding of mixing times beyond the classical polynomial-versus-exponential dichotomy.

In particular, for geometric intersection graphs, our results demonstrate that the Glauber dynamics is as efficient as exact algorithms for computing the partition function and activation probabilities of each vertex, while being significantly more practical to implement. Moreover, our approach does not require knowledge of the geometric representation of the graph.

Acknowledgements. I would like to express my sincere gratitude to Rémi Watrigant for his guidance throughout this work. I would also like to thank Thomas Begin, Anthony Busson, Loïc Chassin de Kergommeaux and Julien Duron for valuable discussions on the subject. I also thank Mark Jerrum for pointing out relevant references. Finally, I greatly thank anonymous reviewers who pointed out the links with the work of Dyer, Greenhill and Müller, which improved significantly the relevance of this work.

References

1. de Berg, M., Bodlaender, H.L., Kisfaludi-Bak, S., Marx, D., van der Zanden, T.C.: A framework for exponential-time-hypothesis-tight algorithms and lower bounds in geometric intersection graphs. SIAM J. Comput. **49**(6), 1291–1331 (2020)
2. de Berg, M., Jansen, B.M., Mukherjee, D.: Independent-set reconfiguration thresholds of hereditary graph classes. Discret. Appl. Math. **250**, 165–182 (2018)
3. de Berg, M., Kisfaludi-Bak, S., Monemizadeh, M., Theocharous, L.: Clique-based separators for geometric intersection graphs. Algorithmica **85**(6), 1652–1678 (2023)
4. Bezáková, I., Sun, W.: Mixing of Markov chains for independent sets on chordal graphs with bounded separators. In: International Computing and Combinatorics Conference, pp. 664–676. Springer, Cham (2020)
5. Bharadwaj, B.S., Chandran, L.S.: Bounds on isoperimetric values of trees. Discret. Math. **309**(4), 834–842 (2009)
6. Bordewich, M., Kang, R.J.: Subset Glauber dynamics on graphs, hypergraphs and matroids of bounded tree-width. Electron. J. Comb. **21**(4), P4-19 (2014)
7. Borgs, C., et al.: Torpid mixing of some Monte Carlo Markov chain algorithms in statistical physics. In: 40th Annual Symposium on Foundations of Computer Science (Cat. No. 99CB37039), pp. 218–229. IEEE (1999)
8. Chen, Z., Liu, K., Vigoda, E.: Optimal mixing of Glauber dynamics: entropy factorization via high-dimensional expansion. SIAM J. Comput. **49**(6), STOC21–104–STOC21–153 (2023)
9. Dallard, C., Milanič, M., Štorgel, K.: Treewidth versus clique number. II. Tree-independence number. J. Comb. Theory Ser. B **164**, 404–442 (2024)
10. Dyer, M., Greenhill, C., Müller, H.: Counting independent sets in graphs with bounded bipartite pathwidth. Random Struct. Algorithms **59**(2), 204–237 (2021)

11. Eppstein, D., Frishberg, D.: Rapid mixing for the Hardcore Glauber dynamics and other Markov chains in bounded-treewidth graphs. In: 34th International Symposium on Algorithms and Computation (ISAAC 2023), pp. 30–1. Schloss Dagstuhl–Leibniz-Zentrum für Informatik (2023)
12. Jerrum, M.: Glauber dynamics for the hard-core model on bounded-degree h-free graphs. arXiv preprint arXiv:2404.07615 (2024)
13. Korach, E., Solel, N.: Tree-width, path-width, and cutwidth. Discret. Appl. Math. **43**(1), 97–101 (1993)
14. Lipton, R.J., Tarjan, R.E.: A separator theorem for planar graphs. SIAM J. Appl. Math. **36**(2), 177–189 (1979)
15. Marx, D., Sidiropoulos, A.: The limited blessing of low dimensionality: when $1-1/d$ is the best possible exponent for d-dimensional geometric problems. In: Proceedings of the Thirtieth Annual Symposium on Computational Geometry, pp. 67–76 (2014)
16. Mossel, E., Weitz, D., Wormald, N.: On the hardness of sampling independent sets beyond the tree threshold. Probab. Theory Relat. Fields **143**(3), 401–439 (2009)
17. Randall, D.: Slow mixing of Glauber dynamics via topological obstructions. In: Symposium on Discrete Algorithms: Proceedings of the Seventeenth Annual ACM-SIAM Symposium on Discrete Algorithms, vol. 22, pp. 870–879. Citeseer (2006)
18. Sinclair, A.: Improved bounds for mixing rates of Markov chains and multicommodity flow. Comb. Probab. Comput. **1**(4), 351–370 (1992)
19. Sly, A.: Computational transition at the uniqueness threshold. In: 2010 IEEE 51st Annual Symposium on Foundations of Computer Science, pp. 287–296. IEEE (2010)
20. Smith, W.D., Wormald, N.C.: Geometric separator theorems and applications. In: Proceedings 39th Annual Symposium on Foundations of Computer Science (Cat. No. 98CB36280), pp. 232–243. IEEE (1998)

Games and Graph Reconfiguration

Complexity and Algorithms
for ARC-KAYLES
and NON-DISCONNECTING ARC-KAYLES

Kyle Burke[1], Antoine Dailly[2,3], and Nacim Oijid[4(✉)]

[1] Florida Southern College, Lakeland, FL 33801, USA
kburke@flsouthern.edu
[2] CNRS, Mines de Saint-Étienne, Clermont-Auvergne-INP, LIMOS, Université
Clermont-Auvergne,
63000 Clermont-Ferrand, France
[3] INRAE, UR TSCF, Université Clermont Auvergne,
63000 Clermont-Ferrand, France
antoine.dailly@inrae.fr
[4] Department of Mathematics and Mathematical Statistics, Umeå University,
Umeå, Sweden
nacim.oijid@umu.se

Abstract. ARC-KAYLES is a game where two players alternate removing two adjacent vertices until no move is left, the winner being the player who played the last move. Introduced in 1978, its computational complexity is still open. More recently, subtraction games, where the players cannot disconnect the graph while removing vertices, were introduced. In particular, ARC-KAYLES admits a non-disconnecting variant that is a subtraction game. We study the computational complexity of subtraction games on graphs, proving that they are PSPACE-complete even on very structured graph classes (split, bipartite of any even girth). We give a quadratic kernel for NON-DISCONNECTING ARC-KAYLES when parameterized by the feedback edge number, as well as polynomial-time algorithms for clique trees and a subclass of threshold graphs. We also show that a sufficient condition for a second player-win on ARC-KAYLES is equivalent to the graph isomorphism problem.

Keywords: Combinatorial games · Parameterized complexity · Vertex deletion games · Subtraction games · Complexity · Arc-Kayles

1 Introduction

ARC-KAYLES is one of the many games introduced by Schaefer in his seminal paper on the computational complexity of games [22]. It is a two-player,

A. Dailly—The research of the second author was supported by the International Research Center "Innovation Transportation and Production Systems" of the I-SITE CAP 20-25 and by the ANR project GRALMECO (ANR-21-CE48-0004).
N. Oijid—The research of the third author was supported by the ANR project P-GASE (ANR- 21-CE48-0001-01) and the Kempe Foundation Grant No. JCSMK24-515 (Sweden).

E. Di Giacomo and D. Mondal (Eds.): WALCOM 2026, LNCS 16444, pp. 293–307, 2026.
https://doi.org/10.1007/978-981-95-7127-7_20

information-perfect, finite vertex deletion game, in which the players alternate removing two adjacent vertices and all their incident edges from a graph. The game ends when the graph is either empty or reduced to an independent graph, *i.e.*, a graph containing no edges, and the first player unable to play loses. Since the game is fully deterministic, we consider that both players play perfectly.

While Schaefer proved the PSPACE-completeness of many games, the complexity of ARC-KAYLES surprisingly remains open. Some FPT and XP algorithms have been proposed, and the game has been studied on some specific graph classes [19], but the results are quite scarce, the game is still generally open even on subdivided stars with three paths.

ARC-KAYLES can also be seen as an *octal game* played on a graph. Octal games are taking-breaking games played on heaps of counters, where the octal code defines how many counters one can take from a heap, as well as how the heap can be left after the move (empty, non-empty, split in two non-empty heaps). An overview of octal games can be found in volume 1 of [4]. In [2], octal games were expanded to be played on graphs, where removing counters means removing a connected subgraph, and splitting a heap means disconnecting the graph. Under this definition, ARC-KAYLES is the octal game **0.07**.

Non-disconnecting octal games, also called *subtraction games*, have also been studied on graphs [10]: for a non-empty set of integers S, the *connected subtraction game on S*, CSG(S), is played by removing connected subgraphs of order k (where $k \in S$) without disconnecting the graph. The non-disconnecting condition of subtraction games allows for better results, compared to ARC-KAYLES, on constrained classes such as trees. Note that ND-ARC-KAYLES (for NON-DISCONNECTING ARC-KAYLES) is CSG($\{2\}$).

The aim of this paper is to further the study of the computational complexity of subtraction games, particularly ND-ARC-KAYLES. We present several PSPACE-completeness results for subtraction games, a sufficient condition for symmetry strategies for ARC-KAYLES, and tractability results for ND-ARC-KAYLES: an FPT algorithm parameterized by the feedback edge number, *i.e.*, the minimum size of a set S of edges such that $G - S$ contains no cycle. as well as several polynomial-time algorithms on structured graph classes.

Combinatorial Game Theory Terminology. ARC-KAYLES and subtraction games are impartial games, a subset of combinatorial games which have been extensively studied. Since both players have the same possible moves, impartial games can have two possible *outcomes*: either the first player has a winning strategy (in which case the game has outcome $\mathcal{N}$) or the second player does (in which case the game has outcome $\mathcal{P}$). A position G' that can be reached by a move from a position G is called an *option* of G, the set of options of G is denoted by opt(G). If every option of G has outcome $\mathcal{N}$, then G has outcome $\mathcal{P}$; conversely, if any option has outcome $\mathcal{P}$, then G has outcome $\mathcal{N}$. A refinement of outcomes are the *Sprague-Grundy values* [15,24], denoted by $\mathcal{G}(G)$ for a given position G. The value can be computed inductively, with $\mathcal{G}(G) = \text{mex}(\text{opt}(G))$, where $\text{mex}(S)$ is the smallest non-negative integer not in S. A position G has outcome $\mathcal{P}$ if and only if $\mathcal{G}(G) = 0$. Furthermore, ARC-KAYLES and connected subtraction games are *vertex deletion games*, a family of games where the players alternate removing vertices from a graph following some fixed constraints. For more information and history about combinatorial games, we refer the reader to [1, 4, 7, 23].

ARC-KAYLES: A Difficult Game. Introduced in 1978 [22], ARC-KAYLES's complexity remains open to this day. However, there have been some algorithmic results. Determining the winner of ARC-KAYLES is FPT when parameterized by the number of rounds [20] and the vertex-cover number, and XP when parameterized by neighborhood diversity [18]. ARC-KAYLES on paths is exactly the octal game **0.07**, and its Grundy sequence is well-known to have period 34 and a preperiod of 68 [16]. ARC-KAYLES on grid graphs is exactly the game CRAM [14], and thus there are winning strategies for the first player on even-by-odd grids and for the second player on even-by-even grids (by an easy symmetry argument). The game remains open on odd-by-odd grids; Sprague-Grundy values for grids with at most 30 squares [25] and of size $3 \times n$ for n up to 21 and a few other odd-by-odd grids [21, 26] have been computed without any regularity appearing. Finally, the game seems to be hard to solve even on trees: an $\mathcal{O}(2^{\frac{n}{2}})$ enumerative algorithm was proposed [18], later improved to $\mathcal{O}(1.3831^n)$ [17]. While the game remains open on subdivided stars with three paths, one of which has size 1 [19]—by fixing the length of the second path and having the third one change—the authors conjectured that the sequence of Sprague-Grundy values thus obtained would ultimately be periodic. However, the Sprague-Grundy values of ARC-KAYLES are unbounded in the general case [9].

Subtraction Games on Graphs. Subtraction games have been extensively studied on heaps of counters, and there are several regularity results on them [1, 4]. It is important to note that these are different rulesets than NIMG or GRAPHNIM, games where NIM-heaps are embedded on the nodes or vertices of the graph [6, 12, 13]. Subtraction games on graphs were introduced in [10]. They are of particular interest since they restrict the types of moves available on sparser classes (such as trees), and because they cannot be solved through a simple symmetry strategy started by cutting the graph in two isomorphic subgraphs. Hence, their analysis differs fundamentally from the analysis of vertex deletion games that allow disconnecting.

In [10], the authors proved that, by fixing G and a vertex u, the Sprague-Grundy values of the sequence of graphs obtained by appending a path of increasing length to u are ultimately periodic. They also found regularity results for the family CSG($\{1, \ldots, N\}$) on subdivided stars with three paths, one of which has size 1, in contrast with ARC-KAYLES, which they used as base cases for subdivided stars with specific values of N. In [2], the authors gave polynomial-time algorithms for the game CSG($\{1, 2\}$) on subdivided stars and bistars.

As for the specific case of ND-ARC-KAYLES[1], which is the subtraction game CSG($\{2\}$), it is known to be polynomial-time solvable on trees, wheels and grids of height at most 3 [8].

---

[1] A playable version is available at https://kyleburke.info/DB/combGames/nonDisconnectingArcKayles.html.

Our Results and Outline of the Paper. Our focus here is the complexity of the problem that takes as input a graph and outputs the outcome of the considered game played on this graph.

We further expand algorithmic results on ARC-KAYLES and subtraction games on graphs. Our first contribution is a general PSPACE-completeness reduction for most subtraction games:

Theorem 1. *If S is finite and $1 \notin S$, then, $CSG(S)$ is PSPACE-complete, even on bipartite graphs of any given even girth.*

We also prove that $CSG(\{k\})$ is PSPACE-complete on split graphs, a very structured graph class. Note that ND-ARC-KAYLES is in this family of games.

Theorem 2. *For $k \geq 2$, $CSG(\{k\})$ is PSPACE-complete, even on split graphs.*

Both of those results are proved in Sect. 2, where we also show that deciding whether a symmetry strategy can be applied for ARC-KAYLES is as hard as the Graph-Isomorphism problem (GI-hard), even on bipartite graphs.

Finally, in Sect. 3, we give polynomial-time algorithms for deciding the winner of ND-ARC-KAYLES on several structured graph classes. Our main result is on parameterized complexity. Note that the parameters used for ARC-KAYLES (vertex cover and neighbourhood diversity) are linked to the number of rounds and do not allow for connectivity conditions, and as such would not be a good fit. On the contrary, since ND-ARC-KAYLES is trivial on trees, the feedback edge number is a natural parameter:

Theorem 3. ND-ARC-KAYLES *has a quadratic kernel parameterized by the feedback edge number of the graph.*

We also study clique trees and subclasses of threshold graphs. Due to space restrictions, proofs of results marked with (∗) are omitted, and can be found in the full version of the paper [5].

2 Hardness Results for Vertex Deletion Games

Before we get to the hardness results, we point out that $CSG(\{1\})$ is a strategy-free game. Every vertex can be removed over the course of a game, so players can count the number of moves remaining.

For our first proof, we need to define NODE-KAYLES, another impartial vertex deletion game. Each turn, the current player selects one vertex on the graph that is not adjacent to an already selected vertex (there is no distinction between the players' selected vertices). Just as in ARC-KAYLES, under normal play the last player to move wins. Note that, while NODE-KAYLES is a vertex deletion game (it can be seen as alternately removing one vertex and all its neighbors), it cannot be expressed as a subtraction or octal game on graphs, since the deletion of vertices is restricted by the graph itself and not by a specific set of integers. Recall that the *girth* of a graph is the length of its smallest cycle.

Theorem 1. *If S is finite and $1 \notin S$, then, $CSG(S)$ is PSPACE-complete, even on bipartite graphs of any given even girth.*

Proof. First, note that $CSG(S)$ is clearly in PSPACE. To prove completeness, we reduce from NODE-KAYLES, which was shown to be PSPACE-complete in [22].

Let $G(V, E)$ be a position of NODE-KAYLES. For a given finite set S such that $1 \notin S$, we construct a position $G'(V', E')$ of $CSG(S)$ such that the outcomes of G for NODE-KAYLES and of G' for $CSG(S)$ are the same.

Let $M = \max(S)$. We construct G' from G as follows:

- Create a so-called *control vertex* c, and attach $M + 1$ leaves to it;
- For every vertex $v \in V$, create a star with center v' and $M - 1$ leaves, and add the edge (cv');
- For every edge $(uv) \in E$, create a vertex e_{uv}, attach M leaves to it, and add the edges $(u'e_{uv})$ and $(v'e_{uv})$.

The reduction is depicted in Fig. 1. It is clearly polynomial, since we have $|V'| = M|V| + (M + 1)|E| + M + 2$.

We now prove that playing NODE-KAYLES on G is exactly the same as playing $CSG(S)$ on G'.

First, note that the only possible moves on G' consist of removing a vertex v' and its attached leaves. Indeed, since $1 \notin S$, it is impossible to play on an induced star of order more than M, except if the graph is reduced to exactly a star of order $M + 1$. Hence, it is always impossible to remove c or any e_{uv} for $(uv) \in E$, leaving the stars centered on vertices v' as the only possible moves.

Furthermore, if a vertex u' has been removed, then it is impossible to remove any vertex v' such that $(uv) \in E$, since we cannot remove the star centered on e_{uv}, as doing so would disconnect the graph.

Hence, the only possible moves for $CSG(S)$ on G' are on vertices equivalent to those of G, and playing on such a vertex (removing the star it is the center of) prevents either player from playing on any of its neighbors. This proves that the moves of $CSG(S)$ on G' are exactly the ones of NODE-KAYLES on G, and thus the game trees (and thus the outcomes and the Grundy values) of those two games are the same.

Note that G' has girth 4, since for any edge (uv) in G there is a 4-cycle $u'cv'e_{uv}$ in G', and that G' is bipartite, since any odd cycle in G now contains twice as many vertices since each edge of G has been subdivided, and no cycle containing c can be of odd length. Furthermore, the girth of G can be increased (but will remain even) by subdividing every edge enough times and adding $M + 1$ leaves to the newly created vertices: a cycle using c will have length $4 + 2i$ if we subdivide the edges incident with c i times, afterwards we can subdivide the edges of any more smaller cycles in G' to increase the girth. This transformation does not affect the proof, since the newly created vertices can never be removed, thus proving that G' can have any even girth.   □

We remark that the non-disconnecting variant of NODE-KAYLES, where the selected vertices and their neighbors must not induce a separating set (*i.e.*, a set of vertices that disconnects the graph), is also PSPACE-complete:

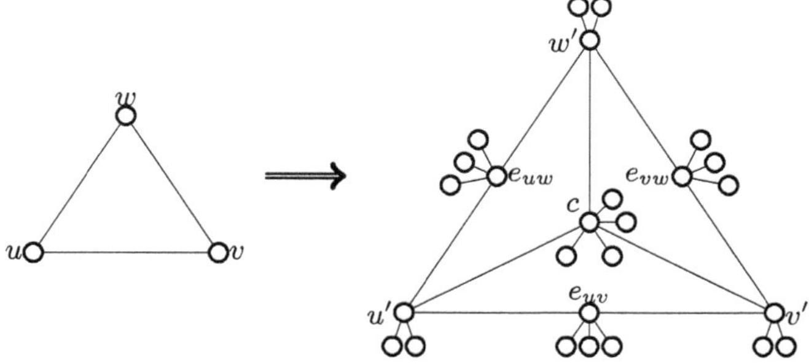

Fig. 1. An illustration of the reduction from NODE-KAYLES to CSG(S) with $S = \{2, 3\}$.

Theorem 4 ($*$). ND-NODE-KAYLES *is PSPACE-complete.*

For the next reduction, we need to define AVOID TRUE, a game played on positive (without negations) boolean formulas and an assignment of all variables such that the formula is false. (The variables are initially set to false.) Each turn the current player selects one false variable and switches the assignment to true. Players may not switch variables that cause the overall formula to evaluate to true. Under normal play, the last player to move wins. AVOID TRUE is known to be PSPACE-complete on formulas in disjunctive normal form [22]. Furthermore, recall that a split graph is a graph where the vertex set can be partitioned into a clique (so every possible edge within the part exists) and an independent set (so no edge exists within this part).

Theorem 2. *For $k \geq 2$, CSG($\{k\}$) is PSPACE-complete, even on split graphs.*

Proof. The proof is obtained by reduction from AVOID TRUE (see Fig. 2).

Let $\varphi = \bigvee_{i=1}^{m} C_i$ be an AVOID TRUE formula in disjunctive normal form over the variables $x_1, \ldots, x_n$. We construct a graph $G = (V, E)$ as follows:

- For any $1 \leq i \leq n$, we include k vertices v_i and $v_i^1, \ldots, v_i^{k-1}$ in V and all the edges $(v_i v_i^j)$ for $1 \leq j \leq k - 1$ in E.
- For any $1 \leq i, j \leq n$ we include the edge $(v_i v_j)$ in E.
- For any $1 \leq j \leq m$, we include a vertex c_j in V, and if $x_i \in C_j$, we add the edge $(v_i c_j)$ to E.

The obtained graph (an example for $k = 2$ is depicted in Fig. 2) is a split graph, as the v_i's are a clique and all the other vertices are an independent set.

First note that the only moves available are of the form $\{v_i, v_i^1, \ldots, v_i^{k-1}\}$. Indeed, as the v_i's are a vertex cover, any move has to contain at least one v_i, and if this move does not contain any of the v_i^j (for $1 \leq j \leq k - 1$), then this vertex becomes isolated and the graph is no longer connected.

We now prove that the first (second resp.) player wins in AVOID TRUE if and only if they win in ND-ARC-KAYLES.

Suppose that the first (second resp.) player has a winning strategy $\mathcal{S}$ in Avoid True. We define their strategy in ND-Arc-Kayles as follows:

- Whenever $\mathcal{S}$ would take a variable x_i, they play the set $\{v_i, v_i^1, \ldots, v_i^{k-1}\}$
- If their opponent plays a set $\{v_i, v_i^1, \ldots, v_i^{k-1}\}$, they consider that the variable x_i has been played against $\mathcal{S}$ in Avoid True.

Following this strategy, as $\mathcal{S}$ is a winning strategy in Avoid True, the second (first resp.) player will be the first to satisfy a clause C_j. The moves satisfying C_j will therefore be removing the last vertex $v_i \in C_j$ and will disconnect the graph. Therefore, the first (second resp.) player wins.

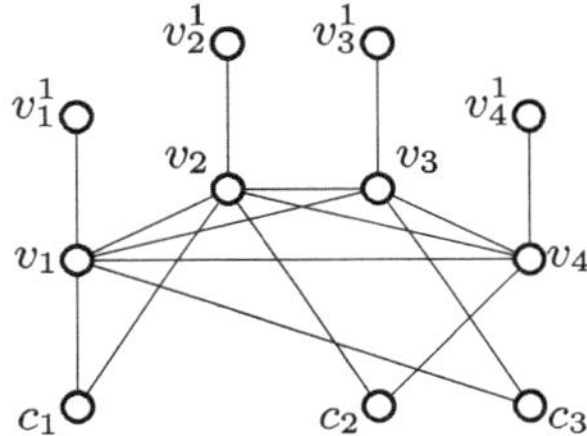

Fig. 2. An illustration of the reduction from Avoid True to CSG(k) with $k = 2$. The clauses are $C_1 = (x_1 \wedge x_2)$, $C_2 = (x_2 \wedge x_4)$ and $C_3 = (x_1 \wedge x_3)$. The v_i's form a clique, and the other vertices an independent set.

Returning to Arc-Kayles, we present a natural and sufficient winning condition for the second player, which is used in the case of grids. When the graph is highly symmetric, the second player may want to apply a symmetry strategy with respect to a given "point" in the graph. We show that, in the general case, deciding whether such a strategy is possible is as hard as the famous Graph Isomorphism problem. Note that such symmetry strategies are often used in games, and are often GI-hard to compute, see [3, 11].

Definition 1. *An automorphism f of a graph G is said to be* edge-disjoint *if for any $e \in E(G)$, we have $e \cap f(e) = \emptyset$.*

Theorem 5. *Let $G = (V, E)$ be a graph such that there exists an edge-disjoint involutive automorphism f of G. Then G has outcome $\mathcal{P}$ in* Arc-Kayles.

Proof. Let G be a graph, and let f be an involutive automorphism of G satisfying the hypothesis of the Theorem. Consider the following strategy for the second player: each time his opponent plays an edge e, he answers by playing $f(e)$. This move is always available as $e \cap f(e) = \emptyset$, and $G \setminus V(e, f(e))$ still satisfies the hypothesis of the theorem.

Unlike Arc-Kayles, in ND-Arc-Kayles, there are graphs with such an f where the symmetry strategy does not give a winner, such as the cycle on six vertices: it has this property, but the graph has outcome $\mathcal{N}$. Indeed, in ND-Arc-Kayles, the move $f(e)$ can be forbidden as it could disconnect the graph.

Theorem 6 (∗). *Verifying that a graph G admits an involutive automorphism f such that, for all e, $e \cap f(e) = \emptyset$ is GI-hard, even in bipartite graphs.*

Proof (Sketch of proof). The proof is a reduction from the graph isomorphism problem. Given two graphs G_1 and G_2, we add first an isolated vertex, then a universal vertex to both of them, resulting in G_1' and G_2', such that G_1' and G_2' are isomorphic if and only if G_1 and G_2 are, with a unique vertex of maximum degree in both of them. Then, we consider the graph G obtained from the disjoint union of G_1' and G_2'. Since, there are only two vertices of maximum degree, one in G_1' and one in G_2', the only nontrivial automorphism φ of G has to exchange them, ensuring an infinite distance between any vertex and its image by φ. Hence, G will admit an involutive automorphism satisfying the statement if and only if G_1 and G_2 are isomorphic. $\square$

Put more simply, deciding whether the second player can win a game of Arc-Kayles by applying a symmetry strategy is GI-hard in the general case.

3   ND-Arc-Kayles on Structured Graph Families

While there are very few polynomial-time algorithms for Arc-Kayles on structured graph classes, its non-disconnecting variant seems more approachable. In this section, we expand on the results in [8], where it was shown that trees are trivial to solve. Indeed, as with $\mathrm{CSG}(\{1\})$, no strategy is required, as all possible moves on a tree will end up being played, and so the Sprague-Grundy value is either 0 or 1. The result can be computed in linear time by maintaining a representation with levels and labeled leaves ensuring that no edge has to be considered more than once, when it is removed.

First, we will show that this result on trees can be extended to tree-like graphs: graphs with bounded feedback edge number and clique trees. Recall that the *feedback edge set* of a graph is a set of edges that, when removed, turn the graph into a forest; the smallest size of a feedback edge set is the *feedback edge number*. Also recall that a graph is a *clique tree* (or *block graph*) if every biconnected component is a clique.

Theorem 3. ND-Arc-Kayles *has a quadratic kernel parameterized by the feedback edge number of the graph.*

To prove Theorem 3, we present the following reduction rules. A reduction rule is called *safe* if it outputs a graph with the same outcome as the original graph. Moreover, since we may have to consider that a move is played, we consider the decision problem "Is the outcome of ND-Arc-Kayles played on G o?" for $o \in \{\mathcal{N}, \mathcal{P}\}$, to handle changes of the current player in the reduction rules.

Reduction Rule 1. *If $v \in G$ has at least three leaves attached to it, remove all of them but three.*

This reduction rule can easily be applied in $\mathcal{O}(|E|)$ since each edge is considered at most twice.

Lemma 1. *Reduction Rule 1 is safe*

Proof. Let v be a vertex of G and let $\ell_1, \ldots, \ell_k$ with $k \geq 3$ be the leaves attached to it. The first time a leaf ℓ_i is played, it must be through the edge $(\ell_i v)$, which would disconnect the other leaves, and at least two other leaves will remain. Thus, playing any of the leaves would not be a legal move. This statement remains true if only three leaves are attached to v. Thus, we can remove all but three of them without changing the legal moves, hence the outcome does not change. $\square$

Let C be a smallest connected subgraph of G containing all the two-connected components of G, and let $T_1, \ldots, T_p$ be the connected components of $G \setminus C$. Note that each T_i is a tree and is connected to C by a single edge. Let $r_1, \ldots, r_p$ be the vertices of $T_1, \ldots, T_p$ connected to C.

Lemma 2. *There are at most $2k - 2$ vertices of degree at least 3 in C.*

Proof. Note that C has minimum degree 2, is connected and has a feedback edge set of size k. Thus, we have $|E| = n - 1 + k$. Then, using the handshaking formula, we have:

$$2|E| = \sum_{v \in C} deg_C(v)$$

$$2(n - 1 + k) = 2n + \sum_{v \in C}(deg_C(v) - 2)$$

$$2n + 2k - 2 \geq 2n + |\{v \in C | deg_C(v) \geq 3\}|$$

$$2k - 2 \geq |\{v \in C | deg_C(v) \geq 3\}|$$

Reduction Rule 2. *Let $e_1 = (ab), e_2 = (cd)$ be two non-adjacent edges of T_{i_1} and T_{i_2} respectively (we may have $i_1 = i_2$), such that the moves on e_1 and e_2 are legal (we may have that the move on e_2 is only legal after the move on e_1). We transform G into $G \setminus \{a, b, c, d\}$.*

C can be computed in time $\mathcal{O}(|E|^2)$ by simply starting with $C = V(G)$ and removing one by one the vertices of degree 1 in C. Then this reduction rule can be applied in time $\mathcal{O}(|E|^2)$ since each time a pair of leaves is considered, one is removed.

Lemma 3. *Reduction Rule 2 is safe.*

Proof. Let G be the original graph and G' be the graph where one move in T_{i_1} and one move in T_{i_2} have been played for some $1 \leq i_1, i_2 \leq p$. Denote by $e_1 = (ab)$ and $e_2 = (cd)$ the two edges such that $G' = G \setminus \{a, b, c, d\}$. Note that it is possible that $i_1 = i_2$. We prove that G and G' have the same outcome by induction on $n = |V(G)|$. If $n \leq 4$, either the graph is $K_{1,3}$ the star with three leaves, or the graph can be emptied and there is nothing to do. Suppose $n \geq 5$.

Assume first that $o(G') = \mathcal{N}$, i.e. the first player wins in G'. Let f_1 be a winning first move in G'.

- If $f_1 \in C$, the first player also plays f_1 in G. Let f_2 be the answer of the second player in G.
  - If f_2 is a possible move in G', by the induction hypothesis, by considering G'' the current graph, G'' and $G'' \setminus \{a, b, c, d\}$ have the same outcome. Thus $o(G'') = \mathcal{N}$.
  - Otherwise, $f_2 \in \{(ab), (cd)\}$. Thus, the first player can play the other edge in $\{(ab), (cd)\}$ and we are left to the game in G' in which the first player has played f_1 and thus wins.
- If $f_1 = (a'b') \notin C$, then play e_1 instead. By the induction hypothesis, applied on $G \setminus \{a, b\}$ with the edges e_2 and f_1 (possible as f_1 is a legal move in G'), and since $G' \setminus \{a', b'\}$ has outcome $\mathcal{P}$, $G \setminus \{a, b\}$ has outcome $\mathcal{P}$, which proves that playing e_1 was a winning move for the first player.

Assume now that $o(G') = \mathcal{P}$. Let $f_1 = (a'b')$ be the first edge played in G.

- If $f_1 \in \{e_1, e_2\}$, then the second player answers by playing the second edge in $\{e_1, e_2\}$, and reaches the position G' which is winning for them by hypothesis.
- If $f_1 \notin \{e_1, e_2\}$, then by the induction hypothesis applied on $G \setminus \{a', b'\}$, with the edges e_1, e_2, $G \setminus \{a', b'\}$ has the same outcome as $G' \setminus \{a'b'\}$, which is winning for the second player by hypothesis.

This proves that the reduction rule preserves the outcome and thus is safe.

After applications of Reduction Rule 2, at most one move can be played in a tree T_i. Up to renaming the trees, suppose it is in T_1. Note that we cannot get rid of the last edge on which a move is possible since it may change the strategy in the rest of the graph, and it might be that the outcome doesn't change by removing an edge in a tree (see Fig. 3).

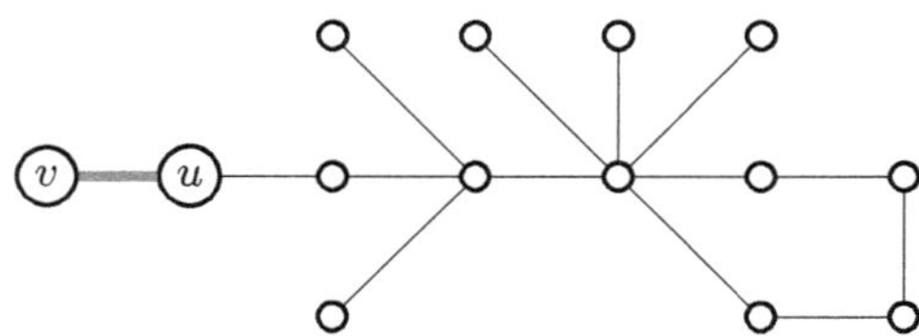

Fig. 3. Example of graph having a pendant edge (uv) such that $o(G) = \mathcal{N}$ and $o(G \setminus \{u, v\}) = \mathcal{N}$.

Reduction Rule 3. *If F is a forest containing at least 2 vertices in which no move can be played, and attached to some vertex u, we replace T_u the subtree induced by u and this forest by a tree T' of Fig. 5 such that $o(T_u) = o(T')$ and $o(T_u - \{u\}) = o(T' - \{u\})$ (or both $T_u - \{u\}$ and $T' - \{u\}$ are disconnected).*

If F is a forest containing at least 4 vertices in which exactly one legal move e can be played, and attached to some vertex u, we proceed the same way considering $F - \{e\}$ and we then add a path of length 2 attached to a leaf of T'.

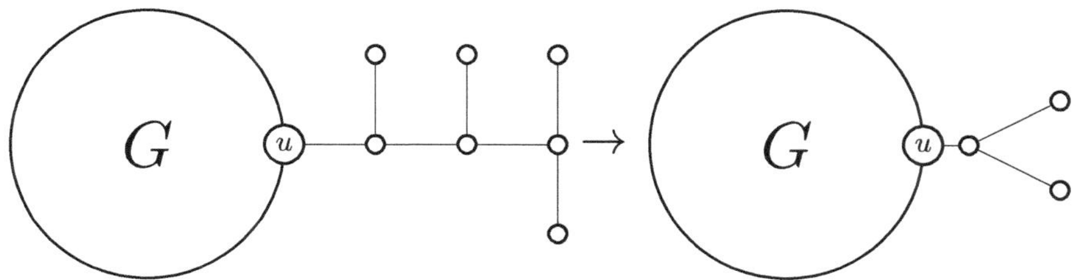

Fig. 4. An example of application of Reduction Rule 3. Here, we have $o(T_u) = \mathcal{P}$ and $o(T_u - \{u\}) = \mathcal{N}$.

An example of application of Reduction Rule 3 is provided in Fig. 4

This reduction rule can be applied in time $\mathcal{O}(|V|)$ since each vertex is in at most one forest and, as mentioned before, the outcome of the game played on a tree can be computed in linear time. If no legal move remains in F, we present in Fig. 5 a set of trees covering all the possible outcomes.

Lemma 4. *Reduction Rule 3 is safe.*

In order to prove Lemma 4, we prove the following stronger lemma (unlike in Reduction Rule 3, we take the root into account in the number of vertices):

Lemma 5. *Let G_0 be a graph, and let (T, r) and (T', r') be two rooted trees of order at least 3 (5 resp.), in which no (exactly one resp.) legal move not containing the root can be made. Let $u \in G$ be a vertex. Suppose $o(T) = o(T')$, $T - \{r\}$ is disconnected if and only if $T' - \{r'\}$ is and $o(T - \{r\}) = o(T' - \{r'\})$ otherwise. Let G and G' be the graphs in which T and T' are attached to G_0 by identifying r and u or r' and u respectively. Then $o(G) = o(G')$.*

Proof. Suppose that the first (second resp.) player has a winning strategy in G. We consider the following strategy in G'. While there is a vertex in G_0, it is not possible to play any edge containing u without disconnecting the remaining vertices of G_0 and the remaining vertices of T or T'. Thus, there is a natural bijection between the moves of G and the moves of G', since there is at most one move outside G_0 that can be made in the bijection if it exists. When the last move is G_0 is made, there are only four possibilities:

- At least three vertices remain in G_0. In this case, the game ends as it is not possible to claim u and a vertex of T nor T', since it would disconnect the remaining vertices of G_0 with the remaining vertices of T or T', which exists since we supposed that they have order at least 5.
- Exactly two vertices remain in G_0. Then, since all the moves have been legal, they are exactly u and a leaf ℓ connected to it. Since the last move of G_0 has been played, playing $(u\ell)$ is not legal, which means that the resulting graph would be $T - \{r\}$, which would be disconnected. But, by the hypothesis this means that $T' - \{r'\}$ is also disconnected and the game is also finished in G'.

– Exactly one vertex remains in G_0. It must be u since, otherwise, G would be disconnected. The game is then equivalent to the game played on T, which has the same outcome as the game played on T'

– No vertex remains in G_0. The game is then equivalent to the game played on $T - \{r\}$, which has the same outcome as the game played on $T' - \{r'\}$. $\square$

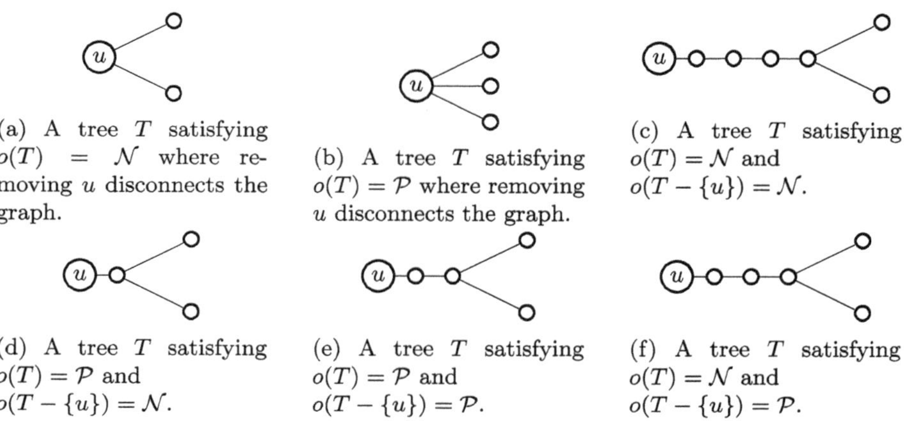

(a) A tree T satisfying $o(T) = \mathcal{N}$ where removing u disconnects the graph.

(b) A tree T satisfying $o(T) = \mathcal{P}$ where removing u disconnects the graph.

(c) A tree T satisfying $o(T) = \mathcal{N}$ and $o(T - \{u\}) = \mathcal{N}$.

(d) A tree T satisfying $o(T) = \mathcal{P}$ and $o(T - \{u\}) = \mathcal{N}$.

(e) A tree T satisfying $o(T) = \mathcal{P}$ and $o(T - \{u\}) = \mathcal{P}$.

(f) A tree T satisfying $o(T) = \mathcal{N}$ and $o(T - \{u\}) = \mathcal{P}$.

Fig. 5. A set of small trees with different outcomes.

After an exhaustive use of Reduction Rule 3, the graph consists in a core C of vertices of degree at least 2 and containing all the 2-connected components of the graph and trees of size at most 10 attached to it. Recall that, using Lemma 2, there are at most $2k - 2$ vertices having at least 3 neighbors in C. Thus, we only have to reduce long paths of vertices having 2 neighbors in C. To achieve this, we give types to the vertices of C as follows:

– type 0: if nothing is attached to it
– type 1: if a leaf is attached to it.
– type M: if a tree in which there is a legal move from a leaf attached to it.
– type B1: if exactly three leaves are attached to it.
– type B2: if any other tree is attached to it.

We also denote by M the set of vertices of type M, and by B the set of vertices of type B1 or B2. Note that, after exhausting Reduction Rule 2, there is at most one vertex of type M. Let $S = \{v \in C | deg_C(v) \geq 3\} \cup M$. Using Lemma 2, we have that $|S| \leq 2k - 1$.

We now prove that there exists a constant α such that we can reduce any path of vertices of type 0, 1 or B to a path of length at most α.

Let P be a path of vertices of type 0, 1 or B, and of length at least 3. Let a, b be the two vertices of S connected to the endpoints of P, denoted by u

and v respectively. For the rest of the proof, we also consider a new outcome $\mathcal{X}$ corresponding to "the move is illegal". Let $\mathcal{O} = \{\mathcal{N}, \mathcal{P}\}$ the classical set of outcome and $\mathcal{O}' = \mathcal{O} \cup \{\mathcal{X}\}$. We define the *extended outcome* $\tilde{o}(P) = (o_1 \times \varepsilon_1, o_2 \times \varepsilon_2, o_3 \times \varepsilon_3, o_4 \times \varepsilon_4, o_5 \times \varepsilon_5, O \times \varepsilon_6)$, with $o_1, o_2, o_3, o_4, o_5 \in \mathcal{O}'$ the outcome of the subtree where $a, b, \{a, u\}, \{b, v\}, \{a, b\}$ are removed respectively, $O \subset 2^{\mathcal{O} \times \mathcal{O}}$ the set of pairs of outcomes possible after a legal move in P on the subtree rooted in v and the one rooted in u, and $\varepsilon_1 \ldots, \varepsilon_6 \in \{0, 1\}$ the parity of the number in the remaining trees. Note that $\tilde{o}(P)$ can only take at most $3^5 * 2^4 * 2^6 = 248832$ different values.

Let $\mathcal{P}$ be a finite set of paths (with a forest attached to them) such that $\tilde{o}(\mathcal{P})$ covers all the possible outcomes of $\tilde{o}$.

Note that, in practice, these paths are paths in C, but are trees in G.

Reduction Rule 4. *While there is a path of vertices of type $0, 1$ or B of length at least 3, that is not in $\mathcal{P}$ in G, replace it with one of $\mathcal{P}$ with the same extended outcome.*

This reduction rule can be computed in time $\mathcal{O}(|E|^2)$ as for each of the graphs, the outcome of two trees has to be computed. The following Lemma can be proved using the same idea as Lemma 4, but with more case analysis.

Lemma 6 ($*$). *Reduction Rule 4 is safe.*

Proof (of Theorem 3). Let α be the maximum size of the trees in $\mathcal{P}$. By exhausting application of Reduction Rule 4, the resulting graph has at most $(2k - 1)$ vertices of degree at least 3 in C, all of them are connected through paths of lengths at most α, and each of these vertices is attached to trees of order at most 10. Overall, after applications of these rules, there are at most $40\alpha k^2 + \mathcal{O}(k)$ vertices in the resulting graph which has the same outcome as G. Thus, it is a quadratic kernel. Moreover, since all the reduction rules can be computed in quadratic time, we can compute this kernel in quadratic time and compute the winner in time $\mathcal{O}(|E|^2 f(k))$.

We now study more structured graph classes to obtain efficient algorithms: we provide polynomial-time algorithms for clique trees and a subclass of threshold graphs.

A clique tree is a tree where every vertex is expanded into a clique (which can be of order 1).

Theorem 7 ($*$). ND-Arc-Kayles *can be solved in polynomial-time on clique trees.*

Threshold graphs are a highly structured subclass of split graphs (*i.e.*, graphs that can be partitioned into a clique and an independent set), where the vertices can be ordered as $v_1, \ldots, v_n$ such that $N(v_1) \subseteq \ldots \subseteq N(v_n)$ (where $N(u)$ denotes the neighborhood of u). Other characterizations of threshold graphs exist, as they arise from the intersection of split graphs and cographs, two important structured graph classes. Recall that ND-Arc-Kayles is PSPACE-complete on split graphs (Theorem 2), so the following result implies that there is a complexity gap between clique-twin-free threshold graphs and split graphs:

Theorem 8 (∗). *Let $T = (K \cup S, E)$ be a threshold graph with $|K| = n$ and such that K is twin-free. If $n \geq 5$, then* ND-ARC-KAYLES *on T is outcome-equivalent to* $SUB(\{1,2\})$ *on a heap of size $n - 1$, and thus its outcome can be computed in polynomial time.*

4   Conclusion

In this paper, we expanded previous PSPACE-hardness results for connected subtraction games, a very large class of vertex deletion games. In particular, we proved that, provided 1 is not in the subtraction set, then the game is PSPACE-complete even on bipartite graphs of any given girth. We also proved that $CSG(k)$ is PSPACE-complete on split graphs, a very restricted class. However, the cases where 1 is in a subtraction set of at least two elements are still open, as well as other strong conditions on the graph (such as planarity, interval graphs, etc.).

We also showed that a natural sufficient winning condition for the second player for ARC-KAYLES is GI-hard, which is one of the few hardness results on this game. While this is not enough to determine the computational complexity of ARC-KAYLES, such an approach might help to understand more about it. A future research direction would be to study more necessary and sufficient conditions for first and second player winning, and determine their complexity.

On the polynomial-time side, we showed that ND-ARC-KAYLES is easy on tree-like graphs, such as clique trees, and gave a polynomial kernel parameterized by feedback edge number. Note that this contrasts with ARC-KAYLES, where even subdivided stars with three paths are open. Future research could consider other tree-like classes with unbounded feedback edge number, such as cacti. Furthermore, our study of clique-twin-free threshold graphs seems to imply that the complexity gap may lie between threshold and split graphs. Another direction would be to look at cographs, since threshold graphs are the intersection of split (for which the game is PSPACE-complete) and cographs.

References

1. Albert, M., Nowakowski, R., Wolfe, D.: Lessons in Play: An Introduction to Combinatorial Game Theory. AK Peters/CRC Press (2019)
2. Beaudou, L., et al.: Octal games on graphs: the game 0.33 on subdivided stars and bistars. Theoret. Comput. Sci. **746**, 19–35 (2018)
3. Bensmail, J., Fioravantes, F., Mc Inerney, F., Nisse, N.: The largest connected subgraph game. Algorithmica **84**(9), 2533–2555 (2022)
4. Berlekamp, E.R., Conway, J.H., Guy, R.K.: Winning Ways for Your Mathematical Plays. AK Peters/CRC Press (2001–2004)
5. Burke, K., Dailly, A., Oijid, N.: Complexity and algorithms for arc-kayles and non-disconnecting arc-kayles. arXiv preprint arXiv:2404.10390 (2024)
6. Burke, K., George, O.: A PSPACE-complete graph nim. Games No Chance **5**, 259–269 (2019)
7. Conway, J.H.: On Numbers and Games. CRC Press (2000)

8. Dailly, A.: Criticalité, identification et jeux de suppression de sommets dans les graphes: Des étoiles plein les jeux. Ph.D. thesis, Université de Lyon (2018)
9. Dailly, A., Gledel, V., Heinrich, M.: A generalization of Arc-Kayles. Int. J. Game Theory **48**, 491–511 (2019)
10. Dailly, A., Moncel, J., Parreau, A.: Connected subtraction games on subdivided stars. Integers: Electron. J. Comb. Number Theory **19**, G3 (2019)
11. Duchêne, E., Oijid, N., Parreau, A.: Bipartite instances of INFLUENCE. Theoret. Comput. Sci. **982**, 114274 (2024)
12. Fukuyama, M.: A Nim game played on graphs. Theoret. Comput. Sci. **304**(1–3), 387–399 (2003)
13. Fukuyama, M.: A Nim game played on graphs II. Theoret. Comput. Sci. **304**(1–3), 401–419 (2003)
14. Gardner, M.: Mathematical games: cram, crosscram and quadraphage: new games having elusive winning strategies. Sci. Am. **230**(2), 106–108 (1974)
15. Grundy, P.M.: Mathematics and games. Eureka **2**, 6–8 (1939)
16. Guy, R.K., Smith, C.A.: The G-values of various games. In: Mathematical Proceedings of the Cambridge Philosophical Society, vol. 52, no. 3, pp. 514–526. Cambridge University Press (1956)
17. Hanaka, T., Kiya, H., Lampis, M., Ono, H., Yoshiwatari, K.: Faster winner determination algorithms for (Colored) Arc Kayles. J. Comput. Syst. Sci. 103716 (2025)
18. Hanaka, T., Kiya, H., Ono, H., Yoshiwatari, K.: Winner determination algorithms for graph games with matching structures. Algorithmica 1–17 (2023)
19. Huggan, M.A., Stevens, B.: Polynomial time graph families for Arc Kayles. Integers **16**, A86 (2016)
20. Lampis, M., Mitsou, V.: The computational complexity of the game of set and its theoretical applications. In: Pardo, A., Viola, A. (eds.) LATIN 2014. LNCS, vol. 8392, pp. 24–34. Springer, Heidelberg (2014). https://doi.org/10.1007/978-3-642-54423-1_3
21. Lemoine, J., Viennot, S.: Computation records of normal and misère cram. http://sprouts.tuxfamily.org/wiki/doku.php?id=records#cram
22. Schaefer, T.J.: On the complexity of some two-person perfect-information games. J. Comput. Syst. Sci. **16**(2), 185–225 (1978)
23. Siegel, A.N.: Combinatorial Game Theory, vol. 146. American Mathematical Society (2013)
24. Sprague, R.: Über mathematische kampfspiele. Tohoku Math. J. First Series **41**, 438–444 (1935)
25. Uiterwijk, J.W.: Construction and investigation of cram endgame databases. ICGA J. **40**(4), 425–437 (2018)
26. Uiterwijk, J.W.: Solving cram using combinatorial game theory. In: Advances in Computer Games, pp. 91–105. Springer, Cham (2019)

Can One Flip Spoil it All?

Pragya Arora[1]([✉])(iD), Palash Dey[2](iD), and Neeldhara Misra[1](iD)

[1] Indian Institute of Technology Gandhinagar, Gandhinagar, India
{pragya.arora,neeldhara.m}@iitgn.ac.in
[2] Indian Institute of Technology Kharagpur, Kharagpur, India
palash@cse.iitkgp.ac.in

Abstract. We study the query complexity of finding tournament solutions in near-transitive tournaments, which are tournaments obtained by flipping a small number of edges in a transitive tournament. While general tournaments require $\Omega(n^2)$ queries for many solution concepts such as Copeland, Top Cycle, and Uncovered Set, we show that we can do better for near-transitive tournaments in several scenarios.

In particular, we introduce three query models: the standard model, where queries return orientations in the input tournament; the partial-flip model, where queries return orientations from the underlying transitive tournament and we have access to edge-flip queries; and the full-flip model, which additionally provides vertex-flip queries that indicate whether a vertex is incident to a flipped edge.

Our main result shows that in the full-flip model, if at most ℓ edges are flipped from a transitive tournament where $\ell \leqslant \sqrt{n \log n}$, then the set of Copeland winners, top cycle, and uncovered set can all be determined in $O(n \log n)$ queries. This represents an improvement from the $\Omega(n^2)$ lower bound in general tournaments. We also show the fine-grained query complexity of obtaining Copeland winners parameterized by the number of flipped edges.

Finally, we study pseudo-transitive tournaments—tournaments that are at most a single flip away from a transitive tournament—at length, and consider the query complexity of finding the Copeland winners, Top cycle, and Uncovered set in all three models. We also study the query complexity of determining if a pseudo-transitive tournament has a Condorcet winner in these models. Surprisingly, in the standard model, we *need* $2n - \lceil \log n \rceil - 2$ queries to determine if a pseudo-transitive tournament has a Condorcet winner: and these many queries are in fact sufficient to address the question on general tournaments, so the restriction to pseudo-transitive tournaments is not helpful. This situation persists in the partial-flip model, although we show an improved bound of $n + \lceil \log n \rceil$ queries in the full-flip model.

Keywords: Tournament solutions · Query complexity · Near-transitive tournaments · Pseudo-transitive tournaments · Condorcet · Copeland · Top cycle · Uncovered set

E. Di Giacomo and D. Mondal (Eds.): WALCOM 2026, LNCS 16444, pp. 308–322, 2026.
https://doi.org/10.1007/978-981-95-7127-7_21

1   Introduction

A tournament is a directed graph where every pair of distinct vertices is connected by exactly one directed edge. Tournaments naturally model scenarios involving pairwise comparisons, such as sports competitions, elections where voters rank candidates, and various decision-making processes [4]. A central objective in these settings is to identify the "best" alternatives, a notion formalized through various tournament solution concepts.

Common solution concepts include the Condorcet winner (an alternative that defeats every other alternative), the Copeland set (alternatives with the maximum number of wins), the Top Cycle (the smallest dominant set of alternatives), and the Uncovered Set (defined in due course). While many of these solutions are efficiently computable once the entire tournament structure is known, the process of learning the tournament structure itself can be resource-intensive.

We study this cost through the lens of *query complexity*. In this model, an algorithm seeks to determine the solution set by making adaptive queries about the orientation of specific edges. The objective is to minimize the number of queries required in the worst case. The query complexity landscape depends significantly on the underlying structure of the tournament. The simplest class is *transitive* tournaments, which admit a linear ordering of the vertices such that all edges align with this order. In a transitive tournament, most standard solution concepts coincide with the unique Condorcet winner. Moreover, the entire structure can be learned efficiently using standard sorting algorithms, requiring $O(n \log n)$ queries (Proposition 1).

However, real-world tournaments rarely exhibit perfect transitivity. In the context of general tournaments, the query complexity changes significantly. For many widely used solution concepts, including Copeland, Top Cycle, and Uncovered Set, any algorithm must query $\Omega(n^2)$ edges in the worst case [6]. This implies that determining the winners is asymptotically as difficult as reading the entire adjacency matrix. A notable exception is finding the Condorcet winner, which can be accomplished with $O(n)$ queries [9].

This dichotomy between the complexity for transitive and general tournaments motivates the central question of this paper: can we obtain improved query complexity bounds if the input tournament is "close" to being transitive? We investigate this by studying *near-transitive tournaments*, defined as tournaments obtained by flipping a limited number (at most ℓ) of edges in an underlying transitive tournament $T^\star$.

In the standard query model, where queries reveal the orientation of edges in the input tournament T, the promise of near-transitivity is difficult to leverage. As we observe, even determining if a tournament is transitive, or identifying the specific flipped edges, requires $\Omega(n^2)$ queries, as an adversary can conceal the flips until the final queries. Consequently, algorithms in this model have no direct access to the structure of the underlying $T^\star$. To explore whether access to $T^\star$ facilitates finding solutions in T, we introduce new query models where we have access to the structure of $T^\star$, and queries that reveal whether or not a pair of players were involved in a flipped edge.

Related Work. The study of query complexity in tournaments has traditionally focused on general tournaments, where structural assumptions are absent.

Condorcet Winner. A foundational result concerns the identification of the Condorcet winner. Balasubramanian et al. [2] showed that the query complexity of determining whether a Condorcet winner exists is at most $2n - \lfloor \log n \rfloor - 2$. Procaccia [9] subsequently proved that this bound is tight, establishing that exactly $2n - \lfloor \log n \rfloor - 2$ queries are necessary and sufficient. The standard algorithmic approach involves conducting a knockout tournament ($n - 1$ queries) to identify a candidate winner, followed by verifying this candidate against others they have not yet faced. Maiti and Dey [8] extended this result, showing that the query complexity of finding the set of Condorcet non-losers is also $2n - \lfloor \log n \rfloor - 2$.

General Tournament Solutions. For more complex tournament solutions, the results in the general setting are predominantly negative. Dey [6] systematically analyzed the worst-case query complexity for several common tournament solutions, establishing that $\Omega(n^2)$ pairwise comparisons are necessary to determine the winners under the Copeland, Slater, Markov, Bipartisan (minimal covering), Uncovered Set, Banks [10], and Top Cycle solution concepts. These lower bounds rely on adversarial constructions where the identity of the winner depends on the orientation of a few hidden edges, compelling any correct algorithm to query nearly the entire tournament. This highlights a stark contrast: while sorting a transitive tournament takes $O(n \log n)$ comparisons, identifying winners under these concepts in an intransitive tournament is asymptotically as hard as reading the entire adjacency matrix.

Structured Tournaments and Ranking. The study of algorithms for tournaments that are "close" to transitive has often been explored in the context of ranking and feedback arc sets (FAS). A tournament is ℓ-close to transitive if it has a FAS of size at most ℓ. The problem of finding a ranking that minimizes the number of upsets (the Slater solution) is equivalent to finding a minimum FAS, which is NP-hard in general tournaments [1]. However, approximation algorithms [7] and heuristics for sorting tournaments with few errors [3] have been developed. The computational complexity has also been studied under parameterizations, showing fixed-parameter tractability when parameterized by the size of the FAS [1].

In the context of query complexity, work has explored scenarios where structural information might reduce the number of required queries. For instance, Dey [6] analyzed the query complexity when the Top Cycle is bounded by size k, showing winner determination requires $O(nk + \frac{n \log n}{\log(1 - 1/k)})$ queries. Other related research explores preference elicitation under structural assumptions on the preferences, such as single-peakedness [5]. Our work contributes to this line of research by focusing on the parameterization by the number of edge flips (FAS size) and by introducing novel query models (partial-flip and full-flip) that provide explicit access to the underlying transitive structure, allowing us to bypass the $\Omega(n^2)$ barrier in several scenarios.

Our Contributions. We consider three query models: the standard model, the partial-flip model, and the full-flip model. Suppose we have a pseudo-transitive tournament T on n players which is obtained by flipping at most one edge in a transitive tournament $T^\star$. In the standard model, when the algorithm queries a pair (u, v), the response is the orientation between u and v in T.

Notice that if we would like to determine whether T is transitive or not; and if not, determine the edge that is flipped in $T^\star$ to obtain T, we need to query all pairs of players, which takes $\binom{n}{2}$ queries. Indeed, let σ be an arbitrary but fixed ordering on $[n]$. An adversary responds to any query (u, v) with $u \to v$ if $\sigma(u) < \sigma(v)$ and $v \to u$ if $\sigma(u) > \sigma(v)$. It can be checked that this will force any correct algorithm to query all pairs of players.

Knowing the edge that is flipped may not be necessary to leverage the pseudo-transitive structure of T, but it does mean that in the standard model, we have no direct access to the structure of $T^\star$. Since we set out to understand if it helps to have some access to the structure of the underlying transitive tournament $T^\star$ to find solutions in T, we also introduce and consider the following extended query models:

- In the *partial-flip model*, when the algorithm queries a pair (u, v), the response is the orientation between u and v in $T^\star$. Additionally, the algorithm has access to a `rev` query, which also takes in a pair (u, v) and returns `true` if the edge (u, v) is flipped in $T^\star$ to obtain T, and `false` otherwise.
- The *full-flip model* is the same as the partial-flip model, but the algorithm additionally has access to a `vrev` query, which takes in a player u and returns `true` if there is some player $v \neq u$ such that the edge (u, v) is flipped in $T^\star$ to obtain T, and `false` otherwise. In other words, `vrev` returns `true` if u is involved in at least one edge reversal.

These models extend beyond near-transitive tournaments: if T is a tournament obtained by flipping at most ℓ edges in a transitive tournament $T^\star$, then we can continue to use these models to learn about T via $T^\star$. Recall that the DAG order of a transitive tournament $T^\star$ can be computed in $O(n \log n)$ time using a sorting algorithm.

Proposition 1 (folklore.). *In a transitive tournament with n players, the top k players of these n players can be identified in $n + (k - 1) \log n$ queries.*

In the full-flip model, if T is obtained by flipping at most ℓ edges on a transitive tournament $T^\star$, it turns out that we can learn the entire tournament structure using only n vertex flip queries, $\binom{2\ell}{2}$ edge flip queries, and $O(n \log n)$ orientation queries:

1. Get the DAG order of $T^\star$ using $O(n \log n)$ queries via Proposition 1.
2. Use n `vrev` queries to identify which vertices are involved in edge reversals.
3. If only ℓ edges are flipped in $T^\star$ to obtain T, they "affect" at most 2ℓ vertices. Thus we can use the `rev` queries on the entire subtournament induced by the affected vertices—which would amount to at most $\binom{2\ell}{2}$ orientation queries—to recover the structure of T.

Contrast this with the standard model, which requires $\binom{n}{2}$ queries to learn the entire tournament structure. If ℓ is at most $\sqrt{n \log n}$, then the full-flip model requires only $O(n \log n)$ queries to learn the entire tournament structure. This implies an improved query complexity for all tournament solutions with an $\Omega(n^2)$ lower bound in the full-flip model.

Theorem 1 (Efficient Tournament Solutions in Full-Flip Model) *In the full-flip model, if T is a tournament obtained by flipping at most ℓ edges from a transitive tournament T^* where ℓ is at most $\sqrt{n \log n}$, then the set of Copeland winners, the top cycle, and the uncovered set can all be determined in $O(n \log n)$ queries.*

We also show the following general upper bound for the query complexity of determining the set of Copeland winners parameterized by ℓ, in the standard and full-flip models.

Theorem 2 (Copeland Winners with ℓ Edge Reversals). *In the full-flip model, if T is a tournament on n players obtained from a transitive tournament T^* by flipping at most ℓ edges, then the set of Copeland winners can be computed using at most: $2n + (\ell + 1) \log n + \binom{2\ell}{2}$ queries. In the standard query model, at most $(4\ell - 2) \cdot n$ queries suffice.*

Beyond this, we shift our focus to the special case of pseudo-transitive tournaments across all three query models. Our results are summarized in Table 1. We describe proofs of Theorems 3 and 4 in Sects. 3 and 4 respectively, while all other results are deferred to a full version of the paper due to lack of space.

For Copeland, Top Cycle, and Uncovered Set, the lower bound for general tournaments is $\Omega(n^2)$ and the upper bound for transitive tournaments—where all these solutions coincide with the Condorcet winner—is $n - 1$. We show linear upper bounds for pseudo-transitive tournaments in all models for all these solutions, suggesting that being "close to transitivity" is indeed useful.

On the other hand—somewhat surprisingly—in the standard model, determining whether a Condorcet winner exists in a pseudo-transitive tournament requires as many queries as it does in a general tournament! This is partly because we can already do very well on general tournaments: $2n - \lceil \log n \rceil - 2$ queries already suffice. We show that these many queries are necessary even when the input is pseudo-transitive in the standard model. However, on the issue of determining if the tournament has a Condorcet winner, we can do slightly better in the full-flip model with $n + \lceil \log n \rceil$ queries, while we are worse off in the partial-flip model with $n + \lceil \log n \rceil$ queries. This case suggests that we may potentially be *worse* off in the partial model compared to the standard model, although we cannot say this for sure because we do not have a matching lower bound.

Table 1. Query complexity bounds for tournament solution concepts in pseudo-transitive tournaments. Here, query models 1, 2, and 3 correspond to the standard model, partial-flip model, and full-flip model, respectively. The colors in the table indicate the nature of the bounds: purple shows improved upper bounds, orange shows improved lower bounds, blue denotes unchanged bounds, and grey represents the trivial lower bound in the partial-flip and full-flip models. The results for Top Cycle and Uncovered Set can be found in a full version of this work.

Solution	Model	General		Pseudo-Transitive		Transitive	
		LB	UB	LB	UB	LB	UB
Condorcet (Theorem 3)	1	$2n - \lceil \log n \rceil - 2$		$2n - \lceil \log n \rceil - 2$		$n - 1$	
	2			$2n - \lceil \log n \rceil - 2$	$2n + \lceil \log n \rceil - 3$		
	3			$n - 1$	$n + \lceil \log n \rceil$		
Copeland (Theorem 4)	1	n^2		$2n$	$4n - \lceil \log n \rceil - 7$	$n - 1$	
	2			$n - 1$	$2n + 2\lceil \log n \rceil - 1$		
	3			$n - 1$	$n + 2\lceil \log n \rceil + 3$		
Top Cycle	1	n^2		$n - 1$	$4n - \lceil \log n \rceil - 7$	$n - 1$	
	2			$n - 1$	$3n - 4$		
	3			$n - 1$	$3n - 4$		
Uncovered Set	1	n^2		$n - 1$	$3n - 5$	$n - 1$	
	2			$n - 1$	$2n + \lceil \log n \rceil - 3$		
	3			$n - 1$	$2n + \lceil \log n \rceil - 3$		

2 Preliminaries

A *tournament* $T = (V, E)$ on n players is a complete directed graph where $V = \{a_1, a_2, \ldots, a_n\}$ represents the set of players and for every pair of distinct players $a_i, a_j \in V$, exactly one of the edges (a_i, a_j) or (a_j, a_i) is present in E. We write $a_i \rightarrow a_j$ to denote that a_i defeats a_j, i.e., $(a_i, a_j) \in E$. A tournament is *transitive* if there exists an ordering σ of the players such that $a_i \rightarrow a_j$ if and only if $\sigma(a_i) < \sigma(a_j)$. Given a tournament T and an arbitrary but fixed order σ on the vertices of T, to conduct a knockout tournament with seed σ involves pairing up the players according to σ for the first round of the knockout tournament and proceeding with the winners to the next round, and so on, until a winner is found. Note that this involves $\lceil \log n \rceil$ rounds and $(n - 1)$ matches.

A tournament T is *pseudo-transitive* if it can be obtained by flipping at most one edge in some transitive tournament $T^\star$. More generally, a tournament T is *near-transitive* (or *ℓ-close to transitivity*) if it can be obtained by flipping at most ℓ edges in some transitive tournament $T^\star$. We refer to $T^\star$ as the underlying transitive tournament.

A player a_i is a *Condorcet winner* in tournament T if a_i defeats every other player, i.e., $a_i \rightarrow a_j$ for all $j \neq i$. Equivalently, a_i has out-degree $n - 1$. A tournament may have at most one Condorcet winner, and some tournaments have no Condorcet winner. The *Copeland score* of player a_i in tournament T

is the number of players that a_i defeats, i.e., $|\{a_j : a_i \to a_j\}|$. The *Copeland winners* (or *Copeland set*) are the players with the maximum Copeland score. In our notation, if $\{a_1, \ldots, a_n\}$ represents the DAG order of the underlying transitive tournament $T^\star$, then a_i has Copeland score $n - i$ in $T^\star$. The *top cycle* of a tournament T is the smallest non-empty set of players S such that every player in S defeats every player outside S. Equivalently, it is the smallest dominant set. In a transitive tournament, the top cycle consists of only the Condorcet winner. A player a_i *covers* player a_j if $a_i \to a_j$ and for every player a_k such that $a_j \to a_k$, we also have $a_i \to a_k$. In other words, a_i covers a_j if a_i defeats a_j and $D(a_i) \supseteq D(a_j)$, where $D(a_x) = \{a_y : a_x \to a_y\}$ denotes the set of players defeated by a_x. We also use $\overline{D}(a_x) = \{a_y : a_y \to a_x\}$ to denote the set of players that defeat a_x. The *uncovered set* consists of all players who are not covered by any other player.

We assume that a query algorithm for a solution is required to output all members in the solution set in question, or correctly determine that the solution set is empty, as may be the case with the Condorcet winner.

3   Condorcet

For general tournaments, it is well-known that a Condorcet winner can be found using $2n - \lceil \log n \rceil - 2$ queries using the following steps after fixing an arbitrary ordering σ on the players:

- Conduct a knockout tournament with seeding σ and let p be the winner of this tournament.
- Query p against every other player $q \neq p$ in the tournament who p has not yet defeated.

If p stands undefeated at the end of this process, then p is the Condorcet winner. Otherwise, there is no Condorcet winner. Note that the first step requires $(n-1)$ queries, and the second step requires $(n - \lceil \log n \rceil - 1)$ queries, for a total of $2n - \lceil \log n \rceil - 2$ queries.

We show that this lower bound extends even to the situation when the input is known to be a pseudo-transitive tournament; both in the standard model as well as in the partial flip model. On the other hand, we are able to do better in the full flip model: the existence of a Condorcet winner can be settled using $n + \lceil \log n \rceil$ queries.

Theorem 3 (Condorcet). *Suppose we are promised that the input to our query algorithm is a pseudo-transitive tournament on n players. Then:*

- *In the standard model, $2n - \lceil \log n \rceil - 2$ queries are necessary and sufficient to determine a Condorcet winner.*
- *In the partial-flip model, a Condorcet winner can be found using at most $2n + \lceil \log n \rceil - 3$ queries, and $2n - \lceil \log n \rceil - 2$ queries are necessary.*
- *In the full-flip model, a Condorcet winner can be found using at most $n + \lceil \log n \rceil$ queries, and at least $n - 1$ queries are necessary.*

Let T denote the input pseudo-transitive tournament obtained by flipping at most one edge of a transitive tournament $T^\star$. Further, let $\{a_1, \ldots, a_n\}$ denote the players in $T^\star$ in the DAG order, with a_1 being the Condorcet winner. While the algorithm is not privy to this ordering, we introduce this notation to enable discussion in the analysis.

Observe that if no edge is flipped or if the edge that is flipped is not incident to a_1, then the Condorcet Winner of T is $\{a_1\}$, which is same as the Condorcet Winner of $T^\star$. Now, suppose the edge that is flipped is (a_1, a_k) for some $k \geq 2$. If $k = 2$, then $\{a_2\}$ is the Condorcet winner of T, and if $k \neq 2$ then T has no Condorcet winner. Therefore, in a pseudo-transitive tournament, the possible Condorcet winner is either a_1 or a_2.

3.1　Standard Model and Partial Flip Models

For the upper bound in partial-flip model, consider the following algorithm. First conduct a knockout tournament with an arbitrary seeding to find a champion p using $(n - 1)$ queries. With another $(n - 1)$ queries, determine if any edge incident to p is flipped. If not, then p is the Condorcet winner. Otherwise, we find the second-best player in $T^\star$ using an additional $\lceil \log n \rceil - 1$ queries. If the edge incident to p and the second-best player is flipped, then the second-best player is the Condorcet winner. Otherwise, the tournament has no Condorcet winner. Note that we used a total of $2n + \lceil \log n \rceil - 3$ queries.

For the lower bound, we adapt the proof by Procaccia [9] demonstrating a lower bound of $2n - \lceil \log n \rceil - 2$ for general tournaments. The key insight is that we design an adversary that maintains a pseudo-transitive tournament throughout the query process.

Proof (Proof of Theorem 3; the lower bound in the standard and partial flip models). We establish that any deterministic algorithm requires at least $2n - \lceil \log n \rceil - 2$ queries to determine if a pseudo-transitive tournament has a Condorcet winner. The theorem trivially holds for $n = 1, 2$ and is easy to verify for $n = 3$. We therefore assume that $n \geq 4$.

We construct an adversary that responds to the algorithm's queries while maintaining the following invariant: the partial tournament revealed so far can be extended to a pseudo-transitive tournament. At time t, let $T^{(t)}$ denote the partial tournament based on the queries up to time t. The adversary maintains two disjoint sets:

- $\text{in}^{(t)}$: the set of players who have not lost any match by time t (potential Condorcet winners).
- $\text{out}^{(t)}$: an ordered sequence of players who have lost at least one match by time t, ordered by the time they first lost.

Initially, $\text{in}^{(0)} = \{1, 2, \ldots, n\}$ and $\text{out}^{(0)} = \emptyset$. We define two notions of score:

- The *score* of player i at time t, denoted $s_i^{(t)}$, is the number of players i has defeated based on the first t queries (i.e., i's outdegree in $T^{(t)}$).

- The *knockout score* (k-score) of player i at time t, denoted $\bar{s}_i^{(t)}$, is the number of other players that i ousted; that is, the number of players j such that at some time $t' < t$ when $j \in \text{in}^{(t')}$, the algorithm queried (i, j) and received $i \to j$.

Adversary Strategy. Given a query (i, j) at time t, the adversary responds as follows:

- If $i, j \in \text{in}^{(t)}$: The adversary responds with $i \to j$ if $\bar{s}_i^{(t)} \leqslant \bar{s}_j^{(t)}$, moves j to $\text{out}^{(t+1)}$ (placing it at the beginning of the sequence), and updates the k-scores. Otherwise, it sets $j \to i$ and moves i to $\text{out}^{(t+1)}$.
- If $i \in \text{in}^{(t)}$ and $j \in \text{out}^{(t)}$: The adversary sets $i \to j$. Conversely, if $j \in \text{in}^{(t)}$ and $i \in \text{out}^{(t)}$, then $j \to i$.
- If $i, j \in \text{out}^{(t)}$: If i appears before j in the sequence $\text{out}^{(t)}$, then $i \to j$; otherwise, $j \to i$.

Claim 1. At any time t, the partial tournament $T^{(t)}$ constructed according to the adversary strategy is transitive and can be extended to either a complete transitive tournament or a complete pseudo-transitive tournament.

Proof (Proof of Claim). We prove by induction that $T^{(t)}$ is always acyclic with a DAG ordering where all players in $\text{in}^{(t)}$ precede all players in $\text{out}^{(t)}$, and within $\text{out}^{(t)}$, players are ordered consistently with the sequence maintained by the algorithm.

Base Case: For $t = 0$, there are no edges, so the partial tournament is trivially transitive. For $t = 1$, there is exactly one edge, which is acyclic.

Inductive Hypothesis: Assume $T^{(t-1)}$ is transitive with a DAG ordering where all players in $\text{in}^{(t-1)}$ precede all players in $\text{out}^{(t-1)}$, and the ordering within $\text{out}^{(t-1)}$ follows the sequence maintained by the adversary.

Inductive Step: We show that $T^{(t)}$ remains transitive after the t-th query. Consider the three cases based on the query (i, j) at time t:

Case 1: Both $i, j \in \text{in}^{(t-1)}$. The adversary orients the edge based on k-scores, and exactly one player (say j if $i \to j$) moves to $\text{out}^{(t)}$. The loser j is placed at the beginning of the $\text{out}^{(t)}$ sequence. Since j had no incoming edges before (being in $\text{in}^{(t-1)}$), and now has exactly one incoming edge from $i \in \text{in}^{(t)}$, the DAG property is preserved: all players in $\text{in}^{(t)}$ still precede all players in $\text{out}^{(t)}$.

Case 2: One player is in $\text{in}^{(t-1)}$, the other in $\text{out}^{(t-1)}$. The edge is always oriented from $\text{in}^{(t-1)}$ to $\text{out}^{(t-1)}$, which is consistent with the DAG ordering where $\text{in}^{(t-1)}$ precedes $\text{out}^{(t-1)}$. No cycles can be created.

Case 3: Both $i, j \in \text{out}^{(t-1)}$. The edge follows the ordering within $\text{out}^{(t-1)}$: if i appears before j in the sequence, then $i \to j$. This maintains the transitive property within $\text{out}^{(t)}$.

In all cases, $T^{(t)}$ remains transitive. The adversary can later complete this to either: a complete transitive tournament (by maintaining the DAG ordering), or a pseudo-transitive tournament (by adding exactly one back edge when completing the tournament).

Counting Argument. Following the analysis in [9], we show that after $t^* = 2n - \lceil \log n \rceil - 3$ queries, no player can be confirmed as a Condorcet winner. Additionally, because of the structure we have maintained, the adversary can always ensure that the reported tournament is a pseudo-transitive tournament.

By the analysis in [9], we know the following: there is always at least one player remaining in $\text{in}^{(t^*)}$, and no player in $\text{in}^{(t^*)}$ can have score $n - 1$ at time t^*.

Now, if exactly one player $i \in N$ is in the in-set at time t^*, we know that $s_i^{(t^*)} < n - 2$, so there is some $j \in N$ with $j \neq i$ where the query (i, j) was not asked. If the algorithm reports that the tournament has no Condorcet winner, the adversary completes the tournament to a transitive tournament led by i. On the other hand, if the algorithm reports that the tournament has a Condorcet winner, the adversary completes the tournament to a pseudo-transitive tournament, which is the same as the transitive tournament in the previous case, except for the edge $j \to i$.

If the in-set has more than one player $i, j \in N$ at time t^*, we can extend the tournament to a transitive tournament where either i or j is in the lead. Thus, the query algorithm cannot pin down the Condorcet winner without at least one additional query between i and j. This argument can be extended to also handle the case when the algorithm is only required to report the existence of a Condorcet winner without identifying one. $\qquad\square$

It is straightforward to check that the adversary algorithm can be extended to the partial flip model by responding to all `rev()` queries as `false`.

3.2   Full Flip Model

Proof (Proof of Theorem 3, the upper bound in the full-flip model. Our algorithm works as follows. We first identify the champion of T^* by running a knockout tournament at a cost of $(n - 1)$ queries. Call this player p. If `vrev(p)` returns false, we are done, p is the Condorcet winner. If not, we find the second-best player of T^*, say q by running a knockout tournament on all players defeated by p at a cost of an additional $\lceil \log n \rceil - 1$ queries. If `vrev(q)` returns true, we are done, q is the Condorcet winner, since in this case we know that it is the (a_1, a_2) edge that was flipped in T^* to obtain T. If not, the tournament has no Condorcet winner. These steps are summarized below.

The correctness of the algorithm follows from the observation that the Condorcet winner in a pseudo-transitive tournament can be either a_1 or a_2. Observe that the worst-case number of queries required by the algorithm is given by:

$$\underbrace{n - 1}_{\text{To find } a_1} + \underbrace{1}_{vrev(a_1)} + \underbrace{\lceil \log n \rceil - 1}_{\text{To find } a_2} + \underbrace{1}_{vrev(a_2)} = n + \lceil \log n \rceil.$$

This concludes the proof of the upper bound in the full-flip model. $\qquad\square$

Algorithm 1. Determining whether the Condorcet winner in Full-Flip Model exist or not.

1: **Input:** Pseudo-transitive tournament T
2: **Output:** Condorcet Winner of T
3:
4: Identify vertex a_1 in the underlying transitive tournament $T^\star$ ▷ Knockout
     Tournament
5:
6: **if** Query $vrev(a_1)$ returns true **then**
7:        Identify vertex a_2 in the underlying transitive tournament $T^\star$
8:
9:        **if** Query $vrev(a_2)$ returns true **then**
10:              **return** $\{a_2\}$ ▷ a_2 has out-degree $n - 1$
11:        **else**
12:              **return** No Condorcet Winner   ▷ no player has out degree $n - 1$
13:        **end if**
14: **else**
15:        **return** $\{a_1\}$ ▷ a_1 retains out-degree $n - 1$
16: **end if**

Remark 1. Note that at least $(n - 1)$ queries are necessary to determine the Condorcet winner in the full-flip model. Assume an adversary always responds in a manner consistent with some arbitrary but fixed DAG order on the players. Indeed, suppose a query algorithm makes fewer than $(n-1)$ queries, and reports either that T has no Condorcet winner or that the Condorcet winner is a_1. In the latter scenario, there exists some a_i with $i > 1$ such that the pair (a_1, a_i) is not queried, and the adversary can make a_1 lose this match to contradict the outcome of the algorithm. In the former scenario, the adversary can extend the partial DAG order to a transitive tournament, again contradicting the outcome of the algorithm.

4   Copeland

We first discuss the algorithm for finding the set of Copeland winners for pseudo-transitive tournaments. As usual, let T denote the input pseudo-transitive tournament obtained by flipping at most one edge of a transitive tournament $T^\star$. Further, let $\{a_1, \ldots, a_n\}$ denote the players in $T^\star$ in the DAG order, with a_1 being the Condorcet winner. We first observe the possibilities for the set of Copeland winners in a pseudo-transitive tournament.

Lemma 1. *In a pseudo-transitive tournament T, the set of Copeland winners can be one of the following: $\{a_1\}$, $\{a_2\}$, $\{a_1, a_2\}$, or $\{a_1, a_2, a_3\}$, where a_1, a_2, and a_3 are the top three players in the DAG order of $T^\star$.*

Proof. If no edge is flipped or if the edge that is flipped is not incident to a_1, then the Copeland set of T is $\{a_1\}$, coinciding with the Copeland set of $T^\star$.

   Now, suppose the flipped edge is (a_1, a_k) for some $k \geqslant 2$:

- If $k = 2$, then a_2 becomes the Condorcet winner of T, and the Copeland set is $\{a_2\}$.
- If $k \geqslant 3$, the out-degree of a_1 decreases by one, that of a_k increases by one, and all other players' out-degrees remain unchanged. If $k = 3$, a_1, a_2, and a_3 all have out-degree $n - 2$, so the Copeland set of T is $\{a_1, a_2, a_3\}$. If $k \geqslant 4$, then only a_1 and a_2 have out-degree $n - 2$, so the Copeland set of T is $\{a_1, a_2\}$.

This completes the proof. $\qquad\qquad\qquad\qquad\qquad\qquad\qquad\qquad\qquad\qquad$ $\square$

We now turn to proving parts of the following theorem. The parts that are not discussed here are available in a full version of this work.

Theorem 4 (Copeland). *Suppose we are promised that the input to our query algorithm is a pseudo-transitive tournament on n players. Then:*

- *In the standard model, $2n$ queries are necessary and $4n - \lceil \log n \rceil - 7$ queries are sufficient to determine the set of Copeland winners.*
- *In the partial-flip model, the set of Copeland winners can be found using at most $2n + 2\lceil \log n \rceil - 1$ queries, and $n - 1$ queries are necessary.*
- *In the full-flip model, the set of Copeland winners can be found using at most $n + 2\lceil \log n \rceil + 3$ queries, and at least $n - 1$ queries are necessary.*

4.1   Standard Model

We now establish the upper bound in the standard model. In general tournaments, Balasubramanian et al. [2] showed how to find a player—if one exists—of score $n - 2$ using at most $4n + o(n)$ queries. We adapt their argument to the pseudo-transitive setting.

Proof (Proof of Theorem 4, the standard model). For the lower bound, we sketch the main idea: notice that a player is a potential Copeland winner until it loses at least two matches. Therefore, unless at least $n - 1$ players have been ruled out at a cost of $2n - 2$ queries, the adversary can always create an ambiguous situation for the algorithm.

By Lemma 1, the maximum out-degree in a pseudo-transitive tournament is either $n - 1$ or $n - 2$. Hence, finding the Copeland winners reduces to finding a player of score $n - 1$ (if one exists), and otherwise finding all the players of score $n - 2$. We proceed as follows:

Step I: Conduct a knockout tournament among the n players of T, and let c be the champion.

Step II: Compare c against all players who were not directly paired with c during the knockout tournament. Then the following three cases arise:

Case I: If c defeats all these players, then c is the Condorcet winner, and the Copeland set is $\{c\}$.

Observe that in this case the worst-case number of queries required by the algorithm is given by:

$$\underbrace{n-1}_{\text{Step I}}+\underbrace{n-\lceil\log n\rceil-1}_{\text{Step II}}=2n-\lfloor\log n\rfloor-2.$$

Case II: Suppose c is defeated by exactly one player, say a. Then the score of c is $n-2$, and no player has score $n-1$. Hence, By Lemma 1, the Copeland set must be either a 2-way tie or a 3-way tie.

To determine the exact form, we examine the score of a:

* If a also has score $n-2$, let b be the player that defeats a. If b also has score $n-2$, then a,b,c form a directed 3-cycle, and the Copeland set is $\{a,b,c\}$. Otherwise, if b has score $< n-2$, then the Copeland set is $\{a,c\}$.

  Observe that in this sub-case the worst-case number of queries required by the algorithm is given by:

$$\underbrace{n-1}_{\text{Step I}}+\underbrace{n-\lceil\log n\rceil-1}_{\text{Step II}}+\underbrace{n-2}_{\text{To find score}(a)}+\underbrace{n-3}_{\text{To find score}(b)}=4n-\lceil\log n\rceil-7.$$

* If a has score $< n-2$, then necessarily c must be the top player in the underlying transitive tournament and the flipped edge is incident on c by a. In this case the Copeland set is $\{c,b\}$, where b is the unique player defeated only by c. The player b can be identified directly from the knockout tournament.

  Observe that in this sub-case the worst-case number of queries required by the algorithm is given by:

$$\underbrace{n-1}_{\text{Step I}}+\underbrace{n-\lceil\log n\rceil-1}_{\text{Step II}}+\underbrace{n-2}_{\text{To find score}(a)}+\underbrace{\lceil\log n\rceil-1}_{\text{To find }(b)}=3n-5.$$

Case III: Suppose c is defeated by at least 2 players. Stop the comparisons against players not paired with c in the knockout tournament as soon as the second loss of c is observed. Let q denote the number of queries made before this point. Then the $q-2$ players defeated by c after the knockout tournament had already lost to someone else in the knockout bracket (since they never faced c there). Hence, their scores are strictly less than $n-2$, and they cannot belong to the Copeland set. Since c itself has at least two losses, it is also excluded. Thus, at most $n-q+1$ players remain, each of whom has been defeated only once so far.

Next, partition these $n-q+1$ candidates into groups based on the round they reached in the knockout tournament. Players in the same group never met during the knockout. Since the depth of the knockout tree is at most $\lceil\log n\rceil$, there are at most that many groups.

Run a knockout tournament within each group using at most $n-q$ additional queries, and retain only the group champions as potential candidates for to be in the Copeland set. All other players in each group now have at least two defeats and thus score $< n-2$.

Algorithm 2. Finding the Copeland set in Partial-Flip Model.

1: **Input:** Pseudo-transitive tournament T
2: **Output:** Copeland set of T
3:
4: Identify vertex a_1 in the underlying transitive tournament $T^\star$ $\triangleright$ Knockout
   Tournament
5: **for** all $k \in \{2, \ldots, n\}$ **do**
6:
7:     **if** Query $rev\langle a_1, a_k \rangle$ returns true **then**
8:         Identify vertex a_2 in the underlying transitive tournament $T^\star$
9:
10:         **if** $a_k = a_2$ **then**
11:             **return** $\{a_2\}$ $\triangleright$ a_2 has out-degree $n - 1$
12:         **else**
13:             Identify vertex a_3 in the underlying transitive tournament $T^\star$
14:
15:             **if** $a_k = a_3$ **then**
16:                 **return** $\{a_1, a_2, a_3\}$ $\triangleright$ By Lemma 1
17:             **else**
18:                 **return** $\{a_1, a_2\}$
19:             **end if**
20:         **end if**
21:     **end if**
22: **end for**
23: **return** $\{a_1\}$ $\triangleright$ a_1 retains out-degree $n - 1$

Finally, query the tournament induced by the $\lceil \log n \rceil$ group champions. Since no player has score $n-1$, Lemma 1 implies that the Copeland set can only be a 2-way or a 3-way tie. However, among these group champions, at most two players can have score $n - 2$, if three did, they would create a 3-cycle, which is impossible because c, the original champion, already has at least two losses. Therefore, a 3-way tie cannot occur, and the Copeland set is precisely these two players.

Observe that in this case the worst-case number of queries required by the algorithm is given by:

$$\underbrace{n - 1}_{\text{Step I}} + \underbrace{q}_{\text{Step II}} + \underbrace{n - q}_{\text{knockout tournament in groups}} + \underbrace{\binom{\lceil \log n \rceil}{2}}_{\text{sub-tournament}} = 2n + \binom{\lceil \log n \rceil}{2} - 1.$$

Thus, in the worse-case the number of queries required by the algorithm is at most $4n - \lceil \log n \rceil - 7$. This concludes the proof of the upper bound in the standard model. $\square$

The algorithm for finding winners in the partial flip model is as shown in Algorithm 2. A detailed proof of Theorem 4 towards the upper bound in the partial-flip and full-flip models is deferred to the full version.

5   Concluding Remarks

Our results demonstrate that modest structural access can break the $\Omega(n^2)$ barrier for several tournament solutions, and that the value of this access differs markedly across solution concepts.

Future directions include tightening bounds in the partial-flip model (closing the gap for Condorcet and sharpening constants for Copeland/Top Cycle/Uncovered Set), relaxing the $\ell \leqslant \sqrt{n \log n}$ threshold in the full-flip model to understand the transition from $O(n \log n)$ to superlinear complexities as ℓ grows, studying randomized algorithms and lower bounds under distributional assumptions, extending to additional solutions (e.g., Banks, Slater, Bipartisan, etc.) and to models with noisy or adversarially corrupted queries, and exploring practical implementations with empirical evaluation on real tournament data.

References

1. Alon, N., Gutin, G., Kim, E.J., Kratsch, S., Yeo, A.: Solving max-at and max-cut in parameterized subexponential time. SIAM J. Comput. **38**(3), 890–905 (2009)
2. Balasubramanian, R., Raman, V., Srinivasaragavan, G.: Finding scores in tournaments. J. Algorithms **24**(2), 380–394 (1997)
3. Bar-Noy, A., Naor, J.: Sorting, minimal feedback sets, and Hamilton paths in tournaments. SIAM J. Discret. Math. **3**, 7–20 (1990)
4. Brandt, F., Brill, M., Harrenstein, P.: The Computational Complexity of Choice Sets, vol. 55. Elsevier (2010)
5. Conitzer, V.: Eliciting single-peaked preferences using comparison queries. In: Proceedings of the 2009 ACM Conference on Electronic Commerce (EC) (2009)
6. Dey, P.: Query complexity of tournament solutions. In: AAAI (2017)
7. Kenyon-Mathieu, C., Schudy, W.: How to rank with few errors. In: Proceedings of the 39th Annual ACM Symposium on Theory of Computing (STOC) (2007)
8. Maiti, A., Dey, P.: Query complexity of tournament solutions. Theoret. Comput. Sci. **991**, 114422 (2024)
9. Procaccia, A.D.: A note on the query complexity of the condorcet winner problem. Inf. Process. Lett. **108**(6), 390–393 (2008)
10. Woeginger, G.J.: Banks winners in tournaments are difficult to recognize. Soc. Choice Welfare **20**(3), 523–528 (2003)

Computing Power Indices in Weighted Majority Games with Formal Power Series

Naonori Kakimura[(✉)][iD] and Yoshihiko Terai

Keio University, Yokohama 223-8522, Japan
`kakimura@math.keio.ac.jp`, `11ty24@keio.jp`

Abstract. In this paper, we propose fast pseudo-polynomial-time algorithms for computing power indices in weighted majority games. We show that we can compute the Banzhaf index for all players in $O(n + q \log(q))$ time, where n is the number of players and q is a given quota. Moreover, we prove that the Shapley–Shubik index for all players can be computed in $O(nq \log(q))$ time. Our algorithms are faster than existing algorithms when $q = 2^{o(n)}$. Our algorithms exploit efficient computation techniques for formal power series.

Keywords: Power index · Weighted majority game · Formal power series · Generating function

1 Introduction

In this paper, we study measuring the power of players in weighted majority games. The weighted majority game (also known as the weighted voting game) is a mathematical model of voting in which each player has a certain number of votes. In the game, each player votes for or against a decision, and the decision is accepted if the sum of players' voting is greater or equal to a fixed quota.

In 1964, Shapley and Shubik [25] proposed how to measure a voting power of each player, called the *Shapley–Shubik index*. The Shapley–Shubik index is a specialization of the Shapley value [24] for cooperative games. Another concept for measuring voting power was introduced by Banzhaf [5], called the *Banzhaf index*. These power indices are used to evaluate political systems such as the European Constitution and the IMF [1,16,17].

It is known that computing the Banzhaf and Shapley–Shubik indices are both known to be NP-hard [10,23]. This paper is thus concerned with calculating these power indices in pseudo polynomial time. To devise fast algorithms, there are various approaches in the literature, that include using dynamic programming [27], generating functions [6,8,20], binary decision diagrams [7], Monte-Carlo methods [4,19,28], and so on. The current best time complexity [17,27] is $O(nq)$ for

---

A full version is available at arXiv [13].

E. Di Giacomo and D. Mondal (Eds.): WALCOM 2026, LNCS 16444, pp. 323–336, 2026.
https://doi.org/10.1007/978-981-95-7127-7_22

computing the Banzhaf index of all n players with a given quota q, and $O(n^2 q)$ for the Shapley-Shubik index of all n players. It should be remarked that we here assume, as previous papers, that the arithmetic operations of n-bit numbers can be performed in constant time. See also a remark in Sect. 5.

The main contribution of this paper is to design efficient algorithms for calculating the Banzhaf and Shapley–Shubik indices for all players, respectively, using techniques in theory of formal power series. Our proposed algorithms run in $O(n + q \log(q))$ time for the Banzhaf index, and $O(nq \log(q))$ time for the Shapley–Shubik index. Our algorithms are faster than existing algorithms when $q = 2^{o(n)}$.

In our algorithms, we first represent power indices with generating functions (i.e., polynomials). In the previous work such as [6,8,20], generating functions are computed by dynamic programming. In this paper, we regard a generating function as a formal power series, which is an infinite sum of monomials, and exploit the fast Fourier transform (FFT) for efficient computation. We remark that similar technique was recently used to solve the subset sum problem [12].

1.1 Related Work

We here describe only algorithmic aspects of power indices in weighted majority games. See also a survey [21]. Practical applications of power indices may be found in, e.g., [1,16,17].

As mentioned before, it is NP-hard to calculate the Banzhaf and Shapley–Shubik indices. In fact, Deng and Papadimitriou [10] showed the problem of computing the Shapley–Shubik index is #P-complete. Prasad and Kelly [23] proved that computing the Banzhaf index is #P-complete, and that a number of decision problems around the Banzhaf and the Shapley–Shubik indices belong to the class of NP-complete problems. See also [22] for other hardness results. It is known that even approximating the Shapley–Shubik index within a constant factor is intractable, unless $P = NP$ [11].

Cantor (as mentioned by [20]) represented the Shapley–Shubik index with a generating function (a polynomial), which was used for calculation [18,20]. See Brams and Affuso [8] for the Banzhaf index. Bilbao et al. [6] proposed an algorithm based on generating functions, running in $O(nq)$ time to compute the Banzhaf index of one fixed player.

Matsui and Matsui [21] designed dynamic programming algorithms for the Banzhaf and Shapley–Shubik indices, respectively, which is a similar approach to the one for counting the number of knapsack solutions. Uno [27] later improved their algorithms, which run in $O(nq)$ and $O(n^2 q)$ time, respectively, for calculating the Banzhaf and Shapley–Shubik indices of all players. See also [17] for slight improvement with other parameters. Klinz and Woeginger [15] proposed an $O(n^2 2^{n/2})$-time algorithm based on the enumeration technique. There is another line of research to calculate the power indices approximately, see, e.g., [4,19,28].

2   Preliminaries

2.1   Power Index

Let us formally introduce our setting. In a weighted majority game, there are n players. Let $N = \{1, 2, \ldots, n\}$ be a set of n players. Each player p in N has a non-negative integer weight w_p. We are also given a non-negative integer q, which is called a *quota*. Then each player votes for or against a decision. A *coalition* is a set $S \subseteq N$ of players who vote for the decision. A coalition S is called *winning* if $w(S) \geq q$, where we define $w(S) = \sum_{p \in S} w_p$, and *losing* otherwise.

For a player p, the *Banzhaf index* Bz_p is defined as

$$\mathrm{Bz}_p := \frac{\#\{S \subseteq N \mid w(S) \geq q,\ w(S \setminus \{p\}) < q\}}{2^n}.$$

Intuitively, for a winning coalition S with $p \in S$, the player p is considered to have a power in S if $S \setminus \{p\}$ is a losing coalition. The Banzhaf index Bz_p is equal to the probability that the player p belongs to a coalition with having a power under the assumption that every coalition occurs uniformly at random.

We next define the Shapley–Shubik index. Consider the situation where, at first, a coalition S is the empty set, and players join the coalition S one by one in an order. Then, at the beginning of this situation, S is a losing coalition, and S becomes a winning one at some point. Let p be the player such that S has changed from a losing coalition to a winning one when p has been added to S. We can naturally think that the player p has a power.

More specifically, let Π be the set of permutations with length n. We note that $|\Pi| = n!$. A permutation $\pi \in \Pi$ indicates an ordering of players to join a coalition, that is, players make coalitions $S_1(\pi), S_2(\pi), \ldots, S_n(\pi)$ in the order, where $S_i(\pi) = \{\pi_1, \pi_2, \ldots, \pi_i\}$ for $i = 1, 2, \ldots, n$. Then there exists a unique player π_k such that $w(S_{k-1}(\pi)) < q$ and $w(S_k(\pi)) \geq q$. The *Shapley–Shubik index* SS_p of player p is defined as

$$\mathrm{SS}_p := \frac{\#\{\pi \in \Pi \mid \exists k \text{ s.t. } w(S_{k-1}(\pi)) < q,\ w(S_k(\pi)) \geq q,\ p = \pi_k\}}{n!}.$$

The Shapley–Shubik index is equal to the probability that the player p has a power under the assumption that every permutation occurs uniformly at random.

2.2   Formal Power Series

Let R be a commutative ring where the zero element and the identity are denoted by 0 and 1, respectively. A *formal power series* f *over* R is an infinite sum of the form

$$f = \sum_{i=0}^{\infty} a_i x^i = a_0 + a_1 x + a_2 x^2 + \cdots,$$

where $a_i \in R$ for every non-negative integer i. We call a_i the *i-th coefficient of* f. The i-th coefficient of f is denoted by $[x^i]f$.

We denote the set of all formal power series over R by $R[[x]]$, that is, $R[[x]] = \{\sum_{i=0}^{\infty} a_i x^i \mid a_i \in R\}$. Then $R[[x]]$ forms a ring. Specifically, the ring $R[[x]]$ defines the following operations for two formal power series $f = \sum_{i=0}^{\infty} a_i x^i$ and $g = \sum_{i=0}^{\infty} b_i x^i$:

1. (Sum)

$$f + g = \sum_{i=0}^{\infty} (a_i + b_i) x^i.$$

2. (Product)

$$f \cdot g = \sum_{k=0}^{\infty} \left(\sum_{i+j=k} a_i \cdot b_j \right) x^k.$$

The *(multiplicative) inverse* of a formal power series f is a formal power series g such that $f \cdot g = 1$. The inverse of f, which is known to be unique (if exists), is denoted by $1/f$ or f^{-1}. For example, the inverse of $1 - x$ is equal to $1 + x + x^2 + \cdots = \sum_{i=0}^{\infty} x^i$, as $(1 - x)(1 + x + x^2 + \cdots) = 1$. Note that f may not necessarily have the inverse. An element having the inverse is called a *unit*.

We observe that a formal power series $f \in R[[x]]$ is a unit if and only if $[x^0]f \in R$ has the inverse in a ring R. In particular, when R is a field, $f \in R[[x]]$ is a unit if and only if $[x^0]f \in R$ is not the zero element 0.

A polynomial is a formal power series with only finitely many non-zero terms. It is well-known that the product of two polynomials over the real field $\mathbb{R}$ can be computed efficiently by the fast Fourier transform (FFT). See e.g., [14].

Lemma 1. *Suppose that we are given two polynomials $f = \sum_{i=0}^{d} a_i x^i$ and $g = \sum_{j=0}^{d} b_j x^j$ over the real field $\mathbb{R}$. Then their product $f \cdot g$ can be computed in $O(d \log(d))$ time. That is, we can compute, in $O(d \log(d))$ time, all the k-th coefficients*

$$[x^k](f \cdot g) = \sum_{i+j=k} a_i \cdot b_j$$

for non-negative integers k with $k \leq 2d$.

By the above lemma, we can efficiently compute the product of two formal power series in the following sense. For two formal power series f and g, if we are given the first t coefficients $[x^i]f$ and $[x^i]g$ for $i = 0, 1, \ldots, t - 1$, then we can obtain the first t coefficients $[x^i](f \cdot g)$ of $f \cdot g$ for $i = 0, 1, \ldots, t - 1$, in $O(t \log(t))$ time. Note that, since a formal power series is an infinite sum, we set a parameter t indicating the number of coefficients we want to compute.

3 Computing the Banzhaf Index

In this section, we present a fast algorithm for computing the Banzhaf index in weighted majority games. Let $N = \{1, 2, \ldots, n\}$ be a set of players where player

p has a non-negative integer weight w_p, and let q be a quota. We may assume that $1 \leq w_p < q$ for any player p, and that there exists at least one winning coalition in the given game, that is, $w(N) \geq q$.

The main result of this section is the following.

Theorem 1. *For a weighted majority game with n players and a quota q, we can compute the Banzhaf index for all players in $\mathrm{O}(n + q\log(q))$ time.*

It is known in [8] that the Banzhaf index Bz_p for a player p can be represented with a polynomial over the real field $\mathbb{R}$. For a player p, we define $f_p \in \mathbb{R}[[x]]$ as

$$f_p = \prod_{\ell \neq p}(1 + x^{w_\ell}).$$

Then we observe that, for every non-negative integer i, the i-th coefficient $[x^i]f_p$ is equal to the number of subsets $S \subseteq N \setminus \{p\}$ with $w(S) = i$.

Lemma 2 (Brams and Affuso [8]). *For a player p, it holds that*

$$2^n \cdot \mathrm{Bz}_p = \sum_{j=q-w_p}^{q-1} [x^j]f_p.$$

In Lemma 3 below, we shall represent the Banzhaf index of all players with one formal power series over the real field $\mathbb{R}$. Define $\hat{f} \in \mathbb{R}[[x]]$ as

$$\hat{f} = \prod_{p=1}^{n}(1 + x^{w_p}).$$

Then, for each player p, we have

$$f_p = \frac{\hat{f}}{1 + x^{w_p}}.$$

Lemma 3. *For each player p, it holds that*

$$2^n \cdot \mathrm{Bz}_p = [x^{q-1}]\frac{f_p}{1-x} - [x^{q-w_p-1}]\frac{f_p}{1-x}$$

$$= [x^{q-1}]\frac{\hat{f}}{(1-x)(1+x^{w_p})} - [x^{q-w_p-1}]\frac{\hat{f}}{(1-x)(1+x^{w_p})}. \tag{1}$$

Proof. Since it holds that $\frac{1}{1-x} = \sum_{i=0}^{\infty} x^i$, we have

$$\frac{f_p}{1-x} = f_p \cdot \left(\sum_{i=0}^{\infty} x^i\right) = \sum_{i=0}^{\infty} \left(\sum_{j=0}^{i} a_{pj}\right) x^i,$$

where we denote $a_{pj} = [x^j]f_p$ for every non-negative integer j. Therefore, it holds that

$$[x^{q-1}]\frac{f_p}{1-x} - [x^{q-w_p-1}]\frac{f_p}{1-x} = \sum_{j=0}^{q-1} a_{pj} - \sum_{j=0}^{q-w_p-1} a_{pj} = \sum_{j=q-w_p}^{q-1} a_{pj},$$

which is equal to $2^n \cdot \mathrm{Bz}_p$ by Lemma 2. Thus the lemma holds. $\qquad\square$

We compute the Banzhaf index for all players using the formal power series in (1). The proposed algorithm is described as follows.

Step 1. Compute the i-th coefficients of $\hat{f}$ for all non-negative integers i with $i < q$.

Step 2. Compute the i-th coefficients of $g = \frac{\hat{f}}{1-x}$ for all non-negative integers i with $i < q$.

Step 3. For each player p with different weights, compute the Banzhaf index Bz_p by (1).

We next evaluate the computational complexity of each step in the proposed algorithm.

Jin and Wu [12] showed as below that a polynomial f in the form of $f = \prod_{i=1}^{n}(1+x^{s_i})$ can be calculated efficiently with the aid of Lemma 1. Thus Step 1 can be performed in $\mathrm{O}(n + q\log(q))$ time.

Theorem 2 (Jin and Wu [12]). *For n positive integers $s_1, s_2, \ldots, s_n$, define a polynomial f over $\mathbb{R}$ to be $f = \prod_{i=1}^{n}(1 + x^{s_i})$. Then, for a positive integer t, the first t coefficients of f can be computed in $\mathrm{O}(n + t\log(t))$ time.*

For Step 2, since

$$g = \hat{f} \cdot \left(\frac{1}{1-x}\right) = \sum_{i=0}^{\infty}\left(\sum_{j=0}^{i}\hat{a}_j\right)x^i,$$

where $\hat{a}_j = [x^j]\hat{f}$ for every non-negative integer j, we have $[x^i]g = \sum_{j=0}^{i}\hat{a}_j$ for $i = 0, 1, \ldots, q - 1$. Hence we can compute $[x^0]g, \ldots, [x^{q-1}]g$ sequentially in $\mathrm{O}(q)$ time.

The following lemma evaluates the time complexity of Step 3.

Lemma 4. *Suppose that we are given the first q coefficients of g, that is, we are given $[x^0]g, [x^1]g, \ldots, [x^{q-1}]g$. For a player p with weight w_p, the Banzhaf index Bz_p can be computed in $\mathrm{O}\left(\frac{q}{w_p}\right)$ time.*

Proof. We use (1) to compute the Banzhaf index Bz_p. It holds that

$$\frac{\hat{f}}{(1-x)(1+x^{w_p})} = \frac{g}{1+x^{w_p}} = \left(\sum_{j=0}^{\infty}(-1)^j x^{jw_p}\right)\cdot g.$$

Then, for $k \leq q - 1$, its k-th coefficient is equal to

$$\sum_{i+jw_p=k}(-1)^j \cdot [x^i]g = \sum_{j=0}^{\lfloor \frac{k}{w_p}\rfloor}(-1)^j \cdot [x^{k-jw_p}]g.$$

Since the right-hand side has at most $\frac{k}{w_p} + 1 = \mathrm{O}\left(\frac{q}{w_p}\right)$ terms, we can compute the k-th coefficient of $\frac{g}{1+x^{w_p}}$ in $\mathrm{O}\left(\frac{q}{w_p}\right)$ time, provided $[x^0]g, [x^1]g, \ldots, [x^{q-1}]g$. Therefore, since the right-hand side of (1) consists of two coefficients of $\frac{g}{1+x^{w_p}}$, Bz_p can be calculated in $\mathrm{O}\left(\frac{q}{w_p}\right)$ time. $\qquad\square$

We are now ready to prove Theorem 1.

Proof (Proof of Theorem 1). As discussed above, Step 1 in the proposed algorithm takes $\mathrm{O}(n + q\log(q))$ time by Theorem 2, and Step 2 takes $\mathrm{O}(q)$ time. It follows from Lemma 4 that Bz_p of a player p can be computed in $\mathrm{O}\left(\frac{q}{w_p}\right)$ time. Since two players p, p' with $w_p = w_{p'}$ satisfy $\mathrm{Bz}_p = \mathrm{Bz}_{p'}$, we can compute the Banzhaf index of all players by computing those of players with different weights. Since weights are integers between 1 and $q - 1$, the complexity for Step 3 can be bounded by

$$\mathrm{O}\left(\sum_{w=1}^{q-1} \frac{q}{w}\right) = \mathrm{O}(q\log(q)).$$

Therefore, the total complexity of the proposed algorithm is $\mathrm{O}(n + q\log(q))$. $\quad\square$

4 Computing the Shapley–Shubik Index

In this section, we show that the Shapley–Shubik index can be computed efficiently.

Theorem 3. *For a weighted majority game with n players and a quota q, we can compute the Shapley–Shubik index of all players in $\mathrm{O}(n + \tilde{n}q\log(q))$ time, where $\tilde{n} = \min\{n, q\}$.*

To prove the theorem, we introduce a polynomial with two variables. For a polynomial $f = \sum_{k=0}^{d} \sum_{j=0}^{d'} b_{jk} x^j y^k$ with two variables, we denote $[y^k x^j] f_p = b_{jk}$ for non-negative integers k, j. We also denote $[x^j] f_p = \sum_{k=0}^{d} b_{jk} y^k$ and $[y^k] f_p = \sum_{j=0}^{d'} b_{jk} x^j$.

4.1 Algorithm

As shown by Cantor (cf. [18,20]), the Shapley–Shubik index can be represented with a polynomial with two variables. For a player p, we define a polynomial f_p in two variables x, y over the real field $\mathbb{R}$ as

$$f_p = \prod_{\ell \neq p}(1 + yx^{w_\ell}).$$

Then $[y^k x^j] f_p$ is equal to the number of subsets $S \subseteq N \setminus \{p\}$ such that $|S| = k$ and $w(S) = j$.

Lemma 5 (Cantor (cf. [18,20])). *For a player p, it holds that*

$$n! \cdot \mathrm{SS}_p = \sum_{k=1}^{n-1} \left(k! \cdot (n-1-k)! \sum_{j=q-w_p-1}^{q-1} [y^k x^j] f_p \right).$$

Since $w_p \geq 1$ for every player p, we see that, if $[y^k x^j] f_p \neq 0$, then $k \leq j$ holds. Hence it suffices to consider $[y^k x^j] f_p$ for $k \leq j$ in the right-hand side, implying that

$$n! \cdot \mathrm{SS}_p = \sum_{k=1}^{\tilde{n}-1} \left(k! \cdot (n-1-k)! \sum_{j=q-w_p-1}^{q-1} [y^k x^j] f_p \right), \tag{2}$$

where $\tilde{n} = \min\{n, q\}$.

Similarly to Lemma 3 in Sect. 3, the Shapley–Shubik index SS_p for a player p can be represented as follows. Define

$$\hat{f} = \prod_{p=1}^{n} (1 + yx^{w_p}) \quad \text{and} \quad g = \frac{\hat{f}}{1-x}.$$

Lemma 6. *For each player p, it holds that*

$$n! \cdot \mathrm{SS}_p = \sum_{k=1}^{\tilde{n}-1} k! \cdot (n-1-k)! \left([y^k x^{q-1}] \frac{g}{1+yx^{w_p}} - [y^k x^{q-w_p-1}] \frac{g}{1+yx^{w_p}} \right). \tag{3}$$

Our proposed algorithm computes the right-hand side of (3) for a player p. By (3), it suffices to compute $[y^k x^j] \frac{g}{1+yx^{w_p}}$ for all $k < \tilde{n}$ and $j < q$. To this end, we regard a two-variable rational function as a formal power series (w.r.t. x) over the formal power series ring (w.r.t. y). Specifically, let $R = \mathbb{R}[[y]]/(y^{\tilde{n}})$, which is the quotient ring of $\mathbb{R}[[y]]$ by the ideal generated by $y^{\tilde{n}}$. That is, an element in R can be written as $\sum_{i=0}^{\tilde{n}-1} b_i y^i$ for $b_i \in \mathbb{R}$. Note that R is isomorphic to the quotient ring $\mathbb{R}[y]/(y^{\tilde{n}})$ of polynomials. Consider the formal power series ring $R[[x]]$. Then, a formal power series f in $R[[x]]$ can be written as

$$f = \sum_{i=0}^{\infty} \left(\sum_{j=0}^{\tilde{n}-1} a_{ij} y^j \right) x^i,$$

where $a_{ij} \in \mathbb{R}$. The *i-th coefficient* $[x^i] f$ is equal to $\sum_{j=0}^{\tilde{n}-1} a_{ij} y^j$. We also denote $[y^j x^i] f = a_{ij}$ for every non-negative integers i and $j < \tilde{n}$.

The proposed algorithm is presented as below.

Step 1. Compute the i-th coefficients of $\hat{f}$ (mod $y^{\tilde{n}}$) for all non-negative integers i with $i < q$.

Step 2. Compute the i-th coefficients of g (mod $y^{\tilde{n}}$) for all non-negative integers i with $i < q$.

Step 3. For each player p with different weights, compute the Shapley–Shubik index SS_p by (3).

To evaluate the time complexity of the proposed algorithm, we first show a variant of Theorem 2 as follows, where the proof is deferred to Sect. 4.2. The theorem implies that Step 1 can be executed in $\mathrm{O}(n + \tilde{n}q \log(\tilde{n}q))$ time.

Theorem 4. *For n positive integers $s_1, s_2, \ldots, s_n$, define a polynomial $f = \prod_{i=1}^{n}(1 + yx^{s_i})$ over the real field $\mathbb{R}$. Then, for two positive integers m, t, we can compute all the coefficients $[y^k x^j] f$ for $j < t$ and $k < m$ in $\mathrm{O}(n + mt \log(mt))$ time.*

Moreover, we have the following lemma for Step 3.

Lemma 7. *Suppose that we are given the first q coefficients of g modulo $y^{\tilde{n}}$, denoted by $a_0, a_1, \ldots, a_{q-1}$. For a player p with weight w_p, the Shapley–Shubik index SS_p can be computed in $\mathrm{O}\left(\frac{\tilde{n}q}{w_p}\right)$ time.*

Proof. We use (3) to compute the Shapley–Shubik index SS_p. It holds that

$$\frac{g}{1 + yx^{w_p}} = \left(\sum_{j=0}^{\infty}(-1)^j y^j x^{jw_p}\right) \cdot g.$$

Then, for $k \leq q - 1$, its k-th coefficient is

$$[x^k]\frac{g}{1 + yx^{w_p}} = \sum_{i+jw_p=k}(-1)^j y^j \cdot [x^i]g = \sum_{j=0}^{\lfloor \frac{k}{w_p} \rfloor}(-1)^j y^j \cdot [x^{k-jw_p}]g.$$

Hence, for $\ell = 1, 2, \ldots, \tilde{n}$, we have

$$[y^\ell x^k]\frac{g}{1 + yx^{w_p}} = \sum_{j=0}^{\lfloor \frac{k}{w_p} \rfloor}(-1)^j \cdot [y^{\ell-j}x^{k-jw_p}]g.$$

Since the right-hand side has at most $\frac{k}{w_p} + 1$ terms, we can compute it in $\mathrm{O}\left(\frac{q}{w_p}\right)$ time, provided $a_0, a_1, \ldots, a_{q-1}$. Therefore, since the right-hand side of (3) can be calculated from $2(\tilde{n} - 1)$ coefficients of $\frac{g}{1+yx^{w_p}}$, SS_p can be obtained in $\mathrm{O}\left(\frac{\tilde{n}q}{w_p}\right)$ time. $\square$

We now prove Theorem 3.

Proof (Proof of Theorem 3). As discussed above, Step 1 in the proposed algorithm takes $\mathrm{O}(n + \tilde{n}q \log(\tilde{n}q))$ time by Theorem 4. Since $\tilde{n} = \min\{n, q\} \leq q$, the time complexity for Step 1 is $\mathrm{O}(n + \tilde{n}q \log(q))$. Moreover, Step 2 takes $\mathrm{O}(\tilde{n}q)$ time, similarly to Sect. 3.

It follows from Lemma 4 that the Shapley–Shubik index SS_p of player p can be computed in $\mathrm{O}\left(\frac{\tilde{n}q}{w_p}\right)$ time. Since two players p, p' with $w_p = w_{p'}$ satisfy $\mathrm{SS}_p = \mathrm{SS}_{p'}$, we can compute the Shapley–Shubik index of all players by computing those of players with different weights. The complexity for Step 3 can be bounded by

$$\mathrm{O}\left(\sum_{w=1}^{q-1} \frac{\tilde{n}q}{w}\right) = \mathrm{O}(\tilde{n}q\log(q)).$$

Therefore, the total complexity is $\mathrm{O}(n + \tilde{n}q\log(q))$, which completes the proof of the theorem. $\qquad\square$

4.2 Two-Variable Formal Power Series

This section is devoted to proving Theorem 4. The proof extends the one of Theorem 2 by [12] for the real field $\mathbb{R}$ to the case when $R = \mathbb{R}[[y]]/(y^m)$, where m is a non-negative integer.

We first show that the product of two-variable polynomials over $\mathbb{R}$ can be computed by Lemma 1.

Lemma 8. *Suppose that we are given two polynomials* $f = \sum_{i=0}^{d_x} \sum_{j=0}^{d_y} a_{ij}x^i y^j$ *and* $g = \sum_{i=0}^{d_x} \sum_{j=0}^{d_y} b_{ij}x^i y^j$. *Then the product* $f \cdot g$ *of* f *and* g *can be computed in* $\mathrm{O}(d_x d_y \log(d_x d_y))$ *time. That is, we can compute, in* $\mathrm{O}(d_x d_y \log(d_x d_y))$ *time, all the coefficients*

$$[y^h x^k](f \cdot g) = \sum_{i_1+i_2=k} \sum_{j_1+j_2=h} f_{i_1 j_1} \cdot g_{i_2 j_2}$$

for two non-negative integers k, h *satisfying* $k \leq 2d_x$ *and* $h \leq 2d_y$.

For a formal power series $f \in R[[x]]$ such that $[x^0]f$ is nilpotent with respect to R, define

$$\exp(f) = \sum_{i=0}^{\infty} \frac{f^i}{i!}.$$

We recall that an element $a \in R$ is *nilpotent* if there exists a positive integer k such that $a^k = 0$. Also, for a formal power series $f \in R[[x]]$ such that $[x^0](f-1)$ is nilpotent with respect to R, define

$$\log(f) = -\sum_{i=1}^{\infty} \frac{(1-f)^i}{i}.$$

When $R = \mathbb{R}$, it is known that $1/f$, $\exp(f)$, and $\log(f)$ can be computed efficiently with the aid of Lemma 1 and the Newton method as below.

Lemma 9 (Brent [9]). *Let* $R = \mathbb{R}$, *and let* f *be a formal power series over* $R[[x]]$ *such that the first* t *coefficients* $a_0, a_1, \ldots, a_{t-1}$, *where* $a_i = [x^i]f$, *are given. Then the following statements hold.*

1. *If f is a unit, the first t coefficients of $1/f$ can be computed in $O(t \log(t))$ time.*
2. *If $a_0 = 1$, then the first t coefficients of $\log(f)$ can be computed in $O(t \log(t))$ time.*
3. *If $a_0 = 0$, then the first t coefficients of $\exp(f)$ can be computed in $O(t \log(t))$ time.*

Using the above lemma, together with the fact that $f = \exp(\log(f))$, Theorem 4 follows. See [12] for the details.

In this section, we show analogous results to Lemma 9 when $R = \mathbb{R}[[y]]/(y^m)$. The proof idea is similar to that of Lemma 9, but the difference is in computing the 0-th coefficients of $\log(f)$ and $\exp(f)$, while it is trivial for the case when $R = \mathbb{R}$. We here show only (2) for computing the logarithm, due to the space limitation.

Lemma 10. *Let $R = \mathbb{R}[[y]]/(y^m)$ for a positive integer m, and let f be a formal power series over $R[[x]]$ such that the first t coefficients $a_0, a_1, \ldots, a_{t-1}$, where $a_i = [x^i]f$ in R, are given. Then the following statements hold.*

1. *If f is a unit, the first t coefficients of $1/f$ can be computed in $O(mt \log(mt))$ time.*
2. *If $a_0 - 1$ is nilpotent, then the first t coefficients of $\log(f)$ can be computed in $O(mt \log(mt))$ time.*
3. *If a_0 is nilpotent, then the first t coefficients of $\exp(f)$ can be computed in $O(mt \log(mt))$ time.*

Proof (Proof of Part (2)). It is known that $\log(f) = \int f'/f$, where f' is the differential of f and $\int$ is an integral (See e.g., [3] for the details). By definition, it holds that $f' = \sum_{i=1}^{\infty} i a_i x^{i-1}$. Hence we know the first $t - 1$ coefficients of f', $[x^0]f', [x]f', \ldots, [x^{t-2}]f'$, from $a_1, \ldots, a_{t-1}$. Also, the first t coefficients of $1/f$ can be computed in $O(mt \log(mt))$ time by the first statement of this lemma.

Let $f' \cdot (1/f) = \sum_{i=0}^{\infty} b_i x^i$. Then it follows from Lemma 8 that the first $t - 1$ coefficients $b_0, \ldots, b_{t-2}$ can be computed in $O(mt \log(mt))$ time. We see that

$$\int \frac{f'}{f} = \sum_{i=1}^{\infty} \frac{b_{i-1}}{i} x^i + [x^0] \log(f).$$

Therefore, since $\log(f) = \int f'/f$, we can obtain $[x^1] \log(f), \ldots, [x^{t-1}] \log(f)$ from $b_0, \ldots, b_{t-2}$. The time complexity of this step is $O(mt \log(mt))$.

It remains to compute $[x^0] \log(f)$. We observe that $[x^0] \log(f) = \log([x^0]f)$. In fact, it holds by definition that

$$[x^0] \log(f) = [x^0]\left(-\sum_{i=1}^{\infty} \frac{(1-f)^i}{i}\right) = -\sum_{i=1}^{\infty} [x^0]\frac{(1-f)^i}{i} = -\sum_{i=1}^{\infty} \frac{(1-a_0)^i}{i},$$

where the last equation follows from $[x^0](1 - f)^i = (1 - a_0)^i$. The most right-hand side, which is $\log(a_0)$, is an element of $\mathbb{R}[[y]]/(y^m)$, and hence it can be computed in $O(m \log(m))$ time by Lemma 9. Thus, $[x^0] \log(f)$ can be obtained in $O(m \log(m))$ time.

Therefore, all the first t coefficients of $\log(f)$ can be calculated in $O(mt\log(mt))$ time in total, which completes the proof. $\square$

We note that the above lemma does not hold for a general commutative ring R, as it might happen that $[x^i]f$ is an infinite sum, and moreover, it is not clear how to compute the 0-th coefficients.

We are now ready to prove Theorem 4.

Proof (Proof of Theorem 4). Recall that $f = \prod_{i=1}^{n}(1 + yx^{s_i})$ is a polynomial over the real field $\mathbb{R}$. We consider f as an element of the formal power series ring $R[[x]]$ where $R = \mathbb{R}[[y]]/(y^m)$, and compute the first t coefficients of f (mod y^m) using the relationship $f = \exp(\log(f))$. We may assume that $s_i \le t$ for every i, as otherwise we may remove $(1 + yx^{s_i})$ for $s_i > t$ from f.

We first compute the first t coefficients of $\log(f)$ as follows. It holds that

$$\log(f) = \log \prod_{i=1}^{n}(1 + yx^{s_i}) = \sum_{i=1}^{n} \log(1 + yx^{s_i}). \tag{4}$$

Thus the first t coefficients of $\log(f)$ can be computed from those of $\log(1+yx^{s_i})$ with different exponent s_i. By the definition of the logarithm, we have

$$\log(1 + yx^{s_i}) = \sum_{j=1}^{\infty}(-1)^j \frac{(yx^{s_i})^j}{j} = \sum_{j=1}^{\infty} \frac{(-1)^j y^j}{j} x^{s_i j}.$$

Then $\log(1 + yx^{s_i})$ has at most $\frac{t}{s_i}$ non-zero coefficients in the first t coefficients, where each coefficient has only one monomial in y. Hence the first t coefficients of $\log(1 + yx^{s_i})$ can be computed in $O\left(\frac{t}{s_i}\right)$ time. Therefore, since $s_i \le t$ for every i, it follows from (4) that the first t coefficients of $\log(f)$ can be computed in time

$$O\left(\sum_{s=1}^{t} \frac{t}{s}\right) = O(t\log(t)).$$

Finally, since $f = \exp(\log(f))$, we can compute the first t coefficients of f in $O(mt\log(mt))$ time by Lemma 10. Thus the theorem holds, as reading the input takes $O(n)$ time. $\square$

5   Conclusion

In this paper, we proposed fast pseudo-polynomial-time algorithms for computing power indices in weighted majority games. Our algorithms calculate the Banzhaf index and the Shapley–Shubik index for all players in $O(n + q\log(q))$ time and $O(nq\log(q))$ time, respectively, where n is the number of players and q is a given quota. Our algorithms are faster than existing algorithms when $q = 2^{o(n)}$.

We remark that our algorithms ignore the time complexity to deal with high-precision integers, following the literature such as [21,27]. As the numerators of the Banzhaf and Shapley–Shubik indices may be $2^{\Omega(n)}$, the time complexity in the standard computation model would increase by a factor of n. On the other hand, we can avoid arithmetic operations of large integers by applying our algorithm $O(n)$ times as follows. Let $\alpha_1, \ldots, \alpha_\ell$ be prime numbers such that $q < \alpha_i \leq n$ for each i and $2^n < \prod_i \alpha_i$, where $\ell \leq n$. Then, for each i, we perform our algorithm over a prime field $\mathbb{F}_{\alpha_i}$ to obtain the output modulo α_i. It follows from the Chinese Remainder Theorem that we can uniquely recover the exact value of the output from these remainders.

In weighted majority games, other indices are proposed to measure the voting power of each player, such as the Deegan–Packel index and the Public Good index. See e.g., [2,17,26] and references therein. It is known that most of the existing approaches for computing the Banzhaf and Shapley–Shubik indices can be adapted to these indices. It would be interesting if our approach can be used to calculate these power indices.

Acknowledgments. We are grateful to anonymous reviewers for their helpful comments. This work was supported by JSPS KAKENHI Grant Numbers 22H05001, 23K21646, and JST ERATO Grant Number JPMJER2301, Japan.

References

1. Algaba, E., Bilbao, J.M., Fernández, J.R.: The distribution of power in the European Constitution. Eur. J. Oper. Res. **176**(3), 1752–1766 (2007). https://doi.org/10.1016/j.ejor.2005.12.002, https://www.sciencedirect.com/science/article/pii/S0377221705008805
2. Alonso-Meijide, J.M., Freixas, J., Molinero, X.: Computation of several power indices by generating functions. Appl. Math. Comput. **219**(8), 3395–3402 (2012). 10. https://doi.org/10.021, https://www.sciencedirect.com/science/article/pii/S0096300312010089
3. Ardila, F.: Algebraic and geometric methods in enumerative combinatorics. Handb. Enumerative Comb. 3–172 (2015)
4. Bachrach, Y., Markakis, E., Resnick, E., Procaccia, A.D., Rosenschein, J.S., Saberi, A.: Approximating power indices: theoretical and empirical analysis. Auton. Agents Multi Agent Syst. **20**(2), 105–122 (2010). https://doi.org/10.1007/s10458-009-9078-9
5. Banzhaf, J.F.: Weighted voting doesn't work: a mathematical analysis. Rutgers Law Rev. **19**(2), 317–343 (1965)
6. Bilbao, J.M., Fernandez, J.R., Jiménez-Losada, A., López, J.: Generating functions for computing power indices efficiently. TOP **8**(2), 191–213 (2000)
7. Bolus, S.: Power indices of simple games and vector-weighted majority games by means of binary decision diagrams. Eur. J. Oper. Res. **210**(2), 258–272 (2011). https://doi.org/10.1016/j.ejor.2010.09.020, https://www.sciencedirect.com/science/article/pii/S0377221710006181
8. Brams, S.J., Affuso, P.J.: Power and size: a new paradox. Theor. Decis. **7**(1), 29–56 (1976)
9. Brent, R.P.: Multiple-precision zero-finding methods and the complexity of elementary function evaluation. In: Traub, J. (ed.) Analytic Computational Complexity, pp. 151–176. Academic Press (1976). https://doi.org/10.1016/B978-0-12-697560-4.50014-9. https://www.sciencedirect.com/science/article/pii/B9780126975604500149
10. Deng, X., Papadimitriou, C.H.: On the complexity of cooperative solution concepts. Math. Oper. Res. **19**(2), 257–266 (1994). https://doi.org/10.1287/moor.19.2.257

11. Elkind, E., Goldberg, L.A., Goldberg, P.W., Wooldridge, M.J.: Computational complexity of weighted threshold games. In: Proceedings of the Twenty-Second AAAI Conference on Artificial Intelligence, pp. 718–723. AAAI Press (2007). http://www.aaai.org/Library/AAAI/2007/aaai07-114.php
12. Jin, C., Wu, H.: A simple near-linear pseudopolynomial time randomized algorithm for subset sum. In: Fineman, J.T., Mitzenmacher, M. (eds.) 2nd Symposium on Simplicity in Algorithms, SOSA 2019. OASIcs, vol. 69, pp. 17:1–17:6. Schloss Dagstuhl - Leibniz-Zentrum für Informatik (2019). https://doi.org/10.4230/OASICS.SOSA.2019.17
13. Kakimura, N., Terai, Y.: Computing power indices in weighted majority games with formal power series (2025). https://arxiv.org/abs/2511.14995
14. Kleinberg, J., Tardos, É.: Algorithm Design. Pearson Education India (2006)
15. Klinz, B., Woeginger, G.J.: Faster algorithms for computing power indices in weighted voting games. Math. Soc. Sci. **49**(1), 111–116 (2005). https://doi.org/10.1016/j.mathsocsci.2004.06.002, https://www.sciencedirect.com/science/article/pii/S016548960400068X
16. Kóczy, L.A.: Beyond Lisbon: demographic trends and voting power in the European Union Council of Ministers. Math. Soc. Sci. **63**(2), 152–158 (2012). https://doi.org/10.1016/j.mathsocsci.2011.08.005. https://www.sciencedirect.com/science/article/pii/S0165489611000916
17. Kurz, S.: Computing the power distribution in the IMF. arXiv preprint arXiv:1603.01443 (2016)
18. Lucas, W.F.: Measuring Power in Weighted Voting Systems, pp. 183–238. Springer, New York (1983). https://doi.org/10.1007/978-1-4612-5430-0_9
19. Mann, I., Shapley, L.S.: Values of Large Games, IV: Evaluating the Electoral College by Montecarlo Techniques. RAND Corporation, Santa Monica (1960)
20. Mann, I., Shapley, L.S.: Values of Large Games, VI: Evaluating the Electoral College Exactly. RAND Corporation, Santa Monica (1962)
21. Matsui, T., Matsui, Y.: A survey of algorithms for calculating power indices of weighted majority games. J. Oper. Res. Soc. Jpn. **43**(1), 71–86 (2000). https://doi.org/10.15807/jorsj.43.71
22. Matsui, Y., Matsui, T.: NP-completeness for calculating power indices of weighted majority games. Theor. Comput. Sci. **263**(1–2), 305–310 (2001). https://doi.org/10.1016/S0304-3975(00)00251-6
23. Prasad, K., Kelly, J.S.: NP-completeness of some problems concerning voting games. Internat. J. Game Theory **19**(1), 1–9 (1990). https://doi.org/10.1007/BF01753703
24. Shapley, L.S.: A value for n-person games. In: Kuhn, H.W., Tucker, A.W. (eds.) Contributions to the Theory of Games II, pp. 307–317. Princeton University Press, Princeton (1953)
25. Shapley, L.S., Shubik, M.: A method for evaluating the distribution of power in a committee system. Am. Polit. Sci. Rev. **48**(3), 787–792 (1954). http://www.jstor.org/stable/1951053
26. Staudacher, J., et al.: Computing power indices for weighted voting games via dynamic programming. Oper. Res. Decis. **31**(2) (2021). https://doi.org/10.37190/ORD210206
27. Uno, T.: Efficient computation of power indices for weighted majority games. In: Chao, K.-M., Hsu, T., Lee, D.-T. (eds.) ISAAC 2012. LNCS, vol. 7676, pp. 679–689. Springer, Heidelberg (2012). https://doi.org/10.1007/978-3-642-35261-4_70
28. Ushioda, Y., Tanaka, M., Matsui, T.: Monte Carlo methods for the Shapley-Shubik power index. Games **13**(3), 44 (2022). https://doi.org/10.3390/g13030044

How to Reconfigure Your Alliances

Henning Fernau and Kevin Mann[(✉)]

Universität Trier, Fachbereich 4 – Abteilung Informatikwissenschaften,
54286 Trier, Germany
`{fernau,mann}@uni-trier.de`

Abstract. Different variations of alliances in graphs have been introduced into the graph-theoretic literature about twenty years ago. More broadly speaking, they can be interpreted as groups that collaborate to achieve a common goal, for instance, defending themselves against possible attacks from outside. In this paper, we initiate the study of reconfiguring alliances. This means that, with the understanding of having an interconnection map given by a graph, we look at two alliances of the same size k and investigate if there is a reconfiguration sequence (of length at most ℓ) formed by alliances of size (at most) k that transfers one alliance into the other one. Here, we consider different (now classical) movements of tokens: sliding, jumping, addition/removal. We link the latter two regimes by introducing the concept of reconfiguration monotonicity. Concerning classical complexity, most of these reconfiguration problems are PSPACE-complete, although some are solvable in LogSPACE. We also consider these reconfiguration questions through the lense of parameterized algorithms and prove various FPT-results, in particular concerning the combined parameter $k + \ell$ or neighborhood diversity together with k or neighborhood diversity together with ℓ.

1 Introduction

Abstractly speaking, the concept of *reconfiguration* addresses the question how different solutions to a problem relate to each other in the sense that it is possible to 'move' from one solution to another one through the space of solutions. For instance, if you do some re-installment of infrastructure, there is a working solution at present and a hopefully working solution in the future, but also all intermediate steps should be planned in a way that the infrastructure is still working for everybody. A concrete instantiation of this setting was investigated in [11] as the POWER SUPPLY RECONFIGURATION problem. Further practically relevant examples can be found in [10,15], to cite just two references, and the current paper will add to this list of relevant problems. Again more abstractly speaking, this type of analysis can be undertaken for any combinatorial problem. For graph problems like INDEPENDENT SET, a reconfiguration instance would consist of a graph G and two solutions I_s and I_t, i.e., independent sets, and the question is whether one can move from I_s to I_t in the solution space. In other

E. Di Giacomo and D. Mondal (Eds.): WALCOM 2026, LNCS 16444, pp. 337–352, 2026.
https://doi.org/10.1007/978-981-95-7127-7_23

words, the question is if there exists a *reconfiguration sequence* from I_s to I_t, formally treated in the next section. This of course depends on the 'connection structure' of the solution space. Typically, an adjacency relation between two solutions is defined based on a notion of 'permitted transformation'. In this context, we imagine a solution as given by a set of tokens placed on the vertices of a graph. For instance, *token sliding* then means that two solutions S, S' are adjacent if $S \triangle S' = \{u, v\}$, $|S| = |S'|$ and u, v are adjacent in the graph, while *token jumping* would not require u, v to be adjacent. Similarly, one can think of *token removal* or *token addition*, to give two more examples of such 'move' operations. Also, apart from the pure reconfigurability question, which is basically the question of reachability within the solution graph, one could also add a time upper bound, or just ask the combinatorial question if the solution graph is connected. A lot of work on various aspects of reconfiguration has been done in recent years; still, a nice introduction in the topic can be found in [17]. It should be mentioned that we assume that only one token can be at a vertex in one point of the sequence. This is not the case for each paper (for example [3]). As most variants of reconfiguration problems that we study in this paper turn out to be computationally hard, we also look at them through the lens of parameterized complexity. As in [16], we can consider the size k of the solutions that we study (or an upper bound on them) and an upper bound ℓ on the length of the reconfiguration sequence as natural parameter choices. Furthermore, we also consider neighborhood diversity as a structural parameter of the underlying graph, as started out with [8] in the context of reconfiguration.

In the present paper, we are going to apply the concept of reconfiguration to different notions of *alliances* that have been defined in the literature, starting with [7,13,14,21,22]. Several surveys have been written on alliances and related notions [6,18,23], and even two chapters of the recent monograph [9] have been devoted to this topic. Possible applications are nicely described in [18], among them also community-detection problems [20]. For instance, given a set of vertices A that should model an alliance, one could think of some $v \in A$ to be a weak spot in the alliance if it has more vertices outside of A (in a sense, enemies) in its neighborhood than allies (situated in A). This idea leads to the notion of a *defensive alliance*, where such weak spots are not permitted. Similarly, an *offensive alliance* is longing for weak spots in the complement of A as possible points of attack. These notions will be defined more formally in the next section. However, the intuition laid so far should suffice to see that reconfiguring alliances makes a lot of sense from a practical perspective. Now, the 'tokens' could be viewed as 'armies' that move around, and 'token sliding' would take care of the geography modeled by the underlying graph. The main results of this paper are the following ones, where we (again) refer to the precise definitions of the problems given below.

- For all variants of alliance reconfiguration problems (defensive, offensive, powerful), we can prove their **PSPACE**-completeness for all variants of token movements. This remains true if the alliances are *global*, i.e., if they also form dominating sets. The picture changes if we require that a (global) offensive alliance is also an independent set; then, the reachability questions are solvable in LogSPACE and therefore much easier. For details, see Table 1.

– We also consider different parameterizations for the (hard) reconfiguration problems. In short, all but few alliance reconfiguration problem variants are proven to be in FPT with the combined parameter $\ell + k$, where ℓ upperbounds the length of the reconfiguration sequence and k denotes the number of tokens. For the powerful or global problem variations, even the parameter k alone suffices to prove membership in FPT. Also neighborhood diversity is a nice starting point for parameterized tractability results, as we show.

– We introduce and discuss the novel notion of *reconfiguration monotonicity* that turns out to be quite helpful in linking token addition and removal together with token jumping. These results could be interesting beyond the reconfiguration of alliances.

2 Definitions and Notations

Let $\mathbb{N}$ denote the set of all nonnegative integers (including 0). For $n \in \mathbb{N}$, we will use the notation $[n] := \{1, \ldots, n\}$. Let $G = (V, E)$ be a graph, i.e., $E \subseteq \binom{V}{2}$. If $X \subseteq V$, then $G[X]$ denotes the subgraph induced by X, i.e., $G[X] := (X, \{e \in E \mid e \subseteq X\})$. $N_G(v)$ describes the *open neighborhood* of $v \in V$ with respect to G. The *closed neighborhood* of $v \in V$ with respect to G is defined by $N[v] := N(v) \cup \{v\}$. For a set $A \subseteq V$, its open neighborhood is defined as $N_G(A) := \bigcup_{v \in A} N_G(v)$. The closed neighborhood of A is given by $N_G[A] := N_G(A) \cup A$. The *degree* of a vertex $v \in V$ with respect to G is denoted $d_G(v) := |N_G(v)|$. The *boundary* of $A \subseteq V$ is defined by $\partial A := N_G(A) \setminus A$. With $N_A(v)$ and $d_A(v)$, we describe the open neighborhood and the degree of v with respect to $G[A \cup \{v\}]$. We suppress the index G if clear from context. A vertex of degree one is called a *leaf*, a vertex of degree zero is an *isolate*. Similarly, an edge connecting two leaves is called an *isolated edge*. A set $C \subseteq V$ is a *clique* in G if $C \subseteq N_G[v]$ for each $v \in C$. A vertex v of G is called *simplicial* in G if $N_G(v)$ is a clique in G. For instance, leaves are always simplicial. The ordering $v_1, \ldots, v_n$ of the vertices of G is a *perfect elimination order* of G if, for all $i \in [n]$, v_i is simplicial in $G[\{v_1, \ldots, v_i\}]$. A graph is *chordal* if it has a perfect elimination ordering. The *neighborhood diversity* $\mathsf{nd}(G)$ of a graph G is defined as the number of equivalence class of the following equivalence relation: vertices $v, u \in V$ are equivalent iff $N(u) \setminus \{v\} = N(v) \setminus \{u\}$. We also say u and v *have the same type*.

Let $A, B \subseteq V$. Then, A can be transformed to B by a *token removal* step if $A \subseteq B$ and $|B \setminus A| = 1$. In this case, B can be transformed by a *token addition* step to A. We say that A can be transformed to B by a *token jumping* step if $|A| = |B|$, $|A \setminus B| = 1$. For $v \in A \setminus B$ and $u \in B \setminus A$, we say the token jumps from v to u. A token jumping step is called a *token sliding* if the vertices in $v \in A \setminus B$ and $u \in B \setminus A$ are neighbors. In this case, we say the token slides from v to u. A sequence $A_1, \ldots, A_\ell$ is a *token addition removal sequence* (or *token jumping sequence* or *token sliding sequence*, respectively) if for each $i \in [\ell - 1]$, A_i can transformed to A_{i+1} by a token addition or removal (or token jumping or token sliding, respectively) step. We also employ the abbreviations TAR (or TJ or TS, respectively) sequence. For $Y \in \{\text{TAR}, \text{TJ}, \text{TS}\}$, such a sequence is named an X-Y (reconfiguration) sequence if all sets A_i in this Y sequence satisfy the property X.

Let $G = (V, E)$ be a graph. A set $I \subseteq V$ is *independent* if $G[I]$ contains only isolates. Graph G is *bipartite* if V can be partitioned into two independent sets. A set $D \subseteq V$ is called *dominating* if $N[D] = V$. A set $A \subseteq V$ is called a *defensive alliance* (or DA for short) if $d_A(v) + 1 \geq d_{V \setminus A}(v)$ for each $v \in A$. $A \subseteq V$ with $d_A(v) \geq d_{V \setminus A}(v) + 1$ for each $v \in \partial A$ is called an *offensive alliance* (or OA for short). If a vertex set is a defensive and an offensive alliance, it is called a *powerful alliance* (or PA for short). For $X \in \{\text{defensive, offensive, powerful}\}$, a *global X* alliance is an X alliance that is also a dominating set. Similarly, an OA which is also an independent set is called an *independent OA*; see [19]. We will now define the decision problems studied in this paper. We differentiate between two versions of reconfiguration problems. We will use X as any alliance version, viewed as a property of vertex sets and abbreviated as D, O, P and sometimes prefixed with G (global) or IDP, while $Y \in \{\text{TAR}, \text{TJ}, \text{TS}\}$.

Problem name: X-ALLIANCE RECONFIGURATION-Y, or XA-RECONF-Y for short.
Given: A graph $G = (V, E)$ and X alliances $A_s, A_t \subseteq V$ (and $k \in \mathbb{N}$ if $Y = \text{TAR}$).
Question: Is there an X-Y reconfiguration sequence $(A_s = A_1, \ldots, A_\ell = A_t)$ (with $|A_i| \leq k$ for $i \in [\ell]$ if $Y = \text{TAR}$)?

Problem name: TIMED X-ALLIANCE RECONFIGURATION-Y, or T-XA-RECONF-Y for short.
Given: A graph $G = (V, E)$, X alliances $A_s, A_t \subseteq V$ and $T \in \mathbb{N}$ ($k \in \mathbb{N}$ if $Y = \text{TAR}$).
Question: Is there an $\ell \in \mathbb{N}$ with $\ell < T$ and an X-Y reconfiguration sequence $(A_s = A_1, \ldots, A_\ell = A_t)$ (with $|A_i| \leq k$ if $Y = \text{TAR}$)?

In these problems, we call A_s the start configuration and A_t the target configuration. The first version asks if there is a reconfiguration sequence between A_s and A_t, while the timed version also gives an upper bound on the number of reconfiguration steps. Sometimes, we also speak of the *underlying combinatorial problem*, referring to: given a graph G and $k \in \mathbb{N}$; is there a set D, $|D| \leq k$, with property X? As usual, we will use the star symbol ($*$) to label theorems etc. whose proofs can be found in the long version [5] due to space constraints.

3   PSPACE-Completeness or Membership in LogSPACE

To motivate our parameterized studies, we will prove PSPACE-completeness for (most of) the alliance reconfiguration problems. These results are not that surprising as there are other PSPACE-complete reconfiguration problems for which the underlying combinatorial problem is NP-complete. (TIMED-)DS-RECONF-TJ is such an example that will be important for us and is hence presented next.

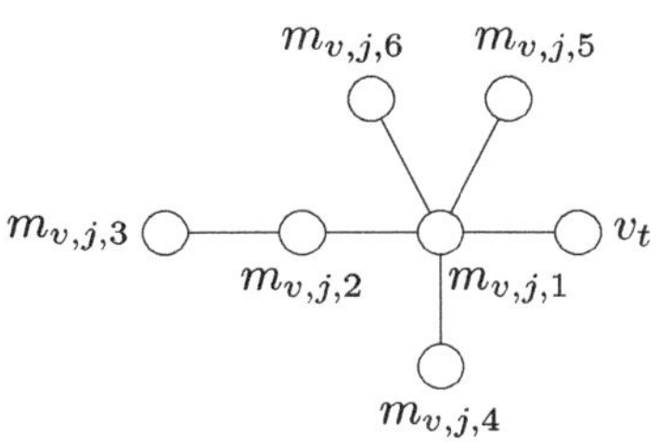

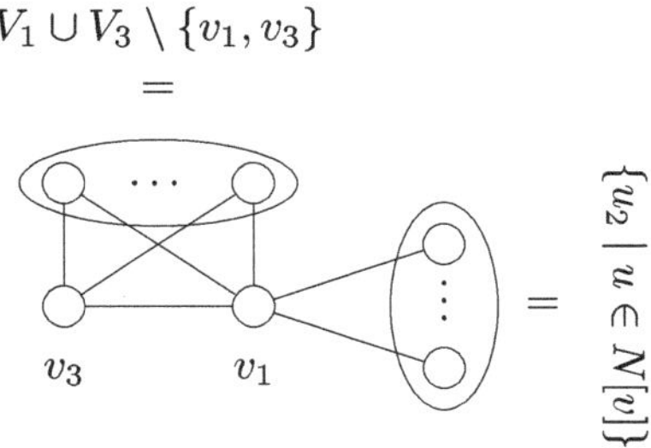

(a) Construction of M_v ($v \in V$, $j \in [d_G(v)]$).
Note that $t = 3$ if $j = d_G(v)$ and $t = 2$, else.

(b) Construction for $\widetilde{G}[V_1 \cup V_2 \cup V_3]$ ($v \in V$).

Fig. 1. Construction for Theorem 3.1

Problem name: (TIMED-)DOMINATING SET RECONFIGURATION-token jumping, or (T-)DS-RECONF-TJ for short.
Given: A graph $G = (V, E)$ and dominating sets $D_s, D_t \subseteq V$ (and $T \in \mathbb{N}$).
Question: Is there an $\ell \in \mathbb{N}$ (with $\ell < T$ and a) dominating set token jumping reconfiguration sequence $(D_s = D_1, \ldots, D_\ell = D_t)$?

We will use this problem to prove the claimed PSPACE-completeness of our problems (see [3]). All the hardness proofs have the same idea: we take a DS-RECONF-TJ instance $(G = (V, E), D_s, D_t)$; we construct a new graph $\widetilde{G} = (\widetilde{V}, \widetilde{E})$ with some copies $V_1, \ldots, V_p$ (p depends on the alliance version that we consider) of the vertex set V, i.e., $V_i = \{v_i \mid v \in V\}$, and some additional vertices. If we consider the timed variant, then the very same time bound T can be also taken for the alliance reconfiguration problem. In every case, the tokens in V_1 represent the tokens in the DS-RECONF-TJ instance. To achieve this, we define, for each $D \subseteq V$, a set $A_D \subseteq \widetilde{V}$, such that for two sets $D, D' \subseteq V$, $A_D \setminus A_{D'} = \{v_1 \mid v \in D \setminus D'\} \subseteq V_1$. Furthermore, D is a dominating set of G iff A_D is an alliance with the right properties of $\widetilde{G}$. This implies one direction of the equivalence. For the other direction, we show that in an alliance reconfiguration sequence $A_{D_s} = A_1, \ldots, A_\ell = A_{D_t}$, there is a D_i with $A_i = A_{D_i}$ for each $i \in [\ell]$. As the reconfiguration sequences on G and $\widetilde{G}$ will have the same length, we will show directly the PSPACE-completeness for the timed versions that are hence not explicitly stated. We will show a detailed proof only in one of the many cases that implement this idea more concretely. Membership in PSPACE immediately follows by guessing the movements in the implicit reconfiguration graph.

3.1 **PSPACE**-Completeness of Token Sliding and Jumping

Theorem 3.1. DA-RECONF-*TS is PSPACE-complete, even on chordal graphs.*

Proof. For PSPACE-hardness, we show a reduction from DS-RECONF-TJ. Hence, let $G = (V, E)$ be a graph and $D_s, D_t \subseteq V$ be dominating sets of G with $k := |D_s| = |D_t|$. Define $\widetilde{G} = (\widetilde{V}, \widetilde{E})$ with $V_q := \{v_q \mid v \in V\}$ for $q \in [3]$ and $M_{v,p} := \{m_{v,j,p} \mid j \in [d_G(v)]\}$ for $p \in [6]$ and $v \in V$. To simplify the notation, denote $M_{v,-} := \bigcup_{p=1}^{6} M_{v,p}$ for $v \in V$ as well as $M_{-,p} := \bigcup_{v \in V} M_{v,p}$ for $p \in [6]$.

$$E_M := \left\{ \{m_{v,j,1}, m_{v,j,p}\}, \{m_{v,j,2}, m_{v,j,3}\} \mid v \in V, j \in [d_G(v)], p \in \{2, 4, 5, 6\} \right\},$$

$$\widetilde{V} := \left(\bigcup_{q=1}^{3} V_q \right) \cup \left(\bigcup_{v \in V} M_{v,-} \right), \text{ and}$$

$$\widetilde{E} := \binom{V_1 \cup V_3}{2} \cup E_M \cup \{\{v_1, u_2\} \mid v, u \in V, u \in N_G[v]\} \cup$$

$$\{\{v_2, m_{v,j,1}\} \mid v \in V, j \in [d_G(v) - 1]\} \cup \{\{v_3, m_{v,d_G(v),1}\} \mid v \in V\}.$$

This construction is illustrated in Fig. 1. $\widetilde{G}$ is chordal as there is a perfect elimination ordering. (1) The vertices in $M_{-,3} \cup M_{-,4} \cup M_{-,5} \cup M_{-,6}$ are leaves. (2) After deleting these vertices, $M_{-,2}$ are leaves. (3) On $\widetilde{G}[V_1 \cup V_2 \cup V_3 \cup M_{-,1}]$, the vertices in $M_{-,1}$ are leaves. (4) V_2 are simplicial vertices on $\widetilde{G}[V_1 \cup V_2 \cup V_3]$ and (5) $V_1 \cup V_3$ is a clique. Construction of $M_{v,-}$ $\qquad\qquad\square$

For $D \subseteq V$, define $D' := \{v_1 \mid v \in D\}$ and $A_D := D' \cup V_2 \cup V_3 \cup M_{-,1} \cup M_{-,2} \subseteq \widetilde{V}$.

Claim 3.2. $(\star)$ *Let* $D \subseteq V$. D *is a dominating set of* G *iff* A_D *is a DA of* $\widetilde{G}$.

This shows that (G, D_s, D_t) is a yes-instance of DS-RECONF-TJ only if $(\widetilde{G}, A_{D_s}, A_{D_t})$ is a yes-instance of DA-RECONF-TS. To see this, we transform each dominating set D_i in our sequence into the DA A_{D_i}.

Now assume there exists a DA TS sequence $A_{D_s} = A_1, \ldots, A_\ell = A_{D_t}$. If we can show that for each $i \in [\ell]$, $V_2 \cup V_3 \cup M_{-,1} \cup M_{-,2} = A_i \setminus V_1$, then the claim implies that there exist dominating sets $D_s = D_1, \ldots, D_\ell = D_t$ with $A_{D_i} = A_i$ for $i \in [\ell]$, so that $D_1, \ldots, D_\ell$ is a DS-RECONF-TJ sequence.

We will show this by contradiction. To this end, let $i \in [\ell]$ $(i \neq 1)$ be the first index such that $V_2 \cup V_3 \cup M_{-,1} \cup M_{-,2} \neq A_i \setminus V_1$. Let $v \in V$ and $j \in [d_G(v)]$. Assume $m_{v,j,1} \notin A_i$. As we consider token sliding, there exists a $p \in \{4, 5, 6\}$ with $m_{v,j,p} \in A_i$. Then $d_{A_i}(m_{v,j,2}) + 1 = 1 < 2 = d_{\widetilde{V} \setminus A_i}(m_{v,j,2})$. Therefore, $M_{-,1} \subseteq A_i$ and as $m_{v,j,1}$ is the only neighbor of $m_{v,j,4}, m_{v,j,5}, m_{v,j,6}$, $(M_{-,4} \cup M_{-,5} \cup M_{-,6}) \cap A_i = \emptyset$. Since $m_{v,j,3}$ is the only neighbor of $m_{v,j,2}$ in $\widetilde{V} \setminus A_{i-1}$ and $m_{v,j,2}$ is the only one of $m_{v,j,3}$ in A_{i-1}, $m_{v,j,3} \in A_i$ iff $m_{v,j,2} \notin A_i$. In this case, $d_{A_i}(m_{v,j,1}) + 1 \leq 2 < 4 = d_{\widetilde{V} \setminus A_i}(m_{v,j,1})$. Thus, $\left(\bigcup_{p=1}^{6} M_{-,p} \right) \cap A_i = M_{-,1} \cup M_{-,2}$. If $v_3 \notin A_i$ for some $v \in V$, $d_{A_i}(m_{v,d_G(v),1}) + 1 = 2 < 4 = d_{\widetilde{V} \setminus A_i}(m_{v,d_G(v),1})$, since $m_{v,d_G(v),4}, m_{v,d_G(v),5}, m_{v,d_G(v),6} \notin N_{\widetilde{G}}(v_3)$. Thus, $V_3 \subseteq A_i$. Analogously, $V_2 \subseteq A_i$. Hence, $V_2 \cup V_3 \cup M_{-,1} \cup M_{-,2} = A_i \setminus V_1$. $\qquad\square$

Observe that each A_{D_i} is also a dominating set: For $i \in [\ell]$, $\widetilde{V} \setminus A_{D_i} \subseteq V_1 \cup M_{-,3} \cup M_{-4} \cup M_{-,5} \cup M_{-,6} \subseteq N(V_3 \cup M_{-,1} \cup M_{-,2}) \subseteq N(A_{D_i})$. For $v_1 \in V_1$, we also know $d_{\widetilde{V} \setminus A_{D_i}}(v_1) + 1 \leq |V_1| = |V_3| \leq d_{A_{D_i}}(v_1)$, as $A_{D_i} \neq \emptyset$. Furthermore, $d_{\widetilde{V} \setminus A_{D_i}}(m_{v,j,p}) + 1 = 1 = d_{A_{D_i}}(m_{v,j,p})$. Hence, the A_{D_i} are even global PAs.

Theorem 3.3. G-DA-Reconf-*TS*, PA-Reconf-*TS*, G-PA-Reconf-*TS* *are* PSPACE-*complete, even on chordal graphs.*

Even if A_{D_i} is an OA, this construction does not provide a proof for the PSPACE-hardness of OA-Reconf-TS, as the DA property is necessary. Namely, we could move the tokens in V_1 as we want. Thus, $(\widetilde{G}, A_{D_s}, A_{D_t})$ is always a yes-instance as an OA-Reconf-TS instance. Even if we bound the number of steps, this is a yes-instance iff $|A_{D_s} \setminus A_{D_t}| \leq \ell$.

With quite similar reductions, yet displaying non-trivial adaptations to the specific cases, we can show the following list of complexity results for TJ and TS.

Theorem 3.4. (∗) *For any* $X \in \{\text{DA,G-DA,OA,G-OA,Idp-OA,PA,G-PA}\}$, X-Reconf-*TJ* *is* PSPACE-*complete, even on bipartite graphs. Moreover,* OA-Reconf-*TS*, G-OA-Reconf-*TS* *and* G-OA-Reconf-*TJ* *are* PSPACE-*complete, even on chordal graphs.*

3.2 LogSPACE Membership Results

Interestingly, some variants of alliance reconfiguration problems are distinctively easier than PSPACE. To prove this, showing membership in LogSPACE suffices, as LogSPACE $\subsetneq$ PSPACE is known by the space hierarchy theorem. We start with a simple combinatorial observation.

Lemma 3.5. *Let* $G = (V, E)$ *be a graph and* $A, B \subseteq V$ *be independent OAs such that* A *can be transformed by one token sliding step into* B. *For* $x \in A \setminus B$ *and* $y \in B \setminus A$, $\{x, y\} \in E$ *is an isolated edge.*

Proof. Since $x \in A \setminus B$ and $y \in B \setminus A$ describe a token sliding step, $\{x, y\} \in E$. As A and B are independent, $N_B(x) \setminus \{y\} = \emptyset = N_A(x)$ and $N_A(y) \setminus \{x\} = \emptyset = N_B(y)$. If $d_G(x) > 1$ or $d_G(y) > 1$, this contradicts B and A being OAs. $\qquad\square$

This lemma has one immediate algorithmic consequence.

Proposition 3.6. (∗) Idp-OA-Reconf-*TS* $\in$ LogSPACE.

With little more effort, one can also show the next algorithmic result.

Lemma 3.7. *For* $Y \in \{TJ, TS\}$, G-Idp-OA-Reconf-Y $\in$ LogSPACE.

Proof. Let $G = (V, E)$ be a graph and let $A_s, A_t \subseteq V$ be global independent OAs with $k := |A_s| = |A_t|$. Let us again take a look at one step of the reconfiguration on G. To this end, let A, B be two global independent OAs such that A can be transformed to B by one token jumping step and $v \in A \setminus B$ and $u \in B \setminus A$. Therefore, $u \in V \setminus A = \partial A$ and there exists a $w \in A$ such that $\{u, w\} \in E$. If $w \neq v$, then this would be contradiction to the independence of B ($w, u \in B$). Hence, u is a leaf, as otherwise, A would not be an OA.

Assume $u \in N(v)$. So $v \in \partial B$. Since B is an OA, v is either a leaf or there exists an $x \in (N(v) \setminus \{u\}) \cap B$. The existence of such an x would contradict the independence of A. Therefore, v is a leaf. So, $\{v, u\} \in E$ is an isolated edge and we can use the same argument as in Proposition 3.6. Since each step is a token sliding step, this algorithm also works for the TS version of this problem. $\qquad\square$

3.3 Token Removal and Addition

Now, we will consider TAR reconfiguration steps. Again, we will derive a number of **PSPACE**-completeness results, but this time, we will provide tight combinatorial links between TAR and TJ to be able to profit from earlier findings. The following proofs are adaptions from Lemma 3 of Bonamy et al. [3] and Theorem 1 of Kaminski et al. [12], but we will treat it more abstractly, based on a novel notion that we introduce now. Let X be a set property. We call X *reconfiguration monotone increasing*, or *rmi* for short (resp. *decreasing*, or *rmd* for short), if for each X-token jumping step A, B (formally, an X-TJ sequence A, B) with $v \in B \setminus A$ (resp. $v \in A \setminus B$), also $A \cup \{v\}$ (resp. $A \setminus \{v\}$) fulfills X. Monotone increasing properties as domination are rmi.

Proposition 3.8. *Let X, Y be rmi (resp. rmd) properties. Then the property that X and Y hold is also rmi (resp. rmd).*

Theorem 3.9. *Let $G = (V, E)$ be a graph, X be a rmi property on vertex sets and $A_s, A_t \subseteq V$ such that $|A_s| = |A_t| = k$ and A_s, A_t have the property X. Then, there exists an X-TJ sequence of length at most ℓ iff there is an X-TAR with threshold $k + 1$ of length at most 2ℓ.*

Proof. Let $A_s = A_1, \ldots, A_\ell = A_t$ be an X-TJ sequence in G. Define $v_i \in A_{i+1} \setminus A_i$ for $i \in [\ell - 1]$. As X is rmi, $A_i' := A_i \cup \{v_i\}$ fulfills the property X for all $i \in [\ell - 1]$. Since A_i can be transformed into A_i' by a token addition step and A_i' into A_{i+1} by a token removal step, $A_s = A_1, A_1', A_2, \ldots, A_{\ell-1}, A_{\ell-1}', A_\ell = A_t$ is an X-TAR sequence of length 2ℓ.

Conversely, let $A_1 = B_1, \ldots, B_{\ell'} = A_t$ be an X-TAR sequence of length $\ell' \leq 2\ell$. We can assume that B_i and B_j are pairwise different for $i, j \in [\ell']$ if $i < j$. Otherwise, delete the $B_i, B_{i+1}, \ldots, B_{j-1}$. The resulting sequence would be an X-TAR sequence of a length at most 2ℓ. Also, observe that ℓ' is odd.

Assume there is an $i \in [\ell']$ with $|B_i| < k$. Clearly, $i \notin \{1, \ell'\}$. Let $i \in [\ell' - 1] \setminus \{1\}$ such that $|B_i|$ is minimum. We now show that there is an X-TAR sequence where we deleted B_i or found a B_i' to replace B_i, with $|B_i| < |B_i'|$. This reduces the number of sets of minimum cardinality in the sequence.

As $|B_i|$ is minimum, B_{i-1} can be transformed into B_i by a token removal step and B_i can be transformed into B_{i+1} by a token addition step. Hence, there is a $v \in B_{i-1} \setminus B_i$ and $u \in B_{i+1} \setminus B_i$. Thus, $B_{i+1} = (B_{i-1} \setminus \{v\}) \cup \{u\}$. If $v = u$ (so $B_{i-1} = B_{i+1}$), we could delete B_i and B_{i+1}. Otherwise, B_{i-1} can be transformed into B_{i+1} by a TJ step. $B_i' := B_{i-1} \cup \{u\}$ fulfills X as X is rmi. Hence, $B_1, \ldots, B_{i-1}, B_i', B_{i+1}, \ldots, B_{\ell'}$ is an X-TAR sequence. We can repeat this, until for each component of the sequence, the cardinality is at least k. Let $C_1, C_2, \ldots, C_{\ell''}$ denote the obtained sequence. This is an X-TAR sequence with threshold $k + 1$ of odd length $\ell'' \leq \ell'$. By construction, TAR-steps always alternate in this sequence. Hence, $|C_i| = k + 1$ for all even $i \in [\ell'']$, and $|C_i| = k$ for all odd $i \in [\ell'']$. As the components of the sequence are pairwise different, $B_1 = C_1, C_3, \ldots, C_{\ell''} = B_{\ell'}$ is an X-TJ sequence of length $\left\lceil \frac{\ell''}{2} \right\rceil \leq \ell$. $\qquad\square$

Proposition 3.10. *The properties* X *and* G-X *are rmi for* $X \in$ $\{\mathrm{DA}, \mathrm{OA}, \mathrm{PA}\}$.

Proof. Let $A, B \subseteq V$ be DAs such that A can be transformed into B by a TJ step with $u \in A \setminus B$ and $v \in B \setminus A$. Define $C := A \cup \{v\} = B \cup \{u\}$. For $x \in A \subseteq C$, $d_C(x) + 1 \geq d_A(x) + 1 \geq d_{V \setminus A}(x) \geq d_{V \setminus C}(x)$. Further $d_C(v) \geq d_B(v) \geq d_{V \setminus B}(v) \geq d_{V \setminus C}(v)$. Hence, the property of being a DA is rmi. The proof for OA works analogously.

A PA is defensive and offensive. By Proposition 3.8, the property PA is rmi. Since domination is a rmi property, G-X is rmi for $X \in \{\mathrm{DA}, \mathrm{OA}, \mathrm{PA}\}$. $\qquad\square$

IDP-OFF is not rmi. A token could jump to a neighbor, disabling Theorem 3.9.

Theorem 3.11. $(*)$ *Let* $G = (V, E)$ *be a graph and* $A_s, A_t \subseteq V$ *be independent OAs s.t.* $|A_s| = |A_t| = k$. *There is an independent OA TJ sequence iff there is an independent OA TAR sequence from* A_s *to* A_t *with threshold* $k + 1$.

Lemma 3.12. $(*)$ *Let* $G = (V, E)$ *be graph. For two global independent OAs* $A_s, A_t \subseteq V$ *of* G, *there exists no global independent OA TAR sequence.*

For a reconfiguration problems Π-RECONF, the reconfiguration graph is often considered. In this graph, the vertices represent a set with the given property and a feasible size. The edges imply that the sets can be transformed into each other by one corresponding transformation step. The previous lemma implies that any reconfiguration graph for G-IDP-OA-RECONF-TAR has no edges. This has the following trivial algorithmic implication.

Corollary 3.13. $(*)$ G-IDP-OA-RECONF-*TAR can be solved in* LogSpace.

These results, together with Theorem 3.4, imply a number of further PSPACE-completeness results for TAR-reconfiguration problems, as summarized next.

Corollary 3.14. *For* $X \in \{\mathrm{DA}, \mathrm{OA}, \mathrm{G\text{-}DA}, \mathrm{G\text{-}OA}, \mathrm{PA}, \mathrm{G\text{-}PA}, \mathrm{IDP\text{-}OA}\}$, X-RECONF-*TAR is* PSPACE-*complete, even on bipartite graphs.* G-OA-RECONF-*TAR is also* PSPACE-*complete on chordal graphs.*

4   FPT-Algorithms: Natural Parameters and Limitations

In this section, we will show that there are FPT-algorithms for DA-RECONF-TJ/TS/TAR and OA-RECONF-TS if the parameter is the number of steps (denoted by ℓ) plus the cardinality of the alliances (denoted by k). The reader might wonder why we look at this combined parameter $k + \ell$. Notice that PSPACE-hardness reductions are also FPT-reductions with respect to ℓ-DS-RECONF-TJ and the corresponding alliance reconfiguration version parameterized by ℓ. By [16], DS-RECONF-TJ is W[2]-hard if parameterized by ℓ.

Corollary 4.1. $(*)$ *For* $X \in \{(\text{G-})\text{DA}, (\text{G-})\text{OA}, (\text{G-})\text{PA}\}$, $Y \in \{TS,\ TJ,\ TAR\}$, X-RECONF-Y *is* W[2]-*hard if parameterized by* ℓ; *this also holds for* IDP-OA-RECONF-TJ/TAR. *The corresponding parameterized problems are in* XP.

In each case, we will provide an $f : \mathbb{N} \to \mathbb{N}$ such that, in each reconfiguration step, there are at most $f(k)$ many choices to consider. Hence, there are at most $f(k)^\ell$ many reconfiguration sequences to be taken into account. Since we can check in polynomial time if such a sequence is feasible, we get FPT-algorithms.

To gain just the information that there exists an FPT-algorithm for each case with this idea, it would be enough to consider TJ, since each TS step is also a TJ step. Nevertheless, we will also present the function f for the TS cases, as f will be significantly smaller. Furthermore, this could be helpful to understand the ideas for the TJ cases. Therefore, we will start with the TS algorithms.

Theorem 4.2. $(k + \ell)$-DA-RECONF-TS, $(k + \ell)$-OA-RECONF-$TS \in$ FPT.

Proof. Let $G = (V, E)$ be a graph. At first we consider TS for DAs. In each step, we can move one of the tokens to one neighbor which is not in the alliance. Since $k \geq d_A(v) + 1 \geq d_{A \setminus S}(v)$ holds for all DAs $A \subseteq V$ with $|A| = k$ and $v \in A$, in each step, we only can put one token of A (k possibilities) on one of at most k neighbors. Thus, in each of the at most ℓ steps, we have $f(k) = k^2$ possibilities. Hence, there are $\sum_{i=1}^{\ell} \left(k^2\right)^i \leq \left(k^2\right)^{\ell+1}$ many possible reconfiguration sequences.

We consider $(k + \ell)$-OA-RECONF-TS next. Let $A, B \subseteq V$ be OAs where A can be transformed into B by one sliding step. Hence, there are unique $v \in A \setminus B$ and $u \in B \setminus A$ with $v \in N(u)$. Thus, $d_{V \setminus B}(v) = d_{V \setminus A}(v) + 1$. As B is an OA, $k \geq d_B(v) \geq d_{V \setminus B}(v) = d_{V \setminus A}(v) + 1$. Hence for each token which we can move, there are at most $k - 1$ possible new positions. As we have k tokens, there are at most $f(k) = k^2 - k$ possibilities of continuation in the next step. Hence, there are at most $\left(k^2 - k\right)^{\ell+1}$ reconfiguration sequences. $\square$

Now, we will consider the token addition/removal and jumping versions. We mostly prove the results directly for token addition/removal. Because of reconfiguration monotonicity, some results transfer directly to token jumping.

For $G = (V, E)$, let $G^{\leq r} = G[V^{\leq r}]$ with $V^{\leq r} = \{v \in V \mid d_G(v) \leq r\}$. If G' is a subgraph of G, then let $\text{dist}_{G'}(x, y)$ denote the shortest-path distance within G', and $\text{dist}_{G'}(x, M) = \min\{\text{dist}_{G'}(x, y) \mid y \in M\}$. The following lemma decreases the number of TAR sequences we need to consider for DA-Reconf-TAR. We use that each DA of size at most k in G needs to be a subset of $V^{\leq 2k}$. Also, we can add at most ℓ tokens. Thus, we only need to consider vertices in $v \in V^{\leq 2k}$, within a distance of ℓ to $A_s \cup A_t$ on $G^{\leq k}$, as the other vertices are irrelevant.

Lemma 4.3. *Let $G = (V, E)$ be a graph and $A_s, A_t \subseteq V$ be DAs, $|A_s| = |A_t| = k$, with a DA TAR sequence $A_s = A_1, \ldots, A_\ell = A_t$ of length ℓ. Then there exists a DA TAR sequence $A_s = A'_1, \ldots, A'_{\ell'} = A_t$, with $\ell' \leq \ell$, such that*

$$\bigcup_{i=1}^{\ell'} A'_i \subseteq \{v \in V^{\leq 2k} \mid dist_{G^{\leq 2k}}(v, A_s \cup A_t) \leq \ell\}.$$

Proof. Clearly, we can assume that the mapping $[\ell] \to 2^V$, $i \mapsto A_i$ is injective. We simplify the notation by setting $A := \bigcup_{i=1}^{\ell} A_i$. Assume there is a $v \in A$ with $dist_{G^{\leq 2k}}(v, A_s \cup A_t) > \ell$. Define $K_j \subseteq A$ for $j \in [\ell] \cup \{0\}$ inductively by $K_0 = \{v\}$ and $K_j = K_{j-1} \cup K'_j$ for $j \in [\ell]$ with

$$K'_j := \{x \in A \setminus K_{j-1} \mid \exists i \in [\ell] : d_{A_i \setminus K_{j-1}}(x) + 1 < d_{V \setminus (A_i \setminus K_{j-1})}(x)\}.$$

In other words, K'_j includes vertices for which there exists an $i \in [\ell]$ such that v violates the DA property for $A_i \setminus K_{j-1}$.

Let $j \in [\ell]$ be fixed. For $x \in V \setminus N[K_j]$, $d_{A_i \setminus K_{j-1}}(x) = d_{A_i}(x)$ as well as $d_{V \setminus (A_i \setminus K_{j-1})}(x) = d_{V \setminus A_i}(x)$. Hence, $K_j \subseteq N(K_{j-1})$. By an inductive argument, $dist_{G^{\leq 2k}}(v, K_j) \leq j \leq \ell < dist_{G^{\leq 2k}}(v, A_s \cup A_t)$. So, $K_j \cap (A_s \cup A_t) = \emptyset$ which implies $A_s \setminus K_j = A_s$ and $A_t \setminus K_j = A_t$.

Assume $K_j \neq K_{j-1}$ for all $j \in [\ell]$. Since K_j is strictly monotone increasing, $|K_\ell| > \ell$. As $K_\ell \cap (A_s \cup A_t) = \emptyset$ and $K_\ell \subseteq A$, this contradicts that $A_s = A_1, \ldots, A_\ell = A_t$ is a TAR sequence, as only one vertex can be added per step.

Hence, we can assume there is a $j \in [\ell]$ such that $K_{j-1} = K_j$, so $K'_j = \emptyset$. Thus, for all $q \in [\ell] \setminus [j-1]$, $K_j = K_q$ and $K'_q = \emptyset$. By the definition of K'_j, $A_1 \setminus K_j, \ldots, A_\ell \setminus K_j$ are DAs. Furthermore, for each $i \in [\ell - 1]$, $A_i \setminus K_j = A_{i+1} \setminus K_j$ or $A_i \setminus K_j$ can be transformed into $A_{i+1} \setminus K_j$ by a TAR step. Hence, we can shorten the $A_1 \setminus K_j, \ldots, A_\ell \setminus K_j$ into a DA TAR reconfiguration sequence $A'_1, \ldots, A'_{\ell'}$ by deleting all but one sets that are the same. Thus, for each $i \in [\ell']$, $v \notin A'_i$. We can use this argument repeatedly to prove the lemma. $\square$

Lemma 4.3 restricts the search space to $\{v \in V^{\leq 2k} \mid dist_{G^{\leq 2k}}(v, A_s \cup A_t) \leq \ell\}$, which gives (together with Theorem 3.9 and Proposition 3.10) the next result.

Theorem 4.4. $(k + \ell)$-DA-Reconf-*TAR*, $(k + \ell)$-DA-Reconf-*TJ* $\in$ *FPT*.

Before considering OA-Reconf-TJ versions, we introduce an auxiliary result. To simplify the notation, we define for a graph $G = (V, E)$ and a set $X \subseteq V$, $Z(X) := \{v \in V \setminus N[X] \mid N(v) \subseteq L \cup \partial X\}$ and $Y(X) := N[X] \cup Z(X) \cup L$, where $L := \{x \in V \mid d_G(x) = 1\}$ is the set of leaves. If X is an OA, $Z(X)$ are vertices $z \in N[X]$ for which $X \cup \{z\}$ is still an OA. Namely, the elements in ∂X already fulfill the degree property for the OAs and leaves fulfill this property if the neighbor is in X. Clearly, $Z(X)$ is an independent set of G for any $X \subseteq V$. $Y(X)$ is the union of the closed neighborhood of X together with the leaves and $Z(X)$. The next lemma gives a combinatorial restriction on our search space.

Lemma 4.5. *Let $G = (V, E)$ be a graph and A, B be OAs of G for which there is an OA TM sequence between both for $TM \in \{\, TAR, TJ \,\}$. Then, $Y(A) = Y(B)$.*

Proof. By Theorem 3.9 and Proposition 3.10 each TJ step can be seen as two TAR steps. So, it is enough to show that this lemma holds if there is a $v \in A$ with $A = B \cup \{v\}$. Using this argument inductively proves the lemma.

Since A is an OA, for each $x \in N(v)$, $x \in \partial A$ with $d_A(x) > d_{V \setminus A}(x)$ or $x \in A$. Thus, if $x \in N(v) \setminus (L \cup A) = N(v) \setminus (L \cup B)$, $(A \setminus \{v\}) \cap N(x) = B \cap N(x)$ is not empty. Thus, $x \in \partial B$ and $N(v) \subseteq N[B] \cup L \subseteq Y(B)$. Therefore, $\partial A \subseteq Y(B)$. Now, we want to show $v \in Y(B)$. If v has a neighbor in B, then $v \in Y(B)$. If vertex v has no neighbor in B, then by the same argument as above, each vertex in $N(v) \setminus (L \cup A) = N(v) \setminus L$ has a neigbor in B. Thus, $v \in Z(B) \subseteq Y(B)$.

This leaves to show $Z(A) \subseteq Y(B)$. Assume there exists a $u \in Z(A)$ such that $u \in Z(A) \setminus Y(B)$. As $u \in Z(A) \subseteq V \setminus N[A] \subseteq V \setminus N[B]$, $u \notin N[B] = B \cup \partial B$. Thus, u has a neighbor $w \in (\partial A) \setminus L$ with $w \notin \partial B$. Therefore, $w \in (N[v] \cap N[u]) \setminus N[B]$. This is a contradiction to $N(v) \subseteq N[B] \cup L$ (see above). Hence, $Y(A) \subseteq Y(B)$.

As $N[B] \subseteq N[A]$, $Z(B) \subseteq Y(A)$ has to be proved. Let $u \in Z(B)$. Then, $u \in N[A] \subseteq Y(A)$ or $N(u) \subseteq L \cup \partial B \subseteq L \cup N[A]$ (i.e., $u \in \partial A \cup Z(A)$). □

As detailed in the appendix, Lemma 4.5 can be helpful to get FPT-algorithms for some versions of OA-RECONF-TJ. We discuss another application next.

Theorem 4.6. k-$(G$-$)PA$-$Reconf$-$Y \in$ FPT *for* $Y \in \{\, TJ, TS, TAR \,\}$.

Proof. Let $G = (V, E)$ be a graph and let $A_s = A_1, \ldots, A_\ell = A_t \subseteq V$ be a PAsTAR sequence with $|A_i| \leq k$, for $i \in [\ell]$. By Lemma 4.5, for all $i \in [\ell]$, $Y(A_s) = Y(A_i)$. As A_s is a DA, $|\partial A_s| \leq k^2$. Furthermore, $|N(\partial A_s) \setminus A_s| \leq k^3$, as A_s is an OA. Consider the case that there is an $i \in [\ell - 1]$ such that A_i can be transformed into A_{i+1} by adding a token to $x \in V \setminus N[\partial A_i]$. As A_{i+1} is a DA, $d_G(x) \leq 1$. If x has a neighbor u, then $d_G(u) = 1$, i.e., we have an isolated edge $\{x, u\}$. Otherwise, A_{i+1} is not an OA. Let M be the set of such vertices. The vertices in M can be also useful for TA steps, but clearly, we only need to consider at most $3k$ of them for reconfiguration, as there is no difference for the sequence differentiating on which vertex in $M \setminus (A_s \cup A_t)$ we put a token: collect these in $M_k \subseteq M$. Define $Y := Y(A_s) \setminus (M \setminus M_k)$. Hence, there are at most $|Y| \leq k^2 + k^3 + 3k$ vertices which are useful for any reconfiguration step. Thus, the number of PAs which are useful for the reconfiguration sequence is at most $r_k = \sum_{j=0}^{k} \binom{k^2 + k^3 + 3k}{j}$. This number is also an upper bound on the number of steps as we can avoid visiting sets twice in the sequence. Therefore, the algorithm runs in $\mathcal{O}^*\left(\left(k^2 + k^3 + 3k \right)^{r_k} \right)$ time. As these are rmi properties, we also get FPT-results for TJ and TS. Additionally checking if the alliances are dominating sets yields FPT-algorithms for the G-variants. □

Quite similarly, one can also attack other alliance reconfiguration problems, even only with the single parameter k, as we can bound the number of steps of a reconfiguration sequence by a computable function in k. We will sketch the proof for offensive alliances and omit the one for defensive alliances here.

Theorem 4.7. $(*)$ k-G-OA-RECONF-$Y \in$ FPT *for* $Y \in \{$ *TJ, TS, TAR* $\}$.

Proof. (Sketch) Let $G = (V, E)$ be a graph and $A_s, A_t \subseteq V$ be global OAs with $k := |A_s| = |A_t|$. Since A_s is global, $V = A_s \cup \partial A_s$. Let $v \in V$ with $d_G(v) > 2k$, then v has to be in any global OA A. Otherwise, $v \in \partial A$ and $d_V(v) \leq k < d_{V \setminus A}(v)$, would contradict that A is an OA. Therefore, $D := \{v \in V \mid d_G(v) > 2k\} \subseteq A$. Define $B = \{v \in V \setminus D \mid d_D(v) > d_{V \setminus D}(v)\}$. We need not check $d_A(v) \geq d_{V \setminus A}(v)$ for $v \in (\partial A) \cap B$ to verify that $A \subseteq V$ with $|A| = k$ is an OA, since $D \subseteq A$ has to hold for such an OA. For $v \in (\partial A_s) \setminus B$, there has to be a $u \in A_s \setminus D$. Since for all $u \in A_s \setminus D$, $d_G(u) \leq 2k$, there are at most $2k^2$ many vertices in $(\partial A_s) \setminus B$. For $W \subseteq V' := V \setminus (D \cup B)$, define $P_W := \{v \in B \mid N_{V'}(v) = W\}$. Clearly, $\mathcal{P} = \{P_W \mid W \subseteq V'\}$ is a partition of B in to 2^{k+2k^2} classes. The following claim will help us to bound the number of global OAs that we need to consider in our sequences. $\qquad \square$

Claim 4.8. $(*)$ *Let the following be given:* $W \subseteq V'$, $v \in P_W$, *a global OA* $A \subseteq V \setminus \{v\}$ *of* G *with* $|A| = k$ *and* $u \in A \cap P_W \neq \emptyset$. *Then* $A' := (A \cup \{v\}) \setminus \{u\}$ *is also a global OA of* G.

By the claim, we now that we need to consider at most $2k$ vertices per partition class ($2k$ as A_s and A_t could have many vertices in one class). We can use this observation as a reduction rule. This implies $|\bigcup_{W \subseteq V'} P_W| \leq 2k \cdot 2^{k+2k^2}$ and $|V| \leq 2k \cdot 2^{k+2k^2} + k + 2k^2$. So, there are at most $\binom{2k \cdot 2^{k+2k^2} + k + 2k^2}{k}$ many global OAs that we need to consider and we can assume that a reconfiguration sequence would have at most this length to avoid repetitions. $\qquad \square$

Theorem 4.9. $(*)$ k-G-DA-RECONF-$Y \in$ FPT *for* $Y \in \{$ *TJ, TS, TAR* $\}$.

These are interesting results as [2] shows that k-DS-RECONF-TJ is XL-complete, k-T-DS-RECONF-TJ is XNL-complete if the maximal number ℓ of steps is given in binary (see [2, Cor. 28]) and XNLP-complete if ℓ is given in unary (see [2, Theorem 36, Corollary 37]) if parameterized by the cardinality k of the dominating sets. For the definition of these classes, we refer to [2]. By definition, XL $\subseteq$ XNL. Chen and Flum [4] proved XNLP $\subseteq$ XNL. By Bodlaender *et al.* [1], XNLP-hardness implies W[t]-hardness for each $t \in \mathbb{N} \setminus \{0\}$. Our results on the reconfiguration of global alliances differ from the results for DS-RECONF.

5   Neighborhood Diversity

We now consider the structural parameter neighborhood diversity nd for our reconfiguration problems. It is known that if the vertex cover number vc is upper-bounded by d, then this also holds for the neighborhood diversity. Thus, an FPT-algorithm with respect to nd would also imply an FPT-algorithm for vc.

Table 1. Survey on PSPACE-completeness results for alliance reconfiguration problems (or membership in LogSPACE when put in parentheses).

	DEF	OFF	POW	IDP-OFF	G-DEF	G-OFF	G-POW	G-IDP-OFF
TS	3.1	3.4	3.3	(3.6)	3.3	3.4	3.3	(3.7)
TJ	3.4	3.4	3.4	3.4	3.4	3.4	3.4	(3.7)
TAR	3.14	3.14	3.14	3.14	3.14	3.14	3.14	(3.13)

Observation 5.1. *Let $G = (V, E)$ be a graph, A be a DA and v, u be vertices of the same type. If $v \in A$ and $u \notin A$, then $(A \setminus \{v\}) \cup \{u\}$ is also a DA. Similar statements hold for (G-)OAs, (G-)PAs, global DAs, and independent OAs.*

Theorem 5.2. *$(\mathsf{nd} + p)$-Z-RECONF-TM $\in$ FPT for TM $\in \{TJ, TS, TAR\}$, $p \in \{k, \ell\}$ and Z $\in \{$ (G-)DA, (G)-OA, (G)-PA, IDP-OA$\}$.*

Proof. Let $G = (V, E)$ be a graph, $k \in \mathbb{N}$ and $A_s, A_t \subseteq V$ be DAs with $|A_s| = |A_t| = k$. Further, $d := \mathsf{nd}(G)$ and let $C_1, \ldots, C_d$ be the neighborhood diversity equivalence classes. By Observation 5.1, we can assume that it is unimportant to which vertex in a class a token moves (unless being in A_s or A_t).

First, consider $p = k$. We only need $2k$ vertices per class to remember. Hence, we have at most $\binom{d \cdot 2k}{k}$ many possible DAs. So, we can go through all possibilities, check these and find a shortest path in this part of the reconfiguration graph.

Now we consider $p = \ell$. We move the token to a vertex in A_t, if possible. In the other cases, it is arbitrary to which vertex we move the token. Hence, there are at most d^2 many possible moves in one token transformation. Thus, there are at most $d^{2\ell}$ many alliance reconfiguration sequences that we need to consider. This argument works analogously for the other versions of alliances. $\square$

Beside the sketched combinatorial algorithm of the last proof, we could also use Integer Linear Programming (ILP) to solve these parameterized problems in FPT-time. Using ILP for solving reconfiguration problems appear to be a new approach. Details are in the appendix. It is still <u>open</u> if we can get FPT-results if we parameterize alliance reconfiguration problems by nd only. We cannot employ the meta-theorem from [8]: alliance problems are not expressible in MSO.

6   Conclusions

We survey our classical complexity results in Table 1. Notice that we alternate between LogSPACE- and PSPACE-results. Admittedly, our FPT-algorithms are not optimized in terms of running times. As most of our arguments are of a combinatorial nature, one could also interpret these results as kernel results. Alternatively, one could construct branching algorithms that make use of our combinatorial findings. The parameterized complexity status of $(k + \ell)$-OA-RECONF-TJ, $(k + \ell)$-IDP-OA-RECONF-TJ, nd-X-RECONF, k-X'-RECONF, where X is any alliance condition and $X' = \{$ DA,OA,IDP-OA $\}$ is still open.

References

1. Bodlaender, H.L., Groenland, C., Nederlof, J., Swennenhuis, C.M.F.: Parameterized problems complete for nondeterministic FPT time and logarithmic space. In: 62nd IEEE Annual Symposium on Foundations of Computer Science, FOCS, pp. 193–204. IEEE (2021)
2. Bodlaender, H.L., Groenland, C., Swennenhuis, C.M.F.: Parameterized complexities of dominating and independent set reconfiguration. Technical report 2106.15907, ArXiv, Cornell University (2021). Revised in 2023
3. Bonamy, M., Dorbec, P., Ouvrard, P.: Dominating sets reconfiguration under token sliding. Discret. Appl. Math. **301**, 6–18 (2021)
4. Chen, Y., Flum, J.: Bounded nondeterminism and alternation in parameterized complexity theory. In: 18th Annual IEEE Conference on Computational Complexity, CCC, pp. 13–29. IEEE Computer Society (2003)
5. Fernau, H., Mann, K.: How to reconfigure your alliances. Technical report 2509.08798, ArXiv, Cornell University, USA (2025)
6. Fernau, H., Rodríguez-Velázquez, J.A.: A survey on alliances and related parameters in graphs. Electron. J. Graph Theory Appl. **2**(1), 70–86 (2014)
7. Fricke, G., Lawson, L., Haynes, T.W., Hedetniemi, S.M., Hedetniemi, S.T.: A note on defensive alliances in graphs. Bull. Inst. Combin. Appl. **38**, 37–41 (2003)
8. Gima, T., Ito, T., Kobayashi, Y., Otachi, Y.: Algorithmic meta-theorems for combinatorial reconfiguration revisited. In: Chechik, S., Navarro, G., Rotenberg, E., Herman, G. (eds.) 30th Annual European Symposium on Algorithms, ESA. LIPIcs, vol. 244, pp. 61:1–61:15. Schloss Dagstuhl - Leibniz-Zentrum für Informatik (2022)
9. Haynes, T.W., Hedetniemi, S.T., Henning, M.A.: Structures of Domination in Graphs. Developments in Mathematics, vol. 66. Springer, Cham (2021)
10. van den Heuvel, J.: The complexity of change. In: Blackburn, S.R., Gerke, S., Wildon, M. (eds.) Surveys in Combinatorics. London Mathematical Society Lecture Note Series, vol. 409, pp. 127–160. Cambridge University Press (2013)
11. Ito, T., et al.: On the complexity of reconfiguration problems. Theoret. Comput. Sci. **412**(12–14), 1054–1065 (2011)
12. Kamiński, M., Medvedev, P., Milanič, M.: Complexity of independent set reconfigurability problems. Theoret. Comput. Sci. **439**, 9–15 (2012)
13. Kim, B.J., Liu, J., Um, J., Lee, S.I.: Instability of defensive alliances in the predator-prey model on complex networks. Phys. Rev. E **72**(4), 041906 (2005)
14. Kristiansen, P., Hedetniemi, S.M., Hedetniemi, S.T.: Alliances in graphs. J. Comb. Math. Comb. Comput. **48**, 157–177 (2004)
15. Mouawad, A.E.: On reconfiguration problems: structure and tractability. Ph.D. thesis, University of Waterloo, Ontario, Canada (2015)
16. Mouawad, A.E., Nishimura, N., Raman, V., Simjour, N., Suzuki, A.: On the parameterized complexity of reconfiguration problems. Algorithmica **78**(1), 274–297 (2017)
17. Nishimura, N.: Introduction to reconfiguration. Algorithms **11**(4), 52:1–25 (2018)
18. Ouazine, K., Slimani, H., Tari, A.: Alliances in graphs: parameters, properties and applications–a survey. AKCE Int. J. Graphs Combin. **15**(2), 115–154 (2018)
19. Rodriguez, J.A., Sigarreta, J.M.: Offensive alliances in cubic graphs. Int. Math. Forum **1**(36), 1773–1782 (2006)
20. Seba, H., Lagraa, S., Kheddouci, H.: Alliance-based clustering scheme for group key management in mobile ad hoc networks. J. Supercomput. **61**(3), 481–501 (2012)

21. Shafique, K.H.: Partitioning a graph in alliances and its application to data clustering. Ph.D. thesis, University of Central Florida, Orlando, USA (2004)
22. Szabó, G., Czárán, T.: Defensive alliances in spatial models of cyclical population interactions. Phys. Rev. E **64**(4), 042902 (2001)
23. Yero, I.G., Rodríguez-Velázquez, J.A.: A survey on alliances in graphs: defensive alliances. Utilitas Math. **105**, 141–172 (2017)

Graph Irregularity via Edge Deletions

Julien Bensmail[1], Noëmie Catherinot[2], Foivos Fioravantes[3(✉)],
Clara Marcille[4], and Nacim Oijid[4,5]

[1] Université Côte d'Azur, CNRS, Inria, I3S, Côte d'Azur, France
[2] DER Informatique, Université Paris-Saclay, ENS Paris-Saclay,
4 Avenue des Sciences, 91190 Gif-sur-Yvette, France
`noemie.catherinot@ens-paris-saclay.fr`
[3] Department of TCS, FIT, Czech Technical University in Prague,
Prague, Czech Republic
`foivos.fioravantes@fit.cvut.cz`
[4] Univ. Bordeaux, CNRS, Bordeaux INP, LaBRI, UMR 5800, 33400 Talence, France
`nacim.oijid@ens-lyon.fr`
[5] Department of Mathematics and Mathematical Statistics, Umeå University,
Umeå, Sweden

Abstract. We pursue the study of edge-irregulators of graphs, which were recently introduced in [Fioravantes et al. Parametrised Distance to Local Irregularity. *IPEC*, 2024]. That is, we are interested in the parameter $I_e(G)$, which, for a given graph G, denotes the smallest $k \geq 0$ such that G can be made locally irregular (*i.e.*, with no two adjacent vertices having the same degree) by deleting k edges. We exhibit notable properties of interest of the parameter I_e, in general and for particular classes of graphs, together with parameterized algorithms for several natural graph parameters.

Despite the computational hardness previously exhibited by this problem (NP-hard, W[1]-hard w.r.t. feedback vertex number, W[1]-hard w.r.t. solution size), we present two FPT algorithms, the first w.r.t. the solution size plus Δ and the second w.r.t. the vertex cover number of the input graph.

Finally, we take important steps towards better understanding the behaviour of this problem in dense graphs. This is crucial when considering some of the parameters whose behaviour is still uncharted in regards to this problem (e.g., neighbourhood diversity, distance to clique). In particular, we identify a sub-family of complete graphs for which we are able to provide the exact value of $I_e(G)$. These investigations lead us to propose a conjecture that $I_e(G)$ should always be at most $\frac{1}{3}m + c$, where m is the number of edges of the graph G and c is some constant. This conjecture is verified for various families of graphs, including trees.

Due to space limitations, many proofs are omitted or just sketched. We refer the reader to the full version [6]. This research was partly funded by Kempe Foundation Grant No. JCSMK24-515 (Sweden) and the International Mobility of Researchers MSCA-F-CZ-III at CTU in Prague, CZ.02.01.01/00/22_010/0008601 Programme Johannes Amos Comenius.

1   Introduction

In graph theory, the family of *regular graphs* is used to capture the most common notion of regularity, that of all vertices having the same degree. A very natural question then, is on possible definitions of graphs that would be sort of antonyms to regular graphs. This led several groups of researchers to investigate various definitions of irregularity in graphs, such as *irregular graphs* (graphs in which all degrees are pairwise distinct [7]), *highly irregular graphs* (graphs in which no vertex has two neighbours with the same degree [1]), and *locally irregular graphs* (graphs in which no two adjacent vertices have the same degree [2]). As we will only focus on the latter notion throughout this work, we encourage the reader to refer to the literature for more details on the former two.

Note that most definitions of (ir)regular graphs do not hold for all simple graphs: a given simple graph can be more or less (ir)regular w.r.t. some notion of (ir)regularity, and exhibit different behaviour for different such notions. This leads to different questions, such as how a given simple graph can be turned into a (ir)regular structure through elementary operations. See, e.g., [4] for an example of a problem where regularity is introduced in the given graph through vertex deletions. We will focus on introducing irregularity. The initiators of this line of research are Chartrand *et al.* [7], who introduced the notion of *irregularity strength*. In brief, this is a parameter that measures how easy/hard it is for a given graph to be turned into an irregular multigraph by increasing the multiplicity of its edges (with the additional objective that the maximum number of parallel edges replacing a given edge should be minimised). This concern was considered as well for the other notions of irregularity, namely in [5] for highly irregular graphs, and in [13] for locally irregular graphs. These investigations resulted in numerous related questions and problems, such as the so-called 1–2-3 Conjecture (introduced by Karoński, Łuczak, and Thomason in [13], and solved recently by Keusch in [14]) for the latter notion.

In all these considerations, a given simple graph is intended to be made irregular by essentially adding edges locally (in a parallel way, upon original edges), and there is an implicit postulate: the number of edges added this way should be minimised. This leads us to the question of whether one could consider making simple graphs somewhat irregular through other elementary operations that would limit even further the eventual number of edges. In particular, could this be achieved without adding any additional edge, by just *moving* edges? We understand this operation as removing said edge and placing a new edge between any two vertices we chose (regardless of whether there already exists an edge between the considered vertices). At this point it is also worth mentioning [3], where the authors consider a similar operation on the edges but with the goal of introducing regularity to the given graph.

Transforming a given graph into a locally irregular graph via edge moves is straightforward, since any multistar on at least 3 vertices is always locally irregular. Things become more interesting when we integrate concerns borrowed from reconfiguration problems, such as minimising the number of edge moves performed. This latter problem is way harder to comprehend. Indeed, when edge moves can be performed at will, it is important to consider the number of edges in a largest (non-induced) locally irregular subgraph of G. This leads us to the study of the (dual) concepts and notions we will focus on in this paper, which are about the minimum number of edges one must remove from a graph to make it locally irregular.

These notions were introduced recently by Fioravantes, Melissinos, and Triommatis [11]. Following their terminology, we define an *edge-irregulator* of a graph G as a set $S \subseteq E(G)$ of edges of G such that $G - S$ is locally irregular. We also set $\mathrm{I}_e(G)$ as the smallest $k \geq 0$ such that G admits edge-irregulators of cardinality k, and say that an edge-irregulator S of G is *optimal* if $|S| = \mathrm{I}_e(G)$. The authors of [11] introduced edge-irregulators as an edge counterpart of *vertex-irregulators* (introduced in [12]), in which vertices, instead of edges, are deleted. It was proven in [11] that the problem of finding optimal edge-irregulators is NP-hard even when restricted to planar bipartite graphs of maximum degree 6, and W[1]-hard when parameterized by the size of the solution, the feedback vertex set of the input graph, or its vertex integrity.

The first contribution of this work is the continuation of the study of the parameterized complexity of computing an optimal edge-irregulator of a given graph. Considering the previously mentioned W[1]-hardness results, we propose adding the maximum degree Δ of the input graph as part of the parameter. This proves to be fruitful, as we show that the problem becomes FPT w.r.t. the solution size plus Δ. We then present an FPT algorithm parameterized by the vertex cover number of the input graph, which is more efficient than the corresponding algorithm that is implied by the work in [11] (which is based on a reduction to an ILP with bounded number of variables).

The behaviour of computing I_e is less understood when it comes to structural parameters capturing dense graphs. In particular, it is unknown whether the problem is FPT when parameterized by parameters such as the neighbourhood diversity or the distance to clique of the input graph. We take an important step towards understanding the case of complete graphs. In particular, we prove that $\mathrm{I}_e(K_{t_k}) = |E(K_{t_k})| - \frac{k(k+1)(k-1)(3k+2)}{24}$, where t_k is any triangular number.

The study of this concept of edge-irregulators leads us to conjecture that there exists a constant c such that any connected graph G has an edge-irregulator of size at most $\frac{|E(G)|}{3} + c$. Note that the connected hypothesis is mandatory–the only edge-irregulator of a graph consisting only in a perfect matching will contain all the edges of the graph. Appart form attacking the case of complete graphs, we also prove this conjecture for various families of simple graphs, including trees, for which we also provide a dedicated polynomial algorithm.

2   Preliminaries, First Results and a Conjecture

We follow standard graph theory notations [9]. Let $G = (V, E)$ be a graph. For any vertex $v \in V$, let $N_G(v) = \{u \in V : uv \in E\}$ denote the *neighbourhood* of v in G, and let $d_G(v) = |N_G(v)|$ denote the *degree* of v in G. We also define

$N_G[v] = N_G(v) \cup \{v\}$. Whenever the graph G is clear from the context, we will omit the subscript and simply write $N(v)$, $d(v)$ and $N[v]$.

Parameterized complexity is a field in algorithm design that takes into consideration additional measures to determine the time complexity. The goal in this paradigm is to design a *Fixed-Parameter Tractable* (FPT) algorithm, i.e., an algorithm that solves the problem in $f(k)|x|^{O(1)}$ time for any arbitrary computable function $f\colon \mathbb{N} \to \mathbb{N}$, where $|x|$ is the size of the input and k is some bounded *parameter*. We say that a problem *is in* FPTif it admits an FPTalgorithm. A parameterized problem is *slicewise polynomial* if it can be determined in $|x|^{f(k)}$ time for a computable function $f\colon \mathbb{N} \to \mathbb{N}$. Such a problem then belongs to the class XP. A problem is presumably not in FPTif it is shown to be W[1]-hard (by a parameterized reduction). We refer the interested reader to classical monographs [8, 10] for a more comprehensive introduction to this topic.

We are now ready to formally define the problem studied in this work. Let $G = (V, E)$ be a graph. We say that G is *locally irregular* if, for every edge $uv \in E$, it holds that $d_G(u) \neq d_G(v)$. Next, let $S \subseteq E$ and consider the graph G' that results from the deletion of the edges of S from G. We will denote this G' as $G - S$. Any set $S \subseteq E$ such that $G - S$ is locally irregular will be called called an *edge-irregulator of* G. Moreover, let $\mathrm{I}_e(G)$ be the minimum order that any edge-irregulator of G can have. We will say that S is an *optimal* edge-irregulator of G if S is an edge-irregulator of G and $|S| = \mathrm{I}_e(G)$. Our goal is to compute an optimal edge-irregulator of a given graph G (optimisation version) or, given an integer $k \geq 1$, decide if $\mathrm{I}_e(G) \leq k$ (decision version).

Irregularity-Deletion
Input: A graph G, and a positive integer $k \geq 1$.
Question: Do we have $\mathrm{I}_e(G) \leq k$?

To begin, observe that checking if a graph G is locally irregular can be done in polynomial time as it suffices to go through its edges once. Thus, checking if $\mathrm{I}_e(G) \leq k$ is (trivially) in XP when parameterized by k.

Intuitively, regular graphs should be the graph having the largest value of I_e. Asymptotically, if G is a cycle with m edges, then all its adjacent vertices are in conflicts and removing an edge resolves at most 3 of these conflicts. Thus, $\mathrm{I}_e(G)$ cannot be too far from $\frac{1}{3}m$. Equality cannot always hold due to divisibility matters; for instance, $\mathrm{I}_e(C_4) = 2$, which is $\frac{1}{2}m$ (where $m = |E(C_4)| = 4$) but also $\lfloor \frac{1}{3}m \rfloor + 1$. Even worse, $\mathrm{I}_e(K_2) = 1$, which is m (since $m = |E(K_2)| = 1$). Perhaps a good way to express our presumption is in the following form:

Conjecture 1. There is an absolute constant $c \geq 1$ such that, for every connected graph G with m edges, we have $\mathrm{I}_e(G) \leq \frac{1}{3}m + c$.

The additive constant c is mandatory to handle some uncontrolled behaviour of small and sparse graphs. In particular, our result on cycles (see Theorem 7) shows that $c \geq 2$ for any cycle C_{3k+2}, where k is an integer.

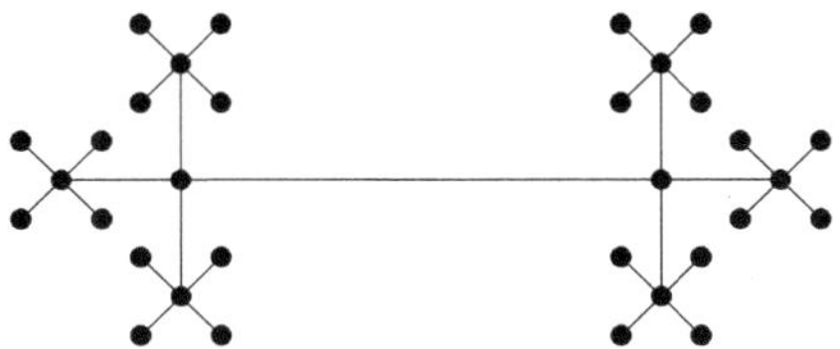

Fig. 1. Example of graph satisfying $I_e(G) = \left\lceil \frac{\text{conf}(G)}{2\Delta-1} \right\rceil$.

We want to insist on the fact that the connectivity requirement in Conjecture 1 is important, as the conjecture would be false otherwise. Consider, e.g., a disjoint union of several copies of K_2.

Throughout Sect. 4, we will prove that this conjecture holds for common classes of graphs. These points apart, Conjecture 1 is also supported by the fact that, when trying to construct connected graphs for which the parameter I_e is a "large" fraction of the number of edges, the best we could achieve is indeed a third of the number of edges. This also shows that Conjecture 1 is tight.

Theorem 1. *Let G be any (connected) graph on m edges obtained from a (connected) graph H by subdividing exactly twice every edge of H. Then, $I_e(G) \geq \frac{1}{3}m$.*

As far as lower bounds are concerned, we observe that upon deleting an edge in a graph, the number of *conflicts*, *i.e.*, of remaining edges uv with $d(u) = d(v)$, can decrease by a limited amount only. Formally, for a graph G, we define $\text{conf}(G)$ as the number of conflicts in G. We stress here the *static* nature of this definition: it only concerns G in its original state.

Remark 1. In any graph G with maximum degree $\Delta \geq 1$, deleting any edge decreases the number of conflicts by at most $2\Delta - 1$. Thus, $I_e(G) \geq \left\lceil \frac{\text{conf}(G)}{2\Delta-1} \right\rceil$.

The bound in Remark 1 is very general, and it can be bad for some graphs. As an illustration, consider $K_{n,n}$: the bound we obtain is $I_e(K_{n,n}) \geq \frac{n^2}{2n-1}$. Later on, we will show this is distant from the optimal by about a factor of 2. There are other contexts, however, in which this bound is not too bad, as we will see later on with the case of paths and cycles. Nevertheless, the bound in Remark 1 can be tight for some graphs, as attested by the graph shown in Fig. 1.

3   FPT Algorithms

In this section we provide our efficient algorithms for computing an optimal edge-irregulator of a given graph.

3.1   Kernelization by Solution Size Plus the Maximum Degree

Theorem 2. IRREGULARITY-DELETION *has a kernel of size $O(k\Delta^{2k+2})$ parameterized by the solution size k plus the maximum degree Δ of the input graph G, and thus an optimal edge-irregulator of G can be computed in FPT-time.*

The kernel that will be used in Theorem 2 will be mostly due to the following lemma, and the fact that we can bound the number of vertices in a graph in function of its diameter and its maximum degree.

Lemma 1. *Let $G = (V, E)$ be a graph, and let S be an optimal edge-irregulator of G. For any edge $e \in S$, there exists an edge $uv \in E$ such that $d(u) = d(v)$, and e is at distance at most $2|S| - 1$ from uv.*

Proof. The proof is by induction on $k = |S|$.

Assume first that $k = 1$. Since $S = \{e\}$ is an optimal edge-irregulator and is not empty, there exists at least one conflict $uv \in E$. Thus, since $G - \{e\}$ is locally irregular, either $e = uv$ or one vertex among u or v has a different degree in $G - S$, which implies that e is either adjacent to u or to v, and thus at distance at most 1 from e.

Suppose now $k \geq 2$. Let uv be a conflict in G. Since in $G - S$, the vertices u and v are not in conflict, there exists one edge e in S that is adjacent to u or to v. Consider now the graph $G - \{e\}$. By the optimality of S, we get that $S - \{e\}$ is an optimal edge-irregulator of $G - \{e\}$. Thus by induction hypothesis, all the edges of S are at distance at most $2|S - \{e\}| - 1 = 2|S| - 3$ from a conflict in $G - \{e\}$. Consider now a conflict ab in $G - \{e\}$. Then either ab was already a conflict in G, or ab has become a conflict due to the deletion of e. This means that either a or b is adjacent to wither u or to v, and thus ab is at distance at most 2 from uv. Finally, using the triangular inequality, any edge of S is at distance at most $2k - 3 + 2 = 2k - 1$ from a conflict of G. $\qquad\square$

3.2   Vertex Cover Number

While it is already known that $I_e(G)$ can be computed in FPT-time parameterized by the vertex cover number using the FPT-algorithm for the more general parameter of *vertex integrity* [11], we present here an algorithm that is tailor-made for the vertex cover number, and, as such, is faster.

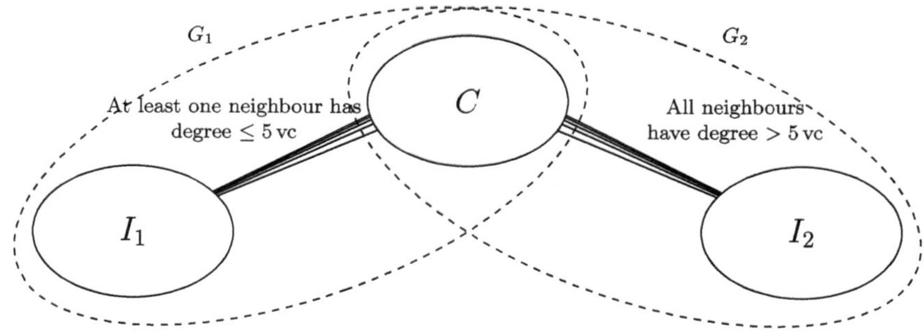

Fig. 2. An abstract depiction of the graph G of Theorem 3.

Theorem 3. *There exists an algorithm to compute $I_e(G)$ in time $2^{O(vc^4)} \cdot n^{O(1)}$, where vc is the size of a minimum vertex cover of G and n is the order of G.*

Proof. We first describe the algorithm and then prove its correctness and time-complexity.

Let $G = (V, E)$ be the given graph. Also, let $C = \{u_1, \ldots, u_{vc}\}$ be a minimum vertex cover of G and $I = V \setminus C$ be an independent set. We partition I into I_1 and I_2 as follows. For every vertex $u \in C$, if $d_I(u) \leq 5vc$, we include all the vertices of $N_I(u)$ into I_1. Once we have finished considering all the vertices of C, we place the remaining vertices of I into I_2. Let $G_1 = G[C \cup I_1]$ and $G_2 = G[C \cup I_2]$ (see Fig. 2). We will show later that this results in $|I_1| \leq 5vc^2$ and $|V(G_1)| \leq 5vc^2 + vc$. We can now guess the projection S_1 of an optimal edge-irregulator S of G onto G_1. That is, $S_1 = S \cap E(G_1)$. Notice that $G_1 - S_1$ is not necessarily locally irregular. All that remains to do is to compute the smallest set $S_2 \subseteq E(G_2)$ such that $G - (S_1 \cup S_2)$ is locally irregular. Notice that such a set S_2 contains only edges between the vertices of C and I_2; let E_2 denote these edges.

Henceforth, we focus on G_2. For every $i \in [vc] = \{1, \ldots, vc\}$, we define a *modifier* $k_i \in [n]$. These modifiers encode the number of edges of E_2 that need to be removed in order to arrive to a locally irregular graph. That is, if k_i edges of E_2 incident to u_i are removed from G_2, then u_i is not in conflict with any of its neighbours (in G). Note that for each i, the modifier k_i does not necessarily have a unique value. We will nevertheless show that it is sufficient to compute a minimum (with respect to summation) tuple $(k_1, \ldots, k_{vc})$ such that $\Sigma_{i=1}^{vc} k_i \leq vc^2$.

Correctness. We first argue about the size of I_1.

Claim 4. *We have that $|I_1| \leq 5vc^2$.*

Proof of the Claim. Consider how I_1 is constructed. By definition of I_1, every vertex of C adjacent to a vertex of I_1 has degree at most $5vc$. Thus, $|I_1| = |\{N[v] | v \in C, d(v) \leq 5vc\}| \leq 5vc \cdot |C| = 5vc^2$.

Claim 5. *Let $S_1 \subset E_1$ be a set of edges. For any optimal edge-irregulator S of G coinciding with S_1 on E_1, we have $S \cap E_2 \leq vc^2$.*

Proof of the Claim. Let S be an optimal edge-irregulator of G coinciding with S_1 on E_1. Assume by contradiction that $S \cap E_2 > vc^2$. Since all the edges of E_2 has one endpoint in C, by the pigeonhole principle, there exists a vertex v in C adjacent to at least $vc+1$ edges of E_2. If at least $2vc+1$ edges of S are adjacent to v, then we first remove vc of them from S. In this case, S may not yet be an edge-irregulator but the only conflicts we may have created are between v and other vertices of C (since all the other neighbours of v are not in C and thus have degree at most vc). Either way, since the other vertices of C can forbid at most $vc - 1$ conflicts, there exist a number $1 \leq k \leq vc$ such that increasing the degree of v by k removes all these conflicts. Since there are at least $vc+1$ vertices of $S \cap E_2$ that are still adjacent to v, we can remove k of them to solve all the conflicts. This leads to a smaller edge-irregulator S' and $S' \cap E_1 = S \cap E_1 = S_1$, which is a contradiction to the optimality of S.

Claim 6. *Let S be an optimal edge-irregulator of G, and u, v_1, v_2 be three vertices with $u \in C$ and $v_1, v_2 \in I_2$. If $uv_1 \in S$, $uv_2 \in E$ and $uv_2 \notin S$, then there exists an optimal edge-irregulator S' of G such that $S' = (S \setminus \{uv_1\}) \cup \{uv_2\}$.*

Proof of the Claim. First, note that the proof of Claim 5 also proves that if S is an optimal edge-irregulator of G, for any vertex $u \in C$ which has neighbours in I_2, we have $d_{G-S}(u) > vc$.

Consider now $S' = (S \setminus \{uv_1\}) \cup \{uv_2\}$ and let $G' = G - S'$. Since $v_1, v_2 \in I_2$, their degrees are at most vc and all their neighbours have a degree of at least $vc + 1$. Thus S' does not create any conflicts for v_1 or for v_2. Moreover, all the other vertices in G have the same degree and the same neighbourhood in $G - S$ and in $G - S'$. Thus, since S was an edge-irregulator, they are not in conflict with each other. Thus, S' is an edge-irregulator.

Running Time. By Claim 4, we have that $|C \cup I_1| \le vc + 5vc^2$. Thus, there are at most $2^{(vc+5vc^2)^2}$ possible guesses of S_1. By Claim 5, we know that an optimal edge-irregulator of G has at most vc^2 edges in E_2, hence $|S_2| \le vc^2$. Due to Claim 6, it suffices to guess for each vertex $u \in C$ how many edges of E_2 incident to u are in S_2, that is, a tuple of integers $(k_1, \ldots, k_{vc})$ such that $k_1 + \cdots + k_{vc} \le vc^2$. There are vc^{vc^2} such possibilities. Recall that checking whether a set $S \subset E(G)$ is an edge-irregulator can be done in running time n^2 where n is the order of G, and since we repeat this step for each guess of S_1, the running time of this algorithm is $2^{(vc+5vc^2)^2} \cdot vc^{vc^2} \cdot n^2 = 2^{O(vc^4)} \cdot n^{O(1)}$.   □

4   Structural Results

In this section we investigate common classes of graphs to determine their optimal edge-irregulators. Our investigation is lead by Conjecture 1, which we verify for most of the considered classes. The results in this section are important in order to gain a better understanding of I_e in regards to structural parameters, but also since general bounds on I_e can be deduced from simpler cases:

Remark 2. Let G be a graph and S be a set of edges of G such that $G - S$ consists of $p \ge 1$ connected components $G_1, \ldots, G_p$. Then, $I_e(G) \le |S| + \sum_{i=1}^{p} I_e(G_i)$.

We start by considering graphs with maximum degree 2 (paths and cycles). For $n \ge 2$, we denote by P_n the path of order n, and, for $n \ge 3$, by C_n the cycle of order n. In this case, we are able to determine I_e completely. Roughly, to prove this result we delete the minimum number of edges to reach as many disjoint P_3s as possible.

Theorem 7. *For every $n \ge 2$, we have:*

- $I_e(P_n) = \frac{n-1}{3}$ *if* $n \equiv 1 \bmod 3$;
- $I_e(P_n) = \lceil \frac{n-1}{3} \rceil$ *if* $n \equiv 2 \bmod 3$;

- $\mathrm{I}_e(P_n) = \lfloor \frac{n-1}{3} \rfloor$ *otherwise.*

Moreover, for every $n \geq 3$, we have $\mathrm{I}_e(C_n) = \mathrm{I}_e(P_n) + 1$.

Next, we deal with complete bipartite graphs. Recall that $K_{n,m}$ denotes the complete bipartite graph with parts of size n and m $(n, m \geq 1)$.

Theorem 8. *For every $n, m \geq 1$, we have:*

- $\mathrm{I}_e(K_{n,m}) = 0$ *if $n \neq m$;*
- $\mathrm{I}_e(K_{n,m}) = n$ *otherwise.*

Proof. The first item is because $K_{n,m}$ is locally irregular whenever $n \neq m$. We now focus on $K_{n,n}$ with $n \geq 2$. Choose any vertex, u, of $K_{n,n}$, and let S be the set containing all edges incident to u. Then, clearly, $K_{n,n} - S$ is locally irregular, and $\mathrm{I}_e(K_{n,n}) \leq n$. To see, now, that $\mathrm{I}_e(K_{n,n}) \geq n$, consider, towards a contradiction, any set S of at most $n-1$ edges such that $K_{n,n} - S$ is locally irregular. Denoting by (U, V) the bipartition of $K_{n,n}$ (where, thus, $|U| = |V| = n$), since $|S| \leq n-1$, there must be at least a $u \in U$ not incident to any edge of S, and similarly a $v \in V$ not incident to any edge of S. Then, $d_{K_{n,n}-S}(u) = d_{K_{n,n}-S}(v) = n$, while u and v are adjacent in $K_{n,n} - S$; this is a contradiction. $\square$

We now move towards more general families of graphs.

Theorem 9. *If $G = (V, E)$ is a connected bipartite graph of order n and minimum degree 1, then $\mathrm{I}_e(G) \leq n - 1$.*

Sketch. Let r be any degree-1 vertex of G, and let $\mathcal{V} = \{V_1, V_2, \ldots V_d\}$ be the partition of $V(G) \setminus \{r\}$ where each V_i contains the vertices of G at distance exactly i from r. We construct a set $S \subseteq E$ such that, in $G - S$, every vertex in V_i has a degree of the same parity as i (except maybe for vertices becoming isolated, and one vertex; see later). We start with S being empty. We consider the vertices of $V(G) \setminus \{r\}$ one by one, following the parts $V_d, V_{d-1}, \ldots, V_1$ in this order. For any vertex $u \neq r$ considered this way, if $d_{G-S}(u)$ does not have the required parity, we add to S a single edge uu' where u' is closer to r than u is. Once all vertices of $\mathcal{V}$ have been treated this way, the only potential issue can come from r having the same degree as its sole neighbour w in $G - S$, which is dealt with easily since r is a degree-1 vertex. $\diamond$

Adapting the above strategy, we can prove that:

Theorem 10. *For every connected graph G of order n, with m edges and maximum degree Δ, we have $\mathrm{I}_e(G) \leq \lfloor \frac{m}{2} \rfloor + n + \Delta - 2$.*

4.1   Trees

We now focus on trees. We first prove that Conjecture 1 holds for this family, and then provide a linear-time algorithm for trees of bounded maximum degree.

We proceed in two steps: we first deal with "star-like" trees, before treating the more general case. Recall that a *bistar* is a tree with exactly two vertices u

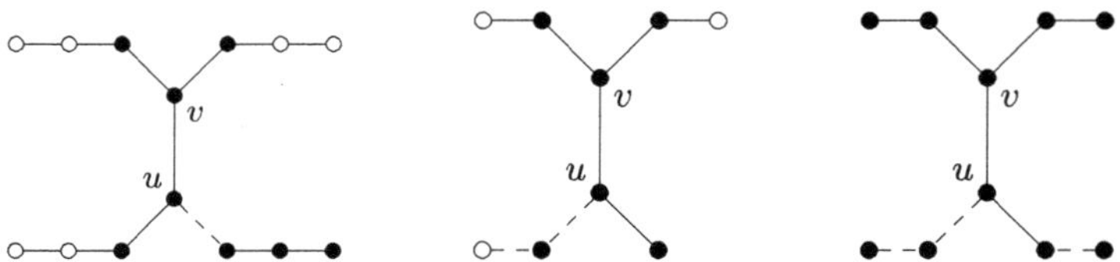

Fig. 3. A representation of the case where $d(u) = d(v) = 3$ and u, v are adjacent in the proof of Lemma 2. A white vertex may or may not exist. The dashed edges form an optimal edge-irregulator. In the leftmost figure, we do not provide an edge-irregulator, but rather fall back in the case of a subdivided star.

and v of degree at least 3 that are adjacent. See Fig. 3 for a representation of a bistar, as well as some intuition on the proof of the following lemma. For the more general case, we also need to observe a lemma concerning bridges.

Lemma 2. *If G is a subdivision of a star or of a bistar, then $I_e(G) \leq \frac{1}{3}|E(G)|$.*

Lemma 3. *Let S and S' be two edge-irregulators of two graphs G, G', respectively, and let $u \in V(G)$ and $v \in V(G')$. Let H be the graph obtained from the disjoint union of G and G' with the addition of the edge uv. Then $I = S \cup S' \cup \{uv\}$ is an edge-irregulator of H.*

Theorem 11. *Let T be a tree. Then either T is a path or $I_e(T) \leq \frac{1}{3}|E(T)|$.*

Proof. Suppose by contradiction that G is a minimum counterexample to this theorem. Then G cannot be a path, and by Lemma 2, G cannot be a subdivision of a star or of a bistar. We root G at an arbitrary vertex r, and we consider u a deepest vertex of degree $k \geq 3$ in G. If several exist, we choose one of the highest degree. Denote by v its parent. Note that $u \neq r$, otherwise the graph would be a subdivided star. If $v = r$, we chose a neighbour of r other than u as the new root. For $w \in V(G)$, we will denote by T_w the subtree induced by w and all its descendant for this rooting.

Claim 12. *Let $P = (p_1, p_2, p_3)$ be a path of G with $d(p_1) = 1$ and $d(p_2) = 2$. Then $d(p_3) \geq 3$.*

Proof of the Claim. Since G is connected, $d(p_3) \neq 1$, otherwise we would have $T = P$, which contradicts that G is not a path. Suppose by contradiction $d(p_3) = 2$, let $G' = G \backslash T_{p_3}$ and let S' be an edge-irregulator of G'. By minimality of G, we have that $|S'| \leq \frac{|G'|}{3}$. Let p_3' be the parent of p_3, and consider $S = S' \cup \{p_3 p_3'\}$. Then S is an edge-irregulator of G of size at most $\frac{|G|}{3}$, a contradiction.

It follows that, by construction, the subtree T_u rooted in u is a subdivided star. Let $P_i = (v_{i,1,}, \ldots, v_{i,k_i})$ for $i \in [k-1]$ be its branches. Without loss of generality, and according to Claim 12, we can suppose for $1 \leq i < j \leq k$, $|P_i| \geq |P_j|$, and thus $2 \geq |P_1| \geq \cdots \geq |P_k| \geq 1$.

- If $k \geq 4$, then by the minimality of G, we get that $G' = G \setminus T_u \cup \{u\}$ has an edge-irregulator S' of size at most $\frac{1}{3}|E(G')|$. If $uv \in S'$, then $S \cup S'$ is an edge-irregulator of the required size, where S is an edge-irregulator of T_u of size at most $\frac{|T_u|}{3}$ obtained from Lemma 2. Now suppose $uv \notin S'$. In $G \setminus S'$, the only remaining conflicts can be the edge uv. Should this happen, all the edges from P_{k-1} form an edge-irregulator of T_u decreasing the degree of u by one, thus resolving the conflict uv, and $|P_i| \leq \frac{|T_u|}{3}$.
- We can now assume $k = 3$. There are thus two induced paths P_1, P_2 rooted at u. We will distinguish cases depending on the lengths of $|P_1|$ and $|P_2|$.
  - Assume $|P_1| = |P_2| = 1$ and let $G' = G \setminus T_u$. By minimality of G, the tree G' is not a path and has an edge-irregulator S' of size $\frac{|G'|}{3}$. By construction, $S = S' \cup \{uv\}$ is an edge-irregulator of G. Moreover, we have $|S| = |S'| + 1$ and $|G| = |G'| + 3$.
  - Assume now $|P_1| = 2$ and $|P_2| \in \{1, 2\}$. Let v' be the parent of v. If $d_G(v) = 1$, then the graph is already locally irregular and we are done. If $d_G(v) = 2$ and G' is a path then we are done by Lemma 2. Otherwise (if $d_G(v) = 2$ and G' is not a path), we consider $G' = G - T_v$ and let S' be an optimal edge-irregulator of G'. Using Lemma 3, the set $S = S' \cup \{vv'\}$ is an edge-irregulator of G and has size less than $\frac{1}{3}|E(G)|$.
    We can now assume $d(v) \geq 3$. By the definition of u, any other descendant of v of degree at least 3 must be a neighbour of v. Moreover, by minimality of G, we can suppose that all its other neighbours of degree 3 cannot fall into the previous cases. Thus, we can suppose that all its other neighbour have degree at most 3, and if they have degree 3, they are adjacent to a path of length two and a path of length at most 2. Moreover, we recall that by minimality of G, there cannot be a path of length at least 3 attached to v.
    - Assume first $d_G(v) \geq 5$ and let $G' = G - T_v$. Consider S' an optimal edge-irregulator of G'. Note that G' can be a path, but a path has an edge-irregulator of size at most $\frac{|G'|}{3} + 1$ which will be sufficient here. Let $S = S' \cup \{vv'\}$. It is easy to check that $G - S$ is indeed locally irregular. Since exactly one edge adjacent to v was removed, its resulting degree is at least 4, all its neighbours have degree at most 3 and the rest of T_v was already locally irregular. Moreover, T_v contains at least 7 edges. Thus $|S| = |S'| + 1 \leq \frac{|G'|}{3} + 2 \leq \frac{|G|-7}{3} + 2 \leq \frac{|G|}{3}$
    - Assume $d(v) = 3$ and let w be the neighbour of v other than u and v'. We consider $G' = G - T_v$. If G' is a path, then $d_g(w) = 3$, as otherwise G would be a bistar. In this case, consider S' an edge-irregulator of G' of size at most $\frac{|G'|}{3} + 1$ and let $S = S' \cup \{vv'\}$. As in the previous case, since there are at least 8 edges in T_v and since T_v is already locally irregular, we are done. Otherwise, G' is not a path and we can consider S' an optimal edge-irregulator of G' of size at most $\frac{|G'|}{3}$. If $d_G(w) \in \{1, 3\}$, then $S = S' \cup \{vv'\}$ is an edge-irregulator of size at most $\frac{|G|}{3}$. Otherwise, $d_G(w) = 2$. Denote by w' the child of w in T_v. Note that w' is a leaf, since we have already proved that there is no

path of length at least 3 adjacent to v. Thus, $S \cup \{vv', ww'\}$ is an edge-irregulator of G of size at most $\frac{|G|}{3}$ as $|T_v| \geq 6$.

- We can finally assume $d(v) = 4$. Let w and x be the two other children of v. W.l.o.g., suppose $d(w) \geq d(x)$. By hypothesis on u, we have $3 = d(u) \geq d(w) \geq d(x)$, and both w and x can only have paths as descendents. In case of equality in their degrees, suppose $|T_u| \geq |T_w| \geq |T_x|$.
  - If $d(w) = 3$, let $G' = G \setminus T_v$. If G' is a path, G is already locally irregular. Indeed, all the vertices of degree 3 or more are v and some of its neighbours. Thus, using Claim 12, and since all the neighbours or v have degree at most 3, there are no conflicts in G. Otherwise, let S' be an edge-irregulator of G', and S_x be an optimal edge-irregulator of T_x. Since G' is not a path and G is a minimum counter-example, we have $|S'| \leq \frac{|G'|}{3}$. Also, $|S_x| \leq \left\lceil \frac{|T_x|}{3} \right\rceil$ since T_x is a path. Consider $S = S' \cup \{vv', vx\} \cup S_x$. By construction, S is an edge-irregulator of G, and $|S| \leq \frac{|G|}{3}$, see Fig. 4(a).
  - If $d_G(w) = 2$ and $d_G(x) = 2$, let $G' = G \setminus (T_u \cup T_w \cup T_x)$ and let S' be an optimal edge-irregulator of G'. If $d_{G'-S'}(v') \neq 4$, since $d_{G'-S'}(v) = 4$, we have that S' is also an edge-irregulator of G and has size at most $\frac{G}{3}$. Otherwise, let S_u be an optimal edge-irregulator of T_u. Note that T_u is either a path on 4 or on 5 vertices, which implies $|S_u| = 1$ or $|S_u| = 2$, respectively. Consider $S = S' \cup S_u \cup \{(vu)\}$. Then S is an edge-regulator of G, and $|S| = |S'| + 1 + |S_u| \leq \frac{|G'|}{3} + 1 + |S_u| \leq \frac{|G|}{3}$ since $|T_v| = 8$ if $|T_u| = 4$ and $|T_v| = 9$ if $|T_u| = 5$, see Fig. 4(b).
  - Finally, suppose $d_G(w) \leq 2$, and $d_G(x) = 1$. Let $G' = G \setminus T_v$. Once again, if G' is a path, G is already locally irregular. Let S' be an edge-irregulator of G', and let $S = S' \cup \{vv', vx\}$. If $d(w) = 2$, we also add to S the edge connecting w to its leaf. In both cases, S is an edge-irregulator of G of size at most $\frac{G}{3}$, see Fig. 4(c).

Thus, a minimal counterexample does not have a vertex of degree at least 3, meaning it is a path, which concludes the proof.

Theorem 11 proves that Conjecture 1 holds for trees. For this family we can also compute the exact value of $I_e(G)$ with an efficient algorithm.

Theorem 13. *If G is a tree of order n and maximum degree Δ, then computing $I_e(G)$ can be done in time $O(n\Delta^3)$.*

4.2   Complete Graphs

For any $n \geq 3$, we denote by K_n the complete graph of order n. Complete graphs are an interested case for our problem, as computing $I_e(K_n)$ is the same as finding the size of locally irregular graphs of order n with the most edges.

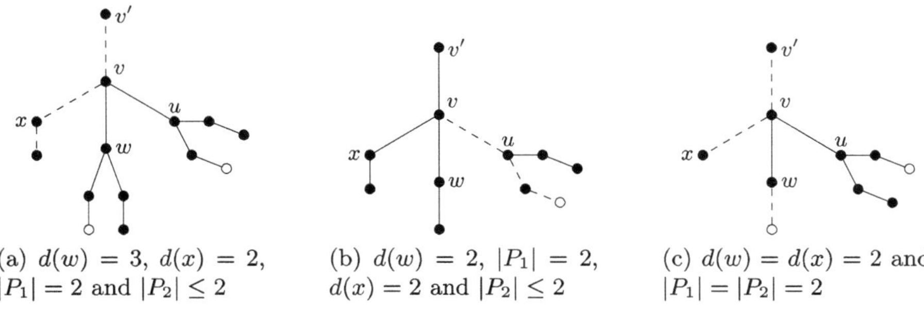

(a) $d(w) = 3$, $d(x) = 2$, $|P_1| = 2$ and $|P_2| \leq 2$

(b) $d(w) = 2$, $|P_1| = 2$, $d(x) = 2$ and $|P_2| \leq 2$

(c) $d(w) = d(x) = 2$ and $|P_1| = |P_2| = 2$

Fig. 4. A representation of some cases of the proof of Theorem 11. The white vertices may or may not exist. The dashed edges are the edges in S.

In what follows, we are solving the question for particular values of n, while we only provide partial results for the remaining ones. Precisely, we focus on the case where n is a *triangular number, i.e.*, of the form $\frac{k(k+1)}{2}$ for some k. For any $k \geq 1$, we denote by t_k the kth triangular number, being the sum of all integers from 1 up to k.

Lemma 4. *If $n = t_k$ is a triangular number, then the maximum number of edges in a locally irregular graph of order n is exactly $m_k := \frac{k(k+1)(k-1)(3k+2)}{24}$.*

Sketch. Let G be a locally irregular graph of order $n = t_k$. Observe that if G is a locally irregular graph of order n, then G has at most k vertices of degree $n - k$ for all $k \in [k]$. Thus, for every $i \in [k]$, there are at most i vertices of degree $n - i$ in G. So, any locally irregular graph of order n having the most edges has at most one vertex of degree $n - 1$, two vertices of degree $n - 2$, and so on. Because n is a triangular number, such a graph T_k can be constructed for all $k \geq 1$; we prove this by induction on k. The graph T_1 is the graph containing a single vertex. Then, for any $i \geq 1$ such that T_i has been constructed, we obtain T_{i+1} from T_i by adding $i + 1$ independent vertices adjacent to all the vertices of T_i. It is easy to see that for any $k \geq 1$, the graph T_k indeed has the claimed degree sequence. To fully prove the claim, it remains to express the number of edges of any T_k, which is achieved through the handshaking lemma.   ◇

Theorem 14. *For every $k \geq 1$ and $n = t_k$, we have $I_e(K_n) = |E(K_n)| - m_k$.*

Even though Theorem 14 is tight, it cannot be generalised easily to complete graphs whose order is not a triangular number. However, since $t_{k+1} - t_k = k = \Theta(\sqrt{t_k})$ for every $k \geq 1$, Lemma 4 yields both a lower and an upper bound on the size of locally irregular graphs of order n having the most edges, when n is not triangular: namely, they have $\frac{n^2}{2} - O(n^{3/2})$ edges. And from this, we proved that Conjecture 1 holds for all complete graphs.

5 Conclusion

In this work, we extended the study of edge-irregulators from [11]. Apart from focusing on the algorithmic aspects of this problem, we also derived general bounds and analysed behaviours of optimal edge-irregulators, leading to Conjecture 1, which we verified for several graph classes.

Determining $I_e(G)$ for a graph G remains challenging due to the intricate properties and elusive structure of maximally edge-dense locally irregular graphs. Let us denote by $\mathcal{L}_n$ (for some $n \geq 3$) the class of all locally irregular graphs of order n with the most edges possible. In light of Lemma 4, a key open question concerns the size and structural properties of $\mathcal{L}_n$ in the case where n is a non-triangular number.

We finally observe that there is room for improvement for many of our results. This could lead to further work on the topic. In particular, we find the following questions appealing.

- Refining the static parameter conf (Sect. 4) to dynamically capture evolving edge conflicts.
- Determining I_e for complete graphs and tightening bounds for cubic graphs.
- Extending Theorem 1 to subcubic graphs. Also, for k-degenerate graphs G of order n, we could get an upper bound on $I_e(G)$ of about $(k-1)n + \lfloor \frac{n}{3} \rfloor$ through exploiting similar ideas to the ones used in this paper.

It is also interesting to further study the algorithmic complexity of determining $I_e(G)$ for a given graph G. In particular, the authors of [12] proved that the vertex counterpart of the problem is in FPT when parameterized by the treewidth and the maximum degree. Their algorithm should adapt in an obvious way to our context. Thus, we prefer to not include such a result here.

We finish by commenting on the FPT algorithm we present here w.r.t. the vertex cover number. First, observe that there is still room for improvement in its running time. Indeed, we are only aware of a $2^{o(vc)}n^{O(1)}$ lower bound (based on the ETH) for this problem (stemming as a direct corollary of the W[1]-hardness w.r.t. the solution size proof presented in [11]). Lastly, we believe that this algorithm could be generalised to work w.r.t. the neighbourhood diversity of the input graph. However, in order to achieve this, it would be mandatory to completely determine I_e for complete graphs.

References

1. Alavi, Y., Chartrand, G., Chung, F., Erdős, P., Graham, R., Oellermann, O.: Highly irregular graph. J. Graph Theory **11**(2), 235–249 (1987)
2. Baudon, O., Bensmail, J., Przybyło, J., Woźniak, M.: On decomposing regular graphs into locally irregular subgraphs. European J. Combin. **49**, 90–104 (2015)
3. Bazgan, C., Cazals, P., Chlebíková, J.: Degree-anonymization using edge rotations. Theoretical Comput. Sci. **873**, 1–15 (2021)
4. Belmonte, R., Sau, I.: On the complexity of finding large odd induced subgraphs and odd colorings. Algorithmica **83**(8), 2351–2373 (2021)

5. Bensmail, J., Li, B., Li, B.: An injective version of the 1-2-3 conjecture. Graphs Combin. **37**, 281–311 (2021)
6. Bensmail, J., Catherinot, N., Fioravantes, F., Marcille, C., Oijid, N.: Graph Irregularity via Edge Deletions (2025). https://hal.science/hal-05084434
7. Chartrand, G., Jacobson, M., Lehel, J., Oellermann, O., Ruiz, S., Saba, F.: Irregular networks. Congr. Numer. **64**, 197–210 (1988)
8. Cygan, M., et al.: Parameterized Algorithms. Springer, Cham (2015). https://doi.org/10.1007/978-3-319-21275-3
9. Diestel, R.: Graph Theory. GTM, vol. 173. Springer, Heidelberg (2017). https://doi.org/10.1007/978-3-662-53622-3
10. Downey, R.G., Fellows, M.R.: Fundamentals of Parameterized Complexity. TCS, Springer, London (2013). https://doi.org/10.1007/978-1-4471-5559-1
11. Fioravantes, F., Melissinos, N., Triommatis, T.: Parameterised distance to local irregularity. In: 19th International Symposium on Parameterized and Exact Computation (IPEC 2024). Leibniz International Proceedings in Informatics (LIPIcs), vol. 321, pp. 18:1–18:15. Schloss Dagstuhl – Leibniz-Zentrum für Informatik (2024)
12. Fioravantes, F., Melissinos, N., Triommatis, T.: Complexity of finding maximum locally irregular induced subgraphs. Discret. Appl. Math. **363**, 168–189 (2025)
13. Karoński, M., Łuczak, T., Thomason, A.: Edge weights and vertex colours. J. Combin. Theory Seri. B **91**, 151–157 (2004)
14. Keusch, R.: A solution to the 1-2-3 conjecture. J. Comb. Theory Seri. B **166**, 183–202 (2024)

Shortest Paths and Minimum Spanning Trees

Forcing a Unique Minimum Spanning Tree and a Unique Shortest Path

Tatsuya Gima[1]([✉])(iD), Yasuaki Kobayashi[1](iD), Yota Otachi[2](iD),
and Takumi Sato[1]

[1] Hokkaido University, Hokkaido, Japan
{gima,koba}@ist.hokudai.ac.jp, sato.takumi.h1@elms.hokudai.ac.jp
[2] Nagoya University, Aichi, Japan
otachi@nagoya-u.jp

Abstract. A *forcing set* S in a combinatorial problem is a set of elements such that there is a unique solution that contains all the elements in S. An *anti-forcing set* is the symmetric concept: a set S of elements is called an anti-forcing set if there is a unique solution disjoint from S. There are extensive studies on the computational complexity of finding a minimum forcing set in various combinatorial problems, and the known results indicate that many problems would be harder than their classical counterparts: the decision version of finding a minimum forcing set for perfect matchings is NP-complete [Adams et al., Discret. Math. 2004] and that of finding a minimum forcing set for satisfying assignments for 3CNF formulas is Σ_2^P-complete [Hatami-Maserrat, DAM 2005]. In this paper, we investigate the complexity of the problems of finding minimum forcing and anti-forcing sets for the shortest s-t path problem and the minimum weight spanning tree problem. We show that, unlike the aforementioned results, these problems are tractable, with the exception of the decision version of finding a minimum anti-forcing set for shortest s-t paths, which is NP-complete.

Keywords: Forcing sets · minimum spanning trees · shortest paths

1 Introduction

This paper focuses on the problem of uniquely determining a solution by prescribing a subset of the solution (or of its complement). This concept has been extensively studied for various combinatorial problems, under several different names, such as (anti-)forcing number [1,2,14,15,24], critical sets [13], and defining sets [16].

Harary, Klein, and Živković [14] formulated the *forcing number* (for perfect matchings) of a graph, which is the smallest cardinality of an edge subset such that there is exactly one perfect matching including it. Prior to this work, the concept has been studied in chemistry as the innate degree of freedom of a Kekulé

This work is partially supported by JSPS KAKENHI Grant Numbers JP22H00513, JP23K28034, JP24H00686, JP24H00697, JP24K23847, JP25K03076, JP25K03077.

Table 1. A summary of known and our results.

	classical setting	forcing model	anti-forcing model
PERFECT MATCHING	P	NP-complete [2]	unknown
3SAT	NP-complete	Σ_2^P-complete [16]	—
2SAT	P	NP-complete [10]	—
VERTEX COLORING	NP-complete	Σ_2^P-complete [16]	—
VERTEX COVER	NP-complete	Σ_2^P-complete [17]	
BIPARTITE VERTEX COVER	P	NP-complete [17]	
SHORTEST s-t PATH	P	P (Theorem 1)	NP-complete (Theorem 2)
MINIMUM SPANNING TREE	P	P (Theorem 5)	

Since the solutions of 3SAT, 2SAT, and VERTEX COLORING do not form subsets, these results cannot be directly regarded as results in the forcing model. However, since these problems deal with unique extensions of partial assignments on subsets of variables or vertices, it is still possible to interpret them as results in the forcing model. (See the discussion in [16].)

structure [20] and gained attention through several studies [1,2,24], including a recent survey article [24] on the forcing number and on its "dual" notion, the anti-forcing number [11]. From the computational perspective, deciding whether the forcing number of an input graph is at most k is known to be NP-complete even on bipartite graphs [2]. Similarly, given a bipartite graph G and its perfect matching M, it is NP-complete to decide whether there is a subset $S \subseteq M$ of size at most k such that every perfect matching M' of G other than M does not include S entirely (i.e., $S \backslash M' \neq \varnothing$) [1]. Deng and Zhang [11] proved that, given a bipartite graph G and its perfect matching, it is NP-complete to decide whether there is a subset $S \subseteq E(G) \setminus M$ of size at most k such that M is the unique perfect matching in $G - S$. In a more general context, this problem can be viewed as finding a smallest partial solution that is uniquely extended to a (complete) solution. Viewed from this broader perspective, the computational complexity of such problems has been investigated for various combinatorial problems, such as graph colorings [16], satisfiability [10,16], puzzles [10,19], and the minimum vertex cover problem [3,17,18]. Table 1 shows several known results. An important observation is that the computational complexity of (most of) these problems would be harder than that of their classical setting.

We study the computational complexity of the problems of finding a minimum (anti-)forcing set for well-known tractable combinatorial problems, namely the shortest s-t path problem and the minimum-weight spanning tree problem. More specifically, let G be a directed graph with $s, t \in V(G)$ and let $w \colon E(G) \to \mathbb{R}_+$ be an edge-weight function. We say that $S \subseteq E(G)$ is a *forcing set for shortest s-t paths* if there is exactly one shortest path P from s to t in (G, w) such that $S \subseteq E(P)$. Similarly, $S \subseteq E(G)$ is an *anti-forcing set for shortest s-t paths* if there is exactly one shortest path P from s to t in (G, w) such that $S \cap E(P) = \varnothing$. Our first two problems are stated formally as follows.

> Problem: FORCINGSP / ANTI-FORCINGSP
> Input: A weighted digraph G with $w\colon E(G) \to \mathbb{R}_+$, $s,t \in V(G)$, and $k \in \mathbb{N}$.
> Goal: Decide whether there is a forcing set / anti-forcing set $S \subseteq E(G)$ with $|S| \leq k$ for shortest s-t paths in (G, w).

Similarly, we can define forcing and anti-forcing sets for minimum weight spanning trees. They are defined as follows.

> Problem: FORCINGMST / ANTI-FORCINGMST
> Input: A weighted graph G with $w\colon E(G) \to \mathbb{R}$.
> Goal: Decide whether there is a forcing set / anti-forcing set $S \subseteq E(G)$ with $|S| \leq k$ for minimum weight spanning trees in (G, w).

Let us note that the weight of an edge can be negative in the latter two problems, while the former two problems require the weight to be positive.

It is not hard to see that these four problems belong to NP, as we can check whether in polynomial time, given an edge set S, there is an optimal solution including or avoiding it. This motivates us to investigate whether these problems are NP-complete or polynomial-time solvable.

For FORCINGMST and ANTI-FORCINGMST, we can naturally generalize these problems to the settings in matroids. This generalization allows us to handle these two problems in a uniform way. In particular, we observe that the problem of finding a minimum forcing set for minimum weight bases in a matroid M is equivalent to that of finding a minimum weight anti-forcing set for maximum weight bases in the dual matroid M^*. We design a polynomial-time algorithm for computing a minimum anti-forcing set for maximum weight bases in a matroid M, assuming that M is given as an independence oracle (see Sect. 2 for details). This yields polynomial-time algorithms for FORCINGMST and ANTI-FORCINGMST (Theorem 5). Our polynomial-time algorithm exploits a well-known greedy algorithm for the maximum weight basis problem.

For FORCINGSP and ANTI-FORCINGSP, we can observe an intriguing distinction between them: FORCINGSP is polynomial-time solvable (Theorem 1), while ANTI-FORCINGSP is NP-complete (Theorem 2). To the best of our knowledge, this is the first example in which there is a complexity gap between the two models (see Table 1). Our polynomial-time algorithm for FORCINGSP is based on the simple reduction that reduces the problem on general weighted directed graphs to that on unweighted directed acyclic graphs. This reduction enables us to design a dynamic programming algorithm. For ANTI-FORCINGSP, we give a polynomial-time reduction from VERTEX COVER in Theorem 2. To overcome this intractability, we show that ANTI-FORCINGSP can be solved in linear time on graphs of bounded treewidth via Courcelle's theorem [8] (Theorem 4).

Due to space limitations, several proofs, marked $\star$, are omitted.

2   Preliminaries

Let E be a finite set. A *property* is a subset $\Pi \subseteq 2^E$. For $X \in \Pi$, a set $S \subseteq E$ is called a *forcing set for* X if it holds that $S \subseteq X$ and $S \not\subseteq X'$ for every $X' \in \Pi \setminus \{X\}$. In other words, S is a forcing set of X if X is the unique set in Π that contains S. Similarly, for $X \in \Pi$, a set $S \subseteq E$ is called an *anti-forcing set for* X if it holds that $S \cap X = \varnothing$ and $S \cap X' \neq \varnothing$ for every $X' \in \Pi \setminus \{X\}$. A set $S \subseteq E$ is called a *forcing set* (resp. *anti-forcing set*) *for property* $\Pi \subseteq 2^E$ if S is a forcing set (resp. anti-forcing set) for some $X \in \Pi$. We sometimes omit the target property Π when it is clear from the context.

For a (directed) graph G, we denote its vertex set by $V(G)$ and its edge set by $E(G)$. For an edge set $X \subseteq E(G)$, $G - X$ denotes the graph obtained from G by deleting all edges in X. Let G be a directed graph. For a vertex $v \in V(G)$, $N^-(v)$ denotes the set of in-neighbors $u : (u, v) \in E(G)$ of v. For a directed edge $e = (u, v) \in E(G)$, head(e) and tail(e) denote v and u, respectively.

Matroids. Let E be a finite set and let $\mathcal{I} \subseteq 2^E$. A pair $M = (E, \mathcal{I})$ is called a *matroid* if $\mathcal{I}$ is nonempty; for $X \in \mathcal{I}$, every subset of X belongs to $\mathcal{I}$; and for $X, Y \in \mathcal{I}$ with $|Y| > |X|$, $Y \setminus X$ has an element e such that $X \cup \{e\} \in \mathcal{I}$. Each set in $\mathcal{I}$ (resp. not in $\mathcal{I}$) is said to be *independent* (resp. *dependent*) in M. A maximal independent set is called a *basis* and a minimal dependent set is called a *circuit* in M. Every matroid has the following *symmetric basis-exchange property* [7]: for any two bases A and B and for $a \in A \setminus B$, there is $b \in B \setminus A$ such that both $(A \cup \{b\}) \setminus \{a\}$ and $(B \cup \{a\}) \setminus \{b\}$ are bases. The collection of all bases of M is denoted by $\mathcal{B}(M)$. A *loop* in M is an element $e \in E$ such that $\{e\}$ is dependent.

Proposition 1 (Corollary 1.2.6 in [22]). *Let $B \in \mathcal{B}(M)$ and $e \notin B$. Then, $B \cup \{e\}$ contains a unique circuit C of M. Moreover, we have $(B \cup \{e\}) \setminus \{e'\} \in \mathcal{B}(M)$ for $e' \in C$.*

Let $M = (E, \mathcal{I})$ be a matroid. It is known that the pair $M^* := (E, \mathcal{I}^*)$ with

$$\mathcal{I}^* = \{X \subseteq E : X \subseteq E \setminus B \text{ for some } B \in \mathcal{B}(M)\}$$

forms a matroid [22], which is called the *dual matroid* of M. Note that $(M^*)^* = M$ for any matroid M. For $X \subseteq E$, we define pairs

$$M \mid X = (X, \{I \subseteq X : I \in \mathcal{I}\}),$$
$$M / X = (E \setminus X, \{I \subseteq E \setminus X : \exists B \in \mathcal{B}(M \mid X) \text{ s.t. } B \cup I \in \mathcal{I}\}).$$

These pairs are called the *restriction* and the *contraction* of M with respect to X, respectively. It is known that these pairs are also matroids [22] for any $X \subseteq E$. In this paper, a matroid M is given as an *independence oracle*, that is, we can query the oracle to decide whether a subset $X \subseteq E$ is independent in M.

Proposition 2 (Proposition 3.1.10 in [22]). *Let $X \subseteq E$. For every circuit C of M with $C \nsubseteq X$, there is a circuit C^* of M / X such that $C^* \subseteq C \setminus X$.*

Let $G = (V, E)$ be a multigraph. A subset of edges is said to be *acyclic* if it induces a forest in G. Let $\mathcal{F} \subseteq 2^E$ be the collection of all acyclic edge subsets of G. Then, it is known that the pair $M = (E, \mathcal{F})$ forms a matroid, which is called a *graphic matroid* [22].

3   Forcing a Unique Shortest Path

In this section, we discuss FORCINGSP and ANTI-FORCINGSP. Recall that in these problems we are given a directed graph G with $s, t \in V(G)$ and an edge-weight function $w \colon E \to \mathbb{R}_+$. The numbers of vertices and edges in G are denoted by n and m, respectively. A forcing (resp. anti-forcing) set $S \subseteq E(G)$ (for shortest s-t paths) is *minimal* if any proper subset of S is not a forcing (resp. anti-forcing) set. It is easy to observe that for any minimal (anti-)forcing set S, every edge in S is contained in a shortest s-t path in G. This would reduce our problems to the cases where G is acyclic. More specifically, we let $d_s \colon V(G) \to \mathbb{R}_+ \cup \{0, \infty\}$ be the distance labeling from s in G, that is, $d_s(v)$ is the (shortest) distance from s to v in (G, w) for $v \in V(G)$. We remove all the edges $e = (u, v)$ that do not satisfy $d_s(v) = d_s(u) + w(e)$, which are edges that do not belong to any shortest path from s to v. We also remove all vertices (and their incident edges) that are not reachable from s or not reachable to t. As $w(e) > 0$ for each $e \in E$, the graph G' obtained in this way is indeed acyclic, which can be computed by a standard shortest path algorithm in $\mathcal{O}(m + n \log n)$ time. The following observation immediately follows from the fact that each shortest s-t path in (G, w) is an s-t path in G', and vice versa.

Observation 1. *Let $S \subseteq E$. Then, S is a minimal forcing set in G if and only if it is a minimal forcing set in G'. Similarly, S is a minimal anti-forcing set in G if and only if it is a minimal anti-forcing set in G'.*

This observation offers several advantages: we can assume that the input directed graph G is acyclic and all the paths from s to t are shortest paths in G, making the subsequent discussions simple.

3.1   Forcing Set

In this subsection, we describe an algorithm for computing a minimum forcing set for shortest s-t paths. By Observation 1, we can assume that the given graph G is acyclic. Moreover, we can ignore the weight of edges, as every path from s to t in G is a shortest s-t path in the original graph.

For $v \in V(G)$, let $\mathcal{P}_v$ be the set of all paths from s to v in G. For vertices $u, v \in V(G)$, we write $u \to_{=1} v$ if there is exactly one path from u to v in G. This relation $\to_{=1}$ is reflexive, i.e., $v \to_{=1} v$ holds for any $v \in V(G)$. We assume that s has out-degree at least 2 since otherwise we can contract the (unique) out-going edge from s without affecting the solution. We define OPT$[e]$ and OPT$[v]$ as

$$\mathrm{OPT}[e] := \min\{|S| : S \text{ is a forcing set for } \mathcal{P}_v \text{ with } e \in S\}$$

for $e = (u, v) \in E(G)$ and

$$\mathrm{OPT}[v] := \min_{u \in N^-(v)} \mathrm{OPT}[(u, v)]$$

for $v \in V(G) \setminus \{s\}$, while $\mathrm{OPT}[s] := 0$. From now on, we show how to compute the values of OPT and then a minimum forcing set for $\mathcal{P}_t$ from these values.

The following lemma gives a characterization of a forcing set for a specific path $P \in \mathcal{P}_v$.

Lemma 1. *Let $S \subseteq E(G)$ and let $P \in \mathcal{P}_v$ be a path from s to v in G such that all edges in S are contained in P. Let $e_1, \ldots, e_k$ be the edges in S appearing in this order on P. For $1 \leq i \leq k$, we let $e_i = (t_i, s_i)$, and let $s_0 = s$ and $t_{k+1} = v$. Then, S is a forcing set for P if and only if $s_i \rightarrow_{=1} t_{i+1}$ for all $0 \leq i \leq k$.*

Proof. Suppose that S is a forcing set for P. If there are at least two paths from s_i to t_{i+1} for some i, we can conclude that there is at least one path $P' \in \mathcal{P}_v$ with $P' \neq P$ that contains all the edges in S, which contradicts the uniqueness of P.

Conversely, suppose that $s_i \rightarrow_{=1} t_{i+1}$ for all $0 \leq i \leq k$. If S is not a forcing set for P, there is another path $P' \in \mathcal{P}_v$ such that all the edges in S are contained in P'. As $P \neq P'$, at least one edge of P is not contained in P'. We can assume that this edge appears on the subpath P_i of P between s_i and t_{i+1}. Since P' also passes though both s_i and t_{i+1}, the subpath of P' between them is distinct from P_i, contradicting $s_i \rightarrow_{=1} t_{i+1}$. □

Corollary 1. *The minimum size of a forcing set for $\mathcal{P}_t$ is equal to*

$$\min_{v \in V(G)} \{\mathrm{OPT}[v] : v \rightarrow_{=1} t\}.$$

We turn to a polynomial-time algorithm to compute the values of OPT. This immediately yields a polynomial-time algorithm for FORCINGSP due to Corollary 1.

Lemma 2. *For every $e = (u, v) \in E(G)$, it holds that*

$$\mathrm{OPT}[e] = \min_{w \in V(G)} \{\mathrm{OPT}[w] : w \rightarrow_{=1} u\} + 1.$$

Proof. Let $S \subseteq E(G)$ be a forcing set for $P \in \mathcal{P}_v$ such that $|S| = \mathrm{OPT}[e]$ and $e \in S$. By Lemma 1, there is a vertex w' in P, which is either s or the head of an edge in $S \setminus \{e\}$ with $w' \rightarrow_{=1} u$. Observe that $S \setminus \{e\}$ is a forcing set for the subpath of P between s and w', as otherwise there are two paths from s to v including all edges in S, contradicting the uniqueness of P. Thus, we have

$$|S| = |S \setminus \{e\}| + 1 \geq \mathrm{OPT}[w'] + 1 \geq \min_{w \in V(G)} \{\mathrm{OPT}[w] : w \rightarrow_{=1} u\} + 1.$$

Conversely, let w be a vertex minimizing $\mathrm{OPT}[w]$ under the condition that $w \rightarrow_{=1} u$ holds. Suppose first that $\mathrm{OPT}[w] = 0$. By definition, we have $w = s$.

Since $s \rightarrow_{=1} u$, there is a unique path from s to u, which is denoted by P_{su}. By concatenating P_{su} and e in this order, we have a path P from s to v in G. Observe that there is exactly one path from s to v passing through e as $s \rightarrow_{=1} u$ and every path containing e must pass through u. Thus, $\{e\}$ is a forcing set for P, implying that $\mathrm{OPT}[e] \leq \mathrm{OPT}[w] + 1$.

Suppose otherwise. In this case, $\mathrm{OPT}[w] = \mathrm{OPT}[e']$ for some edge e' incoming to w. Let S' be a forcing set for $P_{sw} \in \mathcal{P}_w$ such that $|S'| = \mathrm{OPT}[w]$ and $e' \in S'$. Similarly to the above case, the path obtained by concatenating P_{sw}, P_{wu}, and e in this order is the unique path P from s to v containing all edges in $S' \cup \{e\}$. Thus, $S' \cup \{e\}$ is a forcing set for P, implying that $\mathrm{OPT}[e] \leq \mathrm{OPT}[w] + 1$. $\square$

Now, we describe our dynamic programming algorithm to compute the values of OPT. We first decide whether $u \rightarrow_{=1} v$ holds for each pair of vertices $u, v \in V(G)$. This can be done in total time $\mathcal{O}(nm)$ for all vertex pairs in G. We can evaluate $\mathrm{OPT}[e]$ and $\mathrm{OPT}[v]$ for each $e \in E(G)$ and each $v \in V(G)$ in time $\mathcal{O}(n)$, assuming that these values are evaluated in a dynamic programming manner. By Corollary 1, we can compute a minimum forcing set for $\mathcal{P}_t$ in time $\mathcal{O}(nm)$.

It is easy to extend our algorithm to compute a minimum forcing set for a specific path P by only evaluating $\mathrm{OPT}[e]$ and $\mathrm{OPT}[v]$ for each edge e and vertex v on the path P. Therefore, we have the following theorem.

Theorem 1. FORCINGSP *can be solved in time* $\mathcal{O}(nm)$. *Moreover, when additionally given a shortest s-t path P in G, we can compute a minimum forcing set for P in time* $\mathcal{O}(nm)$ *as well.*

It is not hard to see that, applying a standard trace-back technique, we can find a minimum forcing set for shortest s-t paths within the same running time. We would like to mention that our algorithm also works for undirected graphs since Observation 1 also holds for undirected graph G.

3.2 Anti-Forcing Set

We next consider the complexity of ANTI-FORCINGSP. In contrast to FORCINGSP, ANTI-FORCINGSP is NP-complete even for undirected and unweighted graphs.

Theorem 2. ANTI-FORCINGSP *is NP-complete even for undirected and unweighted graphs.*

Proof. We perform a polynomial-time reduction from VERTEX COVER. Recall that VERTEX COVER asks whether, given an undirected graph G and an integer k, there exists a vertex set $S \subseteq V$ of size at most k such that any edge $e \in E(G)$ is incident with at least one vertex in S. VERTEX COVER is known to be NP-complete [12]. Our reduction produces a weighted multigraph, which can easily turn into an unweighted simple graph.

Construction. Let $\langle G, k \rangle$ be an instance of Vertex Cover. Let $n = |V(G)|$. Now we give a construction of the reduction. See Fig. 1a for an example of the construction. Let $V(G) = \{v_1, \ldots, v_n\}$.

For each vertex v_i, we construct a vertex gadget H_i as follows. The gadget H_i contains a path of four vertices t_i, x_i, y_i, s_i, where the vertices have neighbors a_i, w_i, z_i, b_i, respectively. We add four paths of length 2 between a_i and b_i. The four internal vertices are denoted by $c_i^1, \ldots, c_i^4$.

We construct the entire multigraph H as follows. Let $N = 3n + k$. First, we add two special vertices $s = s_0$ and $t = t_{n+1}$. Then, we connect s_i and t_{i+1} by an edge for all $0 \leq i \leq n$. For each $1 \leq i \leq n$, we add $N + 2$ parallel edges between s and w_i and between z_i and t. For each edge $\{v_i, v_j\} \in E(G)$ with $i < j$, we add $N + 2$ parallel edges between y_i and x_j. We may regard and refer to these $N + 2$ parallel edges between two vertices as a single "thick edge". The weights of edges are defined as: $w(\{x_i, y_i\}) = 2$; $w(y_i, x_j) = 5(j-i-1)+3$; $w(s, w_i) = 5(i-1)+1$; and $w(z_i, t) = 5(n-i)+1$ for all $i < j$; all the other edges are of weight 1. The graph constructed in this way is denoted by H. In this construction, every edge belongs to a shortest path (of length $5n + 1$) between s and t. Moreover, every shortest s-t path in H goes from the left to the right in Fig. 1.

We show that G has a vertex cover of cardinality at most k if and only if H has an anti-forcing set of cardinality at most N for shortest s-t paths. To simplify notation, we may refer to an edge $\{p_i, q_i\} \in E(H_i)$ as $(pq)_i$ and to a path with vertices $p_i, q_i, \ldots, r_i \in V(H_i)$ as $(pq \ldots r)_i$.

The Only-if Direction: Assume that G has a vertex cover $S \subseteq V(G)$ such that $|S| \leq k$. We define the edge set F as follows. If $v_i \in S$ then F includes $(xy)_i, (ac^2)_i, (ac^3)_i, (ac^4)_i$, otherwise F includes $(ta)_i, (wx)_i, (yz)_i$. Since $|S| \leq k$, the cardinality of F is at most $4|S| + 3|V(G) \setminus S| \leq 4k + 3(n - k) = N$. Let P be an s-t path in H that includes the subpath $(tac^1bs)_i$ if $v_i \in S$; includes the subpath $(txys)_i$ otherwise. Since P is a shortest s-t path (of length $5n + 1$) that does not contain any edge in F, it suffices to show that F is an anti-forcing set for P. Observe that for each i, either $(xy)_i \in F$ or $\{(wx)_i, (yz)_i\} \subseteq F$. Moreover, for each edge $\{v_i, v_j\} \in E(G)$, at least one of $(xy)_i$ and $(xy)_j$ is contained in F as S is a vertex cover of G. Thus, every s-t path that contains a thick edge is hit by F. This implies that P contains both t_i and s_i for each $1 \leq i \leq n$. For each i, there is exactly one path between t_i and s_i in $H_i - F$, which is the subpath of P. Hence, P is the unique path between s and t in $H - F$.

The If Direction: Let F be an anti-forcing set with cardinality at most N and let P be the unique shortest s-t path in $H - F$. Since $|F| \leq N$ and there are $N + 2$ parallel edges as a thick edge, $H - F$ contains at least two of them: Due to the uniqueness of P, none of those parallel edges are contained in P. This implies that P contains both t_i and s_i for each $1 \leq i \leq n$. Hence, P contains either $(txys)_i$ or $(tac^jbs)_i$ as a subpath for each $1 \leq i \leq n$. Suppose that P contains $(txys)_i$. As $H - F$ contains both thick edges $\{s, w_i\}$ and $\{z_i, t\}$, F contains both $(wx)_i$ and $(yz)_i$. Moreover, due to the uniqueness of P, F contains at least one edge of the path forming $(tac^jbs)_i$ for all j. Thus, F contains at least three edges

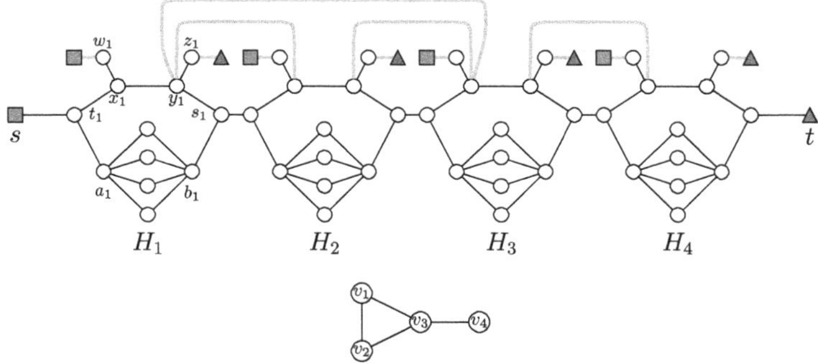

(a) The lower graph is the original instance $\langle G, k \rangle$, and the upper graph is the constructed instance H.

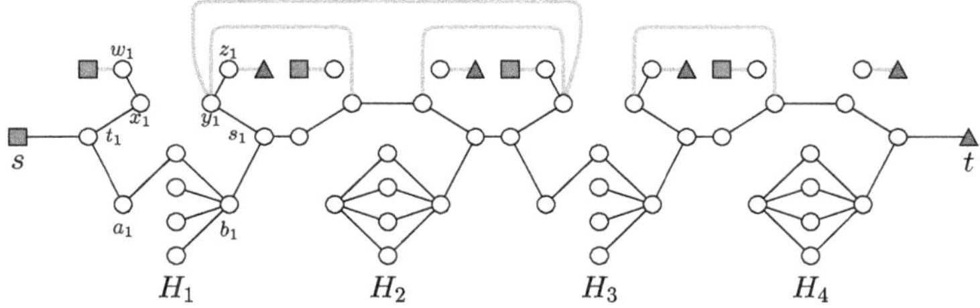

(b) Illustration of $H - F$, where F is the anti-forcing set constructed from a vertex cover $\{v_1, v_3\}$ of G.

Fig. 1. Example of the reduction from VERTEX COVER to ANTI-FORCINGSP in the proof of Theorem 2. For an aesthetic reason, the figure contains multiple copies of vertices representing the same vertex: all red boxes correspond to the vertex s and all blue triangles correspond to the vertex t. Thick orange edges represent "thick edges", which are parallel edges between them.

in H_i. Suppose otherwise that P contains $(tac^j bs)_i$ for some $1 \le j \le 4$. Similarly to the above case, F contains at least one edge on the path $(txys)_i$. Moreover, as there are four edge-disjoint paths between a_i and b_i, F contains at least three edges between them. Thus, F contains at least four edges in H_i in this case. Let $S \subseteq V$ be the set of all vertices v_i such that P contains $(tac^j bs)_i$ as a subpath for some j. Since F contains at most $N = 3n + k$ edges, S contains at most k vertices. For each edge $\{v_i, v_j\} \in E(G)$ with $i < j$, P cannot contain both subpaths $(txys)_i$ and $(txys)_j$, which make a bypass passing through the thick edge $\{y_i, x_j\}$. This yields that S covers all edges in G.

Making H Simple and Unweighted: Finally, let us note that H can be converted into a simple graph with uniform edge weight. We can replace an edge of weight $q \in \mathbb{N}$ with a path of length q. Moreover, we can remove parallel edges by replacing all edges in H (including non-parallel edges) with paths of length 2. It is easy to see that these replacements preserve the size of a smallest anti-forcing set of H. Since the weight of each edge is $\mathcal{O}(n)$ and the multiplicity of each thick edge is $\mathcal{O}(n)$, the size of the resulting graph is upper bounded by a polynomial in n, completing our reduction. $\square$

While ANTI-FORCINGSP is NP-complete, we can find in polynomial time a minimum cost anti-forcing set for a given shortest s-t path P in G by reducing it to the minimum multiway cut problem on directed acyclic graphs, which can be solved in polynomial time [5].

Theorem 3 ($\star$). *Let G be an edge-weighted directed graph with $s, t \in V(G)$ and let P be a shortest path from s to t in G. Then, a minimum anti-forcing set S for P can be computed in polynomial time.*

Finally, we observe that the problem can be solved in linear time on graphs of bounded treewidth. Since it is not necessary to discuss what treewidth is in the proof, we omit its definition in this paper (see e.g. [9, Chapter 7]). We observe that ANTI-FORCINGSP is reduced to one for evaluating a monadic second-order (MSO_2) formula, and thus, by Courcelle's theorem [4,8] and Bodlaender's algorithm [6], it can be solved in linear time on graphs of bounded treewidth.

Theorem 4 ($\star$). *Let G be a directed graph with two specified vertices s and t such that the underlying graph of G has bounded treewidth. Then, a minimum anti-forcing set S for shortest s-t paths on G can be computed in linear time.*

4   Forcing a Unique Minimum Weight Matroid Basis

In this section, we give polynomial-time algorithms for FORCINGMST and ANTI-FORCINGMST. The basic strategy of both algorithms follows Kruskal's algorithm [21] for computing a minimum weight spanning tree. We first briefly explain our algorithms for FORCINGMST and ANTI-FORCINGMST in Sect. 4.1 and then generalize them to the setting of matroids in Sect. 4.2.

4.1   Special Case: Minimum Weight Spanning Trees

First, we sketch an intuition behind our algorithm for ANTI-FORCINGMST. A formal discussion, including the correctness of the algorithm, is deferred to Sect. 4.2. Let G be a connected edge-weighted multigraph. Contracting an edge of G may create parallel edges or self-loops, and we do not remove them in the contraction operation. Let us recall Kruskal's algorithm for computing a minimum weight spanning tree of G: Starting from $T = \varnothing$, we repeatedly choose a minimum weight edge e of G that does not form a cycle with the previously chosen edges T and add it to T until it becomes a spanning tree of G. Instead of adding edges one by one, we can modify this procedure to add, in bulk, a maximal forest F consisting only of minimum weight edges E_{min} that have not been chosen yet. We repeatedly apply this to the graph obtained from G by contracting all edges in E_{min} as long as $E(G)$ is nonempty. It is easy to observe that this procedure also computes a minimum weight spanning tree T of G. Since

---

Algorithm 1: An algorithm for ANTI-FORCINGMST.

Input: A multigraph G with weight function $w : E(G) \to \mathbb{R}$.

Output: A minimum anti-forcing set for minimum weight spanning trees of (G, w).

1 **begin**
2     $S \leftarrow \varnothing$; $T \leftarrow \varnothing$;
3   **while** $E(G) \neq \varnothing$ **do**
4        $w_{\min} \leftarrow \min_{e \in E(G)} w(e)$;
5        $E_{\min} \leftarrow \{e \in E(G) : w(e) = w_{\min}\}$;
6   Let $G_{\min}$ be the subgraph $(V(G), E_{\min})$ of G;
7   Let $F_{\min}$ be a maximal forest of $G_{\min}$;
8        $T \leftarrow T \cup F_{\min}$;
9        $S \leftarrow S \cup (E_{\min} \setminus (F_{\min} \cup \{e \in E_{\min} : e$ is a self-loop in $G_{\min}\}))$;
10   Let G be the graph obtained by contracting all edges in $E_{\min}$;
11   **return** S;

Algorithm 2: An algorithm for FORCINGMST.

Input: A multigraph G with weight function $w : E(G) \to \mathbb{R}$.

Output: A minimum forcing set for minimum weight spanning trees of (G, w).

1 **begin**
2     $S \leftarrow \varnothing$; $T \leftarrow \varnothing$;
3   **while** $E(G) \neq \varnothing$ **do**
4        $w_{\min} \leftarrow \min_{e \in E(G)} w(e)$;
5        $E_{\min} \leftarrow \{e \in E(G) : w(e) = w_{\min}\}$;
6   Let $G_{\min}$ be the subgraph $(V, E_{\min})$ of G;
7   Let $F_{\min}$ be a maximal forest of $G_{\min}$;
8        $T \leftarrow T \cup F_{\min}$;
9        $S \leftarrow S \cup \{e \in F_{\min} : e$ is not a bridge of $G_{\min}\}$;
10   Let G be the graph obtained by contracting all edges in $E_{\min}$;
11   **return** S;

---

the choice of a maximal forest F consisting of edges in $E_{\min}$ is not unique, the solution T obtained by this procedure is not unique as well. Thus, to force the solution to be unique, it is necessary to include all edges in $E_{\min} \setminus F$, except for self-loops, as an anti-forcing set. The algorithm that formalizes this intuition is given in Algorithm 1. For FORCINGMST, we can design a similar algorithm: Instead of including all non-loop edges in $E_{\min} \setminus F$, include all edges in F that are not bridges in $G_{\min}$. The pseudocode of this algorithm is given in Algorithm 2. It can be easily confirmed that both algorithms for ANTI-FORCINGMST and FORCINGMST run in time $\mathcal{O}(m \log n)$ using a standard analysis of Kruskal's algorithm and a linear-time algorithm for enumerating bridges [23]. The proof of their correctness is deferred to the next subsection.

Theorem 5. FORCINGMST *and* ANTI-FORCINGMST *can be solved in time* $\mathcal{O}(m \log n)$, *where* $n = |V(G)|$ *and* $m = |E(G)|$.

4.2   General Case: Matroid Bases

In this subsection, we consider generalized versions of FORCINGMST and ANTI-FORCINGMST, where the goals are to find minimum forcing and anti-forcing sets for minimum weight bases of a matroid M. Clearly, FORCINGMST and ANTI-FORCINGMST are special cases, where M is a graphic matroid.

Theorem 6. *Let* $M = (E, \mathcal{I})$ *be a matroid with weight function* $w \colon E \to \mathbb{R}$. *Assume that* M *is given as an independence oracle. Then, a minimum forcing set for minimum weight bases of* (M, w) *can be computed in polynomial time. Similarly, a minimum anti-forcing set for minimum weight bases of* (M, w) *can be computed in polynomial time as well.*

Let $\mathcal{B}_{\min}(M)$ and $\mathcal{B}_{\max}(M)$ be the collections of minimum and maximum weight bases of M, respectively. At first, we show that the problem of finding a minimum forcing set for $\mathcal{B}_{\min}(M)$ is equivalent to that of finding a minimum anti-forcing set for $\mathcal{B}_{\max}(M^*)$, where M^* is the dual matroid of M.

Observation 2 ($\star$). *A set* $S \subseteq E$ *is a forcing set for* $\mathcal{B}_{\min}(M)$ *if and only if* S *is an anti-forcing set for* $\mathcal{B}_{\max}(M^*)$.

Observation 2 suggests the relations in Fig. 2, enabling us to find a minimum forcing set for $\mathcal{B}_{\min}(M)$ by applying an algorithm for finding a minimum anti-forcing set for $\mathcal{B}_{\max}(M^*)$. Note that a basis B minimizes the value $w(B)$ if and only if it maximizes $-w(B)$. Now, we are ready to describe our algorithm for computing a minimum anti-forcing set for $\mathcal{B}_{\min}(M)$, which is shown in Algorithm 3. The underlying idea of the algorithm is analogous to Algorithm 1. At each iteration of the main loop, we consider the matroid $M_{\min}$ consisting only of the minimum weight elements $E_{\min}$, and select a basis $B_{\min}$ from $M_{\min}$. The crux of its correctness is that it is necessary and sufficient to include all the elements in $E_{\min} \setminus B_{\min}$ except for loops in any anti-forcing set for $\mathcal{B}_{\min}(M)$.

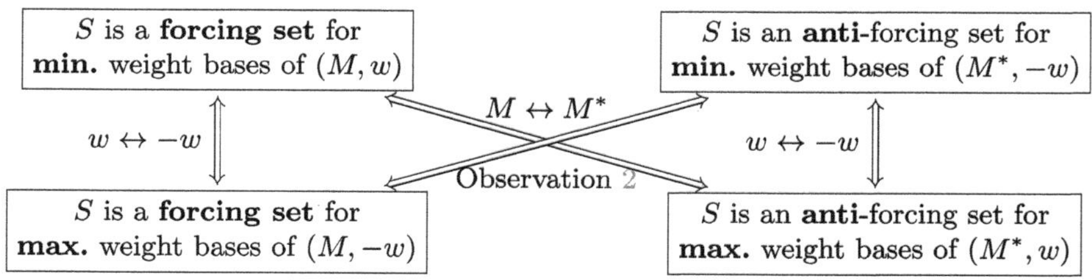

Fig. 2. The equivalence of the problems.

Algorithm 3: An algorithm for a minimum anti-forcing set for $\mathcal{B}_{\min}(M)$

Input: Matroid $M = (E, \mathcal{I})$ with weight function $w \colon E \to \mathbb{R}$.
Output: A minimum anti-forcing set for $\mathcal{B}_{\min}(M)$.

1 **begin**
2 $\quad S \leftarrow \varnothing;\ B \leftarrow \varnothing;$
3 $\quad$ **while** $E \neq \varnothing$ **do**
4 $\quad\quad w_{\min} \leftarrow \min_{e \in E} w(e);$
5 $\quad\quad E_{\min} \leftarrow \{e \in E : w(e) = w_{\min}\};$
6 $\quad\quad M_{\min} \leftarrow M \mid E_{\min};$
7 $\quad\quad$ Let $B_{\min}$ be a basis of $M_{\min};$
8 $\quad\quad B \leftarrow B \cup B_{\min};$
9 $\quad\quad S \leftarrow S \cup (E_{\min} \setminus (B_{\min} \cup \{e \in E_{\min} : e \text{ is a loop in } M_{\min}\}));$
10 $\quad\quad M \leftarrow M / E_{\min};$

11 $\quad$ **return** $S;$

Lemma 3. *Algorithm 3 returns a minimum anti-forcing set for $\mathcal{B}_{\min}(M)$.*

Proof. First, we show that the output S of Algorithm 3 is an anti-forcing set for $\mathcal{B}_{\min}(M)$. From the correctness of the greedy algorithm for computing a minimum weight basis of a matroid, at the end of the main loop, we have $B \in \mathcal{B}_{\min}(M)$. Moreover, we have $B \cap S = \varnothing$.

Let $B' \in \mathcal{B}_{\min}(M)$ with $B' \neq B$. It suffices to show that $B' \cap S \neq \varnothing$. Now, there exists an element $e \in B' \setminus B$ since B and B' are distinct bases of M. We can assume that, at some iteration of the main loop, $e \in E_{\min}$ and $e \notin B_{\min}$ hold. Since e is not a loop in $M_{\min}$ and $e \in E_{\min} \setminus B_{\min}$, we have $e \in S$. Hence, $B' \cap S \neq \varnothing$. This concludes that S is an anti-forcing set for $\mathcal{B}_{\min}(M)$.

Next, we show that S is a minimum anti-forcing set for $\mathcal{B}_{\min}(M)$. Let S' be a minimum anti-forcing set for $\mathcal{B}_{\min}(M)$ and $B' \in \mathcal{B}_{\min}(M)$ be the unique minimum weight basis of M with $B' \cap S' = \varnothing$. Let w_i be the value of $w_{\min}$ in the i-th iteration of the while-loop and let $E_i = \{e \in E : w(e) = w_i\}$. For each i, let $E_{\leq i} = E_1 \cup \cdots \cup E_i$ and let $M_i = M / E_{\leq i-1}$. Note that $M_i \mid E_i = M_{\min}$ in the i-th iteration.

Claim $(\star)$. $|B \cap E_i| = |B' \cap E_i|$ holds for all i.

Suppose to the contrary that $|S'| < |S|$. This implies that there is an index j such that $|S' \cap E_j| < |S \cap E_j|$. By the above claim, we have $|B \cap E_j| = |B' \cap E_j|$. Let $e \in E_j \setminus (B' \cup S')$ that is not a loop in $M_j \mid E_j$. We can choose such an element e since $|E_j| = |B \cap E_j| + |S \cap E_j| + |L_j|$, where L_j is the set of loops in $M_j \mid E_j$, and $|S' \cap E_j| < |S \cap E_j|$. Note that e is not a loop of M_j. Due to Proposition 1, $B' \cup \{e\}$ has a unique circuit C of M. If C contains an element $e' \in E \setminus E_{\leq j}$, then $(B' \cup \{e\}) \setminus \{e'\}$ is a basis of M with weight strictly smaller than B', contradicting $B' \in \mathcal{B}_{\min}(M)$. Thus, C consists of only elements in $E_{\leq j}$. If C contains an element $e' \in E_j$, the basis $(B' \cup \{e\}) \setminus \{e'\}$, which belongs to $\mathcal{B}_{\min}(M)$, avoids S', contradicting the uniqueness of B'. Hence, all the elements of C except for e belong to $E_{\leq j-1}$. As $C \not\subseteq E_{\leq j-1}$, by Proposition 2, there is a circuit C^* of M_j such that $C^* \subseteq C \setminus E_{\leq j-1}$. This circuit is indeed a singleton $C^* = \{e\}$, contradicting the fact that e is not a loop in M_j. $\square$

Finally, we consider the running time of Algorithm 3. Given a matroid $M = (E, \mathcal{I})$ as an independence oracle and $X, Y \subseteq E$, we can decide whether Y is independent in $M \mid X$, in M / X, and in M^* with a polynomial number of oracle calls to M. Thus, each step of Algorithm 3 can be performed in polynomial time. Thus, Theorem 6 holds.

Similarly to Theorem 1, we can compute smallest forcing and anti-forcing sets for a given $B^* \in \mathcal{B}_{\min}(M)$ in polynomial time, by just taking $\mathcal{B}_{\min}$ at line 7 in Algorithm 3 as $\mathcal{B}_{\min} = B^* \cap E_{\min}$.

Corollary 2. *Let $M = (E, \mathcal{I})$ be a matroid with weight function $w : E \to \mathbb{R}$. Assume that M is given as an independence oracle. Given a minimum weight basis $B^* \in \mathcal{B}_{\min}(M)$, a minimum forcing set / a minimum anti-forcing set for B^* can be computed in polynomial time.*

References

1. Adams, P., Mahdian, M., Mahmoodian, E.S.: On the forced matching numbers of bipartite graphs. Discret. Math. **281**(1–3), 1–12 (2004)
2. Afshani, P., Hatami, H., Mahmoodian, E.S.: On the spectrum of the forced matching number of graphs. Australas. J. Comb. **30**, 147–160 (2004)
3. An, S., Chang, Y., Cho, K., Kwon, O., Lee, M., Oh, E., Shin, H.: Pre-assignment problem for unique minimum vertex cover on bounded clique-width graphs. In: 39th AAAI Conference on Artificial Intelligence, AAAI 2025, pp. 26886–26894 (2025)
4. Arnborg, S., Lagergren, J., Seese, D.: Easy problems for tree-decomposable graphs. J. Algorithms **12**(2), 308–340 (1991)
5. Bentz, C.: The maximum integer multiterminal flow problem in directed graphs. Oper. Res. Lett. **35**(2), 195–200 (2007)
6. Bodlaender, H.L.: A linear-time algorithm for finding tree-decompositions of small treewidth. SIAM J. Comput. **25**(6), 1305–1317 (1996)
7. Brualdi, R.A.: Comments on bases in dependence structures. Bull. Aust. Math. Soc. **1**(2), 161–167 (1969)
8. Courcelle, B., Mosbah, M.: Monadic second-order evaluations on tree-decomposable graphs. Theoret. Comput. Sci. **109**(1&2), 49–82 (1993)
9. Cygan, M., et al.: Parameterized Algorithms. Springer, Cham (2015). https://doi.org/10.1007/978-3-319-21275-3_15
10. Demaine, E.D., Ma, F., Schvartzman, A., Waingarten, E., Aaronson, S.: The fewest clues problem. In: 8th International Conference on Fun with Algorithms, FUN 2016. LIPIcs, vol. 49, pp. 12:1–12:12 (2016)
11. Deng, K., Zhang, H.: Anti-forcing spectra of perfect matchings of graphs. J. Comb. Optim. **33**(2), 660–680 (2017)
12. Garey, M.R., Johnson, D.S.: Computers and intractability: a guide to the theory of NP-completeness. Freeman, W. H (1979)
13. Ghandehari, M., Hatami, H., Mahmoodian, E.S.: On the size of the minimum critical set of a latin square. Discret. Math. **293**(1–3), 121–127 (2005)

14. Harary, F., Klein, D.J., Živković, T.P.: Graphical properties of polyhexes: perfect matching vector and forcing. J. Math. Chem. **6**, 295–306 (1991)
15. Harary, F., Slany, W., Verbitsky, O.: On the computational complexity of the forcing chromatic number. SIAM J. Comput. **37**(1), 1–19 (2007)
16. Hatami, H., Maserrat, H.: On the computational complexity of defining sets. Discret. Appl. Math. **149**(1–3), 101–110 (2005)
17. Horiyama, T., Kobayashi, Y., Ono, H., Seto, K., Suzuki, R.: Theoretical aspects of generating instances with unique solutions: Pre-assignment models for unique vertex cover. In: Thirty-Eighth AAAI Conference on Artificial Intelligence, AAAI 2024, pp. 20726–20734 (2024)
18. Horiyama, T., Seto, K., Sakamoto, F., Suzuki, R.: Hardness of pre-assignment problem for unique minimum vertex cover on planar graphs with maximum degree 3. In: 25th International Symposium on Fundamentals of Computation Theory, FCT 2025. Lecture Notes in Computer Science, vol. 16106, pp. 238–251 (2025)
19. Kimura, K., Kamehashi, T., Fujito, T.: The fewest clues problem of picross 3d. In: 9th International Conference on Fun with Algorithms, FUN 2018. LIPIcs, vol. 100, pp. 25:1–25:13 (2018)
20. Klein, D.J., Randić, M.: Innate degree of freedom of a graph. J. Comput. Chem. **8**(4), 516–521 (1987)
21. Kruskal, J.B.: On the shortest spanning subtree of a graph and the traveling salesman problem. Proc. Amer. Math. Soc., 48–50 (1956)
22. Oxley, J.G.: Matroid theory. Oxford University Press, 2nd edn. (2011)
23. Tarjan, R.E.: A note on finding the bridges of a graph. Inf. Process. Lett. **2**(6), 160–161 (1974)
24. Zhang, Y., He, X., Liu, Q., Zhang, H.: Forcing, anti-forcing, global forcing and complete forcing on perfect matchings of graphs - a survey. Discret. Appl. Math. **376**, 318–347 (2025)

On the MST-Ratio: Theoretical Bounds and Complexity of Finding the Maximum

Afrouz Jabal Ameli[1] [ID], Faezeh Motiei[2]([✉]) [ID], and Morteza Saghafian[3] [ID]

[1] Utrecht University, Utrecht, Netherlands
`a.jabalameli@uu.nl`
[2] Eindhoven University of Technology, Eindhoven, Netherlands
`f.motiei@tue.nl`
[3] IST Austria (Institute of Science and Technology Austria), Klosterneuburg, Austria
`morteza.saghafian@ist.ac.at`

Abstract. Given a finite set of red and blue points in $\mathbb{R}^d$, the MST-ratio is defined as the total length of the Euclidean minimum spanning trees of the red points and the blue points, divided by the length of the Euclidean minimum spanning tree of their union. The MST-ratio has recently gained attention due to its direct interpretation in topological models for studying point sets with applications in spatial biology. The maximum MST-ratio of a point set is the maximum MST-ratio over all proper colorings of its points by red and blue. We prove that finding the maximum MST-ratio of a given point set is NP-hard when the dimension is part of the input. Moreover, we present a quadratic-time 3-approximation algorithm for this problem. As part of the proof, we show that in any metric space, the maximum MST-ratio is smaller than 3. Furthermore, we study the average MST-ratio over all colorings of a set of n points. We show that this average is always at least $\frac{n-2}{n-1}$, and for n random points uniformly distributed in a d-dimensional unit cube, the average tends to $\sqrt[d]{2}$ in expectation as n approaches infinity.

Keywords: Minimum Spanning Tree · NP-hardness · Discrete and Computational Geometry · Approximation Algorithms

1 Introduction

Recently, motivated by applications in spatial biology, the interactions between color classes in a colored point set have been studied in [17] from a topological perspective. To this end, they developed a framework based on the *chromatic Delaunay mosaic* and explored its combinatorial and topological properties; see

A. J. Ameli—Supported by the project COALESCE (ERC grant no. 853234).
M. Saghafian—Partially supported by the European Research Council (ERC), grant no. 788183, and by the Wittgenstein Prize, Austrian Science Fund (FWF), grant no. Z 342-N31.

E. Di Giacomo and D. Mondal (Eds.): WALCOM 2026, LNCS 16444, pp. 386–401, 2026.
https://doi.org/10.1007/978-981-95-7127-7_26

also [6]. Moreover, they introduced the concept of *MST-ratio* as one of the measures for the mingling of points with different colors in a colored point set. Investigating this measure opens the door to interesting discrete geometry questions related to Euclidean minimum spanning trees (EMST), which is the main focus of this paper. For a set P of n points in the Euclidean space $\mathbb{R}^d$, an *Euclidean minimum spanning tree* of P, denoted by $\mathrm{EMST}(P)$, is a minimum spanning tree (MST) of the complete geometric graph on P where the weight of each edge is the Euclidean distance between its endpoints. For a partition $P = R \cup B$ of P into red and blue points, the EMST-ratio (sometimes referred to as the MST-ratio) can be defined as follows.

$$\mu(P, R) = \frac{|\mathrm{EMST}(R)| + |\mathrm{EMST}(P \setminus R)|}{|\mathrm{EMST}(P)|},$$

where $|\mathrm{EMST}(X)|$ for a point set X is the length of an Euclidean minimum spanning tree of X. How much longer can EMSTs of two finite sets be compared to an EMST of their union? Given a point set P, we are interested in the maximum ratio, $\gamma(P)$, over all proper partitionings of P into two sets. The upper and lower bounds for γ of certain classes of point sets can be found in [11,16], see Sect. 1.1 for further details. However, limited research has been conducted on the maximum EMST-ratio in higher-dimensional Euclidean spaces. Moreover, the question of whether there is an efficient algorithm to compute γ for a given point set remains unanswered.

The question can also be raised in the abstract setting. Namely, given a weighted complete graph G with positive weights and a bi-partition $V(G) = R \cup B$ of the vertices of G, we define the MST-ratio of this partition as

$$\mu(G, R) = \frac{|\mathrm{MST}(G[R])| + |\mathrm{MST}(G[P \setminus R])|}{|\mathrm{MST}(G)|},$$

where $G[R]$ and $G[B]$ are the induced subgraphs of G on the vertex sets R and B, respectively, and $|H|$ is the total weight of the edges of the graph H. The maximum MST-ratio of G, denoted by $\gamma(G)$, is the maximum ratio over all proper bi-partitions of $V(G)$.

In the *MAX-MST-ratio* problem, we aim to find a partition $V(G) = R \cup B$ of the vertices of G that maximizes the MST-ratio. As for the geometric setting, we introduce the *MAX-EMST-ratio* problem, in which given dimension d and a point set P in $\mathbb{R}^d$, we aim to find the maximum EMST-ratio of P. When d is a fixed number, we call the problem *d-MAX-EMST-ratio*.

Our work establishes new bounds on the maximum and average MST-ratio. We also analyze the computational complexity of MAX-MST-ratio and study it through the lens of approximation algorithms. By an α-approximation ($\alpha \geq 1$) for MAX-MST-ratio, we mean a polynomial-time algorithm that for every weighted graph G provides a bi-partition $V(G) = R \cup B$ satisfying $\alpha \cdot \mu(G, R) \geq \gamma(G)$.

We summarize our main results as follows:

- The MAX-MST-ratio problem is NP-hard and hard to approximate even within a factor of $O(n^{1-\varepsilon})$ (Theorems 1).
- The MAX-EMST-ratio problem is NP-hard (Theorem 2).
- In any metric space, the maximum MST-ratio is smaller than 3 (Theorem 3).
- There exists an $O(n^2)$ time 3-approximation algorithm for the MAX-EMST-ratio problem (Theorem 4).
- For every set of n points in Euclidean space, the average EMST-ratio over all colorings is at least $\frac{n-2}{n-1}$ (Theorem 6).
- The average EMST-ratio for a set of n random points uniformly distributed in $[0,1]^d$ tends to $\sqrt[d]{2}$ as n goes to infinity (Theorem 5).

1.1 Related Work

The maximum EMST-ratio is closely related to the Steiner ratio of the Euclidean space. Recall that the Steiner tree of a point set is a tree that connects the points via segments of minimum total length, allowing the use of some extra points. The Steiner ratio, ρ_d of $\mathbb{R}^d$, is the infimum over all finite point sets in $\mathbb{R}^d$ of the length of the Steiner tree divided by the length of an EMST of the set. Figure 1 shows an example of three points that form an equilateral triangle of side length 1 with Steiner ratio $\sqrt{3}/2$. Gilbert and Pollack [13] conjectured that this is the most extreme example in the plane and therefore $\rho_2 = \sqrt{3}/2 \approx 0.866$. The best-known lower bound for ρ_2 is 0.824..., due to Chung and Graham [7]. It is also important for us to have a universal lower bound on ρ_d that holds for all d. Gilbert and Pollack [13] presented a short proof for $\rho_d \geq 0.5$. Later, Graham and Hwang [14] showed $\rho_d \geq 0.577$ for any d.

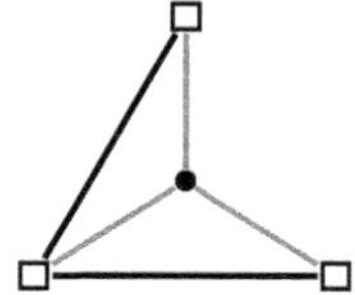

Fig. 1. The point set P shown by square nodes is 3 vertices of an equilateral triangle of side length 1. The black edges form $\mathrm{EMST}(P)$ with length 2. Adding an extra vertex in the center, the green edges form the minimum Steiner tree of P with length $\sqrt{3}$. (Color figure online)

Using the above bounds for ρ_2, the authors in [16] showed that the supremum over all point sets in $\mathbb{R}^2$ of the maximum EMST-ratio is between 2.154 and 2.427. It is not hard to see that the infimum of the maximum EMST-ratio is 1. Their attention then shifted to lattice point sets, where they introduced the MST-ratio (despite the infinite number of points). They subsequently demonstrated that the infimum and supremum of the maximum MST-ratios across all 2-dimensional lattices are 1.25 and 2, respectively.

Dumitrescu, Pach, and Tóth proved that any set P of at least 12 points in $\mathbb{R}^2$ satisfies $\gamma(P) > 1$ [11]. They also showed that for n points sampled uniformly at random in $[0,1]^2$, the maximum EMST-ratio is at least $\sqrt{2} - \varepsilon$, for every $\varepsilon > 0$, with a probability that tends to 1 as n goes to infinity. A tighter analysis shows that in the d-dimensional unit cube, the expected value of the average EMST-ratio over all colorings tends to $\sqrt[d]{2}$ as n goes to infinity; see Theorem 5. The proof relies on the classic result by Beardwood, Halton, and Hammersley [3], saying that the length of an MST of a set P of n points uniformly distributed in $[0,1]^d$ (or any bounded region of volume 1 in $\mathbb{R}^d$) satisfies

$$|\mathrm{EMST}(P)|/n^{1-1/d} \to \beta(d), \tag{1}$$

with probability 1, where $\beta(d) > 0$ is a constant depending only on the dimension. The best known lower and upper bounds for $\beta(2)$ are approximately 0.6 by Avram and Bertsimas [2] and 0.707 by Gilbert [12], respectively.

1.2 Outline

Section 2 focuses on abstract graphs and presents complexity results on the MAX-MST-ratio problem. Section 3 addresses the geometric setting, establishing bounds for the maximum MST-ratio and providing complexity results for MAX-EMST-ratio problem. Section 4 explores the average EMST-ratio across all colorings of a point set. Section 5 concludes with open questions to deepen understanding of the MST-ratio.

2 Maximum MST-Ratio of Abstract Graphs

In this section, we study the complexity of finding the maximum MST-ratio for abstract graphs. Throughout this section and the subsequent sections, we call an edge *colorful* if the two endpoints of it have different colors; otherwise, we call it *monochromatic*.

As the main result of this section, we show that not only MAX-MST-ratio is NP-hard, but it is extremely hard to approximate. In Sect. 3, we will use the NP-hardness of the MAX-MST-ratio to demonstrate that the geometric version of the problem (i.e., the MAX-EMST-ratio) is also NP-hard. Theorem 1 presents a strong inapproximability result for the MAX-MST-ratio problem.

Theorem 1. *For every $0 < \varepsilon \leq 1$, there is no poly-time $O(n^{1-\varepsilon})$-approximation with weights restricted to 1 and n for MAX-MST-ratio problem, unless P=NP.*

Proof. We prove that given an α-approximation for the MAX-MST-ratio, one can obtain a 2α-approximation for the MAX-Clique problem. Let $G = (V, E)$ be a graph on n vertices on which we wish to find the maximum clique. From this, we create an instance of MAX-MST-ratio with the input graph G', where G' is a complete graph with $V(G') = V$ and the weight of any edge $e \in E(G')$ is n if

$e \in E$, and it is 1 otherwise. Given a bi-partition of vertices into sets R and B, we refer to $|\text{MST}(R)| + |\text{MST}(B)|$ as the *value* of this bi-partition.

Given a clique C of size ℓ in G, we can find a coloring of value k for G' in polynomial time, where $\ell \leq \lfloor \frac{k}{n} \rfloor + 2$. This is trivial if $\ell \leq 2$, and for $\ell \geq 3$ it is sufficient to color all but one of the vertices in C by red, and the remaining vertices by blue. In this case, the weight of any MST of the red vertices is $(\ell-2)\cdot n$ and thus $\ell \leq \lfloor \frac{k}{n} \rfloor + 2$.

Conversely, given a coloring Q of value k for G', we show that we can find a clique of size at least $\frac{\lfloor k/n \rfloor}{2} + 1$ in G in polynomial time. Let R_Q and B_Q denote a red MST and a blue MST of G, respectively, in the coloring Q. Given that the weights in G' belong to $\{1, n\}$ and since $R_Q \cup B_Q$ has exactly $n - 2$ edges, then $R_Q \cup B_Q$ has exactly $\lfloor k/n \rfloor$ edges of weight n. W.l.o.g, we can assume that R_Q has at least $\frac{\lfloor k/n \rfloor}{2}$ edges of weight n. Consider $F := (V(R_Q), E'')$, where E'' is the set of edges of weight 1 in R_Q. Let $C_1, \ldots, C_r$ be the connected components of F. Clearly, we have $r \geq \frac{\lfloor k/n \rfloor}{2} + 1$. For every $1 \leq i \leq r$, let u_i be a vertex in C_i. As R_Q is an MST on the set of red vertices, then for every $i < j$, the weight of the edge $u_i u_j$ in G is n. Hence, $\{u_1, u_2, \ldots, u_r\}$ is a clique of size $r \geq \frac{\lfloor k/n \rfloor}{2} + 1$ in G.

Altogether, this shows that one can use an α-approximation algorithm for the MAX-MST-ratio problem to find a 2α-approximation for the MAX-Clique problem. Zuckerman [20] proved that there exists no $O(n^{1-\varepsilon})$-approximation for MAX-Clique problem, unless $P = NP$. Therefore, there exists no $O(n^{1-\varepsilon})$-approximation for MAX-MST-ratio, unless $P = NP$. $\qquad\square$

3   Maximum MST-Ratio of Geometric Point Sets

We show that MAX-EMST-ratio problem is NP-hard. We remark that a weighted complete graph does not necessarily correspond to the distance graph of a point set in $\mathbb{R}^d$. A weighted complete graph G on n vertices is *realizable* in $\mathbb{R}^d$ if there exist n points in $\mathbb{R}^d$ such that the distance graph of these points is isomorphic to G.

We rely on a result by Dekster and Wilker [9], which shows that any weighted complete graph, whose weights belong to a small range, is realizable in Euclidean space.

Lemma 1. *[9, Theorem 2] For every positive integer $n > 0$ there exists a real number $\lambda_n = \sqrt{1 - \frac{1}{\Theta(n)}}$ such that every complete graph G on n vertices whose all edges have weights in $[\lambda_n \cdot \ell, \ell]$, for a positive real number ℓ is realizable in $\mathbb{R}^{n-1}$.*

We use the above lemma to increase all the weights of an instance in Theorem 1 by a fixed number in order to make it realizable in Euclidean space.

Lemma 2. *Let G be a weighted complete graph G on n vertices. Let w_1 and w_2 be the minimum and maximum weights of the edges of G and let λ_n be as defined in Lemma 1. For every N such that $N \geq \max\{|w_1|, \frac{\lambda_n w_2 - w_1}{1 - \lambda_n}\}$, by increasing the weight of all edges in G by N, we obtain a realizable graph in $\mathbb{R}^{n-1}$.*

Proof. Increasing the weights of the edges in G by N results in a graph G' with positive weights in which the minimum and maximum weights are $N + w_1$ and $N + w_2$. As $N > \frac{\lambda_n w_2 - w_1}{1 - \lambda_n}$ then $N + w_1 \geq \lambda_n(N + w_2)$ and hence by Lemma 1, G' is realizable in $\mathbb{R}^{n-1}$. $\qquad\square$

Theorem 2. *The MAX-EMST-ratio problem is NP-hard.*

Proof. By Lemma 2, the instances of the MAX-MST-ratio problem with weights 1 and n can be turned into a geometric graph by increasing the weights of all edges by a fixed amount. Note that this action does not change the optimal coloring of the graph in terms of the maximum MST-ratio. This is because the total number of edges in the MSTs of the two colors is always $n - 2$. Moreover, the amount we add is $O(n^2)$, according to Lemma 2. Thus, from any instance of the MAX-MST-ratio with weights 1 and n, we can construct an instance of the MAX-EMST-ratio problem in polynomial time and hence by Theorem 1. $\qquad\square$

Theorem 1 shows that it is even hard to approximate the maximum MST-ratio for general graphs. However, in Euclidean space, it is not difficult to approximate the maximum EMST-ratio as it cannot be too large. We show a general upper bound for the maximum MST-ratio in any metric space that leads to a simple 3-approximation algorithm for the MAX-EMST-ratio problem. In the proof, we use the following well-known fact about the path double cover of trees.

Lemma 3 ([19]). *Given a tree T with ℓ leaves, there exist ℓ (not necessarily distinct) paths in T such that each edge of T is in precisely two of these paths.*

Theorem 3. *In any metric space, the maximum MST-ratio for any set of $n \geq 5$ points is at most $3 - \frac{4}{n-1}$.*

Proof. Consider an arbitrary point set S in a metric space, and let T be an MST of S. Let P^* be a path with maximum total weight among all paths in T whose endpoints are not leaves of T. Assume w.l.o.g. That one endpoint of P^*, denoted by r, is red, and we hang T from r. We traverse the tree from the root r using Depth-First Search (DFS), prioritizing the vertices of P^* when exploring branches. The starting time of a node refers to the moment it is first visited during the traversal.

To prove the statement, for every coloring Q of S, we build spanning trees R_Q and B_Q on the set of red vertices and the set of blue vertices of Q respectively, such that $|R_Q \cup B_Q|$ is at most $(3 - \frac{4}{n-1})|T|$. We construct R_Q and B_Q in the following way (see Fig. 2).

- We add every edge e of T with both endpoints in red (blue) to R_Q (B_Q).
- Consider any red (blue) vertex v in T and any blue (red) child w of v such that there exists at least one red (blue) vertex in the descendants of w (e.g. $v = r$ in Fig. 2). Let $u_1, u_2, \ldots, u_x$ be the red descendants of w for which all the internal vertices of the wu_i-path in T are blue (red). Assume that these vertices are sorted by starting time (i.e. u_1 has the earliest starting time, and u_x has the latest). We add the edges $u_1u_2, u_2u_3, \ldots, u_{x-1}u_x, u_xv$ to R_Q (to B_Q).

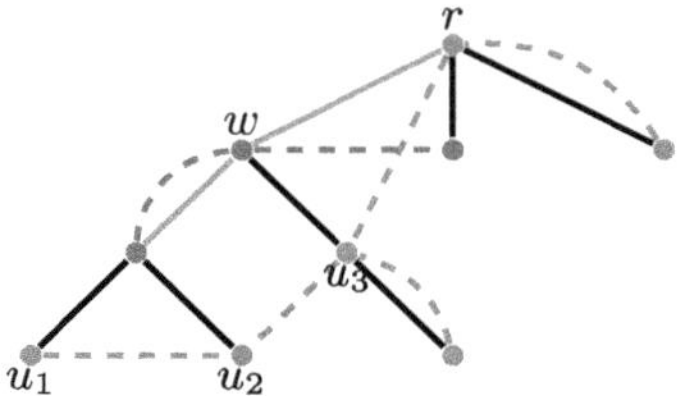

Fig. 2. The construction of R_Q and B_Q (dashed edges) according to the coloring Q of vertices in the proof of Theorem 3. The solid black and green edges form the tree T rooted at r. The children of any vertex are from left to right according to the order of their starting time in DFS. The green edges are the edges of P^*. (Color figure online)

- Let v_1, v_2, ..., v_y be the set of blue vertices that have no blue ancestors, sorted by starting time. We add the edges $v_1 v_2, v_2 v_3, \ldots, v_{y-1} v_y$ to B_Q.

Let $\mathcal{P}$ be the set of all uv-paths P_e in T, where $e = uv$ is an edge in $R_Q \cup B_Q$. Observe that every edge of T belongs to at most 3 paths in $\mathcal{P}$. Namely, by construction, if x is the parent of y and x is blue (similarly red) then the edge xy appears in at most one path P_{e_b} where $e_b \in B_Q$ and at most two paths P_{e_r} where $e_r \in R_Q$. Furthermore, certain edges in T are covered by at most two paths in $\mathcal{P}$, namely the leaf edges (i.e. edges incident to a leaf) and the edges in P^* (i.e. the green path in Fig. 2).

Since the instance is metric, the weight of every edge uv in $R_Q \cup B_Q$ is not more than the total weight of the corresponding uv-path in $\mathcal{P}$. So we have $|R_Q| + |B_Q| \le 3|T| - W_L - W_{P^*}$, where L is the set of leaf edges in T and W_L and W_{P^*} are the total weight of the edges of L and P^*, respectively. It suffices to show that $W_{P^*} + W_L \ge (\frac{4}{n-1})|T|$.

If the diameter of T is less than 4, then $P^* \cup L = T$, and hence $W_{P^*} + W_L = |T| \ge (\frac{4}{n-1})|T|$. Assume that the diameter of T is at least 4. Let T' be the tree obtained from T by removing all the leaves of T, and let ℓ and ℓ' be the number of leaves of T and T', respectively. Note that a vertex in T that is a leaf in T' must be adjacent to at least one leaf in T and hence $\ell' \le \ell$. Furthermore, as the diameter of T is at least 4, then T' has at least one non-leaf vertex which implies that $\ell + \ell' \le n - 1$ and thus, $\ell' \le \frac{n-1}{2}$. Using Lemma 3, there are ℓ' paths in T' such that all edges of T' belong to exactly two such paths. Hence, W_{P^*} is lower bounded by $\frac{2|T'|}{\ell'} = \frac{2(|T|-W_L)}{\ell'} \ge \frac{4(|T|-W_L)}{n-1}$. Therefore, it can be concluded that $W_{P^*} + W_L \ge \frac{4(|T|-W_L)}{n-1} + W_L \ge \frac{4}{n-1}|T|$.   □

Note that the above bound is the best we can achieve in metric spaces. In fact, for every odd n, there exist metric spaces with n points (elements) for which the maximum MST-ratio is arbitrarily close to $3 - \frac{4}{n-1}$. To see this, let $n = 2k + 1$, and define the point set $S = \{v_1, \ldots, v_n\}$. For every $i < n$, we set the weight of $v_n v_i$ to 1. For every $1 \le i \le k$, we set the weight of $v_{2i} v_{2i-1}$ to be ε for a sufficiently small $\varepsilon > 0$. Any other edge has weight 2.

We first argue that the weights are metric. The edges of weight 1 are the edges incident to v_n, and the edges of weight ε form a matching. Therefore, any triangle with edge weights $a \leq b \leq c$ has zero or two edges of weight 1 and has at most one edge of weight ε. Thus, (a, b, c) is in $\{(\varepsilon, 1, 1), (\varepsilon, 2, 2), (1, 1, 2), (2, 2, 2)\}$, and in all cases we have $c \leq a + b$. Observe that $|\mathrm{MST}(S)| = \varepsilon \times k + 1 \times k = k(1 + \varepsilon)$. If we color $v_1, v_3, \ldots, v_n$ by red and $v_2, v_4, \ldots, v_{n-1}$ by blue, then $|\mathrm{MST}(R)| = k$ and $|\mathrm{MST}(B)| = 2(k - 1) = 2k - 2$. Thus, $|\mathrm{MST}(R)| + |\mathrm{MST}(B)| = 3k - 2$ and hence the MST-ratio is $\frac{3k-2}{k(1+\varepsilon)} = (3 - \frac{4}{n-1})\frac{1}{1+\varepsilon}$.

Theorem 4. *There exists an $O(n^2)$-time 3-approximation algorithm for the MAX-EMST-ratio problem.*

Proof. By Theorem 3, the maximum EMST-ratio for an n-element point set is upper bounded by $3 - \frac{4}{n-1}$. On the other hand, one can achieve an EMST-ratio of at least $\frac{n-2}{n-1}$ by first computing a minimum spanning tree of the point set in $O(n^2)$ (see [8, 18]); then, removing the shortest edge; and finally coloring the vertices of the two components of it by red and blue. Finally, as $\frac{n-2}{n-1} \geq \frac{(3 - \frac{4}{n-1})}{3}$, this is an $O(n^2)$ time 3-approximation algorithm. $\square$

We remark that a slightly improved running time for computing an EMST is provided in [1]. Moreover, a weaker approximation factor 3.47 could be achieved in a simpler way using a trivial upper bound $\frac{2}{\rho_d} \leq \frac{2}{0.577} < 3.47$ on the MAX-EMST-ratio. Indeed, if we restrict to $\mathbb{R}^2$ (i.e. 2-MAX-EMST-ratio), the approximation factor in Theorem 4 can be improved to 2.427, because the upper bound for the maximum EMST-ratio in the plane is ≈ 2.427 (see [16]), and for $n > 12$, a coloring with EMST-ratio greater than 1 can be computed in polynomial time (see [11]). As can be observed, the coloring used to present the 3-approximation is fairly straightforward. However, there seems to be room for improvement by identifying more mingled colorings.

We introduce *Bipartite Coloring* as a good candidate to approximate the maximum EMST-ratio of a point set. By a bipartite coloring of a point set, we mean considering one of the Euclidean minimum spanning trees and then partitioning the point set into two subsets labeled by colors in which every edge of the chosen EMST is colorful. Note that since the EMST of a point set is not necessarily unique, it may have more than one bipartite coloring, namely, one for each EMST. Intuitively, bipartite colorings give well-distributed colorings of the point set that can be computed in $O(n^2)$ time ($O(n \log n)$ in the plane; see [4]). Although bipartite coloring seems to be effective, it does not always produce the maximum EMST-ratio. Here is an extreme example: Consider $n = 2k + 1$ points on a zigzag path in the triangular grid and call them a *Triangular Chain* (see Fig. 3). By stretching the chain slightly, one can ensure a unique EMST, which is the green tree depicted in Fig. 3 and hence a unique bipartite coloring. The maximum EMST-ratio of a triangular chain with $2k + 1$ points is $\frac{3k-2}{2k}$ (The proof is provided in Sect. A), while the ratio for a bipartite coloring is $\frac{2k-1}{2k}$. As k increases, the proportion between the maximum EMST-ratio and this bipartite EMST-ratio approaches 1.5.

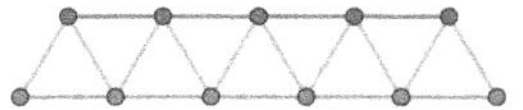

Fig. 3. An example of a point set where the maximum EMST-ratio has a significant gap from the EMST-ratio of a bipartite coloring. The left panel shows a triangular chain for $k = 5$, its EMST in green, and the red and blue EMSTs in a bipartite coloring. The right panel presents the coloring that achieves the maximum EMST-ratio. (Color figure online)

4   Average MST-Ratio

A natural approach to approximating the maximum EMST-ratio is to consider randomized approximation algorithms that assign random colorings to the point set. To analyze the performance guarantees of such algorithms, it is reasonable to study the average EMST-ratio over all possible colorings of a given point set.

The *Average EMST-ratio* of a point set P is the average of EMST-ratio over all proper bi-colorings of P. The average EMST-ratio is automatically a lower bound for the maximum EMST-ratio, and the key question is whether this average provides a good approximation or not. In addition, it could be considered as a measure to see how close a point set is to a typical random point set, as we know the behavior of random point sets. In fact, it is proved that the expected value for the average EMST-ratio of a random point set in a unit square is $\sqrt{2}$ in the limit [10]. We present the proof for the generalized version of this result, i.e. for points within a d-dimensional unit cube (or any bounded region in $\mathbb{R}^d$).

Theorem 5. *For a set of n random points uniformly distributed in $[0,1]^d$, the expected value of the average EMST-ratio tends to $\sqrt[d]{2}$ as n goes to infinity.*

Proof. Let n be large enough and P be a set of n random points uniformly distributed in $[0,1]^d$. Consider all $2^n - 2$ proper bi-partitions $P = R \cup B$ into red and blue points. Clearly, for $1 \leq k \leq n - 1$, in $\binom{n}{k}$ of such bi-partitions $|R| = k, |B| = n - k$. Ignoring the relatively small k or $n - k$, by (1), we know that for every $k_0 \leq k \leq n - k_0$ we have

$$|\mathrm{EMST}(R)| \sim \beta(d).k^{1-1/d}, |\mathrm{EMST}(B)| \sim \beta(d).(n-k)^{1-1/d},$$

$$|\mathrm{EMST}(P)| \sim \beta(d).n^{1-1/d},$$

with probability approaching 1, where $\beta(d)$ is the constant in (1), and k_0 is an appropriate constant. Since n is significantly greater than k_0, and the EMST-ratio is at most 3, the contribution of the EMST-ratios for colorings with $k < k_0$ red points or with $n - k < k_0$ blue points are negligible when computing the average EMST-ratio. Thus, the average EMST-ratio over all proper bi-partitions of P is

$$\sum_{P=R\cup B} \frac{|\mathrm{EMST}(R)| + |\mathrm{EMST}(B)|}{|\mathrm{EMST}(P)|} \sim \sum_{k=k_0}^{n-k_0} \frac{[\beta(d).k^{1-1/d} + \beta(d).(n-k)^{1-1/d}].\binom{n}{k}}{\beta(d).n^{1-1/d}.(2^n - 2)}$$

$$= \sum_{k=k_0}^{n-k_0} \frac{[k^{1-1/d} + (n-k)^{1-1/d}].\binom{n}{k}}{n^{1-1/d}.(2^n - 2)},$$

with probability 1 as n goes to infinity. The right-hand side is a special case of convergence of Bernstein polynomials (see [5,15]), and therefore converges to $\sqrt[d]{2}$ as n goes to infinity. $\qquad\square$

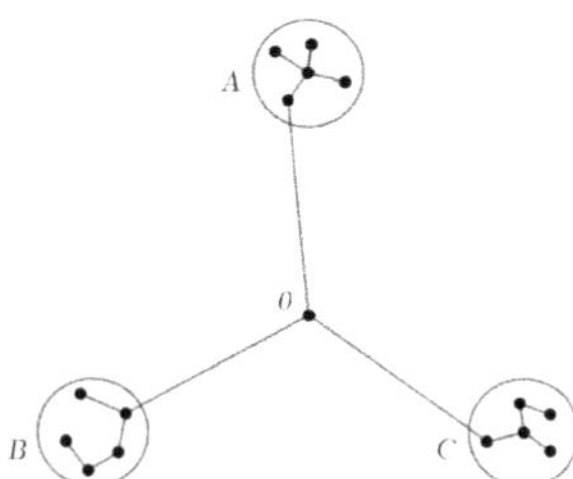

Fig. 4. A point set consisting of a point called the origin and several points clustered near the three vertices of an equilateral triangle centered at the origin. The corresponding minimum spanning tree is depicted. As the number of points increases, the average EMST-ratio approaches approximately 2.154.

Certainly, Theorem 3 implies that 3 is an upper bound for the average EMST-ratio of point sets in Euclidean space. How much can this upper bound be improved? There exist point sets in $\mathbb{R}^2$ with an average EMST-ratio ≈ 2.154. For instance, consider the unit circle centered at the origin in the plane, an equilateral triangle abc on it, and then a point set $P = A \cup B \cup C \cup \{0\}$, where A (resp. B, C) consists of many points that lie within distance ε of a (resp. b, c), for ε being small enough (see Fig. 4). In this case, the average EMST-ratio approaches $\frac{3+2\sqrt{3}}{3} \approx 2.154$ as the number of points increases, because with high probability, in a random bi-partition, each of the sets A, B, and C will contain both colors. Regarding the lower bound, we show that the average EMST-ratio for n points is always at least $\frac{n-2}{n-1}$. This is a stronger result than the first part of Theorem 2 in [11].

Theorem 6. *For every set of n points in Euclidean space, the average EMST-ratio is at least $\frac{n-2}{n-1}$.*

Proof. Consider a set of n points P and let $e_1, ..., e_{n-1}$ be the edges of an EMST of P, called T. Assume that the edge e_i has weight w_i and $w_1 \le w_2 \le ... \le w_{n-1}$. Considering a coloring $P = R \cup B$ of P into red and blue points, let i be the

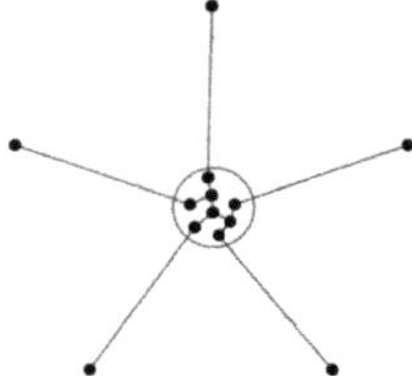

Fig. 5. A point set with average EMST-ratio smaller than 1. The point set consists of $n - 5$ points in a very small circle through the origin (the core), and 5 outer points as vertices of a regular pentagon inscribed in the unit circle around the origin. The corresponding MST is depicted.

smallest index such that the edge e_i is colorful. As $\mathrm{EMST}(R) \cup \mathrm{EMST}(B) \cup \{e_i\}$ is a spanning tree of P, it holds that $|\mathrm{EMST}(R)| + |\mathrm{EMST}(B)| \geq |\mathrm{EMST}(P)| - w_i$.

On the other hand, considering a random coloring of P with two colors, the probability that the edge e_i is colorful is $\frac{1}{2}$, and since these are the edges of a tree, their corresponding events are independent. Thus, for a fixed index j, the probability that in a random coloring, the shortest colorful edge of T is e_j is $(\frac{1}{2})^j$. Therefore, there are precisely 2^{n-j} colorings of P in which j is the smallest index such that the edge e_j is colorful. Let denote the average EMST-ratio by a_r. By the above argument and since there are $2^n - 2$ possible (proper) colorings, we have

$$a_r \geq |\mathrm{EMST}(P)| - \sum_{i=1}^{n-1} \frac{2^{n-i} w_i}{2^n - 2}.$$

As we have $\sum_{i=1}^{n-1} 2^{n-i} = 2^n - 2$ and $w_1 \leq w_2 \leq \ldots \leq w_{n-1}$, it can be concluded that $a_r \geq |\mathrm{EMST}(P)| - \frac{1}{n-1} \cdot |\mathrm{EMST}(P)|$, and hence the claim.   □

Note that Theorem 6 holds also for general graphs. The average EMST-ratio appears to be greater than 1 in many cases, though this is not always true. The only instances we found in $\mathbb{R}^2$ consist of point sets where $n - 6 \leq k \leq n - 1$ points are clustered closely together, forming a core, and $1 \leq n - k \leq 6$ points are sufficiently far from the core such that their pairwise distances are not less than their distances to the core. We believe these are the only point sets with an average EMST-ratio of less than 1.

As an example, we show that the average EMST-ratio for the point set in Fig. 5 is less than 1. The point set consists of $n - 5$ points in a small circle through the origin whose radius is $\varepsilon \in o(\frac{1}{n2^n})$ (the core), along with 5 outer points as vertices of a regular pentagon inscribed in the unit circle around the origin. Observe that each side of this pentagon is $s = 2 \sin 36 \approx 1.176$ and each diagonal is $d = \frac{1+\sqrt{5}}{2} s \approx 1.902$. In this example, clearly we have $|\mathrm{EMST}(P)| \geq 5 - O(n\varepsilon)$. In any bi-coloring of P by red and blue, if an outer point has the same color as one of the core points, then the edge connecting this outer point to a core point appears in one of the monochromatic EMSTs. Therefore, in $2^n - 64$ bi-colorings of P where the core contains both colors, we have $|\text{EMST}(R)| + |\text{EMST}(B)| \leq 5 + O(n\varepsilon)$. There are only $2^6 - 2 = 62$ proper bi-colorings with a mono-chromatic core. Consider one of these 62 colorings and assume that the points in the core are red. In this case, if $0 \leq n_r < 5$ denotes the number of outer points that are red, then $|\text{EMST}(R)| = n_r + O(n\varepsilon)$. Also, if the outer blue points are consecutive vertices of the outer pentagon, $|\text{EMST}(B)| = (4 - n_r)s$ and otherwise $|\text{EMST}(B)| = (3 - n_r)s + d$. So in each of these colorings $|\text{EMST}(R)| + |\text{EMST}(B)| \leq d + s + 2 + O(n\varepsilon) < 5.1$. Moreover, in $62 - 2 \times 5 = 52$ of such colorings, we even have $|\text{EMST}(R)| + |\text{EMST}(B)| \leq d + 3 + O(n\varepsilon) < 4.91$. Thus, the average EMST-ratio for this point set is upper bounded by

$$\frac{(2^n - 64) \cdot (5 + O(n\varepsilon)) + 52 \cdot (4.91) + 10 \cdot (5.1)}{(2^n - 2) \cdot (5 - O(n\varepsilon))} < 1.$$

Question A. Which point sets in $\mathbb{R}^2$ have an average EMST-ratio < 1?

5 Discussion

This section presents questions and conjectures motivated by our earlier results, along with related directions for further study.

Theorem 2 shows that determining the maximum EMST-ratio is NP-hard. We conjecture that this remains true in specific dimensions, particularly in $\mathbb{R}^2$ and $\mathbb{R}^3$ as they are relevant for practical applications.

Conjecture 1. The k-MAX-EMST-ratio problem for $k \in \{2, 3\}$ is NP-hard.

We also conjecture that the computed approximation factor for the maximum EMST-ratio of the bipartite coloring in Sect. 3 is the worst that can be obtained from a bipartite coloring for a point set in the plane. If this is true, this yields a $O(n \log n)$-time algorithm that achieves a $\frac{3}{2}$-approximation 2-MAX-EMST-ratio.

Conjecture 2. For every point set in $\mathbb{R}^2$, the EMST-ratio for any bipartite coloring is a $\frac{3}{2}$-approximation for the maximum EMST-ratio.

Apart from the bipartite coloring, we are interested in any improved approximation algorithm for the MAX-EMST-ratio that can be obtained by a simple mixed coloring of a point set.

Question B. Does there exist a polynomial-time 2-approximation algorithm for MAX-EMST-ratio?

Fig. 6. A point set and a bi-partition with chromatic crossing number 4.

Coloring with Many Crossings. For a point set P in the plane, partitioned as $P = R \cup B$ of P, the *Chromatic Crossing Number* is the number of crossings between pairs of edges, one in $\mathrm{EMST}(R)$ and one in $\mathrm{EMST}(B)$. Intuitively, This measures how mixed the coloring is: more crossings indicate more mingling. In fact, for any point set, it is easy to find colorings with no crossings between the two MSTs. More interesting, however, is the *Maximum Chromatic Crossing Number*, which represents the highest chromatic crossing number achievable across all possible colorings of the point set (Fig. 6).

Question C. How large and how small can the maximum chromatic crossing number of a set of n points in the plane be?

Question D. Are there relations between the chromatic crossing number and the EMST-ratio? In particular, does the coloring that maximizes the chromatic crossing number tend to have a relatively high EMST-ratio?

Conjecture 3. Finding the maximum chromatic crossing number of a point set in the plane is NP-hard.

Maximum MST Length over Subsets. A related question to finding the maximum MST-ratio is the *MAX-EMST-subset* problem. Here, given a point set P, we aim to find the maximum EMST length over all subsets $Q \subseteq P$. This extends to the abstract setting as the *MAX-MST-subset* problem, where, given a weighted graph G, we seek the maximum MST weight over all induced subgraphs $H \subseteq G$.

Question E. Are MAX-MST-subset and MAX-EMST-subset NP-hard?

More Colors. Similar questions arise for point sets with $c > 2$ colors. In this case, the MST-ratio is the total length of the MST of each color divided by the length of an MST of the (uncolored) point set. Theorem 3 can be generalized to obtain $2c - 1$ as an upper bound for the MST-ratio in any metric space, which can be improved in the Euclidean space for $c \geq 4$. In particular, as $\rho_d \geq 0.577$ for every

d, the maximum EMST-ratio is upper bounded by $\frac{c}{0.577}$. It remains unclear how the results in [16] on lattices can be generalized to more than 2 colors. Furthermore, exploring the average EMST-ratio for c colors is an interesting direction to pursue.

A   Maximum EMST-Ratio of Triangular Chain

In this section, we show that the maximum EMST-ratio for the triangular chain, shown in Fig. 3 is $\frac{3k-2}{2k+1}$. Let P be the point set of a triangular chain with $2k+1$ points and let us number the points of P from left to right by $p_1, p_2, \ldots, p_{2k+1}$.

As $|\mathrm{EMST}(P)| = 2k$, we first show that for any proper coloring of the chain to red and blue, we have $|\mathrm{EMST}(R)| + |\mathrm{EMST}(B)| \leq 3k - 2$. We prove this using induction on k. The claim holds trivially for $k = 1$, as any proper coloring has $|\mathrm{EMST}(R)| + |\mathrm{EMST}(B)| = 1 = 3 - 2$.

Assume that the claim holds for the case of chains with $2k + 1$ points. Now, we prove the claim for any chain that consists of $2(k + 1) + 1$ points. Consider one such chain with point set P and let B and R be the set of blue points and red points in a proper coloring of P. Let i be the smallest index such that p_i, p_{i+1} and p_{i+2} have the same color. W.l.o.g., we assume these points belong to R. We prove the claim by distinguishing the following cases.

a. $i = 1$) In this case, $R \setminus \{p_1, p_2\}$ and B is a proper coloring for $P \setminus \{p_1, p_2\}$. Also, $T_R := \mathrm{EMST}(R \setminus \{p_1, p_2\}) \cup \{p_1 p_2, p_2 p_3\}$ is a spanning tree of R. By induction hypothesis $|\mathrm{EMST}(R \setminus \{p_1, p_2\})| + |\mathrm{EMST}(B)| \leq 3k - 2$, which implies that

$$|\mathrm{EMST}(R)| + |\mathrm{EMST}(B)| \leq |T_R| + |\mathrm{EMST}(B)|$$
$$\leq 3k - 2 + |p_1 p_2| + |p_2 p_3| < 3(k + 1) - 2,$$

and hence the claim.

b. $i = 3$) Let P' be the triangular chain obtained from P by removing p_1 and p_2. We define R' and B' as a coloring for P' in the following way.
   - For every $j \geq 5$, $p_j \in R'$ if and only if $p_j \in R$.
   - For every $j \in \{3, 4\}$, $p_j \in R'$ if and only if $p_{j-2} \in R$.

   Note that R' and B' is a proper coloring of P', and hence by induction hypothesis, we can conclude $|\mathrm{EMST}(R')| + |\mathrm{EMST}(B')| \leq 3k - 2$. Since $i \neq 1$ and considering that p_3 is in R, one of p_1 and p_2 must belong to B. We have $|\mathrm{EMST}(R')| = |\mathrm{EMST}(R)| - 2$ and $|\mathrm{EMST}(B')| + 1 \geq |\mathrm{EMST}(B)|$. Therefore, we can conclude that

$$|\mathrm{EMST}(R) + |\mathrm{EMST}(B)| \leq |\mathrm{EMST}(R')| + |\mathrm{EMST}(B')| + 3 \leq 3(k + 1) - 2.$$

c. Otherwise) W.l.o.g. assume $p_3 \in R$. First, assume that p_1 and p_2 are in B. Since we are not in case b, there exists $j \in \{4, 5\}$ such that $p_j \in B$. Thus, R and $B \setminus \{p_1, p_2\}$ is a proper coloring for $P \setminus \{p_1, p_2\}$. The tree $T_B :=

$\text{EMST}(B \setminus \{p_1, p_2\}) \cup \{p_1 p_2, p_2 p_j\}$ is a spanning tree of B. By the induction hypothesis, we have $|\text{EMST}(B \setminus \{p_1, p_2\})| + |\text{EMST}(R)| \leq 3k - 2$, and thus,

$$|\text{EMST}(R)| + |\text{EMST}(B)| \leq |T_B| + |\text{EMST}(R)|$$
$$\leq 3k - 2 + |p_1 p_2| + |p_2 p_j| < 3(k + 1) - 2.$$

Now, we assume that there exists $z \in \{1, 2\}$ such that $p_z \in R$. Since we are not in any of the previous cases, then p_{3-z} is blue and at least one of the points p_4 and p_5 (say p_j) is in B. Therefore, R and $B \setminus \{p_1, p_2\}$ is a proper coloring for $P \setminus \{p_1, p_2\}$. Now, $T_B := \text{EMST}(B \setminus \{p_{3-z}\}) \cup \{p_{3-z} p_j\}$ is a spanning tree of B and $T_R := \text{EMST}(R \setminus \{p_z\}) \cup \{p_z p_3\}$. Note that the induction hypothesis yields $|T_B| + |T_R| \leq 3k - 2 + 1 + 2$ and hence the claim.

Finally, if $R = \{u, v\}$ is the pair of points in P with maximum distance from each other (see Fig. 3), then $|EMST(B)| + |EMST(R)| = 3k - 2$. So, the EMST-ratio of P is $\frac{3k-2}{2k-1}$.

References

1. Agarwal, P.K., Edelsbrunner, H., Schwarzkopf, O., Welzl, E.: Euclidean minimum spanning trees and bichromatic closest pairs. Disc. Comput. Geometry **6**(3), 407–422 (1991). https://doi.org/10.1007/BF02574698
2. Avram, F., Bertsimas, D.: The minimum spanning tree constant in geometrical probability and under the independent model: a unified approach. Ann. Appl. Prob. **2**, 113–130 (1992)
3. Beardwood, J., Halton, J.H., Hammersley, J.M.: The shortest path through many points. Math. Proc. Cambridge Philos. Soc. **55**, 299–327 (1959)
4. de Berg, M., Cheong, O., van Kreveld, M., Overmars, M.: Computational Geometry: algorithms and Applications. Springer, 3 edn. (2008)
5. Bernstein, S.: Démonstration du théorème de weierstrass fondée sur le calcul des probabilités. Commun. Kharkov Math. Soc. **13**, 1–2 (1912)
6. Biswas, R., Cultrera di Montesano, S., Draganov, O., Edelsbrunner, H., Saghafian, M.: On the size of chromatic delaunay mosaics. arXiv:2212.03121 (2022)
7. Chung, F.R.K., Graham, R.L.: A new bound for Euclidean Steiner minimal trees. In: Discrete Geometry and Convexity, pp. 328–346. Annals of the New York Academy of Sciences (1985)
8. Cormen, T., Leiserson, C.E., Rivest, R.L., Stein, C.: Introduction to Algorithms. MIT Press, 3 edn. (2009)
9. Dekster, B.V., Wilker, J.B.: Edge-lengths guaranteed to form a simplex. Arch. Math. **49**, 351–366 (1987)
10. Draganov, O., Saghafian, M.: Private communication (2024)
11. Dumitrescu, A., Pach, J., Tóth, G.: Two trees are better than one. arXiv:2312.09916 (2023)
12. Gilbert, E.N.: Random minimal trees. J. Soc. Ind. Appl. Math. **13**, 376–387 (1965)
13. Gilbert, E.N., Pollack, H.O.: Steiner minimal trees. SIAM J. Appl. Math. **16**, 1–29 (1968)

14. Graham, R.L., Hwang, F.K.: Remarks on Steiner minimal trees. Bull. Inst. Math. Academia Sinica **4**(1), 177–182 (1976)
15. Koralov, L., Sinai, Y.: Theory of probability and random processes. Springer, 2 edn. (2007)
16. Cultrera di Montesano, S., Draganov, O., Edelsbrunner, H., Saghafian, M.: The euclidean mst-ratio for bi-colored lattices. In: Symposium on Graph Drawing and Network Visualization (2024)
17. Cultrera di Montesano, S., Draganov, O., Edelsbrunner, H., Saghafian, M.: Chromatic alpha complexes. Found. Data Sci. (2025)
18. Prim, R.C.: Shortest connection networks and some generalizations. Bell Syst. Tech. J. **36**(6), 1389–1401 (1957)
19. Thiyagarajan, G., Saravanan, M.: Path double covering number of a graph. Int. J. Math. Comput. Res. **2**(8), 565–573 (2014)
20. Zuckerman, D.: Linear degree extractors and the inapproximability of max clique and chromatic number. In: Proceedings of the 38th Annual ACM Symposium on Theory of Computing, pp. 681–690 (2006)

Disjoint Tours and the Price of Diversity

Mark de Berg, Andrés López Martínez[(⊠)], and Frits Spieksma

Department of Mathematics and Computer Science, TU Eindhoven,
Eindhoven, Netherlands
`{m.t.d.berg,a.lopez.martinez,f.c.r.spieksma}@tue.nl`

Abstract. We study a variant of the Traveling Salesman Problem,
where instead of finding a single tour, we want to find a pair of two
edge-disjoint tours whose longer tour is as short as possible. We investigate the *Price of Diversity (PoD)* for this problem, which is the ratio of
the cost of the longer of the two tours and the cost of a single optimal
tour, in the worst case over all possible instances. We prove (almost)
tight bounds on this quantity for a special 1-dimensional scenario and
for general metric spaces. We believe that the Price-of-Diversity framework that we introduce is interesting in its own right, and may lead to
follow-up work on other problems as well.

Keywords: Diverse Solutions · TSP · Path TSP · Price of Diversity

1 Introduction

Traditionally, algorithms for combinatorial optimization problems focus on computing a single, optimal solution. There are many situations, however, where
it is desirable to compute multiple solutions that are sufficiently distinct. For
example, a decision maker who lacks complete information about the costs or
feasibility of candidate solutions may prefer to keep multiple alternatives open
until the missing details are revealed, reducing the risk of committing to a sub-
optimal choice. Likewise, in time-sensitive settings where last-minute constraints
can arise, a diverse solution set increases the likelihood that at least one feasible
option is available.

Typically it is not possible to find multiple distinct solutions that are all
optimal. Thus, requiring a diverse set of solutions comes at a price. In this paper,
we study this phenomenon for the Traveling Salesman Problem (TSP): given an
edge-weighted graph, find a shortest tour visiting all its vertices. Suppose that
instead of computing a single optimal tour, we must produce k tours subject to a
diversity requirement, while minimizing their *bottleneck cost*—the length of the
longest tour. If too few short tours satisfy this requirement, longer tours must
be included, and the longest of the k tours may end up much more expensive
than the optimal single tour.

The full version of the paper, including all omitted proofs, is available at
https://arxiv.org/abs/2507.13026.

To measure the trade-off between solution quality and the desired number of solutions, we analyze the ratio between the length of a longest tour in the best set of k diverse tours and the length of an optimal single tour. Taking the worst-case value of this ratio over all instances provides a measure of how much the demand for diversity impacts the cost of solutions. We call this measure the *Price of Diversity (PoD)*, which we introduce more formally and more generally in Sect. 2.

Our Problems and Results. We study the price of diversity in the context of the TSP in a metric space (i.e., edge costs satisfy the triangle inequality). Our main focus is on the setting where diversity is enforced through edge-disjointness and k, the number of required solutions, equals 2. This is one of the simplest and most natural cases, yet deriving tight bounds is already nontrivial. More precisely, we consider the following problems.

Disjoint-Path TSP_2. Given a weighted graph $G = (V, E)$ and two specified vertices $s, t \in V$, find a pair (H_1, H_2) of edge-disjoint Hamiltonian (s, t)-paths that minimizes $cost(H_1, H_2) := \max(c(H_1), c(H_2))$.

Disjoint-TSP_2. Given a weighted graph $G = (V, E)$, find a pair (T_1, T_2) of edge-disjoint tours (that is, Hamiltonian cycles) that minimizes $cost(T_1, T_2) := \max(c(T_1), c(T_2))$.

We establish the PoD of Disjoint-Path TSP_2 and Disjoint-TSP_2 for different metric spaces. Already in one dimension, these problems exhibit nontrivial behavior. In Sect. 3, we consider Disjoint-Path TSP_2 on the line with uniformly spaced vertices and show that the PoD equals $\frac{13}{7}$ (approaching $\frac{8}{5}$ asymptotically). In Sect. 4, we extend the analysis to Disjoint-TSP_2 on the circle with uniformly spaced vertices.

In Sect. 5, we turn to arbitrary metrics and show that the price of diversity of Disjoint-Path TSP_2 is exactly 3, while the PoD of Disjoint-TSP_2 is exactly 2. Each of these bounds is matched by a constructive algorithm that produces a pair of edge-disjoint solutions whose bottleneck cost achieves the claimed PoD. Note that we are, in general, not interested in the running time of the algorithms; we briefly discuss efficiency considerations in Sect. 6. All missing proofs can be found in the full version of this paper.

Related Work. The generation of multiple solutions has been studied in many areas of computer science, including constraint programming [20,25], mixed integer optimization [2,11,16], k-best enumeration [13,18], genetic algorithms [12,15], social choice [9], and computational economics [17,24]. More recently, attention has shifted to producing not just many but *diverse* solutions [5,14,19,23,27].

There are many ways to specify a diversity requirement for a collection of solutions. Of particular relevance in this work is the case when diversity is specified in terms of disjointness. For problems where the output is a set of edges, such as matching, spanning tree, or minimum s-t cut, edge-disjointness serves as a natural diversity criterion. Recent works have addressed such settings [8,14,19].

In the context of the TSP, several works have considered the problem of finding edge-disjoint tours in specific graph classes. We mention the works of Rowley and Bose [26] on De Bruijn graphs, Lü and Wu [22] on balanced hypercubes, and Alspach [4] on complete graphs. The peripatetic k-TSP [21] similarly requires multiple edge-disjoint tours, aiming to minimize their total length. Another related problem is Two-TSP, where $2n$ points in the plane must be partitioned into two equal-sized sets, and the objective is to minimize the maximum TSP tour length over the two parts [7]. These works focus on computing multiple non-overlapping solutions but do not quantify the cost overhead incurred by the disjointness requirement.

In this work, we introduce the price of diversity (PoD) as a measure of the tradeoff between solution quality and the requirement to compute multiple diverse solutions. The term has appeared before, especially in computational social choice, but usually refers to the cost of enforcing diversity within a *single* solution—for example, in matching, assignment, clustering, or committee selection [3,6,10,28]. In contrast, we analyze the PoD in the setting where a *collection* of diverse solutions is required. To our knowledge, this is the first formal treatment of this perspective.

2　Preliminaries

The PoD Framework. Let Π be a minimization problem—like TSP—where each instance I has an associated set of *feasible solutions* $\Gamma(I)$. Each solution $s \in \Gamma(I)$ has a cost $c(s)$, and the objective is to find an optimal solution; that is, a feasible solution with minimum cost, $\mathrm{OPT}(I)$. Consider now the following *diverse* variant of Π, called simply Π_k, which, given an instance I of problem Π and a fixed integer $k > 0$, is tasked with finding a *minimum-cost* set of k feasible solutions to I that satisfies a specified *diversity requirement*.

PROBLEM Π_k. Given an instance I of Π, find a set S of k feasible solutions that (i) satisfies a *diversity requirement* and (ii) has minimum *cost*.

Depending on the problem, there are many ways to define a diversity requirement. For example, for problems where solutions are subsets of a larger set, one could measure the diversity of a set of solutions as the cardinality of their union, the average size of their pairwise symmetric differences, or the minimum size among these differences. A requirement can then be imposed for the diversity to be at least a given value t; compare this with the edge-disjointness requirement for TSP in our earlier example. Similarly, the cost of a collection of feasible solutions can be defined in different ways. For instance, in TSP, one might measure cost as the length of the longest tour in the collection, or as the average length of all tours.

Returning to problem Π_k, let $\mathcal{S}(I) \in \Gamma(I)^k$ represent the collection of all k-sized sets of feasible solutions for instance I that satisfy the diversity requirement, and let $cost(S)$ denote the cost of a set $S \in \mathcal{S}(I)$ of feasible solutions.

Further, let $\mathcal{I}(\Pi)$ denote the set of all instances of problem Π. We define the Price of Diversity of problem Π_k, abbreviated with $PoD(\Pi_k)$, as

$$PoD(\Pi_k) = \sup_{I \in \mathcal{I}(\Pi)} \left\{ \frac{\min_{S \in \mathcal{S}(I)} cost(S)}{OPT(I)} \right\}. \tag{1}$$

In plain words, $PoD(\Pi_k)$ represents the supremum, taken over all instances of problem Π, on the ratio between the minimum cost among diverse sets of feasible solutions and the cost of a single optimal solution. Note that we have assumed that Π is a minimization problem, but this is without loss of generality as a similar measure to (1) can be defined for maximization problems in a straightforward manner.

In some cases, as we shall see in Sects. 3 and 4, analyzing the *asymptotic* worst-case behavior of $PoD(\Pi_k)$ provides a more meaningful perspective on the trade-off between diversity and solution quality. This is because the supremum of the ratio in (1) may be dictated by a small instance, making $PoD(\Pi_k)$ less representative of its behavior across all instances. To address this, we define the *asymptotic price of diversity* $PoD^\infty(\Pi_k)$ as

$$PoD^\infty(\Pi_k) = \lim_{n \to \infty} \sup_{I \in \mathcal{I}(\Pi),|I|=n} \left\{ \frac{\min_{S \in \mathcal{S}(I)} cost(S)}{OPT(I)} \right\}. \tag{2}$$

While *PoD* and *PoD$^\infty$* are properties corresponding to a particular problem (including a specification of diversity), we also want to be able to express the quality of an algorithm that outputs k diverse solutions in terms of the quality of the set of diverse solutions as used above. Indeed, let A be an algorithm for problem Π_k, and let $A(I)$ denote the set of k solutions that algorithm A produces when given instance I of problem P. We define both the *loss ratio LR_A*, and the *asymptotic loss ratio LR_A^∞*, of algorithm A as

$$LR_A = \sup_{I \in \mathcal{I}(\Pi)} \left\{ \frac{cost(A(I))}{OPT(I)} \right\} \text{ and } LR_A^\infty = \lim_{n \to \infty} \sup_{I \in \mathcal{I}(\Pi),|I|=n} \left\{ \frac{cost(A(I))}{OPT(I)} \right\}.$$

Notation. For a weighted graph $G = (V, E)$, let $n := |V|$ and $m := |E|$ denote the number of vertices and edges, respectively. If G is a complete graph whose nodes correspond to points in a metric space and whose edge weights correspond to the distances between these points, then we call G a *metric graph*. We use $\mathcal{M}$ to denote the class of all metric graphs. Note that this is simply the class of complete weighted graphs whose edge weights satisfy the triangle inequality (and are non-negative). We are especially interested in $\mathcal{R}$, the class of metric graphs in $\mathbb{R}^1$, and in $\mathcal{S}$, the class of metric graphs in $\mathbb{S}^1$, where distances are measured using standard Euclidean geometry. (Here $\mathbb{S}^1$ denotes the circular 1-dimensional space.) We define $R_n^1 \in \mathcal{R}$, the *uniform metric graph on n vertices in $\mathbb{R}^1$*, to be the metric graph whose vertices correspond to the set $[n] \subset \mathbb{R}^1$. Similarly, we define $S_n^1 \in \mathcal{S}$, the *uniform metric graph on n vertices in $\mathbb{S}^1$*, to be the metric graph whose vertices correspond to evenly spaced points in a circle of length n, such that the angle between consecutive vertices is $2\pi/n$ and the distance (arc length) between them is 1. For a metric graph in $\mathcal{R}$ (or $\mathcal{S}$), we refer to the edges that connect consecutive vertices along $\mathbb{R}^1$ (or $\mathbb{S}^1$) as *segments*.

3   The Price of Diversity of DISJOINT-PATH TSP$_2$ in $\mathcal{R}_n^1$

In this section, we establish both the regular and the asymptotic price of diversity of DISJOINT-PATH TSP$_2$ for the class of uniform metric graphs in $\mathbb{R}^1$, specifically when $s = 1$ and $t = n$, thereby proving the following.

Theorem 1. *PoD(DISJOINT-PATH TSP_2)= $\frac{13}{7}$, and PoD$^\infty$(DISJOINT-PATH TSP$_2$) = $\frac{8}{5}$ for the class of uniform metric graphs in $\mathbb{R}^1$ when $s = 1$ and $t = n$.*

Throughout this section, we take the uniform metric graph $R_n^1 = (V, E)$ as the input graph and define P_n as the path graph formed by the vertex set V and the segments of R_n^1. We say that P_n is the path graph induced by R_n^1. We assume that the vertices of P_n are labeled from left to right $(v_1, v_2, \ldots, v_n)$, with $s = v_1$ and $t = v_n$. Since we assume that all Hamiltonian (s, t)-paths in this section have the same endpoints, we omit explicit reference to s and t when referring to a Hamiltonian path. For a subset $X \subseteq V$ of vertices, we use $R_n^1[X]$ to denote the subgraph of R_n^1 induced by the vertex set X. For a Hamiltonian path H in R_n^1, we use $H[X]$ (with a slight abuse of notation) to denote the (possibly disconnected) subgraph induced by X of the path graph defined by H.

We prove Theorem 1 as follows. In Sect. 3.1, we first establish a lower bound of $\frac{8}{5}$ for PoD(DISJOINT-PATH TSP$_2$). Then, in Sect. 3.2, we develop an algorithm whose asymptotic loss ratio matches this bound, thus showing that PoD^∞(DISJOINT-PATH TSP$_2$) = $\frac{8}{5}$. Also in Sect. 3.2, we establish a loss ratio of $\frac{13}{7}$ for the algorithm and argue that this coincides with the price of diversity of DISJOINT-PATH TSP$_2$, thereby completing the proof.

3.1   A Lower Bound

We prove that $\frac{8}{5}$ is a lower bound to PoD(DISJOINT-PATH TSP$_2$) for the class of uniform metric graphs in $\mathbb{R}^1$. First, we introduce some new terminology. Let v_i be a vertex in a path H on R_n^1. We say that v_i is *covered* by H if there is an edge $(v_j, v_k) \in H$ such that $j < i < k$. Conversely, the edge (v_j, v_k) is said to *cover* the vertex v_i. The concept extends naturally to segments. An edge $(v_j, v_k) \in H$ is said to cover the segment $(v_i, v_{i+1}) \in P_n$ if $j \le i$ and $i + 1 \le k$. The *depth* $\mu_H(s)$ of a segment s in P_n w.r.t. path H is the number of edges from H that cover it. Note that the cost of H is equal to the sum of the depths of each segment w.r.t. H.

We now introduce the notions of *ℓ-piece* and *cut-point*, which play a key role in the proof of the lower bound. Any sequence of ℓ consecutive segments in P_n is called an *ℓ-piece*. The depth of a piece (w.r.t. a path H) is the sum of the depths of its constituting segments. A pair of edge-disjoint Hamiltonian paths

(H_1, H_2) is said to have a cut-point v if v is not covered by H_1 nor by H_2.[1] The *total depth* of an ℓ-piece w.r.t. the pair (H_1, H_2) is the sum of depths w.r.t. H_1 and H_2, respectively.

With this terminology in place, we can establish the following connection between cut-points and the total cost of a pair of Hamiltonian paths. Due to space limitations, the proof is deferred to the full version of the paper.

Lemma 1. *Consider a pair of disjoint Hamiltonian paths on the uniform metric graph R_n^1. If they have no cut-point, then the sum of their costs is at least $\frac{16}{5}(n-1)$ for $n \geq 6$.*

In addition, we have verified by computer the following claim for small uniform metric graphs in $\mathbb{R}^1$.[2] Part (i) of the observation is also stated by Ageev *et al.* [1].

Observation 1. *Consider the uniform metric graph R_n^1 for some $n \leq 8$. Then (i) there is no pair of edge-disjoint Hamiltonian paths when $n \leq 5$, and (ii) there is no pair of edge-disjoint Hamiltonian paths with total cost less than $\frac{16(n-1)}{5}$ when $n \in \{6, 7, 8\}$.*

With Lemma 1 and Observation 1, we are ready to prove the section's main result.

Lemma 2. *There is no pair of disjoint Hamiltonian paths in the uniform metric graph R_n^1 such that the sum of their costs is less than $\frac{16}{5}(n-1)$ for $n \geq 6$.*

Proof. By induction on n. For the base case, $n = 6$, Observation 1 shows that the statement is true. Now, assume that the statement holds for any integer k such that $6 < k < n$, with $n > 6$. We will prove that, under this assumption, the statement is also true for the input graph R_n^1 of size n. For the sake of contradiction, suppose that there is a pair (H_1, H_2) of edge-disjoint Hamiltonian paths of total cost less than $16(n-1)/5$. By Lemma 1, the pair (H_1, H_2) has a cutpoint v. We show that this gives a contradiction.

Let P_n denote the path graph induced by R_n^1. If H_1 and H_2 have a cutpoint v, we can partition P_n into the two path graphs $P_{n_1} = (V_1, E_1)$ and $P_{n_2} = (V_2, E_2)$—on n_1 and n_2 vertices, respectively—such that $s \in P_{n_1}$, $t \in P_{n_2}$, and $V_1 \cap V_2 = \{v\}$. Note that $n = n_1 + n_2 - 1$. We define the total cost of a pair of paths (H_1, H_2) as $C_{\text{tot}}(H_1, H_2) := c(H_1) + c(H_2)$, i.e., the sum of the costs of the paths. There are two cases to consider: Either $\min(n_1, n_2) < 6$ or $\min(n_1, n_2) \geq 6$.

- Let $\min(n_1, n_2) < 6$. W.l.o.g., assume that $n_1 < 6$. In other words, we assume the shorter subpath is $H_1[V_1]$. By part (i) of Observation 1, we know that no pair of disjoint Hamiltonian paths exists. This means the subpath

---

[1] Not to be confused with a *cut-vertex*, which, in standard graph theory terminology, refers to a vertex whose removal results in a disconnected graph.

[2] Code available at https://github.com/andoresu47/Disjoint-Tours-and-the-PoDhttps://github.com/andoresu47/Disjoint-Tours-and-the-PoD.

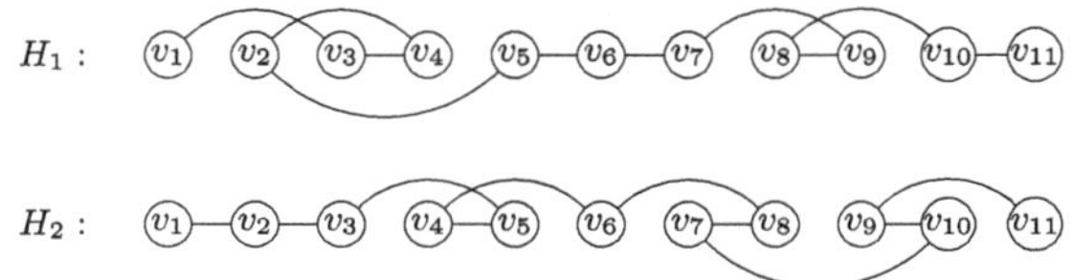

Fig. 1. An optimal pair of edge-disjoint Hamiltonian paths with $cost(H_1, H_2) = 16$ for $n = 11$.

$H_2[V_1]$ cannot be edge-disjoint with $H_1[V_1]$. But we are given that $H_1 = H_1[V_1] \cup H_1[V_2]$ and $H_2 = H_2[V_1] \cup H_2[V_2]$ are edge-disjoint. Hence, we reach a contradiction.

- Let $\min(n_1, n_2) \geq 6$. The inductive hypothesis implies that $C_{\text{tot}}(H_1[V_1], H_2[V_1]) \geq 16(n_1 - 1)/5$ and $C_{\text{tot}}(H_1[V_2], H_2[V_2]) \geq 16(n_2 - 1)/5$. Then $C_{\text{tot}}(H_1, H_2) = C_{\text{tot}}(H_1[V_1], H_2[V_1]) + C_{\text{tot}}(H_1[V_2], H_2[V_2]) \geq 16(n_1 + n_2 - 2)/5 = 16(n - 1)/5$. But this contradicts our initial assumption that the total cost of H_1 and H_2 was strictly less than $16(n - 1)/5$.

Both cases result in a contradiction; therefore, the statement is true for $n > 6$ and the theorem is proven. $\qquad\square$

Notice that Lemma 2 implies that in any pair of Hamiltonian paths on the uniform metric graph R_n^1, the longest of the two paths has a cost of at least $\frac{8}{5}(n - 1)$. This can be restated as the following corollary.

Corollary 1. $PoD(\text{DISJOINT-PATH TSP\_2})) \geq \frac{8}{5}$ *for uniform metric graphs in* $\mathbb{R}^1$ *when* $s = 1$ *and* $t = n$.

3.2   An Algorithm for DISJOINT-PATH TSP$_2$

We now present an algorithm for DISJOINT-PATH TSP$_2$ on the uniform metric graph R_n^1, which achieves a loss ratio of $\frac{13}{7}$ and an asymptotic loss ratio of $\frac{8}{5}$. The main idea is to *concatenate* small optimal building blocks of disjoint paths. The algorithm is detailed below as Algorithm 1. A more formal description can be found in the full version of the paper.

When the number of segments, $n - 1$, is a multiple of 10, a pair of edge-disjoint Hamiltonian paths of optimal bottleneck cost $\frac{8}{5}(n-1)$ can be constructed by concatenating k copies of an optimal solution for $n = 11$. One such pair is shown in Fig. 1. For general n, we extend this idea by attaching suitable smaller optimal solutions for $6 \leq n \leq 10$ (found by computer-aided exhaustive search) or by slightly augmenting the $n = 11$ construction. The illustrations of optimal solutions for $n \in \{6, 7, 8, 9, 10\}$, as well as the augmented constructions for $n \in \{12, 13, 14, 15\}$, are deferred to the full version.

Algorithm 1. Algorithm `Paths`

Input: Uniform metric graph R_n^1
Output: A pair of disjoint Hamiltonian paths.

1: **if** $n \leq 11$ **then**
2:     **return** an optimal pair (precomputed).
3: **else**
4:     Let $(n - 1) = 10k + \ell$.
5:     **if** $\ell = 0$ **then**
6:         **return** k concatenated copies of the $n = 11$ solution of Figure 1.
7:     **else if** $5 \leq \ell \leq 9$ **then**
8:         Concatenate k copies of the $n = 11$ solution with a precomputed
9:         (optimal) solution for $n = \ell + 1$.
10:     **else** $\triangleright \ell \in \{1, 2, 3, 4\}$
11:         concatenate $k - 1$ copies of the $n = 11$ solution with a precomputed solution
12:         for $n = 11 + \ell$.

Analysis. Note that each time the algorithm concatenates pairs of disjoint paths, a cut-point is created, namely, the vertex joining the concatenated paths. Consider the pair of paths returned by the algorithm. By construction, the pair of subpaths induced by the vertices between cutpoints are edge-disjoint. Hence, the paths in the returned pair are also edge-disjoint.

Lemma 3. *Algorithm 1 generates a pair of edge-disjoint Hamiltonian paths.*

We show that its loss ratio is $\frac{13}{7}$ and its asymptotic loss ratio is $\frac{8}{5}$. Due to space constraints, we defer the proofs to the full version, but we sketch the key ideas here. For small instances, the worst case occurs at $n = 8$, where the ratio $\frac{13}{7}$ is attained. For larger n, the ratio decreases, converging to $\frac{8}{5}$ as $n \to \infty$.

Lemma 4. *Algorithm 1 has loss ratios* $LR_{Paths} = \frac{13}{7}$ *and* $LR_{Paths}^{\infty} = \frac{8}{5}$.

The loss ratio of Algorithm 1 provides an upper bound for $PoD(\text{DISJOINT-PATH TSP}_2)$ with $s = 1$ and $t = n$. The proof of Lemma 4 actually shows this bound is tight, because the ratio $\frac{\min_{S \in \mathcal{S}(\mathcal{I})}\{\text{cost}(S)\}}{OPT(I)}$ achieved by the optimal solution at $n = 8$ is already an upper bound on the loss ratio for all other values of n.

4   The Price of Diversity of **DISJOINT-TSP$_2$** in $\mathcal{S}_n^1$

We now extend the results and concepts of the previous section to DISJOINT-TSP$_2$ in $\mathcal{S}_n^1$. We establish the following.

Theorem 2. $PoD(\text{DISJOINT-PATH TSP\_2}) = \frac{13}{7}$, *and* $PoD^{\infty}(\text{DISJOINT-PATH TSP\_2}) = \frac{8}{5}$ *for the class of uniform metric graphs in* $\mathbb{S}^1$.

We prove Theorem 2 as follows. In Sect. 4.1, we extend the lower bound established in Corollary 1 to pairs of edge-disjoint tours. Subsequently, in Sect. 4.2, we present an algorithm that builds on the results of Sect. 3, generalizing the approach from paths to tours while maintaining the loss ratio. Together, these results provide the proof of Theorem 2. Recall that a segment is an edge of G that connects consecutive vertices along $\mathbb{S}^1$. Throughout this section, we use C_n to denote the cycle graph formed by the vertex set V and the segments of G. We refer to C_n as the cycle graph induced by G.

4.1 A Lower Bound

We now extend the lower bound results of Sect. 3.1—which focused on Hamiltonian (s, t)-paths, with $s = 1$ and $t = n$—to tours in uniform metric graphs. To this end, we adopt the notions of *covered* vertices and segments introduced in Sect. 3.1, modifying them by replacing the Hamiltonian path with a tour in each definition. As before, we denote the depth of a segment $s \in C_n$ w.r.t. a tour T by $\mu_T(s)$. Note that the cost of T is equal to the sum of the depths of all segments in C_n w.r.t. T. We begin with the following crucial lemma (see proof in the full version).

Lemma 5. *In any tour of a metric graph in $\mathbb{S}^1$, all segments have odd depth or all segments have even depth.*

Consider the case when the input graph G is the uniform metric graph S_n^1. Lemma 5 implies that any tour that covers some segment of S_n^1 an even number of times has a total cost that is roughly twice the optimal value. (Note that we treat 0 as an even number.)

Corollary 2. *Any pair of disjoint tours T_1 and T_2 in S_n^1 such that one of them has exclusively even depth segments satisfies $\max(c(T_1), c(T_2)) \geq 2 \cdot (n - 1)$.*

Hence, only pairs of tours in S_n^1 with exclusively odd depth segments may have a bottleneck cost smaller than $2 \cdot (n-1)$. We call such tours *odd-depth* tours. Similarly, *even-depth* tours are tours with exclusively even-depth segments. In the rest of the section, we restrict our discussion to pairs of *odd-depth* tours.

We now extend the notions of *cut-point* and *ℓ-piece* from Hamiltonian paths to tours. A pair of edge-disjoint tours (T_1, T_2) is said to have a *cut-point* v if v is not covered by T_1 nor by T_2. Any sequence of ℓ consecutive segments in C_n is called an *ℓ-piece*. The depth of a piece (w.r.t. a tour T) is the sum of the depths of its constituting segments. The *total depth* of an ℓ-piece w.r.t. the pair (T_1, T_2) is the sum of depths w.r.t. T_1 and T_2, respectively.

With these definitions, we establish the following analogue of Lemma 1 for odd-depth tours (see proof in the full version).

Lemma 6. *Consider a pair of odd-depth disjoint tours on the uniform metric graph S_n^1. If they have no cut-point, then the sum of their costs is at least $\frac{16}{5}n$ for $n \geq 5$.*

With Lemma 6, we have the necessary ingredients to extend the result of Lemma 2 from pairs of Hamiltonian paths to pairs of odd-depth tours.

Lemma 7. *There is no pair of disjoint odd-depth tours in the uniform metric graph S_n^1 such that the sum of their costs is less than $\frac{16}{5}n$ for $n \geq 5$.*

Proof. Consider an arbitrary pair of odd-depth tours in S_n^1. By contrapositive of Lemma 6, if the sum of their costs is less than $16n/5$, then either they are not disjoint, or they have a cut-point, or both. We are interested in low-cost pairs of disjoint tours, so we may only consider the case of the tours having a cutpoint. If such a pair (T_1, T_2) of disjoint tours existed, then we could construct two disjoint Hamiltonian (s,t)-paths H_1 and H_2 of total cost less than $16n/5$ as follows. Let v denote the cutpoint of T_1 and T_2. We construct H_1 by creating two copies, s and t, of vertex v and replacing the edges $(u_1, v), (v, w_1) \in T_1$ with (u_1, t) and (s, w_1), respectively. We proceed similarly for path H_2 and edges $(u_2, v), (v, w_2) \in T_2$. The (s, t)-paths H_1 and H_2 just created are edge-disjoint since the edges we replaced from T_1 and T_2 were already disjoint. Further, the new edges have the same length as the ones replaced. Hence, the total cost of the paths we created is the same as the total cost of the original tours: less than $16n/5$. This is a pair of Hamiltonian paths on the uniform metric graph R_n^1 whose total cost is $16n/5$. But, by Theorem 2, we know that no such pair of Hamiltonian paths exists. We arrive at a contradiction. Hence, the theorem is proved. $\qquad\square$

Putting Corollary 2 and Lemma 7 together, we get the following result for pairs of disjoint tours, regardless of their depth.

Corollary 3. *There is no pair of disjoint tours in the uniform metric graph S_n^1 such that the cost of the longer tour is less than $\frac{8}{5}n$ for $n \geq 5$.*

Proof. Let (T_1, T_2) be an arbitrary pair of disjoint tours in S_n^1. If at least one of T_1 or T_2 is an even-depth tour, then by Corollary 2, we have $cost(T_1, T_2) \geq 2(n-1) \geq \frac{8}{5}n$ for $n \geq 5$. If both T_1 and T_2 are odd-depth tours, Lemma 7 implies $cost(T_1, T_2) \geq \frac{16}{10}n = \frac{8}{5}n$ for for $n \geq 5$ as well. $\qquad\square$

Finally, we get the following as an immediate consequence of Corollary 3.

Corollary 4. $PoD(\textsc{Disjoint-Path TSP\_2})) \geq \frac{8}{5}$ *for the class of uniform metric graphs in* $\mathbb{S}^1$.

4.2   An Algorithm for DISJOINT-TSP$_2$

We now describe an algorithm, referred to as **Tours**, that generates two disjoint tours such that the cost of the longer tour is always within a factor $\frac{13}{7}$ of the cost of an optimal tour, and asymptotically within a factor $\frac{8}{5}$. A more detailed description of the algorithm is deferred to the full version.

The algorithm is an extension of Algorithm 1 for DISJOINT-PATH TSP$_2$. It works by first using the earlier algorithm as a subroutine to generate two edge-disjoint Hamiltonian paths, each of length $n \Rightarrow n + 1$. Each path is then transformed into a tour by contracting its first and last vertices into a single new vertex. As a result, the cycles each contain exactly n vertices and n edges. Note that the newly created vertex is a cutpoint, ensuring that the cycles remain edge-disjoint, given that the original paths were edge-disjoint. Furthermore, the loss ratio of the algorithm inherits the bounds established for Algorithm 1.

Theorem 3. *Algorithm* **Tours** *has loss ratio* $LR_{Tours} = \frac{13}{7}$ *and asymptotic loss ratio* $LR^{\infty}_{Tours} = \frac{8}{5}$.

Proof. Replace n with $n+1$ in the proof of Lemma 4.

5 The PoD of DISJOINT-PATH TSP$_2$ and DISJOINT-TSP$_2$ in $\mathcal{M}$

We conclude our investigation of the PoD by examining DISJOINT-PATH TSP$_2$ and DISJOINT-TSP$_2$ in general metric graphs. First, in Sect. 5.1, we derive lower bounds of $3-\varepsilon$ and $2-\varepsilon$, for any $\epsilon > 0$, on the price of diversity for these problems, respectively. Next, in Sect. 5.2, we present two simple algorithms whose loss ratios essentially match these lower bounds, thus establishing the price of diversity of the problems.

Theorem 4. $PoD(\text{DISJOINT-PATH } TSP\_2)) = 3$ *and* $PoD(\text{DISJOINT-}TSP\_2))$ $= 2$ *in* $\mathcal{M}$.

5.1 Lower Bounds

In this section, we present lower bounds for the PoD of both DISJOINT-PATH TSP$_2$ (Sect. 5.1) and DISJOINT-TSP$_2$ (Sect. 5.1) for general metric graphs.

DISJOINT-PATH TSP$_2$. Consider a pair of edge-disjoint Hamiltonian (s, t)-paths H_1 and H_2 in a (not necessarily uniform) metric graph $G = (V, E)$ in $\mathbb{R}^1$, where s and t are the first and last vertices of G along the line $\mathbb{R}^1$. Like in Sect. 3, we use P_n to denote the path graph induced by G and assume that its vertices are labeled from left to right $(v_1, v_2, \ldots, v_n)$, with $s = v_1$ and $t = v_n$. We make the following claim (see proof in the full version).

Claim 1. The depth of the segment (v_2, v_3) (resp. (v_{n-2}, v_{n-1})) w.r.t. at least one of H_1 and H_2 is at least 3.

We now use this claim to derive a lower bound for the PoD of DISJOINT-PATH TSP$_2$ in a general metric. We show that there is a metric graph on n points in $\mathbb{R}^1$ where the longest path of an arbitrary pair of disjoint Hamiltonian paths has a cost of almost three times the optimal value.

Lemma 8. *For any* $\varepsilon > 0$, *the PoD of* DISJOINT-PATH TSP$_2$ *in a general metric is at least* $3 - \varepsilon$.

Proof. Consider the metric graph $G = (V, E)$ in $\mathbb{R}^1$ of size $n > 5$, where $V = \{v_1, \ldots, v_n\}$ and edge weights are defined as follows: $w(v_i, v_{i+1}) = 1$ for all $i \neq 2$, and $w(v_i, v_{i+1}) = W$ for $i = 2$. Now, let H_1 and H_2 be a pair of edge-disjoint Hamiltonian (s, t)-paths in G, with $s = v_1$ and $t = v_n$. By Claim 1, the depth of segment (v_2, v_3) is at least 3 in one of the two paths, which we assume w.l.o.g. to be H_1. Then, $c(H_1) \geq 3 \cdot W + (n - 2)$, while $OPT = W + (n - 2)$. Hence

$$PoD(\text{Disjoint-Path TSP}_2) \geq \frac{3W + (n - 2)}{W + (n - 2)} = 3 - \frac{2n - 4}{W + n - 2}.$$

Thus, for any $\varepsilon > 0$, we can obtain a PoD of $3 - \varepsilon$ by setting $W := \frac{(2-\varepsilon)(n-2)}{\varepsilon}$. $\square$

Corollary 5. *There is no algorithm for* Disjoint-Path TSP$_2$ *in metric graphs with a loss ratio strictly less than 3.*

Disjoint-TSP$_2$. We now turn our attention to Disjoint-TSP$_2$ in general metric graphs. We show that there is a metric graph in $\mathbb{S}^1$ where the longest tour of an arbitrary pair of disjoint tours has a cost that is (almost) twice the optimal value. The proof is long and technical, so we provide only a sketch here; the full argument appears in the full version of the paper.

Lemma 9. *For any $\varepsilon > 0$, the PoD of* Disjoint-Path TSP$_2$ *in a general metric is at least $2 - \varepsilon$.*

Proof sketch. The construction is similar in spirit to Lemma 8 for the path case. We consider a cycle graph on n vertices where two specific edges are made very expensive, of weight W, while all other edges have unit weight. Any single optimal tour has cost roughly $2W$, but when two edge-disjoint tours are required, one of them is forced to traverse both expensive edges, incurring a cost of about $4W$. This yields a ratio approaching 2 as W grows.

The main technical challenge is to show that, regardless of whether the tours are odd- or even-depth, one of the two tours must indeed pick up both heavy edges. This is established through a series of structural claims about depth patterns of consecutive segments. $\square$

Corollary 6. *There is no algorithm for* Disjoint-Path TSP$_2$ *in $\mathcal{M}$ with a loss ratio strictly less than 2.*

5.2   Algorithms

We now present two simple algorithms, one for Disjoint-Path TSP$_2$ and another for Disjoint-TSP$_2$, that achieve loss ratios of 3 and 2, respectively, in any weighted graph.

Algorithm for Disjoint-Path TSP$_2$. Let us first consider Disjoint-Path TSP$_2$, where the goal is to find a pair of edge-disjoint Hamiltonian (s, t)-paths in an (arbitrarily) weighted graph G. Let $H_{\text{opt}} = (v_1, v_2, \ldots, v_n)$ be an optimal single Hamiltonian (s, t)-path in G. Consider now a metric graph G' in $\mathbb{R}^1$ with the same vertex set as G, where the vertices are placed in the order given by H_{opt} and spaced so that the distance between v_i and v_{i+1} in G' is equal to the weight of the segment (v_i, v_{i+1}) in H_{opt}. Using Algorithm 1, we can obtain a pair of edge-disjoint Hamiltonian (s, t)-paths in G'.

What happens if we use the same pair of solutions but with the edge weight of the original graph G? We claim that the loss ratio of this algorithm is 3. Indeed, this follows from the fact that the solutions generated by Algorithm 1 never cover any segment more than three times. Since the segments are defined by the optimal solution H_{opt}, each of our edge-disjoint paths will have a cost at most three times the cost of H_{opt}.

Lemma 10. *The described algorithm has loss ratio 3.*

An algorithm for DISJOINT-PATH TSP$_2$. We now present a simple algorithm for DISJOINT-TSP$_2$ that achieves a loss ratio of 2. This algorithm is listed below as Algorithm 2. We omit a detailed description as the provided pseudocode is self-explanatory. We defer the proof of Lemma 11 below to the full version of the paper.

Algorithm 2. Naïve 2-loss ratio algorithm

Input: Metric graph G on $n > 4$ vertices.
Output: A pair (T_1, T_2) of tours of minimum cost $\max(c(T_1), c(T_2))$.

1: Compute an optimal cost tour $H_{\mathrm{opt}} = (v_1, v_2, \ldots, v_n, v_1)$ in G.
2: **if** n is odd **then**
3:     $T_1 \leftarrow H_{\mathrm{opt}}$
4:     $T_2 \leftarrow (v_1, v_3, \ldots, v_{n-2}, v_n, v_2, v_4, \ldots, v_{n-1}, v_1)$
5: **if** n is even **then**
6:     $T_1 \leftarrow (v_1, v_n, v_2, v_3, \ldots, v_{n-1}, v_1)$
7:     $T_2 \leftarrow (v_1, v_3, \ldots, v_{n-1}, v_n, v_{n-2}, \ldots, v_2, v_1)$
8: Return the pair of tours (T_1, T_2)

Lemma 11. *Algorithm 2 has a loss ratio of 2.*

The loss ratios achieved by the algorithms described in this section, together with the lower bounds of Sect. 6, establish the PoD of DISJOINT-PATH TSP$_2$ and DISJOINT-TSP$_2$ in general metric graphs.

6   Concluding Remarks

We introduced two problems DISJOINT-PATH TSP$_2$ and DISJOINT-TSP$_2$, and studied them on metric graphs. For the classes of uniform metric graphs in $\mathbb{R}^1$

and $\mathbb{S}^1$, respectively, we presented $\frac{13}{7}$-loss ratio algorithms for both problems and proved a lower bound of $8/5$ for their PoD. Both of our algorithms are asymptotically tight, as they achieve an asymptotic loss ratio of $8/5$ when $n \to \infty$. For the class of general metric graphs, for any $\varepsilon > 0$, we show lower bounds of 3 and 2 for the PoD of DISJOINT-PATH TSP$_2$ and DISJOINT-TSP$_2$, respectively. We then provide two algorithms: Algorithm 1 for DISJOINT-PATH TSP$_2$ and Algorithm 2 for DISJOINT-TSP$_2$, whose loss ratios match these lower bounds, respectively. Hence, the PoD of DISJOINT-PATH TSP$_2$ and DISJOINT-TSP$_2$ in a general metric are 3 and 2, respectively.

In the case of DISJOINT-TSP$_2$, the naive algorithm relies on the computation of an optimal TSP solution, making it impractical in terms of running time. The algorithm can be made efficient (running in polynomial time); however, if one is willing to settle for a starting tour that is only approximate with respect to the optimal tour. For instance, starting the algorithm with a sub-optimal tour, namely one obtained via the Christofides $3/2$-approximation algorithm, results in a naive solution to DISJOINT-TSP$_2$ with approximation ratio 3. The question of whether one can design a polynomial time algorithm with loss-ratio $2 < \alpha < 3$ remains open. We have only presented results for DISJOINT-PATH TSP$_k$ and DISJOINT-TSP$_k$ with $k = 2$. Hence, it is also natural to ask what the price of diversity is for these problems when considering $k > 2$.

Acknowledgements. This research was supported by the European Union's Horizon 2020 research and innovation programme under the Marie Skłodowska-Curie grant agreement no. 945045, and by the NWO Gravitation project NETWORKS under grant no. 024.002.003. We are grateful to the reviewers for their valuable comments and suggestions.

References

1. Ageev, A.A., Pyatkin, A.V.: A 2-approximation algorithm for the metric 2-peripatetic salesman problem. In: International Workshop on Approximation and Online Algorithms, pp. 103–115. Springer (2007)
2. Ahanor, I., Medal, H., Trapp, A.C.: Diversitree: a new method to efficiently compute diverse sets of near-optimal solutions to mixed-integer optimization problems. INFORMS J. Comput. **36**(1), 61–77 (2024)
3. Ahmed, F., Dickerson, J.P., Fuge, M.: Diverse weighted bipartite b-matching. In: Proceedings of the Twenty-Sixth International Joint Conference on Artificial Intelligence, IJCAI-17, pp. 35–41 (2017)
4. Alspach, B.: The wonderful Walecki construction. Bull. Inst. Combin. Appl. **52**(52), 7–20 (2008)
5. Baste, J., Jaffke, L., Masařík, T., Philip, G., Rote, G.: FPT algorithms for diverse collections of hitting sets. Algorithms **12**(12), 254 (2019)
6. Benabbou, N., Chakraborty, M., Ho, X.V., Sliwinski, J., Zick, Y.: The price of quota-based diversity in assignment problems. ACM Trans. Econ. Comput. (TEAC) **8**(3), 1–32 (2020)

7. Bereg, S., et al.: The two-squirrel problem and its relatives. In: Japanese Conference on Discrete and Computational Geometry, Graphs, and Games. pp. 104–120. Springer (2022)

8. de Berg, M., López Martínez, A., Spieksma, F.: Finding diverse minimum ST cuts. In: 34th International Symposium on Algorithms and Computation (ISAAC 2023), pp. 24–1. Schloss Dagstuhl–Leibniz-Zentrum für Informatik (2023)

9. Boehmer, N., Niedermeier, R.: Broadening the research agenda for computational social choice: Multiple preference profiles and multiple solutions. In: Proceedings of the 20th International Conference on Autonomous Agents and MultiAgent Systems, pp. 1–5 (2021)

10. Bredereck, R., Faliszewski, P., Igarashi, A., Lackner, M., Skowron, P.: Multiwinner elections with diversity constraints. In: Proceedings of the AAAI Conference on Artificial Intelligence. vol. 32 (2018)

11. Danna, E., Woodruff, D.L.: How to select a small set of diverse solutions to mixed integer programming problems. Oper. Res. Lett. **37**(4), 255–260 (2009)

12. Do, A.V., Guo, M., Neumann, A., Neumann, F.: Niching-based evolutionary diversity optimization for the traveling salesperson problem. In: Proceedings of the Genetic and Evolutionary Computation Conference, pp. 684–693 (2022)

13. Eppstein, D.: k-best enumeration. Encycl. Algorithms, 1003–1006 (2016)

14. Fomin, F.V., Golovach, P.A., Jaffke, L., Philip, G., Sagunov, D.: Diverse pairs of matchings. Algorithmica **86**(6), 2026–2040 (2024)

15. Gabor, T., Belzner, L., Phan, T., Schmid, K.: Preparing for the unexpected: diversity improves planning resilience in evolutionary algorithms. In: 2018 IEEE International Conference on Autonomic Computing (ICAC), pp. 131–140. IEEE (2018)

16. Glover, F., Løkketangen, A., Woodruff, D.L.: Scatter search to generate diverse mip solutions. In: Computing Tools for Modeling, Optimization and Simulation: Interfaces in Computer Science and Operations Research, pp. 299–317. Springer (2000)

17. Gupta, S., Moondra, J., Singh, M.: Balancing notions of equity: trade-offs between fair portfolio sizes and achievable guarantees. In: Proceedings of the 2025 Annual ACM-SIAM Symposium on Discrete Algorithms (SODA), pp. 1136–1165. SIAM (2025)

18. Hamacher, H.W., Queyranne, M.: K best solutions to combinatorial optimization problems. Ann. Oper. Res. **4**(1), 123–143 (1985)

19. Hanaka, T., Kobayashi, Y., Kurita, K., Otachi, Y.: Finding diverse trees, paths, and more. In: Proceedings of the AAAI Conference on Artificial Intelligence. vol. 35, pp. 3778–3786 (2021)

20. Hebrard, E., Hnich, B., O'Sullivan, B., Walsh, T.: Finding diverse and similar solutions in constraint programming. In: AAAI. vol. 5, pp. 372–377 (2005)

21. Krarup, J.: The peripatetic salesman and some related unsolved problems. In: Combinatorial Programming: Methods and Applications: Proceedings of the NATO Advanced Study Institute held at the Palais des Congrès, Versailles, France, 2–13 September, 1974, pp. 173–178. Springer (1995)

22. Lü, H., Wu, T.: Edge-disjoint hamiltonian cycles of balanced hypercubes. Inf. Process. Lett. **144**, 25–30 (2019)

23. Misra, N., Mittal, H., Rai, A.: On the parameterized complexity of diverse sat. In: 35th International Symposium on Algorithms and Computation (ISAAC 2024), pp. 50–1. Schloss Dagstuhl–Leibniz-Zentrum für Informatik (2024)

24. Paris, Q.: Dealing with Multiple Optimal Solutions, pp. 295–307. Palgrave Macmillan US, New York (2016)

25. Petit, T., Trapp, A.C.: Finding diverse solutions of high quality to constraint optimization problems. In: IJCAI. International Joint Conference on Artificial Intelligence (2015)
26. Rowley, R., Bose, B.: Edge-disjoint hamiltonian cycles in de bruijn networks. In: Proceeding 6th Distributed Memory Computing Conference, pp. 707–709 (1991)
27. Shida, Y., Punzi, G., Kobayashi, Y., Uno, T., Arimura, H.: Finding diverse strings and longest common subsequences in a graph. In: 35th Annual Symposium on Combinatorial Pattern Matching (CPM 2024), pp. 27–1. Schloss Dagstuhl–Leibniz-Zentrum für Informatik (2024)
28. Thejaswi, S., Ordozgoiti, B., Gionis, A.: Diversity-aware k-median: clustering with fair center representation. In: Machine Learning and Knowledge Discovery in Databases. Research Track: European Conference, ECML PKDD 2021, Bilbao, Spain, September 13–17, 2021, Proceedings, Part II 21, pp. 765–780. Springer (2021)

Shortcutting the Diameter of a Polygon

Taekang Eom[1] ⓘ, Taehoon Ahn[2] ⓘ, Minju Song[3], and Hee-Kap Ahn[1,3](✉) ⓘ

[1] Department of Computer Science and Engineering, Pohang University of Science and Technology, Pohang, Republic of Korea
{tkeom0114,heekap}@postech.ac.kr
[2] Department of Computer Science, Sookmyung Women's University, Seoul, Republic of Korea
taehoon@sookmyung.ac.kr
[3] Graduate School of Artificial Intelligence, Pohang University of Science and Technology, Pohang, Republic of Korea
mjsong@postech.ac.kr

Abstract. We study the problem of minimizing the diameter of a polygon in the plane by attaching a segment to the polygon. Both endpoints of the segment must lie on the boundary of the polygon, while its relative interior must be disjoint from the polygon. Such a segment is considered an addition to the polygon, and thus it (or part of it) can be used in geodesic paths connecting pairs of points in the union of the polygon and itself. This problem can be considered a geometric analogue to the problem of augmenting graphs by inserting edges to minimize the diameter. We present an $O(n)$-time algorithm to determine whether the diameter of an x-monotone polygon with n vertices can be reduced by attaching a horizontal segment to the polygon under the L_1 metric. We also present an $O(n \log n)$-time algorithm using $O(n)$ space for finding a horizontal segment that minimizes the diameter.

Keywords: Shortcuts · Diameter · Polygons

1 Introduction

We study the problem of minimizing the diameter of a polygon P in the plane by attaching a segment to the polygon. The geodesic distance of two points in a figure X is the length of a shortest path that is contained in X and that connects the points. The diameter of X is the largest one among all geodesic distances defined over pairs of points in X. Suppose that we attach (or augment) a segment σ to P such that both endpoints of σ lie on the boundary of P and the relative interior of σ, which is σ without its endpoints, is disjoint from P. It is considered an addition to the polygon. Then σ (or part of it) can be used as a shortcut to geodesic paths connecting pairs of points in $P \cup \sigma$. We call such a segment a *shortcut* for P. Thus, our problem is to find a shortcut σ for P minimizing the diameter of $P \cup \sigma$. This problem can be considered as a geometric analogue to the problem of augmenting graphs by inserting edges to minimize the diameter.

© The Author(s), under exclusive license to Springer Nature Singapore Pte Ltd. 2026
E. Di Giacomo and D. Mondal (Eds.): WALCOM 2026, LNCS 16444, pp. 418–431, 2026.
https://doi.org/10.1007/978-981-95-7127-7_28

One motivation of the problem is from the *transport facility location* problem [7,8,14] that asks for locating a facility to a region to minimize the travel time or distance of customers. In our problem, the facility is a shortcut (a line segment) connecting two boundary points of the region such that customers can travel along it to reduce their travel time or distance.

In this paper, we consider an x-monotone polygon P in the plane as the domain and place a *horizontal* shortcut σ to minimize the diameter of P with σ under the L_1 metric. A polygon is x-monotone if for any line ℓ orthogonal to the x-axis, the intersection of the polygon with ℓ is connected. Recall that both endpoints of a shortcut σ lie on the boundary of P and the relative interior of σ is disjoint from P. Thus, there is no horizontal shortcut if P is convex. In the remainder of the paper, we assume that P is not convex.

Since a shortcut σ for P is considered an addition to P, the diameter of P with σ is the diameter of $P \cup \sigma$, that is, the largest geodesic distance between any pair of points in $P \cup \sigma$. We call a pair of points in $P \cup \sigma$ that realizes the diameter of $P \cup \sigma$ a *diametral pair* of $P \cup \sigma$. Note that the diameter may increase after augmenting a segment to a polygon. See Fig. 1(a). We say that a shortcut σ is *useful* for P if the diameter of $P \cup \sigma$ is strictly smaller than the diameter of P. A useful shortcut σ is an *optimal* shortcut for P if the diameter of $P \cup \sigma$ is the minimum diameter of $P \cup \sigma'$ for all useful shortcuts σ' for P. See Fig. 1(b) for an illustration of such optimal shortcuts for an x-monotone polygon.

Interestingly, no optimal shortcut may exist for some x-monotone polygons even though they have useful shortcuts. See Fig. 1(c) for an illustration. In the figure, the diameter of $P \cup \sigma$ is the length of a geodesic path connecting α and β using σ. Since the diameter of $P \cup \sigma$ is smaller than the diameter of P, σ is a useful shortcut for P. Observe that the diameter of $P \cup \sigma$ gets smaller as the shortcut σ goes closer to σ'. Observe also that the diameter of $P \cup \sigma'$ is strictly smaller than the diameter of $P \cup \sigma$ for any useful shortcut $\sigma \neq \sigma'$. But σ' is not a 'shortcut' for P as its relative interior intersects P. Thus, P has no optimal shortcut while P has useful shortcuts. In this case, we call such a segment σ' an *infimum* shortcut for P.

Thus, our problem can be rephrased as follows: Given an x-monotone polygon P, compute a horizontal optimal shortcut for P under the L_1 metric. If there is a useful shortcut but no optimal shortcut exists for P, compute an infimum shortcut for P.

Related Work. In the transport facility location problem, customers may choose whether they use the transport facility or not in minimizing their maximum travel time. Most of previous studies on the transport facility location problem assume that the customers are given as a set of points in the plane. Cardinal and Langerman [4] considered three variants of the transport facility location problem, including one for placing a highway (line segment) of infinite speed to minimize the maximum travel time between any two input points. Later, Ahn et al. [1] studied the problem of placing a highway of a positive constant speed. In contrast, we assume that the customers are everywhere in the input polygon.

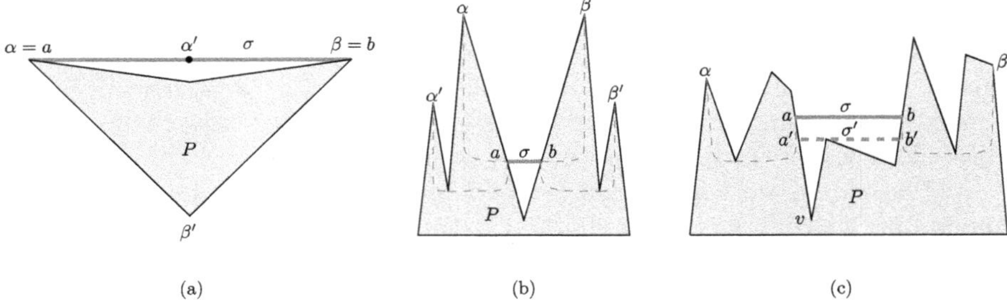

(a)   (b)   (c)

Fig. 1. We use pq to denote the line segment connecting two points p and q in the plane, and $|pq|$ the length of pq. If $p, q \in P$, $d(p, q)$ denotes the length of a geodesic path from p to q in P. (a) (α, β) is the diametral pair of P, and (α', β') is the diametral pair of $P \cup \sigma$ with $\sigma = ab$. We have $d(\alpha, \beta) < |\alpha'a| + d(a, \beta') = |\alpha'b| + d(b, \beta')$. (b) (α, β) is the diametral pair of P. Both (α, β) and (α', β') are diametral pairs of $P \cup \sigma$ with $\sigma = ab$ such that $d(\alpha, a) + |ab| + d(b, \beta) = d(\alpha', a) + |ab| + d(b, \beta')$. σ is an optimal shortcut for P. (c) (α, β) with distance $d(\alpha, \beta) = d(\alpha, v) + d(v, \beta)$ is the diametral pair of P. The distance $d(\alpha, a) + |ab| + d(b, \beta)$ between α and β in $P \cup \sigma$ with $\sigma = ab$ decreases linearly as σ moves towards σ', and the diameter of $P \cup \sigma$ converges to $d(\alpha, a') + |a'b'| + d(b', \beta)$.

Our problem can be considered a geometric analogue of augmenting graphs by inserting edges to minimize the diameter. There has been a fair amount of work on minimizing the diameter or radius of graphs embedded in the Euclidean plane by inserting *segments* (edges) connecting two points in graphs [9]. Große et al. [11] gave an $O(n \log^3 n)$-time algorithm for inserting an edge connecting two vertices of an input tree that minimizes the diameter of the resulting graph. De Carufel et al. [6] gave an $O(n \log n)$-time algorithm for inserting a segment connecting any two points of edges of an input tree that minimizes the diameter of the resulting graph. Also, there have been a fair amount of work on inserting a segment to a planar graph embedded in the Euclidean space to minimize the diameter of the resulting graph [3,10,17]. See also [12,13] for inserting a segment that minimizes the radius of the resulting graph.

In the geometric domain, most of previous studies focus on inserting a *bridge*, a line segment, connecting two disjoint objects that minimizes the diameter of their union. Wang [16] gave an algorithm for inserting an optimal bridge connecting two disjoint rectilinear simple polygons of total n vertices under the L_1 metric in $O(n)$ time. Tan [15] gave an algorithm for inserting an optimal bridge connecting two disjoint simple polygons of total n vertices in $O(n \log^3 n)$ time. Later, Bhosle and Gonzalez [2] gave an algorithm for inserting a bridge connecting two disjoint simple polygons of total n vertices that minimizes the geodesic diameter in their union in $O(n^2 \log n)$ time, where the bridge may intersect the polygon interiors. However, little is known for inserting an optimal shortcut connecting two boundary points of the same object, except those algorithms for augmenting graphs.

Our Results. For an x-monotone polygon P with n vertices, we present a linear time algorithm for determining the existence of a shortcut σ that reduces the diameter under the L_1 metric, that is, the diameter of $P \cup \sigma$ is strictly smaller than the diameter of P. We also present an $O(n \log n)$-time algorithm using $O(n)$ space for finding an optimal shortcut or an infimum shortcut for P that reduces the diameter under the L_1 metric, if one exists. To the best of our knowledge, our algorithms are the first results for finding an optimal shortcut for a polygon with respect to diameter.

Outline. In Sect. 3, we present an $O(n)$-time algorithm to determine whether P has a useful shortcut. As mentioned earlier, a shortcut σ is *useful* if the diameter of $P \cup \sigma$ is strictly smaller than the diameter of P under the L_1 metric. We call a vertex v an *anchor* of P if, for any diametral pair (α, β) of P, every geodesic path connecting α and β passes through v. We show that P has a useful shortcut if and only if P has an anchor. Then we show how to find the anchors of P in $O(n)$ time.

In Sect. 4, we present an algorithm that computes an optimal shortcut for P. We say a shortcut σ is *valid* if σ contains an anchor of P in the range of its x-coordinates. Let S denote the union of all valid shortcuts. We partition S into trapezoids and find the trapezoids that contain optimal shortcuts using a binary-search-like algorithm. For the binary-search-like algorithm, we define two functions f and g on shortcuts such that for a fixed shortcut σ, the diameter of $P \cup \sigma$ is $\max\{f(\sigma), g(\sigma)\}$. Here, we can compute $f(\sigma)$ in $O(n)$ time in the worst case, and $g(\sigma)$ in $O(\log n)$ time with help of appropriate query data structures. We show that $f(\sigma)$ satisfies a monotonicity such that $f(\sigma) \leq f(\sigma')$ holds for shortcuts σ and σ' satisfying a certain ordering; the projection of σ onto the x-axis is strictly contained in the projection of σ'. We also show a unimodality of g in Sect. 4.3.

In Sect. 4.4, we find the trapezoids that contain optimal shortcuts efficiently by using $O(n)$ segments of S. Such a segment is referred to as a *critical* shortcut. We compute $g(\sigma)$ for every critical shortcut σ and the local minimum of g in each trapezoid in $O(n \log n)$ time. Then we compute $f(\sigma)$ for critical shortcuts σ with a binary-search-like algorithm to find those trapezoids containing optimal shortcuts. This can be done in $O(n \log n)$ time in total.

Then, for each such trapezoid, we find a shortcut minimizing the diameter. Using the monotonicity of f and the unimodality of g, we apply binary search in each trapezoid and find such a shortcut. In total, this takes $O(n \log n)$ time and $O(n)$ space. Finally, we return the one achieving the minimum diameter.

Due to page limit, descriptions for data structures, proofs of lemmas, and some details of the algorithm for computing f and g are omitted. They are given in the full version.

2   Preliminaries

We use P to denote an x-monotone polygon with n vertices in the plane. We assume that P is given as a list of vertices sorted along its boundary in clockwise order. For ease of the description, we assume that the vertices of P have distinct x-coordinates and y-coordinates. This assumption can be removed with little modification to the algorithms without affecting their running times.

A polygon Q is a subpolygon of P if $Q \subseteq P$. For a subpolygon Q of P, we denote by $|Q|$ the number of vertices of Q. Also, we use $\mathsf{bd}(Q)$ and $\int(Q)$ to denote the boundary and the interior of Q, respectively. For any two points $p, q \in \mathsf{bd}(Q)$, we use $Q[p, q]$ to denote the subchain of $\mathsf{bd}(Q)$ from p to q in clockwise order, and $Q(p, q)$ to denote $Q[p, q] \setminus \{p, q\}$. We use p_l and p_r to denote the leftmost vertex of P and the rightmost vertex of P, respectively. We use $P[p_l, p_r]$ and $P[p_r, p_l]$ to denote the *upper chain* and the *lower chain* of P, respectively.

We often omit 'x-monotone' and use subpolygons to refer to x-monotone subpolygons of P as every subpolygon of P we consider is x-monotone. We call an x-monotone polygon a *terrain polygon* if its upper or lower boundary chain is a single horizontal edge. We call the horizontal edge the *base* of the terrain polygon. Polygons in Fig. 1(b) and 1(c) are terrain polygons.

For a point $p \in P$, we use $x(p)$ and $y(p)$ to denote the x-coordinate and the y-coordinate of p, respectively. We denote by $\eta(p)$ and $\upsilon(p)$ the maximal horizontal segment and the maximal vertical segment through p that are contained in P, respectively. We use $\overleftarrow{p}$ and $\overrightarrow{p}$ to denote the left and the right endpoint of $\eta(p)$, respectively. A vertex v of P is a *peak* if $\overleftarrow{v} = \overrightarrow{v} = v$, and a *pit* if $\overleftarrow{v} \neq v$ and $\overrightarrow{v} \neq v$. For any point $p \in \mathsf{bd}(P)$, we use $\overline{p}$ to denote the vertical projection of p onto $\mathsf{bd}(P)$. We denote by $L(\upsilon(p))$ and $R(\upsilon(p))$ the subpolygons of P induced by $\upsilon(p)$, $L(\upsilon(p))$ lying left to $\upsilon(p)$ and $R(\upsilon(p))$ lying right to $\upsilon(p)$.

We use pq to denote the line segment connecting two points p and q in the plane, and $|pq|$ the length of pq. For a horizontal segment s, we use $y(s)$ to denote the y-coordinate of an endpoint of s.

Let $\mathsf{M}(P)$ denote the trapezoidal map of P, which is a partition of P into trapezoids by adding $\eta(v)$ for each vertex v of P. The map can be computed in $O(n)$ time [5]. For a horizontal edge $e = uv$ of $\mathsf{M}(P)$, let $U(e)$ and $D(e)$ be the subpolygon of P, $U(e)$ lying above $D(e)$ locally along e, induced by e.

For any two points $p, q \in \mathbb{R}^2$, we use $d_1(p, q) = \|p-q\|_1 = |x(p)-x(q)|+|y(p)-y(q)|$ to denote their L_1 distance. A *geodesic path* connecting two points $p, q \in P$ is a shortest path connecting them that consists of line segments contained in P. There can be one or more geodesic paths connecting them. We use $\pi(p, q)$ to denote a fixed geodesic path from p to q in P. We use $d(p, q)$ to denote the length of $\pi(p, q)$, which is the sum of $d_1(u, v)$ for each segment uv in $\pi(p, q)$. We denote by $\mathsf{diam} = \mathsf{diam}(P) = \max_{p,q \in P} d(p, q)$ the diameter of P. A pair of points (α, β) is a *diametral pair* of P if $d(\alpha, \beta) = \mathsf{diam}$.

A *shortcut* σ of P is a horizontal segment ab such that $ab \cap \int(P) = \emptyset$ and $ab \cap \mathsf{bd}(P) = \{a, b\}$. To ease the description, we assume that every shortcut $\sigma = ab$ we consider has $a, b \in P[p_l, p_r]$ with $x(a) < x(b)$, unless stated otherwise. We abuse the notation so that $P[\sigma]$ denotes $P[a, b] \subseteq P[p_l, p_r]$, and $P(\sigma)$ denotes $P(a, b) \subset P[p_l, p_r]$.

For any two points $p', q' \in P \cup \sigma$, we use $\pi_\sigma(p', q')$ to denote a fixed geodesic path in $P \cup \sigma$ from p' to q'. We use $d_\sigma(p', q')$ to denote the length of $\pi_\sigma(p', q')$.

For any $p, q \in P$, we use $\pi_{\bar{\sigma}}(p, q)$ to denote a fixed geodesic path from p to q that uses σ, and use $d_{\bar{\sigma}}(p, q)$ to denote the length of $\pi_{\bar{\sigma}}(p, q)$. Thus, $d_\sigma(p, q) = \min\{d(p, q), d_{\bar{\sigma}}(p, q)\}$ for $p, q \in P$.

For a set A, the handle of A is the point $p \in A$ with the smallest x-coordinate among points $\arg\max_{q \in A}(y(q) - x(q))$.

3   Useful Shortcuts

We say a shortcut σ is *useful* for a pair of points (p, q) in P if $d_\sigma(p, q) < d(p, q)$. We call a pit v of P an *anchor* of (p, q) if every geodesic path connecting p and q passes through v. In this section, we show a few properties of useful shortcuts and anchors, and derive a necessary condition for a shortcut σ to satisfy $\mathsf{diam}(P \cup \sigma) < \mathsf{diam}(P)$.

Lemma 1. $P(\sigma)$ *contains an anchor of a pair* (p, q) *of points in* P *if* σ *is useful for* (p, q).

Lemma 2. *If a shortcut* ab *is useful for a diametral pair* (α, β) *of* P, *neither* α *nor* β *lies on* $P(ab)$ *and every anchor of* (α, β) *in* $P(ab)$ *is an anchor of* (a, b).

We say that a shortcut σ is *useful* for P if $\mathsf{diam}(P \cup \sigma) < \mathsf{diam}(P)$. By Lemmas 1 and 2, we have the following lemma.

Lemma 3. *A shortcut* ab *is useful for* P *if an anchor of* (a, b) *in* $P(ab)$ *is a common anchor of all diametral pairs of* P.

We say that a pit is an anchor of P if it is an anchor of every diametral pair of P. If P has an anchor, a shortcut arbitrarily close to the anchor is useful for every diametral pair of P, and thus it is useful for P. Moreover, we can observe that neither p nor q can lie on the shortcut for any diametral pair (p, q) of P when the shortcut is added. Therefore, by Lemma 3, P has a useful shortcut if and only if P has an anchor.

Theorem 1. *We can determine whether* P *has a useful shortcut in* $O(n)$ *time. If one exists, we can compute all anchors of* P *in the same time.*

4   Optimal Shortcut for x-monotone Polygons

Let $\mathsf{diam}_\sigma = \max_{p, q \in P \cup \sigma} d_\sigma(p, q)$. We find a shortcut σ that minimizes diam_σ, which we call an *optimal* shortcut. We consider shortcuts with endpoints on $P[p_l, p_r]$. Recall that $\mathsf{diam} = \mathsf{diam}(P)$ denotes the diameter of P. If P has a shortcut σ with $\mathsf{diam}_\sigma < \mathsf{diam}$, but no optimal shortcut exists, as shown in Fig. 1(b), we return an infimum shortcut.

We say a shortcut σ is *valid* if $P[\sigma]$ contains an anchor of P. Every useful shortcut is valid. But not every valid shortcut is useful. Let S denote the union of all valid shortcuts. Figure 2 illustrates S of P. We will define two functions $f(\sigma)$ and $g(\sigma)$ for shortcuts σ of P such that $\mathsf{diam}_\sigma = \max\{f(\sigma), g(\sigma)\}$. We present two algorithms for a given shortcut σ, one to compute $f(\sigma)$ and the other to compute $g(\sigma)$ efficiently. Using the two algorithms, we find an optimal shortcut or an infimum shortcut in S in binary-search-like manner, which takes $O(n \log n)$ time in total.

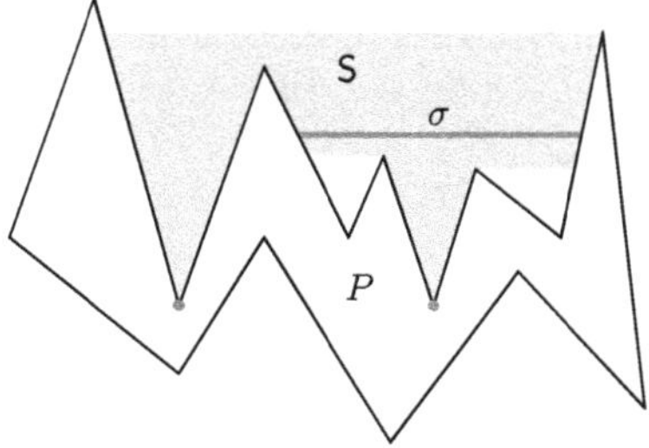

Fig. 2. Anchors of P are shown in red disks. $\mathsf{S}(Color\,figure\,online)$ (gray) is the union of all valid shortcuts.

4.1   Partition of an x-monotone Polygon

Before defining $f(\sigma)$ and $g(\sigma)$ for a given valid shortcut σ, we partition P into subpolygons using two polygonal chains γ_l and γ_r such that each subpolygon is either a terrain polygon or an xy-monotone polygon. The chain $\gamma_l = p_0 p_1 \ldots$ starts from $p_0 = p_l$. Let $p_1 := \vec{p}_0$ if $p_0 \neq \vec{p}_0$, and let p_1 be the right endpoint v of the edge e such that $\vec{p}_0 \in e \setminus \{v\}$ if $p_0 = \vec{p}_0$. Repeat this process with p_1 and find p_2, and so on, until we reach p_r. Observe that γ_l eventually reaches p_r since it always goes rightwards. Similarly, the chain $\gamma_r = p'_0 p'_1 \ldots$ starts from $p'_0 = p_r$, and goes leftwards. Let $p'_1 := \overleftarrow{p}'_0$ if $p'_0 \neq \overleftarrow{p}'_0$, and let p_1 be the left endpoint v of the edge e such that $\overleftarrow{p}'_0 \in e \setminus \{v\}$ if $p'_0 = \overleftarrow{p}'_0$. Repeat this process with p'_1 and find p'_2, and so on, until we reach p_l. Observe that γ_l and γ_r intersect each other at p_l and p_r, and at every anchor of (p_l, p_r). Moreover, the subchains of γ_l or γ_r lying in between any two consecutive intersections are xy-monotone. Let $\mathcal{P}$ be the resulting partition. See Fig. 3 for an illustration.

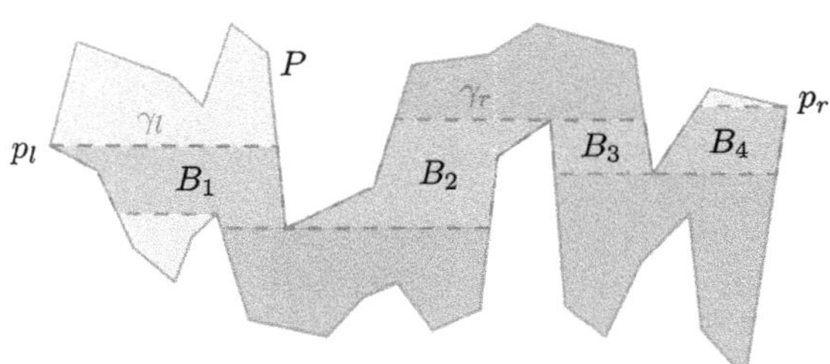

Fig. 3. Two polygonal chains γ_l and γ_r that connect p_l and p_r. Each red subpolygon is a B-gon; B_1 and B_3 are downward, and B_2 and B_4 are upward. Each blue subpolygon is a primary T-gon. Each gray subpolygon is a secondary T-gon. (Color figure online)

Backbone Polygons. If a polygon in $\mathcal{P}$ is enclosed by γ_l and γ_r, it is called a *backbone polygon,* or simply a B-gon. We call the leftmost and rightmost vertices of a B-gon the *joints* of the polygon. Each joint is an intersection point of γ_l

and γ_r. Observe that every B-gon is an xy-monotone polygon because both the upper and lower chains are xy-monotone. A B-gon is called *upward* if the left joint has the smallest y-coordinate among the vertices of the polygon, and *downward* otherwise.

Terrain Polygons. We show that every polygon in $\mathcal{P}$ that is not a B-gon is a terrain polygon. Let H be a polygon in $\mathcal{P}$ that is not a B-gon. By definition, H is incident to γ_l, γ_r, or an anchor c of (p_l, p_r), and thus either the upper chain or the lower chain of H is a segment of γ_l or γ_r , or $\eta(c)$. Hence, H is a terrain polygon whose base is contained in $\gamma_l \cup \gamma_r$. We call a terrain polygon in $\mathcal{P}$ a T-gon. If a T-gon is incident to an anchor and it is adjacent to two B-gons, then we call it a *primary* T-gon. If a T-gon is adjacent to one B-gon, then we call it a *secondary* T-gon.

4.2   Two Functions for Shortcuts

We define two functions $f(\sigma)$ and $g(\sigma)$ for shortcuts $\sigma = ab$ such that $\mathsf{diam}_\sigma = \max\{f(\sigma), g(\sigma)\}$. To do this, we define subpolygons $Q(\sigma)$ for shortcuts σ of P such that for any two shortcuts σ, σ' satisfying $P[\sigma] \subset P[\sigma']$, $\mathsf{diam}_\sigma(Q(\sigma)) \leq \mathsf{diam}_{\sigma'}(Q(\sigma'))$, where $\mathsf{diam}_\sigma(Q(\sigma)) = \max_{p,q \in Q(\sigma)} d_\sigma(p,q)$.

In the following, we define two subpolygons $W_l(\sigma)$ and $W_r(\sigma)$ such that $Q(\sigma) = P \setminus (W_l(\sigma) \cup W_r(\sigma))$. Recall that $\pi_{\bar\sigma}(p,q)$ denotes a fixed geodesic path from p to q using σ. For a point $p \in P$, let $S_p(\sigma) := \{q \in P \mid d_{\bar\sigma}(p,q) < d(p,q)\}$. To make the description easier, we may omit σ if it is understood in context.

We define two subpolygons W_1 and W_2 of P such that $S_p = S_{p'}$ for any two points $p, p' \in W_1$. There are a few cases. If a is incident to a B-gon of $\mathcal{P}$, let $W_1 := U(\bar{a}a)$ and $W_2 = \emptyset$ (Fig. 4(a)). If $a \in \mathsf{bd}(H)$ for a T-gon $H \in \mathcal{P}$, let u be the left endpoint of the base of H, and let h denote the handle of $P(u,a)$. Let $W_2 := \{p \in H \mid x(p) \leq x(a)$ and $\min_{t \in \pi(p,a)} y(t) \geq \min_{t \in \pi(h,a)} y(t)\}$. Then $d(p,a) \leq d(h,a)$ for all $p \in W_2(\sigma)$. If u is incident to an upward B-gon of $\mathcal{P}$, let W_1 denote the left subpolygon in the partition of P induced by $v(u)$ (Fig. 4(b)). If u is incident to a downward B-gon of $\mathcal{P}$, let $W_1 := U(\bar{u}u)$ (Fig. 4(c)).

Let $W_l := W_1 \cup W_2$. Observe that $W_2 \subseteq H$. We define W_r with respect to b analogously. Then let $Q(\sigma) = P \setminus (W_l(\sigma) \cup W_r(\sigma))$. Lemmas 4 and 6 are the properties of $Q(\sigma)$ we will use in computing an optimal shortcut.

Lemma 4. *For any two shortcuts σ and σ' satisfying $P[\sigma] \subset P[\sigma']$, we have* $\mathsf{diam}_\sigma(Q(\sigma)) \leq \mathsf{diam}_{\sigma'}(Q(\sigma'))$.

Before defining f, we introduce a property of a diametral pair of $P \cup \sigma$. Let P_σ be the portion of P lying between $x = x(a)$ and $x = x(b)$.

Lemma 5. *Let (α, β) be a diametral pair of $P \cup \sigma$ with $\beta \in \int(\sigma)$. Then $\alpha \in P_\sigma$.*

Let $\Pi_1(A) := \{(p,q) \mid p,q \in A, x(p) \leq x(q)\}$ for a set $A \subset \mathbb{R}^2$. For any two sets $A_0, A_1 \subset \mathbb{R}^2$, let $\Pi_2(A_0, A_1) := \{(p,q) \mid p \in A_i, q \in A_{1-i}$, and $x(p) \leq x(q)$ for $i = 0,1\}$. Let $\mathsf{diam}_\sigma(\Pi') := \max_{(p,q) \in \Pi'} d_\sigma(p,q)$ for $\Pi' \subseteq \Pi_1(P \cup \sigma)$. By

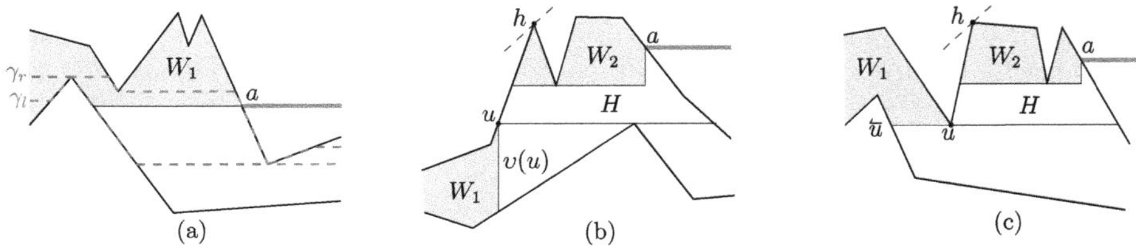

Fig. 4. W_1 and W_2 for a shortcut $\sigma = ab$. (a) a is incident to a B-gon. (b) a is incident to a T-gon H, and the left endpoint u of the base of H is incident to an upward B-gon. (c) a is incident to a T-gon H, and u is incident to a downward B-gon.

Lemma 5, $\mathsf{diam}_\sigma = \mathsf{diam}_\sigma(\Pi_1(P) \cup \Pi_2(P_\sigma, \sigma))$. Let $f(\sigma) := \mathsf{diam}_\sigma(\Pi_1(Q(\sigma)) \cup \Pi_2(P_\sigma, \sigma))$. Since $\mathsf{diam}_\sigma(Q(\sigma)) = \mathsf{diam}_\sigma(\Pi_1(Q(\sigma)))$, we obtain the following property of f by Lemma 4.

Lemma 6. *For any two shortcuts σ and σ' satisfying $P[\sigma] \subset P[\sigma']$, $f(\sigma) \leq f(\sigma')$.*

Let g be a function on a shortcut $\sigma = ab$ such that $\mathsf{diam}_\sigma = \max\{f(\sigma), g(\sigma)\}$. Let $L_\sigma = L(v(c_1))$ and $R_\sigma = R(v(c_k))$, where c_1 and c_k are the leftmost anchor and the rightmost anchor of (a, b), respectively (Fig. 5). Since $W_l(\sigma) \subset L_\sigma$, $\Pi_2(W_l(\sigma), P \setminus W_l(\sigma)) \subset \Pi_1(L_\sigma) \cup \Pi_2(W_l(\sigma), P \setminus L_\sigma)$.

Let $\Pi_l(\sigma) := \Pi_1(L_\sigma) \cup \Pi_2(W_l(\sigma), P \setminus L_\sigma)$ and $\Pi_r(\sigma) := \Pi_1(R_\sigma) \cup \Pi_2(P \setminus R_\sigma, W_r(\sigma))$. Let $\Pi_g(\sigma) := \Pi_l(\sigma) \cup \Pi_r(\sigma)$. Then $\Pi_1(Q(\sigma)) \cup \Pi_g(\sigma) = \Pi_1(P)$. Let $g(\sigma) := \mathsf{diam}_\sigma(\Pi_g(\sigma))$. Therefore, $\mathsf{diam}_\sigma = \max\{f(\sigma), g(\sigma)\}$.

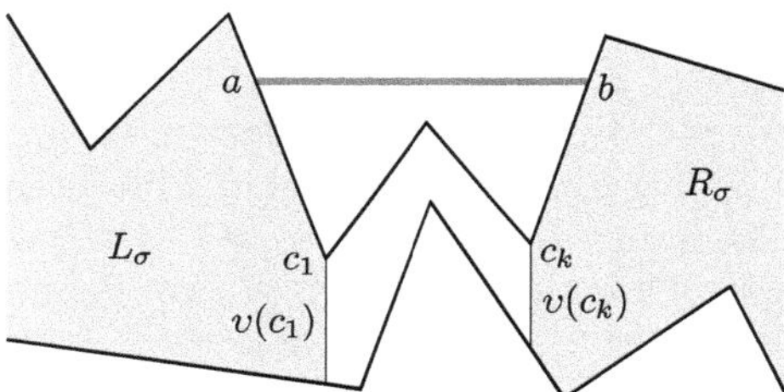

Fig. 5. $L_\sigma = L(v(c_1))$ and $R_\sigma = L(v(c_k))$ for the leftmost anchor c_1 and the rightmost anchor c_k of (a, b).

We can compute $f(\sigma)$ in $O(|Q(\sigma)|)$ time and $g(\sigma)$ in $O(\log n)$ time with the dynamic farthest point query data structure.

4.3   Construction of a Forest

Let S denote the union of all valid shortcuts ab of P, and let Φ denote the set of anchors of P. A valid shortcut in S is *critical* if it is incident to a vertex of

$M(P)$. Let $S_c \subseteq S$ denote the set of all critical shortcuts of S. The *thread* of a critical shortcut $\sigma \in S_c$ is the maximal horizontal line segment containing σ that does not intersect $\int(P)$. Let S_t denote the set of threads of critical shortcuts of S_c. Let $\Lambda := S \cup S_t \cup \Phi$ and $\Lambda_c := S_c \cup S_t \cup \Phi$.

For any two line segments ab and $a'b'$ in Λ, we use $ab \preceq a'b'$ if $P[a,b] \subseteq P[a',b']$, and $ab \prec a'b'$ if $P[a,b] \subsetneq P[a',b']$. We can make a proper ordering of the segments in Λ in $O(n)$ time as follows. Let M be the decomposition of S by $S_c \cup S_t$. Consider the dual forest $F = (V, E)$ of Λ_c such that each node w of F corresponds to a segment $s(w)$ of Λ_c. Two nodes w, w' of F are connected by an edge of Λ_c if $s(w)$ and $s(w')$ bound the same trapezoid of M, one from above and one from below, or $s(w')$ is the thread of $s(w)$. We assign a direction to each edge ww' of F such that it is directed from w to w' if and only if $y(s(w)) < y(s(w'))$ or $s(w')$ is the thread of $s(w)$. Then F is a directed acyclic graph, and we can get a topological ordering of the segments in Λ_c such that $s(w) \prec s(w')$ if and only if there is a directed path from w to w'. This ordering can be computed in $O(n)$ time. See Fig. 6 for an illustration of F.

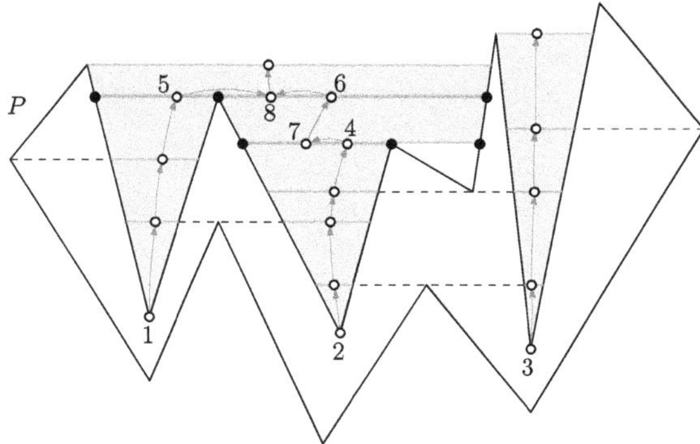

Fig. 6. Illustration of the dual forest F of Λ_c. The gray region is S. Each node has a corresponding element of Λ_c. Each leaf node, labeled 1, 2, and 3 of F, corresponds to an anchor of P. Each node, labeled 4, 5, and 6, corresponds to a critical shortcut in S_c. Node 7 corresponds to the thread of $s(4)$, and node 8 corresponds to the thread of both $s(5)$ and $s(6)$.

Observe that threads in S_t and anchors in Φ are not shortcuts. We extend the definitions of f and g so that they are defined for all elements of Λ. We can compute $f(s(u))$ for each $u \in V$ in $O(|Q(s(u))|)$ time, which takes $O(n^2)$ time for all vertices in V. In Sect. 4.4, we show how to find an optimal shortcut using $f(s(u))$ for u in a subset of V, and reduce the total running time to $O(n \log n)$. We compute $g(s(u))$ for each $u \in V$ in $O(\log n)$ time in bottom-up manner while maintaining the dynamic farthest point query data structure in total $O(n \log n)$ time.

For any edge $e = (u, w) \in E$ with $s(u) \prec s(w)$, let $\Lambda(e)$ be the set consisting of all $s \in \Lambda$ that satisfy $s(u) \preceq s \preceq s(w)$. If $s(w)$ is not the thread of $s(u)$, observe that the function f is monotone for $\Lambda(e)$ with respect to $\prec$ by Lemma 6 and continuous by definition. Similarly, the function g is continuous for $\Lambda(e)$ with respect to $\prec$ by definition. However, g is not necessarily monotone for $\Lambda(e)$ with respect to $\prec$. Instead, g satisfies a weak unimodality, that is, there exists a segment $\omega(e) \in \Lambda(e)$ with $s(u) \preceq \omega(e) \preceq s(w)$ such that $g(s)$ is decreasing linearly with respect to the y-coordinate of s satisfying $s(u) \preceq s \preceq \omega(e)$ and weakly monotonically increasing for s satisfying $\omega(e) \preceq s \preceq s(w)$. We can find $\omega(e)$ and compute $g(\omega(e))$ in $O(\log n)$ time, after computing $g(s(u))$ and $g(s(w))$. Let $\mathsf{S_p}$ be the set of $\omega(e)$ for all edges e of F. We can compute $\mathsf{S_p}$ and $g(s)$ for all $s \in \mathsf{S_p}$ in $O(n \log n)$ time.

Let $\Lambda_\mathsf{p} := \mathsf{S_c} \cup \mathsf{S_p} \cup \mathsf{S_t} \cup \Phi$. For $u \in \mathsf{V}$, we compute $s^*(u) \in \Lambda_\mathsf{p}$ and $g(s^*(u))$, where $s^*(u)$ is the segment achieving the minimum $g(s)$ value among segments $s \preceq s(u)$. We can compute $s^*(u)$ for all $u \in \mathsf{V}$ in $O(n)$ time in bottom-up manner if $g(s)$ values for all $s \in \Lambda_\mathsf{p}$ are computed.

4.4 Pruning F and Finding an Optimal Shortcut

We show how to find an optimal shortcut using F efficiently. We assume that we have computed $g(s(u))$ and $s^*(u)$ for all $u \in \mathsf{V}$ in total $O(n \log n)$ time. We will compute $f(s(u))$ for u in a subset of V in $O(n \log n)$ time that will be sufficient for finding an optimal shortcut.

We say two edges $e_1, e_2 \in \mathsf{E}$ are x-*disjoint* if any $s_1 \in \Lambda(e_1)$ and $s_2 \in \Lambda(e_2)$ are disjoint along the x-axis. By Lemma 6 and the definition of $g(s^*(u))$ for a node $u \in \mathsf{V}$, we have the following lemma.

Lemma 7. *For any two nodes u and w such that u is an ancestor of w, we have $f(s(u)) \geq f(s(w))$ and $g(s^*(u)) \leq g(s^*(w))$.*

We start pruning with $\mathsf{F}_1 := \mathsf{F}$ and get F_i from F_{i-1} for i increasing from 2 until every edge (u, w) of the resulting forest satisfies $f(s(u)) \leq g(s^*(u))$ and $f(s(w)) \geq g(s^*(w))$. We obtain F_i from F_{i-1} as follows. Consider a (rooted) tree T of F_{i-1} such that the root of T is at level 0. Let $V_{\lceil h/2 \rceil}$ be the set of nodes at level $\lceil h/2 \rceil$, where h is the height of T. For each $u \in V_{\lceil h/2 \rceil}$, we compute $f(s(u))$. If $f(s(u)) \leq g(s^*(u))$, we remove all descendants of u from T. If $f(s(u)) > g(s^*(u))$, we remove every edge in the path from u to the root of T. After doing this for every node in $V_{\lceil h/2 \rceil}$, we are left with trees that were parts of T. We remove singleton trees (height 0) from F_{i-1}. Note that the remaining part of T consists of x-disjoint rooted trees with level at most $\lceil h/2 \rceil$. After doing this for every tree of F_{i-1}, we obtain F_i. See Fig. 7 for an illustration.

During the pruning, F_i remains to be a forest that consists of x-disjoint rooted trees. We repeat this until every edge (u, w) of the resulting forest satisfies $f(s(u)) \leq g(s^*(u))$ and $f(s(w)) \geq g(s^*(w))$. Let $\mathsf{F}_k = (\mathsf{V}_k, \mathsf{E}_k)$ be the final forest from the pruning. By Lemma 7, F_k consists of trees with height one. Since F consists of $O(n)$ nodes, the pruning procedure terminates in $O(\log n)$ iterations.

Let V'_i be the set of nodes u such that $f(s(u))$ is computed while obtaining F_i from F_{i-1}. Then for any two nodes $u, w \in \mathsf{V}'_i$, $s(u)$ and $s(w)$ are disjoint

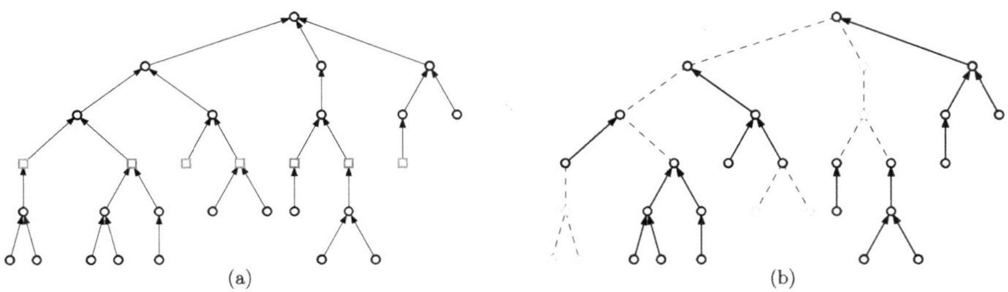

Fig. 7. A pruning step of a tree T in F_{i-1}. (a) Each node u with $f(s(u)) > g(s^*(u))$ is marked as a blue square. Each node w with $f(s(w)) \le g(s^*(w))$ is marked as a red square. (b) Remaining part of T in F_i after pruning.

along the x-axis. Note that $f(s(u))$ can be computed in $O(|Q(s(u))|)$ time as mentioned in Sect. 4.2. Now we bound the total complexity of $Q(\sigma)$ for a set of valid shortcuts that are disjoint each other along the x-axis to show that we can compute F_i from F_{i-1} in $O(n)$ time.

Lemma 8. *For any set S consisting of valid shortcuts that are disjoint to each other along the x-axis, $\sum_{\sigma \in S} |Q(\sigma)| = O(n)$.*

By Lemma 8, we can compute F_i from F_{i-1} in $O(n)$ time. Consider an edge $e = (u, w)$ of $\mathsf{F}_k = (\mathsf{V}_k, \mathsf{E}_k)$ with w being the parent of u. Then $\mathsf{diam}(s(w)) = f(s(w))$, and for any shortcut σ satisfying $s(w) \prec \sigma$, $\mathsf{diam}_\sigma = f(\sigma) \ge f(s(w))$ by Lemma 7. Hence, we consider shortcuts σ with $\sigma \preceq s(w)$. Let s_e^* denote an element of $\arg\min\{\mathsf{diam}(s) \mid s \in \Lambda \text{ and } s \preceq s(w)\}$. Observe that an optimal shortcut or an infimum shortcut for P is s_e^* for some edge $e \in \mathsf{E}_k$. Thus, we compute s_e^* for all edges $e \in \mathsf{E}_k$ and return s^* among them satisfying $\mathsf{diam}(P \cup s^*) \le \mathsf{diam}(P \cup s_e^*)$ for all edges $e \in \mathsf{E}_k$ as an optimal shortcut or an infimum shortcut for P. Recall that $g(s^*(u)) = \min\{g(s) \mid s \in \Lambda \text{ and } s \preceq s(u)\}$ for $u \in \mathsf{V}_k$.

We show how to compute s_e^* for an edge $e = (u, w) \in \mathsf{E}_k$ with w being the parent of u. If $g(s^*(u)) = g(s^*(w))$, then $\mathsf{diam}(s^*(w)) = g(s^*(w)) \le g(s) \le \mathsf{diam}(s)$ for any $s \preceq s(w)$, and thus $s^*(w)$ is s_e^*. If $g(s^*(u)) > g(s^*(w))$, then $g(s^*(w)) = g(s(w))$ and $g(s(u)) > g(s(w))$. In this case, $f(s(u)) \le g(s(u))$ and $f(s(w)) \ge g(s(w))$. Then $s_e^* = s^*(u)$ or $s(u) \preceq s_e^* \preceq s(w)$. There are two subcases. If one endpoint of $s(u)$ is a peak of a T-gon, $s(w)$ is the thread of $s(u)$, and thus no shortcut $\sigma \in \Lambda$ satisfies $s(u) \prec \sigma \prec s(w)$. Then s_e^* is $s^*(u)$ or $s(w)$, which can be determined by comparing $\mathsf{diam}(s^*(u))$ and $\mathsf{diam}(s(w))$. If no endpoint of $s(u)$ is a peak of P, there is $s_e' \in \Lambda$ such that $s(u) \preceq s_e' \preceq w(e)$ and $f(s_e') = g(s_e')$, since g satisfies the weak unimodality in Sect. 4.3. We find s_e' in $O(|Q(s(u))| \log n)$ time using a binary search. Then s_e^* is $s^*(u)$ or s_e'.

Observe that any two shortcuts $s(u)$ and $s(u')$ are disjoint along the x-axis for two distinct edges $e = (u, w)$ and $e' = (u', w')$, u being a child of w and u' being a child of w', of F_k. Thus, we can compute s_e^* for edges $e \in \mathsf{E}_k$ in $O(n \log n)$

time in total by Lemma 8. We return s^* among them satisfying $\mathsf{diam}(P \cup s^*) \leq \mathsf{diam}(P \cup s_e^*)$ for all s_e^*'s. The algorithm uses $O(n)$ space in total.

Theorem 2. *For an x-monotone polygon P with n vertices, we can find a shortcut or an infimum shortcut of P that minimizes the diameter in $O(n \log n)$ time and $O(n)$ space, if it exists.*

Concluding remarks. The problem has generalizations and variants, including shortcuts under Euclidean metric, shortcuts of arbitrary orientation, shortcuts over simple polygons, and k shortcuts for $k > 1$.

References

1. Ahn, H.K., et al.: Constructing optimal highways. Int. J. Found. Comput. Sci. **20**(01), 3–23 (2009)
2. Bhosle, A.M., Gonzalez, T.F.: Exact and approximation algorithms for finding an optimal bridge connecting two simple polygons. Int. J. Comput. Geometry Appl. **15**(06), 609–630 (2005)
3. Cáceres, J., Garijo, D., González, A., Márquez, A., Puertas, M.L., Ribeiro, P.: Shortcut sets for the locus of plane Euclidean networks. Appl. Math. Comput. **334**, 192–205 (2018)
4. Cardinal, J., Langerman, S.: Min-max-min geometric facility location problems. In: Proceedings of the 22nd European Workshop on Computational Geometry (EWCG 2006), pp. 149–152 (2006)
5. Chazelle, B.: Triangulating a simple polygon in linear time. Disc. Comput. Geometry **6**(3), 485–524 (1991). https://doi.org/10.1007/BF02574703
6. de Carufel, J.L., Grimm, C., Maheshwari, A., Schirra, S., Smid, M.: Minimizing the continuous diameter when augmenting a geometric tree with a shortcut. Comput. Geom. **89**, 101631 (2020)
7. Drezner, Z.: Facility location: a survey of applications and methods. Springer Series in Operations Research and Financial Engineering, Springer-Verlag (1996)
8. Drezner, Z., Hamacher, H.W.: Facility location: applications and theory. Springer-Verlag (2004)
9. Frati, F., Gaspers, S., Gudmundsson, J., Mathieson, L.: Augmenting graphs to minimize the diameter. Algorithmica **72**, 995–1010 (2015)
10. Garijo, D., Márquez, A., Rodríguez, N., Silveira, R.I.: Computing optimal shortcuts for networks. Eur. J. Oper. Res. **279**(1), 26–37 (2019)
11. Große, U., Gudmundsson, J., Knauer, C., Smid, M., Stehn, F.: Fast algorithms for diameter-optimally augmenting paths. In: Halldórsson, M.M., Iwama, K., Kobayashi, N., Speckmann, B. (eds.) ICALP 2015. LNCS, vol. 9134, pp. 678–688. Springer, Heidelberg (2015). https://doi.org/10.1007/978-3-662-47672-7_55
12. Gudmundsson, J., Sha, Y.: Algorithms for radius-optimally augmenting trees in a metric space. Comput. Geom. **114**, 102018 (2023)
13. Gudmundsson, J., Sha, Y.: Augmenting graphs to minimize the radius. Comput. Geom. **113**, 101996 (2023)
14. Robert, J.M., Toussaint, G.T.: Computational geometry and facility location. Technical Report, McGill University. School Comput. Sci. (1990)
15. Tan, X.: Finding an optimal bridge between two polygons. Int. J. Comput. Geometry Appl. **12**(03), 249–261 (2002)

16. Wang, D.P.: An optimal algorithm for constructing an optimal bridge between two simple rectilinear polygons. Inf. Process. Lett. **79**(5), 229–236 (2001)
17. Yang, B.: Euclidean chains and their shortcuts. Theoret. Comput. Sci. **497**, 55–67 (2013)

Parameterized Complexity of Reconfiguring Vertex-Disjoint Shortest Paths

Rin Saito[(✉)] and Takehiro Ito

Graduate School of Information Sciences, Tohoku University, Sendai, Japan
`rin.saito@dc.tohoku.ac.jp`, `takehiro@tohoku.ac.jp`

Abstract. Given k terminal pairs and two tuples of (internally) vertex-disjoint shortest paths in an unweighted graph, we study the problem of determining whether one tuple can be transformed into the other by repeatedly replacing a single vertex in one shortest path, while maintaining a tuple of vertex-disjoint shortest paths at every step. In this paper, we analyze the problem from the viewpoint of parameterized complexity. We first provide a tight complexity classification based on the maximum terminal distance D, defined as the largest distance among the terminal pairs: the problem is (para-)PSPACE-complete for any fixed $D \geq 3$, while it is solvable in polynomial time when $D \leq 2$. We then show that the problem is fixed-parameter tractable when parameterized by $\ell + \Delta$, where ℓ is the number of transformation steps and Δ is the maximum degree of an input graph. Finally, we examine the effect of structural graph parameters, by showing that the problem is fixed-parameter tractable when parameterized by k plus one of the modular-width, cluster deletion number, and tree-depth of an input graph.

Keywords: Graph algorithm · combinatorial reconfiguration · fixed-parameter tractability · vertex-disjoint paths

1 Introduction

Combinatorial reconfiguration problems explore the solution spaces of various combinatorial (search) problems under prescribed reconfiguration rules. Given two feasible solutions, the task is to determine whether one solution can be transformed into the other by repeatedly applying a reconfiguration rule while maintaining feasibility at each intermediate step. A wide range of classical graph problems has been studied under this framework, including independent sets, dominating sets, colorings, and shortest paths; see the surveys by van den Heuvel [11] and Nishimura [17].

Among these problems, the SHORTEST PATH RECONFIGURATION (SPR) problem, introduced by Kamiński et al. [13], has been extensively studied in

This work was partially supported by JST SPRING Grant Number JPMJSP2114, and JSPS KAKENHI Grant Numbers JP24H00686 and JP24H00690.

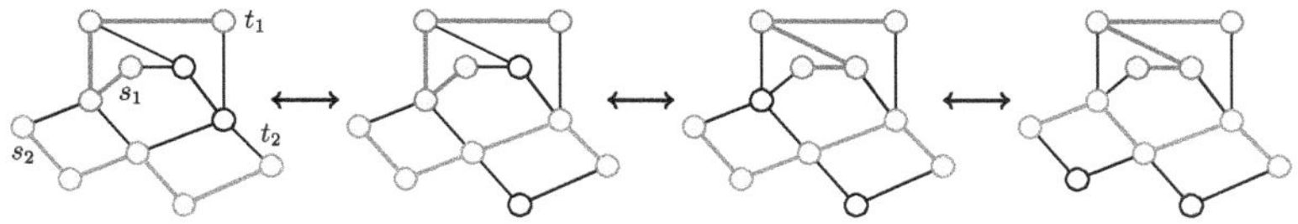

Fig. 1. A sequence of tuples of (internally) vertex-disjoint shortest paths for terminal pairs (s_1, t_1) and (s_2, t_2).

this field. In SPR, we are given an unweighted graph G and two shortest paths P and Q between two specified vertices s and t. The goal is to determine whether P can be transformed into Q by changing a single vertex on the current shortest path at each step, such that all intermediate results remain shortest paths between s and t. This problem is known to be PSPACE-complete [2], and its complexity has been further investigated in terms of graph classes [1–3,7] and parameterized complexity [4,21].

In this paper, we study a generalized problem of SPR, called the REACH-ABILITY OF VERTEX-DISJOINT SHORTEST PATHS (RVDSP) problem, introduced by Saito et al. [19]. In RVDSP, we are given a set of k terminal pairs $(s_1, t_1), (s_2, t_2), \ldots, (s_k, t_k)$ in an unweighted graph G, along with two tuples of shortest paths, $\mathcal{P} = (P_1, P_2, \ldots, P_k)$ and $\mathcal{Q} = (Q_1, Q_2, \ldots, Q_k)$, where P_i and Q_i are shortest paths between s_i and t_i for each $i \in \{1, 2, \ldots, k\}$. In each of the tuples $\mathcal{P}$ and $\mathcal{Q}$, the paths must be *internally vertex-disjoint*, namely, no two paths in the tuple share an internal vertex, and no internal vertex coincides with any terminal. Then, RVDSP asks whether $\mathcal{P}$ can be transformed into $\mathcal{Q}$ by iteratively changing a single vertex of one shortest path in the current tuple at a time, such that all intermediate tuples remain internally vertex-disjoint shortest paths. (See Fig. 1.) Note that RVDSP for $k = 1$ is equivalent to SPR.

1.1 Known Results and Related Work

Saito et al. [19] introduced RVDSP, and analyzed the polynomial-time solvability of RVDSP with respect to graph classes. They showed that RVDSP is PSPACE-complete for any fixed k, even on bipartite graphs, while the problem is solvable in polynomial time for split graphs and for distance-hereditary graphs even if k is part of the input. Furthermore, since the computational hardness of SPR extends to RVDSP, we here note that SPR (namely, RVDSP for $k = 1$) is known to be PSPACE-complete for bipartite graphs [2], bounded bandwidth graphs [21], graph powers, and line graphs [7]. On the other hand, we omit the details of existing work developing polynomial-time algorithms for SPR [1–3,7], because they do not directly yield ones for RVDSP when k is part of the input.

SPR has also been studied from the viewpoint of parameterized complexity. Bousquet et al. [4] showed that SPR (and hence RVDSP) is W[1]-hard for bounded degeneracy graphs when parameterized by $\ell + D$, where ℓ is the number of transformation steps and D is the distance between a given terminal pair. On the positive side, they developed fixed-parameter algorithms for SPR when

Table 1. Summary of the complexity of RVDSP when parameterized by the length ℓ of a reconfiguration sequence, the maximum terminal distance D, the maximum degree Δ of an input graph G, and the number k of terminal pairs.

	ℓ: reconf. length		D: max terminal distance	Δ: max degree	k: # terminal pairs
ℓ	W[1]-h [4]				
	XP (Proposition 2)				
D	W[1]-h [4]		PSPACE-c (Theorem 2)		
	XP (Proposition 2)				
Δ	FPT (Theorem 3)		PSPACE-c (Theorem 2)	PSPACE-c [19,21] (Theorem 2)	
k	W[1]-h [4]		W[1]-h [4]	PSPACE-c [19,21]	PSPACE-c [2,19,21]
	XP (Proposition 2)		XP (Proposition 1)		

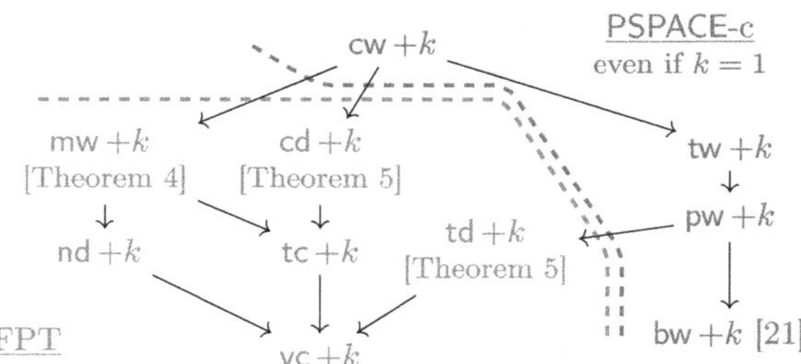

Fig. 2. Relationship between some structural graph parameters, and the parameterized complexity of RVDSP. The parameters are vc (vertex cover number), td (treedepth), tc (twin cover), nd (neighborhood diversity), mw (modular-width), cd (cluster deletion number), bw (bandwidth), pw (pathwidth), tw (treewidth), and cw (clique-width). An arrow from parameter a to parameter b indicates that a can be bounded by some function of b.

parameterized by structural graph parameters, including modular-width, cluster deletion number, and treedepth of an input graph (Table 1).

1.2 Our Contribution

To the best of our knowledge, RVDSP has not yet been investigated from the perspective of parameterized complexity. In this paper, we thus study the parameterized complexity of RVDSP, as follows.

We first analyze the role of the maximum terminal distance D, defined as the largest distance between any terminal pair. The parameter D denotes simply the distance between two given terminals when $k = 1$, and hence this is a natural generalization of the known parameter by Bousquet et al. [4]. We will prove that RVDSP is PSPACE-complete for any fixed constant $D \geq 3$, and hence the problem is unlikely to admit even an XP algorithm with the parameter D. We complement this hardness by providing a polynomial-time algorithm when $D \leq 2$. As a result, we provide a tight complexity classification based on D.

We next focus on the number ℓ of transformation steps, which is one of the most well-studied parameters for combinatorial reconfiguration problems [14,15].

As shown by Bousquet et al. [4], SPR (and hence RVDSP) is W[1]-hard for bounded degeneracy graphs when parameterized by $\ell + D$. In contrast, we will prove that RVDSP is fixed-parameter tractable when parameterized by $\ell + \Delta$, where Δ is the maximum degree of an input graph. Note that RVDSP remains PSPACE-complete even for $\Delta = 5$ (Theorem 2), which implies that the problem is unlikely to admit even an XP algorithm with the single parameter Δ.

Finally, we examine the effect of structural graph parameters (see Fig. 2). Bousquet et al. [4] showed that SPR is fixed-parameter tractable when parameterized by modular-width mw, cluster deletion number cd, or treedepth td. By taking k as an additional parameter, we generalize these results to RVDSP: namely, we will prove that RVDSP is fixed-parameter tractable when parameterized by $\mathsf{mw} + k$, $\mathsf{cd} + k$, or $\mathsf{td} + k$.

Due to space limitation, we omit the proofs of the claims marked with ($*$).

2   Preliminaries

For a positive integer r, let $[r] := \{1, 2, \ldots, r\}$. For two sets A and B, let $A \triangle B$ denote their *symmetric difference*, i.e., $A \triangle B := (A \setminus B) \cup (B \setminus A)$.

Graph Notation. Throughout the paper, a graph G is assumed to be connected, unweighted and undirected. We write $V(G)$ and $E(G)$ for the vertex and edge sets of G, respectively. For each $v \in V(G)$, we denote by $N_G(v)$ the set of all vertices that are adjacent to v in G. For a vertex subset U of G, let $G[U]$ denote the subgraph of G induced by U. For any two vertices $u, v \in V(G)$, a *uv-path* is a path in G joining u and v. The *distance between two vertices* $u, v \in V(G)$, denoted by $\mathsf{dist}(u, v)$, is the minimum number of edges in any uv-path. The *diameter of* G is the maximum value of $\mathsf{dist}(u, v)$ over all pairs of vertices $u, v \in V(G)$.

For a graph G, the *width of a bijection* $\pi \colon V(G) \to [|V(G)|]$ is the maximum value of $|\pi(u) - \pi(v)|$ over all $uv \in E(G)$. The *bandwidth of* G is the minimum width over all such bijections. For an integer $d \geq 0$, a graph G is *d-degenerate* if every subgraph of G contains a vertex of degree at most d. The *degeneracy of* G is the minimum integer d such that G is d-degenerate.

Definitions of Our Problem. Let k be a positive integer, and let (s_i, t_i) be a pair of vertices in G, called *terminals*, for each $i \in [k]$. Then, k paths $P_1, P_2, \ldots, P_k$ in G are said to be *internally vertex-disjoint* if P_i is an $s_i t_i$-path in G for every $i \in [k]$, and the internal vertices of these paths are all distinct and do not include any terminals. Note that internally vertex-disjoint paths may share terminal vertices. Hereafter, we refer to internally vertex-disjoint paths simply as *vertex-disjoint* paths. For a tuple $\mathcal{P} = (P_1, P_2, \ldots, P_k)$ of vertex-disjoint paths, we define $V(\mathcal{P}) := \bigcup_{i=1}^{k} V(P_i)$ as the set of all vertices (including terminals) that are passed through by some P_i in $\mathcal{P}$, $i \in [k]$.

In this paper, we only consider *shortest $s_i t_i$-paths* in G for each $i \in [k]$. Let $\mathcal{P} = (P_1, P_2, \ldots, P_k)$ and $\mathcal{P}' = (P_1', P_2', \ldots, P_k')$ be two tuples of vertex-disjoint shortest paths. We say that $\mathcal{P}$ and $\mathcal{P}'$ are *adjacent*, denoted by $\mathcal{P} \leftrightarrow \mathcal{P}'$, if $\sum_{i=1}^{k} |V(P_i) \triangle V(P_i')| = 2$; that is, $\mathcal{P}'$ can be obtained from $\mathcal{P}$ by replacing a single (internal) vertex in some shortest path P_i with a vertex which is not contained in $V(\mathcal{P})$. A sequence $\langle \mathcal{P}_0, \mathcal{P}_1, \ldots, \mathcal{P}_\ell \rangle$ of tuples of vertex-disjoint shortest paths is called a *reconfiguration sequence between* $\mathcal{P}_0$ and $\mathcal{P}_\ell$ if $\mathcal{P}_{r-1} \leftrightarrow \mathcal{P}_r$ for all $r \in [\ell]$. We call ℓ the *length* of the sequence. We write $\mathcal{P} \rightsquigarrow \mathcal{Q}$ if there exists a reconfiguration sequence between $\mathcal{P}$ and $\mathcal{Q}$. Note that any reconfiguration sequence is reversible; therefore, $\mathcal{P}$ is reconfigurable to $\mathcal{Q}$ if and only if $\mathcal{Q}$ is reconfigurable to $\mathcal{P}$. We now formally define the problem.

REACHABILITY OF VERTEX-DISJOINT SHORTEST PATHS (RVDSP)

> **Input:** An unweighted graph G, and two tuples $\mathcal{P}$ and $\mathcal{Q}$ of vertex-disjoint shortest paths for k terminal pairs (s_i, t_i).
>
> **Question:** Determine whether $\mathcal{P} \rightsquigarrow \mathcal{Q}$.

Note that RVDSP is a decision problem, and does not require an actual reconfiguration sequence as part of the output. For a set of k terminal pairs (s_i, t_i), we define the *maximum terminal distance* as the maximum distance over all terminal pairs, that is, $\max_{i=1}^{k} \mathsf{dist}(s_i, t_i)$.

3   Maximum Terminal Distance

In this section, we analyze the complexity of RVDSP from the viewpoint of the maximum terminal distance. We first obtain the following polynomial-time solvability result by utilizing [19, Theorem 2].

Theorem 1 ($*$). RVDSP *can be solved in polynomial time when the maximum terminal distance is at most two.*

3.1   PSPACE-Completeness

In contrast to Theorem 1, we then give the following hardness result. We note that our analysis is tight with respect to the maximum terminal distance.

Theorem 2. *For any fixed integer $D \geq 3$,* RVDSP *is* PSPACE-*complete for graphs of bounded bandwidth with maximum degree five, even if the maximum terminal distance is D.*

We prove Theorem 2 in this subsection. Observe that RVDSP belongs to the class PSPACE, and hence it suffices to prove that the problem is PSPACE-hard. To do so, we give a polynomial-time reduction from a PSPACE-complete problem, called NONDETERMINISTIC CONSTRAINT LOGIC [10,22], to our problem.

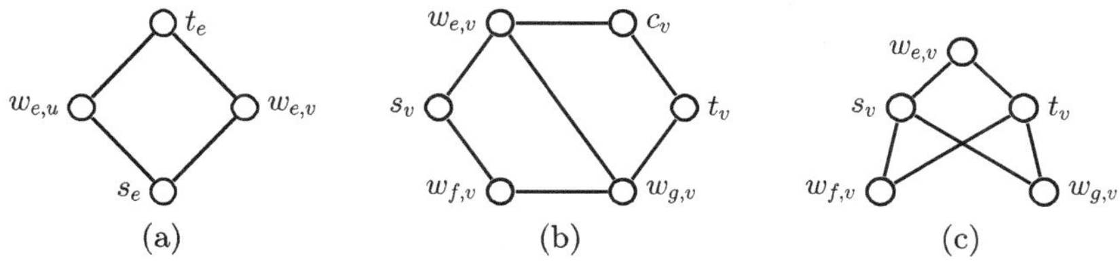

Fig. 3. (a) The edge gadget G_e for an edge $e = uv \in E(M)$, (b) the AND gadget G_v for an AND vertex $v \in V(M)$, and (c) the OR gadget G_v for an OR vertex $v \in V(M)$.

Nondeterministic Constraint Logic. We here define the NONDETERMINISTIC CONSTRAINT LOGIC problem [10,22]. An *NCL machine* is an undirected and edge-weighted graph such that the weight of each edge is either 1 or 2. Every vertex in an NCL machine is of degree three, and is either an AND vertex or an OR vertex, defined as follows: An *AND vertex* is a vertex such that exactly one edge has weight 2 and the other two edges have weight 1, and an *OR vertex* is a vertex such that all the three edges have weight 2.

A *configuration of an NCL machine* is an orientation of the edges such that, for any vertex v in the NCL machine, the sum of weights of the incoming arcs to v is at least 2. Two configurations of an NCL machine are *adjacent* if one can be obtained from the other by flipping the direction of a single edge. Given an NCL machine M and two configurations σ and τ of M, the NONDETERMINISTIC CONSTRAINT LOGIC problem is to determine whether there exists a sequence of adjacent configurations of M that transforms σ into τ. We sometimes refer to such a sequence as a *reconfiguration sequence between σ and τ*. NONDETERMINISTIC CONSTRAINT LOGIC is known to be **PSPACE**-complete even when the input NCL machine is planar and has bounded bandwidth [22].

Reduction. Let $I' = (M, \sigma, \tau)$ be an instance of NONDETERMINISTIC CONSTRAINT LOGIC such that M has bounded bandwidth.[1] In our reduction, we first introduce three types of gadgets, called *edge gadgets*, *AND gadgets* and *OR gadgets*, and then construct the instance $I = (G, \mathcal{P}_\sigma, \mathcal{P}_\tau)$ of RVDSP corresponding to $I' = (M, \sigma, \tau)$.

We first define the *edge gadget G_e* for each edge $e = uv \in E(M)$, as follows (see also Fig. 3a):

$$V(G_e) = \{s_e,\ t_e,\ w_{e,u},\ w_{e,v}\}, \quad E(G_e) = \{s_e w_{e,u},\ s_e w_{e,v},\ t_e w_{e,u},\ t_e w_{e,v}\}.$$

Each edge gadget G_e has one terminal pair (s_e, t_e). For each $x \in \{u, v\}$, the vertex $w_{e,x}$ is referred to as the *connector vertex of $e = uv$ for x*. Connector vertices also appear in AND and OR gadgets and serve to connect gadgets.

We then define the AND gadget G_v for each AND vertex $v \in V(M)$. Let $e, f, g \in E(M)$ be the three edges incident to v, where e has weight 2 and each

---

[1] Our reduction does not preserve the planarity, so M is not required to be planar.

Fig. 4. (a) An edge $e = uv \in E(M)$ oriented from u to v, and (b) the corresponding shortest $s_e t_e$-path $s_e w_{e,u} t_e$, highlighted with red, in the edge gadget G_e.

of f and g has weight 1. Then, the *AND gadget* G_v for v is defined as follows (see also Fig. 3b):

$$V(G_v) = \{s_v,\ t_v,\ w_{e,v},\ w_{f,v},\ w_{g,v},\ c_v\},$$
$$E(G_v) = \{s_v w_{e,v},\ s_v w_{f,v},\ w_{e,v} c_v,\ w_{e,v} w_{g,v},\ w_{f,v} w_{g,v},\ c_v t_v,\ w_{g,v} t_v\}.$$

Each AND gadget G_v has one terminal pair (s_v, t_v). For each $x \in \{e, f, g\}$, the vertex $w_{x,v}$ is referred to as the *connector vertex of x for v*, and it will be identified with the connector vertex of the same label in the edge gadget G_x.

We then define the OR gadget G_v for each OR vertex $v \in V(M)$. Let $e, f, g \in E(M)$ be the three edges incident to v, where all the three edges are of weight 2. Then, the *OR gadget* G_v is defined as follows (see also Fig. 3c):

$$V(G_v) = \{s_v,\ t_v,\ w_{e,v},\ w_{f,v},\ w_{g,v}\},$$
$$E(G_v) = \{s_v w_{e,v},\ s_v w_{f,v},\ s_v w_{g,v},\ t_v w_{e,v},\ t_v w_{f,v},\ t_v w_{g,v}\}.$$

Each OR gadget G_v has a terminal pair (s_v, t_v). For each $x \in \{e, f, g\}$, the vertex $w_{x,v}$ is referred to as the *connector vertex of x for v*.

We are now ready to construct the graph G. For each AND (resp., OR) vertex v in $V(M)$, we prepare the corresponding AND (resp., OR) gadget G_v. Similarly, for each edge e in $E(M)$, we prepare the corresponding edge gadget G_e. Then, we connect these gadgets by identifying the connector vertices having the same label; there are exactly two connector vertices having the same label. Let G be the resulting graph. Then, the maximum degree of G is at most five, and the bandwidth of G is at most a constant factor larger than that of M.

To complete the construction, we now define a tuple $\mathcal{P}_\sigma$ of vertex-disjoint shortest paths in G which corresponds to a given configuration σ of M; it is the same way to construct $\mathcal{P}_\tau$ corresponding to τ. By the construction of G, for every terminal pair (s, t) in G, we note that any shortest st-path in G passes through only vertices inside the gadget containing (s, t).

We first define a shortest $s_e t_e$-path P_e on the edge gadget G_e for each edge $e = uv \in E(M)$. If e is oriented from u to v by σ, we set $P_e = s_e w_{e,u} t_e$ (see also Fig. 4); otherwise, it is oriented from v to u, and we set $P_e = s_e w_{e,v} t_e$.

We next define a shortest $s_v t_v$-path P_v on the AND gadget G_v for each AND vertex $v \in V(M)$. Let $e, f, g \in E(M)$ be the three edges incident to v, where e has weight 2 and each of f and g has weight 1. If the weight-2 edge e is oriented inward for v, we set $P_v = s_v w_{e,v} c_v t_v$; otherwise, both f and g must be oriented inward for v, and we set $P_v = s_v w_{f,v} w_{g,v} t_v$.

We then define a shortest $s_v t_v$-path P_v on the OR gadget G_v for each OR vertex $v \in V(M)$. Let $e, f, g \in E(M)$ be the three edges incident to v, where all edges have weight 2. Since σ is a configuration of M, at least one of the edges e, f, g must be oriented inward for v; let x be such an edge chosen arbitrarily. Then, we set $P_v = s_v w_{x,v} t_v$.

Let $\mathcal{P}_\sigma$ be the tuple of the shortest paths defined above for all terminal pairs in G. By the construction, the shortest paths in $\mathcal{P}_\sigma$ do not pass through the same connector vertex $w_{e,v}$ shared between an edge gadget G_e and an AND/OR gadget G_v. Thus, $\mathcal{P}_\sigma$ forms a tuple of vertex-disjoint shortest paths.

This completes the construction of the corresponding instance $I = (G, \mathcal{P}_\sigma, \mathcal{P}_\tau)$ of RVDSP. Then, the following lemma completes the proof of Theorem 2.

Lemma 1 (∗). (M, σ, τ) *is a yes-instance of* NONDETERMINISTIC CONSTRAINT LOGIC *if and only if* $(G, \mathcal{P}_\sigma, \mathcal{P}_\tau)$ *is a yes-instance of* RVDSP.

3.2 Parameterization by the Maximum Terminal Distance

From the viewpoint of parameterized complexity, Theorem 2 says that RVDSP is para-PSPACE-complete with respect to the parameter D. In addition, the problem is known to be para-PSPACE-complete with respect to k [19, 21]. Thus, RVDSP is unlikely to admit even an XP algorithm with a single parameter D or k. On the other hand, we here point out that the combination of the two parameters gives the following result.

Proposition 1 (∗). RVDSP *admits an XP algorithm with the parameter* $k + D$.

A natural question to Proposition 1 is the fixed-parameter tractability of RVDSP when parameterized by $k + D$. However, we observe from the result by Bousquet et al. [4] that the problem is W[1]-hard even if we take the length ℓ of a reconfiguration sequence as an additional parameter.

Observation 1. *For any fixed* $k \geq 1$, RVDSP *is* W[1]*-hard for 6-degenerate graphs when parameterized by* $\ell + D$, *where* ℓ *is the length of a reconfiguration sequence and* D *is the maximum terminal distance.*

4 Parameterization by Reconfiguration Length

Throughout this section, we denote by ℓ the length of a reconfiguration sequence, and regard it as a parameter. We first point out that a simple enumeration gives the following result.

Proposition 2 (∗). RVDSP *admits an XP algorithm with the parameter* ℓ.

Observation 1 says that RVDSP is unlikely to admit a fixed-parameter algorithm with $\ell + D$ (and hence with the single parameter ℓ). Therefore, to obtain a fixed-parameter algorithm for RVDSP, we need an additional parameter other than D. In this section, we take the maximum degree Δ of an input graph as an additional parameter, and give the following theorem.

Theorem 3. RVDSP *is fixed-parameter tractable when parameterized by* $\ell + \Delta$.

We note that Δ itself is not so powerful parameter in the following sense: RVDSP is PSPACE-complete for $\Delta = 5$ (Theorem 2), and hence the problem is unlikely to admit even an XP algorithm with the single parameter Δ.

Definitions and Notation. To prove Theorem 3, we provide a fixed-parameter algorithm for RVDSP parameterized by $\ell + \Delta$. We first introduce some notation. For a vertex v in a graph G and an integer $d \geq 0$, let $N^d[v]$ denote the set of vertices within distance d from v, that is, $N^d[v] := \{u \in V(G) \mid \mathsf{dist}(u, v) \leq d\}$. Note that $|N^d[v]| \leq \Delta^d$ holds for any vertex $v \in V(G)$. For a vertex subset $S \subseteq V(G)$, we define $N^d[S] := \bigcup_{v \in S} N^d[v]$.

Let $I = (G, \mathcal{P}, \mathcal{Q})$ be an instance of RVDSP, where $\mathcal{P} = (P_1, P_2, \ldots, P_k)$ and $\mathcal{Q} = (Q_1, Q_2, \ldots, Q_k)$ are tuples of vertex-disjoint shortest paths with $\mathcal{P} \neq \mathcal{Q}$. We define $V_{\mathrm{diff}}(\mathcal{P}, \mathcal{Q}) := \bigcup_{i \in [k]} V(P_i) \setminus V(Q_i)$. This means that $V_{\mathrm{diff}}(\mathcal{P}, \mathcal{Q})$ is the set of all vertices v in $V(\mathcal{P})$ that belong to some path P_i in $\mathcal{P}$, but do not belong to the corresponding path Q_i in $\mathcal{Q}$. Note that, however, such a vertex v may belong to another path $Q_{i'}$ in $\mathcal{Q}$. By the definition, $V_{\mathrm{diff}}(\mathcal{P}, \mathcal{Q}) \subseteq V(\mathcal{P})$ and $V_{\mathrm{diff}}(\mathcal{Q}, \mathcal{P}) \subseteq V(\mathcal{Q})$, and it does not necessarily hold that $V_{\mathrm{diff}}(\mathcal{P}, \mathcal{Q}) = V_{\mathrm{diff}}(\mathcal{Q}, \mathcal{P})$.

Assume that there is a reconfiguration sequence $\mathcal{R} = \langle \mathcal{P}_0, \mathcal{P}_1, \ldots, \mathcal{P}_{\ell'} \rangle$ between $\mathcal{P} (= \mathcal{P}_0)$ and $\mathcal{Q} (= \mathcal{P}_{\ell'})$, where $\ell' \leq \ell$. We say that a vertex $u \in V(G)$ is *deleted during* $\mathcal{R}$ if there is some $r \in [\ell']$ such that $\{u\} = V(\mathcal{P}_{r-1}) \setminus V(\mathcal{P}_r)$. Conversely, we say that a vertex $v \in V(G)$ is *added during* $\mathcal{R}$ if there is some $r \in [\ell']$ such that $\{v\} = V(\mathcal{P}_r) \setminus V(\mathcal{P}_{r-1})$. Note that $\mathcal{R}$ may both delete and add the same vertex, and may do so multiple times. Let $D_{\mathcal{R}} := \bigcup_{r \in [\ell']} V(\mathcal{P}_{r-1}) \setminus V(\mathcal{P}_r)$ and $A_{\mathcal{R}} := \bigcup_{r \in [\ell']} V(\mathcal{P}_r) \setminus V(\mathcal{P}_{r-1})$. Then, it always holds that $V_{\mathrm{diff}}(\mathcal{P}, \mathcal{Q}) \subseteq D_{\mathcal{R}}$ and $V_{\mathrm{diff}}(\mathcal{Q}, \mathcal{P}) \subseteq A_{\mathcal{R}}$.

Key Lemma and Algorithm. The key observation underlying our algorithm is the following: In each reconfiguration step, the deleted and added vertices are within distance at most two. Therefore, to find a reconfiguration sequence between $\mathcal{P}$ and $\mathcal{Q}$ of length at most ℓ, it suffices to consider deletions and additions involving only vertices in $N^{2\ell}[V_{\mathrm{diff}}(\mathcal{P}, \mathcal{Q})]$. Formally, we have the following lemma.

Lemma 2. *Let* $I = (G, \mathcal{P}, \mathcal{Q})$ *be an instance of* RVDSP *such that there exists a reconfiguration sequence* $\mathcal{R}$ *between* $\mathcal{P}$ *and* $\mathcal{Q}$ *of length at most* ℓ. *Then, there exists a reconfiguration sequence* $\mathcal{R}^*$ *between* $\mathcal{P}$ *and* $\mathcal{Q}$ *of length at most* ℓ *such that* $D_{\mathcal{R}^*} \cup A_{\mathcal{R}^*} \subseteq N^{2\ell}[V_{\mathrm{diff}}(\mathcal{P}, \mathcal{Q})]$.

Proof. For convenience, we introduce the following notation: let $S_0 := V_{\mathrm{diff}}(\mathcal{P}, \mathcal{Q})$, and $S_d := N^{2d}[V_{\mathrm{diff}}(\mathcal{P}, \mathcal{Q})] \setminus N^{2d-2}[V_{\mathrm{diff}}(\mathcal{P}, \mathcal{Q})]$ for each integer $d \geq 1$. By the definition, the sets $S_0, S_1, \dots$ are pairwise disjoint, in particular, $S_0, S_1, \dots, S_\ell$ form a partition of $N^{2\ell}[V_{\mathrm{diff}}(\mathcal{P}, \mathcal{Q})]$.

Let $\mathcal{R} = \langle \mathcal{P}_0, \mathcal{P}_1, \dots, \mathcal{P}_{\ell'} \rangle$ be any reconfiguration sequence from $\mathcal{P}$ ($= \mathcal{P}_0$) to $\mathcal{Q}$ ($= \mathcal{P}_{\ell'}$) of length ℓ', where $\ell' \leq \ell$. Since $|D_{\mathcal{R}}| \leq \ell$ and $S_0, S_1, \dots, S_\ell$ are pairwise disjoint, there exists an index z such that $z \leq \ell$ and $S_z \cap D_{\mathcal{R}} = \emptyset$. We indeed know that $z \geq 1$ since $S_0 = V_{\mathrm{diff}}(\mathcal{P}, \mathcal{Q}) \subseteq D_{\mathcal{R}}$.

Claim 1 (*). *For each $r \in [\ell']$, let $u \in V(\mathcal{P}_{r-1}) \setminus V(\mathcal{P}_r)$ and $v \in V(\mathcal{P}_r) \setminus V(\mathcal{P}_{r-1})$. Then, the following (1)–(3) hold:*

(1) if $u \in N^{2z-2}[V_{\mathrm{diff}}(\mathcal{P}, \mathcal{Q})]$, then $v \in N^{2z}[V_{\mathrm{diff}}(\mathcal{P}, \mathcal{Q})]$;
(2) $u \notin S_z = N^{2z}[V_{\mathrm{diff}}(\mathcal{P}, \mathcal{Q})] \setminus N^{2z-2}[V_{\mathrm{diff}}(\mathcal{P}, \mathcal{Q})]$; and
(3) if $u \in V(G) \setminus N^{2z}[V_{\mathrm{diff}}(\mathcal{P}, \mathcal{Q})]$, then $v \in V(G) \setminus N^{2z}[V_{\mathrm{diff}}(\mathcal{P}, \mathcal{Q})]$.

Consider any tuple $\mathcal{P}'$ of vertex-disjoint shortest paths for terminal pairs (s_i, t_i), $i \in [k]$. Then, for the tuple $\mathcal{P}'$ and each vertex $v \in V(G)$, we define

$$
\phi(\mathcal{P}', v) := \begin{cases} i & \text{if } v \text{ belongs to the shortest } s_i t_i\text{-path in } \mathcal{P}'; \\ \mathsf{null} & \text{otherwise.} \end{cases}
$$

Note that, if $\phi(\mathcal{P}', v) = i$, then v appears at the position $\mathrm{dist}(s_i, v)$ from the terminal s_i on the shortest $s_i t_i$-path in $\mathcal{P}'$. Therefore, $\phi(\mathcal{P}', v)$ provides sufficient information to identify the location of v on $\mathcal{P}'$ whenever $v \in V(\mathcal{P}')$.

Claim 2 (*). $\phi(\mathcal{P}, v) = \phi(\mathcal{Q}, v)$ *for each $v \in V(G) \setminus N^{2z}[V_{\mathrm{diff}}(\mathcal{P}, \mathcal{Q})]$.*

We now construct a new sequence $\mathcal{R}'$ from $\mathcal{R} = \langle \mathcal{P}_0, \mathcal{P}_1, \dots, \mathcal{P}_{\ell'} \rangle$ by ignoring any reconfiguration step that deletes or adds a vertex in $V(G) \setminus N^{2z}[V_{\mathrm{diff}}(\mathcal{P}, \mathcal{Q})]$. Formally, for each $r \in \{0, 1, \dots, \ell'\}$, we define a tuple $(V'_{r,1}, V'_{r,2}, \dots, V'_{r,k})$ of k vertex subsets, as follows: for each $i \in [k]$,

$$
\begin{aligned}
V'_{r,i} = &\{v \in N^{2z}[V_{\mathrm{diff}}(\mathcal{P}, \mathcal{Q})] \mid \phi(\mathcal{P}_r, v) = i\} \\
&\cup \{v \in V(G) \setminus N^{2z}[V_{\mathrm{diff}}(\mathcal{P}, \mathcal{Q})] \mid \phi(\mathcal{P}_0, v) = i\}.
\end{aligned}
$$

Let $\mathcal{P}'_r := (G[V'_{r,1}], G[V'_{r,2}], \dots, G[V'_{r,k}])$. Then, we have the following claim.

Claim 3 (*). *For each $r \in \{0, 1, \dots, \ell'\}$ and each $i \in [k]$, the subgraph $G[V'_{r,i}]$ is a shortest $s_i t_i$-path, and $\mathcal{P}'_r$ forms a tuple of vertex-disjoint shortest paths. In particular, $\mathcal{P}'_0 = \mathcal{P}_0 = \mathcal{P}$ and $\mathcal{P}'_{\ell'} = \mathcal{P}_{\ell'} = \mathcal{Q}$.*

It may hold that $\mathcal{P}'_{r-1} = \mathcal{P}'_r$ for some $r \in [\ell']$, because we ignored any reconfiguration step that involves a vertex in $V(G) \setminus N^{2z}[V_{\mathrm{diff}}(\mathcal{P}, \mathcal{Q})]$. We observe that, if $\mathcal{P}'_r \neq \mathcal{P}'_{r-1}$, then the tuple $\mathcal{P}'_r$ can be obtained from $\mathcal{P}'_{r-1}$ by replacing a single vertex. From the sequence $\langle \mathcal{P}'_0, \mathcal{P}'_1, \dots, \mathcal{P}'_{\ell'} \rangle$, we delete duplicated tuples and obtain the sequence $\mathcal{R}^*$. Then, $\mathcal{R}^*$ is a reconfiguration sequence between $\mathcal{P}$ and $\mathcal{Q}$ of length at most ℓ such that $D_{\mathcal{R}^*} \cup A_{\mathcal{R}^*} \subseteq N^{2\ell}[V_{\mathrm{diff}}(\mathcal{P}, \mathcal{Q})]$. This completes the proof of the lemma. $\qquad\square$

Proof of Theorem 3. For an instance $I = (G, \mathcal{P}, \mathcal{Q})$ of RVDSP, we can assume that $|V_{\text{diff}}(\mathcal{P}, \mathcal{Q})| \leq \ell$, because otherwise we can answer "no" immediately. By Lemma 2, to find a reconfiguration sequence between $\mathcal{P}$ and $\mathcal{Q}$ of length at most ℓ, it suffices to consider the case where only vertices in $N^{2\ell}[V_{\text{diff}}(\mathcal{P}, \mathcal{Q})]$ are involved. Since $|N^{2\ell}[v]| \leq \Delta^{2\ell}$ holds for any vertex $v \in V(G)$, we have

$$\left| N^{2\ell}[V_{\text{diff}}(\mathcal{P}, \mathcal{Q})] \right| \leq |V_{\text{diff}}(\mathcal{P}, \mathcal{Q})| \cdot \Delta^{2\ell} \leq \ell \cdot \Delta^{2\ell}.$$

Therefore, $|N^{2\ell}[V_{\text{diff}}(\mathcal{P}, \mathcal{Q})]|$ is bounded by a function of $\ell + \Delta$, and hence we can enumerate all possible reconfiguration sequences in fixed-parameter time with respect to $\ell + \Delta$. This completes the proof of Theorem 3. $\qquad\square$

5   Structural Graph Parameters

In this section, we analyze RVDSP parameterized by structural graph parameters: modular-width, cluster deletion number, and tree-depth. (See also Fig. 2.)

5.1   Modular-Width

In a graph $G = (V, E)$, a *module of* G is a subset M of vertices such that all vertices in M have the same neighborhoods outside the module, namely, for all $x, y \in M$ and any $z \in V \setminus M$, if $xz \in E$, then $yz \in E$. A graph G is said to have *modular-width at most* w if it satisfies at least one of the following conditions: (i) $|V| \leq w$, or (ii) V can be partitioned into at most w non-empty subsets $V_1, V_2, \ldots, V_{w'}$, where $w' \leq w$, such that for each $j \in [w']$, the subgraph $G[V_j]$ of G induced by V_j has modular-width at most w and V_j forms a module of G. We refer to such a partition as a *modular partition of* G. We denote by $\mathsf{mw}(G)$ the smallest w such that G has modular-width at most w. It is known that there is a polynomial-time algorithm which computes a modular partition of a given graph G into at most $\mathsf{mw}(G)$ modules [6,9,20]. We show the following lemma, which plays an important role in designing our algorithm.

Lemma 3 ($*$). *Let G be a connected graph such that $|V(G)| > \mathsf{mw}(G)$, and let $V_1, V_2, \ldots, V_w$ be a modular partition of G, where $w \leq \mathsf{mw}(G)$. Let s, t be any two vertices in G, and let P be any shortest st-path in G. Then, every module V_j, $j \in [w]$, intersects with at most one internal vertex of P, that is, $|V_j \cap (V(P) \setminus \{s, t\})| \leq 1$.*

In this subsection, we consider RVDSP parameterized by $k + \mathsf{mw}$, where mw is an upper bound on the modular-width of an input graph.

Theorem 4. RVDSP *is fixed-parameter tractable when parameterized by $k + \mathsf{mw}$. In particular, RVDSP admits a kernel with at most $\mathsf{mw} \cdot (k^2 + 5k)$ vertices.*

Let $I = (G, \mathcal{P}, \mathcal{Q})$ be an instance of RVDSP such that G has modular-width at most mw. To prove Theorem 4, we propose a kernelization algorithm based on two reduction rules, each of which produces a smaller instance $(G', \mathcal{P}', \mathcal{Q}')$. We say that a reduction rule is *safe* if the following holds: the instance $(G', \mathcal{P}', \mathcal{Q}')$ obtained by applying the reduction rule is a yes-instance if and only if the original instance $(G, \mathcal{P}, \mathcal{Q})$ is a yes-instance.

In the remainder of this subsection, we may assume without loss of generality that the input graph G satisfies $|V(G)| > \mathsf{mw}(G)$; otherwise G already forms a desired kernel. Then, any modular partition $V_1, V_2, \ldots, V_w$ of G, with $w \leq \mathsf{mw}(G)$, satisfies $w \geq 2$ and $V_j \subsetneq V(G)$ for all $j \in [w]$. There may exist more than one such modular partition of G, but we choose one of them arbitrarily and will apply our reduction rules based on it. Furthermore, we can *ignore* any terminal pair (s, t) such that $\mathsf{dist}(s, t) = 1$, because any shortest st-path for such a terminal pair is always a single edge st which cannot be used by any other vertex-disjoint paths. Although we do not delete such terminal pairs from I for preserving the connectedness of the graph, we may assume in the following that $\mathsf{dist}(s, t) \geq 2$ for each terminal pair (s, t).

We categorize each module V_j for $j \in [w]$ into one of the following two types:

- *Type 1*: there is no terminal pair (s, t) in I such that $s, t \in V_j$; and
- *Type 2*: there exists a terminal pair (s, t) in I such that $s, t \in V_j$.

The first reduction rule is based on the following observation. Consider any module V_j of Type 1. Then, for each terminal pair (s, t) in I, Lemma 3 implies that any shortest st-path passes through at most one vertex in V_j. Therefore, to preserve the reconfigurability between $\mathcal{P}$ and $\mathcal{Q}$, it suffices that at most k vertices in V_j are involved. The following rule formalizes this idea and provides the condition under which excess vertices in V_j can be safely removed.

Rule 1. *If there is a module V_j of Type 1 such that $|V_j \setminus (V(\mathcal{P}) \cup V(\mathcal{Q}))| > k$, then arbitrarily choose a vertex $v \in V_j \setminus (V(\mathcal{P}) \cup V(\mathcal{Q}))$ and remove v from G.*

Since the removed vertex v is not contained in $V(\mathcal{P}) \cup V(\mathcal{Q})$, the resulting instance $(G - v, \mathcal{P}, \mathcal{Q})$ remains an instance of RVDSP, where $G - v$ denotes the graph obtained from G by deleting v. By the choice of v, we note that $G - v$ remains connected. From the modular partition $V_1, V_2, \ldots, V_j, \ldots, V_w$ of G, we obtain the modular partition $V_1, V_2, \ldots, V_j \setminus \{v\}, \ldots, V_w$ of $G - v$. Note that deleting vertices does not increase the modular-width of the graph.

Proposition 3 ($*$). *Rule 1 is safe.*

If Rule 1 is no longer applicable to $(G, \mathcal{P}, \mathcal{Q})$, then $|V_j \setminus (V(\mathcal{P}) \cup V(\mathcal{Q}))| \leq k$ for each module V_j of Type 1. Since V_j is of Type 1, at most one of the terminals s_i and t_i is contained in V_j for each $i \in [k]$. Furthermore, Lemma 3 implies that $|V_j \cap (V(P) \setminus \{s, t\})| \leq 1$ holds for each of $2k$ shortest st-paths P in $\mathcal{P}$ and $\mathcal{Q}$. As a result, we have $|V_j \cap (V(\mathcal{P}) \cup V(\mathcal{Q}))| \leq 3k$. Therefore, if Rule 1 is no longer applicable to G, then it holds that $|V_j| \leq 4k$ for each module V_j of Type 1.

We next consider a module V_j of Type 2. Then, there is a terminal pair (s_i, t_i) such that $s_i, t_i \in V_j$. Since G is connected and admits a modular partition $V_1, V_2, \ldots, V_w$ with $w \geq 2$, it follows that $\mathsf{dist}(s_i, t_i) = 2$. Let $L^i := N_G(s_i) \cap N_G(t_i)$ for each $i \in [k]$, that is, L^i is the set of common neighbors of s_i and t_i. In other words, vertices in L^i may be used as internal vertices of shortest $s_i t_i$-paths during reconfiguration. We also define

$$L_{\bar{j}}^{\leq k} := \bigcup \{L^i \cap V_j \mid i \in [k] \text{ such that } s_i, t_i \in V_j \text{ and } |L^i \cap V_j| \leq k\}.$$

Notice that $|L_{\bar{j}}^{\leq k}| \leq k^2$. We then introduce the following reduction rule.

Rule 2. *If there is a module V_j of Type 2 such that $|V_j \setminus (V(\mathcal{P}) \cup V(\mathcal{Q}) \cup L_{\bar{j}}^{\leq k})| > k$, then arbitrarily choose a vertex $v \in V_j \setminus (V(\mathcal{P}) \cup V(\mathcal{Q}) \cup L_{\bar{j}}^{\leq k})$ and remove v from G.*

Proposition 4 ($*$). *Rule 2 is safe.*

If Rule 2 is no longer applicable to $(G, \mathcal{P}, \mathcal{Q})$, then we claim that $|V_j| \leq k^2 + 5k$ holds for any module V_j of Type 2, as follows. Since V_j is of Type 2, we have $|V_j \cap \{s_i, t_i\}| \leq 2$ for each $i \in [k]$. Furthermore, Lemma 3 implies that $|V_j \cap (V(P) \setminus \{s, t\})| \leq 1$ holds for each of $2k$ shortest st-paths P in $\mathcal{P}$ and $\mathcal{Q}$. As a result, we have $|V_j \cap (V(\mathcal{P}) \cup V(\mathcal{Q}))| \leq 4k$. Since $|L_{\bar{j}}^{\leq k}| \leq k^2$ and Rule 2 is no longer applicable to $(G, \mathcal{P}, \mathcal{Q})$, we thus have

$$|V_j| \leq |V_j \setminus (V(\mathcal{P}) \cup V(\mathcal{Q}) \cup L_{\bar{j}}^{\leq k})| + |V_j \cap (V(\mathcal{P}) \cup V(\mathcal{Q}))| + |L_{\bar{j}}^{\leq k}|$$
$$\leq k + 4k + k^2 = k^2 + 5k.$$

Proof of Theorem 4. For a given graph G with $\mathsf{mw}(G) \leq \mathsf{mw}$, we can obtain a modular partition $V_1, V_2, \ldots, V_w$ of G, with $w \leq \mathsf{mw}$, in polynomial time [6,9,20]. Based on the modular partition, each reduction rule can be applied in polynomial time, and hence our kernelization algorithm terminates in polynomial time. As we have discussed, if neither Rule 1 nor Rule 2 is applicable to $(G, \mathcal{P}, \mathcal{Q})$, then it holds that $|V_j| \leq k^2 + 5k$ for any module V_j, $j \in [w]$. Since there are at most mw modules, we have $|V(G)| \leq \mathsf{mw} \cdot (k^2 + 5k)$, as claimed. In this way, RVDSP is fixed-parameter tractable when parameterized by $k + \mathsf{mw}$. $\square$

5.2   Other Parameters

In this subsection, we show that RVDSP is fixed-parameter tractable when parameterized by k plus cluster deletion number or k plus tree-depth.

For an integer $d \geq 0$, a graph G has *tree-depth at most d* if there is a rooted tree T spanning $V(G)$, with depth at most d, such that for every edge $uv \in E(G)$, either u is an ancestor of v in T, or v is an ancestor of u. It is known that finding such a tree T, if it exists, is fixed-parameter tractable when parameterized by d [18]. The *tree-depth of G*, denoted by $\mathsf{td}(G)$, is the smallest such d. The tree-depth provides an upper bound on the diameter: for any graph G, the diameter of G is less than $2^{\mathsf{td}(G)}$ [16].

A *cluster deletion set of a graph G* is a subset $S \subseteq V(G)$ such that every connected component of $G - S$ is a complete graph, where $G - S$ is the graph obtained from G by removing the vertices in S. The *cluster deletion number of G*, denoted by $\mathsf{cd}(G)$, is the size of a minimum cluster deletion set. It is known that finding a minimum cluster deletion set of a given graph G is fixed-parameter tractable when parameterized by $\mathsf{cd}(G)$ [12]. Moreover, the cluster deletion number provides an upper bound on the diameter: for any graph G, the diameter of G is less than $3 \cdot \mathsf{cd}(G)$ [5, Proposition 25].

We give the following theorem in this subsection, where td and cd denote upper bounds on $\mathsf{td}(G)$ and $\mathsf{cd}(G)$ of an input graph G, respectively.

Theorem 5 (∗). RVDSP *is fixed-parameter tractable when parameterized by* $k + \mathsf{td}$ *or* $k + \mathsf{cd}$.

Proof sketch. We utilize the algorithmic meta-theorem for combinatorial reconfiguration problems by Gima et al. [8]. To do so, we observe that RVDSP can be expressed as a first-order logic (FO) formula whose length depends on k plus the diameter of an input graph, because any tuple of k vertex-disjoint shortest paths consists of at most k times the diameter of vertices. By the upper bounds on the diameter by td [16] or cd [5], this implies that the length of the FO formula depends on $k + \mathsf{td}$ or $k + \mathsf{cd}$. Then, we can apply the algorithmic meta-theorems [8, Theorem 4.1] for td, and [8, Corollary 4.6] for cd.[2]   □

6   Concluding Remarks

We conclude this paper with two open questions. The first question is to clarify the complexity status of RVDSP for planar graphs. In the proof of Theorem 2, even though NONDETERMINISTIC CONSTRAINT LOGIC remains PSPACE-complete for planar graphs of bounded bandwidth [22], our reduction does not preserve the planarity. We note that SPR (i.e., RVDSP for $k = 1$) is solvable in polynomial time for planar graphs [3].

The second question is to determine the parameterized complexity of RVDSP with respect to structural graph parameters alone: Can we drop the dependence on k from Theorems 4 and 5, and establish the fixed-parameter tractability only with mw, td, or cd? Note that RVDSP deals with *internally* vertex-disjoint paths, and hence terminals may share the same vertices. Therefore, it is not straightforward to bound k by structural graph parameters alone.

References

1. Asplund, J., Edoh, K.D., Haas, R., Hristova, Y., Novick, B., Werner, B.: Reconfiguration graphs of shortest paths. Discret. Math. **341**(10), 2938–2948 (2018). https://doi.org/10.1016/j.disc.2018.07.007
2. Bonsma, P.S.: The complexity of rerouting shortest paths. Theoret. Comput. Sci. **510**, 1–12 (2013). https://doi.org/10.1016/j.tcs.2013.09.012

---

[2] More precisely, Gima et al. [8] established fixed-parameter tractability of reconfiguration problems definable in MSO_2 for td plus the solution size, and in MSO_1 for cd plus the solution size, both of them are more expressive logic than FO.

3. Bonsma, P.S.: Rerouting shortest paths in planar graphs. Discret. Appl. Math. **231**, 95–112 (2017). https://doi.org/10.1016/j.dam.2016.05.024

4. Bousquet, N., Gajjar, K., Lahiri, A., Mouawad, A.E.: Parameterized shortest path reconfiguration. In: Proceedings of IPEC 2024. Leibniz International Proceedings in Informatics, vol. 321, pp. 23:1–23:14 (2024). https://doi.org/10.4230/LIPICS.IPEC.2024.23

5. Bousquet, N., Gajjar, K., Lahiri, A., Mouawad, A.E.: Parameterized shortest path reconfiguration. CoRR abs/2406.12717 (2024). https://doi.org/10.48550/ARXIV.2406.12717

6. Cournier, A., Habib, M.: A new linear algorithm for modular decomposition. In: Tison, S. (ed.) CAAP 1994. LNCS, vol. 787, pp. 68–84. Springer, Heidelberg (1994). https://doi.org/10.1007/BFB0017474

7. Gajjar, K., Jha, A.V., Kumar, M., Lahiri, A.: Reconfiguring shortest paths in graphs. Algorithmica **86**(10), 3309–3338 (2024). https://doi.org/10.1007/S00453-024-01263-Y

8. Gima, T., Ito, T., Kobayashi, Y., Otachi, Y.: Algorithmic meta-theorems for combinatorial reconfiguration revisited. Algorithmica **86**(11), 3395–3424 (2024). https://doi.org/10.1007/S00453-024-01261-0

9. Habib, M., Paul, C.: A survey of the algorithmic aspects of modular decomposition. Comput. Sci. Rev. **4**(1), 41–59 (2010). https://doi.org/10.1016/J.COSREV.2010.01.001

10. Hearn, R.A., Demaine, E.D.: Games, Puzzles and Computation. A K Peters (2009)

11. van den Heuvel, J.: The complexity of change. In: Surveys in Combinatorics 2013. London Mathematical Society Lecture Note Series, vol. 409, pp. 127–160. Cambridge University Press (2013). https://doi.org/10.1017/CBO9781139506748.005

12. Hüffner, F., Komusiewicz, C., Moser, H., Niedermeier, R.: Fixed-parameter algorithms for cluster vertex deletion. Theory Comput. Syst. **47**(1), 196–217 (2010). https://doi.org/10.1007/S00224-008-9150-X

13. Kamiński, M., Medvedev, P., Milanič, M.: Shortest paths between shortest paths. Theoret. Comput. Sci. **412**(39), 5205–5210 (2011). https://doi.org/10.1016/j.tcs.2011.05.021

14. Mouawad, A.E., Nishimura, N., Raman, V., Simjour, N., Suzuki, A.: On the parameterized complexity of reconfiguration problems. Algorithmica **78**(1), 274–297 (2017). https://doi.org/10.1007/S00453-016-0159-2

15. Mouawad, A.E., Nishimura, N., Raman, V., Wrochna, M.: Reconfiguration over tree decompositions. In: Cygan, M., Heggernes, P. (eds.) IPEC 2014. LNCS, vol. 8894, pp. 246–257. Springer, Cham (2014). https://doi.org/10.1007/978-3-319-13524-3_21

16. Nešetril, J., Ossona de Mendez, P.: Sparsity - Graphs, Structures, and Algorithms. Algorithms and Combinatorics, vol. 28. Springer, Heidelberg (2012) https://doi.org/10.1007/978-3-642-27875-4

17. Nishimura, N.: Introduction to reconfiguration. Algorithms **11**(4), 52 (2018). https://doi.org/10.3390/a11040052

18. Reidl, F., Rossmanith, P., Villaamil, F.S., Sikdar, S.: A faster parameterized algorithm for treedepth. In: Esparza, J., Fraigniaud, P., Husfeldt, T., Koutsoupias, E. (eds.) ICALP 2014. LNCS, vol. 8572, pp. 931–942. Springer, Heidelberg (2014). https://doi.org/10.1007/978-3-662-43948-7_77

19. Saito, R., Eto, H., Ito, T., Uehara, R.: Reconfiguration of vertex-disjoint shortest paths on graphs. J. Graph Algorithms Appl. **28**(3), 87–101 (2024). https://doi.org/10.7155/JGAA.V28I3.2973

20. Tedder, M., Corneil, D., Habib, M., Paul, C.: Simpler linear-time modular decomposition via recursive factorizing permutations. In: Aceto, L., Damgård, I., Goldberg, L.A., Halldórsson, M.M., Ingólfsdóttir, A., Walukiewicz, I. (eds.) ICALP 2008. LNCS, vol. 5125, pp. 634–645. Springer, Heidelberg (2008). https://doi.org/10.1007/978-3-540-70575-8_52
21. Wrochna, M.: Reconfiguration in bounded bandwidth and tree-depth. J. Comput. Syst. Sci. **93**, 1–10 (2018). https://doi.org/10.1016/j.jcss.2017.11.003
22. van der Zanden, T.C.: Parameterized complexity of graph constraint logic. In: Proceedings of IPEC 2015, pp. 282–293 (2015). https://doi.org/10.4230/LIPICS.IPEC.2015.282

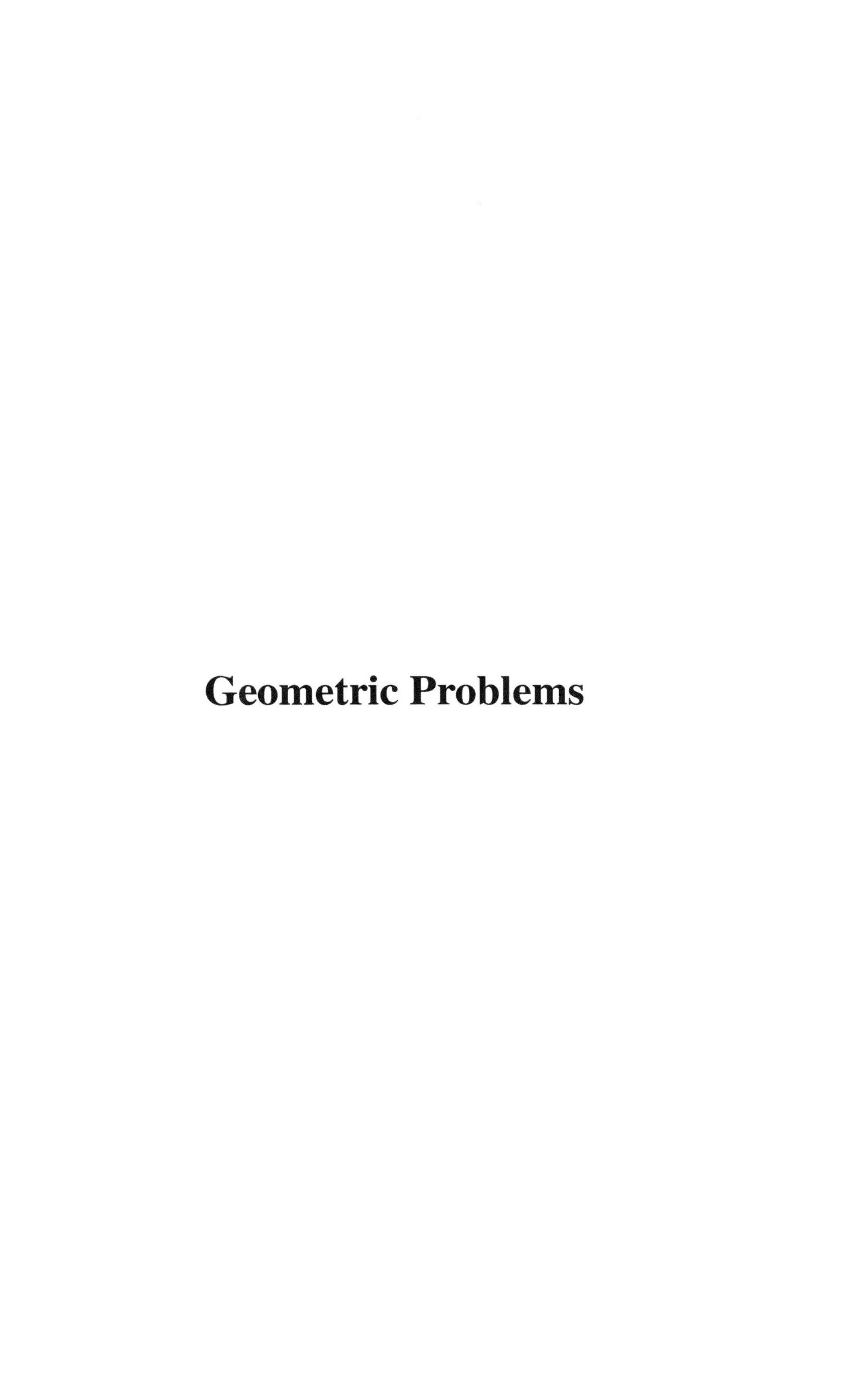

Geometric Problems

Further Results on Rendering Geometric Intersection Graphs Sparse by Dispersion

Nicolás Honorato-Droguett[1]([✉]) [iD], Kazuhiro Kurita[2] [iD], Tesshu Hanaka[3] [iD], Hirotaka Ono[1] [iD], and Alexander Wolff[4] [iD]

[1] Nagoya University, Nagoya, Japan
`honorato.droguett.nicolas.n7@s.mail.nagoya-u.ac.jp`, `ono@i.nagoya-u.ac.jp`
[2] Okayama University, Okayama, Japan
`k-kurita@okayama-u.ac.jp`
[3] Kyushu University, Fukuoka, Japan
`hanaka@inf.kyushu-u.ac.jp`
[4] Universität Würzburg, Würzburg, Germany

Abstract. Removing overlaps is a central task in domains such as scheduling, visibility, and map labelling. This can be modelled using graphs, where overlap removals correspond to enforcing a certain sparsity constraint on the graph structure. We continue the study of the problem GEOMETRIC GRAPH EDIT DISTANCE (GGED), where the aim is to minimise the total cost of editing a geometric intersection graph to obtain a graph contained in a specific graph class. For us, the edit operation is the movement of objects, and the cost is the movement distance.

We present an algorithm for rendering the intersection graph of a set of unit circular arcs (i) edgeless, (ii) acyclic, and (iii) k-clique-free in $O(n \log n)$ time, where n is the number of arcs. We also show that GGED remains strongly NP-hard on *unweighted* interval graphs, solving an open problem of Honorato-Droguett et al. [WADS 2025]. We complement this result by showing that GGED is strongly NP-hard on sets of d-balls and d-cubes, for any $d \geq 2$. Finally, we present an XP algorithm (parameterised by the number of maximal cliques) that removes all edges from the intersection graph of a set of *weighted unit* intervals.

Keywords: Graph modification · Overlap removal · Intersection graphs

1 Introduction

Optimally displacing geometric objects to reduce overlaps arises in several domains including visibility, scheduling, and interference. In scheduling, start times of jobs on a machine are shifted (either delay or advance) to minimise the

This work is partially supported by JSPS KAKENHI Grant Numbers: JP22H03549, JP25K21273, JP25K03080, JP25K00136, JP20H05967, JP21K19765, JP22H00513, JP21K17707, JP23H04388, and JST CRONOS Grant Number JPMJCS24K2.

average deviation from the original schedule. In map labelling, labels are relocated to avoid overlaps while keeping the modification cost small and ensuring visual clarity. In interference, overlapping ranges of sensors or antennas must be reduced in order to improve connectivity while preserving specific global network properties. For a specific graph class Π, consider the following problem. Let $\mathcal{S}$ be a finite collection of (possibly weighted) geometric objects in $\mathbb{R}^d$ and let $G(\mathcal{S})$ be the intersection graph of $\mathcal{S}$. We want to move the objects of $\mathcal{S}$ to obtain a new collection $\mathcal{S}'$ such that (i) $G(\mathcal{S}')$ is in Π and (ii) the sum of the (weighted) distances of the objects in $\mathcal{S}$ is minimised. This problem was introduced in [17,18] under the name GEOMETRIC GRAPH EDIT DISTANCE for Π (GGED(Π) for short), as a way to address graph modification from a geometric perspective, following previous work [10–12]. Through the choice of Π, GGED models diverse problems, such as overlap removal and enhancing connectivity.

The works [17,18] present several polynomial-time algorithms for GGED for obtaining dense and sparse graphs. A complete graph can be optimally obtained in linear time given (i) weighted intervals and (ii) unit squares. Edgeless, acyclic, and k-clique-free graphs and graphs containing a k-clique can be obtained in $O(n \log n)$ time, given n unweighted unit intervals. In contrast, when generalised to arbitrary lengths and weights, the problem becomes strongly NP-hard for obtaining the same sparse graphs. For the minimax variant, it is also strongly NP-hard to obtain edgeless graphs given weighted unit disks.

Given the above context, our motivation is to address unsolved cases of GGED and extending known algorithms to generalisations of intervals. The problem is tractable on intervals with unit length and weight, but becomes strongly NP-hard when both are arbitrary. It is therefore pertinent to analyse cases when only one restriction is relaxed. We also study its parameterised complexity, aligning with recent work on geometric graph modification. While [18] focused on one-dimensional objects, we extend the study to higher dimensions, providing hardness results and structural insights for open cases. On the application side, GGED appears as an abstract model for tasks such as scheduling and map labelling, where the goal is to minimise the overall distortion required to remove overlaps while preserving the original structure. We employ geometric intersection graphs to capture these applications within a unified model.

Next, we briefly summarise results regarding geometric graph modification problems and list applications of overlap removal.

Related Work in Geometric Graph Modification. Researchers [10–12] have studied the problem of minimising the maximum cost of geometric edit operations, specifically *moving* and *scaling*, to make the intersection graph of a given set of disks connected or edgeless. For some cases, they have provided FPT results, for others W[1]-hardness results. Minimising the maximum moving distance is strongly NP-hard given *weighted* unit disks [18]. Other geometric graph modification problems have been considered [3,27], exhibiting the current trend of studying graph modification in geometric domains. A recent survey by Xue and Zehavi [30] gives an overview over this subject.

Applications of Overlap Removal. Scheduling is central in computer science and operations research. Given a set of time lapses (jobs or tasks), the goal is to shift the lapses without overlaps under given constraints. A fundamental example is minimising the total weighted completion time of n given jobs on a single machine, solvable in $O(n \log n)$ [15] but strongly NP-hard with release dates even for unit weights [23]. Greedy approximation algorithms have since been proposed for hard cases [14]. Another variant is to minimise the maximum completion time under precedence constraints, which can be solved in $O(n^2)$ time [22]. For more related results, we refer to surveys and textbooks on scheduling [1,26,28].

Other problems in optimising a parameter under non-overlap constraints appear in the literature. In boundary labelling [2], the labels must be disjointly placed on the boundary of a given figure and connected to features inside the image with curves of minimum length. Various boundary shapes have been considered [2,4,9,20]. Other displacement problems arise in barrier coverage [6], interference [21], visibility [25] and point dispersion [24].

Table 1. Overview and comparison between previous work and our three main results for obtaining graphs in Π_{edgeless}. In this table, L_1 and L_2 are the Manhattan and Euclidean distances, respectively. In the third row, k is the number of maximal cliques.

Object type	Distance	Weight	Complexity	Reference
Unit interval	$L_2(= L_1)$	Yes	Open	[18]
Unit interval	$L_2(= L_1)$	Yes	$\left(1 + \frac{n}{k}\right)^k \cdot \text{poly}(n)$	Theorem 5
Unit interval	$L_2(= L_1)$	No	$O(n \log n)$	[18, Cor. 2]
Unit circular arc	$L_2(= L_1)$	No	$O(n \log n)$	Theorem 2
Interval	$L_2(= L_1)$	Yes	strongly NP-hard	[18, Thm. 2]
d-ball, d-cube	L_2, L_1	No	strongly NP-hard for any $d \geq 1$	Theorem 1, Corollary 2

These examples illustrate that overlap removal is a recurring task across several domains, where the common difficulty lies in achieving disjointness while optimising a specific cost function. This motivates a general framework for overlap removal, which we aim to achieve through GEOMETRIC GRAPH EDIT DISTANCE.

Our Contribution. Table 1 outlines our results for the graph class Π_{edgeless}. For unweighted intervals, we resolve an open problem of Honorato-Droguett et al. [18], showing that $\text{GGED}(\Pi)$ is strongly NP-hard for (i) $\Pi = \Pi_{\text{edgeless}}$, (ii) $\Pi = \Pi_{\text{acyc}}$ and (iii) $\Pi = \overline{\Pi_{k\text{-clique}}}$ (Sect. 3). We extend this reduction to d-dimensional objects, showing that for $d = 2$, the problem remains strongly NP-hard on d-balls and d-cubes for the same graph classes. Subsequently, we show that GGED is solvable in $O(n \log n)$ time on unweighted unit circular arcs for the same graph classes (Sect. 4). Lastly, we show that given weighted unit intervals, $\text{GGED}(\Pi_{\text{edgeless}})$ is solvable in $O(n^4)$ time if the intervals are pairwise intersecting, and provide an XP algorithm when $\text{GGED}(\Pi_{\text{edgeless}})$ is parameterised by the number of maximal cliques (Sect. 5). Proofs of statements marked by ($\star$) appear in the full version of the paper [19].

2 Preliminaries

We provide most of the definitions used throughout the paper, referencing terminology from [5,7,29]. Given $n \in \mathbb{Z}^+$, we use $[n]$ as shorthand for $\{1, 2, \ldots, n\}$ and $[n]_0$ for $\{0, 1, \ldots, n\}$. For two points $p = (p_1, \ldots, p_d)$ and $q = (q_1, \ldots, q_d)$ in $\mathbb{R}^d$, the L_m *distance* of p and q is $\|p, q\|_m = (\sum_{i=1}^{d}(p_d - q_d)^m)^{1/m}$ for an integer $m \geq 1$. We use the L_1 (Manhattan) distance and the L_2 (Euclidean) distance.

Throughout the paper, we assume that all objects presented are open, unless otherwise stated. An *interval* $I = [\ell(I), r(I)]$, $\ell(I) < r(I)$ is the subset $\{x \in \mathbb{R} : \ell(I) \leq x \leq r(I)\}$ of $\mathbb{R}$ where $\ell(I)$ and $r(I)$ are the *left endpoint* and the *right endpoint* of I, respectively. The *centre* of I is the point $c(I) = (r(I) + \ell(I))/2$ and its length is $\mathrm{len}(I) = r(I) - \ell(I)$. The interval I is called *unit interval* when $\mathrm{len}(I) = 1$ and *open* when $\ell(I), r(I) \notin I$ (denoted by $(\ell(I), r(I))$). The *leftmost* and *rightmost endpoints* of a set of intervals $\mathcal{I}$ are defined as $\ell(\mathcal{I}) = \min_{I \in \mathcal{I}} \ell(I)$ and $r(\mathcal{I}) = \max_{I \in \mathcal{I}} r(I)$, respectively. For two intervals I and I', we say that $I \preceq I'$ whenever $c(I) \leq c(I')$. Throughout the paper, we assume that the indices of an n-tuple of intervals $(I_1, \ldots, I_n)$ follow the order $I_i \preceq I_{i+1}$ for all $i \in [n-1]$, unless otherwise stated. However, no order is assumed on $\mathcal{I}$ given as input.

Let C be a circle of radius $r > 0$ centred at the origin. We set $\mathrm{len}(C)$ to the circumference of C, that is, to $2\pi r$. We identify every point p on C with the angle from the positive x-axis and the ray from the origin through p. For $0 \leq p, q < 2\pi$, the *circular arc* $A = [p, q]$ is the subset of C that starts at the point corresponding to p and goes to the point corresponding to q in counterclockwise direction. Note that $[p, q] \cup [q, p] = C$. We let $\mathrm{len}(A)$ denote the length of A. If A is open (i.e. $p, q \notin A$), A is denoted by (p, q). If $\mathrm{len}(A) = 1$, we say that A is a *unit circular arc*. For two points p and q on C, we define $d(p, q) = \min\{\mathrm{len}([p, q]), \mathrm{len}([q, p])\}$ to be the *distance* of p and q. Observe that $d(p, q) \leq C/2$.

A *hypercube* or *d-cube* S centred at $p \in \mathbb{R}^d$ with edge length s is the set $S = \{x \in \mathbb{R}^d : \max_{i \in [d]} |p_i - x_i| \leq s/2\}$. We let $\mathrm{len}(S)$ denote s. An *open* d-cube is a d-cube without its boundary. A 2-cube is called a *square*; it is a *unit square* if $\mathrm{len}(S) = 1$. Given a radius $r > 0$ and $p \in \mathbb{R}^d$, a *d-ball* B centred at p is the set $B = \{x \in \mathbb{R}^d : \|x, p\|_2 \leq r\}$. An *open* d-ball is $B = \{x \in \mathbb{R}^d : \|x, p\|_2 < r\}$. A 2-ball is called a *disk* and a *unit disk* if also $r = 1/2$ (diameter 1). When $d = 1$, d-cubes and d-balls degenerate to intervals.

Graphs. A graph $G = (V, E)$ is assumed to be a simple, finite, and undirected graph with vertex set V and edge set E. A graph G is *edgeless* when $E = \emptyset$. A *k-clique* of G is a subset $W \subseteq V$ such that $|W| = k$ and for all $u, v \in W$, $u \neq v$, $\{u, v\} \in E$, for $k \leq n$. If such W exists in V, we say that G *contains a k-clique*. Given n geometric objects $\mathcal{S} = (S_1, \ldots, S_n)$ in $\mathbb{R}^d$, the *geometric intersection graph of $\mathcal{S}$* is a graph $G(\mathcal{S}) = (V, E)$ where the ith vertex of $V = \{v_1, \ldots, v_n\}$

corresponds to S_i and an edge $\{v_i, v_j\} \in E$ exists if and only if $S_i \cap S_j \neq \emptyset$, for any $i, j \in [n]$, $i \neq j$. We differentiate $G(\mathcal{S})$ according to the type of objects in $\mathcal{S}$. If $\mathcal{S}$ consists of (unit) intervals, then $G(\mathcal{S})$ is a *(unit) interval graph*. Similarly, $G(\mathcal{S})$ is a *(unit) circular arc graph* (provided with a circle C of radius r), *square graph*, *disk graph*, *d-cube graph* and *d-ball graph*. An (infinite) set of graphs Π is a *graph class* (or simply a class), and we say that *G is in Π* if $G \in \Pi$. A graph class Π is *non-trivial* if infinitely many graphs belong to Π and infinitely many graphs do not belong to Π. We consider the following non-trivial classes: (i) $\Pi_{\text{edgeless}} = \{G : G \text{ is edgeless}\}$, (ii) $\Pi_{\text{acyc}} = \{G : G \text{ is acyclic}\}$, (iii) $\Pi_{k\text{-clique}} = \{G : G \text{ has a } k\text{-clique}\}$ and (iv) $\overline{\Pi_{k\text{-clique}}} = \{G : G \notin \Pi_{k\text{-clique}}\}$.

Let $D = (d_{i,j})_{i\in[n], j\in[d]} \in \mathbb{R}^{n\times d}$ be a matrix representing distances, where $d_{i,*} = (d_{i,j})_{j\in[d]}$ is a d-vector. When $d = 1$, D is a vector in $\mathbb{R}^n$. We sometimes call D a *dispersal*. For a geometric object $S \subset \mathbb{R}^d$ and $v \in \mathbb{R}^d$, the object $S + v$ is $\{x + v : x \in S\}$. For an n-tuple $\mathcal{S}$ of geometric objects in $\mathbb{R}^d$, the n-tuple $\mathcal{S} + D$ is $(S_1 + d_{1,*}, \ldots, S_n + d_{n,*})$. We say that a dispersal D is *contiguous* if the n-tuple $\mathcal{I} + D = (I_1', \ldots, I_n')$ ordered by centres satisfies $I_{i+1}' = I_i' + 1$ for every $i \in [n-1]$. For an $n \times m$-matrix $M = (m_{i,j})_{i\in[n], j\in[m]} \in \mathbb{R}^{n\times m}$, $\|M\|_1 = \sum_{i\in[n], j\in[m]} |m_{i,j}|$ is the *1-norm* of M. Let $v \in \mathbb{R}^n$ and $M \in \mathbb{R}^{n\times d}$. The matrix $vM = \text{diag}(v) \cdot M$ is the *matrix M scaled by v*, where $\text{diag}(v) \in \mathbb{R}^{n\times n}$ and $\text{diag}(v)_{i,j} = v_i$ if $i = j$, and 0 otherwise. Let $\mathbf{w} = (w_1, \ldots, w_n) \in \mathbb{R}^n_{>0}$ be a weight vector. Given D as described above, the *total moving distance* is defined as $\|\mathbf{w}D\|_1$.

Now we can formally define our main problem, $\text{GGED}(\Pi)$ for short.

Problem: GEOMETRIC GRAPH EDIT DISTANCE w.r.t. graph class Π

Input: An n-tuple $\mathcal{S}$ of geometric objects in $\mathbb{R}^d$, a weight vector $\mathbf{w} \in \mathbb{R}^n_{>0}$.

Output: A distance matrix D that minimises $\|\mathbf{w}D\|_1$ among all matrices D with $G(\mathcal{S} + D) \in \Pi$.

Parameterised Complexity. We define the two most important notions for this paper; for more background, refer to textbooks [5]. A parameterised problem L is called *fixed-parameter tractable* (FPT) with respect to the parameter k if there is an algorithm $\mathcal{A}$, a computable function $f \colon \mathbb{N} \to \mathbb{N}$ and a constant c such that, given an instance (x, k) of L, $\mathcal{A}$ decides whether $(x, k) \in L$ in $f(k) \cdot |(x, k)|^c$ time. Similarly, L is called *slice-wise polynomial* (XP) if for an additional computable function $g \colon \mathbb{N} \to \mathbb{N}$, the algorithm $\mathcal{A}$ correctly decides whether $(x, k) \in L$ in $f(k) \cdot |(x, k)|^{g(k)}$ time. Correspondingly, we call $\mathcal{A}$ an FPT or an XP algorithm.

3 Hardness of GGEDon Unweighted Intervals and in Higher Dimensions

Recall that $\text{GGED}(\Pi_{\text{edgeless}})$ is strongly NP-hard for weighted intervals. We show that the problem remains strongly NP-hard on unweighted intervals by a reduction from 3-PARTITION and extend it to two generalisations of intervals, namely d-cubes and d-balls. We start by showing the reduction for intervals.

3.1   Dispersing Tuples of Unweighted Intervals Is Strongly NP-hard

We extend the reduction of [18, Thm. 16] to the unweighted case. Recall the definition of 3-PARTITION [13]: Given set A of $3m$ integers and a bound $L \in \mathbb{Z}^+$ with each $a \in A$ satisfying $L/4 < a < L/2$ and $\sum_{a \in A} a = mL$, decide whether A can be partitioned into m disjoint sets $A_1, \ldots, A_m$ such that $|A_i| = 3$ and $\sum_{a \in A_i} a = L$ for each $i \in [m]$. We build a multiset of intervals $\mathcal{I}_A$ and show that its intersection graph can be rendered edgeless with total moving distance of at most T if and only if A can be partitioned as above, for a threshold T to be defined later. The reduction structure is commonly used to simulate partitions on the real line (see e.g. [6] for a reduction of this type in barrier coverage). In particular, $\mathcal{I}_A$ is defined as $\mathcal{I}_A = \mathcal{I} \cup \mathcal{B}$ such that $\mathcal{I} = (I_1, \ldots, I_{3m})$ is a $3m$-tuple of intervals representing A, where $\mathrm{len}(I_i) = a_i$ and $c(I_i) = -a_i/2$ for each $i \in [3m]$ and, $\mathcal{B} = \mathcal{B}_0 \cup \cdots \cup \mathcal{B}_m$, is a family of *barriers*, where $\mathrm{len}(B) = 1/T$ for all $B \in \mathcal{B}$ and for each $i \in [m-1]$, the barrier $\mathcal{B}_i$ is a T-tuple of intervals $\mathcal{B}_i = (B_1^i, \ldots, B_T^i)$. We set $c(B_j^i) = iL + (i-1) + (2j-1)/(2T)$ for all $i \in [m-1]$ and $j \in [T]$. The 0th and mth barriers are defined as $\mathcal{B}_0 = (B_1^0, \ldots, B_{T(T+L)}^0)$ and $\mathcal{B}_m = (B_1^m, \ldots, B_{T(T+L)}^m)$. We set $c(B_i^0) = (1-2i)/(2T)$ for each $i \in [T(T+L)]$. Similarly, we set $c(B_i^m) = mL + (m-1) + (2i-1)/(2T)$ for each $i \in [T(T+L)]$.

The skeleton of $\mathcal{I}_A$ is illustrated in Fig. 1. For $i \in [m]$, the *ith free space* is the interval $[r(\mathcal{B}_{i-1}), \ell(\mathcal{B}_i)]$ of size L. By the strong NP-hardness of 3-PARTITION, we have $L \in \mathrm{poly}(n)$. Hence, the reduction takes polynomial time even if $\mathrm{poly}(L)$ elements are added to $\mathcal{I}_A$. We will define T as a polynomial in n.

Let $A_1, \ldots, A_m$ be a partition of A into disjoint triplets $A_i = \{a_1^i, a_2^i, a_3^i\}$ for all $i \in [m]$. Without loss of generality, suppose that the corresponding intervals of A_i are moved to the ith free space of $\mathcal{I}_A$. The moving distance is given by $a_1^i + (L+1)(i-1)$, $a_1^i + a_2^i + (L+1)(i-1)$ and $a_1^i + a_2^i + a_3^i + (L+1)(i-1)$ for the first, second and last interval, respectively. This gives a moving distance of $3(L+1)(i-1) + 3a_1^i + 2a_2^i + a_3^i$ and consequently the total moving distance of $\mathcal{I}$ is given by $3(L+1)\sum_{i=1}^{m}(i-1) + \sum_{i=1}^{m} 3a_1^i + 2a_2^i + a_3^i$. We show a bound for this value using the strict upper bound of elements in A:

Lemma 1 $(\star)$. $3(L+1)\sum_{i=1}^{m}(i-1) + \sum_{i=1}^{m} 3a_1^i + 2a_2^i + a_3^i < \frac{3m}{2}(mL + m + L - 1)$.

Hence, we set $T = 3m(mL + m + L - 1)/2$ and use this value to prove the correctness of the reduction.

Lemma 2. *An instance* (A, L) *of* 3-PARTITION *is a yes-instance if and only if there is a distance vector* D *such that* $G(\mathcal{I}_A + D) \in \Pi_{\text{edgeless}}$ *and* $\|D\|_1 < T$.

Proof. We label $\mathcal{I} = (I_1, \ldots, I_{3m})$ as $(I_1^1, I_2^1, I_3^1, \ldots, I_1^m, I_2^m, I_3^m)$, where the ith interval corresponds to the ith value of the vector $(a_1^1, a_2^1, a_3^1, \ldots, a_1^m, a_2^m, a_3^m)$.

First, assume that (A, L) is a yes-instance of 3-PARTITION. We define the distance vector $D_{\mathcal{I}} = (d_1^1, \ldots, d_3^m)$, where $d_j^i = (L+1)(i-1) + \sum_{k=1}^{j} a_k^i$. It follows that $\|D\|_1 < T$ by Lemma 1, thus we only need to prove that $G((\mathcal{I} + D) \cup \mathcal{B}) \in \Pi_{\text{edgeless}}$. Two arbitrary intervals I and I' with $I \preceq I'$ are disjoint if $r(I') - r(I) \geq \mathrm{len}(I')$. Equivalently, I and I' are disjoint if $\ell(I') - \ell(I) \geq \mathrm{len}(I)$. Since $r(I) = 0$

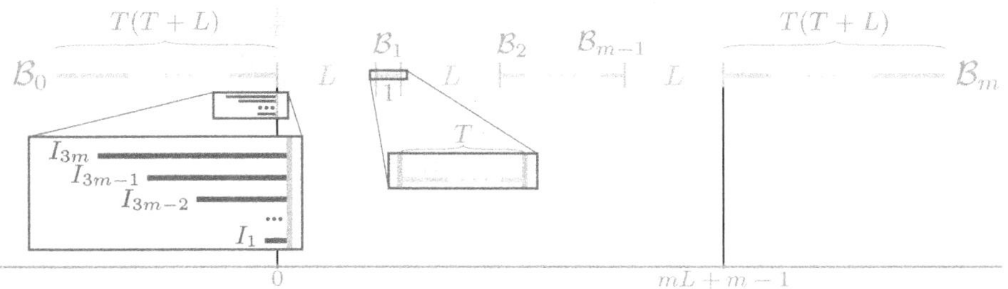

Fig. 1. Reduction overview: Skeleton of the reduction $\mathcal{I}_A = \mathcal{I} \cup \mathcal{B}$.

for all $I \in \mathcal{I}$, we have that $r(I_j^i) = d_j^i$ for all $i \in [m]$ and $j \in [3]$. Then, $r(I_{j+1}^i) - r(I_j^i) = d_{j+1}^i - d_j^i = a_{j+1}^i = \mathrm{len}(I_{j+1}^i)$. Hence, the intervals in $\mathcal{I} + D_{\mathcal{I}}$ are disjoint. Similarly, for all $i \in [m]$, it holds that $r(I_1^i + d_1^i) - r(\mathcal{B}_{i-1}) = \mathrm{len}(I_1^i)$ and $\ell(\mathcal{B}_i) - \ell(I_3^i) = \mathrm{len}(I_3^i)$. In other words, the tuple $\mathcal{I} + D_{\mathcal{I}}$ does not intersect with $\mathcal{B}$. By the above, if we set $\mathcal{I}_A + D = (I + D_{\mathcal{I}}) \cup \mathcal{B}$, then we conclude that $G(\mathcal{I}_A + D) \in \Pi_{\mathrm{edgeless}}$ and $\|D\|_1 \leq T$.

Conversely, assume that there is a vector D such that $G(\mathcal{I}_A + D) \in \Pi_{\mathrm{edgeless}}$ and $\|D\|_1 < T$. We divide D into two vectors $D_{\mathcal{I}}$ and $D_{\mathcal{B}}$. Suppose $\|D_{\mathcal{B}}\|_1 = 0$ and that there is an interval $I \in \mathcal{I} + D_{\mathcal{I}}$ such that $c(I)$ is not in any free space. Then $c(I) \leq \ell(\mathcal{B}_0)$ or $r(\mathcal{B}_m) \leq c(I)$, which implies that $\|D_{\mathcal{I}}\|_1 \geq T$. Hence all intervals in $\mathcal{I} + D_{\mathcal{I}}$ are in the free spaces delimited by $\mathcal{B}_0$ and $\mathcal{B}_m$. Suppose that there is an $i \in [m]$ for which $\mathrm{len}(I_1^i) + \mathrm{len}(I_2^i) + \mathrm{len}(I_3^i) = L + 1$. To move I_1^i, I_2^i and I_3^i to a free space, we need to shift intervals in $\mathcal{B}$ so there is an extra one-unit space. Since $\mathrm{len}(B) = 1/T$ for all $B \in \mathcal{B}$, we need to move at least T intervals by at least one unit, which implies that $\|D_{\mathcal{B}}\|_1 \geq T$. On the other hand, $\sum_{i=1}^{m} \mathrm{len}(I_1^i) + \mathrm{len}(I_2^i) + \mathrm{len}(I_3^i) \geq L$ by the definition of (A, L). Consequently, the m free spaces delimited by $\mathcal{B}$ yield m triplets of intervals whose lengths sum to L, giving a yes-instance of 3-PARTITION. This completes the proof. $\square$

Theorem 1 ($\star$). GGED$(\Pi_{\mathrm{edgeless}})$ *is strongly* NP-*hard on unweighted intervals.*

Theorem 1 can be extended to show that GGED is strongly NP-hard for Π_{acyc} and $\overline{\Pi_{k\text{-clique}}}$. By the chordality of interval graphs, it is sufficient to obtain a graph in $\overline{\Pi_{k\text{-clique}}}$ when $k = 3$ for obtaining a graph in Π_{acyc}. For the general case of k, we add the intervals of $\mathcal{B}$ to $\mathcal{I}_A$ $k - 1$ times. We also add $k - 2$ intervals $(r(\mathcal{B}_i), \ell(\mathcal{B}_{i+1}))$ for all $i \in [m - 1]_0$. Note that each interval of $\mathcal{I}$ forms a k-clique when intersecting with $\mathcal{B}$. Thus, moving the intervals to the free spaces is equivalent to removing all k-cliques from $\mathcal{I}_A$. This yields the following.

Corollary 1. GGED(Π_{acyc}) *and* GGED$(\overline{\Pi_{k\text{-clique}}})$ *are strongly* NP-*hard on unweighted intervals.*

3.2　Extending the Reduction to Higher Dimensions

In this section, we show that Theorem 1 can be extended to d-dimensional objects. Let (A, L) be an instance of 3-PARTITION. We assume that $d = 2$ and show a reduction that works for squares and disks. We give a multiset of squares and disks based on the structure of $\mathcal{I}_A$. To maintain simplicity, we give the definitions using squares. In particular, $\mathcal{S}_A = \mathcal{S} \cup \mathcal{B}$ are squares such that $\mathcal{S} = (S_1, \ldots, S_{3m})$ is a tuple of squares representing A where $\mathrm{len}(S_i) = \Delta + a_i$ and $c(S_i) = (-a_i/2, 0)$ for each $i \in [3m]$ and, $\mathcal{B} = \mathcal{B}_0 \cup \cdots \cup \mathcal{B}_m \cup \mathcal{B}_\uparrow \cup \mathcal{B}_\downarrow$, $\mathrm{len}(B) = 1/T$ for all $B \in \mathcal{B}$, is the family of barriers (see also Fig. 2).

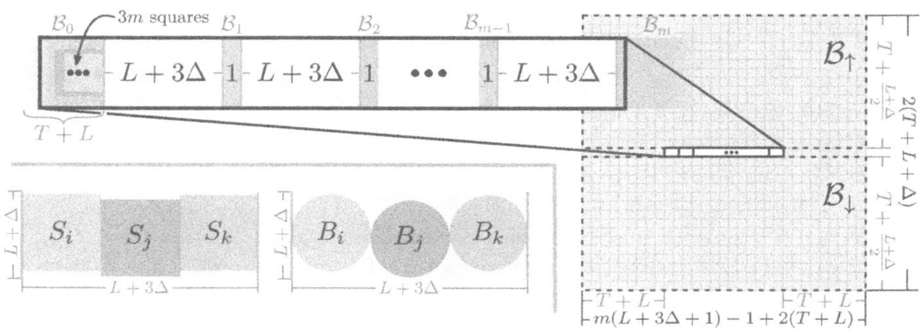

Fig. 2. Skeleton of the reduction of Corollary 2 (squares).

We defer the exact numbers, positions and weights of the objects in $\mathcal{B}$ to the appendix [19]. For disks, we construct an analogous tuple where $r(D_i) = (a_i + \Delta)/2$ for $D_i \in \mathcal{S}$ and $r(D) = 1/(2T)$ for $D \in \mathcal{B}$. The value $\Delta \in \mathbb{N}$ is a sufficiently large number used to compensate for differences among the lengths of the squares. Our reduction is analogous to the reduction in the proof of Theorem 1 except for: (i) the value Δ and (ii) the tuples $\mathcal{B}_\uparrow$ and $\mathcal{B}_\downarrow$.

Lemma 3 ($\star$). *For sufficiently large $\Delta \in \mathbb{N}$, every free space can contain at most three squares (disks) without overlap.*

By Lemma 3, we can choose any $\Delta \in \mathrm{poly}(n)$ such that the values of A and L do not add a significant amount to the area of the free spaces, causing the reduction to fail. Analogously to Lemma 1, we define the bound of the total moving distance as $T = \frac{3m}{2}(m(L+1) + L - 1) + (9\Delta/2)(m(m-1))$. The details on the correctness are given in the appendix [19].

Corollary 2. GGED$(\Pi_{\mathrm{edgeless}})$ *is strongly* NP-*hard for unweighted d-balls and d-cubes for any $d \geq 2$ under the L_1 and L_2 distances.*

4 Rendering Unit Circular Arc Graphs Edgeless in $O(n \log n)$ Time

Given an n-tuple $\mathcal{A} = (A_1, \ldots, A_n)$ of n unit circular arcs on a circle C, we show that finding a vector D with $G(\mathcal{A} + D) \in \Pi_{\text{edgeless}}$ and minimum $\|D\|_1$ can be done in $O(n \log n)$ time. Circular arcs generalise intervals and also appear in overlap removal [4]. We say that a circular arc A *is moved by* d if its endpoints are replaced by (p', q') such that $p', q' \in [0, 2\pi)$ and $d(p, p') = d(q, q') = d$. We assume that $\mathcal{A}$ is *normalised*; that is, $\sum_{A \in \mathcal{A}} \text{len}(A) \leq \text{len}(C)$. Any non-normalised instance can be rejected immediately, as no movement of arcs yields $G(\mathcal{A} + D)$ edgeless. We also assume that no circular arc intersects the point at angle 0. Since $\mathcal{A}$ is normalised, there exists region of C not intersected by any arc (even when $G(\mathcal{A}) \in \Pi_{\text{edgeless}}$ and $\sum_{A \in \mathcal{A}} \text{len}(A) = \text{len}(C)$). By rotating C, we place this region at angle 0 without changing the optimal solution.

We notice that unlike intervals, the elements of $\mathcal{A}$ follow a *cyclic order* induced by their positions in C. That is, whenever $A \preceq A' \preceq A''$ holds, then $A'' \preceq A \preceq A'$ and $A' \preceq A'' \preceq A$ also hold, for any $A, A', A'' \in \mathcal{A}$. Then, the edgeless condition for $\mathcal{A}$ can be adapted as follows: The graph $G(\mathcal{A})$ is in Π_{edgeless} if and only if $A \cap A' = \emptyset$ and $A' \cap A'' = \emptyset$ for all triplets $A, A', A'' \in \mathcal{A}$ such that $A \preceq A' \preceq A''$. With this, we make the following observation:

Observation 1. *For any labelling $\mathcal{A} = (A_1, \ldots, A_n)$ following the cyclic order of $\mathcal{A}$, $G(\mathcal{A})$ is in Π_{edgeless} if and only if $A_{i+1} \cap A_i = \emptyset$, for all $i \in [n]$, where indices are taken modulo n.*

Observation 1 implies that a vector D such that $G(\mathcal{A} + D) \in \Pi_{\text{edgeless}}$ can be obtained from any pair of labellings $(A_1, \ldots, A_n)$ and $(A_i, \ldots, A_n, A_1, \ldots, A_{i-1})$, $i \in [n]$. We initially consider the n-tuple $\mathcal{A} = (A_1, \ldots, A_n)$ labelled by sweeping C from angle 0 in counterclockwise order. We treat $\mathcal{A}$ as a tuple of intervals on the real line by *unwrapping* the circle C: we map C to $[0, \text{len}(C))$ and for each $i \in [n]$, the arc $A_i = (p, q)$ corresponds to the interval $I_i = (p \cdot \text{len}(C)/(2\pi), q \cdot \text{len}(C)/(2\pi))$. The resulting n-tuple of unit intervals is denoted by $\mathcal{I} = (I_1, \ldots, I_n)$. Analogously, we *wrap* $\mathcal{I}$ into C by mapping each $I = (x_1, x_2) \in \mathcal{I}$ to the arc $(p \cdot (2\pi/\text{len}(C)), q \cdot (2\pi/\text{len}(C)))$ taken modulo 2π.

Let UUID be the $O(n \log n)$-time algorithm of [18, Alg. 1] that disperses unweighted unit intervals at minimum total cost. Below, we give a high-level description of UUID. First, we recall the following statement.

Lemma 4 (Lem. 6 in [18]). *If two tuples of intervals, $\mathcal{I}$ and $\mathcal{I}'$, have optimal contiguous dispersals, say D and D', then $\mathcal{I} \cup \mathcal{I}'$ has an optimal contiguous dispersal if and only if an interval in $\mathcal{I} + D$ overlaps an interval in $\mathcal{I}' + D'$.*

Given an n-tuple of unit intervals $\mathcal{I}$, UUID sorts and partitions $\mathcal{I}$ into $m \leq n$ disjoint tuples $\mathcal{I}_{a_1, b_1}, \ldots, \mathcal{I}_{a_m, b_m}$ such that, for every $i \in [m]$, $\mathcal{I}_{a_i, b_i} = (I_{a_i}, \ldots, I_{b_i})$ has an optimal contiguous dispersal. Whenever two tuples $\mathcal{I}_{a_i, b_j}$ and $\mathcal{I}_{a_k, b_\ell}$ with $i \leq j < k \leq \ell$ satisfy Lemma 4, the algorithm recursively merges both tuples into a single tuple $\mathcal{I}_{a_i, b_\ell} = (I_{a_i}, \ldots, I_{b_j}, I_{a_k}, \ldots, I_{b_\ell})$. After all such merges, the remaining tuples have optimal contiguous dispersals and do not satisfy Lemma 4. Deciding whether two tuples satisfy Lemma 4 can be done without explicitly computing D and D' [18, Lem. 5]. Finally, UUID outputs a dispersal D for $\mathcal{I}$ that is a concatenation of the optimal distance vectors of the individual tuples.

Let D be the solution given by UUID for $\mathcal{I}$. To disperse $\mathcal{A}$, we (i) unwrap C, (ii) run UUID on $\mathcal{I}$ and (iii) wrap $\mathcal{I} + D$ in C (also identified by $\mathcal{A} + D$). At this point, $G(\mathcal{A} + D)$ may be in Π_{edgeless}. We distinguish two cases: If $I_n + d_n \preceq I_1 + d_1 + \text{len}(C)$, then we return D as the solution (case 1). If $I_n + d_n \succ I_1 + d_1 + \text{len}(C)$, then $G(\mathcal{A} + D)$ is not edgeless. We relabel $\mathcal{I}$ by shifting intersecting intervals by $\text{len}(C)$ distance (case 2). We execute a modified version of UUID considering the above cases. This leads to the following result.

Lemma 5. *By going through the above cases at most $n - 1$ times, we obtain an optimal distance vector from the set $\{D \in \mathbb{R}^n : G(\mathcal{A} + D) \in \Pi_{\text{edgeless}}\}$.*

Proof. We first show how to obtain D by considering both cases. Given n unweighted unit intervals $\mathcal{I}$ such that $c(I_{i+1}) - c(I_i) \leq 1$ for all $i \in [n]$, it is known that there is a contiguous D for $\mathcal{I}$ that is optimal [18]. Assume that the two tuples containing I_1 and I_n are $\mathcal{I}_{1,i}$ and $\mathcal{I}_{j,n}$ for $i < j$, respectively (equivalently we define $\mathcal{A}_{j,n}$ and $\mathcal{A}_{1,i}$). The case 1 means that for $i < j$, it is not necessary to check whether the arcs in $\mathcal{A}_{j,n}$ and $\mathcal{A}_{1,i}$ intersect, as the positions of their corresponding intervals imply that $A_n + d_n \cap A_1 + d_1 = \emptyset$. Hence we only focus on the second case. The case 2 implies that $\mathcal{A}_{j,n}$ and $\mathcal{A}_{1,i}$ intersect when dispersed. We construct a new n-tuple of unit intervals $\mathcal{I}'$ by shifting $\mathcal{I}_{1,i}$ to the right of $\mathcal{I}_{j,n}$ by $\text{len}(C)$ distance. That is, $\mathcal{I}' = (I_1', \ldots, I_n') = (I_{i+1}, \ldots, I_j, \ldots, I_n, I_1 + \text{len}(C), \ldots, I_i + \text{len}(C))$. Note that this is equivalent to label $\mathcal{A}$ as $(A_{i+1}, \ldots, A_j, \ldots, A_n, A_1, \ldots, A_i)$. By Observation 1, the translation of $\mathcal{I}_{1,i}$ does not change the optimal solution for $\mathcal{A}$. We then continue the procedure of UUID on $\mathcal{I}'$, where $\mathcal{I}'_{k,n} = (I_j, \ldots, I_i + \text{len}(C))$, $k = n - (|\mathcal{I}_{1,i}| + |\mathcal{I}_{j,n}| - 1)$, is the tuple obtained by merging $\mathcal{I}_{j,n}$ and $\mathcal{I}_{1,i}$ translated by $\text{len}(C)$. Hence, by checking both cases, we find a distance vector D such that $G(\mathcal{A} + D)$ is edgeless and $\|D\|_1$ is minimum. It suffices to show that the number of repetitions is at most $n - 1$. Suppose that case 2 occurs $n - 1$ times in $\mathcal{I}$. Then, the first tuple with an optimal contiguous dispersal of $\mathcal{I}$ is now the last tuple. Moreover, there is one tuple $\mathcal{I}' = (I_n, \ldots, I_1)$ with an optimal contiguous dispersal, which is exactly $\mathcal{I}$ in reverse order. If case 2 occurs again, then $\mathcal{I}'$ intersects with itself when dispersed, contradicting the normalisation of $\mathcal{A}$. The lemma follows. $\square$

Theorem 2 ($\star$). *Given an n-tuple of unit circular arcs, $\mathrm{GGED}(\Pi_{\text{edgeless}})$ can be solved in $O(n \log n)$ time.*

5 A Parameterised Algorithm for $\mathbf{GGED}(\boldsymbol{\Pi}_{\mathbf{edgeless}})$ for Weighted Unit Interval Graphs

This section is about $\mathrm{GGED}(\Pi_{\text{edgeless}})$ on weighted unit intervals. We present an $O(n^3)$-algorithm for assigning n weighted unit intervals to $m \in O(n)$ slots on the real line, and then show that this leads to an $O(n^4)$-time algorithm for dispersing cliques. An n-tuple $\mathcal{I} = (I_1, \ldots, I_n)$ is a *clique* if $G(\mathcal{I})$ is a complete graph, that is, if $I_n \prec I_1 + 1$. Finally, we generalise the approach for cliques and show that, given an n-tuple $\mathcal{I}$ of weighted unit intervals with k maximal cliques, $\mathrm{GGED}(\Pi_{\mathrm{edgeless}})$ can be solved in $(1 + n/k)^k \cdot \mathrm{poly}(n)$ time. We first show that $\mathrm{GGED}(\Pi_{\mathrm{edgeless}})$ admits an optimal solution with at least one interval fixed.

Lemma 6. *Given an n-tuple $\mathcal{I}$ of unit intervals with weights $\mathbf{w} \in \mathbb{R}^n_{>0}$, there is a vector $D \in \mathbb{R}^n$ such that (i) $G(\mathcal{I} + D) \in \Pi_{\mathrm{edgeless}}$, (ii) $\|\mathbf{w}D\|_1$ is minimum among all vectors D that fulfil (i), and (iii) at least one component of D is zero.*

Proof. Let D be such that $G(\mathcal{I} + D) \in \Pi_{\mathrm{edgeless}}$ and $\|\mathbf{w}D\|_1$ is minimum among all such vectors D. If there is a $k \in [n]$ such that $d_k = 0$, we are done. Otherwise, let $L = \{i \in [n] : d_i < 0\}$, and let $R = \{i \in [n] : d_i > 0\}$. Note that $L \cup R = [n]$. Let $W_L = \sum_{i \in L} w_i$ and $W_R = \sum_{i \in R} w_i$ be the sum of the weights of the intervals that D moves to the left and right, respectively. Assume, without loss of generality, that $W_L \leq W_R$. Let $k = \arg\min_{i \in R}\{d_i\}$ be the smallest of the distances that D moves an interval to the right. Let $D' = (d'_1, \ldots, d'_n)$, where $d'_i = d_i - d_k$ for every $i \in [n]$. Clearly, $d'_k = 0$, $d_i > d'_i \geq 0$ for every $i \in R$, and $d'_i < d_i < 0$ for every $i \in L$. Moreover, $\|\mathbf{w}D'\|_1 = \sum_{i=1}^n w_i \cdot |d_i - d_k| = \sum_{i=1}^n w_i|d_i| + \sum_{i \in L} w_i d_k - \sum_{i \in R} w_i d_k = \|\mathbf{w}D\|_1 + W_L d_k - W_R d_k$, which yields that $\|\mathbf{w}D'\|_1 \leq \|\mathbf{w}D\|_1$ since $W_L \leq W_R$. $\qquad\square$

5.1   Assigning Slots to Weighted Intervals in Polynomial Time

For $m \geq n$, let $S = (s_1, \ldots, s_m) \in \mathbb{R}^m$ be an m-tuple describing a sequence of *slots* such that $s_{i+1} - s_i \geq 1$ for all $i \in [m-1]$. Let $X = (x_i)_{i \in [n]} \in [m]^n$ be an n-vector such that $x_i \neq x_j$ whenever $i \neq j$ and let $f : \mathbb{R}^n \to \mathbb{R}^n$ be a function defined as $f(X) = (s_{x_i} - c(I_i))_{i \in [n]}$. We find a distance vector D^* such that $\|\mathbf{w}D^*\|_1 = \min_{X \in [m]^n} \|\mathbf{w}f(X)\|_1$.

We use a *one-sided perfect matching* approach. First, we construct a complete bipartite graph $G = (U, V, E)$ where (i) there is a unique vertex $u_i \in U$ for the interval I_i, (ii) U contains a subset U' of $m - n$ *dummy* vertices, (iii) there is a unique vertex $v_i \in V$ for the slot $s_i \in S$ and (iv) E contains an edge $\{u, v\}$ for each pair of the form $\{u, v\} \in U \times V$. Notice that $|U| = |V| = m$. Let $C : U \times V \to \mathbb{R}_{\geq 0}$ be a cost function. For an edge $\{u_i, v_j\}$, $u_i \in U$ and $v_j \in V$, the value of $C(\{u_i, v_j\})$ is equal to $w_i|c(I_i) - s_j|$ if $u_i \notin U'$ and 0 otherwise.

By the definition of C, we have $\sum_{e \in M} C(e) = \|\mathbf{w}f(X)\|_1$ for some $X \in [m]^n$. We compute a perfect matching $M \subset E$ for G that minimises $\sum_{e \in M} C(e)$. Given such M, the set $M' = \{\{u, v\} \in M : u \notin U'\}$ describes a valid solution to assign a unique slot to each interval such that the total weighted distance is minimised. We use M' and define the n-vector $X = (x_i)_{i \in [n]}$ where $x_i = j$ if $\{u_i, v_j\} \in M'$. Since $\sum_{e \in M} C(e)$ is minimum and $\sum_{e \in \{(u,v) \in M : u \in U'\}} C(e) = 0$, the value of $\|\mathbf{w}f(X)\|_1$ is also minimum. Hence we set $D^* = f(X)$. A min-cost perfect matching can be computed in cubic time [8]. This yields the following.

Theorem 3. *Given an n-tuple of weighted unit intervals and $m \geq n$ slots, in $O(m^3)$ time, we can assign a unique slot to each interval such that the total moving distance of the intervals to the slots is minimised.*

5.2   One-Sided Perfect Matching for Cliques

Suppose $\mathcal{I}$ as described above is a clique. We find a distance vector D such that $G(\mathcal{I} + D) \in \Pi_{\text{edgeless}}$ and $\|\mathbf{w}D\|_1$ minimum in $O(n^4)$ time using Theorem 3. First, we prove the following.

Lemma 7. *Every clique of intervals has a contiguous dispersal that is optimal.*

Proof. By Lemma 6, we can assume that an interval I_k with $k \in [n]$ is fixed in an optimal dispersal of $\mathcal{I}$. Let $\mathcal{I}_\ell = (L_i)_{i \in [n_\ell]}$ and $\mathcal{I}_r = (R_i)_{i \in [n_r]}$ be the n_ℓ- and n_r-tuple of intervals, $n_\ell + n_r = n - 1$, moved to the left and right of I_k, respectively. Let $\mathbf{w}_\ell$ and $\mathbf{w}_r$ be their corresponding weight vectors. Let D_ℓ and D_r be the vectors such that $\|\mathbf{w}_\ell D_\ell\|_1$ and $\|\mathbf{w}_r D_r\|_1$ are minimum. Also, suppose that there exists an $i \in [n_\ell - 1]$ such that $(L_{i+1} + d^\ell_{i+1}) - (L_i + d^\ell_i) > 1$ and let $\delta = (L_{i+1} + d^\ell_{i+1}) - (L_i + d^\ell_i) - 1$. Since $\mathcal{I}$ is a clique, all components of D_ℓ are negative. We define a new vector $D' = (d'_1, \ldots, d'_{n_\ell})$ such that $d'_j = d^\ell_j$ if $j \leq i$ and $d'_j = d^\ell_j + \delta$ if $j \geq i + 1$. Since $\delta > 0$, $\|\mathbf{w}_\ell D'\|_1 < \|\mathbf{w}_\ell D_\ell\|_1$ holds. Hence, D' is a better dispersal for $\mathcal{I}$. Moreover, $(L_{i+1} + d'_{i+1}) - (L_i + d'_i) = 1$, which implies that D' is contiguous. The same argument applies if D_r is non-contiguous, contradicting the existence of an optimal non-contiguous dispersal. $\square$

Algorithm for Weighted Cliques. By Lemmas 6 and 7, an optimal contiguous dispersal always exists. For each $i \in [n]$, do the following. Assume that I_i is fixed. Let $\mathcal{I}_i = (I_j)_{j \in [n] - \{i\}}$. Define a sequence of $2(n-1)$ slots $S = (c(I_i) \pm j)_{j \in [n-1]}$, $n - 1$ on each side of I_i. Then, compute the minimum total moving distance D_i of $\mathcal{I}_i$ to S using Theorem 3. Finally, select the solution corresponding to the step i that minimised D_i. We now summarise the above.

Theorem 4 ($\star$). *If $\mathcal{I}$ is a clique,* $\mathrm{GGED}(\Pi_{\text{edgeless}})$ *is solvable in $O(n^4)$ time.*

5.3   Dispersing Weighted Unit Intervals Parameterised by the Number of Maximal Cliques

We can bound the number of fixed intervals for a tuple $\mathcal{I}$ of intervals such that $G(\mathcal{I})$ has k maximal cliques.

Lemma 8 ($\star$). *If $G(\mathcal{I})$ has k maximal cliques, then at most k intervals are fixed in an optimal dispersal of $\mathcal{I}$.*

We apply Lemma 8 assuming at most k fixed intervals, one per maximal clique. For two consecutive fixed intervals F and F', $F \prec F'$, we say that $g(F, F') = r(F') - \ell(F)$ is the *gap* between F and F'. Assuming that $i \leq k$ intervals are fixed, we have at most $i - 1$ gaps between the fixed intervals. An interval can be moved to a gap if there is sufficient space. Depending on the size of $g(F, F')$, we add slots in the gaps:

(i) $g(F, F') < 1$: No interval can be placed in $g(F, F')$. We discard this gap.

(ii) $1 \leq g(F, F') < 2(n - i)$: We consider $n' = \lfloor g(F, F') \rfloor$ cases. In each case, we contiguously place some $r \in [n']_0$ slots immediately to the right of F and $n' - r$ slots immediately to the left of F'.

(iii) $g(F, F') \geq 2(n - i)$: We contiguously place $n - i$ slots immediately to the left of F and immediately to the right of F'.

Lemma 9 ($\star$). *Given an optimal distance vector D, the set $\{c(I): I \in \mathcal{I} + D\}$ is a subset of a set of slots as described above.*

By Lemma 9, we can test all possible slot sets using the algorithm of Theorem 3. We select the set of slots that yields the minimum total moving distance as the optimal solution for $(\mathcal{I}, k)$. This result is summarised in the following.

Theorem 5 ($\star$). *For n weighted unit intervals with k maximal cliques, the problem* $\mathrm{GGED}(\Pi_{\mathrm{edgeless}})$ *can be solved in* $(1 + n/k)^k \cdot \mathrm{poly}(n)$ *time.*

We conclude this section by observing that $\mathrm{GGED}(\Pi_{\mathrm{edgeless}})$ is in NP on tuples of weighted unit intervals, under the word-RAM model. As a decision problem, we ask whether there exists a distance vector D such that $G(\mathcal{I} + D) \in \Pi_{\mathrm{edgeless}}$ and $\|\mathbf{w}D\|_1 \leq T$, for a given threshold T. Since unit interval graphs have at most n maximal cliques, the algorithm for Theorem 5 also solves the non-parameterised problem in $O(2^n)$ time. Although trivial, the algorithm provides a certificate that can be verified in polynomial time. This claim was not possible before, since a certificate consisting of the n intervals and their moving distances could have involved arbitrary reals. Now we know that the slots are obtained by adding/subtracting integers up to $2n$ to/from the interval centres. Assuming that the input $(\mathcal{I}, \mathbf{w}, T)$ is encoded by w-bit words, each slot can be encoded using bit-length that is polynomial with respect to w. We summarise this observation.

Corollary 3 ($\star$). *For weighted unit intervals,* $\mathrm{GGED}(\Pi_{\mathrm{edgeless}})$ *is in NP under the word-RAM model.*

6 Open Problems and Further Research

The main open question is whether $\mathrm{GGED}(\Pi_{\mathrm{edgeless}})$ is NP-hard on weighted unit intervals. Because intervals cannot be distinguished by their length, we cannot adapt the reduction used in Sect. 3.1. On the other hand, arbitrary weights make it hard to design a polynomial-time algorithm. Solving this case would make the intractability border of $\mathrm{GGED}(\Pi_{\mathrm{edgeless}})$ much clearer.

Instances where lengths are proportional to weights are also noteworthy. This models scenarios where movement cost scales with size (e.g., scheduling when processing time equals weight [16]). Such restrictions complicate reductions such as the one for Theorem 1 since heavy intervals cannot be simulated using a number of short intervals. They may also yield useful properties, such as order-preservation [18] or convex-cost formulations that can be exploited by polynomial-time algorithms. Another promising restriction assumes that two-dimensional objects are given with ordering constraints in one dimension [25].

Considering other geometric transformations (scaling or rotation) and parameterisations of strongly NP-hard cases also remains open. Parameters such as the number of objects moved, maximal cliques, or distinct lengths/weights can be considered. Our observations from Sect. 5 may help design a parameterised algorithm when both length and weight are arbitrary. Finally, heuristics and approximation algorithms are also natural extensions for hard cases.

References

1. Agnetis, A., Billaut, J.C., Pinedo, M., Shabtay, D.: Fifty years of research in scheduling - theory and applications. Europ. J. Oper. Res. (2025). https://doi.org/10.1016/j.ejor.2025.01.034
2. Bekos, M.A., Kaufmann, M., Symvonis, A., Wolff, A.: Boundary labeling: models and efficient algorithms for rectangular maps. Comput. Geom. **36**(3), 215–236 (2007). https://doi.org/10.1016/j.comgeo.2006.05.003
3. de Berg, M., Kisfaludi-Bak, S., Woeginger, G.J.: The complexity of dominating set in geometric intersection graphs. Theoret. Comput. Sci. **769**, 18–31 (2019). https://doi.org/10.1016/j.tcs.2018.10.007
4. Bonerath, A., Nöllenburg, M., Terziadis, S., Wallinger, M., Wulms, J.: Boundary labeling in a circular orbit. In: GD 2024. Schloss Dagstuhl – LZI (2024). https://doi.org/10.4230/LIPICS.GD.2024.22
5. Cygan, M., et al.: Parameterized Algorithms. Springer, Cham (2015). https://doi.org/10.1007/978-3-319-21275-3
6. Czyzowicz, J., Kranakis, E., Krizanc, D., Lambadaris, I., Narayanan, L., Opatrny, J., Stacho, L., Urrutia, J., Yazdani, M.: On minimizing the sum of sensor movements for barrier coverage of a line segment. In: Nikolaidis, I., Wu, K. (eds.) ADHOC-NOW 2010. LNCS, vol. 6288, pp. 29–42. Springer, Heidelberg (2010). https://doi.org/10.1007/978-3-642-14785-2_3
7. Diestel, R.: Graph Theory. Graduate texts in mathematics, 5th edn. Springer, Cham (2017). https://doi.org/10.1007/978-3-662-53622-3
8. Edmonds, J., Karp, R.M.: Theoretical improvements in algorithmic efficiency for network flow problems. J. ACM **19**(2), 248–264 (1972). https://doi.org/10.1145/321694.321699
9. Fink, M., Haunert, J.H., Schulz, A., Spoerhase, J., Wolff, A.: Algorithms for labeling focus regions. IEEE Trans. Vis. Comput. Graphics **18**(12), 2583–2592 (2012). https://doi.org/10.1109/TVCG.2012.193
10. Fomin, F.V., Golovach, P.A., Inamdar, T., Saurabh, S., Zehavi, M.: Kernelization for spreading points. In: ESA. LIPIcs, vol. 274, pp. 48:1–48:16. Schloss Dagstuhl – LZI (2023). https://doi.org/10.4230/LIPICS.ESA.2023.48
11. Fomin, F.V., Golovach, P.A., Inamdar, T., Saurabh, S., Zehavi, M.: (re)packing equal disks into rectangle. Discrete Comput. Geom. **72**(4), 1596–1629 (2024). https://doi.org/10.1007/s00454-024-00633-1
12. Fomin, F.V., Golovach, P.A., Inamdar, T., Saurabh, S., Zehavi, M.: Parameterized geometric graph modification with disk scaling. In: ITCS. LIPIcs, vol. 325, pp. 51:1–51:17. Schloss Dagstuhl – LZI (2025). https://doi.org/10.4230/LIPICS.ITCS.2025.51

13. Garey, M.R., Johnson, D.S.: Computers and Intractability: A Guide to the Theory of NP-Completeness. W.H. Freeman (1979)

14. Goemans, M.X., Queyranne, M., Schulz, A.S., Skutella, M., Wang, Y.: Single machine scheduling with release dates. SIAM J. Discrete Math. **15**(2), 165–192 (2002). https://doi.org/10.1137/s089548019936223x

15. Graham, R.L., Lawler, E.L., Lenstra, J.K., Rinnooy Kan, A.H.G.: Optimization and approximation in deterministic sequencing and scheduling: a survey. Ann. Discrete Math. **5**, 287–326 (1979). https://doi.org/10.1016/s0167-5060(08)70356-x

16. Györgyi, P., Kis, T.: Minimizing total weighted completion time on a single machine subject to non-renewable resource constraints. J. Schedul. **22**(6), 623–634 (2019). https://doi.org/10.1007/s10951-019-00601-1

17. Honorato-Droguett, N., Kurita, K., Hanaka, T., Ono, H.: Algorithms for optimally shifting intervals under intersection graph models. In: Li, B., Li, M., Sun, X. (eds.) IJTCS-FAW 2024. LNCS, vol. 14752, pp. 66–78. Springer, Cham (2024). https://doi.org/10.1007/978-981-97-7752-5_5

18. Honorato-Droguett, N., Kurita, K., Hanaka, T., Ono, H.: On the complexity of minimising the moving distance for dispersing objects. In: WADS, vol. 349, pp. 36:1–36:14. Schloss Dagstuhl – LZI (2025). https://doi.org/10.4230/LIPICS.WADS.2025.36

19. Honorato-Droguett, N., Kurita, K., Hanaka, T., Ono, H., Wolff, A.: Further results on rendering geometric intersection graphs sparse by dispersion (2025). https://arxiv.org/abs/2509.20903

20. Kindermann, P., Niedermann, B., Rutter, I., Schaefer, M., Schulz, A., Wolff, A.: Multi-sided boundary labeling. Algorithmica **76**(1), 225–258 (2015). https://doi.org/10.1007/s00453-015-0028-4

21. Kranakis, E., Shaikhet, G.: Sensor allocation problems on the real line. J. Appl. Prob. **53**(3), 667–687 (2016). https://doi.org/10.1017/jpr.2016.33

22. Lawler, E.L.: Optimal sequencing of a single machine subject to precedence constraints. Manag. Sci. **19**(5), 544–546 (1973). https://doi.org/10.1287/mnsc.19.5.544

23. Lenstra, J.K., Rinnooy Kan, A.H., Brucker, P.: Complexity of machine scheduling problems. Ann. Discrete Math. **1**, 343–362 (1977). https://doi.org/10.1016/s0167-5060(08)70743-x

24. Li, S., Wang, H.: Algorithms for minimizing the movements of spreading points in linear domains. Comput. Geom. Topol. **4**(1), 1:1–1:15 (2025). https://doi.org/10.57717/CGT.V4I1.21

25. Meulemans, W.: Efficient optimal overlap removal: algorithms and experiments. Comput. Graph. Forum **38**(3), 713–723 (2019). https://doi.org/10.1111/cgf.13722

26. Adamu, M.O., Adewumi, A.O.: A survey of single machine scheduling to minimize weighted number of tardy jobs. J. Indust. Manag. Optim. **10**(1), 219–241 (2014). https://doi.org/10.3934/jimo.2014.10.219

27. Panolan, F., Saurabh, S., Zehavi, M.: Contraction decomposition in unit disk graphs and algorithmic applications in parameterized complexity. ACM Trans. Algorithms **20**(2), 15 (2024). https://doi.org/10.1145/3648594

28. Pinedo, M.L.: Scheduling: Theory, Algorithms, and Systems. Springer, Cham (2022). https://doi.org/10.1007/978-3-031-05921-6
29. Preparata, F.P., Shamos, M.I.: Computational Geometry - An Introduction. Springer, Cham (1985). https://doi.org/10.1007/978-1-4612-1098-6
30. Xue, J., Zehavi, M.: Parameterized algorithms on geometric intersection graphs. Comput. Sci. Rev. **58**, 100796:1–11 (2025). https://doi.org/10.1016/j.cosrev.2025.100796

Fundamentals of Computing Continuous Dynamic Time Warping in 2D Under Different Norms

Kevin Buchin[1], Maike Buchin[2], Jan Erik Swiadek[2]([✉]),
and Sampson Wong[3]

[1] Technical University Dortmund, Dortmund, Germany
kevin.buchin@tu-dortmund.de
[2] Ruhr University Bochum, Bochum, Germany
{maike.buchin,jan.swiadek}@rub.de
[3] University of Copenhagen, Copenhagen, Denmark

Abstract. Continuous Dynamic Time Warping (CDTW) measures the similarity of polygonal curves robustly to outliers and to sampling rates, but the design and analysis of CDTW algorithms face multiple challenges. We show that CDTW cannot be computed exactly under the Euclidean 2-norm using only algebraic operations, and we give an exact algorithm for CDTW under norms approximating the 2-norm. The latter result relies on technical fundamentals that we establish, and which generalise to any norm and to related measures such as the partial Fréchet similarity.

Keywords: Continuous Dynamic Time Warping · curve similarity · geometric optimisation · integral calculus

1 Introduction

The similarity of two curves can be measured in various ways, where the choice of technique is a trade-off depending on the application [29, 30]. Discrete similarity measures consider only the vertices of polygonal curves, often given as sampling points of some motion. In particular, Dynamic Time Warping (DTW) matches the vertices of the two input curves monotonically such that the sum of their distances is minimised [31]. The appeal of DTW lies in its simplicity, and in the fact that the curves can be detected as similar even if their underlying motions have different speeds. Its main disadvantage is that it may yield poor results if the curves' sampling rates are too different or not sufficiently high [10, 30].

This is due to the fact that DTW disregards information by not interpreting the polygonal curves as continuous objects, given by the linear interpolations of their vertex sequences. Therefore, several continuous versions of DTW have been proposed in the literature [5, 11, 16, 19, 24, 27, 28], some of which are equivalent or nearly equivalent. (See Sect. 1.1 for a brief overview.) Each of these Continuous Dynamic Time Warping (CDTW) variants in some way utilises continuous and monotone matchings of arbitrary curve points, as illustrated in Fig. 1.

E. Di Giacomo and D. Mondal (Eds.): WALCOM 2026, LNCS 16444, pp. 467–482, 2026.
https://doi.org/10.1007/978-981-95-7127-7_31

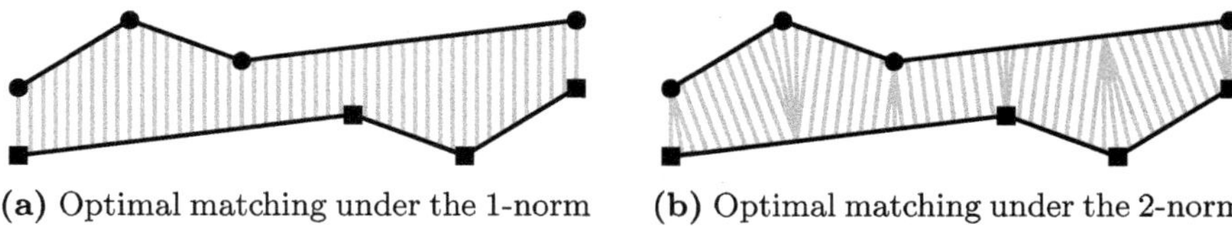

(a) Optimal matching under the 1-norm   (b) Optimal matching under the 2-norm

Fig. 1. Simple example of optimal matchings for our CDTW variant

Efficiently computing any CDTW variant exactly or with strong approximation guarantees has turned out to be challenging. The strongest results so far are a pseudo-polynomial-time $(1+\varepsilon)$-approximation for 2D curves under the Euclidean 2-norm, due to Maheshwari, Sack and Scheffer [23], and a polynomial-time exact algorithm in 1D, first described by Klaren [21], then analysed by Buchin, Nusser and Wong [10]. The difficulty of computing CDTW has multiple reasons, which we address in this paper. Our goals are to advance the understanding of related computational challenges and to establish approaches for handling these.

First of all, the fundamental properties of CDTW are still not well understood, especially in 2D and under different norms. Like in the above works [10,21,23], our formulation is based on a definition introduced by Buchin [11] as a summed version of the Fréchet distance, which minimises the maximum distance of continuously and monotonically matched curve points [1,17]. The CDTW definition at hand replaces the outlier-sensitive maximum with a path integral, aiming to overcome the drawbacks of the Fréchet distance and of DTW [7,10]. As our first contribution, presented in Sect. 2, we show that our path integral formulation admits locally optimal matchings that are advantageous from an algorithmic perspective.

Another challenge for integral-based measures is the integration itself, which typically complicates exact computations. Many existing approaches for CDTW resort to heuristics or approximations by discretising the input curves [5,7,16,19, 23]. This confines the integration to individual steps, but causes the solution quality or the running time to be dependent on the discretisation's resolution. In contrast, the exact 1D algorithm utilises the 1D setting's piecewise linear integrands for a dynamic program that propagates piecewise quadratic functions [10,21].

Brankovic [6] describes a 2D generalisation of this algorithm under the 1-norm, which gives piecewise linear integrands too. The 2-norm, however, gives integrands with a square root and thus seems unsuited to the function propagation approach. We give a new and deeper insight into the difficulty of CDTW under the 2-norm, in that the numbers involved may be transcendental and hence not computable exactly using only algebraic operations. This result is presented in Sect. 3 and further motivates the usage of approximations for the 2-norm [2,14,15].

In Sect. 4 we then develop a $(1+\varepsilon)$-approximation without discretising the curves by generalising the exact algorithm to a large class of norms with polygons as level sets. The final remaining challenge is the running time analysis, i.e. the combinatorial problem of bounding the number of propagated pieces [10,24,27]. A technical analysis of the exact 1D algorithm has shown an $O(n^5)$ running time for curves with n segments [10]. Its 2D generalisation has not been analysed yet. While the techniques of the 1D analysis do not readily facilitate a polynomial-time result in 2D, we highlight the main issue that would need to be resolved for such a result, and we prove technical properties that we deem helpful for this.

1.1 Related Work

Standard algorithms compute DTW in $O(n^2)$ time [31] and the Fréchet distance in $O(n^2 \log(n))$ time [1]. There also exist improved upper bounds [8,13,18]. The first continuous version of DTW was due to Serra and Berthod [27]. It matches curve points continuously, but sums over a finite quantity instead of using integration. An equivalent definition was later analysed by Munich and Perona [24].

Serra and Berthod [28] further introduced an integral-based measure that considers the change of distance vectors instead of the distances themselves, so it is a translation-invariant relative of CDTW. Efrat, Fan and Venkatasubramanian [16] proposed a more general integral-based variant of CDTW, which is rather similar to our definition; remaining differences are explained in [23, Section 2.3].

CDTW has applications in signature verification [16], map matching [5] and clustering [7]. Several practical works [5,7,19] use CDTW formulations equivalent to some definition by Buchin [11], which are moreover related to the lexicographic Fréchet distance [25] and the partial Fréchet similarity [9]: When restricted to one pair of curve segments, optimal matchings for all these measures are realised via local optimality [23, Section 5], upon which our results in Sect. 2 build.

1.2 Preliminaries

A polygonal curve P in $\mathbb{R}^2$ is composed of $n \in \mathbb{N}$ consecutive line segments and represented by its vertex sequence $\langle p_0, \ldots, p_n \rangle$ with $p_i \neq p_{i-1}$ for $i \in \{1, \ldots, n\}$. The arc length of P under a norm $\| \cdot \|$ on $\mathbb{R}^2$ is $\|P\| := \sum_{i=1}^{n} \|p_i - p_{i-1}\|$, and we denote the corresponding arc length parametrisation with constant speed 1 under $\| \cdot \|$ by $P_{\|\cdot\|} \colon [0, \|P\|] \to \mathbb{R}^2$. Moreover, we define the set $\Pi(P)$ containing all functions $f \colon [0, 1] \to \mathbb{R}^2$ that parametrise P monotonically and are piecewise continuously differentiable. Note that $\int_0^1 \|f'(t)\| \, dt = \|P\|$ for all $f \in \Pi(P)$.

Definition 1 ([11, Chap. 6]). *Let P, Q be polygonal curves, let $\| \cdot \|$ be a norm and let $d \colon \mathbb{R}^2 \times \mathbb{R}^2 \to \mathbb{R}_{\geq 0}$ be a continuous function. The measure* Continuous Dynamic Time Warping (CDTW) *of P, Q in $(\mathbb{R}^2, \| \cdot \|)$ under d is defined by*

$$\mathrm{cdtw}_{\|\cdot\|,d}(P, Q) := \inf_{(f,g) \in \Pi(P) \times \Pi(Q)} \int_0^1 d(f(t), g(t)) \cdot \left\| \begin{pmatrix} \|f'(t)\| \\ \|g'(t)\| \end{pmatrix} \right\|_1 dt.$$

We set $\mathrm{cdtw}_{\|\cdot\|}(P, Q) := \mathrm{cdtw}_{\|\cdot\|,(p,q) \mapsto \|p-q\|}(P, Q)$ *for CDTW of P, Q under $\| \cdot \|$.*

So far, only the vector space $(\mathbb{R}^2, \| \cdot \|_2)$ equipped with the Euclidean 2-norm has been used for 2D CDTW, where the $\tilde{p}$-norm $\| \cdot \|_{\tilde{p}}$ on $\mathbb{R}^2$ is, as usual, given by $\|x\|_{\tilde{p}} = \|(x_1, x_2)^{\mathsf{T}}\|_{\tilde{p}} := (|x_1|^{\tilde{p}} + |x_2|^{\tilde{p}})^{1/\tilde{p}}$ for $\tilde{p} \geq 1$. Brankovic [6, Section 5.2] considered CDTW in $(\mathbb{R}^2, \|\cdot\|_2)$ under the distance function $(p, q) \mapsto \|p - q\|_1$ induced by the 1-norm, which differs from the equipped norm. Our formulation of CDTW under a norm employs the chosen norm for both arc lengths of curves and distances between curve points. This not only yields an intuitive generalisation of other definitions (cf. [11, Section 6.2.2] and [19, Section 5.5.1]), but also comes with useful geometric properties, as shown in Sect. 2. The dependence on the chosen norm is likewise reflected by our definition of the parameter space.

Definition 2 ([10, Sect. 2]). *Let $P = \langle p_0, \ldots, p_n \rangle$ and $Q = \langle q_0, \ldots, q_m \rangle$ be polygonal curves, and let $\|\cdot\|$ be a norm. The* parameter space *of P, Q under $\|\cdot\|$ is $[0, \|P\|] \times [0, \|Q\|] \subseteq \mathbb{R}^2$. Each pair $(i, j) \in \{1, \ldots, n\} \times \{1, \ldots, m\}$ is associated with a* cell $[\|\langle p_0, \ldots, p_{i-1} \rangle\|, \|\langle p_0, \ldots, p_i \rangle\|] \times [\|\langle q_0, \ldots, q_{j-1} \rangle\|, \|\langle q_0, \ldots, q_j \rangle\|]$.

As usual, we represent monotone matchings of P, Q by monotone paths in their parameter space under $\|\cdot\|$. Since Definition 1 combines parametrisation speeds by fixing the 1-norm, the arc length of each path is $\sigma := \|P\| + \|Q\|$ [11, Section 6.2.4]. This gives a simple *regularisation constraint*[1] for paths: We let $\Gamma_{\|\cdot\|}(P, Q)$ contain all functions $\gamma \colon [0, \sigma] \to [0, \|P\|] \times [0, \|Q\|]$ such that there are $(f, g) \in \Pi(P) \times \Pi(Q)$ with $\gamma_1(s) = 1/\sigma \cdot \int_0^s \|f'(t/\sigma)\| \, dt$ and $\gamma_2(s) = 1/\sigma \cdot \int_0^s \|g'(t/\sigma)\| \, dt$, subject to the constraint $\|\gamma'(s)\|_1 = 1/\sigma \cdot (\|f'(s/\sigma)\| + \|g'(s/\sigma)\|) = 1$, for all $s \in [0, \sigma]$.

Definition 3 ([10, Definition 8]). *Let P, Q be polygonal curves, let $\|\cdot\|$ be a norm, and let $x, y \in [0, \|P\|] \times [0, \|Q\|]$ with $(x_1 \leq y_1) \wedge (x_2 \leq y_2)$ be two points in parameter space. An (x, y)-path is a restriction $\widehat{\gamma}$ of any $\gamma \in \Gamma_{\|\cdot\|}(P, Q)$ that goes through x and y, i.e. $\gamma(\|x\|_1) = x$ and $\gamma(\|y\|_1) = y$, to the domain $[\|x\|_1, \|y\|_1]$. We further say that an (x, y)-path $\widehat{\gamma}$ is* optimal *for P, Q under $\|\cdot\|$ if its assigned cost $\int_{\|x\|_1}^{\|y\|_1} \|P_{\|\cdot\|}(\widehat{\gamma}_1(t)) - Q_{\|\cdot\|}(\widehat{\gamma}_2(t))\| \, dt$ is minimum among all (x, y)-paths.*

Thus, the cost of an optimal $(\mathbf{0}, (\|P\|, \|Q\|)^{\mathsf{T}})$-path for P, Q under $\|\cdot\|$, which goes from the bottom left $\mathbf{0} := (0, 0)^{\mathsf{T}}$ to the top right of the parameter space, is equal to the value $\mathrm{cdtw}_{\|\cdot\|}(P, Q)$. This is easy to verify (cf. [10, Lemma 1]).

2 Geometry of Parameter Space Cells

To construct a dynamic program propagating optimal path costs through the grid of parameter space cells (see Sect. 4), we need to characterise optimal (x, y)-paths within a single cell. For now, we represent each cell (i, j) by corresponding *polygonal segments*, i.e. single-segment curves $\overline{P} := \langle p_{i-1}, p_i \rangle$ and $\overline{Q} := \langle q_{j-1}, q_j \rangle$ whose arc length parametrisations $\overline{P}_{\|\cdot\|}, \overline{Q}_{\|\cdot\|}$ under a norm $\|\cdot\|$ have domains extended to $\mathbb{R}$. Section 2.1 states some known basics. In Sect. 2.2 we establish a robust characterisation of optimal (x, y)-paths for $\overline{P}, \overline{Q}$ under arbitrary $\|\cdot\|$. Note that this is also useful for the lexicographic Fréchet distance [25] and the partial Fréchet similarity [9]. Omitted proofs are given in the full version.

---

[1] Some constraint is needed to prevent infinite speed. Due to its simplicity, using the 1-norm is popular [6,10,23]. Another natural choice is the ∞-norm [21,25].

2.1   Optimal Paths Through Cell Terrain

As we regularise path speeds to the value 1 via the 1-norm, all $\gamma \in \Gamma_{\|\cdot\|}(\overline{P}, \overline{Q})$ and $s \in [0, \sigma]$ satisfy $\|\gamma(s)\|_1 = \gamma_1(s) + \gamma_2(s) = s$. Intuitively, the points in parameter space reachable after time exactly s are those on the line of slope -1 through $\gamma(s)$. Hence, terrain minima on lines of slope -1 may induce optimal (x, y)-paths by means of steepest monotone descent [25]. If such minima are themselves attained on a line, known as a *valley* [10], there is an optimal (x, y)-path travelling as long as possible on the valley or as close as possible to the valley, see Fig. 2a.

Definition 4 ([6, Definition 2]). *A* valley *for polygonal segments $\overline{P}, \overline{Q}$ under a norm $\|\cdot\|$ is a line $\ell \subseteq \mathbb{R}^2$ not of slope -1 such that every point $z \in \ell$ is a sink, i.e. the function $t \mapsto \|\overline{P}_{\|\cdot\|}(z_1 + t) - \overline{Q}_{\|\cdot\|}(z_2 - t)\|$ that evaluates along the line of slope -1 through z is non-increasing on $\mathbb{R}_{\leq 0}$ and non-decreasing on $\mathbb{R}_{\geq 0}$.*

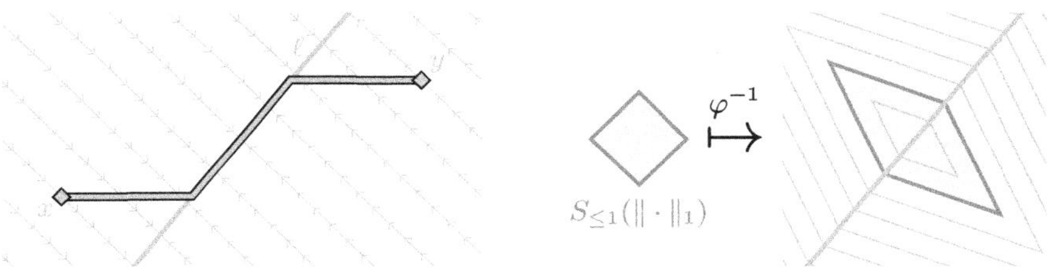

(a) Valley-induced optimal (x, y)-path   **(b)** Terrain and valley under the 1-norm

Fig. 2. Connection between optimal paths through cell terrain and geometric shape of cell terrain, as provided by valleys

Theorem 5 ([6, Sect. 5.5]). *Let $\overline{P}, \overline{Q}$ be polygonal segments with a valley ℓ of positive slope under a norm $\|\cdot\|$, and let $x, y \in \mathbb{R}^2$ with $\Xi := [x_1, y_1] \times [x_2, y_2] \neq \varnothing$ be two points in parameter space. If $\ell \cap \Xi \neq \varnothing$, then the (x, y)-path tracing line segments from x to $\widehat{x}$ to $\widehat{y}$ to y is optimal for $\overline{P}, \overline{Q}$, where $\widehat{x}, \widehat{y} \in \ell \cap \Xi$ share a coordinate with x, y respectively. Else, the (x, y)-path tracing line segments from x to ξ to y is optimal, where $\xi \in \{(x_1, y_2)^{\mathsf{T}}, (y_1, x_2)^{\mathsf{T}}\}$ is closest to ℓ in Ξ.*

The terrain that paths travel through in a cell consists of affinely transformed versions of the (sub)level sets of $\|\cdot\|$, see Fig. 2b. For $\mu \in \mathbb{R}$ the *μ-sublevel set* of a function $h \colon \mathrm{dom}(h) \to \mathbb{R}$ is the set $S_{\leq \mu}(h) := \{z \in \mathrm{dom}(h) \mid h(z) \leq \mu\}$.

Lemma 6 ([1, Lemma 3]). *Let $\overline{P}, \overline{Q}$ be polygonal segments, let $\|\cdot\|$ be a norm, and let $\varphi \colon \mathbb{R}^2 \to \mathbb{R}^2$ be the affine map defined through $\varphi(z) := \overline{P}_{\|\cdot\|}(z_1) - \overline{Q}_{\|\cdot\|}(z_2)$. If $\overline{P}, \overline{Q}$ are parallel, then φ is constant either on every line of slope 1 or on every line of slope -1 in the extended parameter space $\mathbb{R}^2$. Else, φ has an affine inverse map $\varphi^{-1} \colon \mathbb{R}^2 \to \mathbb{R}^2$ satisfying $S_{\leq \mu}(\|\cdot\| \circ \varphi) = \varphi^{-1}(S_{\leq \mu}(\|\cdot\|))$ for all $\mu \geq 0$.*

2.2   Existence and Computation of Valleys

Every cell has a valley under the 2-norm [23, Lemma 4] and 1-norm [6, Lemma 24], but for CDTW in $(\mathbb{R}^2, \|\cdot\|_2)$ under $(p, q) \mapsto \|p - q\|_1$ there are cells whose single valley has negative slope. Because monotone paths cannot travel on such valleys, computing optimal (x, y)-paths then deviates from Theorem 5 (see [6, Lemma 27]), and handling their costs algorithmically becomes more difficult. We generalise and improve upon previous results by showing that every norm $\|\cdot\|$ guarantees valleys of positive slope for our formulation of CDTW under $\|\cdot\|$ from Definition 1.

To this end, we employ the geometric definition of norms: Given an absorbing set $K \subseteq \mathbb{R}^2$, i.e. every $z \in \mathbb{R}^2$ satisfies $z \in \lambda K$ for all $\lambda \in \mathbb{R}$ with large enough $|\lambda|$, let its *gauge* be denoted by $\mathcal{G}_K \colon \mathbb{R}^2 \to \mathbb{R}_{\geq 0}$, where $\mathcal{G}_K(z) := \inf\{\lambda \geq 0 \mid z \in \lambda K\}$. If K is balanced, i.e. $\lambda K \subseteq K$ for all $\lambda \in [-1, 1]$, as well as closed and bounded, then the sublevel sets of $\mathcal{G}_K$ are $S_{\leq \mu}(\mathcal{G}_K) = \mu K$ for all $\mu \geq 0$. If K is also convex, i.e. $\lambda z + (1 - \lambda) z' \in K$ for all $z, z' \in K$ and $\lambda \in [0, 1]$, then its gauge $\mathcal{G}_K$ is a norm on $\mathbb{R}^2$. Conversely, any norm $\|\cdot\|$ is equal to the gauge of its absorbing, balanced, closed, bounded and convex 1-sublevel set $S_{\leq 1}(\|\cdot\|)$. (See [26, pp. 39–40].)

To find sinks for valleys, we minimise a norm $\mathcal{G}_K$ on a line L by duality (cf. [4, Section 5.1.6]). The intuition is: We scale K by $\mu^* \geq 0$ such that L is a *supporting line* of $\mu^* K$, i.e. $L \cap \mu^* K \neq \varnothing$ and $\mu^* K$ lies within a closed half-plane bounded by L. The minimum of $\mathcal{G}_K$ on L is then attained on $L \cap \mu^* K$, see Fig. 3a.

Lemma 7. *Let $L = \{z \in \mathbb{R}^2 \mid u^\mathsf{T} \cdot z = t\}$ be a line with $u \in \mathbb{R}^2 \setminus \{\mathbf{0}\}$ and $t \in \mathbb{R}$. The gauge $\mathcal{G}_K$ of an absorbing, balanced, closed and bounded $K \subseteq \mathbb{R}^2$ attains its minimum on L at $(t/t^*)v^*$, where $v^* \in \arg\max\{u^\mathsf{T} \cdot v \mid v \in K\}$ and $t^* := u^\mathsf{T} \cdot v^*$.*

Proof. We have $(t/t^*)v^* \in L$ due to $u^\mathsf{T} \cdot (t/t^*)v^* = t$ by definition of t^*. If $t = 0$, then $(t/t^*)v^* = \mathbf{0}$ minimises the gauge $\mathcal{G}_K$. Otherwise, it is $\mathbf{0} \notin L$, so $\mathcal{G}_K(z) > 0$ and $z/\mathcal{G}_K(z) \in K$ hold for all $z \in L$, as K is bounded and closed. This yields

$$\mathcal{G}_K(z) = \frac{|t|}{|t|} \cdot \mathcal{G}_K(z) = \frac{|t|}{|u^\mathsf{T} \cdot z|} \cdot \mathcal{G}_K(z) = \frac{|t|}{|u^\mathsf{T} \cdot (z/\mathcal{G}_K(z))|} \geq \frac{|t|}{|u^\mathsf{T} \cdot v^*|} = \frac{|t|}{|t^*|}$$

for all $z \in L$, where the inequality follows from v^* maximising $v \mapsto u^\mathsf{T} \cdot v$ on K and thereby also maximising $v \mapsto |u^\mathsf{T} \cdot v|$ on K, as K is balanced. On top of that, $v^* \in K$ and K being balanced imply $|t/t^*| \geq |t/t^*|\mathcal{G}_K(v^*) = \mathcal{G}_K((t/t^*)v^*)$. $\square$

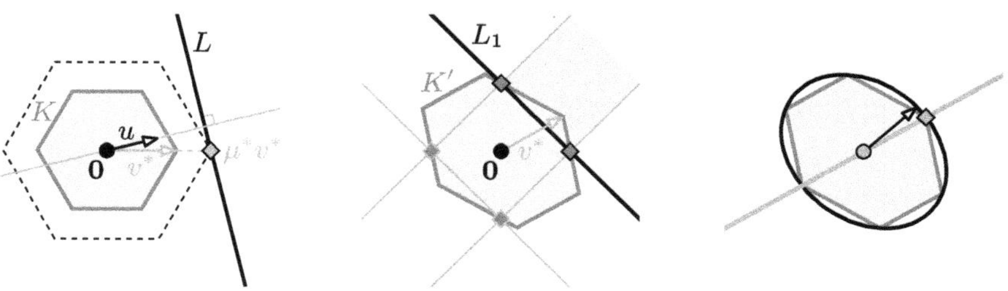

(a) Minimising the norm   (b) Finding positive slope   (c) Computing a valley

Fig. 3. Valley characterisation under a norm with regular hexagons as sublevel sets

We next prove the main result of this section, which additionally uses the affine map φ from Lemma 6. Note that polygonal segments $\overline{P}, \overline{Q}$ give $\varphi(\mathbf{0}) = p_{i-1} - q_{j-1}$, while the translational part of an analogously defined affine map for a parameter space cell (i, j) of curves P, Q may depend on all of $p_0, \dots, p_i$ and $q_0, \dots, q_j$.

Theorem 8. *Let $\overline{P} = \langle p_{i-1}, p_i \rangle$ and $\overline{Q} = \langle q_{j-1}, q_j \rangle$ be polygonal segments, and let $\| \cdot \|$ be a norm. There exists a valley ℓ of positive slope for $\overline{P}, \overline{Q}$ under $\| \cdot \|$. Computing such a valley reduces to finding a sink via Lemma 7, where we use $K := S_{\leq 1}(\| \cdot \|)$ or $K' := \varphi^{-1}(K) - \varphi^{-1}(\mathbf{0})$ with $\varphi \colon \mathbb{R}^2 \to \mathbb{R}^2$ from Lemma 6.*

Proof. By definition of K and φ, we have $\mathcal{G}_K(\varphi(z)) = \|\overline{P}_{\|\cdot\|}(z_1) - \overline{Q}_{\|\cdot\|}(z_2)\|$ for all $z \in \mathbb{R}^2$. Moreover, K is absorbing, balanced, closed, bounded and convex.

Assume first that $\overline{P}, \overline{Q}$ are parallel, so Lemma 6 says that φ and $\mathcal{G}_K \circ \varphi$ are constant either on every line of slope 1 or -1. In the latter case every line not of slope -1 is a valley by Definition 4. In the former case it suffices to find one sink. For this, consider $L := \{\varphi(s, -s) \mid s \in \mathbb{R}\}$, where $\varphi(s, -s) = 2s \cdot \frac{p_i - p_{i-1}}{\|p_i - p_{i-1}\|} + \varphi(\mathbf{0})$ holds for all $s \in \mathbb{R}$. We obtain an $s^* \in \mathbb{R}$ such that $\mathcal{G}_K$ attains its minimum on L at $\varphi(s^*, -s^*)$ by applying Lemma 7 for any $u \in \mathbb{R}^2 \setminus \{\mathbf{0}\}$ with $u^\mathsf{T} \cdot (p_i - p_{i-1}) = 0$ and $t := u^\mathsf{T} \cdot \varphi(\mathbf{0})$. Also, the properties of K imply that $L \cap S_{\leq \mu}(\mathcal{G}_K) = L \cap \mu K$ is a convex subset of L for each $\mu \geq 0$. It follows that $(s^*, -s^*)^\mathsf{T}$ is a sink and that the line $\ell := \{(\lambda + s^*, \lambda - s^*)^\mathsf{T} \mid \lambda \in \mathbb{R}\}$ is a valley of slope 1 for $\overline{P}, \overline{Q}$.

Now assume that $\overline{P}, \overline{Q}$ are not parallel. Then Lemma 6 says $S_{\leq \mu}(\mathcal{G}_K \circ \varphi) = \varphi^{-1}(S_{\leq \mu}(\mathcal{G}_K)) = \varphi^{-1}(\mu K)$ for all $\mu \geq 0$. Because φ^{-1} is a bijective affine map, the set $K' = \varphi^{-1}(K) - \varphi^{-1}(\mathbf{0})$ inherits the properties of K and further satisfies $\mu K' = \varphi^{-1}(\mu K) - \varphi^{-1}(\mathbf{0})$ for all $\mu \geq 0$. Thus, $\mathcal{G}_{K'} = (\mathcal{G}_K \circ \varphi) \circ (z \mapsto z + \varphi^{-1}(\mathbf{0}))$ is fulfilled. Given some line $L_t := \{z \in \mathbb{R}^2 \mid z_1 + z_2 = t\}$ of slope -1 with $t \in \mathbb{R}$, we apply Lemma 7 for K' and $u' = (1, 1)^\mathsf{T}$ to see that $\mathcal{G}_{K'}$ attains its minimum on L_t at a $(t/t^*)v^*$, and hence that $(t/t^*)v^* + \varphi^{-1}(\mathbf{0})$ is a sink similar to before. By fixing any $v^* \in \arg\max\{(1, 1) \cdot v \mid v \in K'\}$ and $t^* := (1, 1) \cdot v^*$, we have that each point on $\ell := \{\lambda v^* + \varphi^{-1}(\mathbf{0}) \mid \lambda \in \mathbb{R}\}$ is a sink, so ℓ is a valley for $\overline{P}, \overline{Q}$.

It remains to show that there is such a valley ℓ of positive slope by establishing that an appropriate v^* with positive components exists. To this end, observe that φ gives unit vectors $\varphi(\pm 1, 0) - \varphi(\mathbf{0}) = \pm \frac{p_i - p_{i-1}}{\|p_i - p_{i-1}\|}$ and $\varphi(0, \pm 1) - \varphi(\mathbf{0}) = \mp \frac{q_j - q_{j-1}}{\|q_j - q_{j-1}\|}$ that lie in the boundary of the unit sublevel set $K = S_{\leq 1}(\| \cdot \|)$. In consequence, the boundary of the linearly transformed set $K' = \varphi^{-1}(K) - \varphi^{-1}(\mathbf{0})$ contains the inverse elements $(\pm 1, 0)^\mathsf{T}$ and $(0, \pm 1)^\mathsf{T}$. Because K' is closed and convex, it is

$$L_1 \cap \mathbb{R}^2_{\geq 0} \subseteq \{v \in K' \mid (1, 1) \cdot v \geq 1\} \subseteq L_1 \cup (L_1 \cap \mathbb{R}^2_{\geq 0} + \{(\lambda, \lambda)^\mathsf{T} \mid \lambda > 0\})$$

with $L_1 = \{z \in \mathbb{R}^2 \mid z_1 + z_2 = 1\}$. These inclusions ensure a $v^* \in \mathbb{R}^2_{>0}$ as desired, see Fig. 3b. The latter inclusion holds since K' has supporting lines through all of its boundary points due to convexity (see [4, Section 2.5.2]) and each such line through $(1, 0)^\mathsf{T} \in L_1$ or $(0, 1)^\mathsf{T} \in L_1$ is constrained by $(0, \pm 1)^\mathsf{T}, (\pm 1, 0)^\mathsf{T} \in K'$. $\square$

To approximate the 2-norm, we can use norms with polygons as sublevel sets. Given a suitable k-gon K, in particular we have that K is convex and k is even,

Theorem 8 allows computing valleys in $O(\log(k))$ time via binary search on the vertices of K or K' (cf. [12, Section 2.2]). If K is regular, we achieve $O(1)$ time by maximising $v \mapsto (1,1) \cdot v$ on the ellipse circumscribing K', see Fig. 3c. More generally, this works if $K = \psi(R_k)$ is a linearly transformed regular k-gon.

Corollary 9. *Let $\|\cdot\|$ be a norm with $S_{\leq 1}(\|\cdot\|) = \psi(R_k)$ for a regular k-gon R_k and a linear map $\psi \colon \mathbb{R}^2 \to \mathbb{R}^2$. Computing a valley ℓ of positive slope under $\|\cdot\|$ for arbitrary $\overline{P}, \overline{Q}$ is possible in $O(1)$ time, independent of $k \in \{4, 6, \dots\}$.*

3   Unsolvability Within ACMℚ Under the 2-Norm

It has been shown that the partial Fréchet similarity, a measure related to CDTW, cannot be computed exactly under the 2-norm by radicals over $\mathbb{Q}$ [14]. Thus, its computational problem is unsolvable within ACMℚ, the *Algebraic Computation Model over* $\mathbb{Q}$ that only allows consecutive applications of the four basic arithmetic operations and the extraction of roots $\sqrt[c]{\cdot}$ for all $c \in \mathbb{N}$. Such algebraic operations are sufficient for many problems from computational geometry, e.g. the Fréchet distance can be computed using only basic arithmetic and square roots [1].

The abovementioned result and others of its kind [2,15] utilise Galois theory. They provide a constraint for potential exact algorithms and hence a motivation for approximation algorithms. Note that, from a numerical perspective, radicals are also approximations, but they enable symbolic computations [15, Remark 1]. Moreover, there is a recent development [32] that solves general polynomials of arbitrary degree by power series, which might start to compete with radicals.

That does not affect our ACMℚ unsolvability result for CDTW under $\|\cdot\|_2$, as we show that the numbers involved may not even be algebraic, i.e. they do not solve any polynomial equation over $\mathbb{Q}$. This stronger result uses the following consequence of Baker's Theorem from transcendental number theory.

Lemma 10 ([3, Theorem 2.2]). *Let $\alpha_1, \dots, \alpha_r \in \mathbb{R}_{>0}$ and $\beta_0, \beta_1, \dots, \beta_r \in \mathbb{R}$ be algebraic numbers. The equality $\beta_0 + \beta_1 \ln(\alpha_1) + \cdots + \beta_r \ln(\alpha_r) = 0$ holds if and only if we have $\beta_0 = 0 = \beta_1 \ln(\alpha_1) + \cdots + \beta_r \ln(\alpha_r)$.*

A single example that gives a transcendental CDTW value under the 2-norm already establishes the unsolvability of the general case. This of course does not rule out practical numerical computations, but we demonstrate another obstacle: Optimal paths can involve transcendental numbers too, so that approximations of their costs could introduce cumulative errors when propagated.

Theorem 11. *CDTW under $\|\cdot\|_2$ is unsolvable within ACMℚ because polygonal curves P, Q may exhibit the following, even with integer vertices.*

(a) The number $\mathrm{cdtw}_{\|\cdot\|_2}(P, Q)$ is transcendental.
(b) Every optimal $(\mathbf{0}, (\|P\|_2, \|Q\|_2)^{\mathsf{T}})$-path for P, Q under $\|\cdot\|_2$ has bending points whose coordinates are transcendental numbers.

Proof (Sketch). Showing (b) requires more effort than showing (a). For the sake of clarity, we defer some details from the proof of (b) to the full version.

(a) Let $P := \langle (1,2)^\mathsf{T}, (1,-4)^\mathsf{T} \rangle$ and $Q := \langle (0,0)^\mathsf{T}, (5,0)^\mathsf{T} \rangle$, so that $\|P\|_2 + \|Q\|_2 = 6 + 5 = 11$ holds and the parameter space has only one cell, which corresponds to the right cell in Fig. 4. The map $z \mapsto z_1 \cdot (0,-1)^\mathsf{T} - z_2 \cdot (1,0)^\mathsf{T} + (1,2)^\mathsf{T}$ realises φ from Lemma 6 and attains $\mathbf{0}$ at $(2,1)^\mathsf{T}$. The line $\ell := \{z \in \mathbb{R}^2 \mid z_1 - z_2 = 1\}$ through that point is a valley of the cell since the 2-norm always gives valleys of slope 1 along ellipse axes in cell terrain (see [23, 25]).

By Theorem 5, the valley ℓ induces an optimal path $\gamma^* \in \Gamma_{\|\cdot\|_2}(P,Q)$ tracing line segments from $\mathbf{0}$ to $(1,0)^\mathsf{T}$ to $(6,5)^\mathsf{T}$. Evaluating its cost yields

$$\mathrm{cdtw}_{\|\cdot\|_2}(P,Q) = \int_0^1 \|\varphi(t,0)\|_2 \, dt + \int_1^{11} \left\| \varphi\left(\tfrac{t+1}{2}, \tfrac{t-1}{2}\right) \right\|_2 \, dt$$

$$= \int_0^1 \sqrt{t^2 - 4t + 5} \, dt + \int_1^{11} \frac{|t-3|}{\sqrt{2}} \, dt$$

$$= \left[\tfrac{1}{2} \cdot \ln(\alpha_1) - \tfrac{1}{\sqrt{2}} - \tfrac{1}{2} \cdot \ln(\alpha_2) + \sqrt{5} \right] + 17 \cdot \sqrt{2}$$

with $\alpha_1 := \sqrt{2} - 1$ and $\alpha_2 := \sqrt{5} - 2$. The left integral makes use of $\int h(t) \, dt = \frac{4\lambda_0 - \lambda_1^2}{8} \cdot \ln(|2 \cdot h(s) + 2s + \lambda_1|) + \frac{2s + \lambda_1}{4} \cdot h(s) + C$ for $h(s) := \sqrt{s^2 + \lambda_1 s + \lambda_0}$ with $\lambda_0, \lambda_1 \in \mathbb{R}$ (see [20, Equation 4.3.4.1.5]). It is $\tfrac{1}{2} \cdot \ln(\alpha_1) - \tfrac{1}{2} \cdot \ln(\alpha_2) \neq 0$ due to $\alpha_1 \neq \alpha_2$. Thus, the numbers $\tfrac{1}{2} \cdot \ln(\alpha_1) - \tfrac{1}{2} \cdot \ln(\alpha_2)$ and $\mathrm{cdtw}_{\|\cdot\|_2}(P,Q)$ are transcendental, as assuming otherwise contradicts Lemma 10.

(b) Now let $P := \langle (4,-2)^\mathsf{T}, (1,2)^\mathsf{T}, (1,-4)^\mathsf{T} \rangle$ and $Q := \langle (0,0)^\mathsf{T}, (5,0)^\mathsf{T} \rangle$, so that this time $\|P\|_2 + \|Q\|_2 = 11 + 5 = 16$ holds and there are two cells with valleys $\ell_1 := \{z \in \mathbb{R}^2 \mid z_1 - z_2 = 0\}$ and $\ell_2 := \{z \in \mathbb{R}^2 \mid z_1 - z_2 = 6\}$ respectively. Optimal path candidates are all $\gamma_s \in \Gamma_{\|\cdot\|_2}(P,Q)$ tracing line segments from $\mathbf{0}$ to $(s/2, s/2)^\mathsf{T}$ to $(s/2+6, s/2)^\mathsf{T}$ to $(11,5)^\mathsf{T}$ for an $s \in [0,10]$, where $\gamma_s(s) = (s/2, s/2)^\mathsf{T}$ is the bending point of γ_s on ℓ_1 like in Fig. 4. This is because any path can be transformed into this shape without increasing its cost, by splitting it at the inter-cell boundary and using Theorem 5 in each cell.

Claim (B1). Let $\mathcal{C}: [0,10] \to \mathbb{R}_{\geq 0}$ be defined by $\mathcal{C}(s)$ being the cost of the optimal path candidate γ_s for $s \in [0,10]$. We have that $\mathcal{C}$ is continuous.

By continuity, $\mathcal{C}$ attains its minimum on the closed interval $[0,10]$, so there is a candidate γ_{s^*} that is optimal and has cost equal to $\mathrm{cdtw}_{\|\cdot\|_2}(P,Q)$.

Claim (B2). The function $\mathcal{C}$ is differentiable on $[0,10]$, and its derivative at s is $\mathcal{C}'(s) = -\tfrac{1}{5} \cdot \int_0^{5/4 \cdot (2 - \kappa_s)} \frac{t}{\sqrt{(t + 2\kappa_s)^2 + \kappa_s^2}} \, dt - \tfrac{1}{2} \cdot \int_{-(\nu_s + 2)}^0 \frac{t}{\sqrt{(t + \nu_s)^2 + \nu_s^2}} \, dt$ for $\kappa_s := 2/5 \cdot (s - 5)$ and $\nu_s := 1/2 \cdot (s - 2)$. It is $\mathcal{C}'(0) < 0$ and $\mathcal{C}'(10) > 0$.

The bounds for $\mathcal{C}'$ at values 0 and 10 imply that any s^* minimising $\mathcal{C}$ lies in the open interval $(0,10)$, which means $\mathcal{C}'(s^*) = 0$ by differentiability.

Claim (B3). Some algebraic functions $\alpha_1, \alpha_2, \beta_0, \beta_1, \beta_2 \colon [0,10] \setminus \{2,5\} \to \mathbb{R}$ satisfying $\mathcal{C}'(s) = \beta_0(s) + \beta_1(s) \cdot \ln(\alpha_1(s)) + \beta_2(s) \cdot \ln(\alpha_2(s))$ exist. We also have $\mathcal{C}'(2) \neq 0 \neq \mathcal{C}'(5)$, and $\beta_0(s_0^*) = 0$ is fulfilled on $(0,10)$ only by $s_0^* := 1/929 \cdot (1880 \cdot \sqrt{10} - 3070) + 60 \cdot \left(\sqrt{16 \cdot \sqrt{10} + 61} \right) / (80 \cdot \sqrt{10} + 269) \in (4, 4.5)$.

Due to Lemma 10, s_0^* is the only algebraic candidate that could minimise $\mathcal{C}$. Numerically, we obtain $s_0^* \approx 4.31$ with $\mathcal{C}'(s_0^*) \approx 0.80$, whereas $\mathcal{C}'(s^*) = 0$ is fulfilled on $(0, 10)$ only by $s^* \approx 2.08$. We can prove $\mathcal{C}'(s_0^*) \neq 0$ as follows.

Claim (B4). It is $\beta_1(s) \cdot \ln(\alpha_1(s)) + \beta_2(s) \cdot \ln(\alpha_2(s)) > 0$ for all $s \in (4, 4.5)$.

So the minimum of $\mathcal{C}$ is not attained at an algebraic value. All s^* minimising $\mathcal{C}$, along with the bending points of all optimal γ_{s^*}, are thus transcendental. $\square$

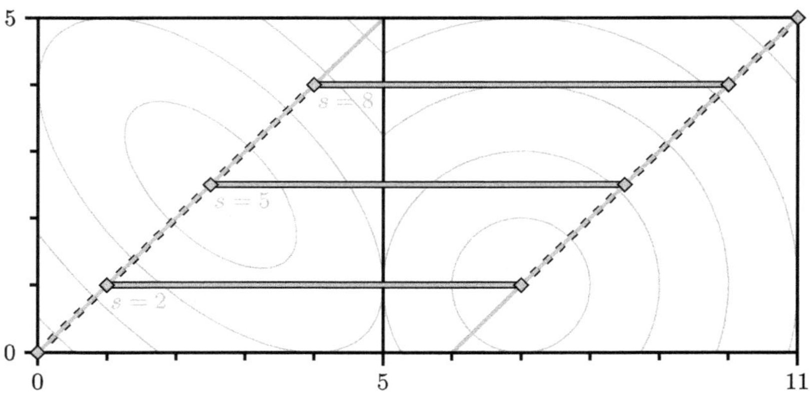

Fig. 4. Parameter space whose optimal path candidates switch valleys

4   An Exact Algorithm Under Polygonal Norms

This section uses a parameter $k \in \{4, 6, \dots\}$, a bijective linear map $\psi \colon \mathbb{R}^2 \to \mathbb{R}^2$, and a norm $\|\cdot\|$ on $\mathbb{R}^2$ as follows: Whenever we write $\|\cdot\|$, the respective definition or result applies to every possible norm. For the algorithm we assume a 1-sublevel set $S_{\leq 1}(\|\cdot\|) = \psi(R_k)$, where R_k denotes the convex regular k-gon with vertices $v_1, \dots, v_k$ defined by $v_r := (\cos(\theta_r), \sin(\theta_r))^\mathsf{T}$ for $\theta_r := r \cdot 2\pi/k$ and $r \in \{1, \dots, k\}$. In that case we use the gauge notation $\mathcal{G}_{\psi(R_k)}$, as specified in Sect. 2.2.

This setting covers all norms with regular polygons as sublevel sets, including the 1-norm and the ∞-norm, and allows to approximate the 2-norm. Instead of discretising the input curves like in previous approaches [5,7,16,19,23], we hence essentially discretise the norm. Omitted proofs are given in the full version.

We apply a dynamic programming scheme known from other exact algorithms for CDTW and related measures [6,9,10,21,24,27,28]. It propagates optimal path costs through the grid of parameter space cells in the form of functions.

Definition 12 ([10, Definition 5]). *Let P, Q be polygonal curves. For every parameter space cell $[a_1, b_1] \times [a_2, b_2] \subseteq [0, \|P\|] \times [0, \|Q\|]$ we define its borders as parametrisations along the top/right/bottom/left side of its boundary, e.g. its top border is $[a_1, b_1] \to \mathbb{R}^2, t \mapsto (t, b_2)^\mathsf{T}$. A cell border $\mathcal{B} \colon \mathrm{dom}(\mathcal{B}) \to \mathbb{R}^2$ has an associated* optimum function *denoted by* $\mathrm{opt}_{0,\mathcal{B}} \colon \mathrm{dom}(\mathcal{B}) \to \mathbb{R}_{\geq 0}$, *where* $\mathrm{opt}_{0,\mathcal{B}}(t)$ *is the cost of an optimal* $(\mathbf{0}, \mathcal{B}(t))$-path *under* $\|\cdot\|$ *for each* $t \in \mathrm{dom}(\mathcal{B})$.

In the dynamic program's base case we compute optimum functions of borders on the parameter space's bottom and left side via costs of straight paths.

4.1   Propagation Procedure

Our propagation procedure builds upon [21, Section 5.2] and [6, Section 5.3].

Definition 13. *Let $\mathcal{A}$ be the bottom or left border and let $\mathcal{B}$ be the top or right border of one cell. The propagation space of $\mathcal{A}, \mathcal{B}$ is defined by $\Sigma_{\mathcal{A},\mathcal{B}} := \{(s,t)^\mathsf{T} \in$ $\mathrm{dom}(\mathcal{A}) \times \mathrm{dom}(\mathcal{B}) \mid (\mathcal{A}_1(s) \leq \mathcal{B}_1(t)) \wedge (\mathcal{A}_2(s) \leq \mathcal{B}_2(t))\}$. If $\mathcal{A}, \mathcal{B}$ are a pair of left and top border or a pair of bottom and right border, we call them* adjoining borders *and write* $\mathrm{adj}(\mathcal{B}) := \mathcal{A}$. *Else, we call $\mathcal{A}, \mathcal{B}$* opposing *and write* $\mathrm{opp}(\mathcal{B}) := \mathcal{A}$.

We equip $\mathrm{opt}_{\mathcal{A},\mathcal{B}} \colon \Sigma_{\mathcal{A},\mathcal{B}} \to \mathbb{R}_{\geq 0}$ on the propagation space, where $\mathrm{opt}_{\mathcal{A},\mathcal{B}}(s,t)$ is the cost of an optimal $(\mathcal{A}(s), \mathcal{B}(t))$-path under $\|\cdot\|$ for each $(s,t)^\mathsf{T} \in \Sigma_{\mathcal{A},\mathcal{B}}$, so

$$\mathrm{opt}_{0,\mathcal{B}}(t) = \min_{\mathcal{A}\in\{\mathrm{adj}(\mathcal{B}),\mathrm{opp}(\mathcal{B})\}} \inf\left\{\mathrm{opt}_{0,\mathcal{A}}(s) + \mathrm{opt}_{\mathcal{A},\mathcal{B}}(s,t) \mid (s,t)^\mathsf{T} \in \Sigma_{\mathcal{A},\mathcal{B}}\right\} \quad (\star)$$

holds for all $t \in \mathrm{dom}(\mathcal{B})$. After determining optimum functions of $\mathrm{adj}(\mathcal{B}), \mathrm{opp}(\mathcal{B})$, we propagate costs to $\mathcal{B}$ by implementing Equation $(\star)$ with the help of Theorem 5. Our implementation PROPAGATE and its correctness are outlined below.

Lemma 14. *(a) We can evaluate $\mathcal{G}_{\psi(R_k)}$ in $O(1)$ time. Its restriction to each cone $\psi(\{\lambda v_{r-1} + \lambda' v_r \mid \lambda, \lambda' \geq 0\})$ is linear, where $r \in \{1,\ldots,k\}$ and $v_0 := v_k$.*
(b) Let $\Sigma_{\mathcal{A},\mathcal{B}}$ be the propagation space of borders $\mathcal{A}, \mathcal{B}$ under $\mathcal{G}_{\psi(R_k)}$. There is an arrangement (V, E, F) of $O(k)$ lines with $\Sigma_{\mathcal{A},\mathcal{B}} = \bigcup F'$ for a subset $F' \subseteq F$ of closed faces such that each restriction of $\mathrm{opt}_{\mathcal{A},\mathcal{B}}$ to an $f \in F'$ is quadratic.
(c) The function $\mathrm{opt}_{0,\mathcal{B}}$ is piecewise quadratic for every border $\mathcal{B}$ under $\mathcal{G}_{\psi(R_k)}$.

PROPAGATE$(\overline{P}, \overline{Q}, \mathcal{B}, \mathrm{opt}_{0,\mathrm{adj}(\mathcal{B})}, \mathrm{opt}_{0,\mathrm{opp}(\mathcal{B})})$

Input: Polygonal segments $\overline{P}, \overline{Q}$ and a top/right border $\mathcal{B}$ of their cell under $\mathcal{G}_{\psi(R_k)}$.
Optimum functions $\mathrm{opt}_{0,\mathcal{A}}$ of $\mathcal{A} \in \{\mathrm{adj}(\mathcal{B}), \mathrm{opp}(\mathcal{B})\}$ as stacks of quadratic pieces.
Output: Optimum function $\mathrm{opt}_{0,\mathcal{B}}$ of $\mathcal{B}$ as a stack of quadratic pieces.

---

1. $\mathcal{S} \leftarrow$ empty stack; $H \leftarrow \varnothing$; $\ell \leftarrow$ valley of positive slope for $\overline{P}, \overline{Q}$ under $\mathcal{G}_{\psi(R_k)}$
2. **foreach** $\mathcal{A}$ **in** $\langle \mathrm{adj}(\mathcal{B}), \mathrm{opp}(\mathcal{B})\rangle$ **do**   // start with adjoining border
3. $(V, E, F) \leftarrow$ arrangement as in Lemma 14b, based on ℓ and Theorem 5
4.   overlay (V, E, F) with vertical lines arranged at breakpoints of $\mathrm{opt}_{0,\mathcal{A}}$
5. $\mathcal{I} \leftarrow$ ascending list of closed intervals split by vertical lines of (V, E, F)
6.   **if** $\mathcal{A} = \mathrm{adj}(\mathcal{B})$ **then** reverse $\mathcal{I}$ // choose order depending on $\mathcal{A}$
7.   **foreach** interval I **in** $\mathcal{I}$ **do**   // process intervals in order
8.   **foreach** edge e **in** $\{e \in E \mid e \subseteq (I \times \mathbb{R}) \cap \Sigma_{\mathcal{A},\mathcal{B}}\}$ **do**
9.   add $[t \mapsto \mathrm{opt}_{0,\mathcal{A}}(s_e(t)) + \mathrm{opt}_{\mathcal{A},\mathcal{B}}(s_e(t), t)]$ to H, where $(s_e(t), t)^\mathsf{T} \in e$
10.   **foreach** face f **in** $\{f \in F \mid f \subseteq (I \times \mathbb{R}) \cap \Sigma_{\mathcal{A},\mathcal{B}}\}$ **do**
11. $e_f \leftarrow \{(s,t)^\mathsf{T} \in f \mid \partial_s[\mathrm{opt}_{0,\mathcal{A}}(s) + \mathrm{opt}_{\mathcal{A},\mathcal{B}}(s,t)] = 0\}$ // extremal edge
12.   add $[t \mapsto \mathrm{opt}_{0,\mathcal{A}}(s_f(t)) + \mathrm{opt}_{\mathcal{A},\mathcal{B}}(s_f(t), t)]$ to H, where $(s_f(t), t)^\mathsf{T} \in e_f$
13. $\mathcal{S}_I \leftarrow$ lower envelope of H as a stack of quadratic pieces; $H \leftarrow \varnothing$
14.   update top of $\mathcal{S}$ with suffix of smaller-valued pieces on top of $\mathcal{S}_I$
15. **return** $\mathcal{S}$

Due to Lemma 14c, PROPAGATE exclusively deals with quadratic pieces, and its method to compute these is based on arrangements as in Lemma 14b. The use of stack data structures (see [10, Lemma 12]) speeds up computations and relies on the following property, which is a stronger version of [10, Lemma 11].

Lemma 15. *Let $\mathcal{B}$ be a top/right border under $\|\cdot\|$, let $\mathcal{A}\colon [a,b] \to \mathbb{R}^2$ be a parametrisation in direction $(1,0)^\mathsf{T}$ or $(0,1)^\mathsf{T}$, and let $[s \triangleright t] := \mathrm{opt}_{0,\mathcal{A}}(s) + \mathrm{opt}_{\mathcal{A},\mathcal{B}}(s,t)$. If $[s \triangleright t] < [s' \triangleright t]$ and $[s \triangleright t'] \geq [s' \triangleright t']$ for $t < t'$, then $s < s'$ if and only if $\mathcal{A}$ and $\mathcal{B}$ have the same direction. This yields a propagation order for $\mathcal{A} \in \{\mathrm{adj}(\mathcal{B}), \mathrm{opp}(\mathcal{B})\}$.*

Theorem 16. PROPAGATE *correctly computes optimum functions under* $\mathcal{G}_{\psi(R_k)}$.

Proof. We obtain $\ell, \mathrm{opt}_{\mathcal{A},\mathcal{B}}, (V,E,F)$ via Corollary 9, Theorem 5 and Lemma 14b. Lines 4 & 5 ensure that the loops cover all of $\Sigma_{\mathcal{A},\mathcal{B}}$ and that lines 9 & 12 deal with quadratic restrictions of $\mathrm{opt}_{0,\mathcal{A}}$ and $\mathrm{opt}_{\mathcal{A},\mathcal{B}}$. Minima of $s \mapsto \mathrm{opt}_{0,\mathcal{A}}(s) + \mathrm{opt}_{\mathcal{A},\mathcal{B}}(s,t)$ are attained on edges and face interiors. The latter minima come with derivative value 0, which yields a linear relation of s and t determined in line 11. To realise Equation $(\star)$, lines 13 & 14 process lower envelopes. Here, Lemma 15 and the order of $\mathcal{I}$ imply that if a value on $\mathcal{S}$ is improved by $\mathcal{S}_I$, all values above it are too. $\qquad\square$

Corollary 17. *We can compute CDTW exactly under* $\mathcal{G}_{\psi(R_k)}$, *yielding a $(1+\varepsilon)$-approximation for CDTW under $\|\cdot\|_2$ using $\psi := \mathrm{id}$ and some $k \in O(\varepsilon^{-1/2})$.*

4.2 Complexity and Properties

We consider the ideas from the 1D running time analysis [10] in our setting.

Proposition 18. *The running time of our algorithm for CDTW under $\mathcal{G}_{\psi(R_k)}$ is bounded by $O(N \cdot k^2 \log(k)\alpha(k))$, where N denotes the total number of quadratic pieces over all optimum functions and α is the inverse Ackermann function.*

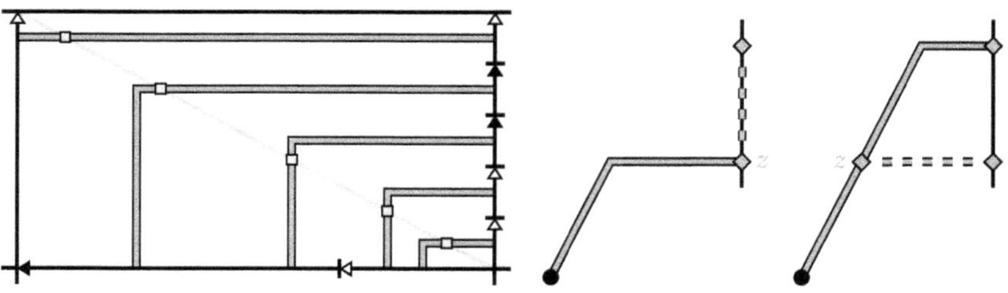

Fig. 5. Doubling of adjoining pieces   **Fig. 6.** Prefix-based path creation

An essential ingredient to showing $N \in O(n^5)$ in 1D is an inductive bound on the number of distinct pairs $(\lambda_1, \lambda_2)^{\mathsf{T}} \in \mathbb{R}^2$ such that $s \mapsto \sum_{i=0}^{2} \lambda_i s^i$ is a quadratic piece [10, Lemma 14]. It is not clear if such a bound holds in 2D, as dealing with the arrangements from Lemma 14b poses a greater challenge in this setting.

That is, the cones from Lemma 14a can induce lines of negative slope, which do not occur in 1D. Even optimal paths in the order from Lemma 15 may have bending points that repeatedly switch faces via these lines, see Fig. 5. It is an unresolved issue whether this could cause an exponential doubling behaviour.

Still, we next generalise another important property, the continuity of optimum functions [10, Lemma 7]. We use $\overline{\lim}$ and $\underline{\lim}$ to denote $\limsup$ and $\liminf$.

Lemma 19. *Consider the parameter space of polygonal curves P, Q under $\|\cdot\|$. Let $(\gamma_r)_{r \in \mathbb{N}}$ be a sequence of (x_r, y_r)-paths that converge to an (x, y)-path γ, i.e. it is $\lim_{r \to \infty}(x_r, y_r) = (x, y)$ and $\lim_{r \to \infty} \gamma_r(s) = \gamma(s)$ for all $s \in (\|x\|_1, \|y\|_1)$. Then the associated sequence $(\mathrm{cost}(\gamma_r))_{r \in \mathbb{N}}$ of path costs converges to $\mathrm{cost}(\gamma)$.*

Proof. First, assume $(x_r, y_r) = (x, y)$ for all $r \in \mathbb{N}$. Then the path costs are integrals over $[\|x\|_1, \|y\|_1]$ by Definition 3. Their integrands converge and are uniformly bounded by $M := \max\{\|P_{\|\cdot\|}(z_1) - Q_{\|\cdot\|}(z_2)\| \mid z \in [0, \|P\|] \times [0, \|Q\|]\}$. The Dominated Convergence Theorem (see [22]) implies $\lim_{r \to \infty} \mathrm{cost}(\gamma_r) = \mathrm{cost}(\gamma)$.

Now consider general paths $\gamma_r \colon [\|x_r\|_1, \|y_r\|_1] \to [0, \|P\|] \times [0, \|Q\|]$. We can normalise their domains to $[\|x\|_1, \|y\|_1]$ by removing excess prefix/suffix paths and adding (arbitrary) missing prefix/suffix paths. The costs of these prefix/suffix paths converge to 0 because of $\lim_{r \to \infty}(x_r, y_r) = (x, y)$ and the uniform bound M. Consequently, the normalisation does not change the limit of $(\mathrm{cost}(\gamma_r))_{r \in \mathbb{N}}$. $\square$

Theorem 20. *The function $\mathrm{opt}_{0,\mathcal{B}}$ is continuous for every border $\mathcal{B}$ under $\|\cdot\|$.*

Proof. We prove that $\overline{\lim}_{t \to t_0} \mathrm{opt}_{0,\mathcal{B}}(t) \le \mathrm{opt}_{0,\mathcal{B}}(t_0) \le \underline{\lim}_{t \to t_0} \mathrm{opt}_{0,\mathcal{B}}(t)$ holds for $t_0 \in \mathrm{dom}(\mathcal{B})$, which implies the result. Let γ_0^* be an optimal $(\mathbf{0}, \mathcal{B}(t_0))$-path, and create $(\mathbf{0}, \mathcal{B}(t))$-paths γ_t for all $t \in \mathrm{dom}(\mathcal{B})$ by concatenating the prefix path of γ_0^* up to its final point $z \in [0, \mathcal{B}_1(t)] \times [0, \mathcal{B}_2(t)]$ with the straight $(z, \mathcal{B}(t))$-path, as in Fig. 6. It is $\mathrm{opt}_{0,\mathcal{B}}(t) \le \mathrm{cost}(\gamma_t)$ for $t \in \mathrm{dom}(\mathcal{B})$, while $(\gamma_t)_t$ converges to γ_0^* for $t \to t_0$, so Lemma 19 gives $\overline{\lim}_{t \to t_0} \mathrm{opt}_{0,\mathcal{B}}(t) \le \lim_{t \to t_0} \mathrm{cost}(\gamma_t) = \mathrm{opt}_{0,\mathcal{B}}(t_0)$.

Now consider optimal $(\mathbf{0}, \mathcal{B}(t))$-paths γ_t^* for $t \in \mathrm{dom}(\mathcal{B})$, and create $(\mathbf{0}, \mathcal{B}(t_0))$-paths γ_0^t with $\mathrm{opt}_{0,\mathcal{B}}(t_0) \le \mathrm{cost}(\gamma_0^t)$ as above. Hence, $\mathrm{opt}_{0,\mathcal{B}}(t_0) \le \underline{\lim}_{t \to t_0} \mathrm{cost}(\gamma_0^t)$. By construction, γ_0^t and γ_t^* share a prefix path up to some point z_t. For $t \nearrow t_0$ we have $(z_t)_t \to \mathcal{B}(t_0)$, so that the suffix costs of $(\gamma_0^t)_t$ converge to 0, while for $t \searrow t_0$ we may assume by Lemma 15 that there is a z with $(z_t)_t \to z$. Then Lemma 19 implies that the suffix costs of $(\gamma_0^t)_t$ and of $(\gamma_t^*)_t$ both converge to the cost of the straight $(z, \mathcal{B}(t_0))$-path. It thus follows $\underline{\lim}_{t \to t_0} \mathrm{cost}(\gamma_0^t) = \underline{\lim}_{t \to t_0} \mathrm{opt}_{0,\mathcal{B}}(t)$. $\square$

5   Conclusion

We have contributed several fundamentals for the computation of CDTW in 2D, tackling the challenges associated with an integral-based measure. To facilitate algorithm design and analysis, our results comprise technical properties that hold under all norms and attest to the robustness of our CDTW formulation. Moreover, we have shown that CDTW under the 2-norm can involve transcendental numbers. Finding a tight bound on our algorithm's complexity remains an open problem.

References

1. Alt, H., Godau, M.: Computing the Fréchet distance between two polygonal curves. Int. J. Comput. Geom. Appl. **5**(1–2), 75–91 (1995). https://doi.org/10.1142/S0218195995000064
2. Bajaj, C.: The algebraic degree of geometric optimization problems. Discrete Comput. Geom. **3**(2), 177–191 (1988). https://doi.org/10.1007/BF02187906
3. Baker, A.: Transcendental Number Theory. Cambridge University Press (1990). https://doi.org/10.1017/CBO9780511565977. Reissue with updated material
4. Boyd, S., Vandenberghe, L.: Convex Optimization, 7th edn. Cambridge University Press (2009). https://web.stanford.edu/~boyd/cvxbook/
5. Brakatsoulas, S., Pfoser, D., Salas, R., Wenk, C.: On map-matching vehicle tracking data. In: Proceedings of the 31st International Conference on Very Large Data Bases, pp. 853–864. VLDB Endowment (2005). https://dl.acm.org/doi/10.5555/1083592.1083691
6. Brankovic, M.: Graphs and trajectories in practical geometric problems. Ph.D. thesis, University of Sydney (2022)
7. Brankovic, M., Buchin, K., Klaren, K., Nusser, A., Popov, A., Wong, S.: (k, l)-medians clustering of trajectories using Continuous Dynamic Time Warping. In: Proceedings of the 28th International Conference on Advances in Geographic Information Systems, pp. 99–110. ACM (2020). https://doi.org/10.1145/3397536.3422245
8. Buchin, K., Buchin, M., Meulemans, W., Mulzer, W.: Four soviets walk the dog: Improved bounds for computing the Fréchet distance. Discrete Comput. Geom. **58**(1), 180–216 (2017). https://doi.org/10.1007/S00454-017-9878-7
9. Buchin, K., Buchin, M., Wang, Y.: Exact algorithms for partial curve matching via the Fréchet distance. In: Proceedings of the Twentieth Annual ACM-SIAM Symposium on Discrete Algorithms, pp. 645–654. SIAM (2009). https://doi.org/10.1137/1.9781611973068.71
10. Buchin, K., Nusser, A., Wong, S.: Computing continuous dynamic time warping of time series in polynomial time. In: 38th International Symposium on Computational Geometry, pp. 22:1–22:16. Schloss Dagstuhl – Leibniz-Zentrum für Informatik (2022). https://doi.org/10.4230/LIPICS.SOCG.2022.22
11. Buchin, M.: On the Computability of the Fréchet distance between triangulated surfaces. Ph.D. thesis, Freie Universität Berlin (2007). https://refubium.fu-berlin.de/handle/fub188/1909
12. Chazelle, B., Dobkin, D.P.: Intersection of convex objects in two and three dimensions. J. ACM **34**(1), 1–27 (1987). https://doi.org/10.1145/7531.24036

13. Cheng, S.W., Huang, H.: Fréchet distance in subquadratic time. In: Proceedings of the 2025 Annual ACM-SIAM Symposium on Discrete Algorithms, pp. 5100–5113. SIAM (2025). https://doi.org/10.1137/1.9781611978322.173

14. De Carufel, J.L., Gheibi, A., Maheshwari, A., Sack, J.R., Scheffer, C.: Similarity of polygonal curves in the presence of outliers. Comput. Geom. **47**(5), 625–641 (2014). https://doi.org/10.1016/J.COMGEO.2014.01.002

15. De Carufel, J.L., Grimm, C., Maheshwari, A., Owen, M., Smid, M.: A note on the unsolvability of the weighted region shortest path problem. Comput. Geom. **47**(7), 724–727 (2014). https://doi.org/10.1016/J.COMGEO.2014.02.004

16. Efrat, A., Fan, Q., Venkatasubramanian, S.: Curve matching, time warping, and light fields: new algorithms for computing similarity between curves. J. Math. Imaging Vision **27**(3), 203–216 (2007). https://doi.org/10.1007/S10851-006-0647-0

17. Fréchet, M.: Sur quelques points du calcul fonctionnel. Rendiconti Circolo Matematico Palermo **22**(1), 1–72 (1906). https://doi.org/10.1007/BF03018603

18. Gold, O., Sharir, M.: Dynamic time warping and geometric edit distance: breaking the quadratic barrier. ACM Trans. Algorithms **14**(4), 50:1–50:17 (2018). https://doi.org/10.1145/3230734

19. Har-Peled, S., Raichel, B., Robson, E.W.: The Fréchet distance unleashed: approximating a dog with a frog. In: 41st International Symposium on Computational Geometry, pp. 54:1–54:13. Schloss Dagstuhl – Leibniz-Zentrum für Informatik (2025). https://doi.org/10.4230/LIPIcs.SoCG.2025.54

20. Jeffrey, A., Dai, H.H.: Handbook of Mathematical Formulas and Integrals, 4th edn. Elsevier (2008)

21. Klaren, K.: Continuous dynamic time warping for clustering curves. Master's thesis, Eindhoven University of Technology (2020). https://research.tue.nl/en/studentTheses/continuous-dynamic-time-warping-for-clustering-curves

22. Luxemburg, W.A.J.: Arzelà's Dominated convergence theorem for the Riemann integral. Am. Math. Monthly **78**(9), 970–979 (1971). https://doi.org/10.1080/00029890.1971.11992915

23. Maheshwari, A., Sack, J.R., Scheffer, C.: Approximating the integral Fréchet distance. Comput. Geom. **70–71**, 13–30 (2018). https://doi.org/10.1016/J.COMGEO.2018.01.001

24. Munich, M.E., Perona, P.: Continuous dynamic time warping for translation-invariant curve alignment with applications to signature verification. In: Proceedings of the Seventh IEEE International Conference on Computer Vision, pp. 108–115. IEEE (1999). https://doi.org/10.1109/ICCV.1999.791205

25. Rote, G.: Lexicographic Fréchet matchings. In: Extended abstract from the 30th European Workshop on Computational Geometry (2014). https://refubium.fu-berlin.de/handle/fub188/15658

26. Schaefer, H.H., Wolff, M.P.: Topological Vector Spaces, 2nd edn. Springer, Cham (1999). https://doi.org/10.1007/978-1-4612-1468-7

27. Serra, B., Berthod, M.: Subpixel contour matching using continuous dynamic programming. In: 1994 Proceedings of IEEE Conference on Computer Vision and Pattern Recognition, pp. 202–207. IEEE (1994). https://doi.org/10.1109/CVPR.1994.323830

28. Serra, B., Berthod, M.: Optimal subpixel matching of contour chains and segments. In: Proceedings of IEEE International Conference on Computer Vision, pp. 402–407. IEEE (1995). https://doi.org/10.1109/ICCV.1995.466911

29. Su, H., Liu, S., Zheng, B., Zhou, X., Zheng, K.: A survey of trajectory distance measures and performance evaluation. VLDB J. **29**(1), 3–32 (2020). https://doi.org/10.1007/S00778-019-00574-9
30. Tao, Y., et al.: A comparative analysis of trajectory similarity measures. GISci. Remote Sens. **58**(5), 643–669 (2021). https://doi.org/10.1080/15481603.2021.1908927
31. Vintsyuk, T.K.: Speech discrimination by dynamic programming. Cybern. Syst. Anal. **4**(1), 52–57 (1968). https://doi.org/10.1007/BF01074755
32. Wildberger, N.J., Rubine, D.: A hyper-Catalan series solution to polynomial equations, and the Geode. Am. Math. Monthly **132**(5), 383–402 (2025). https://doi.org/10.1080/00029890.2025.2460966

The Gate-Cover Problem

Esther M. Arkin[1], Alon Efrat[2], Omrit Filtser[3], Stephen Kobourov[4],
Jan Kratochvíl[5], Joseph S. B. Mitchell[1], and Ariel Rosenberg[3(✉)]

[1] Stony Brook University, New York, USA
{esther.arkin,joseph.mitchell}@stonybrook.edu
[2] University of Arizona, Arizona, USA
[3] The Open University of Israel, Ra'anana, Israel
ariel.rosenberg.84@gmail.com
[4] Technical University Munich, Munich, Germany
stephen.kobourov@tum.de
[5] Faculty of Mathematics and Physics, Charles University, Prague, Czech Republic
honza@kam.mff.cuni.cz

Abstract. We introduce a new variant of the art gallery problem,
namely, the *Gate-Cover Problem* for thin polyominoes. The problem is
equivalently formulated as guarding streets in a city by placing cameras in (some) intersections. We study two variants - of bounded and
unbounded range of visibility. We show that both variants are NP-hard,
the unbounded variant even for polyominoes in which all junctions are
T-junctions. In the case of bounded-range visibility we observe that the
problem admits a PTAS, and for the unbounded variant we present a
polynomial time 2-approximation algorithm. However, our experimental results indicate that the geometric setting may allow for a better
approximation ratio. In particular, the greedy algorithm yields $\ll 1.5$-
approximate results for randomly generated instances. Though we construct examples of instances for which the guaranteed approximation
ratio of the greedy algorithm is not better than $2 - \varepsilon$, we raise the question if more sophisticated tie-breaking rules may result in a provably
better performance.

Keywords: Geometric Covering and Guarding · Segment Intersection

1 Introduction

The ART GALLERY PROBLEM (AGP) is a classic problem in computational
geometry: given a polygon P, find the smallest set of points to guard P, where
a point $p \in P$ is guarded by a point $g \in P$ if and only if the line segment
$\overline{gp}$ is contained in P [16,19,21]. The motivation for AGP, as implied by its
name, is placing a small set of cameras that allow detecting an intruder in an
art gallery. However, sometimes guarding the entire gallery is unnecessary, and
it is enough to guard a certain set of "gates" in P, so that one can output
an approximate location of an intruder, by following it from the moment it
enters the gallery. For example, suppose we have a city where a thief roams the

E. Di Giacomo and D. Mondal (Eds.): WALCOM 2026, LNCS 16444, pp. 483–498, 2026.
https://doi.org/10.1007/978-981-95-7127-7_32

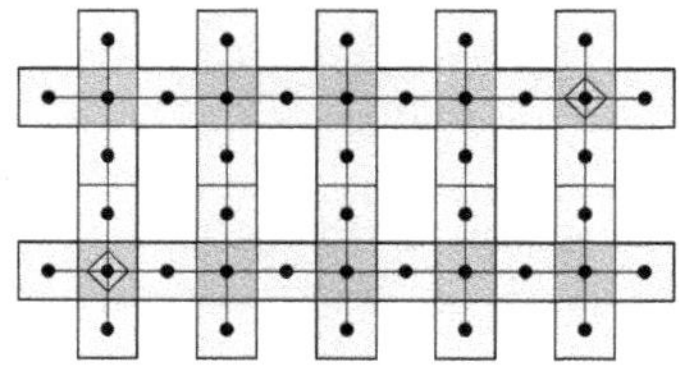

Fig. 1. A thin polyomino P and its dual graph G_P. It has two horizontal streets and five vertical streets. The gates are the orange unit squares, and the two vertices marked with a green diamond are an optimal solution for the gate-cover problem. Note that to cover the entire polyomino (as in the art gallery problem) one needs a linear number of cameras. (Color figure online)

streets, and we would like to delimit his location to a certain street, even without knowing his exact location. To represent the city, we use a thin polyomino P. A polyomino P is a polygon formed by joining together $|P| = m$ unit squares on the square lattice, and a polyomino is thin if it does not contain a 2×2 block of unit squares. In this analogy, a "street" would be a maximal sequence of unit squares in P such that every pair of consecutive unit squares share a horizontal (resp. vertical) edge. Thus, to be able to tell in which street the thief is located at every moment in time, we simply need to guard all the "junctions" of P, i.e., unit squares that separate two streets (see Fig. 1). In other words, our motivation is path monitoring in road networks, using a small number of sensors. This problem received a fair amount of attention in the last few years, since it solves resource allocation in new architectures of vehicle-to-vehicle (V2V) communication networks, which are a key component in autonomous and semi-autonomous vehicles [9,20].

We therefore introduce a new variant of AGP in thin polyominoes, which we call *The Gate-Cover Problem*, and where the goal is to guard a given set $\mathcal{X}$ of gates (unit squares) in a thin polyomino P. We assume that cameras are placed on the center points of unit squares (e.g. on traffic lights), and that a camera c can see a gate $x \in \mathcal{X}$ only if x belongs to a unit square that is completely visible to c, i.e., it must be in the same horizontal or vertical street as c. We will consider both unbounded range of sight and bounded range of sight for the cameras. Note that gates are not necessarily placed on junctions of P, and can be placed on any square of the polyomino.

There is a large body of work on the different variants of the art gallery problem, including many papers that study guarding in orthogonal polygons and polyominoes under different models of vision (see, e.g., [3,5,8,12,14,17]). We discuss the most relevant related work in Sect. 1.3 below. The art gallery problem and many of its variants are known to be NP-hard and even APX-hard; thus, it is natural to study special variants motivated by real-world applications, such as the Gate-Cover Problem, which can be solved with efficient approximation algorithms.

1.1 Definitions and Problem Statement

Let P be a *thin polyomino* (a polyomino which does not contain a 2×2 block of unit squares) with m unit squares. The dual graph G_P of P has a vertex at the center point of each unit square of P, and an edge between two center points whose respective unit squares share an edge. Note that each vertex of G_P has at most four neighbors.

Every maximal sequence of vertical (resp. horizontal) edges in G_P is referred to as vertical (resp. horizontal) street. Notice that the set of streets of P can be represented as set of vertical and horizontal segments in the plane, whose endpoints have integer coordinates, and such that any two segments in the same direction are disjoint (two non-disjoint vertical and horizontal segments can either cross or touch each other). For a vertex u in G_P, denote by $S_h(u)$ (resp. $S_v(u)$) the horizontal (resp. vertical) street that contains u. If one of $S_h(u)$ or $S_v(u)$ do not exist, then we set it to NULL.

Denote by $\mathcal{X}$ the set of gates – a subset of the vertices of G_P. We say that a camera placed on a vertex u of G_P sees a gate $x \in \mathcal{X}$ if x belongs to either $S_h(u)$ or $S_v(u)$. In other words, u and x are on the same street. We will also consider the case when the range of sight is bounded. In this case, given an integer k, a camera on u sees a gate x if u and x are on the same street and the hop-distance between them in G_P is at most k. The gate-cover problem in thin polyominoes is defined as follows.

> **The gate-cover problem:** Given a thin polyomino P and a set of gates $\mathcal{X}$, find a minimum set C of cameras such that every gate is seen by at least one camera.

Throughout the paper, sometimes it will be more convenient to use the representation of a thin polyomino as a set of vertical and horizontal streets, with endpoints on the integer grid. Accordingly, gates and cameras are also points on the integer grid.

1.2 Our Results

We show that the gate-cover problem in thin-polyominoes is NP-hard, both under bounded and unbounded range of vision, and present approximation algorithms for both variants.

In Sect. 2 we show that the unbounded-range variant is equivalent to edge domination (also referred to as minimum maximal matching) in grid intersection graphs (i.e., intersection graphs of vertical and horizontal segments in the plane). By showing that edge domination is NP-hard for planar bipartite graphs of maximum degree three, we get that the gate-cover problem is NP-hard even for polyominoes in which all junctions are T-junctions (i.e. no unit square of the polyomino is adjacent to more than three other squares), and even when the set of gates equals the set of junctions. We then present a near-linear time 2-approximation algorithm for this variant.

In Sect. 3 we consider the case of bounded-range k. We observe that the problem is related (and for bound 1 equivalent) to distance-domination in grid graphs (i.e., induced subgraphs of the infinite unit-square grid), and show that the problem is NP-hard for any value of k. For $k = 1$ we reduce from dominating set in planar cubic graphs, while for $k > 1$ we reduce from Planar-(2or3)-SAT. We then present a PTAS based on local search for this variant.

Finally, in Sect. 4, we explore greedy algorithms for the gate-cover problem. Though we construct examples of instances for which the guaranteed approximation ratio of the greedy algorithm is not better than $2 - \epsilon$, our experimental results indicate that the geometric setting may allow for a better approximation ratio. In particular, we present several greedy heuristics that yield $\ll 1.5$-approximate results for randomly generated instances. This raises the question weather more sophisticated tie-breaking rules may result in a provably better performance.

1.3   Related Work

The most closely related work to our problem is by Biedl and Mehrabi [5], who consider r-guarding in thin orthogonal polygons. In this setting, a point p guards a point q if and only if the minimum axis-aligned rectangle spanned by p and q is inside the polygon. They show that r-guarding thin polyominoes (with holes) is NP-hard, and present a linear time algorithm for guarding simple thin polyominoes. Their algorithm generalizes to polygons with h holes (which makes the problem fixed-parameter tractable in h), and also allow to specify subsets of points to guard and guards to use. Our problem can be regarded as a special case of r-guarding in thin polyominoes, restricted to a predefined set of candidate guards. Motivated by modeling realistic city layouts, we impose additional constraints on guard placement, allowing cameras only at centers of squares (e.g., to model cameras that are placed on traffic lights). Consequently, each camera is placed at the center of a unit square, and covers exactly one vertical and one horizontal street (see Fig. 2).

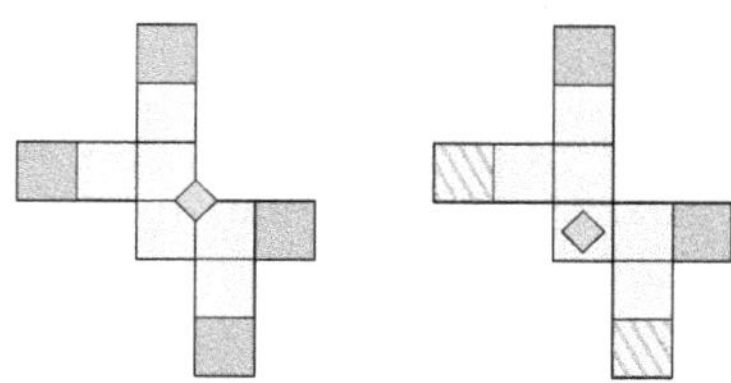

Fig. 2. In r-guarding (left), a point on the boundary may see gates on more than two streets. In our problem (right), a camera positioned at the center can see only the streets that contain it (it does not see the striped squares).

Another closely related work is by Alpert and Roldán [1], considering the Art Gallery Problem with rook vision. Rook vision in thin polyominoes is equivalent to our definition of vision. They show that guarding a (non-thin) polyomino using a minimum number of rook guards is NP-hard.

Biedl et al. [4] study guarding problems in polyominoes. They provide tight combinatorial bounds on the number of point guards (a guard located at any point inside the whole polyomino) and pixel guards (a guard located exactly at the center of a polyomino's unit square) required, and establish NP-hardness of computing optimal guard placements. Filtser et al. [12] considered k-hop visibility in polyominoes. In this problem, two squares are visible to each other if they are at hop-distance at most k. They prove NP-hardness for several cases, give bounds on the VC-dimension of visibility systems, and design approximation algorithms for guarding thin polyomino trees.

As mentioned above, a thin polyomino can be represented as a set of orthogonal line segments in the plane (a line for every street of the polyomino). In the case when the cameras have an unbounded range of sight, and the set of gates is exactly the set of intersection points, the gate-cover problem is equivalent to finding a minimum set C of points to "cover" all the intersection points between horizontal and vertical segments, i.e. for any such intersection point x there exists a point $c \in C$ such that x and c are on the same horizontal or vertical segment. To the best of our knowledge, this natural problem was not investigated before. Related problems include, e.g., hitting a set of orthogonal segments (see [10]), where the entire segments needs to be covered (in Fig. 1, the size of the optimal solution for hitting segments is linear).

2   Unbounded Vision

In this section we consider the case when the cameras have an unbounded range of vision, i.e. a camera u guards any gate x in $S_h(u)$ or $S_v(u)$. Clearly, a gate x is guarded if and only if a camera is placed on either $S_h(x)$ or $S_v(x)$.

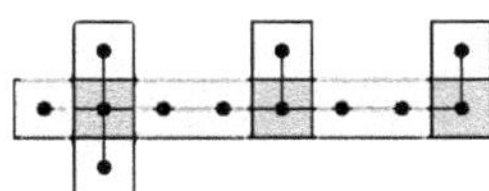

Fig. 3. A thin polyomino with one horizontal street and three vertical streets that create three types of junctions (in orange). The middle one is a T-junction. (Color figure online)

Denote by H (resp. V) the set of horizontal (resp. vertical) streets of P. Let $\mathcal{J}$ be the set of "junctions" (see Fig. 3): all the vertices of G_P that correspond to intersection points between two streets. Observe that in the case of unbounded vision, we may assume that all the cameras are placed on junctions: if a camera is not placed on a junction, we can simply move it to the nearest junction on the same street (the problem is trivial if there is no such junction).

Consider the bipartite intersection graph between horizontal and vertical streets: $\mathcal{G} = (H \cup V, E)$ where $v, h \in E$ if the segments v and h intersect. Clearly, the set E of edges corresponds to the set of junctions $\mathcal{J}$ (there are no other intersection points between streets because every two streets of P in the same direction are disjoint).

2.1 NP-Hardness

We show that gate-cover in thin polyominoes is NP-hard, even if no unit square of the polyomino is adjacent to more than three other ones (the maximum degree of G_P is 3), and the set $\mathcal{X}$ of gates equals the set $\mathcal{J}$ of junctions.

The bipartite graph $\mathcal{G} = (H \cup V, E)$ belongs to a family of graphs called *grid intersection graphs*—these are intersection graphs of vertical and horizontal segments in the plane such that segments in the same direction are disjoint. Every grid intersection graph is then bipartite. A representation in which no two segments cross, i.e., segments that are not disjoint only touch each other, is called a *grid contact representation*. Graphs which allow a grid contact representation are called *grid contact graphs*. Clearly, every grid contact graph is planar. The reverse is also true: De Fraysseix, Ossona de Mendez and Pach [13] proved that every planar bipartite graph is a grid contact graph. A proof for the following lemma is given in the full version of the paper.

Lemma 1. *The gate-cover problem in thin polyominoes with $\mathcal{X} = \mathcal{J}$ is polynomially equivalent to edge domination in grid intersection graphs.*

Edge domination is the problem of selecting a set of edges in a graph such that every edge in the graph shares at least one endpoint with at least one selected edge. By Lemma 1, to prove that the gate-cover problem in thin polyominoes is NP-hard, it is enough to prove that the edge domination problem is NP-hard in grid intersection graphs. We will prove that is NP-hard even in grid contact graphs, which translates to NP-hardness for polyominoes in which every junction is a T-shaped junction (and $\mathcal{X} = \mathcal{J}$).

Yannakakis and Gavril [22] showed that edge domination is NP-hard for bipartite graphs and for planar graphs. We will demonstrate that their reduction actually shows NP-hardness for planar bipartite graphs. Then we use the theorem of de Fraysseix et al. [13] to conclude that edge domination is NP-hard in grid contact graphs and we will be done. A complete proof of Lemma 2 is provided in the full version of the paper.

Lemma 2. *Edge domination is NP-hard for planar bipartite graphs of maximum degree 3.*

This concludes the proof of the following theorem.

Theorem 1. *The gate-cover problem in thin polyominoes is NP-hard, even if no unit square of the polyomino is adjacent to more than three other ones, and the set of gates equals the set of junctions.*

2.2   A 2-Approximation Algorithm

Our goal is to cover gates by cameras—find a minimum set C of unit squares (cameras) in P that see all the gates (every gate in $\mathcal{X}$ is on the same street with a camera from C). Consider the problem of covering gates by streets—find a minimum set S of streets in P that "cover" all the gates (every gate in $\mathcal{X}$ belongs to some street in S). An optimal solution to the latter problem is a factor-2 approximation to our problem: Let OPT_S be an optimal solution for covering gates by streets, and OPT_C an optimal solution for covering gates by cameras. Clearly, $|\mathrm{OPT}_C| \leq |\mathrm{OPT}_S|$, because in every street from OPT_S we can place one camera that will see all the gates in this street. On the other hand, $|\mathrm{OPT}_S| \leq 2|\mathrm{OPT}_C|$ because we can replace each camera in OPT_C by at most two streets (horizontal and vertical) that cover the same set of gates.

When $\mathcal{X} = \mathcal{J}$, a minimum set S of streets in P to cover all the gates is exactly a *minimum vertex cover* in the graph $\mathcal{G}$: every edge (h, v) in E corresponds to a gate in $\mathcal{X}$, which can be covered by either h or v. To accommodate the case when $\mathcal{X} \neq \mathcal{J}$, i.e., gates can be placed on any unit square of P, we need to slightly modify the construction of $\mathcal{G}$.

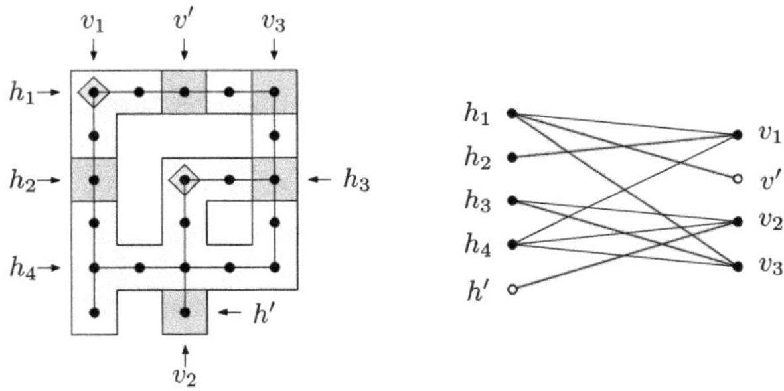

Fig. 4. The orange edges belong to the bipartite graph $\mathcal{G}^{\mathcal{X}}$. The optimal solution for gate-cover has size 2, while the optimal vertex cover has size 4. (Color figure online)

We denote the new graph by $\mathcal{G}^{\mathcal{X}} = ((H \cup \hat{H}) \cup (V \cup \hat{V}), E^{\mathcal{X}})$, and construct it as follows (see Fig. 4). The sets H, V are all the horizontal and vertical streets of P, as before. For any pair of streets $h \in H$, $v \in V$, if they intersect at a gate, then we add the edge (h, v) to $E^{\mathcal{X}}$ (we do not add edges that correspond to junctions that are not gates). The sets of vertices $\hat{H}$ and $\hat{V}$ correspond to "dummy streets" that we add to include edges that correspond to gates that do not lie on junctions: let $x \in \mathcal{X} \setminus \mathcal{J}$. If x is on a horizontal street ($S_h(x)$ is not empty), then add a new vertex v_x to $\hat{V}$ and the edge $(S_h(x), v_x)$ to $E^{\mathcal{X}}$. Otherwise, x is on a vertical street and we add a new vertex h_x to $\hat{H}$ and the edge $(h_x, S_v(x))$ to $E^{\mathcal{X}}$.

Now, the edges of $\mathcal{G}^{\mathcal{X}}$ correspond to gates in P. Some vertices of $\mathcal{G}^{\mathcal{X}}$ are "dummy vertices" that correspond to dummy streets, which are not streets in P.

However, by our construction, these dummy vertices will always have degree 1, and thus a minimum vertex cover would not contain both a dummy vertex and its neighbor. Therefore, if a vertex cover contains a dummy vertex, then we can replace it by its only neighbor. We conclude that a minimum set S of streets in P to cover all the gates corresponds to a *minimum vertex cover* in $\mathcal{G}^{\mathcal{X}}$.

Lemma 3. *Let S be a set of streets in P. Then, S covers all the gates in $\mathcal{X}$ if and only if the corresponding set of vertices in $\mathcal{G}^{\mathcal{X}}$ is a vertex cover for $\mathcal{G}^{\mathcal{X}}$.*

Since $\mathcal{G}^{\mathcal{X}}$ is a bipartite graph with $|\mathcal{X}|$ edges, a minimum vertex cover can be computed in $O(|\mathcal{X}|^{1+o(1)})$ time [7]. We thus obtain the following theorem.

Theorem 2. *Given a thin polyomino P with m unit squares and a set $\mathcal{X}$ of gates in P, one can compute in $O(m + |\mathcal{X}|^{1+o(1)})$ time a set C of cameras that cover $\mathcal{X}$, such that $|C| \leq 2|OPT_C|$ where OPT_C is an optimal solution to the gate-cover problem.*

3   Bounded Range of Vision

In this section we consider cameras that have bounded vision, i.e., a camera can only see gates that are within hop-distance at most k from it, for a given integer $k \geq 1$. For a given grid graph G (i.e., an induced subgraph of the integer grid, given with its plane embedding with all edges of unit length and all pairs of vertices placed in grid points of distance one being adjacent), a subset C of vertices is *k-range dominating* if every vertex not in C has a vertical or horizontal path of length at most k to a vertex of C. Note the difference from k-distance domination, which requires that every vertex is at distance at most k from some vertex of the dominating set; in this case the path from a vertex to a dominating one is not required to be straight in the underlying grid. If all cells of a polyomino P are to be guarded, then a minimum set of cameras guarding all gates is a minimum k-range dominating set of the associated grid graph G_P, and vice versa. This implies the following observation.

Observation 1. *The gate-cover problem with bounded range k in thin polyominoes is equivalent to the k-range domination problem in grid graphs of girth greater than 4.*

Recall that the *girth* of a graph is the length of the shortest cycle in the graph. Thin polyominoes precisely satisfy the case of girth > 4 because they forbid 2×2 blocks, which would otherwise form 4-cycles.

3.1   NP-Hardness

We show that bounded-range gate-cover is NP-hard for any value of k.

Theorem 3. *The gate-cover problem with bounded range k is NP-hard on thin polyominoes for every $k \geq 1$.*

The proof for $k = 1$ is by a reduction from the domination problem in cubic planar graphs, and it is provided in the full version of the paper. For $k > 1$ we use a reduction from a variant of 3-SAT, as follows.

Lemma 4. *The gate-cover problem with bounded range k is NP-hard on thin polyominoes for every $k > 1$.*

Proof. We reduce from Planar-(2or3)-SAT, a variant of SAT in which the input formula is planar, every variable occurs twice negated and once positive, and every clause has either 2 or 3 literals. A truth-assignment satisfies the formula if every clause contains at least one positively valued literal. This satisfiability problem is NP-hard [11].

Given a formula, we start with an orthogonal drawing of its graph, as in the previous case. We replace variable vertices by variable gadgets, clause vertices by clause gadgets, and edges by linking gadgets. In order to do this, we blow up the representation so that all above-mentioned coordinates are divisible by $t = 2k + 1$, and are sufficiently far apart from each other.

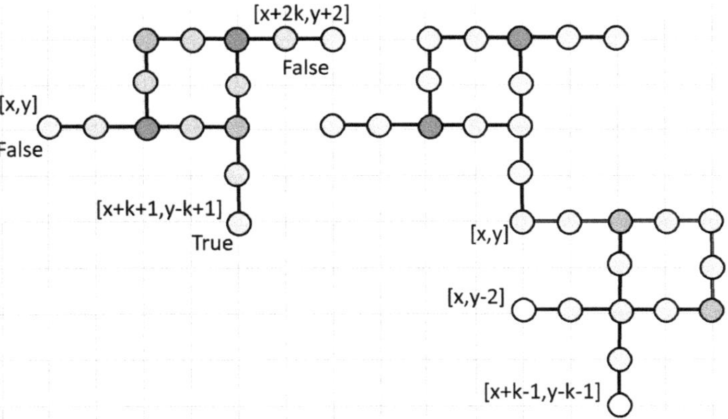

Fig. 5. The variable gadget (left) and an illustration of its adjustment to attain the coordinates of the ports divisible by t.

The variable gadget is depicted in Fig. 5 left. We refer to the octagon consisting of the green, blue, and red vertices as the *core* of the variable gadget. Each gray vertex represents a path consisting of $k - 2$ vertices.

Suppose C is a k-range dominating set. We argue that within the core of each variable gadget, we need at least two vertices to be in C, to dominate all four green vertices. In order to minimize the size of the presumed k-range dominating set, we need to use only two dominating vertices, and then it must be either the pair of the blue ones, or the pair of the red ones. In the former case, the two pink vertices remain not dominated, in the latter case the light blue one is not dominated. These vertices are called the "ports". The left and right will be connected to clauses which contain this literal in negation, the bottom one to the clause which contains this variable positive. Hence, putting the blue vertices in C corresponds to a true assignment of this variable and the negative ports remain not dominated. The placement of the red vertices in C corresponds to a false assignment of the variable and leaves the positive port not dominated.

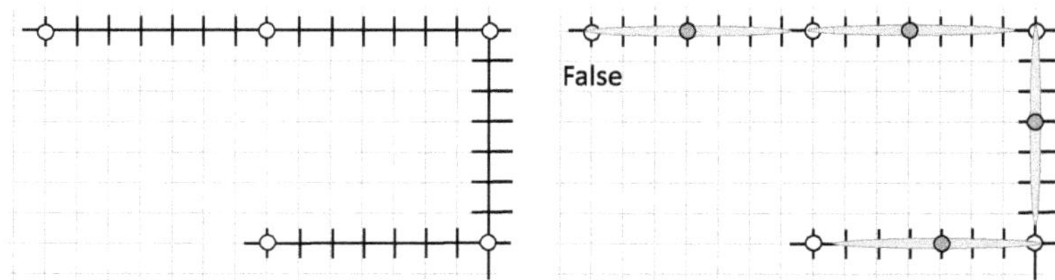

Fig. 6. The linking gadget (left) and how it is used to propagate truth values (right).

The linking gadget in Fig. 6 (left) is a piecewise linear path with the length of every piece divisible by t. This gadget propagates the value "false" in the sense that if one terminal point is not dominated from elsewhere (and has to be dominated from the linking gadget), the placement of the vertices of the linking gadget in C is forced and leaves the other terminal vertex not dominated if we are to use the minimal possible number of dominating vertices, which is the length of the path divided by t; see Fig. 6 (right). Note that if one terminal point has both coordinates divisible by t, so does every white vertex of the linking gadget, and in particular the other terminal point as well.

The clause gadget is simple. It is a path of length t if the clause has two literals, and of length $2t$ if the clause has 3 literals. The terminal vertices of this path, plus the middle vertex in case of a 3-literal clause, are identified with terminal vertices of the linking gadgets arriving from the corresponding variables. We refer to them as the *ports* of the clause gadget. Each clause gadget of the first type must contain at least one dominating vertex, and each gadget of the clause of the second type at least two. These numbers suffice if and only if at least one of the ports is dominated from outside of this gadget, i.e., from the linking gadget, i.e., from a variable which is not evaluated to "false" in this clause. If one of the ports has both coordinates divisible by t, so do the other(s).

The only problem is caused by the variable gadgets, and in particular by the coordinates of their ports. Have a closer look at them. If the left port has coordinates $[x, y]$, the right one has coordinates $[x + 2k, y + 2]$ and the bottom one $[x + k + 1, y - k + 1]$. So if both x, y are divisible by t, the right port is $[-1, 2] \bmod t$ and the bottom one is $[(t + 1)/2, (-t + 3)/2] \bmod t$. However, we have two tools to fix this. The first one is to add a path of length t, i.e., one segment of a linking gadget. This propagates the value "false" and keeps the coordinates of the same residues mod t. We mainly use this tool to gain space.

The other tool is a reduced copy of the core; see Fig. 5 (right). It propagates the "false" value from the vertex $[x, y]$ to a neighbor $[x, y - 2]$ and diagonally to vertex $[x + k - 1, y - k - 1]$.

Looking at the variable gadget in Fig. 5 (left) we note that the pink vertices have the same parity of their coordinates (if x, y are even, so are $x+2k$ and $y+2$). If k is odd, the same is true for the negative port. If k is even, we apply the diagonal propagation to what now becomes vertex $[x + 2k, y - 2k]$ and again the parity of the coordinates is preserved. Once all ports have all coordinates even, we make them all divisible by $2t$ by using the close-by false value propagation from Fig. 5 (right). (If the x coordinate needs to be adjusted, we use a 90-degree rotation of the tool.)

Now we put everything together. Suppose the formula Φ we started with contained v variables, c_2 clauses of size 2 and c_3 clauses of size 3, and suppose that in the construction we used ℓ linking gadgets (counting both those in the variable-clause connections and in adjusting the variable gadgets) and h propagation tools from Fig. 5 (right). Then every k-range dominating set in the constructed graph G_Φ contains at least $2v + \ell + c_2 + 2c_3 + 2h$ vertices, and this minimum possible size is attained if and only if Φ is satisfiable. Note finally that the girth of G_Φ is greater than 4, and so it corresponds to a thin polyomino.

3.2   PTAS via Local Search for Constant k

Local search is a heuristic method for solving NP-hard optimization problems. A local-search-based framework for obtaining PTASs for NP-hard geometric optimization problems was introduced, independently, by Chan and Har-Peled [6] and Mustafa and Ray [15]. Aschner et al. [2] generalize this framework by extending its analysis to additional families of graphs, beyond the family of planar graphs, and one their applications is guarding a polygon under some shallow-ness assumption. More precisely, their approach implies the following. Let P be polygon, G a set of potential guards (points) in P, and X a set of points in P that needs to be guarded. If every point in X is only seen by a constant number of points from G, then the local search framework yields a PTAS for finding the minimum number of points from G to guard X. In our case, every gate can be seen by $O(k)$ cameras, and thus we obtain the following theorem.

Theorem 4. *There exists a PTAS for the gate-cover problem in thin polyomi-noes with range k, for any constant k.*

4   Greedy Heuristics and Experimental Results

Consider the following greedy algorithm for solving the gate-cover problem. Given a polyomimo P with m unit squares, a set $\mathcal{X}$ of gates, and a camera (square) c in P, denote by $\mathrm{Vis}(c)$ the set of gates from $\mathcal{X}$ that are seen by the camera c. Initialize an empty set S. While $\mathcal{X}$ is not empty, find a camera c that maximizes $|\text{Vis}(c)|$ (if there is more than one such camera, choose one arbitrarily). Add c to the set S, and remove $\text{Vis}(c)$ from $\mathcal{X}$. The algorithm stops when all gates are covered, and thus S is a gate-cover for P and $\mathcal{X}$.

Running Time. To bound the running time, we first observe that not all the unit squares have to be considered for placing cameras. More precisely, we can consider only "dominating" cameras (a camera c is dominating if there is no camera c' for which $\text{Vis}(c) \subseteq \text{Vis}(c')$). Dominating cameras are either gates, or junctions that cover at least two gates, and therefore the running time of the algorithm is $O(m + |\mathcal{X}| \log |\mathcal{X}|)$.

Quality of Approximation. Figure 7 shows a polyomino in which the above algorithm outputs a set of three cameras, while the optimal solution has two cameras. In this special case, no tie-breaking is needed in the first step and the greedy algorithm consistently yields a 3/2-approximation. Next, we provide another special example showing that the greedy algorithm, under arbitrary tie-breaking, may yield solutions with $(2 - \varepsilon)$-approximation ratio. We then propose several tie-breaking heuristics that guide the algorithm toward better choices. On various types of cities, randomly generated instances, all heuristics achieve performance strictly better than a 3/2-approximation.

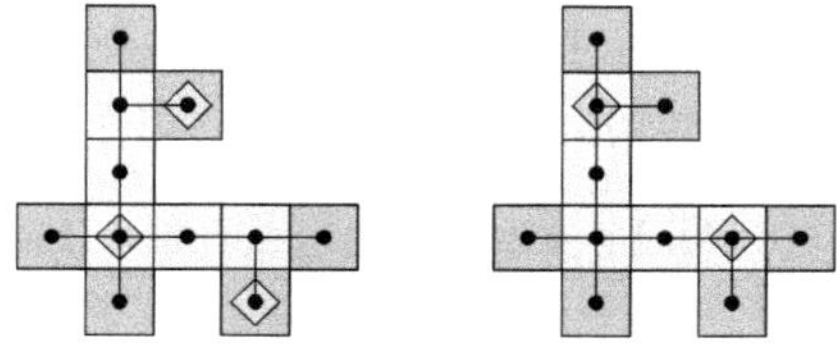

Fig. 7. Left: a simple greedy heuristic. Right: an optimal solution.

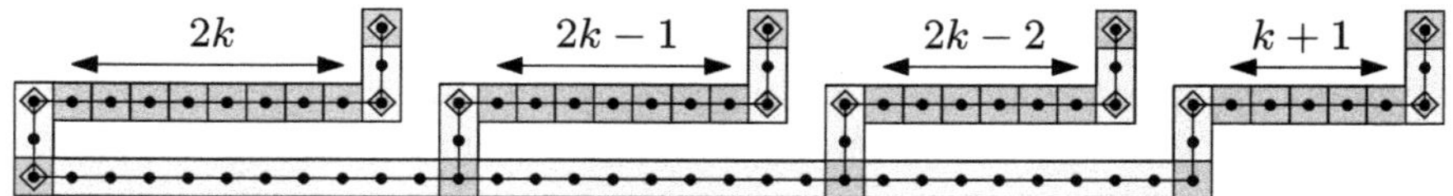

Fig. 8. An example in which greedy is not too good, for $k = 4$.

Consider the polyomino in Fig. 8. This special instance consists of one long horizontal street at the bottom with k components attached to it. In each such component there are two junctions (one is in blue and the other green) that see exactly the same number of gates. Thus in each step of the algorithm, there is a tie-breaking option: In the first step, the maximum number of unguarded gates is $2k + 1$ for both the blue and green intersections in the first component. In the second step, this value becomes $2k + 2$ for similar intersections in the second component, and so on. If the greedy algorithm selects the blue option in each of these steps, then it has to add the other blue junctions in the components (they are unguarded gates after all components are processed). Thus greedy ends up with $2k$ cameras, while the green options in the components together with a single green junction on the bottom street yields an optimal solution of $k + 1$ cameras. Thus the ratio is $\frac{2k}{k+1} = 2 - \frac{2}{k+1}$. With k large enough, it gets arbitrarily close to 2.

4.1 Experimental Results

We have implemented several greedy heuristics, addressing the question of whether greedy can be improved by a tie-breaking rule. In the full version of our paper, we describe our implementation which uses the representation of the city streets as a bipartite graphs (as illustrated in Fig. 4), where the edges correspond to possible locations for the cameras, and a subset of the edges (in orange) are gates that need to be guarded. Five greedy heuristics were tested: (1) random edge selection, (2) greedy selection of edges with maximum coverage (tie breaking randomly), (3) selection prioritizing leaf edges, (4) selection prioritizing gates that are visible by the fewest number of cameras, and (5) minimum adjacent-to-adjacent selection minimizing second-order adjacency (We focus on edges with the smallest second-order coverage, aiming to detect the "rare" edges that have the fewest opportunities to be covered by other edges). Each algorithm repeatedly removes chosen edges and their incident vertices until no gates remain.

Experiments were performed on synthetic cities modeled as thin polyominoes of varying structures: sparse or dense cities (the number of junctions), cities with many holes or tree-like, cities with no dead-ends, cities with only T-junctions, randomly generated layouts, and trees. For each setup, we considered both the case when gates are placed only at junctions, and the case when gates are placed randomly. We ran each algorithm thousands of times and record the best and worst solutions. Results were compared against 0–1 Linear Programming optima, with detailed ratios reported in Tables 1 and 2.

The findings show that while naive random selection performs poorly, structured tie-breaking heuristics (especially Algorithms 3âĂŞ5) consistently produce near-optimal solutions across diverse city types. Despite theoretical weaknesses in worst-case cases, greedy approaches with heuristics prove highly practical for realistic city instances. Experiments can be conducted using the following web interface at this link [18].

Table 1. Results for cities with gates at junctions

	Algo 1	Algo 2	Algo 3	Algo 4	Algo 5
Controlled number of intersections	1.3500	1.2000	1.1250	1.1250	1.1250
Controlled number of holes	1.3158	1.1875	1.1875	1.1875	1.1875
Polyominoes composed solely of squares	1.2857	1.2143	1.2143	1.2143	1.2143
Polyominoes composed solely of T-junctions	1.3519	1.1739	1.1282	1.1282	1.1282
Regular randomly generated polyominoes	1.2857	1.2000	1.1429	1.1429	1.1429
Trees	1.2000	1.2000	1.2000	1.0000	1.0000

Table 2. Results for cities with randomly distributed gates

	Algo 1	Algo 2	Algo 3	Algo 4	Algo 5
Controlled number of intersections	1.7931	1.1154	1.1154	1.1154	1.1154
Controlled number of holes	1.8571	1.1429	1.1429	1.0930	1.0930
Polyominoes composed solely of squares	1.8571	1.2500	1.0714	1.0500	1.0455
Polyominoes composed solely of T-junctions	1.6438	1.1429	1.1429	1.0833	1.0833
Regular randomly generated polyominoes	1.6364	1.0909	1.0909	1.0870	1.0870
Trees	1.5714	1.1250	1.1250	1.0625	1.0625

Acknowledgments. The third author acknowledges support of the Israel Science Foundation (grant No. 2135/24). The fifth author acknowledges support of Czech Science Foundation by Research grant No. GAČR 23-04949X.

Disclosure of Interests. The authors have no competing interests to declare that are relevant to the content of this article.

References

1. Alpert, H., Roldán, É.: Art gallery problem with rook and queen vision. Graphs Comb. **37**(2), 621–642 (2021). https://doi.org/10.1007/S00373-020-02272-8
2. Aschner, R., Katz, M.J., Morgenstern, G., Yuditsky, Y.: Approximation schemes for covering and packing. In: WALCOM: Algorithms and Computation (2013). https://doi.org/10.1007/978-3-642-36065-7_10
3. Biedl, T., Chan, T.M., Lee, S., Mehrabi, S., Montecchiani, F., Vosoughpour, H.: On guarding orthogonal polygons with sliding cameras. In: Poon, S.-H., Rahman,

M.S., Yen, H.-C. (eds.) WALCOM 2017. LNCS, vol. 10167, pp. 54–65. Springer, Cham (2017). https://doi.org/10.1007/978-3-319-53925-6_5

4. Biedl, T., Irfan, M.T., Iwerks, J., Kim, J., Mitchell, J.S.: Guarding polyominoes. In: Proceedings of the 27th Annual Symposium on Computational Geometry (SoCG), pp. 387–396. Association for Computing Machinery (2011). https://doi.org/10.1145/1998196.1998261

5. Biedl, T., Mehrabi, S.: On r-Guarding thin orthogonal polygons. In: 27th International Symposium on Algorithms and Computation (ISAAC 2016), pp. 17:1–17:13. Leibniz International Proceedings in Informatics (LIPIcs) (2016)

6. Chan, T.M., Har-Peled, S.: Approximation algorithms for maximum independent set of pseudo-disks. Discret. Comput. Geom. **48**(2), 373–392 (2012). https://doi.org/10.1007/S00454-012-9417-5

7. Chen, L., Kyng, R., Liu, Y.P., Peng, R., Gutenberg, M.P., Sachdeva, S.: Maximum flow and minimum-cost flow in almost-linear time. In: 63rd IEEE Annual Symposium on Foundations of Computer Science, FOCS, pp. 612–623. IEEE (2022). https://doi.org/10.1109/FOCS54457.2022.00064

8. Daescu, O., Malik, H.: New bounds on guarding problems for orthogonal polygons in the plane using vertex guards with halfplane vision. Theoret. Comput. Sci. **882**, 63–76 (2021). https://doi.org/10.1016/j.tcs.2021.06.012

9. Fabris, M., Ceccato, R., Zanella, A.: Efficient sensors selection for traffic flow monitoring: An overview of model-based techniques leveraging network observability. Sensors **25**(5) (2025). https://doi.org/10.3390/s25051416, https://www.mdpi.com/1424-8220/25/5/1416

10. Fekete, S.P., Huang, K., Mitchell, J.S.B., Parekh, O., Phillips, C.A.: Geometric hitting set for segments of few orientations. Theory Comput. Syst. **62**(2), 268–303 (2018). https://doi.org/10.1007/S00224-016-9744-7

11. Fellows, M.R., Kratochvíl, J., Middendorf, M., Pfeiffer, F.: The complexity of induced minors and related problems. Algorithmica **13**(3), 266–282 (1995). https://doi.org/10.1007/BF01190507

12. Filtser, O., Krohn, E., Nilsson, B.J., Rieck, C., Schmidt, C.: Guarding polyominoes under k-hop visibility. Algorithmica **87**(4), 572–593 (2025). https://doi.org/10.1007/S00453-024-01292-7

13. de Fraysseix, H., de Mendez, P.O., Pach, J.: Representation of planar graphs by segments. Intuitive Geom. **63**, 109–117 (1991)

14. Györi, E., Mezei, T.R.: Partitioning orthogonal polygons into ≤ 8-vertex pieces, with application to an art gallery theorem. Comput. Geom. **59**, 13–25 (2016). https://doi.org/10.1016/J.COMGEO.2016.07.003

15. Mustafa, N.H., Ray, S.: Improved results on geometric hitting set problems. Discret. Comput. Geom. **44**(4), 883–895 (2010). https://doi.org/10.1007/S00454-010-9285-9

16. O'Rourke, J.: Art Gallery Theorems and Algorithms. Oxford University Press Inc, USA (1987)

17. Rieck, C., Scheffer, C.: The dispersive art gallery problem. Comput. Geom. **117**, 102054 (2024)

18. Rosenberg, A.: The gate cover problem in thin polyomino simulator. https://arikrosen.pythonanywhere.com/

19. Shermer, T.C.: Recent results in art galleries (geometry). Proc. IEEE **80**(9), 1384–1399 (1992). https://doi.org/10.1109/5.163407

20. Thabit, A.S., Kerrache, C.A., Calafate, C.T.: A survey on monitoring and management techniques for road traffic congestion in vehicular networks. ICT Express **10**(6), 1186–1198 (2024). https://doi.org/10.1016/j.icte.2024.10.007

21. Urrutia, J.: Art gallery and illumination problems. In: Sack, J.R., Urrutia, J. (eds.) Handbook of Computational Geometry, pp. 973–1027. North-Holland, Amsterdam (2000). https://doi.org/10.1016/B978-044482537-7/50023-1
22. Yannakakis, M., Gavril, F.: Edge dominating sets in graphs. SIAM J. Appl. Math. **38**(3), 364–372 (1980)

Trajectory Visibility at First Sight

Mohammad Ali Abam[1], Mohammad Ghodsi[1,2],
and Seyed Mohammad Hussein Kazemi[1,3(✉)]

[1] Department of Computer Engineering, Sharif University of Technology,
Tehran, Iran
{abam,ghodsi}@sharif.edu
[2] School of Computer Science, Institute for Research in Fundamental Sciences (IPM),
Tehran, Iran
[3] LaBRI, University of Bordeaux, Bordeaux, France
skazemi@u-bordeaux.fr

Abstract. Let P be a simple polygon with n vertices, and let two moving entities $q(t)$ and $r(t)$ travel at constant (possibly distinct) speeds v_q and v_r along line-segment trajectories τ_q and τ_r inside P. We study the exact first-visibility time $t^* = \min\{\, t \geq 0 \mid \overline{q(t)\,r(t)} \subseteq P \,\}$, the earliest moment at which the segment joining $q(t)$ and $r(t)$ lies entirely within P.

Prior work by Eades et al. [9, 10] focused on this question in the setting of a simple polygon. They gave a one-shot decision algorithm running in $\mathcal{O}(n)$ time. For a stationary entity and a moving one, they suggested a structure that, after $\mathcal{O}(n \log n)$ pre-processing, answers the decision query in $\mathcal{O}(\log n)$ time, requiring $\mathcal{O}(n)$ space. In addition, for moving entities, after preprocessing time of $\mathcal{O}(n \log^5 n)$, they construct a data structure with $\mathcal{O}(n^{3/4} \log^3 n)$ query time and $\mathcal{O}(n \log^5 n)$ space. Variants for polygonal domains with holes or when entities cross the boundary of P lie beyond our scope.

In this work, we go beyond the decision to compute t^* exactly under three models for a simple polygon P. When both trajectories are known in advance, we preprocess P in $\mathcal{O}(n)$ time and space and thereafter answer each query in $\mathcal{O}(\log n)$ time. If one trajectory τ_r is fixed while τ_q is given as query, we build a structure in $\mathcal{O}(n \log n)$ time and space that computes t^* in $\mathcal{O}(\log^2 n)$ time per query. In a setting where the trajectories are not known in advance, we develop a randomized structure with $\mathcal{O}(n^{1+\varepsilon})$ expected pre-processing time and $\mathcal{O}(n)$ space, achieving an $\mathcal{O}(n^{1/2} \operatorname{polylog}(n))$ expected query time for any fixed $\varepsilon > 0$.

Keywords: trajectories · visibility · semi-algebraic range searching

1 Introduction

The analysis of moving entities constitutes a broad research area, ranging from robotics and geographic information science (GIScience) to meteorology and

---

M. A. Abam: Authors' names are sorted alphabetically.

E. Di Giacomo and D. Mondal (Eds.): WALCOM 2026, LNCS 16444, pp. 499–511, 2026.
https://doi.org/10.1007/978-981-95-7127-7_33

ecology [4,8,15,19,22]. A body of work focuses on extracting features from historical trajectory data, such as identifying when entities have passed closely together or detecting groups that move in formation [13]. However, a class of problems emerges when the goal is not to analyse the past, but to predict future events for entities on known or planned paths. This is especially true in environments with obstacles or defined boundaries, which mirrors the constrained polygonal setting of our work.

Practical applications of such predictive line-of-sight analysis are numerous. In robotics and automated systems, for example, determining when two micro-robots can establish direct communication [21] or when an autonomous vehicle will gain a clear view of a target is a fundamental task. Similar challenges arise in biology when tracking potential interactions within animal colonies in complex habitats [3] and in optimising sensor-based systems, such as dual-axis solar trackers that must maintain an unobstructed path to the sun [16]. As the complexity and autonomy of these systems grow, the demand for a robust theory of trajectory visibility and the development of efficient, provable algorithms naturally escalates.

1.1 Preliminaries

Let P be a simple polygon with n vertices, and let ∂P denote the boundary of P, which consists of its vertices and edges. A point $p \in P$ is considered to be *weakly visible* from a segment $s \subset P$ if there exists a point $q \in s$ such that the segment $\overline{pq}$ is entirely contained within P. Similarly, two segments s_1 and $s_2 \subset P$ are weakly visible to each other if, for every point in one segment, there exists a corresponding point in the other segment such that the connecting segment is also contained within P [5,12]. For a given query point $p \in P$, the *visibility polygon* of p, denoted by $V(p)$, is defined as

$$V(p) = \{q \in P \mid \overline{pq} \subseteq P\}.$$

It is well known that $V(p)$ can be computed in $O(n)$ time using a rotational sweep algorithm [18]. Given a segment $s \subset P$, the *weak visibility polygon* of s is defined as

$$W(s) = \{q \in P \mid \exists r \in s \text{ such that } \overline{qr} \subseteq P\}.$$

Using similar techniques to those used for the visibility polygon, $W(s)$ can also be computed in $O(n)$ time [12]. Additionally, a *ray shooting* query is defined as follows: given a query ray r with origin $p \in P$, the goal is to determine the first intersection point between r and the boundary ∂P. After a linear-time preprocessing phase, such queries can be answered in $\mathcal{O}(\log n)$ time [6].

Splinegons (also known as curved polygons) are considered extensions of traditional polygons. A splinegon $\mathcal{S}$ is created from a polygon P by substituting one or more of its edges with curved edges, ensuring that the area enclosed by each curved edge and the line segment connecting its endpoints remains convex [7]. Provided a simple splinegon S with n edges, there exists a data structure requiring $\mathcal{O}(n)$ preprocessing time and $\mathcal{O}(n)$ space such that for any query point p and a ray $\overrightarrow{r}$ emanating from p, the first intersection of $\overrightarrow{r}$ with S can be reported in $\mathcal{O}(\log n)$ time [11,20].

Consider a point $p \in P$. For every vertex v of the polygon P, let $\pi(p, v)$ denote the shortest Euclidean path inside P connecting p and v. The shortest-path tree rooted at p, denoted $T(p)$, is the union of these paths:

$$T(p) = \bigcup_{v \in P} \pi(p, v).$$

Equivalently, $T(p)$ is the tree formed by taking, for every polygon vertex v, the unique polygonal path of minimum length from p to v, where all intermediate vertices on each $\pi(p, v)$ are reflex vertices of P [2]. Let us assume all paths in $T(p)$ are stored as *two-way linked lists*.

For two line segments $s_1, s_2 \subset P$ let $L(s_1, s_2)$ be their visibility glass. The $L(s_1, s_2)$ comprises the (potentially empty) collection of all line segments that lie between s_1 and s_2 within P. When $L(s_1, s_2)$ is not empty, there exist segments $s_1' \subset s_1$ and $s_2' \subset s_2$ such that $L(s_1, s_2)$ corresponds to the hourglass [14] formed by s_1' and s_2'. The segments s_1' and s_2' are constrained by two bi-tangents along the shortest paths connecting the endpoints of s_1 and s_2. The total running time of constructing $L(s_1, s_2)$ is proportional to $\mathcal{O}(n)$ [9]. For every pair of segments s_1 and s_2, denote by $p_{s_1}^+$, $p_{s_1}^-$ and $p_{s_2}^+$, $p_{s_2}^-$ the upper and lower endpoints of s_1 and s_2, respectively. Let $\gamma^+ = \pi(p_{s_1}^+, p_{s_2}^+)$ be the *upper chain* and $\gamma^- = \pi(p_{s_1}^-, p_{s_2}^-)$ the *lower chain* of the hourglass of s_1 and s_2.

In their work on semi-algebraic range searching, Agarwal et al. [1] fixed constants d, Δ, s and an arbitrary $\varepsilon > 0$, and showed that an arbitrary n-point set in $\mathbb{R}^d$ admits a data structure with *expected* preprocessing time $\mathcal{O}(n^{1+\varepsilon})$, and the space of $\mathcal{O}(n)$, that answers any constant complexity semi-algebraic-range query in $\mathcal{O}\left(n^{1-1/d} \log^B n\right)$ s.t. $B = B(d, \Delta, s, \varepsilon)^1$.

Let a trajectory be a sequence of time-stamped locations in $\mathbb{R}^d$, which models the movement of an entity in a polygon. The problem of trajectory visibility was explored by P. Eades et al. [9]. Informally, given a simple polygon or a polygonal domain P', and the paths of two moving entities q and r, determine if there is ever a time at which q and r can see each other. Certainly, there are various scenarios based on whether P' is a simple polygon or a polygonal area, and whether the trajectories cross $\partial P'$ or not. Throughout this paper, we assume that the entities are moving in line segments trajectories at possibly different constant speeds and cannot see through the edges of P'. For the sake of notation, suppose that two entities q and r move on trajectories $\tau_q, \tau_r \subset P'$. Eades et al. emphasized identifying whether there is ever a time when the two entities can see each other. That enables a temporal breakdown of the problem: a conclusive answer of *no* is determined if it is *no* for all pairs of successive timestamps [9].

1.2   Our Contributions

From now on, we only consider a simple polygon P, and that $\tau_q, \tau_r \subset P$, that is, they never cross ∂P. Specifically, our work aims to solve some of the directions

---

[1] This can be $\mathcal{O}(n^{1-1/d+\varepsilon})$ if we assume D_0-general position for every point.

Table 1. Preprocessing, storage, and query-time complexities for the trajectory visibility problem in a simple polygon with line-segment trajectories. The first two rows list the best-known data-structure bounds from P. Eades et al. [9] for deciding whether two moving entities ever become visible (referred to as **Decision**). The last three rows summarise our new results (Theorems 1, Theorems 2, Theorems 3), which determine the earliest moment at which mutual visibility occurs (referred to as **Finding** t^*).

Reference	Scenario	Preprocessing	Space	Query time
[9] (Sect. 5)	Stationary & moving (**Decision**)	$\mathcal{O}(n \log n)$	$\mathcal{O}(n)$	$\mathcal{O}(\log n)$
[9] (Sect. 4)	Query trajectories (**Decision**)	$\mathcal{O}(n \log^5 n)$	$\mathcal{O}(n \log^5 n)$	$\mathcal{O}(n^{3/4} \log^3 n)$
Theorem 1	Fixed trajectories (**Finding** t^*)	$\mathcal{O}(n)$	$\mathcal{O}(n)$	$\mathcal{O}(\log n)$
Theorem 2	One trajectory as query (**Finding** t^*)	$\mathcal{O}(n \log n)$	$\mathcal{O}(n \log n)$	$\mathcal{O}(\log^2 n)$
Theorem 3	Query trajectories (**Finding** t^*)	$\mathcal{O}(n^{1+\varepsilon})$ (expected)	$\mathcal{O}(n)$	$\mathcal{O}(n^{1/2} \log^B n)$ (expected)

left for future explorations in [9] (see also [10]). We aim to determine the first time t^* at which q and r can see each other. We emphasize that in the *most general case* explored in this paper, the constant speeds of q and r, denoted as v_q and v_r, together with the trajectories of q and r, are not given alongside P, but provided as *queriers* (see also Table 1 in [9]). More specifically, we address the following (the readers may refer to Table 1 as well):

1. Suppose that two moving entities q and r with fixed line-segment trajectories are given. For query speeds v_q and v_r, determine the first time t^* when they become mutually visible. We achieve a query time of $\mathcal{O}(\log n)$ after $\mathcal{O}(n)$ preprocessing. Moreover, our approach can naturally extend to reporting the entire set of time intervals during which the entities are mutually visible.
2. We extend the above case by assuming that q's trajectory is not known in advance and is provided only as a query, along with v_q and v_r. We develop a data structure that, after $\mathcal{O}(n \log n)$ preprocessing, answers queries in $\mathcal{O}(\log^2 n)$ time, using $\mathcal{O}(n)$ space.
3. We finally remove all the above relaxations, meaning none of q's and r's trajectories are known in advance, same as v_q and v_r. We develop a data structure that, after $\mathcal{O}(n^{1+\varepsilon})$ *expected* pre-processing time, answers queries in $\mathcal{O}(n^{1/2} \log^B n)$ *expected* time, where B is a constant. Our data structure requires $\mathcal{O}(n)$ space.

We note a recent similar attempt [17], but our approach does not overlap with theirs. While they have explored a similar problem, our investigations suggest that our approach resolve serious issues in their analysis.

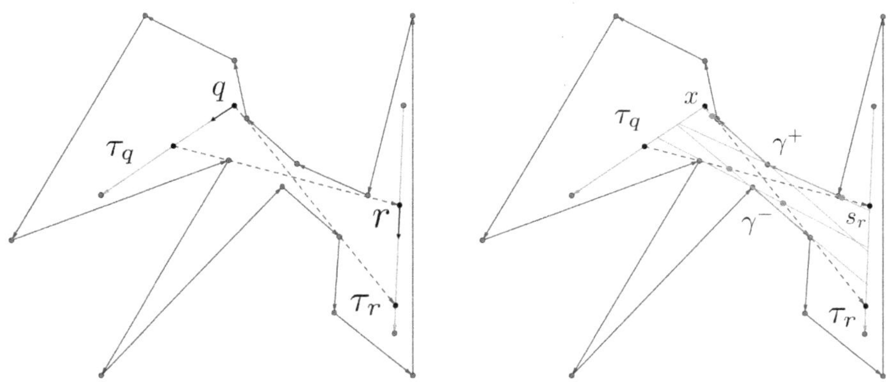

Fig. 1. (Left) Given two moving entities q and r with their trajectories τ_r and τ_q in red and blue within a simple polygon, the dotted lines indicate the bi-tangents of their visibility glass. (Right) Extensions of segments between consecutive reflex vertices (in green) (known as critical constraints [2]) intersect their bi-tangents (red dots). For every $x \in I_q$, there is a corresponding subsegment $s_r \in \tau_r$ such that every point in s_r is visible to x within P. (Color figure online)

2   Fixed Trajectories

Consider the $L(\tau_q, \tau_r)$. For every pair of consecutive reflex vertices v_i and v_{i+1} on its upper chain γ^+ (and similarly on γ^-), let $l(v_i, v_{i+1})$ be the line through them. Let the ray emitting from v_i to v_{i+1} on $l(v_i, v_{i+1})$ refer to the ray with origin at v_i directed along $l(v_i, v_{i+1})$ toward v_{i+1} (similarly for the ray from v_{i+1} to v_i). Let each of these rays be followed until its first intersection with ∂P (excluding v_i and v_{i+1}). The closed segment of $l(v_i, v_{i+1})$ between those two intersection points on ∂P is denoted as a *critical constraint* by Aronov et al. [2] and others before them. Define:

$$I_q = \bigcup_i \big(l(v_i, v_{i+1}) \cap \tau_q\big) \quad \text{and} \quad I_r = \bigcup_i \big(l(v_i, v_{i+1}) \cap \tau_r\big)$$

These intersection points partition τ_q and τ_r into subsegments. For each intersection point $x \in I_q$, there is a corresponding subsegment $s_r \subset \tau_r$ such that every point in s_r is visible to x within P (see Fig. 1).

Next, we define a mapping to map positions along τ_q and τ_r into a two-dimensional diagram $D \subset \mathbb{R}^2$. Let $\varphi_q : \tau_q \to [0, \text{len}(\tau_q)]$ and $\varphi_r : \tau_r \to [0, \text{len}(\tau_r)]$ be the arc-length parametrizations of τ_q and τ_r, respectively. Note that we map the starting point of the trajectories to 0. In D, the x-coordinate is determined by φ_q and the y-coordinate by φ_r. Each $x \in I_q$ corresponds to a vertical segment in D that represents a segment along τ_r on which every point is visible to x (see Fig. 3).

Observe that the mapping from positions on τ_q to positions on τ_r is non-linear. For example, an entity moving along a segment s_1 at constant speed v

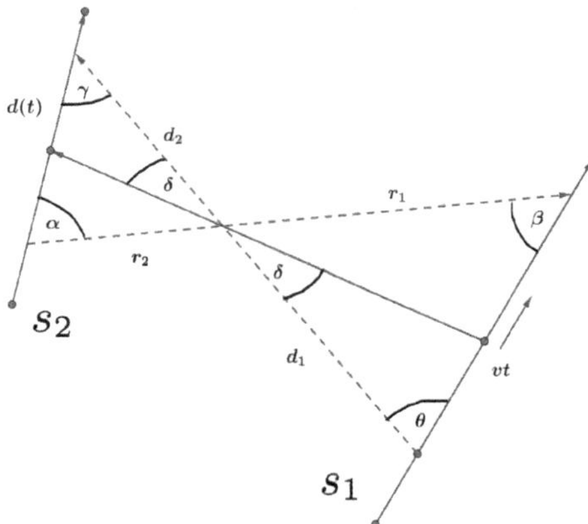

Fig. 2. An entity travels along segment s_1 at constant speed v, covering distance vt from its start. The ray from this entity to segment s_2 meets the hourglass bi-tangent at angle δ, while the fixed angles between the bi-tangent and the upper and lower chains are θ and γ, respectively. Distances d_1 and d_2 are measured from the endpoints of s_1 and s_2 to the corresponding tangency points.

maps to a point on another segment s_2 via a nonlinear function $f(t)$, where the distance $d(t)$ from a fixed endpoint of s_2 is strictly convex.

Corollary 1. *Let $s_1, s_2 \subset P$ be two line segments. Suppose an entity moves with constant speed v on s_1. Construct $L(s_1, s_2)$ and consider its intersection points. Define a mapping of the entity's position as $f(t)$ to be a point on s_2 at distance $d(t)$ from a certain endpoint of s_2 at time t. Then, as the entity moves, the mapping $f(t)$ transitions between the intersection points, and $d(t)$ is strictly convex and non-linear in t.*

Proof. Let $d(t)$ be denoted as x. Referring to Fig. 2 and applying the sine law on the two relevant triangles, we obtain: $vt = \dfrac{d_1 \sin \delta}{\sin(\delta + \theta)}$ and $x = \dfrac{d_2 \sin \delta}{\sin(\gamma + \delta)}$.

Eliminating δ gives the closed-form: $x = \dfrac{d_2 \, vt \, \sin \theta}{d_1 \sin \gamma + vt \sin(\theta - \gamma)}$.

Now set $A = \dfrac{v \, \sin(\theta - \gamma)}{d_1}$, $B = \dfrac{d_1 \sin \gamma}{v \sin \theta}$, so that $x = d_2 \cdot \dfrac{t}{B} - d_2 \cdot \dfrac{At^2}{B(At + B)}$

A straightforward differentiation shows that the second derivative of x with respect to t is strictly positive (or negative) under non-degeneracy conditions (e.g., $d_1 > vt \cos \theta$ and non-zero angles), which establishes that $d(t)$ is strictly convex.

Given query speeds v_q and v_r for q and r, their positions become $\varphi_q(t) = v_q t$ and $\varphi_r(t) = v_r t$. In D, consider the ray $\overrightarrow{r}$ from the origin with slope $\tan \alpha = \frac{v_q}{v_r}$. The first time t^* of mutual visibility occurs when the point $(v_q t, v_r t)$ lies on (or above) the curve boundary (according to Corollary 1) of D. By Corollary 1, the image in D of any constant-speed traversal of a straight subsegment of τ_q or τ_r is an algebraic arc (coordinates given by rational functions of t of constant degree). Together with the vertical segments arising from the mapped critical-constraint intersections, these arcs form the boundary of the mutual-visibility region in D. Each arc endpoint coincides with a mapped critical-constraint intersection, and there are only $\mathcal{O}(n)$ such intersections overall; hence the boundary is a closed chain of $O(n)$ piecewise-algebraic arcs. This boundary is therefore a *splinegon*. Ray shooting on the splinegon in D after a preprocessing in $\mathcal{O}(n)$ time yields t^* in $\mathcal{O}(\log n)$ time. Therefore, the following theorem holds:

Theorem 1. *Constructing D in $\mathcal{O}(n)$ time, using $\mathcal{O}(n)$ space, one can process any query (v_q, v_r) to determine in $\mathcal{O}(\log n)$ time the smallest $t^* \geq 0$ for which q and r become mutually visible.*

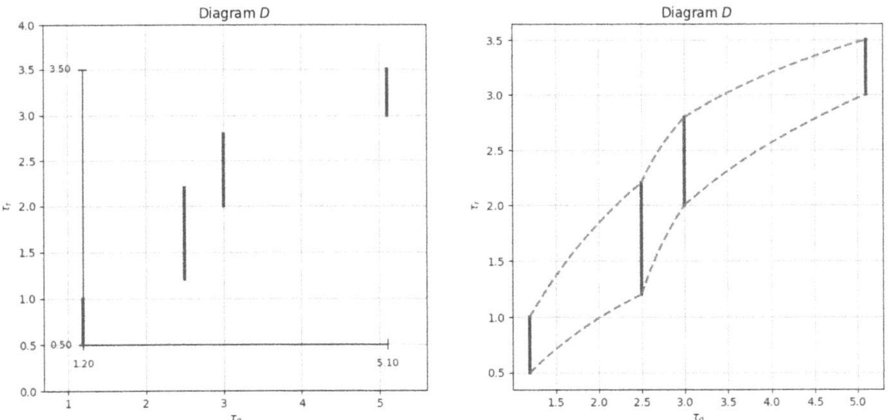

Fig. 3. (Left) Vertical segments in the diagram D induced by the intersection points I_q on τ_q: each $x = \varphi_q(x_i)$ spans the interval of φ_r-values for which points on τ_r are visible to x_i. (Right) The curved boundary (splinegon) obtained by the nonlinear mapping discussed in (Corollary 1 at) Sect. 2.

3   One Query-Provided Trajectory

In contrast to the previous section, where both trajectories are fixed, here the trajectory τ_q of entity q is provided as part of the query, while the trajectory τ_r of entity r is known a priori. The goal is to determine the first time t^* at which q and r become mutually visible within P. More specifically, suppose the trajectory $\tau_r \subset P$ of r is given. At query time, a trajectory $\tau_q \subset P$ for q is provided, as well as the query speeds v_q and v_r. For convenience, we scale the speeds by assuming that the speed of r is *one*, while the speed of q becomes $v = v_q/v_r$.

Since τ_q is provided only at the time of the query, we cannot directly construct $L(\tau_q, \tau_r)$. We instead begin with pre-computing $W(\tau_r)$ and process it to support ray-shooting queries. This allows us to identify only the portion of τ_q that intersects with $W(\tau_r)$ and to discard the remainder. Observe that τ_q enters $W(\tau_r)$ only once and then leaves it. That is, there will not be more than one segment on τ_q that intersects with $W(\tau_r)$. This can be verified by noting that first, the entities are not allowed to intersect P, and second, the trajectories are line segments. So, if we assume τ_q might enter $W(\tau_r)$ more than exactly once, we reach a contradiction. With a slight violation of notation, *we denote the intersection of τ_q with $W(\tau_r)$ as τ_q only during the rest of this section.*

Another step before processing a query is computing $T(p_r^+)$ and $T(p_r^-)$, restricted to vertices of $W(\tau_r)$. For each vertex $v \in T(p_r^+)$ (similarly for $T(p_r^-)$), we pre-compute the distance of v to the root, and up_j, that is a pointer to its 2^j-th ancestor on the root path, for $0 \leq j \leq \lfloor \log_2 \text{depth}(v) \rfloor$.

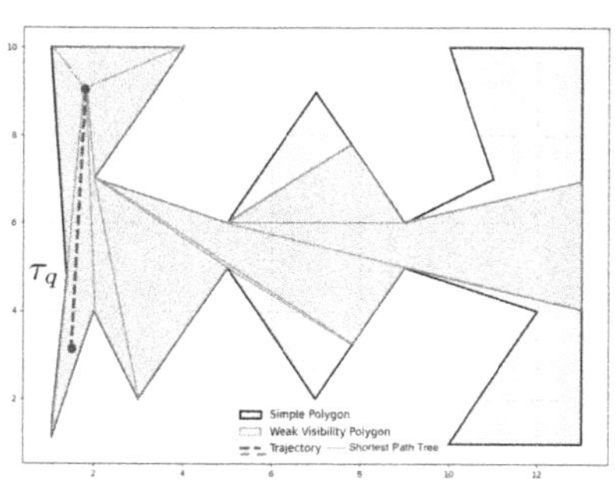

(a) Given trajectory τ_q (blue) together with $T(p_q^+)$ (purple) within $W(\tau_q)$

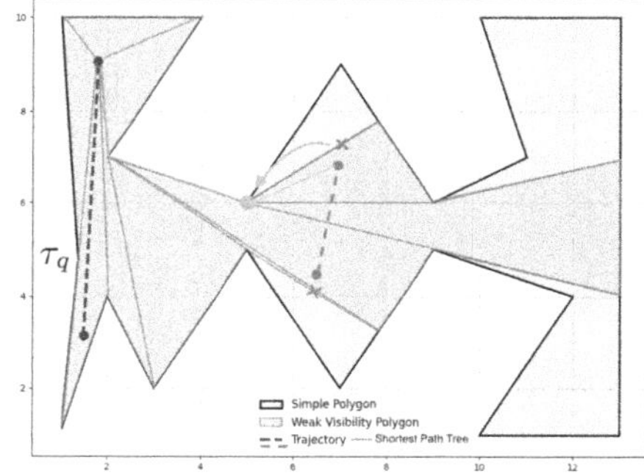

(b) Identifying the visibility glass for the query trajectory τ_r

Fig. 4. An illustration of finding the visibility glass when one trajectory is provided as a query. (a) The $T(\tau_q)$ (purple) is computed within $W(\tau_q)$. (b) This tree is used to find the boundaries of the visibility glass $L(\tau_q, \tau_r)$.

Upon receiving a query we first locate two vertices of $W(\tau_r)$ by ray shooting: we shoot rays in the directions $\overrightarrow{p_{\tau_q}^+ p_{\tau_q}^-}$ and $\overrightarrow{p_{\tau_q}^- p_{\tau_q}^+}$, let each ray hit the boundary of $W(\tau_r)$, and choose an endpoint of the intersected edge. From these endpoints the precomputed shortest-path trees $T(p_r^+)$ and $T(p_r^-)$ immediately provide pointers into the corresponding shortest paths toward $p_{\tau_r}^+$ and $p_{\tau_r}^-$ (See Fig. 4). Next, we find, on each such shortest path, the first vertex (in the path order) that is weakly visible to either $p_{\tau_q}^+$ or $p_{\tau_q}^-$. Because consecutive vertices along a shortest path are pairwise weakly visible, this vertex can be found by binary searching the path while testing weak visibility with an additional ray-shooting query. This yields pointers to $\pi(p_{\tau_q}^+, p_{\tau_r}^+)$ and $\pi(p_{\tau_q}^-, p_{\tau_r}^-)$, and hence to the visibility glass $L(\tau_q, \tau_r)$. It remains to discuss the way we eventually find t^*.

There can be two cases in this step to find t^*. Either the entities are moving in a *similar direction*, by which we mean each entity starts from p^+ (similarly for p^-) of its trajectory, or in a *different direction*, by which we mean one of them begins at p^+ and the other at p^- of its trajectory. We will suggest a general way of handling both cases.

Corollary 2. *If q's and r's positions become co-linear with a vertex $v \in \gamma^+$ (similarly for γ^-), while the line crossing q, r, and v does not intersect ∂P, obviously except at v, then a candidate for t^* is found.*

Proof. Let $v = (x_v, y_v)$, $q(t) = (x_q(t), y_q(t)) = (x_{q,0} + v_{qx}t,\ y_{q,0} + v_{qy}t)$, and $r(t) = (x_r(t), y_r(t)) = (x_{r,0} + v_{rx}t,\ y_{r,0} + v_{ry}t)$, where $(x_{q,0}, y_{q,0})$ and $(x_{r,0}, y_{r,0})$ are the starting coordinates (at $t = 0$) on τ_q and τ_r respectively. The pairs (v_{qx}, v_{qy}) and (v_{rx}, v_{ry}) are the constant components of the speed vectors for q and r. For $q(t)$, v, and $r(t)$ to be co-linear, the z-component of the cross product of the vector from v to $q(t)$ and the vector from v to $r(t)$ must be zero. This can be expressed as: $(x_{q(t)} - x_v)(y_{r(t)} - y_v) - (y_{q(t)} - y_v)(x_{r(t)} - x_v) = 0$. Let $\Delta x_{q0} = x_{q,0} - x_v$, $\Delta y_{q0} = y_{q,0} - y_v$, $\Delta x_{r0} = x_{r,0} - x_v$, and $\Delta y_{r0} = y_{r,0} - y_v$. Substituting the expressions for $q(t)$ and $r(t)$: $(\Delta x_{q0} + v_{qx}t)(\Delta y_{r0} + v_{ry}t) - (\Delta y_{q0} + v_{qy}t)(\Delta x_{r0} + v_{rx}t) = 0$. Expanding this expression yields a quadratic equation in t of the form $At^2 + Bt + C = 0$, where $A = v_{qx}v_{ry} - v_{qy}v_{rx}^2$, $B = \Delta x_{q0}v_{ry} - \Delta y_{q0}v_{rx} + \Delta y_{r0}v_{qx} - \Delta x_{r0}v_{qy}$, and $C = \Delta x_{q0}\Delta y_{r0} - \Delta y_{q0}\Delta x_{r0}$. If $A \neq 0$: $t = \frac{-B \pm \sqrt{B^2 - 4AC}}{2A}$, assuming $B^2 - 4AC \geq 0$ for real solutions. Otherwise, if $A = 0$ (meaning the speed vectors of q and r are parallel), there can be two cases. First, If $B \neq 0$: $t = -\frac{C}{B}$. This occurs if the initial displacement vectors are not co-linear in the same way as the speed vectors. The second is when $B = 0$. In this case, if $C = 0$, the points q_0, r_0, v are already co-linear, and since their speed vectors are parallel and related by the condition that also makes $B = 0$, they remain co-linear for all t. Otherwise, if $C \neq 0$, the initial points are not co-linear. Since the speed vectors are parallel but do not satisfy the condition for $B = 0$, no co-linearity occurs at any time t (unless by a non-physical negative t).

Every real $t \geq 0$, if found, is then a candidate for t^*, if the line crossing $q(t^*)$, $r(t^*)$, and the vertex v intersects ∂P only at that vertex.

To find t^*, we perform a binary search over the vertices of *upper chain* $\gamma^+ = (v_1, \ldots, v_k)$ of $L(\tau_q, \tau_r)$ and similarly over the *lower chain* γ^-. The binary search aims to find the vertex that allows the earliest time $t \geq 0$ where $q(t)$, $r(t)$, and the vertex are collinear, and the entities become weakly visible. The binary search begins with indices *low* set to 1 and *high* set to k. It continues as long as *low* is less than *high*. In each iteration, a middle index *mid* is computed as $low + \lfloor(high - low)/2\rfloor$. To guide this search, the earliest time that v_{mid}

---

² Note that A is the z-component of the cross product of the speed vector of q and the speed vector of r.

might allow visibility occurrence is compared to that of its successor, $v_{\text{mid}+1}$. The determination of this specific time, t_j, for any given vertex v_j proceeds as follows: The collinearity of $q(t), r(t)$, and v_j will be examined. If, for v_j, the coefficients A, B, and C are all zero (see Corollary 2), this vertex v_j does not define a *discrete moment* at which it establishes collinearity. Such a vertex is *discarded* from providing a t_j value. Note that if another vertex becomes co-linear with it, together with the entities, we might still be able to conclude the visibility as shown in Corollary 2. That holds if the segment connecting these four never intersects another edge or vertex of P. Otherwise, if not all of A, B, C are zero, solving $At^2 + Bt + C = 0$ yields potential non-negative times t, up to two distinct, real roots. For each such t, an $\mathcal{O}(\log n)$ ray-shooting query within P verifies if $q(t)$ and $r(t)$ are weakly visible. The t_j for this case is the minimum t from this set that also satisfies weak visibility. If no such t is found, then for comparison purposes, t_j is treated as ∞. We then compute t_{mid} and $t_{\text{mid}+1}$. If $t_{\text{mid}} \leq t_{\text{mid}+1}$, it is inferred that the vertex producing the earliest time is v_{mid} or to its left, so *high* becomes *mid*. Otherwise, *low* becomes *mid* + 1. Upon termination, the minimum time is t_{low}.

The overall t^* is the minimum of times from γ^+ and γ^-. The $\mathcal{O}(\log k)$ binary search iterations, each costing $\mathcal{O}(\log n)$ for calculating these times, yield $\mathcal{O}(\log^2 n)$ complexity per chain, as $k \in \mathcal{O}(n)$. Therefore, theorem 2 follows:

Theorem 2. *Given the fixed trajectory $\tau_r \subset P$ of entity r, for any query-provided trajectory $\tau_q \subset P$ and query speeds v_q and v_r, t^* can be computed in $\mathcal{O}(\log^2 n)$ time, requiring the pre-processing time of $\mathcal{O}(n \log n)$ and space of $\mathcal{O}(n \log n)$.*

4   Query Trajectories

In this section, we remove all the relaxations in the previous sections, meaning none of q's and r's trajectories are known in advance, along with v_q and v_r. Before processing any query, we must pre-process P so that we can perform ray-shooting queries. To handle any constant complexity semi-algebraic-range query [1] on the *vertices of P* as given points in $\mathbb{R}^2$, we must yet pre-process the vertices of P another time.

Once we receive τ_q and τ_r as the query, together with v_q and v_r, we consider the *area* which the *leash* between q and r may sweep in P. That is, a line segment connecting q and r that moves forward as q and r move forward in their trajectories. Denote the leash as $\ell_{qr}(t)$, and the area swept by $\ell_{qr}(t)$ as S (See Fig. 5).

Corollary 3. *The area swept by $\ell_{qr}(t)$ is a constant complexity semi-algebraic range.*

Proof. Let $q(t) = (x_q(t), y_q(t))$ and $r(t) = (x_r(t), y_r(t))$. Also, let t_{qf} be the time for q to traverse τ_q, and t_{rf} be the time for r to traverse τ_r. The leash $\ell_{qr}(t)$ is defined for $t \in [0, t_f]$, where $t_f = \max(t_{qf}, t_{rf})$. The area $S = \bigcup_{t \in [0, t_f]} \ell_{qr}(t)$

swept by this leash has a boundary that includes $\ell_{qr}(0)$, $\ell_{qr}(t_f)$, τ_q, and τ_r. If the line containing $\ell_{qr}(t)$ generates an envelope, portions of this envelope to which interior points of $\ell_{qr}(t)$ are tangent can form part of the boundary of S.

The line through $q(t)$ and $r(t)$ is: $At^2 + B(x,y)t + C(x,y) = 0$, where $A = v_{qx}v_{ry} - v_{qy}v_{rx}$, $B(x,y) = (v_{qy} - v_{ry})x - (v_{qx} - v_{rx})y + (x_{q0}v_{ry} + v_{qx}y_{r0} - y_{q0}v_{rx} - v_{qy}x_{r0})$, and $C(x,y) = (y_{q0} - y_{r0})x - (x_{q0} - x_{r0})y + (x_{q0}y_{r0} - y_{q0}x_{r0})$. If the speed vectors are not parallel, then $A \neq 0$, and the envelope is defined by the quadratic discriminant $B(x,y)^2 - 4AC(x,y) = 0$. Since $B(x,y)$ is linear, its square is quadratic, and $C(x,y)$ is also linear, the resulting equation describes a degree-two algebraic curve in x and y that may contribute to the boundary of region S. If the speed vectors are parallel ($A = 0$), the line becomes $B(x,y)t + C(x,y) = 0$, linear in t. In both cases, S's boundary consists of a constant number of algebraic curves, namely at most four line segments and possibly at most two degree-two curves.

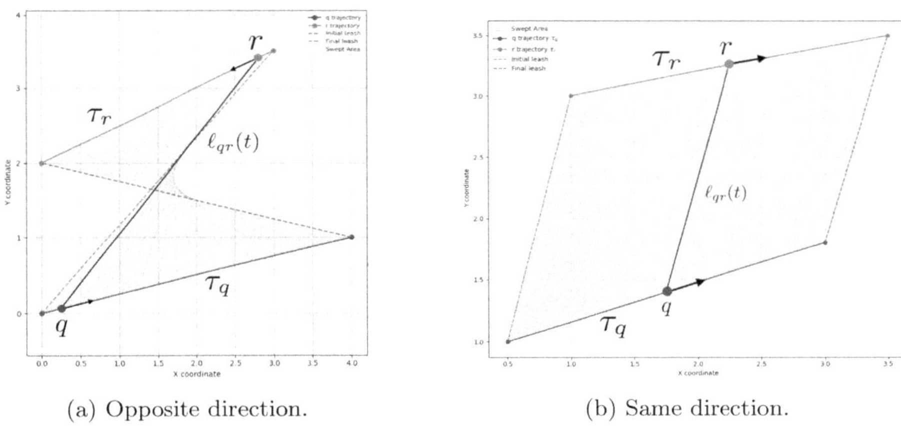

(a) Opposite direction.   (b) Same direction.

Fig. 5. The area S (yellow) swept by the leash $\ell_{qr}(t)$ between entities q (trajectory τ_q, blue) and r (trajectory τ_r, red). The boundary of S is formed by the initial leash (green), final leash (magenta), and portions of τ_q and τ_r. An example intermediate leash $\ell_{qr}(t)$ is shown in black. As discussed in Corollary 3 within Sect. 4, S is a constant complexity semi-algebraic range. (Color figure online)

Clearly, one can compute S in constant time. It is also clear how to support every constant complexity semi-algebraic-range query on the vertices of P as given points in $\mathbb{R}^2$, which includes S, namely our query of interest. So, at this stage, we perform the semi-algebraic range searching query and isolate the vertices of P that fall within S, which takes $\mathcal{O}(n^{1/2}\log^B n)$ time [1]. Once we isolate the intersecting vertices[3] of P with S, we need to find a way to reuse Corollary

---

[3] Indeed, P is a sequence of vertices in a certain *order*. We isolate those intersecting with S.

2 in Sect. 3. Observe that since τ_q and τ_r are provided at the query time, it is not clear how one may construct $L(\tau_q, \tau_r)$ faster than $\mathcal{O}(n)$ running time. To circumvent this issue, we adopt a simple randomised approach: We uniformly at random pick a vertex among the isolated ones by the semi-algebraic range search. Same as Corollary 2, we check if q, r, and the vertex ever become co-linear. Recall that we can perform ray-shooting queries once we need to check if the segment connecting q, r, and the vertex we picked intersects any other edge or vertex of P. While we have not yet found t^*, we keep updating *low*, *mid*, and *high*, same as Sect. 3. As such, it is trivial to see that *in expectation*, we perform $\mathcal{O}(\log n)$ steps in our randomised binary search. Thus, we find t^* in $\mathcal{O}(\log^2 n)$ expected time, as we perform a ray-shooting at every step of the binary search. Therefore, Theorem 3 holds:

Theorem 3. *Given P, after $\mathcal{O}(n^{1+\varepsilon})$ expected pre-processing time, one can find t^* once receiving τ_q and τ_r, as well as v_q and v_r, as queries, in $\mathcal{O}(n^{1/2} \log^B n)$ expected time, where B is a constant. This requires $\mathcal{O}(n)$ space.*

5   Concluding Remarks and Open Problems

We presented data structures for the first-visibility time t^* for two entities moving on line-segment trajectories inside a simple polygon. Two immediate open problems can be explored in the future. First, extending the results to polygonal domains with $h > 0$ holes. It is unknown whether the same strategy used in this paper can be maintained. Second, allowing one or both trajectories to be polylines with m segments. It is open to determine tight trade-offs between m and n for preprocessing and query costs (e.g., whether near-linear preprocessing in $n + m$ with polylogarithmic queries is achievable). Other directions include multiple-entity visibility, and letting the entities to cross ∂P (with or without the presence of holes in P).

References

1. Agarwal, P.K., Matousek, J., Sharir, M.: On range searching with semialgebraic sets. ii. SIAM J. Comput. **42**(6), 2039–2062 (2013)
2. Aronov, B., Guibas, L.J., Teichmann, M., Zhang, L.: Visibility queries and maintenance in simple polygons. Discrete Comput. Geom. **27**, 461–483 (2002)
3. Bozek, K., Hebert, L., Portugal, Y., Mikheyev, A.S., Stephens, G.J.: Markerless tracking of an entire honey bee colony. Nat. Commun. **12**(1), 1733 (2021)
4. Calenge, C., Dray, S., Royer-Carenzi, M.: The concept of animals' trajectories from a data analysis perspective. Eco. Inform. **4**(1), 34–41 (2009)
5. Chazelle, B.: Triangulating a simple polygon in linear time. In: Proceedings of the 14th Annual ACM Symposium on Theory of Computing, pp. 270–282 (1983)
6. Chazelle, B., et al.: Ray shooting in polygons using geodesic triangulations. Algorithmica **12**(1), 54–68 (1994)
7. Dobkin, D.P., Souvaine, D.L.: Computational geometry in a curved world. Algorithmica **5**(1), 421–457 (1990)

8. Dodge, S., Weibel, R., Forootan, E.: Revealing the physics of movement: comparing the similarity of movement characteristics of different types of moving objects. Comput. Environ. Urban Syst. **33**(6), 419–434 (2009)

9. Eades, P., van der Hoog, I., Löffler, M., Staals, F.: Trajectory visibility. In: 17th Scandinavian Symposium and Workshops on Algorithm Theory (SWAT 2020). Schloss-Dagstuhl-Leibniz Zentrum für Informatik (2020)

10. Eades, P.F.: Uncertainty Models in Computational Geometry. Ph.D. thesis (2020). https://hdl.handle.net/2123/23909

11. Ghodsi, M., Maheshwari, A., Nouri-Baygi, M., Sack, J.R., Zarrabi-Zadeh, H.: alpha-visibility. Comput. Geom. **47**(3, Part A), 435–446 (2014). https://doi.org/10.1016/j.comgeo.2013.10.004, https://www.sciencedirect.com/science/article/pii/S0925772211300117X

12. Ghosh, S.K.: Visibility Algorithms in the Plane. Cambridge University Press, Cambridge (2007)

13. Gudmundsson, J., Laube, P., Wolle, T.: Movement patterns in spatio-temporal data, pp. 1362–1370. Springer, Cham (2017).https://doi.org/10.1007/978-3-319-17885-1_823, https://digitalcollection.zhaw.ch/handle/11475/15060, 2nd edition

14. Guibas, L.J., Hershberger, J.: Optimal shortest path queries in a simple polygon. J. Comput. Syst. Sci. **39**(2), 126–152 (1989)

15. Gurarie, E., Andrews, R.D., Laidre, K.L.: A novel method for identifying behavioural changes in animal movement data. Ecol. Lett. **12**(5), 395–408 (2009)

16. Jamroen, C., Fongkerd, C., Krongpha, W., Komkum, P., Pirayawaraporn, A., Chindakham, N.: A novel UV sensor-based dual-axis solar tracking system: implementation and performance analysis. Appl. Energy **299**, 117295 (2021)

17. Kazemi, S.M.H., Vaezi, A., Abam, M.A., Ghodsi, M.: Trajectory range visibility. arXiv preprint arXiv:2209.04013 (2022)

18. Lee, D.T., Preparata, F.P.: Computing visibility graphs. In: Proceedings of the 3rd Annual Symposium on Computational Geometry, pp. 165–174 (1986)

19. Li, X., Li, X., Tang, D., Xu, X.: Deriving features of traffic flow around an intersection from trajectories of vehicles. In: 2010 18th International Conference on Geoinformatics, pp. 1–5. IEEE (2010)

20. Melissaratos, E.A., Souvaine, D.L.: Shortest paths help solve geometric optimization problems in planar regions. SIAM J. Comput. **21**(4), 601–638 (1992)

21. Pane, S., Iacovacci, V., Sinibaldi, E., Menciassi, A.: Real-time imaging and tracking of microrobots in tissues using ultrasound phase analysis. Appl. Phys. Lett. **118**(1), 014102 (2021)

22. Stohl, A.: Computation, accuracy and applications of trajectories–a review and bibliography. Atmos. Environ. **32**(6), 947–966 (1998)

Tile Reconfiguration by a Finite Automaton

Jonas Friemel[1(✉)] , David Liedtke[2] , and Christian Scheffer[1]

[1] Bochum University of Applied Sciences, Bochum, Germany
`{jonas.friemel,christian.scheffer}@hs-bochum.de`
[2] Paderborn University, Paderborn, Germany
`liedtke@mail.upb.de`

Abstract. Shape formation is one of the most thoroughly studied problems in programmable matter and swarm robotics. However, in many models, the class of shapes that can be formed is highly restricted due to the particles' limited memory. In the hybrid model, an active agent with the computational power of a deterministic finite automaton can form shapes by lifting and placing passive tiles on the triangular lattice. We study the shape reconfiguration problem where the agent additionally has the ability to distinguish so-called target nodes from non-target nodes and needs to form a target shape from the initial tile configuration. We present a worst-case optimal $\mathcal{O}(mn)$ algorithm for simply connected target shapes, where m is the initial number of unoccupied target nodes and n is the total number of tiles. Furthermore, we show how an agent can reconfigure a large class of target shapes with holes in $\mathcal{O}(n^4)$ steps.

Keywords: Programmable matter · Reconfiguration · Automaton

1 Introduction

In the field of programmable matter, small (possibly nano-scale) particles are envisioned to solve tasks like self-assembling into desired shapes, making coordinated movements, or coating objects [36]. The particles may be controlled by external stimuli or act on their own with limited computational capabilities. In the future, programmable matter could become relevant for targeted medical treatments [1] or as self-assembling structures in environments that are inaccessible by humans such as in space [25]. There are multiple computational models of programmable matter that differ in the types of particles, their abilities, and the underlying graph structure. We focus on the *hybrid model* of programmable matter where passive hexagonal tiles and an active agent with the computational power of a deterministic finite automaton populate the triangular lattice [19,20].

A central research problem in programmable matter is shape formation [6]. In the hybrid model, the active agent has to rearrange passive tiles from an arbitrary initial configuration into a shape [20] such as a line or a triangle. Once the shape is formed, it is typically assumed that the work is finished and the shape will remain intact. Thus, existing shape formation algorithms are not designed to repair small shape defects such as individually misplaced tiles and may deconstruct the entire structure to rebuild the desired shape.

E. Di Giacomo and D. Mondal (Eds.): WALCOM 2026, LNCS 16444, pp. 512–526, 2026.
https://doi.org/10.1007/978-981-95-7127-7_34

Our Contributions. We provide first results for shape reconfiguration in the hybrid model with the assumption that the agent is able to recognize nodes that belong to the target shape. We present a worst-case optimal algorithm for reconfiguring simply connected target shapes within $\mathcal{O}(mn)$ steps, where n is the configuration size and m is the number of tiles initially located outside of the target shape (Sect. 3). We also describe how an agent equipped with a single counter can reconfigure arbitrary shapes with holes in $\mathcal{O}(mn^2)$ steps and we present an $\mathcal{O}(n^4)$ algorithm for scaled shapes, i.e., target shapes that do not contain any nodes with multiple disconnected non-target neighbors, that does not require the use of counters or other auxiliary means (Sect. 4). This is particularly interesting as it is impossible for finite automata to visit all nodes if they are unable to modify their environment [3]. Due to limited space, some results and many proof details are deferred to the full version [18].

Related Work. Shape formation has been studied in a multitude of models for programmable matter. The particles in each model generally fall into one of the following two categories: *active* agents that can perform (typically limited) computations and move on the underlying graph by themselves, and *passive* entities that do not move or act without external influence.

Models with distributed active agents where shape formation has been studied include: the geometric *amoebot* model [9,10,12,29], where particles with restricted vision occupy either one node (contracted) or two adjacent nodes (expanded) and have the computational capabilities of finite automata [6–8]; the *silbot* model [32], where the particles additionally lack persistent memory and means of computation [5]; the *nubot* model [4,38], where particles (*monomers*) may form bonds with their neighbors and can move and interact according to predefined rules; or the *sliding square* model [14,24,27] where square particles can move via slides and convex transitions.

A well-known model of passive matter is the *abstract tile assembly model (aTAM)* [13,37], where tiles with different types of glue with varying strengths occupy a square grid graph. Starting from a seed configuration, the shape self-assembles with new tiles attaching themselves to compatible existing tiles. For a survey of the aTAM and variants such as the *kinetic tile assembly model (kTAM)* and the *two-handed assembly model (2HAM)*, we refer to Patitz [35].

In this paper, we are working with the *hybrid model* introduced by Gmyr et al. [20]. Here, active finite automata move on the triangular lattice and can lift and place passive tiles. The authors present shape formation algorithms for simple shapes such as triangles. On square grid graphs, finite automata can construct bounding boxes and scale polyominoes using markers [15,17]. In the three-dimensional variant of the hybrid model, an agent can coat objects [26] and construct lines [22] as well as hole-free intermediate shapes called "icicles" [21].

There has been some research into repairing simple shapes such as lines. In the amoebot model, Di Luna et al. [11] and Nokhanji and Santoro [34] separately study the problem of repairing a line when some of the particles that are part of the shape malfunction. Nokhanji et al. [33] also explore a similar problem in the hybrid model where tiles in a line become faulty and need to be removed.

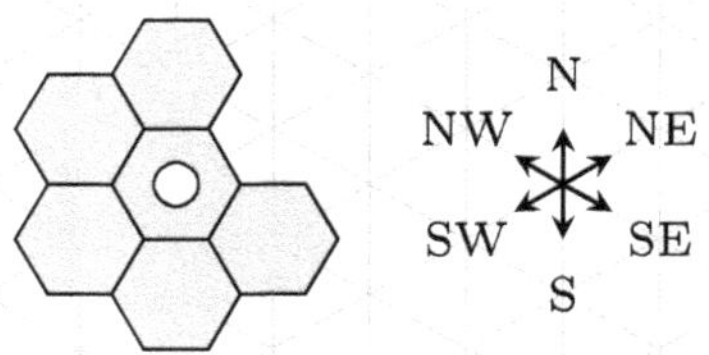

Fig. 1. An agent on tiles and global compass directions on the triangular lattice.

Finally, Kostitsyna et al. [28] propose a linear-time (in expectation) algorithm to reconfigure simply connected amoebot shapes where particles are pulled to their target positions along shortest path trees.

2 Preliminaries

In our setting, the triangular lattice $G = (V, E)$ is occupied by a single active agent r with limited computational capabilities and a finite number of passive hexagonal tiles, see Fig. 1. We call a node $v \in V$ *tiled* if it is occupied by a tile and denote the set of tiled nodes with T. At any time, a node may be occupied by at most one tile and each tile may only occupy a single node. Similarly, the agent may only occupy a single (tiled or untiled) node at a time. Tiles are uniform, i.e., r cannot distinguish any two tiles from one another, and r can carry up to one tile. When carrying a tile, r can still enter tiled nodes. A tuple $C = (T, p)$, where $p \in V$ is the node occupied by r, is called a *configuration*. It has *size* $|T|$. All initial and target configurations in this paper have size n.

We call a node set $S \subseteq V$ *connected* if the induced subgraph $G[S]$ is connected and we call S *simply connected* if the set $V \setminus S$ is connected. A configuration $C = (T, p)$ is *connected* if T is connected or if $T \cup \{p\}$ is connected and r is carrying a tile. Similarly, C is *simply connected* if T is simply connected, i.e., $V \setminus T$ is connected, or if $T \cup \{p\}$ is simply connected and r is carrying a tile. We require configurations to be connected to ensure that the tile system does not drift apart in practical implementations, e.g., in liquid domains.

The agent has an internal compass to differentiate between the six edge directions on the graph G (N, NE, SE, S, SW, NW). For ease of presentation, we assume that this compass aligns with the global directions on the triangular lattice shown in Fig. 1. We denote the set of compass directions by $\mathcal{D} := \{\text{N}, \text{NE}, \text{SE}, \text{S}, \text{SW}, \text{NW}\}$ and define $\mathcal{D}$ to be isomorphic to the ring of integers modulo six $\mathbb{Z}/6\mathbb{Z}$ with N $\equiv 0$, NE $\equiv 1$, and so on, up to NW $\equiv 5$. With a slight abuse of notation, this allows us to perform simple additions on directions, e.g., NE $+ 2 = $ S. Intuitively, by adding $k \in \mathbb{Z}$ to a direction $d \in \mathcal{D}$, we obtain the next direction from d after k clockwise turns of $60°$ around the compass shown in Fig. 1.

Each node $v \in V$ is uniquely identified by a coordinate pair $(x, y) \in \mathbb{Z} \times \mathbb{Z}$ where $x + y$ is even. We write $v = (x, y)$. The x-coordinate grows from west to east, and the y-coordinate grows from north to south. By this convention, the six compass directions correspond to the following directional vectors.

$$\vec{\mathrm{N}} = (0, -2),\ \vec{\mathrm{NE}} = (1, -1),\ \vec{\mathrm{SE}} = (1, 1),\ \vec{\mathrm{S}} = (0, 2),\ \vec{\mathrm{SW}} = (-1, 1),\ \vec{\mathrm{NW}} = -(1, 1).$$

For a node $v \in V$ and a direction $d \in \mathcal{D}$, the node $v + \vec{d}$ is called a *neighbor of v* (in direction d) and $N(v) := \{v + \vec{d} \mid d \in \mathcal{D}\}$ is called the *neighborhood of v*.

The agent acts in *look-compute-move* cycles. In the *look* phase, it observes its surroundings. The agent's "vision" is limited to neighboring nodes, i.e., it can only see tiles within unit hop-distance to its node in G. Next, r enters the *compute* phase where it uses the gathered information to determine its next internal state and its action on the graph. The agent has the computational capabilities of a deterministic finite-state automaton. Consequently, it has only constant memory and cannot store a map of the tile configuration. Finally, r enters the *move* phase, where it may perform any of the following actions, provided that connectivity is maintained: (i) move to an adjacent (tiled or untiled) node (regardless of whether it is carrying a tile), (ii) lift the tile at its position if it is not carrying a tile, and (iii) place a tile at its position if it is carrying a tile and the node is untiled.

2.1   Problem Statement

Consider two connected sets of nodes $\mathcal{I}, \mathcal{T} \subseteq V$ with $|\mathcal{I}| = |\mathcal{T}| = n$ and a non-empty and connected intersection $\mathcal{I} \cap \mathcal{T}$, and an initial position $p^0 \in \mathcal{I}$ for the agent. We refer to $\mathcal{I}$ and $\mathcal{T}$ as the *initial* and *target shape*, respectively, with the corresponding nodes being referred to as *initial* and *target nodes*. Note that a shape is defined by its exact coordinates on the lattice, i.e., two shapes that are translationally or rotationally symmetrical are generally not identical.

An agent solves the SHAPE RECONFIGURATION PROBLEM by executing an algorithm that results in a sequence of connected configurations $C^0 = (T^0, p^0), \dots, C^\ell = (T^\ell, p^\ell)$ for some $p^\ell \in V$ with $T^0 = \mathcal{I}$ and $T^\ell = \mathcal{T}$ such that each configuration C^t results from configuration C^{t-1} by applying the agent's legal move actions (i)–(iii) to p^{t-1} for $0 < t \leq \ell$. When the time step is clear from the context or not relevant, we drop the superscripts. At any time t, a node $v \in T^t \setminus \mathcal{T}$ is called a *supply node* and a node $w \in \mathcal{T} \setminus T^t$ is called a *demand node*. We denote the initial number of supply and demand nodes by $m := |\mathcal{I} \setminus \mathcal{T}| = |\mathcal{T} \setminus \mathcal{I}|$. Finally, tiles on target nodes are called *target tiles* and tiles on supply nodes are called *supply tiles*. Thus, to solve the SHAPE RECONFIGURATION PROBLEM, an agent needs to move all m supply tiles to demand nodes, see Fig. 2.

At any time t, the agent can determine whether $p^t \in \mathcal{T}$ when it is in the *look* phase of a look-compute-move cycle. In a practical implementation, this node distinguishability could be realized with a simple binary signal from the outside, e.g., from a light source. We also assume that the agent can see which of its neighboring nodes are target nodes. This assumption does not make our agent more powerful than an agent r' that can only query $\mathcal{T}$ for its own position as r' could simply visit all six adjacent nodes within a constant number of steps. Some adjacent nodes may be unreachable without violating the connectivity requirement, but our agent ignores these nodes in the presented algorithms.

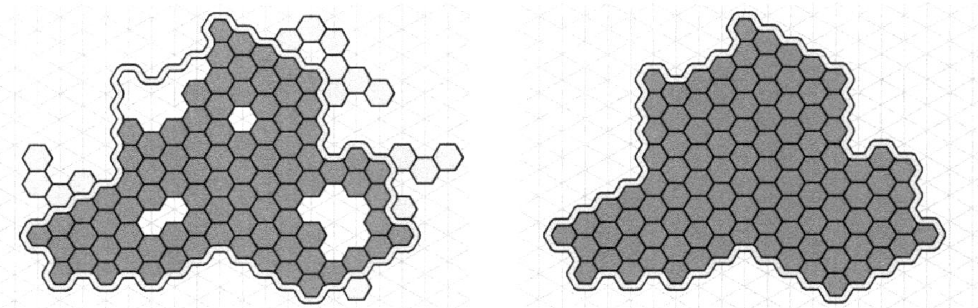

Fig. 2. An example instance of the SHAPE RECONFIGURATION PROBLEM. The light blue line encircles the target shape $\mathcal{T}$. The blue tiles are target tiles. The yellow tiles are supply tiles and need to be moved to untiled target nodes (demand nodes). Left: The positions of the tiles in the initial shape $\mathcal{I}$. Right: The final shape after all supply tiles have been moved to the target shape $\mathcal{T}$. In this example, the target shape $\mathcal{T}$ is simply connected while the initial shape $\mathcal{I}$ is not.

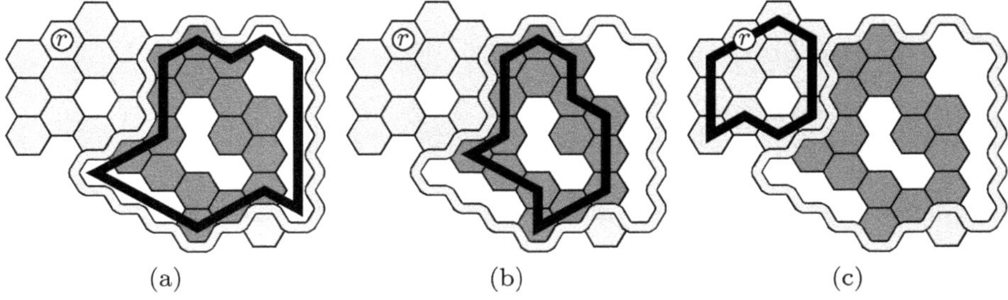

(a)   (b)   (c)

Fig. 3. Boundaries (a) $B(\mathcal{T})$, (b) $B(T \cap \mathcal{T})$, and (c) $B_p(T \setminus \mathcal{T})$. Agent r is on node p.

2.2 Boundaries and Boundary Traversal

Let $S \subseteq V$ be a finite subset of nodes (a *shape*). Consider the node sets $M_1, \ldots, M_k \subseteq V \setminus S$ of all connected components of $G[V \setminus S]$. If S is simply connected, then $k = 1$. We define the *boundary* of M_i as $B^S(M_i) := \bigcup_{v \in M_i} N(v) \cap S$ and refer to M_i as the *outside* of the boundary $B^S(M_i)$. Let M^* be the unique set of infinite size among the M_i. We refer to M^* as the *outside* of the shape S, to $B^S(M^*)$ as the *main boundary* of S and to $B^S(M_i)$ as an *inner boundary* for any $M_i \neq M^*$. For ease of presentation, we write $B(S) := B^S(M^*)$. Let $w \in S$ and let M_w^* be a set of maximum size (not necessarily infinite) among all M_i that contain a node adjacent to w in G. Then $B_w(S) := B^S(M_w^*)$ is called a w-*boundary* of S. Note that $B_w(S)$ is not uniquely defined in general. We refer to $B(T)$ as the *tile boundary*, to $B(\mathcal{T})$ as the *target boundary*, to $B(T \cap \mathcal{T})$ as the *target tile boundary*, and to $B_w(T \setminus \mathcal{T})$ as the *boundary of a supply component* for a supply node $w \in T \setminus \mathcal{T}$. See Fig. 3 for examples of boundaries.

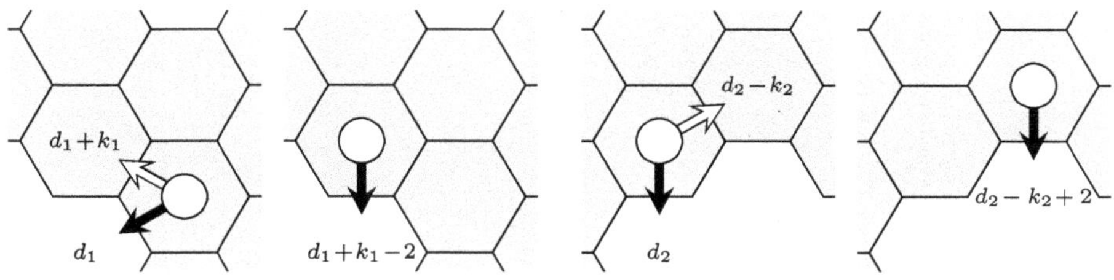

Fig. 4. An agent traversing the tile boundary $B(T)$. Left: LHR. Right: RHR.

The algorithms presented in this paper rely on the agent r's ability to traverse boundaries by the so-called *left-hand rule* (LHR): For some boundary $B \subseteq S$ of a shape S with outside M, r moves along nodes in B by always keeping the outside M to its left. Concretely, r encodes an "outside pointer" $d_{\text{out}} \in \mathcal{D}$ in its internal states that points to an outside node $w \in M$. To compute the next movement direction, r picks the smallest positive integer $k \in \mathbb{N}$ such that $p + \overrightarrow{d_{\text{out}} + k} \in S$, moves in direction $d_{\text{out}} + k$, and updates d_{out} to $d_{\text{out}} + k - 2$ to ensure that d_{out} continues to point to an outside node $\tilde{w} \in M$, which is not necessarily equal to w. The *right-hand rule* (RHR) works analogously; see Fig. 4.

For simplicity, we do not explicitly mention the outside pointer $d_{\text{out}} \in \mathcal{D}$ in our algorithm descriptions and instead just write that the agent traverses a boundary by the LHR (or RHR). We give more details in proofs when necessary.

3   Simply Connected Target Shapes

We first present a worst-case optimal algorithm to solve the SHAPE RECONFIGURATION PROBLEM for simply connected target shapes. Note that the initial shape $\mathcal{I}$ may contain holes. For simplicity, we assume that r is initially located at the target tile boundary, i.e., $p^0 \in B(T \cap \mathcal{T})$, and is equipped with a pointer $d_{\text{out}} \in \mathcal{D}$ to the outside of $B(T \cap \mathcal{T})$, i.e., $p^0 + \overrightarrow{d_{\text{out}}} \notin T \cap \mathcal{T}$. We prove the following lemma in the full version [18] to show that this assumption is not required.

Lemma 1. *The agent can find the target tile boundary in $\mathcal{O}(mn)$ time steps on instances of the* SHAPE RECONFIGURATION PROBLEM *with simply connected target shapes, maintaining connectivity of the target tile shape.*

On a high level, the algorithm works as follows: The agent first traverses the target tile boundary $B(T \cap \mathcal{T})$ by the LHR until it finds a connected component of supply tiles (a *supply component*), see Fig. 5. This happens because the target shape $\mathcal{T}$ is simply connected, which ensures that every supply component is adjacent to $B(T \cap \mathcal{T})$. The agent moves to the component and traverses it until it finds a safely removable supply tile, i.e., a supply tile that can be lifted and carried away without violating connectivity. A deterministic finite automaton cannot always find safely removable tiles on tile shapes with holes whereas finding tiles that can safely be moved to adjacent nodes is possible [20]. Therefore,

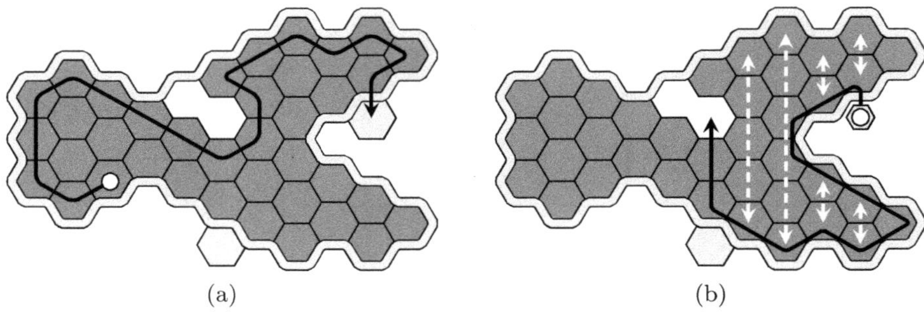

Fig. 5. (a): The agent traverses the target tile boundary $B(T \cap \mathcal{T})$ by the LHR until it finds a supply tile. (b): After lifting a supply tile, the agent traverses the target boundary $B(\mathcal{T})$ and the target shape's columns in phases TRAVERSEBOUNDARY and TRAVERSECOLUMN until it reaches a demand node. Here, $d_{\mathrm{col}} = \mathrm{N}$.

$$\text{FINDSUPPLY} \xleftrightarrow{} \text{COMPACTSUPPLY} \longrightarrow \text{TRAVERSECOLUMN} \longleftrightarrow \text{TRAVERSEBOUNDARY}$$

Fig. 6. Phase transitions of the algorithm for simply connected target shapes.

we require the agent to reconfigure the supply component while looking for removable tiles. In particular, it compacts the supply component by moving supply tiles away from the outside of the component's boundary $B_p(T \setminus \mathcal{T})$ whenever possible. This way, it "creates" safely removable tiles which can always be lifted without disconnecting the supply component, see Fig. 7. A supply node $v \in T \setminus \mathcal{T}$ is called *free* if the set of supply nodes adjacent to v is connected. A tile on a free supply node is called a *free supply tile*.

After the agent lifts such a tile, it returns to the target shape and traverses it until it reaches a demand node where it can place its carried tile. Traversing the boundary of the target shape is not sufficient as some components of demand nodes may be fully enclosed by tiled target nodes. To find them, the agent traverses all *columns* of the target shape, which are defined as follows: For a direction d and a shape $S \subseteq V$, a d-column is a path of nodes $(v_1, \ldots, v_k)$ with $v_i \in S$ such that $v_{i+1} = v_i + \vec{d}$ for every $1 \leq i < k$ and $v_1 + \vec{d+3}, v_k + \vec{d} \notin S$. The node v_1 is called the *start* and the node v_k is called the *end* of the d-column. Since $\mathcal{T}$ is simply connected, all column start nodes lie on the target boundary $B(\mathcal{T})$. Thus, the agent can simply traverse $B(\mathcal{T})$ by the LHR and traverse a column whenever it enters the column's start node until it eventually finds a demand node where it can place its carried tile, see Fig. 5. After a demand node is found, the agent repeats the above steps to move the next supply tiles to demand nodes.

We now give a more detailed description of the algorithm, divided into four phases. The agent starts in phase FINDSUPPLY with $p^0 \in B(T \cap \mathcal{T})$ and is equipped with a pointer to the boundary's outside. See Fig. 6 for an illustration of the algorithm's phase transitions.

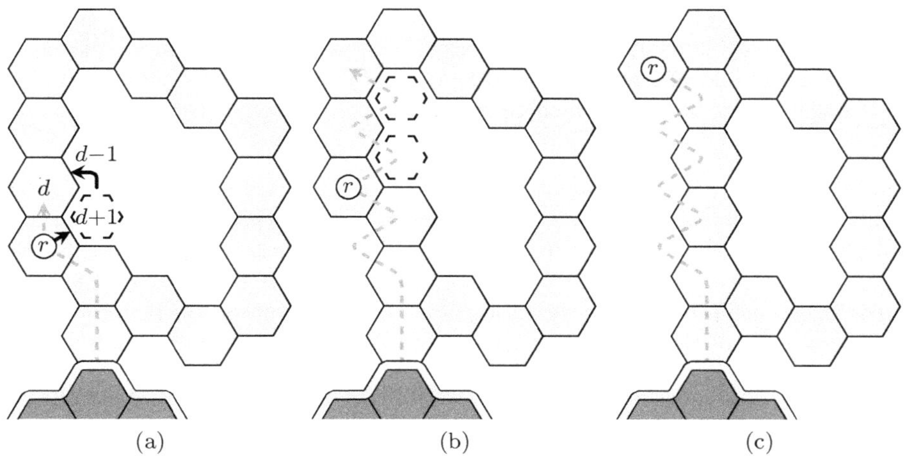

Fig. 7. Phase COMPACTSUPPLY. The agent r traverses $B_p(T \setminus \mathcal{T})$ (gray dashed line) until it enters a supply tile in (a) that can be moved inward (node v_i in the proof sketch of Lemma 3). Direction $d = $ N is the next LHR movement direction, but the highlighted node in direction $d + 1 = $ NE is untiled, so r moves its current tile there and then continues its traversal by moving in direction $d - 1 = $ NW. In (b), the agent continues the same process for the following two tiles. Finally, r ends on a free supply node in (c) (node v_j in the proof sketch of Lemma 3).

- FINDSUPPLY: The agent traverses the target tile boundary $B(T \cap \mathcal{T})$ by the LHR until it is adjacent to a supply node $v \in T \setminus \mathcal{T}$. It moves to v and enters phase COMPACTSUPPLY.

- COMPACTSUPPLY: The agent r moves one step along the boundary $B_p(T \setminus \mathcal{T})$ of the supply component by the LHR and computes the direction d of the next node of the boundary traversal. If the node in direction $d + 1$ is neither tiled nor a target node, the node in direction $d + 2$ holds a supply tile, and none of the nodes in directions $d - 1, \ldots, d - 3$ hold a supply tile, r lifts the tile at its current position, moves in direction $d + 1$, places its carried tile, and moves in direction $d - 1$. Otherwise, r just moves in direction d. As soon as r enters a free supply node, it lifts the free supply tile, moves to an adjacent supply node (if one exists), and traverses the boundary of the supply component $B_p(T \setminus \mathcal{T})$ by the LHR until it is adjacent to a tiled target node $v \in T \cap \mathcal{T}$. Then, it stores the direction to v as d_{col}, moves to v and enters phase TRAVERSECOLUMN.

- TRAVERSECOLUMN: The agent r moves in direction d_{col} until it either enters an untiled node, in which case r places the tile it is holding, or the node in direction d_{col} is not a target node. Then, r moves in direction $d_{\mathrm{col}} + 3$, i.e., the opposite direction of d_{col}, until there is no tiled target node in that direction. If r carries a tile, it enters phase TRAVERSEBOUNDARY. Otherwise, it enters phase FINDSUPPLY.

- TRAVERSEBOUNDARY: The agent r traverses the target boundary $B(\mathcal{T})$ (at least one step) until it either enters an untiled node, in which case r places the carried tile and enters phase FINDSUPPLY, or the node in direction $d_{\text{col}} + 3$ is not a target node, in which case r enters phase TRAVERSECOLUMN.

Analysis. We consider an agent r executing the algorithm given above on an arbitrary instance of the SHAPE RECONFIGURATION PROBLEM with a connected initial shape $\mathcal{I}$ and a simply connected target shape $\mathcal{T}$, assuming that the number of supply tiles is $m > 0$; otherwise, the problem is already solved. We observe r's behavior in each phase in a series of lemmas before combining the results to show the correctness of the algorithm. See the full version for deferred proofs [18].

Lemma 2. *If the agent is in phase FINDSUPPLY and the number of supply tiles is $m > 0$, it switches to phase COMPACTSUPPLY within $\mathcal{O}(n)$ time steps.*

Proof. The agent is initially in phase FINDSUPPLY and it only re-enters the phase from phases TRAVERSECOLUMN or TRAVERSEBOUNDARY. In all three cases, r is at the target tile boundary $B(T \cap \mathcal{T})$. The first case is by assumption. For the other two cases, see the proof sketch of Lemma 4. As all supply components are adjacent to $B(T \cap \mathcal{T})$, r finds a supply node during its LHR traversal. $\square$

Lemma 3. *If the agent is in phase COMPACTSUPPLY, it switches to phase TRAVERSECOLUMN carrying a tile within $\mathcal{O}(n)$ time steps.*

Proof (Sketch). It suffices to show that r encounters a safely removable supply tile during phase COMPACTSUPPLY. If the supply component is simply connected, this happens within an LHR traversal of $B_p(T \setminus \mathcal{T})$. An LHR traversal is possible since r begins this phase adjacent to $\mathcal{T}$, which is on the outside of $B_p(T \setminus \mathcal{T})$, allowing r to set d_{out} accordingly. We thus focus on the case where the component has holes and no safely removable supply tile is encountered during the traversal.

Let $P = (v_1, \ldots, v_k)$ with $v_1 = p$ and $k = \mathcal{O}(n)$ be the path of a full LHR traversal around $B_p(T \setminus \mathcal{T})$, i.e., the traversal repeats after v_k. We can show that r encounters a safely removable tile while traversing P and moving tiles as specified above. To do so, let $\alpha_i \in (-180, 180]$ be the degree by which r needs to turn to the right to move from v_i to v_{i+1}. Note that the tile at v_i is safely removable if $|\alpha_i| > 60$.

Assume $|\alpha_i| \leq 60$ for all i. Since P is a clockwise circular path, $\sum_i \alpha_i = 360$. Thus, there must be a pair (i, j) with $\alpha_i = \alpha_j = 60$ and $\alpha_{i'} = 0$ for $i < i' < j$. This is the case on the path between r's positions in Figs. 7a and 7c. One can now show that the tiles at all nodes $v_{i'}$ for $i \leq i' < j$ are moved in direction $d + 1$. After the tile at node v_{j-1} is moved, it is not hard to show that lifting the tile at v_j does not disconnect the configuration, see Fig. 7c, and that r can return to $B(T \cap \mathcal{T})$ by continuing to apply the LHR on $B_p(T \setminus \mathcal{T})$. $\square$

Lemma 4. *If the agent is in phase TRAVERSECOLUMN or TRAVERSE-BOUNDARY, it switches to phase FINDSUPPLY within $\mathcal{O}(n)$ time steps.*

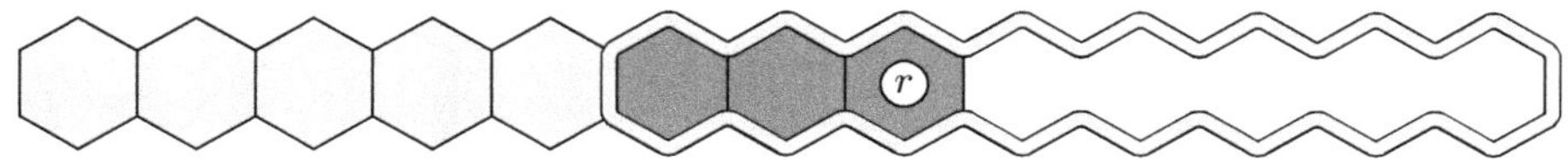

Fig. 8. The agent r requires $\Omega(mn)$ steps to move all supply tiles to demand nodes.

Proof (Sketch). By traversing all columns of $\mathcal{T}$ within $\mathcal{O}(n)$ time steps, r eventually finds a demand node $w \in \mathcal{T} \setminus T$ where it can place its carried tile. Then, it returns to the start of its current column and switches to phase FINDSUPPLY. $\square$

The correctness and runtime of the algorithm follow.

Theorem 1. *The agent can solve an instance of the SHAPE RECONFIGURATION PROBLEM with a simply connected target shape in $\mathcal{O}(mn)$ time steps.*

Existing shape formation algorithms typically have runtimes of $\mathcal{O}(n^2)$ or $\mathcal{O}(nD)$ where D is the diameter of the tile shape [20, 21]. Thus, due to the agent's ability to distinguish target from non-target nodes, our algorithm is faster than existing algorithms for $m = o(n)$, i.e., if the initial and target shape largely overlap. Furthermore, the algorithm is worst-case optimal: Translating a line of n tiles by m positions requires $\Omega(mn)$ steps, see Fig. 8.

Theorem 2. *The agent requires $\Omega(mn)$ time steps to solve the SHAPE RECONFIGURATION PROBLEM in general.*

4   Target Shapes with Holes

If the target shape $\mathcal{T}$ has holes, our previous algorithm fails: In such a shape, some columns' start nodes are on an inner boundary $B \neq B(\mathcal{T})$, so they may not be found by an agent executing the previous traversal phases. This failure is not surprising: In a related setting, a finite automaton cannot visit all cells in a grid maze with holes when it is equipped with fewer than two pebbles that can be used to mark cells [3, 23]. We conjecture that this impossibility extends to the SHAPE RECONFIGURATION PROBLEM for arbitrary target shapes with holes.

We first sketch a reconfiguration algorithm for an agent that can place pebbles on up to two tiles at a time. Then, we argue how an agent without this ability can emulate pebbles on sufficiently scaled target shapes and present a reconfiguration algorithm using this approach. Due to limited space, we only give high-level ideas here and refer to the full version for missing details [18, Section 4].

4.1   Exploration and Reconfiguration with Pebbles

A finite automaton equipped with a single counter can explore arbitrary orthogonal mazes of size n in $\mathcal{O}(n^2)$ steps as shown by Blum and Kozen [2, Theorem 3.1].

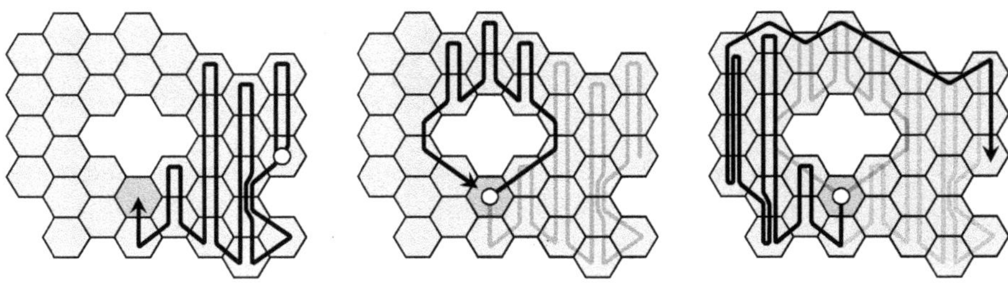

Fig. 9. The traversal path of an agent executing *maze exploration* (without the unique point detection subroutine) on a tile shape with a hole. The unique point of the inner boundary is highlighted in green. Left: The agent traverses the outer boundary and its columns until it enters the unique point of an inner boundary. Middle: It traverses the inner boundary and its columns until it re-enters the boundary's unique point. Right: It continues its traversal of the outer boundary's columns, eventually visiting all tiles.

If the agent is instead equipped with two placeable markers whose distance in the maze emulates the counter, the execution time is $\mathcal{O}(n^3)$. This *maze exploration* algorithm can be adapted for the hybrid model where an agent equipped with a counter—or two pebbles that can be placed on tiles—can visit every tile in a connected configuration [30]. Here, similarly to the exploration phases presented in Sect. 3, the agent traverses boundaries by the LHR and columns whose start nodes lie on the boundaries. To ensure that all columns are traversed, the agent must traverse all boundaries. The agent begins traversing an inner boundary when it reaches the westernmost node of the boundary's southernmost nodes (its *unique point*) and it stops the traversal once this unique point is reached again, see Fig. 9. The counter is used to detect unique points.

By making use of *maze exploration* to find supply components and demand nodes, and phase COMPACTSUPPLY from Sect. 3 to lift supply tiles, we get:

Theorem 3. *An agent equipped with a counter or two pebbles can solve an instance of the SHAPE RECONFIGURATION PROBLEM and terminate in $\mathcal{O}(mn^2)$ or $\mathcal{O}(mn^3)$ time steps, respectively.*

To the best of our knowledge, no published maze exploration algorithm for finite automata achieves execution times below $\mathcal{O}(n^2)$ (with a counter) or $\mathcal{O}(n^3)$ (with pebbles) on mazes with holes. Therefore, we do not expect faster shape reconfiguration algorithms in the hybrid model without new traversal strategies.

4.2   Reconfiguring Scaled Shapes with Holes

The main building block in this section is a pebble emulation subroutine that allows r to execute *maze exploration* without pebbles or counters. To emulate pebbles, we use r's ability to distinguish target from non-target nodes. On sections of $\mathcal{T}$ that are completely tiled, r can emulate placing a pebble by displacing a tile and creating a demand node, see Fig. 10.

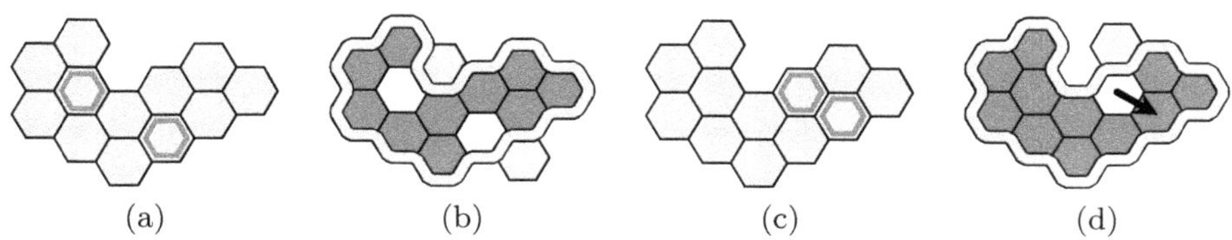

(a)   (b)   (c)   (d)

Fig. 10. A tile shape with two pebbles visualized as gray inner hexagons. The pebbles are emulated by moving tiles at the pebble positions to the outside of the target shape. If two pebbles are adjacent, the position of the second pebble is encoded relative to the first pebble's position as visualized by the black arrow. (Color figure online)

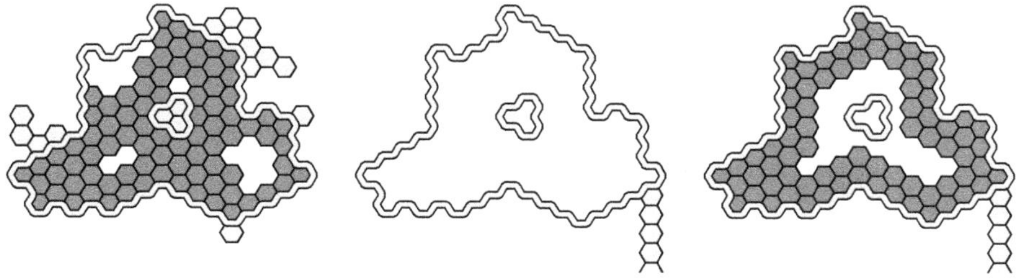

Fig. 11. Left: An instance with a bottleneck-free target shape. Middle: The tile depot is built. Right: The main target boundary $B(\mathcal{T})$ is filled, enabling pebble emulation.

To be able to displace tiles without risking disconnection, the shape needs to be "thick enough". More formally, we require $\mathcal{T}$ to contain no *bottlenecks*, which are defined as follows: A node $v \in \mathcal{T}$ is a *bottleneck* if $N(v) \setminus \mathcal{T}$ is not connected. Upscaling arbitrary shapes by a small constant always results in bottleneck-free shapes. Such scaling assumptions are common in the literature, e.g., [16,31].

Now, to reconfigure bottleneck-free shapes, r first creates a tile depot on the outside of $\mathcal{T}$ using an existing algorithm [20] and fills $B(\mathcal{T})$ with tiles, allowing pebble emulation along $B(\mathcal{T})$. This enables r to execute *maze exploration* to repeatedly move tiles from the depot to demand nodes. Whenever it encounters an untiled boundary, the boundary is filled with tiles from the depot before exploration continues. Since *maze exploration*, which takes $\mathcal{O}(n^3)$ steps, is executed twice per tile (once to find a demand node and once to return to the depot), r terminates within $\mathcal{O}(n^4)$ time steps. See Fig. 11 for a high-level overview.

Theorem 4. *The agent can solve an instance of the* SHAPE RECONFIGURATION PROBLEM *with a bottleneck-free target shape and terminate in* $\mathcal{O}(n^4)$ *time steps.*

5   Discussion and Future Work

We have shown that a single agent can solve the SHAPE RECONFIGURATION PROBLEM for simply connected target shapes in worst-case optimal $\mathcal{O}(mn)$

steps (Theorems 1 and 2) and for bottleneck-free shapes in $\mathcal{O}(n^4)$ steps (Theorem 4). Additionally, an agent can solve the problem for arbitrary target shapes without constructing an intermediate tile depot in $\mathcal{O}(mn^2)$ steps if equipped with a counter or $\mathcal{O}(mn^3)$ steps if equipped with two pebbles (Theorems 3). It remains an open question whether shape reconfiguration is possible on arbitrary shapes without the use of pebbles or counters. We believe this is not the case, just like visiting every cell in a grid maze is impossible for a deterministic finite automaton as shown by Budach [3]. However, proving this conjecture may not be straightforward as the agent in our setting is more powerful than the automaton in Budach's proof due to its ability to reconfigure its environment.

An agent executing the first algorithm for simply connected target shapes does not terminate once the problem is solved. We can easily adjust the algorithm to enable termination in $\mathcal{O}(n^2)$ steps by making use of the tile depot formation algorithm presented in the full version [18, Section 4.2], but it would be interesting to see whether faster termination is possible.

Finally, a natural follow-up is to examine the Shape Reconfiguration Problem for multiple cooperating agents. As a first step omitted from this paper, we developed and implemented an algorithm for simply connected initial and target shapes and agents with a shared sense of direction. However, for more general settings, the ideas presented here cannot easily be adapted since agents lifting tiles at different positions can unknowingly disconnect the configuration.

Acknowledgments. This work was supported by the Deutsche Forschungsgemeinschaft (German Research Foundation, DFG) – SCHE 1931/4-1; SCHE 1592/10-1. We thank the anonymous reviewers for their valuable comments and suggestions, which helped improve the presentation of our results.

References

1. Becker, A.T., et al.: Targeted drug delivery: Algorithmic methods for collecting a swarm of particles with uniform, external forces. In: ICRA, pp. 2508–2514 (2020). https://doi.org/10.1109/ICRA40945.2020.9196551
2. Blum, M., Kozen, D.: On the power of the compass (or, why mazes are easier to search than graphs). In: SFCS, pp. 132–142 (1978). https://doi.org/10.1109/SFCS.1978.30
3. Budach, L.: Automata and labyrinths. Math. Nachr. **86**(1), 195–282 (1978). https://doi.org/10.1002/mana.19780860120
4. Chen, H.-L., Doty, D., Holden, D., Thachuk, C., Woods, D., Yang, C.-T.: Fast algorithmic self-assembly of simple shapes using random agitation. In: Murata, S., Kobayashi, S. (eds.) DNA 2014. LNCS, vol. 8727, pp. 20–36. Springer, Cham (2014). https://doi.org/10.1007/978-3-319-11295-4_2
5. D'Angelo, G., D'Emidio, M., Das, S., Navarra, A., Prencipe, G.: Asynchronous silent programmable matter achieves leader election and compaction. IEEE Access **8**, 207619–207634 (2020). https://doi.org/10.1109/ACCESS.2020.3038174
6. Daymude, J.J., Hinnenthal, K., Richa, A.W., Scheideler, C.: Computing by programmable particles. Distrib. Comput. Mob. Entities: Curr. Res. Moving Comput., 615–681 (2019). https://doi.org/10.1007/978-3-030-11072-7_22

7. Daymude, J.J., Richa, A.W., Scheideler, C.: The canonical Amoebot model: algorithms and concurrency control. Distrib. Comput. **36**(2), 159–192 (2023). https://doi.org/10.1007/s00446-023-00443-3

8. Derakhshandeh, Z., Dolev, S., Gmyr, R., Richa, A.W., Scheideler, C., Strothmann, T.: Amoebot–a new model for programmable matter. In: SPAA, pp. 220–222 (2014). https://doi.org/10.1145/2612669.2612712

9. Derakhshandeh, Z., Gmyr, R., Richa, A.W., Scheideler, C., Strothmann, T.: Universal shape formation for programmable matter. In: SPAA, pp. 289–299 (2016). https://doi.org/10.1145/2935764.2935784

10. Derakhshandeh, Z., Gmyr, R., Strothmann, T., Bazzi, R., Richa, A.W., Scheideler, C.: Leader election and shape formation with self-organizing programmable matter. In: DNA, pp. 117–132 (2015). https://doi.org/10.1007/978-3-319-21999-8_8

11. Di Luna, G.A., Flocchini, P., Prencipe, G., Santoro, N., Viglietta, G.: Line recovery by programmable particles. In: ICDCN, pp. 1–10 (2018). https://doi.org/10.1145/3154273.3154309

12. Di Luna, G.A., Flocchini, P., Santoro, N., Viglietta, G., Yamauchi, Y.: Shape formation by programmable particles. Distrib. Comput. **33**(1), 69–101 (2019). https://doi.org/10.1007/s00446-019-00350-6

13. Doty, D.: Theory of algorithmic self-assembly. CACM **55**(12), 78–88 (2012). https://doi.org/10.1145/2380656.2380675

14. Dumitrescu, A., Suzuki, I., Yamashita, M.: Motion planning for metamorphic systems: feasibility, decidability, and distributed reconfiguration. IEEE Trans. Robot. **20**(3), 409–418 (2004). https://doi.org/10.1109/TRA.2004.824936

15. Fekete, S.P., Gmyr, R., Hugo, S., Keldenich, P., Scheffer, C., Schmidt, A.: CADbots: algorithmic aspects of manipulating programmable matter with finite automata. Algorithmica **83**(1), 387–412 (2021). https://doi.org/10.1007/S00453-020-00761-Z

16. Fekete, S.P., Keldenich, P., Kosfeld, R., Rieck, C., Scheffer, C.: Connected coordinated motion planning with bounded stretch. Auton. Agents Multi-Agent Syst. **37**(2), 43 (2023). https://doi.org/10.1007/S10458-023-09626-5

17. Fekete, S.P., Niehs, E., Scheffer, C., Schmidt, A.: Connected reconfiguration of lattice-based cellular structures by finite-memory robots. Algorithmica **84**(10), 2954–2986 (2022). https://doi.org/10.1007/S00453-022-00995-Z

18. Friemel, J., Liedtke, D., Scheffer, C.: Tile reconfiguration by a finite automaton. arXiv:2501.08663 (2025)

19. Gmyr, R., Hinnenthal, K., Kostitsyna, I., Kuhn, F., Rudolph, D., Scheideler, C.: Shape recognition by a finite automaton robot. In: MFCS, pp. 1–15 (2018). https://doi.org/10.4230/LIPIcs.MFCS.2018.52

20. Gmyr, R., et al.: Forming tile shapes with simple robots. Nat. Comput. **19**(2), 375–390 (2019). https://doi.org/10.1007/s11047-019-09774-2

21. Hinnenthal, K., Liedtke, D., Scheideler, C.: Efficient shape formation by 3D hybrid programmable matter: an algorithm for low diameter intermediate structures. Theor. Comput. Sci. **1057**, 115552 (2025). https://doi.org/10.1016/j.tcs.2025.115552

22. Hinnenthal, K., Rudolph, D., Scheideler, C.: Shape formation in a three-dimensional model for hybrid programmable matter. In: EuroCG, pp. 1–9 (2020). https://www1.pub.informatik.uni-wuerzburg.de/eurocg2020/data/uploads/papers/eurocg20_paper_50.pdf

23. Hoffmann, F.: One pebble does not suffice to search plane labyrinths. In: Gécseg, F. (ed.) FCT 1981. LNCS, vol. 117, pp. 433–444. Springer, Heidelberg (1981). https://doi.org/10.1007/3-540-10854-8_47

24. Hurtado, F., Molina, E., Ramaswami, S., Sacristán, V.: Distributed reconfiguration of 2D lattice-based modular robotic systems. Auton. Robot. **38**(4), 383–413 (2015). https://doi.org/10.1007/s10514-015-9421-8
25. Jenett, B., Gregg, C., Cellucci, D., Cheung, K.: Design of multifunctional hierarchical space structures. In: Aerosp. Conf., pp. 1–10 (2017). https://doi.org/10.1109/AERO.2017.7943913
26. Kostitsyna, I., Liedtke, D., Scheideler, C.: Universal coating by 3D hybrid programmable matter. In: SIROCCO, pp. 384–401 (2024). https://doi.org/10.1007/978-3-031-60603-8_21
27. Kostitsyna, I., Liedtke, D., Scheideler, C.: Distributed rhombus formation of sliding squares. In: SSS, pp. 325–342 (2025). https://doi.org/10.1007/978-3-032-11127-2_26
28. Kostitsyna, I., Peters, T., Speckmann, B.: Fast reconfiguration for programmable matter. In: DISC, pp. 1–21 (2023). https://doi.org/10.4230/LIPIcs.DISC.2023.27
29. Kostitsyna, I., Scheideler, C., Warner, D.: Fault-tolerant shape formation in the Amoebot model. In: DNA, pp. 1–22 (2022). https://doi.org/10.4230/LIPICS.DNA.28.9
30. Liedtke, D.: Exploration and convex hull construction in the three-dimensional hybrid model. Master's thesis, Paderborn University, Germany (2021)
31. Luchsinger, A., Schweller, R., Wylie, T.: Self-assembly of shapes at constant scale using repulsive forces. Nat. Comput. **18**(1), 93–105 (2018). https://doi.org/10.1007/s11047-018-9707-9
32. Navarra, A., Piselli, F., Prencipe, G.: Line formation and scattering in silent programmable matter. J. Parallel Distrib. Comput. **204**, 105129 (2025). https://doi.org/10.1016/j.jpdc.2025.105129
33. Nokhanji, N., Flocchini, P., Santoro, N.: Dynamic line maintenance by hybrid programmable matter. Int. J. Networking Comput. 13(1), 18–47 (2023). https://doi.org/10.15803/ijnc.13.1_18
34. Nokhanji, N., Santoro, N.: Line reconfiguration by programmable particles maintaining connectivity. In: Martín-Vide, C., Vega-Rodríguez, M.A., Yang, M.-S. (eds.) TPNC 2020. LNCS, vol. 12494, pp. 157–169. Springer, Cham (2020). https://doi.org/10.1007/978-3-030-63000-3_13
35. Patitz, M.J.: An introduction to tile-based self-assembly and a survey of recent results. Nat. Comput. **13**(2), 195–224 (2013). https://doi.org/10.1007/s11047-013-9379-4
36. Toffoli, T., Margolus, N.: Programmable matter: concepts and realization. Physica D **47**(1), 263–272 (1991). https://doi.org/10.1016/0167-2789(91)90296-L
37. Winfree, E.: Algorithmic self-assembly of DNA. Ph.D. thesis, California Institute of Technology, Pasadena, CA, USA (1998). https://doi.org/10.7907/HBBV-PF79
38. Woods, D., Chen, H.L., Goodfriend, S., Dabby, N., Winfree, E., Yin, P.: Active self-assembly of algorithmic shapes and patterns in polylogarithmic time. In: ITCS, pp. 353–354 (2013). https://doi.org/10.1145/2422436.2422476

Enumeration Problems

Enumerating All Graph Colorings Using Zero-Suppressed Binary Decision Diagrams

Ryohei Okuda[ID], Jun Kawahara[ID], and Shin-ichi Minato[(✉)][ID]

Graduate School of Informatics, Kyoto University, Kyoto, Japan
`minato@i.kyoto-u.ac.jp`

Abstract. The graph coloring problem is a well-known combinatorial problem that seeks a coloring pattern such that no two adjacent vertices share the same color. In this paper, we propose an efficient method not only for finding one solution but also for efficiently enumerating all solutions that satisfy the graph coloring constraints. Our method uses ZDDs (Zero-suppressed Binary Decision Diagrams) for efficiently indexing sets of graph coloring patterns. In addition, we propose an algorithm to generate all coloring patterns that do not split any cell of a given set partition. Our experimental results demonstrate that the proposed method significantly outperforms existing solvers. Compared to the leading ASP (Answer Set Programming) solver, our approach generates all coloring patterns more than 1,000 times faster. Our method is applicable to many practical graph instances with up to around 25 vertices, in a feasible computation time.

Keywords: Graph coloring problem · Zero-suppressed Binary Decision Diagrams · Enumeration algorithm · Set partition

1 Introduction

The graph coloring problem is the problem of determining, given an undirected graph $G = (V, E)$ and the number of colors, c, whether it is possible to assign a color from the set $\{1, \ldots, c\}$ to every vertex in V such that no two adjacent vertices share the same color [6]. This is a well-known problem in discrete mathematics, and it is known that for $c \geq 3$, the problem of determining whether a general graph is colorable is NP-complete [7], and that any planar graph can be colored using at most four colors [5].

This problem has applications in areas such as the frequency assignment problem [11] and scheduling problems [1], where it can model and solve problems of partitioning potentially conflicting elements into different sets. Previous studies have proposed methods for finding a valid graph coloring using heuristics and genetic algorithms [2,3]. However, most methods presented in previous studies focus on finding a single coloring solution, and to the best of the authors' knowledge, no research has been conducted on efficiently enumerating and indexing

E. Di Giacomo and D. Mondal (Eds.): WALCOM 2026, LNCS 16444, pp. 529–544, 2026.
https://doi.org/10.1007/978-981-95-7127-7_35

all coloring solutions. By enumerating and indexing multiple coloring solutions and presenting them to the user, the user is enabled to select a desired solution. This is particularly useful when the user has requirements or preferences that are difficult to formally define.

When enumerating graph colorings, the number of possible solutions increases exponentially with the size of the graph, and a brute-force enumeration of all solutions would require exponential computation time. Therefore, this paper attempts to enumerate all possible graph colorings while reducing computation time by using a data structure called Zero-suppressed Binary Decision Diagrams (ZDDs) [9]. A ZDD is a directed acyclic graph data structure that can compactly represent a family of combinations and is efficient for performing set operations on large families of combinations such as union and intersection [10]. Takahashi et al. [12] proposed a method for representing a family of set partitions as a ZDD, which we call a *backward labelling ZDD*. In their method, a partition consists of pairs (x, i), where x is an element and i is the number of the cell to which the element belongs, and a backward labelling ZDD represents a family of sets of such pairs.

Graph coloring can be regarded as a type of set partition, where vertices assigned the same color belong to the same cell. This paper proposes a method to construct a backward labelling ZDD that represents all vertex partitions corresponding to all the valid graph colorings of an input graph. Once the ZDD is constructed, it is easy to enumerate all vertex partitions. Furthermore, it becomes possible to count the total number of partitions and to quickly perform set operations such as the union and intersection using ZDD operations [10]. In the proposed method, distinct coloring solutions that can be obtained by permuting the color labels are considered identical.

Furthermore, since it is practically difficult to compare and examine an exponential number of coloring solutions, an operation to filter for desired solutions is necessary. Therefore, this paper proposes a method, called *All-Coarser*, for constructing a backward labelling ZDD representing all partitions that do not separate the elements within any cell of a given partition P into different cells. By taking the intersection (using the ZDD operation intersection) of the ZDD obtained by All-Coarser and the ZDD representing the set of coloring solutions from our proposed method, we can narrow down the solutions to only those graph colorings that do not separate the elements within each cell of P. In other words, we can specify multiple vertices and impose the condition that they must have the same color, thereby extracting only the applicable coloring solutions. For example, in a frequency assignment application, if we provide a vertex partition P representing the current frequency assignments, we can obtain a ZDD representing all vertex partitions that do not separate base stations assigned to the same cell into different frequency bands. Since All-Coarser is a general operation applicable to any backward labelling ZDD, it has other potential applications, such as in electoral districting to narrow down partitions where desired regions are kept within the same district.

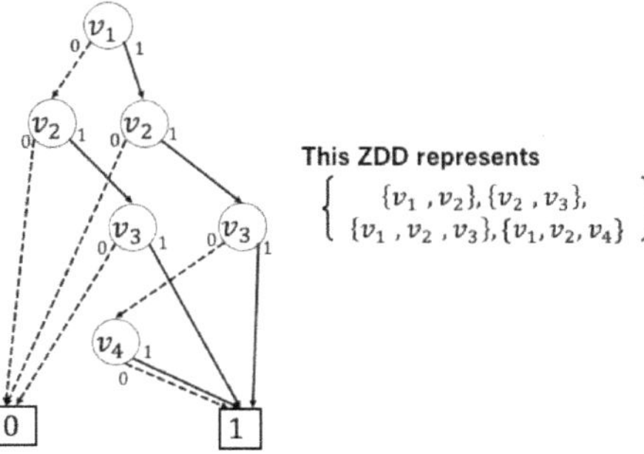

Fig. 1. Example of ZDD. The ZDD on the left in the figure represents the set of combinations on the right. For example, the path that traverses the 1-arc from the node for v_1, the 1-arc from v_2, the 0-arc from v_3, and the 1-arc from v_4 represents the combination $\{v_1, v_2, v_4\}$.

The organization of the paper is as follows. Section 2 defines ZDDs, set partitions, and backward labelling ZDDs. In Sect. 3, we propose a method for enumerating graph colorings using backward labelling ZDDs and the All-Coarser method. Section 4 evaluates the performance of these methods through computational experiments. Finally we conclude this paper in Sect. 5.

2 Preliminaries

2.1 Zero-Suppressed Binary Decision Diagrams

A Zero-suppressed Binary Decision Diagram (ZDD) [9] is a data structure that represents a set of combinations (or family of combinations) of elements from a base set, say $U = \{v_1, \ldots, v_n\}$, using a directed acyclic graph. A ZDD consists of *terminal nodes* whose out-degree is 0, called the *0-terminal* and *1-terminal*, *non-terminal nodes*, and directed arcs originating from non-terminal nodes. Each non-terminal node has two arcs, called *0-arc* and *1-arc*. A ZDD has a single root node, whose in-degree is 0. The root node is the only node from which all other nodes are reachable. Each non-terminal node has an element of the base set, which we call an *item*. Any path from the root node to the 1-terminal in a ZDD corresponds to a combination in the base set U in the following way: if the path contains a node with item v_i and its 1-arc, the corresponding combination includes item v_i. A ZDD represents a set of combinations through the set of all such paths. It is necessary to define a total order of U. The items of the non-terminal nodes on any path from the root to 1-terminal node must appear according to this total order. The same item is not allowed to appear more than once on a path, but some items may not appear on a path. Figure 1 shows an example of a ZDD.

Various set operations have been proposed for families of combinations represented by ZDDs, such as union and intersection (see Table 1) [10]. Taking the

Table 1. Basic operations of ZDDs. "Constructs the ZDD representing" is added to the beginning of each definition. Note that symbols $\mathcal{F}$ and $\mathcal{G}$ represent families and their corresponding ZDDs, and x is an item of the base set.

Operation	Definition
$\emptyset$	The empty family
$\{\emptyset\}$	The family consisting of the empty combination
$\mathcal{F}.\mathrm{onset}(x)$	The family consisting of the combinations in $\mathcal{F}$ including x.
$\mathcal{F}.\mathrm{offset}(x)$	The family consisting of the combinations in $\mathcal{F}$ not including x
$\mathcal{F} \cup \mathcal{G}$	The union of $\mathcal{F}$ and $\mathcal{G}$
$\mathcal{F} \cap \mathcal{G}$	The intersection of $\mathcal{F}$ and $\mathcal{G}$

union of two ZDDs, say $\mathcal{F}$ and $\mathcal{G}$, means generating a new ZDD that represents the set of combinations contained in at least one of the families represented by $\mathcal{F}$ and $\mathcal{G}$. These set operations are defined as manipulations of the ZDD structure itself and can be executed without explicitly enumerating the individual combinations.

In this paper, we also define an operation that adds an element $x \in U$ to all the combinations in the family represented by $\mathcal{F}$, which we denote by $\mathcal{F}.\mathrm{add\_item}(x)$. The add_item operation can only be performed if the set of items used to generate the ZDD $\mathcal{F}$ does not contain x or any item that precedes x in the total order of U. When executed, it creates a new root node with the label x, connects its 0-arc to the 0-terminal, and its 1-arc to the root node of $\mathcal{F}$, and returns the resulting ZDD. This can be executed in $O(1)$ time.

Henceforth, ZDDs will be denoted by calligraphic letters such as $\mathcal{F}$ and $\mathcal{G}$. Furthermore, in pseudocode, a set of combinations and its corresponding ZDD will be treated as interchangeable. For example, $\mathcal{F} \leftarrow \{\emptyset\}$ means that $\mathcal{F}$ is the ZDD representing the set containing only the empty set.

2.2 Set Partitions

A set partition is a way of dividing the elements of a set into multiple cells. For a set V, a set of its non-empty subsets $\{C_1, \ldots, C_k\}$ is called a set partition of V if any two distinct subsets C_i and C_j have no elements in common (i.e., $C_i \cap C_j = \emptyset$ for $i \neq j$) and the union of all subsets covers the entire set V (i.e., $\bigcup_{i=1}^{k} C_i = V$). We call C_i a *cell*. The number of cells, k, is called the partition number. For example, for the set $V = \{a, b, c, d\}$, the collection of subsets $\{C_1, C_2, C_3\} = \{\{a\}, \{b, d\}, \{c\}\}$, where $C_1 = \{a\}, C_2 = \{b, d\}$, and $C_3 = \{c\}$, is a set partition of V. For simplicity in this paper, such a set partition is represented as $(a|bd|c)$ by arranging the elements in the order of their respective cell numbers. Following this notation, we can obtain partitions such as $(a|bd|c), (a|c|bd)$, and $(bd|c|a)$. However, these partitions are essentially the same, merely differing by a reordering of the cells. Therefore, this definition alone cannot uniquely represent what is fundamentally the same partition. To ensure a

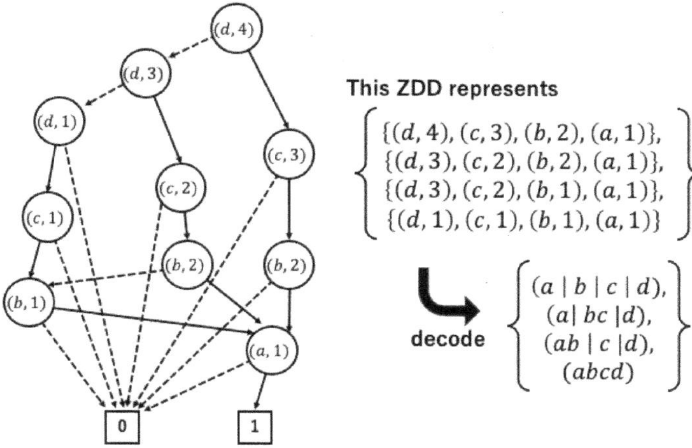

Fig. 2. An example of backward labelling ZDD.

unique representation, we introduce the following constraints. We define a total order on a set V to be partitioned and sort the elements within each cell C_i according to this order. A representation of a set partition is called a canonical form if, after this sorting, the first elements of each cell are arranged in ascending order from C_1 onwards. For instance, consider the set $V = \{a, b, c, d\}$ with the total order defined by the sequence $a < b < c < d$. $(a|bd|c)$ is a canonical form of set partition because the first elements of each cell—a, b, and c—are in ascending order according to the order of the set. In contrast, $(a|c|bd)$ and $(bd|c|a)$ are not in canonical form.

2.3   Backward Labelling ZDDs

Takahashi et al. [12] proposed a method for representing sets of set partitions using ZDDs (although their work was published as an unrefereed Japanese paper). In this method, for a finite set U, the assignment of an element $x \in U$ into a cell C_i is denoted by a pair (x, i), which we call an *assignment pair*. A partition is represented by the combination of these assignment pairs for all $x \in U$. For example, for a set $U = \{a, b, c, d\}$, the set partition $(a|bd|c)$ is represented by the combination of assignment pairs $\{(a, 1), (b, 2), (c, 3), (d, 2)\}$. A backward labelling ZDD is a ZDD that represents a family of partitions as a family of combinations of items, where each item is an assignment pair and the base set of the ZDD is $\{(x, i) \mid x \in U, i \in \{1, 2, \ldots, |U|\}\}$. For $U = \{x_1, \ldots, x_n\}$, we determine that the order of ZDD items is the reverse lexicographical order $(x_n, n), (x_n, n-1), \ldots, (x_n, 1), (x_{n-1}, n), \ldots, (x_1, 1)$. Figure 2 shows an example of a backward labelling ZDD. In the original paper, if none of $(x, 1), \ldots, (x, n)$ are included in a combination, it is interpreted that x belongs to cell 1, and $(x, 1)$ is omitted. However, in this paper, we do not omit it.

Takahashi et al. also proposed a method for constructing a backward labelling ZDD that represents all partitions on a given set. This algorithm constructs a ZDD that enumerates all canonical forms of set partitions by modeling the sequential assignment of elements to cells as the addition of items on the ZDD. Let us consider assigning the fourth element, d, to some cell for the partition $\{(a,1),(b,2),(c,2)\} = (a|bc)$. (Although we should not call $(a|bc)$ a partition of $\{a,b,c,d\}$, we abuse the term.) In this case, the possible assignments that maintain the canonical form are $(d,1), (d,2)$, and $(d,3)$. It is only possible to either add d to an existing cell or to create a new, third cell and place d in it. Generalizing this, we can see that to assign a new element to a partition with partition number k while preserving the canonical form, the cell number for the new element must be in the range $1, \ldots, k+1$.

Let $\mathcal{H}_j^i$ be the family consisting of the encodings of all partitions of $\{x_1, \ldots, x_i\}$ with a partition number of exactly j. Then, the following recurrence relations hold:

$$\mathcal{H}_1^1 = \{\{(x_1,1)\}\}, \quad \mathcal{H}_j^1 = \emptyset \ (j \geq 2), \tag{1}$$

$$\mathcal{H}_j^i = \mathcal{H}_{j-1}^{i-1}.\text{add\_item}((x_i,j)) \cup \left(\bigcup_{l=1,\ldots,j} \mathcal{H}_j^{i-1}.\text{add\_item}((x_i,l)) \right) \tag{2}$$

Equation (1) represents the base case. The only set partition for the set $\{x_1\}$ is $\{(x_1,1)\}$, which has a partition number of 1. Therefore, we set $\mathcal{H}_1^1 = \{\{(x_1,1)\}\}$. Equation (2) represents the operation to obtain $\mathcal{H}_j^i$ by adding an assignment pair for x_i to the partitions of $\{x_1, \ldots, x_{i-1}\}$. The term $\mathcal{H}_{j-1}^{i-1}.\text{add\_item}((x_i,j))$ adds (x_i,j) to all combinations in $\mathcal{H}_{j-1}^{i-1}$, which indicates that we create a new cell, j, and place x_i in it for all partitions that had a partition number of $j-1$. The term $\mathcal{H}_j^{i-1}.\text{add\_item}((x_i,l))$ corresponds to placing x_i into cell l for partitions that already have a partition number of j. Since x_i can be placed in any cell from 1 to j, we take the union of the resulting sets of partitions for all possible values of $l = 1, \ldots, j$.

According to the recurrence relations above, by calculating $\mathcal{H}_j^i$ in order for $i = 1, \ldots, |U|$, we can generate all partitions for the set U categorized by their partition number, resulting in $\mathcal{H}_1^{|U|}, \ldots, \mathcal{H}_{|U|}^{|U|}$. The set operations add_item and $\cup$ used in these relations can be executed as ZDD operations. Among them, add_item is a lightweight operation that can be performed in $O(1)$ time as described in Sect. 2.1.

3 Proposed Method

3.1 Enumerating All Graph Colorings

In this section, we propose a method to enumerate families of vertex partitions representing graph colorings using backward labelling ZDDs. The proposed algorithm takes an undirected graph $G = (V, E)$ as input and enumerates, in the form of a backward labelling ZDD, the family of set partitions of V such that no two adjacent vertices are assigned to the same cell. Since a backward labelling ZDD represents a family of partitions, a graph coloring is expressed not as an assignment to specific colors, but as a partition of vertices into multiple cells of the same color. Furthermore, following the definition of a backward labelling ZDD, the item order is the reverse of lexicographical order, i.e., $(v_{|V|}, c), \ldots, (v_{|V|}, 1), (v_{|V|-1}, c), \ldots, (v_1, 1)$, where c is the upper limit on the number of colors provided by the user.

Let $\mathcal{H}_j^i$ represent the set of all vertex partitions of $\{v_1, \ldots, v_i\}$ that have a partition number of exactly j and do not violate the coloring constraint. Then, the following recurrence relations hold:

$$\mathcal{H}_1^1 = \{\{(v_1, 1)\}\}, \quad \mathcal{H}_j^1 = \emptyset \ (j \geq 2), \tag{3}$$

$$\mathcal{H}_j^i = \mathcal{H}_{j-1}^{i-1}.\text{add\_item}((v_i, j)) \cup \left(\bigcup_{l=1,\ldots,j} \mathcal{R}(\mathcal{H}_j^{i-1}.\text{add\_item}((v_i, l)), v_i, l) \right) \tag{4}$$

$$\mathcal{R}(\mathcal{I}, v_i, l) = \mathcal{I} \cap \left(\bigcap_{v_m : (v_m, v_i) \in E} \mathcal{I}.\text{offset}((v_m, l)) \right) \tag{5}$$

Regarding (3) and (4), the initial value and update rule for $\mathcal{H}_j^i$ are largely consistent with the previous section, but the rightmost term on the right-hand side of (4) differs. This term corresponds to the operation of adding v_i to an existing cell for the family of partitions $\mathcal{H}_j^{i-1}$ (with partition number j) over the elements up to v_{i-1}. However, simply adding the assignment pair (v_i, l) as in the previous section could generate patterns that violate the coloring constraint. Therefore, we use $\mathcal{R}(\mathcal{H}_j^{i-1}.\text{add\_item}((v_i, l)), v_i, l)$ to represent the family of partitions obtained by adding the assignment (v_i, l) to all partitions in $\mathcal{H}_j^{i-1}$ while removing all partitions that violate the coloring constraint. The definition of $\mathcal{R}(\mathcal{I}, v_i, l)$ is given in Eq. (5). The arguments of $\mathcal{R}(\mathcal{I}, v_i, l)$ are the ZDD $\mathcal{I}$, the vertex v_i being assigned, and its target cell l. For each vertex v_m adjacent to v_i, the operation $\mathcal{I}.\text{offset}((v_m, l))$ (shown in Sec. 2.1) excludes partitions where v_m is also assigned to cell l. By taking the intersection of these results for all such adjacent vertices, we obtain $\mathcal{R}(\mathcal{I}, v_i, l)$, which is the family of partitions where no vertex adjacent to v_i is assigned to cell l. By repeatedly applying these recurrence relations, $\mathcal{H}_1^{|V|}, \ldots, \mathcal{H}_{|V|}^{|V|}$ are generated. The algorithm proposed in this section executes these recurrence relations using set operations on ZDDs.

Algorithm 1. Enumerating all graph colorings

Input: A graph $G = (V(= \{v_1, \ldots, v_{|V|}\}), E)$.
Output: A backward labelling ZDD $\mathcal{F}$ that represents all graph colorings of G.
1: $\mathcal{H}_1^1 \leftarrow \{\emptyset\}$.add_item$((v_1, 1))$
2: **for** $i = 2, \ldots, |V|$ **do**
3:     **for** $j = 1, \ldots, i$ **do**
4:         **if** $j > 1$ **then** $\mathcal{H}_j^i \leftarrow \mathcal{H}_{j-1}^{i-1}$.add_item$((v_i, j))$
5:         **for** $k = j, \ldots, i - 1$ **do**
6:             $\mathcal{H}_k^i \leftarrow \mathcal{H}_k^i \cup \text{ColoringRestriction}(\mathcal{H}_k^{i-1}.\text{add\_item}((v_i, j)), v_i, j)$
7: $\mathcal{F} \leftarrow \emptyset$
8: **for** $i = 1, \ldots, |V|$ **do**
9:     $\mathcal{F} \leftarrow \mathcal{F} \cup \mathcal{H}_i^{|V|}$
10: **return** $\mathcal{F}$

Algorithm 2. ColoringRestriction

Input: A backward labelling ZDD $\mathcal{I}$, a vertex $v_i \in V$, a cell number j.
Output: A backward labelling ZDD $\mathcal{R}$ representing partitions where no vertex adjacent to v_i is assigned to cell j.
1: $\mathcal{R} \leftarrow \mathcal{I}$
2: **for** $m = 1, \ldots, i - 1$ **do**
3:     **if** $(v_m, v_i) \in E$ **then**
4:         $\mathcal{R} \leftarrow \mathcal{R} \cap \mathcal{I}.\text{offset}((v_m, j))$
5: **return** $\mathcal{R}$

Algorithm 1 takes a graph G as input and constructs a backward labelling ZDD that enumerates all its graph colorings. It internally uses an operation called ColoringRestriction (Algorithm 2). This algorithm is constructed by executing the add_item operations for each assignment pair (v_i, j) collectively. In Algorithm 1, Line 4 corresponds to the left term on the right-hand side of (4), adding (v_i, j) to $\mathcal{H}_{j-1}^{i-1}$ as an assignment that creates a new cell j. The only other case for adding (v_i, j) is when assigning v_i to an existing cell j for partition sets $\mathcal{H}_k^{i-1}$ that already have a partition number $k \geq j$. This case corresponds to the process in Lines 5 and 6. Here, ColoringRestriction is used to generate $\mathcal{R}(\mathcal{H}_k^{i-1}.\text{add\_item}((v_i, j)), v_i, j)$, thereby filtering the family to include only partitions that satisfy the coloring constraint.

This algorithm can also be used to find the set of graph colorings that use exactly c colors. When doing so, we can ignore two types of patterns during the construction phase: (i) patterns that already exceed c partitions (colors), and (ii) patterns that will not reach c partitions even if all remaining elements are assigned to new, separate cells. By pruning the search for these patterns during construction, we can build the set of coloring solutions much more quickly. The specific algorithm is described in the full version of this paper.

3.2   Extracting Coarser Partitions

In this section, we define a "coarser" and "finer" relationship for families of set partitions. A set partition P' is coarser than another set partition P if P' does not break the elements of any cell in P into different cells. Formally, for two set partitions P and P', we say that P' *is coarser than P* or P *is finer than P'* if for any cell C_i in P, there exists a cell C'_j in P' such that $C_i \subseteq C'_j$. For example, if a partition is $P_1 = (a|bc|d)$, then $P_2 = (ad|bc)$ is a partition that is coarser than P_1. In this section, for a given partition P of the vertex set $V = \{v_1, \ldots, v_{|V|}\}$ of a given graph $G = (V, E)$, we will describe a method to construct a ZDD, denoted as $\mathcal{F}(P)$, that enumerates the family of vertex partitions coarser than P. It should be noted that the set of partitions represented by $\mathcal{F}(P)$ includes P itself, and the vertex partitions contained in $\mathcal{F}(P)$ are not necessarily valid graph colorings.

For the purpose of explanation, we will call the smallest element in each cell of P a *leader*. Here, we define a function $\psi : V \to V$ such that $\psi(v_i)$ returns the leader element of the cell to which v_i belongs. A partition coarser than P is one that does not split elements that were in the same cell in P into different cells. If the cell to which v_i belongs in P has a leader $v_s(= \psi(v_i))$ other than v_i, and v_s has already been placed in cell j, then v_i must also be placed in cell j. Therefore, if we let $\mathcal{H}^i_j$ represent the set of all vertex partitions of $\{v_1, \ldots, v_i\}$ that have a partition number of j and do not split any cell of P, the following recurrence relation holds.

$$\mathcal{H}^1_1 = \{\{(v_1, 1)\}\}, \quad \mathcal{H}^1_j = \emptyset \ (j \geq 2), \tag{6}$$

$$\mathcal{H}^i_j = \begin{cases} \mathcal{H}^{i-1}_{j-1}.\text{add\_item}((v_i, j)) \cup \left(\bigcup_{l=1,\ldots,j} \mathcal{H}^{i-1}_j.\text{add\_item}((v_i, l)) \right) \\ \qquad\qquad\qquad\qquad\qquad\qquad\qquad \text{if } v_i \text{ is a leader of } P \\ \bigcup_{l=1,\ldots,j} \mathcal{H}^{i-1}_j.\text{onset}((\psi(v_i), l)).\text{add\_item}((v_i, l)) \qquad \text{otherwise} \end{cases} \tag{7}$$

Regarding Eq. (7), if v_i is a leader in P, there are no additional constraints on its assignment, so the process is the same as in Eq. (2). Otherwise, the cell to which v_i belongs in P has a leader $\psi(v_i)$ other than v_i. In this case, only assignments where v_i is assigned to the same cell as $\psi(v_i)$ are valid. Therefore, when adding the assignment (v_i, l), it is necessary to use the "onset" operation (shown in Sect. 2.1) to filter for patterns that already include the assignment $(\psi(v_i), l)$.

By executing these set operations on a ZDD, we can construct the ZDD $\mathcal{F}(P)$ that enumerates the family of vertex partitions coarser than P. We call this construction method All-Coarser. The concrete algorithm is presented in Algorithm 3.

Algorithm 3. All-Coarser

Input: The base set $V = \{v_1, \ldots, v_{|V|}\}$, a set partition P.
Output: A backward labelling ZDD $\mathcal{F}$ that represents all set partitions coarser than P.

1: $\mathcal{H}_1^1 \leftarrow \{\emptyset\}.\text{add\_item}((v_1, 1))$
2: **for** $i = 2, \ldots, |V|$ **do**
3:     **if** v_i is a leader of P **then**
4:         **for** $j = 1, \ldots, i$ **do**
5:             **if** $j > 1$ **then** $\mathcal{H}_j^i \leftarrow \mathcal{H}_{j-1}^{i-1}.\text{add\_item}((v_i, j))$
6:             **for** $k = j, \ldots, i - 1$ **do**
7:                 $\mathcal{H}_k^i \leftarrow \mathcal{H}_k^i \cup \mathcal{H}_k^{i-1}.\text{add\_item}((v_i, j))$
8:     **else**
9:         **for** $j = 1, \ldots, i$ **do**
10:             **for** $k = j, \ldots, |V|$ **do**
11:                 $\mathcal{H}_k^i \leftarrow \mathcal{H}_k^i \cup \mathcal{H}_k^{i-1}.\text{onset}((\psi(v_i), j)).\text{add\_item}((v_i, j))$
12: $\mathcal{F} \leftarrow \emptyset$
13: **for** $i = 1, \ldots, |V|$ **do**
14:     $\mathcal{F} \leftarrow \mathcal{F} \cup \mathcal{H}_i^{|V|}$
15: **return** $\mathcal{F}$

The process from Lines 3 to 7 handles the case where v_i is a leader in P, assigning v_i without any additional constraints. Lines 8 to 11 handle the case where v_i is not a leader. Here, the algorithm finds the leader $\psi(v_i)$ and constructs only the patterns where v_i is assigned to the same cell. It utilizes the ZDD operation "onset" so that when assigning (v_i, j), it extracts only those patterns where the leader is also assigned to the same cell number, as in $(\psi(v_i), j)$.

By taking the intersection (using the ZDD "intersection" operation) of this ZDD with a ZDD enumerating valid graph colorings (generated by the proposed methods), we can obtain the intersection of the families of set partitions represented by these ZDDs. This allows us to filter for coloring solutions that do not split any cell of P and have a smaller number of cells.

4   Experiments

4.1   Enumerating Graph Colorings

In this section, we report on an experiment that enumerated graph colorings for a specific number of partitions, using adjacency graphs of Japan's electoral districts as input (Table 2).

We compared our proposed method against the Answer Set Programming (ASP) [8] solver clingo (version 5.8.0) [4], which was also tasked with enumerating all valid graph colorings as canonical set partitions. To ensure a fair comparison, the solutions found by ASP were kept in memory and not written to a file. The experiments were run on an Intel Xeon CPU E5-2637 v3 (3.50GHz) with 1.5TB of memory (Ubuntu 16.04.7 LTS). We set a two-hour timeout for all computations.

In the following tables, "TO" marks a timeout and "MO" marks a memory-out error.

Table 2 shows the results for enumerating all graph colorings using exactly four colors. Our method was over 1,000 times faster than ASP, confirming its efficiency. While ASP's runtime increases almost linearly with the number of solutions, our method enumerates them using a compact ZDD, which drastically cuts down the execution time.

We also ran experiments on a 5×5 grid graph, where we varied the number of colors, c (Table 3). ASP could find solutions when the number of colors was very small or very large (since there are fewer solutions), but it timed out for moderate numbers of colors. In contrast, our method finished for all numbers of colors without timing out, taking at most about 100 s. Interestingly, the longest runtime occurred not when the number of solutions was largest, but when the ZDD size peaked. We believe the ZDD size grows as the number of colors increases from small to moderate because it requires a greater variety of ZDD items (in the form of (v_i, j) pairs). For a very large number of colors, the solution set is small, leading to a quick enumeration. Furthermore, the high-speed construction for cases with small or large partition numbers is thought to be due to the algorithm ignoring partitions with extremely large or small partition numbers during the construction phase.

Table 4 shows a similar experiment on a 6×6 grid. Again, our method was overwhelmingly faster than ASP, but it did run out of memory for moderate numbers of colors.

Based on these results, we believe our method can enumerate coloring solutions in a practical timeframe for graphs with up to about 25 vertices, regardless of the number of colors used.

Table 2. Enumerating all graph colorings using exactly four colors

G	$|V|$	$|E|$	ASP		Proposed Method		
			solutions	time[s]	ZDD size	solutions	time[s]
04_miyagi	39	86	1,607,565,312	6004.47	3,454	1,607,565,312	<0.01
07_fukushima	58	142	–	TO	4,295,450	29,379,580,416	8.67
09_tochigi	25	56	85,440	0.19	2,737	85,440	<0.01
15_niigata	38	81	1,091,045,376	3515.96	5,622	1,091,045,376	0.01
16_toyama	15	27	2,700	0.01	212	2,700	<0.01
29_nara	39	94	28,763,520	81.07	46,596	28,763,520	0.07
34_hiroshima	31	65	20,030,976	45.93	6,365	20,030,976	0.01
39_kochi	34	66	905,969,664	2646.51	900	905,969,664	<0.01

Table 3. Enumerating all graph coloring using exactly c colors(grid_5×5, $|V| = 25$, $|E| = 40$)

c	ASP		Proposed Method		
	solutions	time[s]	ZDD size	solutions	time[s]
2	1	0.005	25	1	0.0014
4	851,690,368	1690.49	12,420	851,690,368	0.0096
6	–	TO	247,202	30,101,741,134,000	0.073
8	–	TO	1,717,710	4,115,701,715,748,344	0.87
10	–	TO	7,104,745	25,532,302,725,012,108	6.23
12	–	TO	20,542,457	19,078,293,494,699,101	25.34
14	–	TO	43,698,664	2,751,178,554,344,595	63.61
16	–	TO	67,215,353	98,149,860,962,962	103.32
18	–	TO	53,525,309	977,800,966,888	86.15
20	–	TO	4,800,532	2,778,496,778	11.75
22	2,031,846	15.46	46,028	2,031,846	0.13
24	260	0.11	306	260	0.0051

4.2   Extracting Coarser Graph Colorings

In this section, we conduct an experiment using the All-Coarser operation to enumerate all graph colorings that are coarser than a specific graph coloring. This experiment takes as input a graph G and a partition P representing a graph coloring of G. First, we enumerate all graph colorings of G using Algorithm 1, and let the output ZDD be $\mathcal{F}$. Simultaneously, we use the All-Coarser algorithm (Algorithm 3) to enumerate all vertex partitions coarser than P, and let the output ZDD be $\mathcal{G}$. Finally, by finding the vertex partitions contained in both $\mathcal{F}$ and $\mathcal{G}$ using the ZDD intersection operation $\mathcal{F} \cap \mathcal{G}$, we can enumerate all graph colorings that are coarser than P. In this experiment, we used graphs for Fukui Prefecture (17 vertices, 26 edges), Akita Prefecture (25 vertices, 52 edges), and Kochi Prefecture (34 vertices, 66 edges) as the input graph G. For the input coloring P, we randomly generated and used 10 samples for each possible partition number for the respective graphs. The experimental environment is the same as in the previous section. The timeout was set to two hours.

To evaluate the performance of the filtering using All-Coarser, Table 5 shows the number of nodes of the backward labelling ZDD, the number of coloring solutions, and the time required for enumerating the coloring solutions before the refinement operation. The results for the Akita Prefecture (25 vertices, 52 edges) graph data, obtained by running All-Coarser for each sample P and then enumerating all graph colorings coarser than P via the intersection operation, are presented in Table 6. In Table 6, c represents the partition number of P, and s represents the sample number of P. Due to space constraints, only the cases for sample numbers 0 and 1 are shown. Regarding the Akita results (Table 6),

Table 4. Enumerating all graph coloring using exactly c colors(grid_6 × 6, $|V| = 36$, $|E| = 60$)

c	ASP		Proposed Method		
	solutions	time[s]	ZDD size		solutions	time[s]
2	1	0.01	36	1	0.0026
4	–	TO	57,594	15,835,535,879,933	0.02
6	–	TO	1,923,494	310,113,550,733,507,673,523	0.62
8	–	TO	20,097,278	2,812,947,903,505,726,935,338,659	15.29
10	–	TO	115,741,391	458,309,349,445,743,593,082,410,973	124.84
12	–	TO	468,144,666	5,786,886,681,267,934,647,424,016,489	640.01
14	–	TO	MO	–	–
30	–	TO	MO	–	–
32	–	TO	6,280,302	2,683,428,582	69.48
34	149,925	4.03	10,090	149,925	0.086
36	1	0.61	36	1	0.0011

Table 5. Enumerating all graph colorings for the experiment of All-coarser

G	$|V|$	$|E|$	ZDD size	solutions	time[s]
18_fukui	17	26	5,910	2,665,291,733	<0.01
05_akita	25	52	17,120,627	26,722,553,935,962,130	27.57
39_kochi	34	66	24,671,373	117,807,697,427,785,337,806,880,852	67.85

although several steps took longer than 100 s for examples with a moderate partition number c, all samples were executed without timing out. In particular, for cases where the value of c was less than 10, the execution time was at most approximately 1 s. The number of final solutions in these cases is less than 100, indicating that we successfully filtered the solutions down to a number that a user could practically compare and examine for application purposes. Furthermore, it was confirmed that for the All-Coarser step, the longest construction time was required for the example with the largest ZDD size. In some examples where the size of the finally constructed ZDD became bloated, the intersection operation was observed to take significantly longer than the other steps. The results for Fukui Prefecture (17 vertices, 26 edges) and Kochi Prefecture (34 vertices, 66 edges) are described in the full version of this paper. As a general summary of the results, we confirmed that for Fukui, all steps could be executed in less than 0.05 s regardless of the value of the partition number c. For Kochi, we confirmed that timeouts and memory-outs occurred due to bloated ZDD sizes at moderate partition numbers. From these results, it can be concluded that for graphs of up to about 25 vertices, it is possible to filter solutions in a practical amount of time.

Table 6. The executing results of All-Coarser and intersection operations for Akita Prefecture (05_akita, $|V| = 25$, $|E| = 52$)

c	s	All-Coarser			intersection		
		ZDD size	solutions	time[s]	ZDD size	solutions	time[s]
6	0	3,108	203	<0.01	25	1	<0.01
6	1	2,792	203	<0.01	25	1	<0.01
8	0	48,285	4,140	0.12	206	12	<0.01
8	1	19,450	4,140	0.05	28	2	<0.01
10	0	114,712	115,975	0.48	365	91	<0.01
10	1	602,599	115,975	1.01	959	68	0.02
12	0	383,353	4,213,597	2.20	4,304	2,401	0.06
12	1	1,439,502	4,213,597	4.57	9,236	5,490	0.08
14	0	510,048	190,899,322	2.69	66,302	256,721	0.19
14	1	369,473	190,899,322	2.18	13,262	251,585	0.06
16	0	1,095,062	10,480,142,147	4.67	146,494	13,972,082	1.48
16	1	68,655,631	10,480,142,147	203.48	2,724,002	16,982,928	81.10
18	0	20,324,799	682,076,806,159	67.59	17,778,228	1,581,314,944	596.36
18	1	1,767,162	682,076,806,159	7.80	5,658,282	1,896,224,368	19.89
20	0	958,724	5.17×10^{13}	2.87	62,509,426	142,427,686,704	512.42
20	1	1,463,890	5.17×10^{13}	3.88	7,375,691	192,068,883,720	127.35
22	0	812,690	4.51×10^{15}	1.99	8,728,734	1.89×10^{13}	85.47
22	1	384,725	4.51×10^{15}	1.40	19,954,480	1.58×10^{13}	372.33
24	0	25,635	4.46×10^{17}	0.06	31,212,624	2.23×10^{15}	16.15
24	1	15,840	4.46×10^{17}	0.07	29,506,368	2.23×10^{15}	26.01

5   Conclusion

In this paper, we proposed a method to enumerate all vertex partitions representing graph colorings using a data structure called the backward labelling ZDD, as well as a method to filter for partitions coarser than a given partition. We then evaluated these methods through computational experiments.

Through comparison experiments with an ASP solver, we confirmed that the proposed graph coloring enumeration method can execute more than 1,000 times faster than the ASP solver. It was also confirmed to operate particularly fast for cases with small or extremely large partition numbers by appropriately pruning the search. Furthermore, in the experiment on filtering solutions with All-Coarser, we were able to narrow down the solutions to a degree that a user could compare by using a partition P with a small partition number. It was found that for small to moderate partition numbers, the size of the generated ZDD increases and the computation time grows as the partition number becomes larger. While we confirmed that for the 25-vertex example (Table 6), nearly all steps were executable within a practical timeframe, there were cases for graphs with more vertices where the computation exceeded practical time limits when P had a moderate partition number. The graph coloring enumeration and solution filtering via All-Coarser proposed in this paper can be executed quickly within a practical range for input graphs of up to about 25 vertices.

The proposed method enables conditioning on cell numbers between elements, as seen in All-Coarser, by enumerating partitions for graph colorings in the form of a backward labelling ZDD and explicitly representing them with pairs of elements and cell numbers, such as (v_i, j).

Future work includes comparative experiments with enumeration methods other than ASP, experiments using more practical graph classes, and investigations into the effect of input graph vertex order on performance. Furthermore, it is a significant challenge to propose and investigate enumeration methods for graph coloring that can handle larger graph sizes and variations in the number of partitions, as well as other types of operations to narrow down the solution set of graph colorings.

Acknowledgement. The authors are grateful to Professor Keisuke Hotta for providing Japan's electoral districts data. This work was partly supported by JSPS KAKENHI Grant Numbers JP24H00690, JP24H00697, and JP25H01114.

References

1. Aletà, A., Codina, J.M., Sánchez, J., González, A.: Graph-partitioning based instruction scheduling for clustered processors. In: Proceedings. 34th ACM/IEEE International Symposium on Microarchitecture, MICRO-34, pp. 150–159 (2001). https://api.semanticscholar.org/CorpusID:14320061
2. Brélaz, D.: New methods to color the vertices of a graph. Commun. ACM **22**, 251–256 (1979). https://api.semanticscholar.org/CorpusID:14838769
3. Fleurent, C., Ferland, J.A.: Genetic and hybrid algorithms for graph coloring. Ann. Oper. Res. **63**, 437–461 (1996). https://api.semanticscholar.org/CorpusID:14020472
4. Gebser, M., Kaminski, R., Kaufmann, B., Schaub, T.: Clingo: a grounder and solver for logic programs. https://potassco.org/clingo/. Accessed 20 Sep 2025
5. Gonthier, G.: Formal proof—the four- color theorem. In: Notices AMS 2008 (2008). https://api.semanticscholar.org/CorpusID:12620754
6. Jensen, T.R., Toft, B.: Graph Coloring Problems. Wiley, Hoboken (1994)
7. Karp, R.M.: Reducibility among combinatorial problems. In: 50 Years of Integer Programming (1972). https://api.semanticscholar.org/CorpusID:33509266
8. Marek, V.W., Truszczynski, M.: Stable models and an alternative logic programming paradigm. In: The Logic Programming Paradigm (1998). https://api.semanticscholar.org/CorpusID:353601
9. Minato, S.: Zero-suppressed BDDs for set manipulation in combinatorial problems. In: 30th ACM/IEEE Design Automation Conference, pp. 272–277 (1993). https://api.semanticscholar.org/CorpusID:11096308
10. Minato, S.: Zero-suppressed BDDs and their applications. Int. J. Softw. Tools Technol. Transfer **3**, 156–170 (2001). https://api.semanticscholar.org/CorpusID:42919336

11. Smith, D.H., Hurley, S.: Bounds for the frequency assignment problem. Discret. Math. **167-168**, 571–582 (1997). https://api.semanticscholar.org/CorpusID: 205877697
12. Takahashi, S., Yoshioka, M.: A proposal of a method for enumerating and indexing set partitions using ZDDs and its experimental evaluation. In: Proceedings of the JSAI Special Interest Group on Fundamental Problems in Artificial Intelligence (SIG-FPAI) (March 2020), no. 112, pp. 08. The Japanese Society for Artificial Intelligence (JSAI)(In Japanese) (2020)

Engineering Algorithms for ℓ-Isolated Maximal Clique Enumeration

Marco D'Elia[1]([✉]) [ID], Irene Finocchi[2] [ID], and Maurizio Patrignani[1] [ID]

[1] Roma Tre University, Rome, Italy
{marco.delia,maurizio.patrignani}@uniroma3.it
[2] Luiss Guido Carli, Rome, Italy
finocchi@luiss.it

Abstract. Maximal cliques play a fundamental role in numerous application domains, where their enumeration can prove extremely useful. Yet their sheer number, even in sparse real-world graphs, can make them impractical to be exploited effectively. To address this issue, one approach is to enumerate ℓ-isolated maximal cliques, whose vertices have (on average) less than ℓ edges toward the rest of the graph. By tuning parameter ℓ, the degree of isolation can be controlled, and cliques that are overly connected to the outside are filtered out. Building on Tomita *et al.*'s very practical recursive algorithm for maximal clique enumeration, we propose four pruning heuristics, applicable individually or in combination, that discard recursive search branches that are guaranteed not to yield ℓ-isolated maximal cliques. Besides proving correctness, we characterize both the pruning power and the computational cost of these heuristics, and we conduct an extensive experimental study comparing our methods with Tomita's baseline and with a state-of-the-art approach. Results show that two of our heuristics offer substantial efficiency improvements, especially on real-world graphs with social network properties.

Keywords: Maximal cliques · Isolated cliques · Enumeration algorithms · Clique summarization · Algorithm engineering

1 Introduction

Cliques capture dense, highly-connected network structures. Their enumeration is important in many application domains, such as Web mining, bioinformatics, computational chemistry, social and financial network analysis. In particular, enumerating *maximal* cliques—those that cannot be extended by another vertex—helps reveal the graph's structure by uncovering tightly connected groups of vertices that share meaningful relationships.

From both theoretical and practical perspectives, the problem presents significant challenges. The set of maximal cliques includes indeed also *maximum* cliques, but computing a maximum clique is a well-known NP-hard problem [13],

This research was supported in part by MUR PRIN Projects EXPAND and NextGRAAL [grant numbers 2022TS4Y3N and 2022ME9Z78].

as well as W[1]-hard [5], and hard to approximate within a factor of $n^{1-\epsilon}$ [10]. In general, the number of maximal cliques in an n-vertex graph can be exponential in n [17]. Hence, not only their enumeration requires considerable time, but navigating such a large number of structures is also difficult. Imposing additional constraints may pinpoint a subset of domain-relevant structures, substantially reduce the number of enumerated cliques, and enable more effective utilization. We refer to [6] for a taxonomic survey of different summarization approaches.

In contexts where cliques with weak connections to the rest of the graph are more meaningful than those strongly intertwined with it, the concept of ℓ-*isolation* may prove particularly useful. First introduced in [12], ℓ-isolation requires that the number of edges leaving a clique of size k does not exceed $\ell \cdot k$. In the same works, the authors proposed an FPT algorithm in the isolation factor ℓ to enumerate all ℓ-isolated maximal cliques. It was also shown that ℓ must be constant to achieve linear-time enumeration, since otherwise one can construct instances admitting a superlinear number of ℓ-isolated maximal cliques.

In [14], two additional isolation principles were introduced, namely max- and min-isolation. A systematic comparison of different isolation notions, complemented by an experimental evaluation on synthetic and financial networks, was presented in [11], underscoring the importance of isolated clique enumeration for network analysis. Isolation concepts have been also explored in temporal networks [9, 16], and applied to practical scenarios such as web structure mining [23], hierarchical representations of scale-free networks [19], and network contraction via isolated cliques [21]. Alternative isolation models have also been proposed, including τ-isolation based on coreness [18], as well as variants for stars [22], bicliques [1], and even approaches inspired by quantum algorithms [8].

Our Contribution. In this paper, building upon the well-known Bron-Kerbosch algorithm [2] for maximal clique enumeration, improved by a pivoting strategy introduced by Tomita *et al.* [20], we first design four pruning heuristics aimed at reducing the search space, proving their correctness. The heuristics, that can be applied individually or in combination, allow the enumeration algorithm to focus only on promising parts of the search tree, discarding branches that are guaranteed not to yield ℓ-isolated maximal cliques. Following a theoretical characterization of the heuristics' pruning power and computational cost, which turn out to be inversely proportional, we conduct an extensive experimental study assessing their practical impact. Our analysis contributes along several dimensions:

- It sheds light on the properties of ℓ-isolated cliques relative to the full set of maximal cliques in real-world graphs. This can be particularly useful for guiding the choice of parameter ℓ.
- It gives insights into the relative power of our pruning strategies and of their combinations, identifying the most effective ones.
- It compares our approach with a carefully tuned baseline algorithm [20] and with a top-down state-of-the-art approach to the problem [11]. Overall, two of our variants offer substantial efficiency improvements over the state of the art, by as much as a factor of 4, especially when used on real-world graphs that exhibit social network properties.

Paper Organization. The rest of this paper is organized as follows. We provide preliminary definitions and a description of the baseline enumeration algorithm in Sect. 2. The pruning strategies are introduced and theoretically analyzed in Sect. 3. Our experimental setup and the main outcomes of our analysis are discussed in Sect. 4 and Sect. 5, respectively. We summarize our results and outline open problems in Sect. 6.

2 Preliminaries

In this section we first introduce preliminary definitions and properties of ℓ-isolation (Sect. 2.1), and then describe the maximal clique enumeration algorithm underlying our approach (Sect. 2.2). Throughout the paper we consider finite simple graphs. For a graph G, let $V = V(G)$ and $E = E(G)$ denote its vertex and edge sets, with $n = |V|$ and $m = |E|$ their respective sizes. For a vertex v, let $\delta(v)$ denote its degree and $N(v) = \{u \mid (u, v) \in E\}$ its set of neighbors. Clearly, $\delta(v) = |N(v)|$. For any vertex subset $I \subseteq V$, let $G[I]$ denote the subgraph of G induced by I, with vertex set $V[I] = I$ and edge set $E[I] = \{(u, v) \in E \mid u, v \in I\}$ consisting of edges with both endpoints in I. For a vertex $v \in I$, let $\delta_e(v, I)$ denote its *external degree with respect to I*, i.e., $\delta_e(v, I) = |\{(u, v) \in E \mid u \notin I\}|$. The *external degree of I*, denoted as $\delta_e(I)$, is then the number of edges with one endpoint in I and the other in $V \setminus I$, i.e.,

$$\delta_e(I) = \sum_{v \in I} \delta_e(v, I) = |\{(u, v) \in E \mid v \in I \text{ and } u \in V \setminus I\}|$$

Moreover, given $I, L \subseteq V$ with $I \cap L = \varnothing$, the *external degree of I neglecting L*, denoted as $\delta_e(I, L)$, is the number of edges with one endpoint in I and the other in $V \setminus (I \cup L)$.

2.1 ℓ-isolated Cliques

A *clique* is a complete subgraph of G, and is *maximal* if it is not contained in any larger clique. The concept of ℓ-isolation has been first introduced in [12]: for any integer value $\ell > 0$, a set $I \subseteq V$ of vertices G of size $|I| = k$ is ℓ-*isolated* in G if it has less than $\ell \cdot k$ outgoing edges, i.e., $\delta_e(I) < \ell \cdot k$. In particular, we are interested into maximal cliques that are also ℓ-isolated (ℓ-isolated maximal cliques). We notice that I may contain vertices with more than ℓ external edges: the property of ℓ-isolation holds on average over the vertices of I, implying that the average external degree of vertices in I is less than ℓ^1.

We remark that the property of ℓ-isolation is monotone. Namely, if $\ell_1 > \ell_2 > 0$, then any ℓ_2-isolated set is also ℓ_1-isolated. This follows immediately from the definition of ℓ-isolation, since for an ℓ_2-isolated set I we have $\delta_e(I) < \ell_2 \cdot k < \ell_1 \cdot k$

---

[1] A stronger notion of *max-ℓ-isolation* requires every $v \in I$ to have fewer than ℓ external edges, but will not be considered in this paper. Algorithmic techniques similar to those presented here could nevertheless be applied.

Algorithm 1. TomitaTanakaTakahashi

1: **Input:** Graph $G = (V, E)$
2: **Output:** The set $\mathcal{M}(G)$ of all maximal cliques of G
3: ProcMCE($\varnothing, V, \varnothing$)
4:
5: **procedure** ProcMCE(C, P, X)
6:      **if** $P = \varnothing$ and $X = \varnothing$ **then**
7:          **yield** C
8:          **return**
9:      Choose a pivot vertex v_p in $P \cup X$ with highest $|N(v_p) \cap P|$
10:     **for all** $v \in P \setminus N(v_p)$ **do**
11:         ProcMCE($C \cup \{v\}, P \cap N(v), X \cap N(v)$)
12:         $P \leftarrow P \setminus \{v\}$
13:         $X \leftarrow X \cup \{v\}$

---

when $\ell_1 > \ell_2$. Hence, when searching for ℓ-isolated maximal cliques, the larger is the value of ℓ, the weaker is the isolation of the cliques relative to the rest of the graph, and the larger is the number of ℓ-isolated cliques in the graph. If ℓ is high enough, the set of ℓ-isolated maximal cliques coincides with the set of all maximal cliques of G.

2.2   Maximal Clique Enumeration with Pivoting

Our approach hinges upon the recursive backtracking procedure for maximal clique enumeration outlined in Algorithm 1. The procedure extends the well-known algorithm by Bron and Kerbosch [2] with a pivoting strategy introduced in [20], aimed at reducing the time spent on non-maximal cliques. Three sets of vertices are maintained by procedure ProcMCE: C represents the clique under construction, P contains candidate vertices that can potentially extend C, and X holds vertices that should no longer be considered, as they are already part of maximal cliques obtained from C or its subsets. As an invariant property, vertices in P and X are adjacent to all vertices in C. The idea is to add at each step a vertex, selected from the candidates in P, expanding the current clique C until it becomes maximal. At the first invocation, $C = X = \varnothing$ and $P = V$. In procedure ProcMCE, base cases occur when P is empty: then C cannot be further expanded and is returned only if $X = \varnothing$, ensuring its maximality. Otherwise, a pivot vertex v_p is chosen so as to minimize recursive calls: ProcMCE is called on each $v \in P$ not adjacent to v_p (possibly including v_p). Correctness follows since, for each neighbor w of v_p, any maximal clique containing $C \cup \{w\}$ will still be reported by recursing on either $C \cup \{v_p\}$ or $C \cup \{u\}$, where $u \in P$ is a non-neighbor of v_p. Recursive calls update (C, P, X) to $(C \cup \{v\}, P \cap N(v), X \cap N(v))$, ensuring vertices in $P \cup X$ to remain adjacent to C. Afterwards, v is removed from further consideration.

When invoked on a graph G, Algorithm 1 implicitly explores a search tree $\mathcal{T}_G$. Each node of $\mathcal{T}_G$ corresponds to a recursive call and is uniquely associated

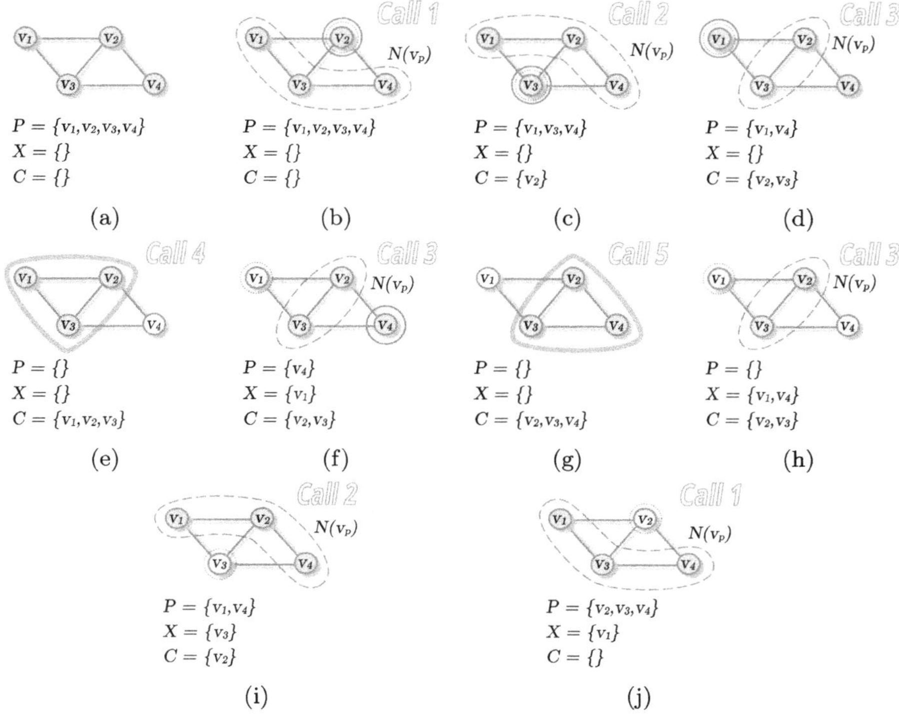

Fig. 1. A running example of Algorithm 1 on a small graph. Green, blue, and red vertices belong to P, C, and X, respectively. The current pivot vertex v_p is circled with a dotted curve. The neighbourhood of v_p is circled with a dashed curve, the current vertex v is circled with a red solid curve.

with a triple (C, P, X). The leaves of $\mathcal{T}_G$ are nodes of the form $(C, \varnothing, X)$. If $X = \varnothing$, the leaf corresponds to a maximal clique that is enumerated. Otherwise, the maximality condition is not satisfied (vertices in X could be added to C), the clique is discarded, but no recursive call needs to be done.

A running example of Algorithm 1 on a small graph is depicted in Fig. 1. When the first call of the procedure `ProcMCE` is launched, all the vertices are in P, while $X = C = \varnothing$ (Fig. 1b). The pivot vertex is $v_p = v_2$ (line 9 of Algorithm 1) and the procedure iterates on vertex v_2 only, since all other vertices in P are neighbors of v_p (line 10 of Algorithm 1). Hence, in the second call of `ProcMCE` (Fig. 1c) vertex v_2 has been moved to C, $v_p = v_3$, and line 10 of Algorithm 1 only iterates on v_3. Instead, Call 3 iterates on both v_1 and v_4: the iteration on v_1 (Fig. 1d) launches Call 4 (Fig. 1e) while the iteration on v_4 (Fig. 1f) launches Call 5 (Fig. 1g). Call 4 and Call 5 detect each a maximal clique, since both P and X are empty (line 6 of Algorithm 1). When Call 5 terminates, the control returns to Call 3 (Fig. 1h) and then to Call 2 (Fig. 1i) and Call 1 (Fig. 1j), terminating the computation.

Algorithm 1 requires $\mathcal{O}(3^{n/3})$ time [20]. While being optimal in the worst-case (matching the number of maximal cliques in Moon-Moser graphs [17]), it has $\Omega(3^{n/6})$ delay and no polynomial-delay variant can exist, even with different pivoting, unless $\mathsf{P} = \mathsf{NP}$ [3]. Despite these worst-case bounds, Bron-Kerbosch-based approaches are known to perform very efficiently in practice [7,20].

3   Pruning for ℓ-Isolated Maximal Clique Enumeration

In this section we show how to adapt Algorithm 1 to enumerate only ℓ-isolated maximal cliques, introducing different pruning strategies to cut uninteresting branches of the search tree. We also analyze the pruning capabilities and computational cost of each strategy, showing an inverse relation that suggests a natural order in which the heuristics may be applied in combination.

Growing ℓ-isolated Cliques. We recall from Sect. 2.2 that in Algorithm 1, during the enumeration, C and P are the current candidate clique and the perspective set, respectively. Let T be any maximal clique in the subgraph $G[P]$ induced by P. Thanks to the invariant property that P is always fully connected to C during the execution, it is not difficult to see that $C \cup T$ is a clique in G, though not necessarily ℓ-isolated. Our approach, when processing a node r of the search tree $\mathcal{T}_G$ associated with a triple (C, P, X), is to avoid recursive calls whenever the subtree rooted at r cannot yield ℓ-isolated maximal cliques.

First, we introduce a necessary condition for a clique grown from C and P to be ℓ-isolated. Let T be any maximal clique in $G[P]$ and let t be its size. By the invariant property maintained by Algorithm 1, $C \cup T$ is a clique in G. Such a clique has at least $\delta_e(C, P) + |C|(|P| - t)$ external edges: each of the $|P| - t$ vertices not included in C is indeed connected through an external edge with all the vertices in C. By definition, the number of external edges in an ℓ-isolated clique must be strictly less than ℓ times the clique size. Hence, if

$$\delta_e(C, P) + |C|(|P| - t) \geq \ell(|C| + t) \tag{1}$$

then $C \cup T$ is not ℓ-isolated. We now introduce a preliminary result that will be useful to prove the correctness of our pruning strategies.

Lemma 1. *Let r be a node in $\mathcal{T}_G$ associated to a candidate set C and a perspective set P. Let P_1 and P_2 be any two subsets of P such that $|P_1| \leq |P_2|$. If $\delta_e(C, P) + |C|(|P| - |P_2|) \geq \ell(|C| + |P_2|)$, then both $C \cup P_2$ and $C \cup P_1$ are not ℓ-isolated.*

Proof. As observed when we introduced Eq. (1), if $\delta_e(C, P) + |C|(|P| - |P_2|) \geq \ell(|C| + |P_2|)$, then $C \cup P_2$ is not ℓ-isolated. The relationship $|P_1| \leq |P_2|$ yields the following chain of inequalities: $\delta_e(C, P) + |C|(|P| - |P_1|) \geq \delta_e(C, P) + |C|(|P| - |P_2|) \geq \ell(|C| + |P_2|) \geq \ell(|C| + |P_1|)$. Hence, $C \cup P_1$ is also not ℓ-isolated. $\square$

Pruning by Maximum Cliques. The size of the largest clique in the subgraph $G[P]$ induced by P, i.e., its clique number $\omega(G[P])$, can be used to prune subtrees of $\mathcal{T}_G$. Namely, we prove in Lemma 2 that it is correct to prune the subtree rooted at a node labelled (C, P, X) whenever

$$\delta_e(C, P) + |C||P| - \ell|C| \geq \omega(G[P]) \cdot (\ell + |C|) \tag{2}$$

Lemma 2. *Let r be a node in $\mathcal{T}_G$ associated to a candidate set C and a perspective set P. If $\delta_e(C, P) + |C||P| - \ell|C| \geq \omega(G[P]) \cdot (\ell + |C|)$, then the subtree of $\mathcal{T}_G$ rooted at r contains no leaf corresponding to an ℓ-isolated maximal clique.*

Proof. We use Lemma 1, choosing P_2 as a maximum clique in $G[P]$, so that $|P_2| = \omega(G[P])$. Notice that the inequality $\delta_e(C, P) + |C||P| - \ell|C| \geq \omega(G[P]) \cdot (\ell + |C|)$ in the statement can be rewritten as $\delta_e(C, P) + |C|(|P| - \omega(G[P])) \geq \ell(|C| + \omega(G[P]))$. When it is satisfied, by Eq. (1) the subgraph induced by C extended by a maximum clique of $G[P]$ cannot be ℓ-isolated. By Lemma 1, if we extend C by any other clique of $G[P]$, the resulting subgraph is also not ℓ-isolated. Any clique P_1 of $G[P]$ is indeed smaller than or equal to a maximum clique, i.e., $|P_1| \leq |P_2| = \omega(G[P])$ and thus Lemma 1 can be applied.

Pruning by Upper Bounding the Maximum Clique Size. Lemma 2 implies that a check on $\omega(G[P])$ would be sufficient to prune at node r. Unfortunately, computing $\omega(G[P])$ is difficult—max-clique being NP-hard. Hence, we exploit upper bounds to $\omega(G[P])$ to guide pruning, inspired by the approach suggested in [24] for the computation of τ-visible summaries. We will prove in Theorem 1 that this maintains correctness with respect to our ℓ-isolation problem. In particular, $\omega(G[P])$ cannot be larger than:

1. the number $|P|$ of vertices in $G[P]$;
2. the maximum degree of $G[P]$, denoted by $\Delta(G[P])$, augmented by 1;
3. the maximum value of k such that $G[P]$ contains at least k vertices with degree $\geq k - 1$, denoted by $\tau(G[P])$;
4. the degeneracy of $G[P]$, denoted by $\kappa(G[P])$, augmented by 1. The degeneracy is the largest value of k for which $G[P]$ has a k-core (i.e., a maximal subgraph in which each vertex has degree at least k).

The first upper bound $\omega(G[P]) \leq |P|$ is the weakest, while the fourth one $\omega(G[P]) \leq \kappa(G[P])$ is the strictest. In particular, the following chain of inequalities holds:

$$\omega(G[P]) \leq \kappa(G[P]) + 1 \leq \tau(G[P]) \leq \Delta(G[P]) + 1 \leq |P| \tag{3}$$

This can be intuitively explained as follows. The fact that $\omega(G[P]) \leq \kappa(G[P]) + 1$ is well-known in the literature: in every clique of size t, every vertex has indeed degree $t - 1$ and is thus part of a $(t - 1)$-core, proving that the degeneracy is at least $t - 1$. The value $\tau(G[P])$ represents a relaxed version of degeneracy, and thus $\kappa(G[P]) + 1 \leq \tau(G[P])$. Indeed, the degeneracy $\kappa(G[P])$ can be computed by repeatedly removing a vertex of minimum degree from the graph, and updating the degree of remaining vertices along the way. Its relaxation $\tau(G[P])$ can be instead obtained by removing vertices based on their original degrees, without updates. Hence, it may be the case that more vertices satisfy the degree condition at point 3. Consider, for instance, a complete binary tree on four levels: $\tau(G[P]) = 4$, as the maximum degree 3 is attained by all the 6 vertices in the two internal layers, while $\kappa(G[P]) + 1 = 2$, as the degeneracy of a tree is 1. Finally, it can be easily seen that $\tau(G[P]) \leq \Delta(G[P]) + 1$, because we cannot have k vertices of degree larger than or equal to $k - 1$ if $k \geq \Delta(G[P]) + 2$.

We use the upper bounds on $\omega(G[P])$ to obtain lower bounds on the total number of external edges that any maximal clique T, grown from the current clique C and the candidate set P, can have. Namely, our algorithm prunes the computation at node (C, P, X) whenever

$$\delta_e(C, P) + |C||P| - \ell|C| \geq \overline{\omega}(G[P]) \cdot (\ell + |C|) \tag{4}$$

where $\overline{\omega}(G[P])$ can be any of the quantities involved in Eq. (3) (i.e., either $\omega(G[P])$ or any of its upper bounds). In the following we discuss the correctness of the pruning condition expressed by Eq. (4), proving that it is safe to prune using upper bounds $\overline{\omega}$ on ω, instead of ω itself.

Theorem 1. *Given a graph G and a positive integer ℓ, let $\mathcal{T}_G$ be the search tree of Algorithm 1 executed on G, let r be a node in $\mathcal{T}_G$ associated to a candidate set C and a perspective set P, and let $\overline{\omega}(G[P])$ be any upper bound on $\omega(G[P])$. If $\delta_e(C, P) + |C||P| - \ell|C| \geq \overline{\omega}(G[P]) \cdot (\ell + |C|)$, then the subtree of $\mathcal{T}_G$ rooted at r contains no leaf corresponding to an ℓ-isolated maximal clique.*

Proof. We exploit Lemma 1, choosing P_1 as a maximum clique in $G[P]$ and choosing P_2 as follows: if the upper bound $\overline{\omega}(G[P])$ in Eq. (3) is $|P|$, then P_2 is P itself; if instead it is $\Delta(G[P]) + 1$, then P_2 is formed by any vertex of maximum degree in P together with its neighbors; if it is $\tau(G[P])$, then P_2 is formed by any $\tau(G[P])$ vertices of P each having degree at least $\tau(G[P]) - 1$; and when it is $\kappa(G[P]) + 1$, then P_2 is the $\kappa(G[P])$-core of P. Hence, $|P_1| = \omega(G[P]) \leq \overline{\omega}(G[P]) = |P_2|$. If the test in Eq. (4) is verified, by Lemma 1 we have that the subgraph induced by C extended by a maximum clique of $G[P]$ is not ℓ-isolated. In particular, from the proof of Lemma 1 we have $\delta_e(C, P) + |C|(|P| - \omega(G[P])) \geq \ell(|C| + \omega(G[P]))$ (Eq. (2)). This in turn implies, by Lemma 2, that any other clique in $G[P]$ is not ℓ-isolated and that pruning at r does not entail discarding any ℓ-isolated clique, thereby proving the statement.

We conclude with a remark on the running time: the quantities $|P|$, $\Delta(G[P])$, $\tau(G[P])$, and $\kappa(G[P])$ involved in Eq. (3) can be computed in $O(1)$, $O(1)$, $O(|P|)$, and $O(|E(G[P])|)$ time, respectively, yielding an inverse relation between running time and pruning power.

Table 1. Statistics of the real-world benchmarks. Δ is the maximum degree, and d is the degeneracy.

Graph	ID	Category	Vertices	Edges	Max. Cliques	Δ	d
brightkite	bk	social	58,228	214,078	290,004	1,134	52
livemocha	lm	social	104,103	2,193,083	3,711,569	2,980	92
gowalla	gw	social	196,591	950,327	1,212,679	14,730	51
twitter	ws	social	465,017	833,540	805,170	677	30
youtube	cy	social	1,134,890	2,987,624	3,265,956	28,754	51
hyves	hy	social	1,402,673	2,777,419	2,590,217	31,883	39
ca-astroph	ap	co-authorship	18,771	198,050	36,427	504	56
dblp	cd	co-authorship	317,080	1,049,866	257,551	343	113
cfinder-google	gc	hyperlink	15,763	148,585	75,258	11,401	102
stanford	sf	hyperlink	281,903	1,992,636	1,055,936	38,625	71
italian-cnr	ic	hyperlink	325,557	2,738,969	1,425,378	18,236	83
notre-dame	nd	hyperlink	325,729	1,090,108	495,947	10,721	155
baidu	bar	hyperlink	415,624	2,374,044	4,596,143	127,066	228
web-google	go	hyperlink	875,713	4,322,051	1,417,580	6,332	44
cit-hepph	phc	citation	34,546	420,877	412,493	846	30
citeseer	cs	citation	384,054	1,736,145	1,232,951	1,739	15

4 Experimental Setup

Algorithms Under Evaluation. The four different heuristics discussed in Sect. 3 are denoted in our analysis by `size`, `degree`, `softcore`, and `degeneracy`, respectively. We also consider a combined approach, called `combo`, that orderly integrates the `size` and `softcore` heuristics: we tested many different combinations of the four basic heuristics, and `combo` consistently proved to be the most effective one. For the sake of comparison, we include the algorithm proposed in [11], denoted by `HKMN`. We relied on the implementation used in [11]. Finally, we consider a baseline algorithm, derived from Algorithm 1 [20] and denoted by `TTT`, that filters each non-ℓ-isolated maximal clique instead of reporting it.

Benchmarks. We conducted experiments on a variety of real-world networks, which are taken from the KONECT repository (http://konect.cc) [15]. Throughout this paper we focus on a representative subset of 16 graphs, whose main properties are summarized in Table 1. Our selection was guided by diversity criteria, taking into account *graph size, number of maximal cliques, maximum degree Δ*, and *degeneracy d*. To further ensure heterogeneity, our benchmarks include different types of networks, namely social, hyperlink, co-authorship, and citation graphs. We also used synthetic instances as described in [4].

Performance Metrics. In addition to running times, we profiled the number of recursive calls and contrasted these values with the number of ℓ-isolated cliques.

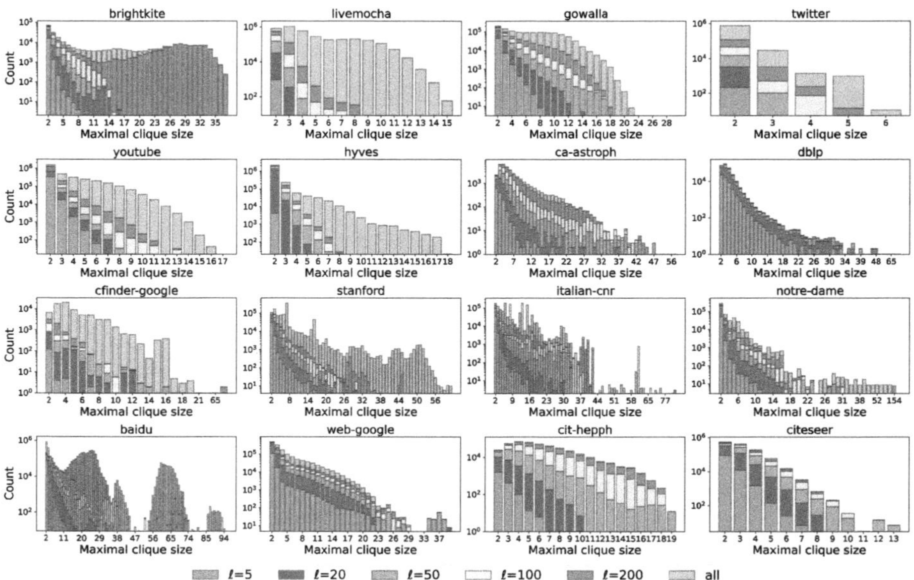

Fig. 2. ℓ-isolated clique size distribution for different values of ℓ.

The running time of each heuristic depends indeed both on its pruning capabilities and on the overhead introduced by the pruning test in Eq. (4): the more complex the pruning strategy is, the larger the test overhead.

Experimental Platform. All the experiments were run on a laptop equipped with an Intel®Core™i5-1135G7 CPU (4 cores, 4.2 GHz) and 8 GB of RAM. Our algorithms were implemented in OCaml (https://ocaml.org) to ensure a fair comparison with the available implementation used in [11].

5   Experimental Results

In this section we present the key findings of our experimental analysis. Section 5.1 examines the properties of ℓ-isolated cliques in real-world benchmarks. We evaluate our pruning strategies in Sect. 5.2, identifying the most effective ones, which are compared against state-of-the-art methods in Sect. 5.3.

5.1   On the Distribution of ℓ-Isolated Maximal Cliques

In Fig. 2 we summarize the distribution of maximal cliques by size, together with their breakdown into ℓ-isolated cliques for different values of ℓ.

The upper profile of the stacked bars shows the total number of maximal cliques with a given number of vertices. In all benchmarks, most cliques consist of a few vertices: the number of cliques grows quickly for small sizes, then drops

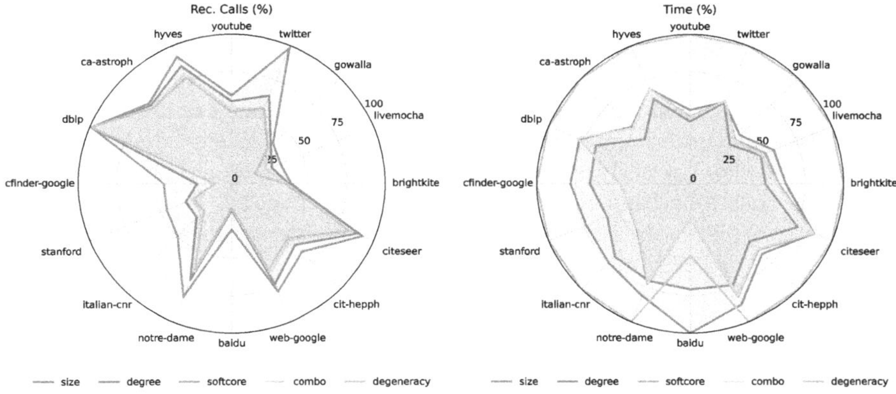

Fig. 3. Comparison of the recursive calls (%) and the running time (% with respect to the slowest execution) of different pruning strategies for $\ell = 50$.

sharply. The maximum clique size varies substantially across graph types (cfr. Table 1), reflecting different connectivity patterns: the first six social networks form only small cliques (typically less than 20 vertices), the two co-authorship networks and the two citation networks reach moderate sizes, while the six Web graphs contain the largest cliques, sometimes exceeding 150 vertices. The difference can be best appreciated on graphs of different types but similar size: compare, e.g., the maximum clique size in `gowalla`, `dblp`, and `notre-dame`, which have comparable numbers of edges.

This analysis can be refined by considering the color segments within each bar of Fig. 2, that show how many maximal cliques are ℓ-isolated for several values of ℓ. For small values of ℓ (e.g., $\ell \leq 20$), the isolation requirement is very strict and only a small fraction of small cliques qualify: large cliques are typically excluded, as they tend to be embedded in denser regions of the graph with many external connections. The distributions for small ℓ are thus sharply truncated and dominated by cliques with just a few vertices. As ℓ increases, the number of ℓ-isolated cliques grows, being the property monotone (see Sect. 2.1). In addition, larger values of ℓ allow increasingly larger cliques to appear in the summary, extending the tail of the distribution. For large thresholds, such as $\ell = 200$, many more cliques are admitted: in many networks (e.g., `brightkite`, `web-google`, `cit-hepph`) the profile for 200-isolated cliques approaches that of the total distribution, making the isolation criterion almost ineffective.

5.2 Assessment of Different Pruning Strategies

We now assess the effectiveness of different pruning strategies considering two different metrics: number of recursive calls and running time. Because the results are largely consistent across different values of ℓ, Fig. 3 shows only the case $\ell = 50$. The left radar plot in Fig. 3 reports the percentage of recursive calls with respect to the number one would have without pruning. This gives a quantitative view of the pruning power of the heuristics: the larger the area, the weaker the pruning. In spite of a large variability across benchmarks, `size` is always the weakest, as expected from the theoretical analysis, producing the largest number of calls. On some benchmarks, e.g., `twitter`, this number can be much larger than the other strategies and very close to the baseline. As expected, `combo` and `softcore` overlap perfectly, since `combo` is `size` followed by `softcore` whenever `size` fails, thus inheriting the pruning power of `softcore`. While `degeneracy` consistently achieves the strongest pruning, yielding the smallest areas across all datasets, it is rather similar to `softcore` and `combo`.

Interestingly, the picture is reversed in the right radar plot, which reports the running time normalized with respect to the slowest execution: `degeneracy`, which is the most powerful pruning strategy, is also almost always the slowest, with the exception of `baidu`, where `degree` performs worst. The heuristics with weaker pruning, most notably `size`, `softcore` and `combo`, are instead the fastest. Among these, `softcore` and `combo` behave very similarly, with `combo` being slightly preferable in some cases. `degree` exhibits an intermediate performance across benchmarks. Overall, the results highlight a clear trade-off between pruning power and efficiency: the running times observed in theory to perform the pruning tests (see Sect. 3) appear to have a stronger impact on performance than the actual amount of pruned nodes. Since running time is ultimately the main performance metric, in the remainder of this section we will focus on `size` and `combo`, which appear to be the most effective strategies.

5.3   Comparison with State-of-the-Art Competitors

In Fig. 4 we compare our best heuristics, `size` and `combo`, against HKMN and TTT, also analyzing how ℓ affects the running times. We chose a range of values for ℓ based on the observations in Sect. 5.1, covering both the lower and the upper spectrum of the isolation factor. A few trends emerge clearly from the charts. First, the running time of TTT is independent of ℓ, thus remaining constant. The two heuristics `size` and `combo` typically deliver the best performance over a large range of values of ℓ, with runtimes that remain low and scale smoothly as ℓ increases, when larger portions of T_G need to be explored. Their curves are often close (see, e.g., `livemocha` or `cfinder-google`), `combo` is marginally faster than `size` on a few instances (most notably, `stanford` and `italian-cnr`), but `size` seems preferable in most of the other ones. On the other side, HKMN is by far the slowest algorithm when ℓ gets large, with runtimes that grow quickly with ℓ: this is due to an ℓ-parameterized subroutine for finding minimal vertex covers. It can be instead competitive when ℓ is very small. TTT is only occasionally competitive with `size` and `combo`, but still outperforms HKMN for large ℓ except on `web-google` and `dblp`. The observed relative differences between the algorithms persist across the datasets. Even weak pruning heuristics that add a small test overhead on nodes of T_G translate into significant speedups, which can be as large as $4\times$ and $2\times$ with respect to HKMN and TTT, respectively. Experiments on synthetic graphs overall confirm the above analysis (see [4]).

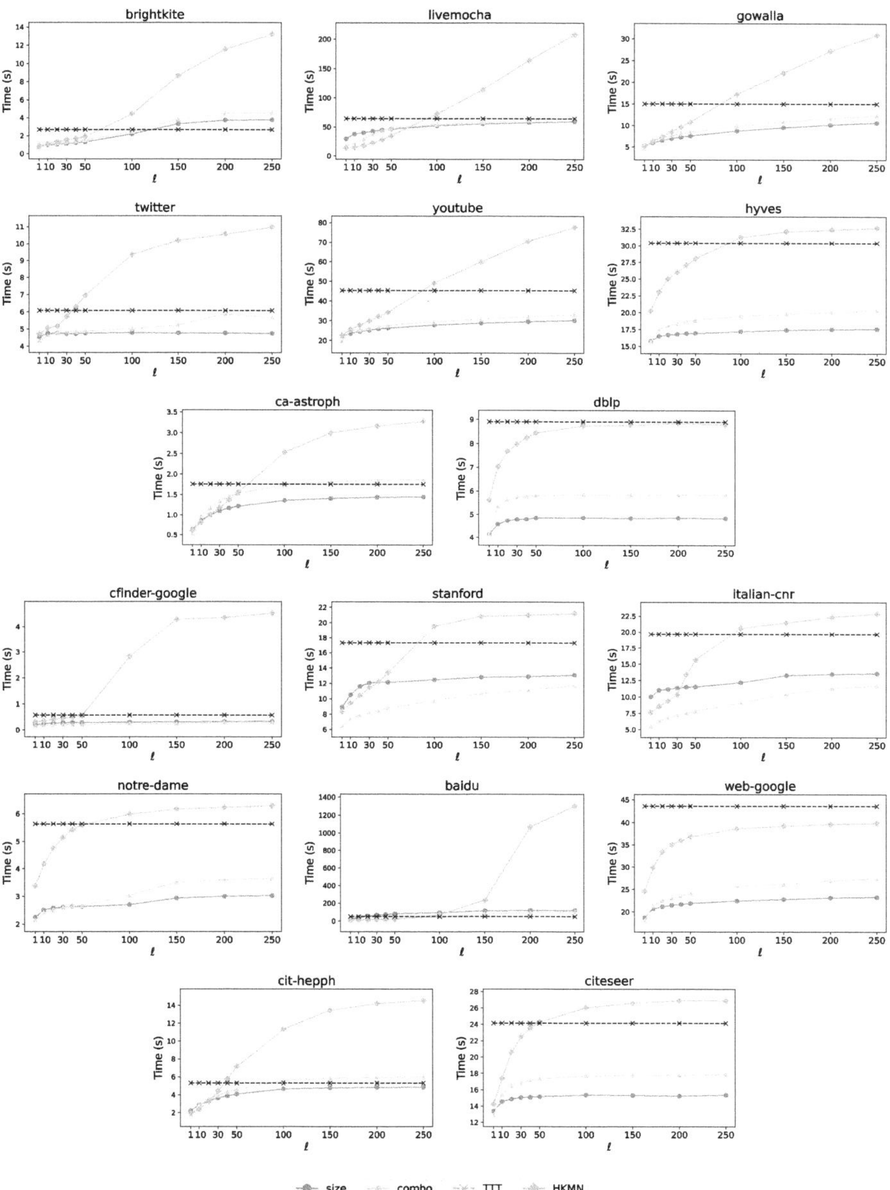

Fig. 4. Comparison of **size**, **combo**, HKMN, and **TTT** on real-world graphs.

6   Conclusions and Open Problems

This paper addressed the problem of enumerating isolated cliques, introducing pruning heuristics that can be applied on top of classical maximal cliques enumeration algorithms. We have theoretically proved correctness and conducted an extensive experimental analysis on a variety of benchmarks, proving the effectiveness of two of our variants against state-of-the-art competitors. Overall, ℓ-isolation appears to be a useful lens for filtering cliques according to how strongly they are separated from the rest of the network. Small values of ℓ expose only the most tightly isolated structures, at the cost of excluding larger cliques. At the other extreme, too large values of ℓ may contribute little beyond the raw clique distribution, as almost every clique qualifies. The most informative patterns emerge at intermediate thresholds, such as $\ell = 50$ or 100, which strike a balance between selectivity and inclusiveness: they admit a substantial fraction of cliques, even larger ones, while still preserving the notion of relative isolation. A natural way to address the tradeoff between selectivity and clique size, which is closely tied to the choice of ℓ, could be to introduce alternative metrics in which the isolation factor is defined relative to clique size, rather than as an absolute threshold.

Isolation, and clique summarization more broadly, pose many other open problems (see also [6]). In particular, it would be worthwhile to investigate whether our approach can be extended to the computation of ℓ-isolated cliques by size (Problem 3 in [6]) and to the stronger notion of max-ℓ-isolation (see Sect. 2.1). Another promising direction is to examine whether the problem admits fixed-parameter tractable algorithms under alternative parameters, such as degeneracy, either by refining the techniques of [12] or by focusing on specific graph families with distinctive structural properties.

Acknowledgements. We thank Robert Bredereck and Manuel Sorge for providing the code used in [11]. We thank Tommaso Caiazzi and Simone Romanelli for interesting conversations.

References

1. Alamgir, Z., Karim, S., Husnine, S.: Linear-time algorithm for generating c-isolated bicliques. Int. J. Comput. Math. **94**(8), 1574–1590 (2017)
2. Bron, C., Kerbosch, J.: Finding all cliques of an undirected graph (algorithm 457). Commun. ACM **16**(9), 575–576 (1973)
3. Conte, A., Tomita, E.: On the overall and delay complexity of the CLIQUES and Bron-Kerbosch algorithms. Theor. Comput. Sci. **899**, 1–24 (2022)
4. D'Elia, M., Finocchi, I., Patrignani, M.: Engineering algorithms for ℓ-isolated maximal clique enumeration. arXiv preprint arXiv:2511.03525 (2025)
5. Downey, R.G., Fellows, M.R.: Parameterized complexity. Springer Science and Business Media (2012)
6. D'Elia, M., Finocchi, I., Patrignani, M.: Maximal cliques summarization: principles, problem classification, and algorithmic approaches. Comput. Sci. Rev. **58**, 100784 (2025)

7. Eppstein, D., Löffler, M., Strash, D.: Listing all maximal cliques in large sparse real-world graphs. ACM J. Exp. Algorithmics **18** (2013)

8. Gall, F.L., Nadler, O., Nishimura, H., Oshman, R.: Quantum simultaneous protocols without public coins using modified equality queries. In: 28th International Conference on Principles of Distributed Systems. LIPIcs, vol. 324, pp. 34:1–34:20 (2024)

9. Gao, J., Hao, F., Yang, E., Yang, Y., Min, G.: Concept stability based isolated maximal cliques detection in dynamic social networks. In: Yu, Z., Becker, C., Xing, G. (eds.) GPC 2020. LNCS, vol. 12398, pp. 131–144. Springer, Cham (2020). https://doi.org/10.1007/978-3-030-64243-3_11

10. Hastad, J.: Clique is hard to approximate within $n^{1-\varepsilon}$. Acta Math. **182**(1), 105–142 (1999)

11. Hüffner, F., Komusiewicz, C., Moser, H., Niedermeier, R.: Isolation concepts for clique enumeration: comparison and computational experiments. Theor. Comput. Sci. **410**(52), 5384–5397 (2009)

12. Ito, H., Iwama, K.: Enumeration of isolated cliques and pseudo-cliques. ACM Trans. Algorithms **5**(4), 40:1–40:21 (2009)

13. Karp, R.M.: Reducibility among combinatorial problems. In: Proc. Symp. Compl. Comp. Computat., pp. 85–103. Plenum Press (1972)

14. Komusiewicz, C., Hüffner, F., Moser, H., Niedermeier, R.: Isolation concepts for efficiently enumerating dense subgraphs. Theor. Comput. Sci. **410**(38–40), 3640–3654 (2009)

15. Kunegis, J.: KONECT: the Koblenz network collection. In: World Wide Web Conference, WWW, pp. 1343–1350 (2013)

16. Molter, H., Niedermeier, R., Renken, M.: Isolation concepts applied to temporal clique enumeration. Netw. Sci. **9**(S1), S83–S105 (2021)

17. Moon, J., Moser, L.: On cliques in graphs. Israel J. Math. **3**, 23–28 (1965)

18. Okubo, Y., Haraguchi, M., Tomita, E.: Enumerating maximal isolated cliques based on vertex-dependent connection lower bound. In: GPC 2020. LNCS (LNAI), vol. 12398, pp. 569–583. Springer, Cham (2016). https://doi.org/10.1007/978-3-319-41920-6_45

19. Shigezumi, T., Uno, Y., Watanabe, O.: A new model for a scale-free hierarchical structure of isolated cliques. J. Graph Algorithms Appl. **15**(5), 661–682 (2011)

20. Tomita, E., Tanaka, A., Takahashi, H.: The worst-case time complexity for generating all maximal cliques and computational experiments. Theor. Comput. Sci. **363**(1), 28–42 (2006)

21. Uno, Y., Oguri, F.: Contracted webgraphs - scale-freeness and structure mining -. IEICE Trans. Commun. **96-B**(11), 2766–2773 (2013)

22. Uno, Y., Ota, Y., Uemichi, A.: Web structure mining by isolated stars. In: Algorithms and Models for the Web-Graph. LNCS, vol. 4936, pp. 149–156 (2006)

23. Uno, Y., Ota, Y., Uemichi, A.: Web structure mining by isolated cliques. IEICE Trans. Inf. Syst. **90-D**(12), 1998–2006 (2007)

24. Wang, J., Cheng, J., Fu, A.W.: Redundancy-aware maximal cliques. In: Knowledge Discovery and Data Mining, KDD, pp. 122–130 (2013)

On the Complexity of Hyperpath and Minimal Separator Enumeration in Directed Hypergraphs

Kazuhiro Kurita[1]([⊠])[iD] and Kevin Mann[2][iD]

[1] Okayama University, Okayama, Japan
`k-kurita@okayama-u.ac.jp`
[2] Universität Trier, Trier, Germany
`mann@uni-trier.de`

Abstract. In this paper, we address the enumeration of (induced) s-t paths and minimal s-t separators. These problems are some of the most famous classical enumeration problems that can be solved in polynomial delay by simple backtracking for a (un)directed graph. As a generalization of these problems, we consider the (induced) s-t hyperpath and minimal s-t separator enumeration in a *directed hypergraph*. We show that extending these classical enumeration problems to directed hypergraphs drastically changes their complexity. More precisely, there are no output-polynomial time algorithms for the enumeration of induced s-t hyperpaths and minimal s-t separators unless $\mathsf{P} = \mathsf{NP}$, and the s-t hyperpath enumeration is at least as hard as the minimal transversal enumeration, even if a directed hypergraph is BF-hypergraph. As a positive result, s-t hyperpath enumeration for a B-hypergraph can be solved in polynomial delay by backtracking.

Keywords: Output-sensitive enumeration · Directed hypergraph · s-t hyperpath · Minimal s-t separator · NP-hardness · Polynomial-delay

1 Introduction

Enumeration problems on graphs have been widely studied, and in particular, the complexity of enumerating minimal subsets that satisfy connectivity constraints has been widely studied, i.e., (induced) s-t paths [29,33], minimal s-t cuts and multiway cuts [24,26], minimal s-t separators [10,30], minimal strongly connected spanning subgraphs [22], and minimal Steiner trees and its variants [23,24]. When evaluating the complexity of enumeration algorithms, we sometimes use an *output-sensitive* manner [21]. An enumeration algorithm is called an *output-polynomial time* algorithm if the total running time is bounded by $O(\mathrm{poly}(n + N))$ time, where n is the size of an input and N is the number of outputs. In enumeration algorithms, *delay* is used as an efficiency measure in addition to the total computation time. The delay is defined as the maximum time until the first solution is output, the time between outputting the i-th and $i+1$-th solutions, and the

E. Di Giacomo and D. Mondal (Eds.): WALCOM 2026, LNCS 16444, pp. 560–575, 2026.
https://doi.org/10.1007/978-981-95-7127-7_37

time between outputting the last solution and termination. If the delay is bounded by $O(\text{poly}(n))$, we call it a *polynomial-delay algorithm*.

Directed hypergraphs are a generalization of directed graphs [7,19]. In a directed graph, an arc consists of a single tail and a single head. In contrast, in a directed hypergraph, a hyperarc is defined by a disjoint pair of sets of vertices. One set of vertices represents the *tails*, and the other set of vertices represents the *heads*. A hyperarc whose tail consists of a single vertex is called a *backward hyperarc* (or B-hyperarc), and a hyperarc whose head consists of a single vertex is called a *forward hyperarc* (or F-hyperarc). In particular, a directed hypergraph in which every hyperarc is B-hyperarc is called a *B-hypergraph*, and a directed hypergraph in which every hyperarc is either B-hyperarc or F-hyperarc is called a *BF-hypergraph*. In addition, as a generalization of reachability in directed graphs, the concept of B-connectivity is defined for directed hypergraphs. As a reachability in directed graphs, B-connectivity from a vertex s to a vertex t is defined as follows. First, a vertex s is B-connected from s. If there is a hyperarc A such that all tails of A is B-connected from s, then all heads of A are B-connected from s. B-connectivity is defined recursively as descrived above.

A B-connectivity on a directed hypergraph is a mathematical model that can represent a broader range of problems than connectivity on a directed graph. It is used to model various applications such as chemical reaction networks [26], the formulation of the maximum Horn SAT problem [18,19], transportation networks [20,34], and conflict-free Petri nets [1–3]. By Alimonti et al. [3], there is a connection between reachability in conflict-free Petri nets and paths on B-hypergraphs[1].

While directed hypergraphs can model a wider variety of phenomena than directed graphs, optimization problems on directed hypergraphs become NP-complete, even for problems that can be solved in polynomial time on directed graphs. Examples of problems that become NP-complete on directed hypergraphs include finding a directed cycle [14],[2] a minimum s-t cut [18], the strongly connected components [4], and a shortest path [6,7]. See the following survey for more details [7].

Given these facts, it is evident that when extending many optimization problems to directed hypergraphs, they tend to be intractable. We aim to investigate how the complexity of fundamental enumeration problems changes when extending from directed graphs to directed hypergraphs.

Enumeration of (induced) s-t paths, minimal s-t cuts, and minimal separators in a (directed) graph are fundamental problems in the field of enumeration algorithms, and many theoretically efficient enumeration algorithms have been developed [11, 28–30, 32, 33]. It is well known that these problems can be enumerated with polynomial delay using a simple backtracking approach (also called the binary partition and flashlight approach).

---

[1] The objects called hypergraphs in [3], are called B-hypergraphs in this paper.

[2] Note that there are multiple definitions of a directed cycle in a directed hypergraph. The definition used in the paper [14] is given in Sect. 2.

The goal of this paper is to show that enumerating induced s-t hyper-paths and minimal s-t separators in directed hypergraphs cannot be solved in output-polynomial time unless $\mathsf{P} = \mathsf{NP}$. Furthermore, we prove that if there is an output-polynomial time algorithm for the s-t hyperpath enumeration for BF-hypergraphs, then the minimal transversal can be solved in output-polynomial time. The existence of an output-polynomial time algorithm for minimal transversal enumeration has remained an open problem for over 45 years and is one of the most famous open problems in the field of enumeration algorithms [16]. Finally, as a positive result, we show that when directed hypergraphs are restricted to B-hypergraphs, s-t hyperpaths can be enumerated with polynomial delay using a backtracking approach. Our algorithm solves a slightly more general problem than the s-t hyperpath enumeration, and it can be applied to minimal directed Steiner tree enumeration. Moreover, this positive result shows that we can enumerate all minimal unsatisfiable subformulas of a Horn SAT formula in polynomial delay.

Due to space limitations, we omit several proofs marked with $\bigstar$.

2   Preliminaries

Let V be a set of elements. We define $\mathcal{A}$ as a set of pairs of disjoint subsets of V. We call $\mathcal{D} = (V, \mathcal{A})$ a *directed hypergraph*. An element of V and $\mathcal{A}$ is called a *vertex* and a *hyperarc* of $\mathcal{D}$, respectively. We denote the set of vertices and hyperarcs in $\mathcal{D}$ as $V(\mathcal{D})$ and $\mathcal{A}(\mathcal{D})$. Moreover, we denote $\sum_{A \in \mathcal{A}(\mathcal{D})} |A|$ as $\|\mathcal{A}\|$. For a hyperarc $A = (T, H)$, T and H are called the *tails* of A and *heads* of A, respectively. It is denoted by $T(A)$ and $H(A)$, respectively. A hyperarc A is called a B-*hyperarc* (a F-*hyperarc*) if $|H(A)| = 1$ ($|T(A)| = 1$). Notice that B and F are abbreviations of "backward" and "forward", respectively. A directed hypergraph $\mathcal{D} = (V, \mathcal{A})$ is called a B-*hypergraph* (a BF-*hypergraph*) if any hyperarc in $\mathcal{A}$ is a B-hyperarc (a B-hyperarc or a F-hyperarc). For a directed hypergraph $\mathcal{D} = (V, \mathcal{A})$, a directed hypergraph $\mathcal{F} = (U, \mathcal{B})$ is a *subhypergraph of* $\mathcal{D}$ if $U \subseteq V$ and $\mathcal{B} \subseteq \mathcal{A}$. For a set of hyperarcs $\mathcal{B} \subseteq \mathcal{A}$, a directed hypergraph $\mathcal{F} = (U, \mathcal{B})$ is an *edge-induced subhypergraph of* $\mathcal{D}$ induced by $\mathcal{B}$ if $U = \bigcup_{(T,H) \in \mathcal{B}} (T \cup H)$. We denote it as $\mathcal{D}[\mathcal{B}]$. A directed hypergraph $\mathcal{F} = (U, \mathcal{B})$ is an *induced subhypergraph of* $\mathcal{D}$ induced by U if $U \subseteq V$ and $B = \{A \in \mathcal{A} \mid H(A) \subseteq U, T(A) \subseteq U\}$. We denote it as $\mathcal{D}[U]$.

In the theory of directed hypergraphs, there exist multiple definitions of connectivity and reachability. We adopt B-connectivity as the notion of connectivity [7,31].[3] The B-*connection* from a vertex s is defined as follows: (i) s is B-connected from s, and (ii) if there is a hyperarc A such that all the vertices in $T(A)$ are B-connected from s, then $H(A)$ is B-connected from s. A B-*hyperpath* from s to a vertex t is an inclusion-wise minimal set of hyperarcs such that t is B-connected from s in its edge-induced subhypergraph. A B-hyperpath is

---

[3] Gallo et al. [19] provides a characterization for making a vertex t B-connected from a vertex s, but Nielsen et al. points out that it is incorrect [27].

sometimes treated as a set of hyperarcs, and at other times as a subhypergraph. From the minimality of B-hyperpath, the set of vertices in an s-t hyperpath is uniquely defined by a set of hyperarcs. Thus, we refer to a set of hyperarcs satisfying the condition as an s-t hyperpath hereafter. A set of hyperarcs P_{st} is an s-t hyperpath if and only if there is a sequence of hyperarcs $(A_1, \ldots, A_k)$ such that (i) for each A_i, $T(A_i) \subseteq \{s\} \cup \bigcup_{1 \leq j \leq i-1} H(A_j)$, (ii) $t \in H(A_k)$, and (iii) for any proper subhypergraph of P_{st}, t is not B-connected from s. See Sect. 2 in [5] for more details. We give an example of an s-t hyperpath in Fig. 1. A set of vertices P_{st} is an *induced s-t hyperpath* if t is B-connected from s in $\mathcal{D}[P_{st}]$ and t is not B-connected from s in any by a subset of P_{st} induced subhypergraph. A set of vertices X of $\mathcal{D} = (V, \mathcal{A})$ is an *s-t separator* if X does not contain s and t and t is not B-connected from s in $D[V \setminus X]$. An s-t separator X is *minimal* if any proper subset of X is not an s-t separator.

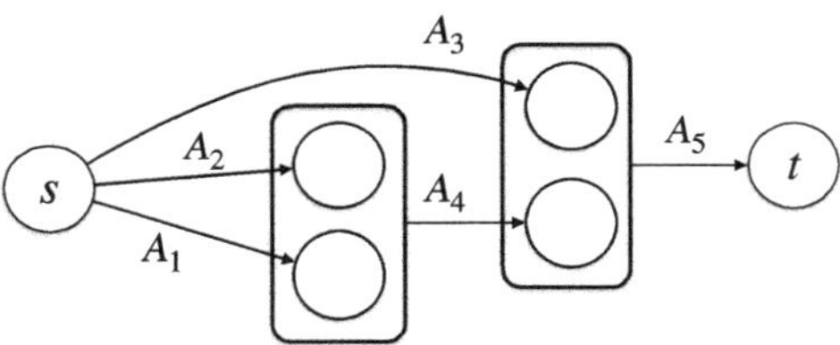

Fig. 1. An example of a directed hypergraph and an s-t hyperpath. If the cardinality of a tail is more than one, we represent tails using a rectangle. It has an s-t hyperpath with an order representation $(A_1, A_2, A_3, A_4, A_5)$.

In this paper, we deal with the following enumeration problems.

Problem 1 ((Induced) s-t Hyperpath Enumeration). Let $\mathcal{D} = (V, \mathcal{A})$ be a directed hypergraph and s and t be vertices in V. The task is to enumerate all (induced) s-t hyperpaths in $\mathcal{D}$.

Problem 2 (Minimal s-t Separator Enumeration). Let $\mathcal{D} = (V, \mathcal{A})$ be a directed hypergraph and s and t be vertices in V. The task is to enumerate all minimal s-t separators in $\mathcal{D}$.

3   Hardness Results

We present three hardness results in this section. In Sect. 3.1 and Sect. 3.2, we show that there are no output-polynomial time algorithms for enumerating all induced s-t hyperpaths and minimal s-t separators on B-hypergraphs, respectively, unless $\mathsf{P} = \mathsf{NP}$. In Sect. 3.3, we show that s-t Hyperpath Enumeration for BF-hypergraphs is at least as hard as Minimal Transversal Enumeration. More formally, if we have an output-polynomial time algorithm for s-t Hyperpath Enumeration, then minimal transversal enumeration can be solved in output polynomial time. We provide a definition of a transversal in a hypergraph $\mathcal{H} = (V, \mathcal{E})$. A set of vertices $U \subseteq V$ is a *transversal* if for any $E \in \mathcal{E}$, $U \cap E \neq \emptyset$. If any subset of U is not transversal, U is a *minimal transversal*. Note that a transversal is also referred to as a hitting set.

To show the hardness results in Sect. 3.1 and Sect. 3.2, we show the NP-hardness of *another solution problem* (also called the finished decision problem [9]). More formally, we consider the following problems.

Problem 3 (ANOTHER INDUCED s-t HYPERPATH). Let $\mathcal{D}$ be a directed hypergraph, s and t be vertices, and $\mathcal{P}_{st}$ be a set of induced s-t hyperpaths. The task is to determine whether there is an induced s-t hyperpath that is not contained in $\mathcal{P}_{st}$.

Problem 4 (ANOTHER MINIMAL s-t SEPARATOR). Let $\mathcal{D}$ be a directed hypergraph, s and t be vertices, and $\mathcal{X}$ be a set of minimal s-t separators. The task is to determine whether there is a minimal s-t separator not contained in $\mathcal{X}$.

If we have an output-polynomial time algorithm for these enumeration problems, these another solution problems can be solved in polynomial time. Such a proof technique has been used as folklore [12,13,21]. For the sake of completeness, we explain why, if these another solution problems are NP-complete, their corresponding enumeration problems cannot be solved in output-polynomial time unless $\mathsf{P} = \mathsf{NP}$.

Suppose that there is an output-polynomial time algorithm ALG for INDUCED s-t HYPERPATH ENUMERATION. We run the algorithm ALG at most $(\|\mathcal{A}\| + |\mathcal{P}_{st}|)^c$ steps, where c is some constant. If there are no induced s-t hyperpaths that are not contained in $\mathcal{P}_{st}$, ALG terminates outputting $\mathcal{P}_{st}$. Otherwise, ALG does not terminate. In other words, if ALG does not terminate, it can be concluded that there exists an induced s-t hyperpath not included in $\mathcal{P}_{st}$. Therefore, if these another solution problems are NP-hard, there is no output-polynomial time enumeration algorithm unless $\mathsf{P} = \mathsf{NP}$. Note that this algorithm can answer only "Yes" or "No", but it cannot find an induced s-t hyperpath not contained in $\mathcal{P}_{st}$. However, since the aim of a decision problem is to answer "Yes" or "No," this algorithm solves these another solution problems. Moreover, this algorithm shows the existence of a polynomial-time algorithm, but it does not allow us to actually construct one. The reason is that the value of c is not explicit.

In Sect. 3.3, we show that s-t hyperpath enumeration is at least as hard as MINIMAL TRANSVERSAL ENUMERATION. More formally, we show that for a hypergraph $\mathcal{H} = (V, \mathcal{E})$, there is a bijection between $\mathrm{Tr}(\mathcal{H})$ and the set of all s-t hyperpaths in a directed hypergraph $\mathcal{D}(\mathcal{H})$, where $\mathrm{Tr}(\mathcal{H})$ is the set of minimal transversals in $\mathcal{H}$ and $\mathcal{D}(\mathcal{H})$ is a directed hypergraph with $O(|\mathcal{E}|)$ vertices and $O(|\mathcal{E}|)$ edges. A set of vertices U is called a *transversal* if for any $E \in \mathcal{E}$, $E \cap U \neq \emptyset$. If for any $U' \subset U$, U' is not a transversal, U is a *minimal transversal*. MINIMAL TRANSVERSAL ENUMERATION can be solved in output-quasi polynomial time [17]. On the other hand, output-polynomial time enumeration of minimal transversals is a longstanding open problem in the field of enumeration

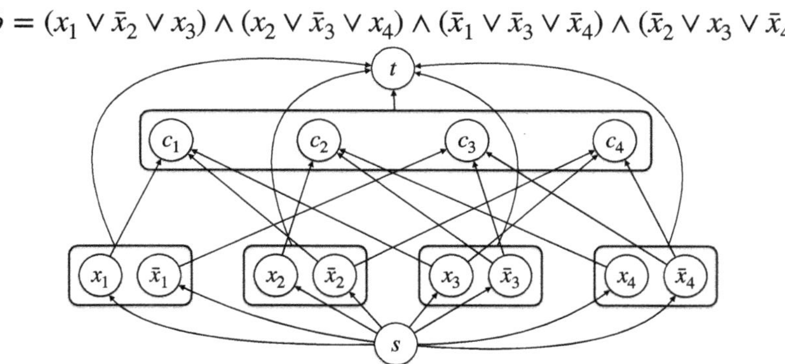

Fig. 2. A reduction from 3-SAT to ANOTHER INDUCED s-t HYPERPATH. A directed hypergraph $\mathcal{D}_\phi$ has an induced s-t hyperpath without $\{s, x_1, \bar{x}_1, t\}$, $\{s, x_2, \bar{x}_2, t\}$, $\{s, x_3, \bar{x}_3, t\}$, and $\{s, x_4, \bar{x}_4, t\}$, if and only if ϕ is satisfiable. In this case, since $x_1 = 1$, $x_2 = 1$, $x_3 = 0$, and $x_4 = 0$ is a satisfying assignment, $\mathcal{D}_\phi$ has an induced s-t hyperpath $\{s, x_1, x_2, \bar{x}_4, c_1, c_2, c_3, c_4, t\}$. If we assign $x_1 = 1$, $x_2 = 1$, and $x_4 = 0$, then ϕ is satisfied regardless of the assignment to x_3. That is, x_3 becomes a "don't care" variable, and therefore neither x_3 nor $\bar{x}_3$ is included in the hyperpath.

algorithms [16]. Thus, our "hardness" result provides evidence that solving s-t HYPERPATH ENUMERATION in output-polynomial time is not easy.

3.1   Another Induced s-t Hyperpath

We give a polynomial-time reduction from 3-SAT. Let $\phi = C_1 \wedge \ldots \wedge C_m$ be a 3-CNF formula. We denote the set of variables in ϕ as $V(\phi)$. We construct a directed hypergraph $\mathcal{D} = (V, \mathcal{A})$ from ϕ as follows. We add two vertices s and t to V. For each variable $x_i \in V(\phi)$, we add two vertices $x_i, \bar{x}_i$ to V. For each clause C_j, we add a vertex c_j to V. We next define the set of hyperarcs. For each x_i and $\bar{x}_i$, we add three hyperarcs $(\{x_i, \bar{x}_i\}, \{t\})$, $(\{s\}, \{x_i\})$, and $(\{s\}, \{\bar{x}_i\})$. For each clause $C_j = \ell_1^j \vee \ell_2^j \vee \ell_3^j$, we add three hyperarcs $(\{\ell_1^j\}, \{c_j\})$, $(\{\ell_2^j\}, \{c_j\})$, and $(\{\ell_3^j\}, \{c_j\})$. Finally, we add a hyperarc $(\{c_1, \ldots, c_m\}, \{t\})$. We denote the resultant directed hypergraph as $\mathcal{D}_\phi$, and $\mathcal{D}_\phi$ is a B-hypergraph. Notice that this reduction can be done in $O(\text{poly}(n + m))$ time, where n is the number of variables in ϕ.

The resultant directed hypergraph $\mathcal{D}_\phi$ has an induced s-t hyperpath with four vertices. For each $1 \leq i \leq n$, $\{s, x_i, \bar{x}_i, t\}$ is an induced s-t hyperpath. We denote the set of such induced s-t hyperpaths as $\mathcal{P}_\phi$. Figure 2 provides an example of our reduction. We show that $\mathcal{D}_\phi$ has an induced s-t hyperpath that is not included in $\mathcal{P}_\phi$ if and only if ϕ has a satisfying assignment.

Lemma 1. *For a 3-CNF formula ϕ, ϕ has a satisfying assignment if and only if $\mathcal{D}_\phi$ has an induced s-t hyperpath that is not contained in $\mathcal{P}_\phi$.*

Proof. Suppose that ϕ has a satisfying assignment $\alpha : V(\phi) \to \{0, 1\}$. We define a set of vertices in $\mathcal{D}_\phi$ using α. We define $U = V(\mathcal{D}_\phi)$. If $\alpha(x_i) = 0$, we remove a vertex x_i from U, otherwise if $\alpha(x_i) = 1$, we remove a vertex $\bar{x}_i$ from U. We show that there is an induced s-t hyperpath contained in U. Since each clause $C_i = \ell_1^i \vee \ell_2^i \vee \ell_3^i$ is satisfied by α, at least one literal ℓ_j^i becomes 1. Therefore, each c_i is B-connected from s in $\mathcal{D}_\phi[U]$. Since $\mathcal{D}_\phi[U]$ has a hyperarc $(\{c_1, \dots, c_m\}, \{t\})$, t is B-connected from s in $\mathcal{D}_\phi[U]$ and there is an induced s-t hyperpath contained in U.

We show the opposite direction. Let P be an induced s-t hyperpath that is not contained in $\mathcal{P}_\phi$. For each pair x_i and $\bar{x}_i$, $x_i \notin P$ or $\bar{x}_i \notin P$. If P contains both x_i and $\bar{x}_i$, it contradicts the minimality since t is B-connected from s in $\mathcal{D}_\phi[\{s, t, x_i, \bar{x}_i\}]$. If there is a vertex $c_j \notin P$, $\mathcal{D}_\phi[P]$ does not have a hyperarc A that contains t as the head. Therefore, it contradicts that P is an induced s-t hyperpath. From the above discussion, $\{c_1, \dots, c_m\} \subseteq P$ and $x_i \notin P$ or $\bar{x}_i \notin P$ for each $1 \leq i \leq n$. We define an assignment α_P as follows: $\alpha_P(x_i) = 1$ if $x_i \in P$ and $\alpha_P(x_i) = 0$ if $x_i \notin P$. We show that α_P satisfies ϕ. For each clause $C_i = \ell_1^i \vee \ell_2^i \vee \ell_3^i$, at least one literal ℓ_j^i becomes 1 by α_P. Otherwise, it contradicts that either P contains $\{c_1, \dots, c_m\}$ or P is an induced s-t hyperpath. Therefore, α is a satisfying assignment of ϕ. $\qquad\square$

Lemma 1 implies there is no output-polynomial time algorithm for INDUCED s-t HYPERPATH ENUMERATION unless $\mathsf{P} = \mathsf{NP}$. By making minor modifications to our reduction, it is possible to show a stronger hardness result. ANOTHER INDUCED s-t HYPERPATH is still NP-hard even if the cardinality of a tail is at most two. Intuitively, since $\mathcal{D}$ contains only one hyperarc with a large tail, we can satisfy this condition by "decomposing" that hyperarc into multiple hyperarcs, each having a tail of cardinality two, in a structure resembling a binary tree.

Let A be the hyperarc $(\{c_1, \dots, c_m\}, \{t\}) \in \mathcal{A}(\mathcal{D}_\phi)$. To simplify the following discussion, we assume that m is a power of two. Notice that if m is not a power of two, we add one dummy variable z and dummy clauses $(z \vee z \vee z)$ to make m a power of two. We add vertices $d_2, \dots, d_{m-1}$ to $\mathcal{D}$, and replace a hyperarc $(\{c_1, \dots, c_m\}, \{t\})$ with hyperarcs $\{(\{w_{2i}, w_{2i+1}\}, \{w_i\}) | 1 \leq i \leq m - 1\}$, where we define w_i as follows. We define $w_1 = t$, if i is at least m and at most $2m - 1$, $w_i = c_{i-m+1}$, and otherwise $w_i = d_i$. We denote the resulting graph as $\mathcal{D}_\phi'$. Notice that the number of dummy clauses is at most m, this reduction works in polynomial time. Any induced s-t hyperpath contains all vertices in $\mathcal{D}'$, there is a bijection between the set of all induced s-t hyperpaths in $\mathcal{D}_\phi$ and $\mathcal{D}_\phi'$. Therefore, Lemma 1 also holds even for $\mathcal{D}_\phi'$ and the following theorem and corollary hold.

Theorem 1. ANOTHER INDUCED s-t HYPERPATH *on a B-hypergraph $\mathcal{D} = (V, \mathcal{A})$ is* NP-*complete even if* $|\mathcal{P}_{st}| \leq |V|$ *and* $|T(A)| \leq 2$ *for any* $A \in \mathcal{A}$. *Therefore, there are no output-polynomial time algorithms for enumerating all induced s-t hyperpaths in a B-hypergraph unless* $\mathsf{P} = \mathsf{NP}$ *even if* $|T(A)| \leq 2$ *for any* $A \in \mathcal{A}$.

3.2   Another Minimal *s-t* Separator

We give a reduction from 3-SAT. This reduction is similar to the reduction in the previous subsection. Let $\phi = C_1 \wedge \ldots \wedge C_m$ be a 3-CNF formula. We construct $\mathcal{D}_\phi = (V_\phi, \mathcal{A}_\phi)$ as follows. We add vertices s, t, and c to V_ϕ. For each variable $x_i \in V(\phi)$, we add three vertices $x_i, \bar{x}_i, y_i$ to V_ϕ.

We next define $\mathcal{A}_\phi$. For each $x_i \in V(\phi)$, we add $(\{x_i\}, \{y_i\})$, $(\{\bar{x}_i\}, \{y_i\})$, $(\{s\}, \{x_i\})$, and $(\{s\}, \{\bar{x}_i\})$. For each clause $C_j = \ell_1^j \vee \ell_2^j \vee \ell_3^j$, we add a hyperarc $(\{\bar{\ell}_1^j, \bar{\ell}_2^j, \bar{\ell}_3^j\}, \{c\})$. Finally, we add a hyperarc $(\{c, y_1, \ldots, y_n\}, \{t\})$, where n is the number of variables in ϕ.

$$\phi = (x_1 \vee \bar{x}_2 \vee x_3) \wedge (x_2 \vee \bar{x}_3 \vee x_4) \wedge (\bar{x}_1 \vee \bar{x}_3 \vee \bar{x}_4) \wedge (\bar{x}_2 \vee x_3 \vee \bar{x}_4)$$

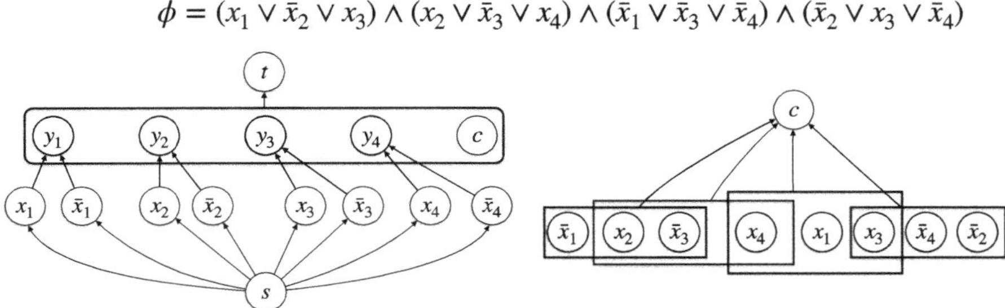

Fig. 3. An example of our reduction from 3-SAT to ANOTHER MINIMAL *s-t* SEPARATOR. The left figure shows the vertices and hyperarcs added for the variables of ϕ, while the right figure shows the hyperarcs added for the clauses. The entire directed hypergraph $\mathcal{D}_\phi$ is the union of the vertices and hyperarcs of these two directed hypergraphs. In this case, $\mathcal{D}_\phi$ has the following minimal *s-t* separators: $\{x_1, \bar{x}_1\}$, $\{x_2, \bar{x}_2\}$, $\{x_3, \bar{x}_3\}$, $\{x_4, \bar{x}_4\}$, $\{y_1\}$, $\{y_2\}$, $\{y_3\}$, $\{y_4\}$, and $\{c\}$. Since ϕ has a satisfying assignment $x_1 = 1$, $x_2 = 1$, $x_3 = 0$, and $x_4 = 0$, $\mathcal{D}_\phi$ has a minimal *s-t* separator $\{x_1, x_2, \bar{x}_4\}$. If we assign $x_1 = 1$, $x_2 = 1$, and $x_4 = 0$, then ϕ is satisfied regardless of the assignment to x_3. That is, x_3 becomes a "don't care" variable, and therefore neither x_3 nor $\bar{x}_3$ is included in the hyperpath.

We next define a set of minimal *s-t* separators $\mathcal{X}(\phi)$. As a set of vertices $\{c\}$ or $\{y_j\}$ for any $1 \leq j \leq n$ is a minimal *s-t* separator, we add such trivial minimal separators to $\mathcal{X}(\phi)$. For each variable $x_i \in V(\phi)$, $\{x_i, \bar{x}_i\}$ is also a minimal *s-t* separator. We define a set of minimal *s-t* separators $\mathcal{X}(\phi)$ as the above $2n + 1$ minimal *s-t* separators. Figure 3 gives an example of our reduction.

Lemma 2. *For a 3-CNF formula ϕ, ϕ has a satisfying assignment if and only if $\mathcal{D}_\phi$ has a minimal s-t separator that is not contained in $\mathcal{X}(\phi)$.*

Proof. Suppose that ϕ has a satisfying assignment α. We define a set of vertices using α. Let x_i be a variable in ϕ. The set of vertices X contains a vertex x_i if $\alpha(x_i) = 1$, otherwise X contains $\bar{x}_i$. From the construction of $\mathcal{D}_\phi$, each y_i is B-connected from s in $\mathcal{D}_\phi[V(\mathcal{D}_\phi) \setminus X]$. Moreover, since α is a satisfying assignment,

c is not B-connected from s in $\mathcal{D}_\phi[V(D_\phi) \setminus X]$. Therefore, X is an s-t separator. Since X an s-t separator, there is a minimal s-t separator $X' \subseteq X$. From the construction of X', $X' \notin \mathcal{X}(\phi)$.

Suppose that $\mathcal{D}_\phi$ has a minimal s-t separator $X \notin \mathcal{X}(\phi)$. From the construction of $\mathcal{X}(\phi)$ and $X \notin \mathcal{X}(\phi)$, for each $1 \le i \le |V(\phi)|$, $x_i \notin X$ or $\bar{x}_i \notin X$. We define an assignment α_X as follows. If $x_i \in X$, $\alpha_X(x_i) = 1$, otherwise $\alpha_X(x_i) = 0$. From the construction of $\mathcal{D}_\phi$, each clause is satisfied. Otherwise, $\mathcal{D}_\phi[V(\mathcal{D}_\phi) \setminus X]$ has an s-t hyperpath and it contradicts that X is an s-t separator. Therefore, α_X is a satisfying assignment of ϕ. $\qquad\square$

We improve our reduction to the cardinality of each tail to two by making a small modification, In our reduction, the maximum cardinality of each tail is $n + 1$ since $\mathcal{D}_\phi$ has a hyperarc $A = (\{y_1, \ldots, y_n, c\}, \{t\})$. In addition, $\mathcal{D}$ has m hyperarcs whose tail has the cardinality three. To simplify the following discussion, we assume that $n + m + 1$ is a power of two. If $n + m + 1$ is not a power of two, we add one dummy variable z and dummy clauses $(z \vee z \vee z)$ to make $n + m + 1$ a power of two.

For each clause $C_j = \ell_1^j \vee \ell_2^j \vee \ell_3^j$, we add one vertex c_j and replace a hyperarc $(\{\bar{\ell}_1^j, \bar{\ell}_2^j, \bar{\ell}_3^j\}, \{c\})$ with two hyperarcs $(\{\bar{\ell}_1^j, \bar{\ell}_2^j\}, \{c_j\})$ and $(\{c_j, \bar{\ell}_3^j\}, \{c\})$. We replace a hyperarc $(\{c, y_1, \ldots, y_n\}, \{t\})$ with $A = (\{c, c_1, \ldots, c_m, y_1, \ldots, y_n\}, \{t\})$. As a result of this modification, all hyperarcs except one have tails with cardinality at most two. This modification adds only $\{c_j\}$ for each $1 \le j \le m$ as new minimal s-t separators since $\{c_j\}$ is a minimal s-t separator. Applying the similar technique as in the previous section, we add vertices $d_2, \ldots, d_{n+m}$, and replace a hyperarc $A = (\{y_1, \ldots, y_n, c_1, \ldots, c_m, c\}, \{t\})$ with a set of hyperarcs $\{(\{w_{2i}, w_{2i+1}\}, \{w_i\}) | 1 \le i \le n+m\}$, where $w_1 = t$, $w_i = d_i$ if i is at least 2 and at most $n+m$, $w_{n+m+1} = c$, $w_i = c_{i-n-m-1}$ if i is at least $n + m + 2$ and at most $n + 2m + 1$, otherwise, $w_i = y_{i-n-2m-1}$. Applying this modification, we obtain new $O(n + m)$ minimal s-t separators. Therefore, by adding such small minimal separators to $\mathcal{X}(\phi)$, the set of minimal separators that is not contained in $\mathcal{X}(\phi)$ does not change.

Theorem 2. *ANOTHER MINIMAL s-t SEPARATOR on a B-hypergraph $\mathcal{D} = (V, \mathcal{A})$ is* NP-*complete even if $|\mathcal{X}| \le |V|$ and $|T(A)| \le 2$ for any $A \in \mathcal{A}$. Therefore, there are no output-polynomial time algorithms for enumerating all minimal s-t separators in a B-hypergraph unless* P $=$ NP *even if $|T(A)| \le 2$ holds for any $A \in \mathcal{A}$.*

3.3 s-t Hyperpath Enumeration

Let $\mathcal{H} = (V, \mathcal{E})$ be a hypergraph. We construct a directed hypergraph $\mathcal{D}_\mathcal{H} = (V_\mathcal{H}, \mathcal{A}_\mathcal{H})$ as follows. The set of vertices $V_\mathcal{H}$ consists of $\mathcal{E} \cup \{s, t\}$. For each $v \in V$, we add a hyperarc $(\{s\}, \mathcal{E}_v)$, where $\mathcal{E}_v$ is the set of hyperedges in $\mathcal{E}$ that contain v. Note that $\mathcal{E}_v$ is the set of hyperedges in $\mathcal{H}$, but it is the set of vertices in $\mathcal{D}_\mathcal{H}$. Thus, $(\{s\}, \mathcal{E}_v)$ is a pair of distinct sets of vertices in $V_\mathcal{H}$. Additionally, we add hyperarc $(\mathcal{E}, \{t\})$. Since each hyperarc $A \in \mathcal{A}_\mathcal{H}$ satisfies either $|T(A)| = 1$ or $|H(A)| = 1$, $\mathcal{D}_\mathcal{H}$ is a BF-hypergraph.

Lemma 3 ($\bigstar$). *Let $\mathcal{H}$ be a hypergraph and $\mathcal{D}_{\mathcal{H}}$ be a BF-hypergraph. There is a bijection between the set of minimal transversals in $\mathcal{H}$ and the set of s-t hyperpaths in $\mathcal{D}_{\mathcal{H}}$.*

Proof. Let $\mathcal{H}$ be a hypergraph and $\mathcal{D}_{\mathcal{H}}$ be a BF-hypergraph obtained by the above procedure. We define a function $f : \mathrm{Tr}(\mathcal{H}) \to 2^{\mathcal{A}_{\mathcal{H}}}$ such that $f(T) = \{(\{s\}, \mathcal{E}_v) | v \in T\} \cup \{(\mathcal{E}, \{t\})\}$, where T is a minimal transversal of $\mathcal{H}$.

We show that for a minimal transversal T, $f(T)$ is an s-t hyperpath. To this end, we show that t is B-connected from s in $\mathcal{D}_{\mathcal{H}}[f(T)]$. Since T is a transversal, any vertex in $V(\mathcal{D}_{\mathcal{H}}) \setminus \{s, t\}$ is connected to s. From the definition of $f(T)$, $f(T)$ contains a hyperarc $(\mathcal{E}, \{t\})$. Since any vertex in $V(\mathcal{D}_{\mathcal{H}}) \setminus \{s, t\}$ is connected to s and $f(T)$ contains $(\mathcal{E}, \{t\})$, t is connected to s. We next show the minimality of $f(T)$ by contradiction. Suppose that there is an s-t hyperpath P_{st} contained in $f(T)$. If $(\mathcal{E}, \{t\}) \notin P_{st}$, P_{st} is not s-t hyperpath. Thus, a hyperarc $A \in f(T) \setminus P_{st}$ satisfies $T(A) = \{s\}$ and $H(A) = \mathcal{E}_v$ for some $v \in V(\mathcal{H})$. However, since T is a minimal transversal, there is a hyperedge $E \in \mathcal{E}_v$ such that $T \cap E = \{v\}$. In this case, E is not connected to s. Therefore, $f(T)$ is an s-t hyperpath.

We next show that f is a bijection between $\mathrm{Tr}(\mathcal{H})$ and the set of all s-t hyperpaths in $\mathcal{D}_{\mathcal{H}}$. We first show that f is injective. Let T_1 and T_2 be two distinct minimal transversals. Since T_1 and T_2 are distinct, there are vertices $v_1 \in T_1 \setminus T_2$ and $v_2 \in T_2 \setminus T_1$. Therefore, $f(T_1) \setminus f(T_2)$ contains a hyperarc $(\{s\}, \mathcal{E}_{v_1})$ and $f(T_2) \setminus f(T_1)$ contains a hyperarc $(\{s\}, \mathcal{E}_{v_2})$, and f is injective.

We next show that f is surjective. Let P_{st} be an s-t hyperpath in $\mathcal{D}_{\mathcal{H}}$. From the construction of $\mathcal{D}_{\mathcal{H}}$, each hyperarc A satisfying $T(A) = \{s\}$ corresponds to a vertex in $\mathcal{H}$. Thus, we consider a set of vertices T corresponds to $P_{st} \setminus \{(\mathcal{E}, \{t\})\}$. Since P_{st} is an s-t hyperpath, each vertex $v \in T$ has a hyperedge E such that $T \cap E = \{v\}$. Therefore, T is a minimal transversal and f is a bijection. $\square$

Theorem 3. *If there is an output-polynomial time algorithm for s-t* HYPER-PATH ENUMERATION *on a BF-hypergraph $\mathcal{D}$, there is an output-polynomial time algorithm for* MINIMAL TRANSVERSAL ENUMERATION.

Proof. Suppose that there is an output-polynomial time algorithm for s-t HYPERPATH ENUMERATION on a BF-hypergraph. For a hypergraph $\mathcal{H} = (V, \mathcal{E})$, we construct $\mathcal{D}_{\mathcal{H}}$ in $O(||\mathcal{H}||)$ time, where $||\mathcal{H}||$ is the sum of the number of vertices and the cardinality of hyperedges. Since we have an output-polynomial time algorithm for enumerating all s-t hyperpaths in $\mathcal{D}_{\mathcal{H}}$, we obtain the set of all s-t hyperpaths in $\mathcal{D}_{\mathcal{H}}$. From Lemma 3, there is a bijection between the set of minimal transversals in $\mathcal{H}$ and the set of all s-t hyperpaths in $\mathcal{D}_{\mathcal{H}}$. Moreover, we can restore $\mathrm{Tr}(\mathcal{H})$ in $O(||\mathcal{H}|| \cdot \mathrm{Tr}(\mathcal{H}))$ time. Therefore, we obtain an output-polynomial time algorithm for MINIMAL TRANSVERSAL ENUMERATION.

4   Polynomial-Delay Enumeration of *s-t* Hyperpaths for *B*-Hypergraphs

We give a positive result of the *s-t* hyperpath enumeration for a B-hypergraph. In this section, a directed hypergraph may have parallel hyperarcs, that is, there may be hyperarcs with the same pair of tail set and head. The proposed algorithm solves a slightly more general problem than the s-t hyperpath enumeration. We first extend the concept of an s-t hyperpath.

Let S be a set of vertices. We define B-*connection from S* as follows: (i) a vertex $s \in S$ is B-connected from S and (ii) if there is a hyperarc A such that all the vertices in $T(A)$ are B-connected from S, then $H(A)$ is B-connected from S. For a set of vertices S and T, an inclusion-wise minimal set of hyperarcs P_{ST} is an *S-T hyperpath* if all vertices in T are B-connected from S. It should be noted that an S-T hyperpath is also a generalization of a minimal directed Steiner tree [25]. Hereafter, we give a polynomial-delay enumeration algorithm for enumerating all S-T hyperpaths. We first introduce a characterization for an S-T hyperpath in a B-hypergraph.

Lemma 4. *For a dictred hypergraph $\mathcal{D}$, a set of hyperarcs P_{ST} is an S-T hyperpath if and only if the followings hold:*

1. *There is a sequence of hyperarcs $(A_1, \ldots, A_{|P_{ST}|})$ such that for each A_i, $T(A_i) \subseteq S \cup \bigcup_{1 \le j \le i-1} H(A_j)$,*
2. *$|\{A \in P_{ST} \mid \{v\} = H(A)\}| = 0$ if $v \in S$, otherwise $|\{A \in P_{ST} \mid \{v\} = H(A)\}| = 1$, and*
3. *$|\{A \in P_{ST} \mid v \in T(A)\}| \ge 1$ if $v \in V(\mathcal{D}[P_{ST}]) \setminus (T \cup S)$.*

Proof. Suppose that P_{ST} satisfies the three conditions and $(A_1, \ldots, A_{|P_{ST}|})$ is an ordering of P_{ST}. From Conditions 1, all vertices in $\mathcal{D}[P_{ST}]$ are B-connected from S. We show the minimality of P_{ST} by contradiction. Suppose that $P'_{ST} \subsetneq P_{ST}$ is an S-T hyperpath. Let A be the largest hyperarc in $P_{ST} \setminus P'_{ST}$ with respect to the order of P_{ST} and a is the head of A. If a is contained in T, a is not B-connected from S since P'_{ST} has no hyperarcs with a as its head. Thus, a is not in T. If a is contained in S, it contradicts Condition 2. Thus, it is contained in $V \setminus (S \cup T)$. From Condition 3, P_{ST} contains a hyperarc A' that contains a as its tail. From Condition 1, A' is larger than A with respect to the order $(A_1, \ldots, A_{|P_{ST}|})$. If A' is contained in P'_{ST}, it contradicts the minimality of P'_{ST} since removing A' from P'_{ST} does not change B-connectivity for other vertices. If A' is not contained in P'_{ST}, it contradicts the condition of A since A' is larger than A with respect to the order of P_{ST}, and P_{ST} is an S-T hyperpath.

Suppose that P_{ST} is an S-T hyperpath. Since P_{ST} is an S-T hyperpath, Conditions 2 and 3 are satisfied, otherwise, it contradicts the minimality of P_{ST}. Therefore, we show that P_{ST} has an order satisfying Condition 1. We order P_{ST} as follows. Let $\mathcal{A}_1$ be the set of hyperarcs such that $T(A) \subseteq S$. We recursively define $\mathcal{A}_j$ as $\{A \in P_{ST} \setminus \bigcup_{1 \le j < i} \mathcal{A}_j \mid T(A) \subseteq S \cup \bigcup_{B \in \bigcup_{1 \le j < i} \mathcal{A}_j} H(B)\}$. Notice that the set of vertices $S \cup \bigcup_{B \in \mathcal{A}_i} H(B)$ is B-connected from S in a directed subhypergraph induced by $\bigcup_{1 \le j \le i} \mathcal{A}_j$. If a hyperarc $A \in P_{ST}$ is not contained in any $\mathcal{A}_i$, it contradicts the minimality of P_{ST}. Moreover, for distinct k_1 and k_2, $\mathcal{A}_{k_1}$ and $\mathcal{A}_{k_2}$ are disjoint. Therefore, $\{\mathcal{A}_1, \ldots, \mathcal{A}_\ell\}$ gives a partition of P_{ST}, where ℓ is the maximum integer satisfying $\mathcal{A}_\ell \ne \emptyset$. We consider an order of P_{ST} as $(\mathcal{A}_1, \ldots, \mathcal{A}_\ell)$. We obtain a desired order of the set of hyperarcs in P_{ST} by ordering the hyperarcs within each $\mathcal{A}_i$ in an arbitrary order.

Our goal is to enumerate all sets of hyperarcs that satisfy the conditions in Lemma 4 by a backtracking approach. To this end, we recursively partition the problem into two subproblems. This approach is inspired by Read and Tarjan's s-t path enumeration algorithm [29]. We describe an overview of our algorithm.

In our enumeration algorithm, we find an S-T hyperpath P_{ST} with an ordering $(A_1, \ldots, A_k = (T_k, h_k))$ in Lemma 4, divide enumeration problems into the enumeration of an S-T hyperpath that contains A_k and does not contain A_k. The S-T hyperpaths outputted by these two enumeration problems are clearly non-duplicate, and their union is equal to the set of S-T hyperpaths of the original problem. One subproblem, enumeration of S-T hyperpaths that do not contain A_k equals the enumeration of S-T hyperpaths in $D[\mathcal{A} \setminus \{A_k\}]$. The key observation is that the enumeration of S-T hyperpaths can also be achieved by modifying $\mathcal{D}$ and T. We give the details of our algorithm in Algorithm 1. In Algorithm 1, P is a set of hyperarcs forming a hyperpath.

Hereafter, we formally define a division of the problem. Let $\mathcal{D} = (V, \mathcal{A})$ be a directed hypergraph and P_{ST} be an S-T hyperpath and $(A_1, \ldots, A_k = (T_k, h_k))$ be an order of P_{ST} satisfying the conditions in Lemma 4. We define $\mathcal{P}_{ST}(\mathcal{D})$ as the set of S-T hyperpaths in $\mathcal{D}$. Furthermore, we introduce $\mathcal{P}^0_{ST}(\mathcal{D}, P_{ST}, A_k)$ and $\mathcal{P}^1_{ST}(\mathcal{D}, P_{ST}, A_k)$ as the set of S-T hyperpaths that do not contain A_k and contain A_k, respectively. Obviously, the set of all S-T hyperpaths on $\mathcal{D}[\mathcal{A} \setminus \{A_k\}]$ corresponds to $\mathcal{P}^0_{ST}(\mathcal{D}, P_{ST}, A_k)$. The remaining part is how to obtain $\mathcal{P}^1_{ST}(\mathcal{D}, P_{ST}, A_k)$ by modifying $\mathcal{D}$, S, and T.

From the definition of $\mathcal{P}^1_{ST}(\mathcal{D}, P_{ST}, A_k)$, any S-T hyperpath in $\mathcal{P}^1_{ST}(\mathcal{D}, P_{ST}, A_k)$ contains A_k. From Lemma 4, the number of hyperarc that has h_k as the head equals one, that is, there are no S-T hyperpaths that contain a hyperarc whose head is h_k without A_k. Thus, removing all hyperarcs that contain h_k as its head by $\mathcal{D}$ does not change the set of S-T hyperpaths containing A_k. Moreover, $u \in T_k$ is B-connected from S in any S-T hyperpath in $\mathcal{P}^1_{ST}(\mathcal{D}, P_{ST}, A_k)$ by Lemma 4. Thus, for any $Q \in \mathcal{P}^1_{ST}(\mathcal{D}, P_{ST}, A_k)$ and $B \in Q$ such that $T(B)$ contains h_k, any vertex in $(T(B) \setminus \{h_k\}) \cup T_k$ is B-connected

---

Algorithm 1: A polynomial-delay and polynomial-space algorithm for enumerating all S-T hyperpaths. When this algorithm is called as EnumHyperpaths($\mathcal{D}, S, T, \emptyset$), it enumerates all S-T hyperpaths.

```
1   Procedure EnumHyperpaths(D = (V, A), S, T, P)
2       if T = ∅ then Output P and return
3       if D has no S-T hyperpaths then return
4       Let (A₁, ..., Aₖ = (Tₖ, hₖ)) be an S-T hyperpath.
5       EnumHyperpaths(D[A \ {Aₖ}], S, T, P)
6       A', T' ← ∅, (T ∪ Tₖ) \ (S ∪ {hₖ})          // Note that A' is a multiset.
7       foreach B ∈ A \ {A'' ∈ A|H(A'') = hₖ} do
8           if hₖ ∈ T(B) then  Add ((T(B) \ {hₖ}) ∪ Tₖ, H(B)) to A'.
9           else Add B to A'.
10      EnumHyperpaths((V \ {hₖ}, A'), S, T', P ∪ {Aₖ})
```

from S in $\mathcal{D}[Q]$. Motivated by these observations, we consider the following modification of hyperarcs in $\mathcal{A} \setminus \{A'' \in \mathcal{A} | H(A) = h_k\}$. For a hyperarc $B \in \mathcal{A} \setminus \{A'' \in \mathcal{A} | H(A'') = h_k\}$, we define a function f as follows. If B does not contain h_k as its tail, $f(B) = B$, otherwise, $f(B) = ((T(B) \setminus \{h_k\}) \cup T_k, H(B))$. Based on f, we define the multiset of hyperarcs as $\mathcal{B} = \{f(B) | B \in \mathcal{A} \setminus \{A'' \in \mathcal{A} | H(A'') = h_k\}\}$. Note that f is a bijection between $\mathcal{A} \setminus \{A'' \in \mathcal{A} | H(A'') = h_k\}$ and $\mathcal{B}$. We show that there is a bijection between the set of all S-T' hyperpaths on $\mathcal{D}'$ and $\mathcal{P}^1_{ST}(\mathcal{D}, P_{ST}, A_k)$, where $\mathcal{D}' = (V \setminus \{h_k\}, \mathcal{B})$ and $T' = (T \cup T(A_k)) \setminus (S \cup \{H(A_k)\})$.

Lemma 5. *There is a bijection between $\mathcal{P}_{ST'}(\mathcal{D}')$ and $\mathcal{P}^1_{ST}(\mathcal{D}, P_{ST}, A_k)$, where $\mathcal{D}' = (V \setminus \{h_k\}, \mathcal{B})$ and $T' = (T \cup T(A_k)) \setminus (S \cup \{H(A_k)\})$.*

Proof. For each $Q \in \mathcal{P}^1_{ST}(\mathcal{D}, P_{ST}, A_k)$, we define a function $g : 2^{\mathcal{A}} \to 2^{\mathcal{A}}$ as follows: $g(Q) = \bigcup_{B \in Q \setminus \{A_k\}} f(B)$. We show that g is a bijection between $\mathcal{P}^1_{ST}(\mathcal{D}, P_{ST}, A_k)$ and $\mathcal{P}_{ST'}(\mathcal{D}')$. We first show that g is injective. Let Q_1 and Q_2 be distinct S-T hyperpaths in $\mathcal{P}^1_{ST}(\mathcal{D}, P_{ST}, A_k)$. Thus, both $Q_1 \setminus Q_2$ and $Q_2 \setminus Q_1$ are non-empty. It implies that $g(Q_1)$ and $g(Q_2)$ are distinct.

We next show that g is surjective. Let Q' be an S-T' hyperpath in $\mathcal{P}_{ST'}(\mathcal{D}')$. Since Q' is an S-T' hyperpath, any vertex in T' is B-connected from S. Thus, h_k is B-connected from S in $\mathcal{D}[Q' \cup \{A_k\}]$ since $T_k \subseteq S \cup T'$. Moreover, any vertex in $\mathcal{D}[\{A_k\} \cup \bigcup_{B' \in Q'} f^{-1}(B')]$ is B-connected from S since any vertex in T_k is B-connected from S in $\mathcal{D}[\{A_k\} \cup \bigcup_{B' \in Q'} f^{-1}(B')]$. Finally, if $\{A_k\} \cup \bigcup_{B' \in Q'} f^{-1}(B')$ is not an S-T hyperpath, then it contradicts the minimality of Q'. Therefore, $\{A_k\} \cup \bigcup_{B' \in Q'} f^{-1}(B')$ is an S-T hyperpath in $\mathcal{P}^1_{ST}(\mathcal{D}, P_{ST}, A_k)$, and g is surjective. $\qquad\square$

Lemma 5 ensures that our branching strategy correctly partitions the set of S-T hyperpaths in $\mathcal{D}$. Moreover, for each S-T' hyperpath $P'_{ST'}$ in $\mathcal{D}'$, it is easy to restore the S-T hyperpath P that corresponds to P'. Therefore, by recursively applying this partitioning, we can enumerate all S-T hyperpaths, and the following theorem holds.

Theorem 4. *Algorithm 1 enumerates all S-T hyperpaths in B-hypergraphs with $O(m^2 \|\mathcal{A}\|)$ delay and $O(m \|\mathcal{A}\|)$ space, where $m = |\mathcal{A}|$ and $\|\mathcal{A}\| = \sum_{A \in \mathcal{A}} (|H(A)| + |T(A)|)$.*

Proof. The correctness follows from Lemma 5, and the space complexity is bounded by $O(m \|\mathcal{A}\|)$ since the depth of this recursion tree is bounded by m, Thus, we give the time complexity analysis.

In each recursion step, we find an S-T hyperpath. It can be done in $O(m \|\mathcal{A}\|)$ time using the B-connectivity checking algorithm in [7]. In addition, finding P_{ST} is a bottleneck of each recursion procedure without line 6 and 10. As the depth of this recursion tree is at most m, the delay of this algorithm is $O(m^2 \|\mathcal{A}\|)$. $\square$

5 Conclusion

In this paper, we show that there are no output-polynomial time algorithms for induced s-t hyperpath enumeration and minimal s-t separator enumeration in a B-hypergraph even if the cardinality of tails is constant, unless $\mathsf{P} = \mathsf{NP}$. Moreover, enumerating s-t hyperpaths in a BF-hypergraph is at least as hard as the problem of enumerating the minimal transversals of a hypergraph. This indicates that enumeration of s-t hyperpaths in a directed hypergraph in output-polynomial time is a challenging problem. Finally, we give a polynomial-delay algorithm for enumerating all S-T hyperpaths. This algorithm is based on a simple backtracking algorithm inspired by [29].

The remaining problem is the enumeration of minimal s-t cuts in a directed hypergraph. This problem is equivalent to enumerating all maximal signatures/minimal unsatisfiable subformula in Horn SAT formulae if an input directed hypergraph is a B-hypergraph. Recently, enumerating (maximal) signatures in tractable SAT formulae have been studied [8, 15]. Creignou et al. showed that in XOR-SAT case, this problem can be solved in incremental polynomial time.

As interesting directions for future research, we consider the improvement of the delay in enumerating S-T hyperpaths and the existence of an output-quasi polynomial time enumeration algorithm, that is, an enumeration algorithm that runs in $(n + N)^{\mathsf{polylog}(n+N)}$ time, for S-T hyperpaths in general directed hypergraphs. The enumeration of s-t paths in undirected graphs is a well-studied topic, and certain optimal algorithms are already known [11]. Furthermore, for the enumeration of minimal Steiner trees, which generalizes directed s-t paths, a linear-delay algorithm has been known [25]. Investigating how much the delay can be improved remains an interesting open problem.

In addition, it remains an open question whether the S-T hyperpath enumeration can also be solved in output-quasi polynomial time since the minimal transversal enumeration can be solved in output-quasi-polynomial time. It would also be interesting to show whether the corresponding another solution problem of s-t HYPERPATH ENUMERATION is NP-complete. Conducting more detailed research on this question would be an interesting direction for future work.

Acknowledgement. This work is partially supported by JSPS KAKENHI Grant Numbers JP22H03549, JP25K21273, JP25K03080, and JP25K00136.

References

1. Alimonti, P., Feuerstein, E.: Petri nets, hypergraphs and conflicts (preliminary version). In: Mayr, E.W. (ed.) WG 1992. LNCS, vol. 657, pp. 293–309. Springer, Heidelberg (1993). https://doi.org/10.1007/3-540-56402-0_55
2. Alimonti, P., Feuerstein, E., Laura, L., Nanni, U.: Linear time analysis of properties of conflict-free and general petri nets. Theor. Comput. Sci. **412**(4), 320–338 (2011)

3. Alimonti, P., Feuerstein, E., Nanni, U.: Linear time algorithms for liveness and boundedness in conflict-free Petri nets. In: Simon, I. (ed.) LATIN 1992. LNCS, vol. 583, pp. 1–14. Springer, Heidelberg (1992). https://doi.org/10.1007/BFb0023812
4. Allamigeon, X.: On the complexity of strongly connected components in directed hypergraphs. Algorithmica **69**(2), 335–369 (2014)
5. Ausiello, G., Franciosa, P.G., Frigioni, D.: Partially dynamic maintenance of minimum weight hyperpaths. J. Discrete Algorithms **3**(1), 27–46 (2005)
6. Ausiello, G., Italiano, G.F., Laura, L., Nanni, U., Sarracco, F.: Structure theorems for optimum hyperpaths in directed hypergraphs. In: Mahjoub, A.R., Markakis, V., Milis, I., Paschos, V.T. (eds.) ISCO 2012. LNCS, vol. 7422, pp. 1–14. Springer, Heidelberg (2012). https://doi.org/10.1007/978-3-642-32147-4_1
7. Ausiello, G., Laura, L.: Directed hypergraphs: introduction and fundamental algorithms–a survey. Theor. Comput. Sci. **658**, 293–306 (2017)
8. Bérczi, K.: Generating clause sequences of a CNF formula. Theor. Comput. Sci. **856**, 68–74 (2021)
9. Bökler, F., Ehrgott, M., Morris, C., Mutzel, P.: Output-sensitive complexity of multiobjective combinatorial optimization. J. Multi-Criteria Decis. Anal. **24**(1–2), 25–36 (2017)
10. Berry, A., Bordat, J.P., Cogis, O.: Generating all the minimal separators of a graph. Int. J. Found. Comput. Sci. **11**(3), 397–403 (2000)
11. Birmelé, E., et al.: Optimal listing of cycles and st-paths in undirected graphs. In: Proceedings of the SODA 2013, pp. 118–128. SIAM (2013)
12. Boros, E., Makino, K.: Generating minimal redundant and maximal irredundant subhypergraphs. Discret. Appl. Math. **358**, 217–229 (2024)
13. Brosse, C.: On the hardness of inclusion-wise minimal separators enumeration. Inf. Process. Lett. **185**, 106469 (2024)
14. Özturan, C.: On finding hypercycles in chemical reaction networks. Appl. Math. Lett. **21**(9), 881–884 (2008)
15. Creignou, N., Defrain, O., Olive, F., Vilmin, S.: On the enumeration of signatures of XOR-CNF's. In: Proceedings of the WADS 2025, vol. 349 of LIPIcs, pp. 19:1–19:14, Dagstuhl, Germany, 2025. Schloss Dagstuhl – Leibniz-Zentrum für Informatik
16. Eiter, T., Makino, K., Gottlob, G.: Computational aspects of monotone dualization: a brief survey. Discrete Appl. Math. **156**(11), 2035–2049 (2008). In Memory of Leonid Khachiyan (1952 - 2005)
17. Fredman, M.L., Khachiyan, L.: On the complexity of dualization of monotone disjunctive normal forms. J. Algorithms **21**(3), 618–628 (1996)
18. Gallo, G., Gentile, C., Pretolani, D., Rago, G.: Max Horn SAT and the minimum cut problem in directed hypergraphs. Math. Program. **80**(2), 213–237 (1998)
19. Gallo, G., Longo, G., Pallottino, S., Nguyen, S.: Directed hypergraphs and applications. Discrete Appl. Math. **42**(2), 177–201 (1993)
20. Gallo, G., Scutellà, M.G.: Directed hypergraphs as a modelling paradigm. Riv. Mat. Sci. Econom. Social. **21**(1), 97–123 (1998)
21. Johnson, D.S., Yannakakis, M., Papadimitriou, C.H.: On generating all maximal independent sets. Inf. Process. Lett. **27**(3), 119–123 (1988)
22. Khachiyan, L., Boros, E., Elbassioni, K.M., Gurvich, V.: On enumerating minimal dicuts and strongly connected subgraphs. Algorithmica **50**(1), 159–172 (2008)
23. Kimelfeld, B., Sagiv, Y.: Efficiently enumerating results of keyword search over data graphs. Inf. Syst. **33**(4), 335–359 (2008)
24. Kobayashi, Y., Kurita, K., Wasa, K.: Linear-delay enumeration for minimal Steiner problems. In: Proceedings of the PODS 2022, PODS '22, pp. 301–313, New York, NY, USA (2022). Association for Computing Machinery

25. Krieger, S., Kececioglu, J.: Shortest hyperpaths in directed hypergraphs for reaction pathway inference. J. Comput. Biol. **30**(11), 1198–1225 (2023). PMID: 37906100

26. Kurita, K., Kobayashi, Y.: Efficient enumerations for minimal multicuts and multiway cuts. In: Proceedings of the MFCS 2020, vol. 170 of LIPIcs, pp. 60:1–60:14. Schloss Dagstuhl - Leibniz-Zentrum für Informatik (2020)

27. Nielsen, L.R., Pretolani, D., Andersen, K.: A remark on the definition of a B-hyperpath. Technical report, Department of Operations Research, University of Aarhus, Technical report (2001)

28. Provan, J.S., Shier, D.R.: A paradigm for listing (s, t)-cuts in graphs. Algorithmica **15**(4), 351–372 (1996)

29. Read, R.C., Tarjan, R.E.: Bounds on backtrack algorithms for listing cycles, paths, and spanning trees. Networks **5**(3), 237–252 (1975)

30. Takata, K.: Space-optimal, backtracking algorithms to list the minimal vertex separators of a graph. Discrete Appl. Math. **158**(15), 1660–1667 (2010)

31. Thakur, M., Tripathi, R.: Linear connectivity problems in directed hypergraphs. Theor. Comput. Sci. **410**(27), 2592–2618 (2009)

32. Tsukiyama, S., Shirakawa, I., Ozaki, H., Ariyoshi, H.: An algorithm to enumerate all cutsets of a graph in linear time per cutset. J. ACM **27**(4), 619–632 (1980)

33. Uno, T., Satoh, H.: An efficient algorithm for enumerating Chordless cycles and Chordless paths. In: Džeroski, S., et al. (eds.) DS 2014. LNCS (LNAI), vol. 8777, pp. 313–324. Springer, Cham (2014). https://doi.org/10.1007/978-3-319-11812-3_27

34. Volpentesta, A.P.: Hypernetworks in a directed hypergraph. Eur. J. Oper. Res. **188**(2), 390–405 (2008)

Enumeration of Bases in Matroid with Exponentially Large Ground Set

Yuki Nishimura and Kazuya Haraguchi[✉]

Graduate School of Informatics, Kyoto University, Kyoto, Japan
{nishimura,haraguchi}@amp.i.kyoto-u.ac.jp

Abstract. When we deal with a matroid $\mathcal{M} = (U, \mathcal{I})$, we usually assume that it is implicitly given by means of the independence (IND) oracle. Time complexity of many existing algorithms is polynomially bounded with respect to $|U|$ and the running time of the IND-oracle. However, they are not efficient anymore when U is exponentially large in some context. In this paper, we propose an algorithm for enumerating minimum-weight bases of a given matroid such that the time complexity does not depend on $|U|$. For some integer L, this algorithm enumerates the first L minimum-weight bases in incremental-polynomial time and the remaining ones in polynomial-delay. To design the algorithm, we assume two oracles other than the IND-oracle: the MinB-oracle that returns a minimum basis and the REL-oracle that returns the relevant elements one by one in non-decreasing order of weight. The proposed algorithm is applicable to enumeration of minimum bases of binary matroids from cycle spaces and cut spaces, both of which have exponentially large U with respect to a given graph. The highlight in this context is that, to design the REL-oracle for a cut space, we develop the first polynomial-delay algorithm that enumerates all relevant cuts of a given graph in non-decreasing order of weight.

1   Introduction

Let $\mathbb{R}, \mathbb{R}_{\geq 0}, \mathbb{R}_+$ denote the sets of reals, nonnegative reals and positive reals, respectively. For a ground set U and a family $\mathcal{I}$ of subsets of U, $\mathcal{M} = (U, \mathcal{I})$ is called a *matroid* if the following axioms are satisfied. (I) $\emptyset \in \mathcal{I}$; (II) for $I, J \subseteq U$ such that $I \subseteq J$, $J \in \mathcal{I}$ implies $I \in \mathcal{I}$; and (III) for $I, J \in \mathcal{I}$, if $|I| < |J|$, then there is $x \in J - I$ such that $I \cup \{x\} \in \mathcal{I}$. A subset $I \in \mathcal{I}$ (resp., $I \notin \mathcal{I}$) is called *independent* (resp., *dependent*), and a maximal independent set is called a *basis*. It is well-known that the cardinality of any basis is equal and called the *rank of* $\mathcal{M}$. Given a weight function $w : U \to \mathbb{R}$, we write $w(S) := \sum_{x \in S} w(x)$ for $S \subseteq U$ and call it the weight of S. A basis is *minimum-weight* (or *minimum* for short) if the weight attains the minimum over all bases. Matroids are among well-studied discrete structures in theoretical computer science. A matroid $\mathcal{M} = (U, \mathcal{I})$ is

---

The preprint of this paper appeared as [18]. This work is partially supported by JSPS KAKENHI Grant Number 25K14993.

E. Di Giacomo and D. Mondal (Eds.): WALCOM 2026, LNCS 16444, pp. 576–590, 2026.
https://doi.org/10.1007/978-981-95-7127-7_38

usually given by means of an independence oracle (a.k.a., a membership oracle) that returns whether $S \in \mathcal{I}$ or $S \notin \mathcal{I}$ to a query subset $S \subseteq U$.

An *enumeration problem* in general asks to list all required solutions without duplication. In general, an enumeration algorithm takes at least the computation time proportional to the output size, where the output size can be exponentially large with respect to the input size. An enumeration algorithm has been evaluated in terms of both input size and output size. According to [10], the algorithm is called

- *output-polynomial* if the overall computation time is bounded by a polynomial with respect to the input size and the output size;
- *incremental-polynomial* (*incremental-poly*) if the computation time for finding the ℓ-th solution is bounded by a polynomial with respect to the input size and ℓ; and
- *polynomial-delay* (*poly-delay*) if the delay (i.e., the computation time between any two consecutive outputs) is bounded by a polynomial with respect to the input size.

Various enumeration problems have been studied in the context of matroids [11,13]. For example, Hamacher et al. [8] presented an algorithm to enumerate k-smallest bases in the sense of weight in $\mathcal{O}(|U| \log |U| + k \cdot |U| \cdot r \cdot \tau_{\mathrm{IND}})$ time, where τ_{IND} denotes the computation time of the independence oracle and r denotes the rank of the matroid. Like this algorithm, many enumeration algorithms on matroids are efficient in the sense that the time complexity is polynomially bounded, where bounding polynomials depend on the cardinality $|U|$ of the ground set U.

In the literature, however, there are matroids such that the ground set consists of an exponential number of elements with respect to the size of a given graph; e.g., binary matroids from cycle spaces and cut spaces [4]. These matroids are never just artificial. For example, a basis in the binary matroid from a cycle space, called a *minimum cycle basis* in the literature, has applications in electric engineering [2] and network analysis [21]. A minimum cycle basis can be found in polynomial time [9,17]. However, we cannot expect the above-mentioned enumeration algorithms to run efficiently for matroids like this.

In this paper, assuming a matroid with an exponentially large ground set, we propose a new enumeration algorithm for all minimum bases. The point is that the polynomials bounding the time complexity do not depend on the ground set size.

After we introduce terminologies, notations and fundamental properties in Sect. 2, we present the main contributions of the paper in Sects. 3 and 4, followed by concluding remarks in Sect. 5. Some proofs and technical details are omitted due to space limitation. See the preprint [18] for details.

Enumeration of Minimum Bases (Sect.3). To design the algorithm for this purpose, we assume two oracles other than the independence oracle: the minimum basis oracle that returns a minimum basis and the relevant oracle that returns the "relevant" elements one by one in non-decreasing order of weight. An element is relevant if there is a minimum basis that contains it. For some integer L, the proposed algorithm generates the ℓ-th solution ($\ell \leq L$) in incremental-poly time with respect to the running times of the three oracles, the matroid rank r and ℓ, and the remaining solutions are enumerated in $O(r)$ delay (i.e., poly-delay).

Application to Binary Matroids (Sect.4). We apply the minimum basis enumeration algorithm in Sect. 3 to the binary matroids from cycle spaces and cut spaces, to obtain an efficient enumeration algorithm. For this, we implement all the oracles by using previous results as subroutines, except the relevant oracle for the binary matroid from a cut space. For this oracle, we develop the first poly-delay algorithm that enumerates all relevant cuts in an edge-weighted graph in non-decreasing order of weight. This is different from a poly-delay algorithm in [23] that enumerates all cuts or s,t-cuts (not necessarily relevant) in non-decreasing order, while ours concentrates on relevant cuts.

2 Preliminaries

For disjoint subsets X, Y of elements, we may denote the disjoint union by $X \sqcup Y$ to emphasize $X \cap Y = \emptyset$. For the minimality under the context, we denote by $\min(X)$ the set of minimum elements in X, where we may use an abuse notation $x = \min(X)$ to indicate $x \in \min(X)$. For integers p, q ($p \leq q$), we denote $[p, q] := \{p, p+1, \ldots, q\}$.

Suppose that a matroid $\mathcal{M} = (U, \mathcal{I})$ is given. We denote by $\mathcal{B}$ the set of all bases of $\mathcal{M}$. For $B \in \mathcal{B}$, we call a pair (x, y) of elements $x \in B$ and $y \notin B$ an *exchange for B* if $(B - \{x\}) \cup \{y\} \in \mathcal{B}$. We will utilize the following property (e.g., see Theorem 13.9 in [14]).

Lemma 1. *For a matroid $\mathcal{M} = (U, \mathcal{I})$, let $B_1, B_2 \in \mathcal{B}$. For any $x \in B_1 - B_2$, there exists $y \in B_2 - B_1$ such that (x, y) is an exchange for B_1. Moreover, for any $y \in B_2 - B_1$, there exists $x \in B_1 - B_2$ such that (x, y) is an exchange for B_1.*

Suppose that a weight function $w : U \to \mathbb{R}$ is given. For a basis $B \in \mathcal{B}$, let $x \in B$ be any element and $y \notin B$ be an element such that (x, y) is an exchange for B. We call $w(x, y) := w(y) - w(x)$ the *weight of exchange (x, y)*. We call (x, y) a *zero-exchange* if $w(x, y) = 0$. For $y \in U - B$, we define $\mathrm{Ex}_{\mathrm{out}}^{\mathrm{zero}}(B; y) := \{x \in B \mid (x, y) \text{ is a zero-exchange for } B\}$. We denote by $\mathcal{B}^* \subseteq \mathcal{B}$ the set of all minimum bases of $\mathcal{M}$. An element $x \in U$ is *relevant* if there is a minimum basis $B^* \in \mathcal{B}^*$ such that $x \in B^*$. We denote by U^* the set of all relevant elements. For subsets $I, O \subseteq U$ ($I \cap O = \emptyset$), let us define $\mathcal{B}^*(I, O)$ to be the family of all minimum bases that contain all elements in I and does not contain any elements in O, that is, $\mathcal{B}^*(I, O) \triangleq \{B^* \in \mathcal{B}^* \mid B^* \supseteq I, B^* \cap O = \emptyset\}$. For $B^* \in \mathcal{B}^*(I, O)$, we say that an exchange (x, y) for B^* is *I, O-preserving* if $(B^* - \{x\}) \cup \{y\}$ belongs to $\mathcal{B}^*(I, O)$.

We introduce a fundamental property of minimum bases. Let $I, O \subseteq U$ be any two disjoint subsets. Lemma 2 shows that, for any minimum bases $P^*, Q^* \in$

$\mathcal{B}^*(I, O)$, there is a sequence of minimum bases $(B_1^* = P^*, B_2^*, \ldots, B_k^* = Q^*)$ such that, for $i = 1, 2, \ldots, k-1$, there is an I, O-preserving zero-exchange (x_i, y_i) for B_i^* that satisfies $B_{i+1}^* = (B_i^* - \{x_i\}) \cup \{y_i\}$.

Lemma 2. *Suppose that a matroid $\mathcal{M} = (U, \mathcal{I})$ with a weight function $w : U \to \mathbb{R}$ is given. For $I, O \subseteq U$ $(I \cap O = \emptyset)$, let $P^*, Q^* \in \mathcal{B}^*(I, O)$. For any minimum element y in $Q^* - P^*$, there exists an element $x \in P^* - Q^*$ such that (x, y) is an I, O-preserving zero-exchange for P^*. Furthermore, x is a minimum element in $P^* - Q^*$.*

Oracles. We assume that a matroid $\mathcal{M} = (U, \mathcal{I})$ with a weight function $w : U \to \mathbb{R}$ is implicitly given by the following oracles: the *independence oracle (IND-oracle)* returns whether a query subset $S \subseteq U$ is independent or not; the *minimum basis oracle (MinB-oracle)* returns a minimum basis; and the *relevant oracle (REL-oracle)* returns the i-th smallest relevant element in the i-th call, where we assume that the oracle is not called more than $b = |U^*|$ times. We denote the time complexities of these oracles by τ_{IND}, τ_{MinB} and τ_{REL}, respectively. In some matroids, we may need to conduct preprocessing before the first call of the REL-oracle. We denote by $\tau_{\mathrm{REL}}^{\mathrm{pre}}$ the time complexity of the preprocessing.

Binary Matroids. We denote by $GF(2)$ the finite field of two elements, that is 0 and 1. A *binary matroid* is a matroid $\mathcal{M} = (U, \mathcal{I})$ such that each element in U is associated with a vector in $GF(2)$, and $I \subseteq U$ is an independent set (i.e., $I \in \mathcal{I}$) if and only if the vectors associated with elements in I are linearly independent in $GF(2)$.

3   An Algorithm for Enumerating All Minimum Bases

In this section, we present an algorithm that enumerates all minimum bases of a matroid $\mathcal{M} = (U, \mathcal{I})$ with a weight function $w : U \to \mathbb{R}$, where $\mathcal{M}$ is implicitly given by an IND-oracle, a MinB-oracle and a REL-oracle, and we denote by r the rank of $\mathcal{M}$.

Naïve Algorithm. We can employ a saturation algorithm [16] to design an incremental-poly algorithm. First, we generate a minimum basis B^* by the MinB-oracle and let $\mathcal{S} := \{B^*\}$. Then we repeat the following for each relevant element y that is generated by the REL-oracle; for each minimum basis B in $\mathcal{S}$, we generate $B' := (B - \{x\}) \cup \{y\}$ for all $x \in B$ such that (x, y) is a zero-exchange. If $B' \in \mathcal{B}^* - \mathcal{S}$, then we add B' to the solution set (i.e., $\mathcal{S} := \mathcal{S} \cup \{B'\}$). We can obtain all the minimum bases by Lemma 2.

Let us analyze the time complexity. Let $\ell := |\mathcal{S}|$. At least one new minimum basis B' will be obtained for each y since it is relevant. To obtain the $(\ell + 1)$-st minimum basis, we call the IND-oracle $\mathcal{O}(\ell r)$ times to identify whether B' is independent. If B' is independent (i.e., a minimum basis), we check whether it has already been generated, where the check can be done in $\mathcal{O}(r)$ time if we store

$\mathcal{S}$ by means of trie. Thus, the computation time to obtain the $(\ell+1)$-st solution is $\mathcal{O}\big((r+\ell)\tau_{\mathrm{REL}} + \sum_{i=1}^{\ell}(ir(\tau_{\mathrm{IND}}+r))\big) = \mathcal{O}\big((r+\ell)\tau_{\mathrm{REL}} + \ell^2 r\tau_{\mathrm{IND}} + \ell^2 r^2\big)$ after $\mathcal{O}(\tau_{\mathrm{MinB}} + \tau_{\mathrm{REL}}^{\mathrm{pre}})$ time preprocessing.

The space complexity is not polynomially bounded since it needs to store all solutions generated so far. As mentioned in [3], it is a major issue to reduce the space complexity of enumeration algorithms of this kind.

Search Tree Based Algorithm. We can reduce the term $\mathcal{O}(\ell^2 r^2)$ in the time bound above if we realize the algorithm by using a search tree. In the search tree algorithm, we do not need to check explicitly whether a minimum basis has been already generated.

The search tree represents a partition structure of the set $\mathcal{B}^*$ of minimum bases. Each node corresponds to a subset of $\mathcal{B}^*$ and has one minimum basis in the subset as the representative. For every minimum basis B^*, there is at least one node that has B^* as the representative. Traversing the search tree, we output the representatives so that duplication does not occur, by which the enumeration task is done.

To be more precise, each node of the search tree is associated with two disjoint subsets $I, O \subseteq U^*$ of relevant elements and denoted by $\langle I, O \rangle$. We call I (resp., O) the *mandatory* (resp., *forbidden*) *subset* of the node $\langle I, O \rangle$. In particular, the root node corresponds to $\langle \emptyset, \emptyset \rangle$. For $\langle I, O \rangle$, the corresponding subset of $\mathcal{B}^*$ is $\mathcal{B}^*(I, O)$, and we choose a minimum basis in $\mathcal{B}^*(I, O)$ as its representative and denote it by $R_{I,O}^*$.

We branch a node in the search tree based on the following lemma.

Lemma 3. *Suppose that a matroid* $\mathcal{M} = (U, \mathcal{I})$ *is given. For disjoint subsets* $I, O \subseteq U^*$ *of relevant elements, let* $B^* \in \mathcal{B}^*(I, O)$ *be a minimum basis and let* $x \in B^* - I$. *(i)* $\mathcal{B}^*(I, O) = \mathcal{B}^*(I \cup \{x\}, O) \sqcup \mathcal{B}^*(I, O \cup \{x\})$. *(ii) Let* $y \in U^* - (B^* \cup O)$. *If* (x, y) *is an* I, O*-preserving zero-exchange for* B^*, *then* $B^* \in \mathcal{B}^*(I \cup \{x\}, O)$ *and* $(B^* - \{x\}) \cup \{y\} \in \mathcal{B}^*(I, O \cup \{x\})$ *hold.*

For a node $\langle I, O \rangle$ in the search tree, let $B^* \in \mathcal{B}^*(I, O)$ denote what we call the representative and $y \in U^* - (B^* \cup O)$ denote what we call the *branching variable*. We explain how we determine B^* and y later. Let $\mathrm{Ex}_{\mathrm{out}}^{\mathrm{zero}}(B^*; y) - I := \{x_1, x_2, \ldots, x_b\}$. That is, for $a = 1, 2, \ldots, b$, (x_a, y) is an I, O-preserving zero-exchange for B^*. By applying Lemma 3(i) repeatedly, $\mathcal{B}^*(I, O)$ can be partitioned as $\mathcal{B}^*(I, O) = \mathcal{B}^*(I \cup \{x_1, x_2, \ldots, x_b\}, O) \sqcup \big(\bigsqcup_{a=1}^{b} \mathcal{B}^*(I \cup \{x_1, x_2, \ldots, x_{a-1}\}, O \cup \{x_a\})\big)$. We branch the node $\langle I, O \rangle$ to take $b+1$ children whose mandatory and forbidden subsets appear in the above equation, that is, $\langle I \cup \{x_1, x_2 \ldots, x_b\}, O \rangle$, $\langle I, O \cup \{x_1\} \rangle, \ldots, \langle I \cup \{x_1, \ldots, x_{b-1}\}, O \cup \{x_b\} \rangle$. The size of the forbidden subset of the first child (i.e., $\langle I \cup \{x_1, x_2 \ldots, x_b\}, O \rangle$) is smaller than others by one. To make the size of forbidden subsets equal, observe that $I \cup \{x_1, x_2, \ldots, x_b\} \cup \{y\}$ contains a dependent set by the definition of $x_1, x_2, \ldots, x_b$. Any basis in $\mathcal{B}^*(I \cup \{x_1, x_2, \ldots, x_b\}, O)$ does not contain y, and hence we regard $\langle I \cup \{x_1, x_2, \ldots, x_b\}, O \cup \{y\} \rangle$ as a child of $\langle I, O \rangle$, instead of $\langle I \cup \{x_1, x_2, \ldots, x_b\}, O \rangle$.

In summary, $\langle I, O \rangle$ has $\langle I_0, O_0 \rangle, \langle I_1, O_1 \rangle, \ldots, \langle I_b, O_b \rangle$ as its children, where $I_0 = I \cup \{x_1, x_2, \ldots, x_b\}$; $O_0 = O \cup \{y\}$; $I_a = I \cup \{x_1, x_2, \ldots, x_{a-1}\}$ and $O_a = O \cup$

$\{x_a\}$ for $a = 1, 2, \ldots, b$. We call $\langle I_0, O_0 \rangle$ the *eldest child of* $\langle I, O \rangle$ to distinguish $\langle I_0, O_0 \rangle$ from the other children. Let r denote the rank of the matroid $\mathcal{M}$. The number of children is at least 1 when $b = 0$ and at most $r + 1$ when $b = r$. The set $\mathcal{B}^*(I, O)$ is partitioned as $\mathcal{B}^*(I, O) = \bigsqcup_{a=0}^{b} \mathcal{B}^*(I_a, O_a)$.

We explain how to determine the representative and the branching variable for node $\langle I, O \rangle$. As the base case, for the root $\langle I, O \rangle = \langle \emptyset, \emptyset \rangle$, we choose any minimum basis as its representative $R^*_{\emptyset, \emptyset}$. Let $U^* - R^*_{\emptyset, \emptyset} := \{y_1, y_2, \ldots, y_{|U^*|-r}\}$, where $w(y_1) \leq w(y_2) \leq \cdots \leq w(y_{|U^*|-r})$. The depth of a node is defined to be the number of edges in the path from the root to that node. The height of the search tree (i.e., the maximum depth) is precisely $h_{\max} := |U^*| - r$. For every node $\langle I, O \rangle$ in the depth $h = 0, 1, \ldots, h_{\max} - 1$, we set the branching variable to y_{h+1}. Let $\langle I_0, O_0 \rangle, \langle I_1, O_1 \rangle, \ldots, \langle I_b, O_b \rangle$ denote the children of $\langle I, O \rangle$. We define the representatives of the children as follows; $R^*_{I_0, O_0} := R^*_{I,O}$; and $R^*_{I_a, O_a} := (R^*_{I,O} - \{x_a\}) \cup \{y_{h+1}\}$ for $a = 1, 2, \ldots, b$. The point is that the eldest child $\langle I_0, O_0 \rangle$ has the same representative as $\langle I, O \rangle$, and that other children have different representatives from $\langle I, O \rangle$. Further, every node in the depth h has a forbidden subset of the size h.

For $h = 0, 1, \ldots, h_{\max}$, let us denote by $\mathcal{N}_h$ the set of all nodes in depth h. Lemma 3 implies that the family of subsets $\mathcal{B}^*(I, O)$ for all $\langle I, O \rangle \in \mathcal{N}_h$ is a partition of $\mathcal{B}^*$.

Let us denote by $\mathcal{R}_h$ the set of all representatives of nodes in depth h. The following lemma shows that, in every depth $h = 1, 2, \ldots, h_{\max}$, there is a representative that does not appear in any lower depth, and that, for every minimum basis $B^* \in \mathcal{B}^*$, there is a leaf whose representative is B^*.

Lemma 4. *Suppose that a matroid $\mathcal{M} = (U, \mathcal{I})$ with a weight function $w : U \to \mathbb{R}$ is given. It holds that $\mathcal{R}_0 \subsetneq \mathcal{R}_1 \subsetneq \cdots \subsetneq \mathcal{R}_{h_{\max}} = \mathcal{B}^*$.*

We can enumerate all minimum bases by traversing the search tree. Specifically, we explore the search tree in the breadth-first manner and output minimum bases in $\mathcal{R}_0$, $\mathcal{R}_1 - \mathcal{R}_0$, $\mathcal{R}_2 - \mathcal{R}_1$, $\ldots$, $\mathcal{R}_{h_{\max}} - \mathcal{R}_{h_{\max}-1}$ in this order.

The algorithm is summarized in Algorithm 1. In this pseudocode, the set $\mathcal{S}_h$, $h = 0, 1, \ldots, h_{\max}$ is introduced to store all nodes in depth h (i.e., $\mathcal{N}_h$).

Theorem 1. *Suppose that a matroid $\mathcal{M} = (U, \mathcal{I})$ with a weight function $w : U \to \mathbb{R}$ is given. Algorithm 1 enumerates all minimum bases in $\mathcal{M}$ in incremental polynomial time with respect to the rank r of $\mathcal{M}$ and the oracle running times. To be more precise, after $\mathcal{O}(\tau_{\mathrm{MinB}} + \tau_{\mathrm{REL}}^{\mathrm{pre}})$-time preprocessing, the ℓ-th minimum basis is output in $\mathcal{O}((r + \ell)\tau_{\mathrm{REL}} + \ell^2 r \tau_{\mathrm{IND}})$ time.*

Proof. (**Correctness**) We claim that $\mathcal{S}_h = \mathcal{N}_h$, $h = 0, 1, \ldots, h_{\max}$ holds upon completion of the algorithm. It is clearly true for $h = 0$ by line 2. Suppose that $\mathcal{S}_h = \mathcal{N}_h$ holds as the assumption of the induction. For every node $\langle I, O \rangle \in \mathcal{S}_h$, the eldest child $\langle I_0, O_0 \rangle$ is added to $\mathcal{S}_{h+1}$ in line 14, and all other children are added to $\mathcal{S}_{h+1}$ in line 11, showing the claim.

No minimum basis is output more than once since a minimum basis $B^* \in \mathcal{R}_h$ is output when and only when it is the representative of a non-eldest child of its

---

Algorithm 1. An algorithm to enumerate all minimum bases in a matroid

---

Require: A matroid $\mathcal{M} = (U, \mathcal{I})$ with a weight function $w : U \to \mathbb{R}$ that is implicitly given by an IND-oracle, a MinB-oracle and a REL-oracle

Ensure: All minimum bases in $\mathcal{M}$

1: $R^*_{\emptyset,\emptyset} \leftarrow$ a minimum basis; output $R^*_{\emptyset,\emptyset}$;

2: $\mathcal{S}_0 \leftarrow \{\langle \emptyset, \emptyset \rangle\}$;

3: **for** $h = 0, 1, \ldots, h_{\max} - 1$ **do**

4:     $\mathcal{S}_{h+1} \leftarrow \emptyset$;

5:     Generate y_{h+1} by the REL-oracle;

6:     **for** $\langle I, O \rangle \in \mathcal{S}_h$ **do**

7:         $\{x_1, x_2, \ldots, x_b\} \leftarrow \mathrm{Ex}^{\mathrm{zero}}_{\mathrm{out}}(R^*_{I,O}; y_{h+1}) - I$;

8:         **for** $a = 1, 2, \ldots, b$ **do**

9:             $I_a \leftarrow I \cup \{x_1, x_2, \ldots, x_{a-1}\}$; $O_a \leftarrow O \cup \{x_a\}$;

10:             $R^*_{I_a, O_a} \leftarrow (R^*_{I,O} - \{x_a\}) \cup \{y_{h+1}\}$; output $R^*_{I_a, O_a}$;

11:             $\mathcal{S}_{h+1} \leftarrow \mathcal{S}_{h+1} \cup \{\langle I_a, O_a \rangle\}$

12:         **end for**;

13:         $I_0 \leftarrow I \cup \{x_1, x_2, \ldots, x_b\}$; $O_0 \leftarrow O \cup \{y_{h+1}\}$;

14:         $R^*_{I_0, O_0} \leftarrow R^*_{I,O}$; $\mathcal{S}_{h+1} \leftarrow \mathcal{S}_{h+1} \cup \{\langle I_0, O_0 \rangle\}$

15:     **end for**

16: **end for**

---

parent; in this case, $B^* \notin \mathcal{R}_0 \cup \mathcal{R}_1 \cup \cdots \cup \mathcal{R}_{h-1}$ holds and B^* is different from any other bases in $\mathcal{R}_h$.

We show by contradiction that all minimum bases are output. Suppose that there is a minimum basis that is not output by the algorithm. By Lemma 4, there is a leaf $\langle I^\clubsuit, O^\clubsuit \rangle$ whose representative $R^*_{I^\clubsuit, O^\clubsuit}$ is not output. Then $\langle I^\clubsuit, O^\clubsuit \rangle$ is the eldest child of the parent, say $\langle I', O' \rangle$. The representative of $\langle I', O' \rangle$ is the same as $\langle I^\clubsuit, O^\clubsuit \rangle$ (i.e., $R^*_{I',O'} = R^*_{I^\clubsuit, O^\clubsuit}$). The representative $R^*_{I',O'}$ is not output and hence $\langle I', O' \rangle$ is the eldest child of the parent. We go upward to the root in this way. However, the representative of the root is output in line 1, a contradiction.

(Complexity) Let $h \in \{0, 1, \ldots, h_{\max} - 1\}$. By Lemma 4, there exists a node $\langle I, O \rangle$ in depth $h + 1$ whose representative $R^*_{I,O}$ is output when this node is visited by the algorithm. Suppose that $R^*_{I,O}$ is the ℓ-th output.

- The MinB-oracle is called once in line 1 to construct a minimum basis as $R^*_{\emptyset,\emptyset}$, which can be done in $\mathcal{O}(\tau_{\mathrm{MinB}})$ time.
- To use the REL-oracle, we conduct preprocessing in $\mathcal{O}(\tau^{\mathrm{pre}}_{\mathrm{REL}})$ time.
- We can determine y_{h+1} that is used in the for-loop from line 6 to 15 in at most $r + h + 1$ calls of the REL-oracle.
- The for-loop from line 6 to 15 is repeated $|\mathcal{S}_h|$ times. In each iteration, the most critical part is determination of $\{x_1, x_2, \ldots, x_b\}$, which can be done in r calls of the IND-oracle.

Finally, $\sum_{i=1}^h |\mathcal{S}_i| = \sum_{i=1}^h |\mathcal{R}_i| \leq h|\mathcal{R}_h| \leq h\ell$ holds by Lemma 4, where $|\mathcal{R}_h|$ equals to the number of representatives that have been output by the depth h.

It holds that $h \leq \ell$ by Lemma 4. After $\mathcal{O}(\tau_{\text{MinB}} + \tau_{\text{REL}}^{\text{pre}})$-time preprocessing, the computation time that the algorithm takes to output $R_{I,O}^*$ is $\mathcal{O}((r+h+1)\tau_{\text{REL}} + \sum_{i=1}^{h+1} |\mathcal{S}_i| r \tau_{\text{IND}}) = \mathcal{O}((r+\ell)\tau_{\text{REL}} + \ell^2 r \tau_{\text{IND}})$. $\qquad\square$

A More Efficient Algorithm. We develop a more efficient enumeration algorithm by using Algorithm 1 as a subroutine. We show that there is an integer L such that the computation time to output the ℓ-th solution is $\mathcal{O}((r+\ell)\tau_{\text{REL}}+\ell^2 r \tau_{\text{IND}})$ for $\ell \leq L$, and that the delay for the remaining solutions is $O(r)$.

Again, let $\mathcal{M} = (U, \mathcal{I})$ denote the given matroid with a weight function $w : U \to \mathbb{R}$. For a subset $X \subseteq U$, we define the *restriction of* $\mathcal{M}$ *to* X to be the system $\mathcal{M}|X := (X, \mathcal{I}|X)$, where $\mathcal{I}|X \triangleq \{I \in \mathcal{I} \mid I \subseteq X\}$. The restriction $\mathcal{M}|X$ is a matroid [20].

Let $P^* = \{x_1, x_2, \ldots, x_r\}$ and $Q^* = \{y_1, y_2, \ldots, y_r\}$ be minimum bases of $\mathcal{M}$ such that $w(x_1) \leq w(x_2) \leq \cdots \leq w(x_r)$ and $w(y_1) \leq w(y_2) \leq \cdots \leq w(y_r)$. By Lemma 2, it is easy to see that $w(x_i) = w(y_i)$ holds for all $i = 1, 2, \ldots, r$. Let $\omega_1, \omega_2, \ldots, \omega_\beta \in \mathbb{R}$ denote the distinct weights that appear in a minimum basis such that $\omega_1 \leq \omega_2 \leq \cdots \leq \omega_\beta$ and $r_1, r_2, \ldots, r_\beta \in \mathbb{Z}_+$ denote their frequencies. In other words, every minimum basis of $\mathcal{M}$ contains r_j elements of weight ω_j, $j = 1, 2, \ldots, \beta$. We define $\mathcal{I}^* \triangleq \{S \subseteq U^* \mid \exists B^* \in \mathcal{B}^*, S \subseteq B^*\}$. The set system $\mathcal{M}^* := (U^*, \mathcal{I}^*)$ is a matroid [15]. For $j \in [1, \beta]$, we also define $U_j^* \triangleq \{e \in U^* \mid w(e) = \omega_j\}$ and $\mathcal{I}_j^* \triangleq \{S \subseteq U_j^* \mid S \in \mathcal{I}^*\}$. Then $\mathcal{M}_j^* := (U_j^*, \mathcal{I}_j^*)$ is a matroid since $\mathcal{M}_j^*$ is a restriction of $\mathcal{M}^*$ (i.e., $\mathcal{M}_j^* = \mathcal{M}^*|U_j^*$).

By the following lemma, we see that the union of any bases of $\mathcal{M}_1^*, \mathcal{M}_2^*, \ldots,$ $\mathcal{M}_\beta^*$ is a minimum basis of $\mathcal{M}$ and that any minimum basis of $\mathcal{M}$ can be partitioned into bases of $\mathcal{M}_1^*, \mathcal{M}_2^*, \ldots, \mathcal{M}_\beta^*$.

Lemma 5. *Suppose that a matroid* $\mathcal{M} = (U, \mathcal{I})$ *with a weight function* $w : U \to$ $\mathbb{R}$ *is given.* (i) *For* $j \in [1, \beta]$, *let* B_j^* *be a basis of the matroid* $\mathcal{M}_j^*$. *The union* $B_1^* \cup B_2^* \cup \cdots \cup B_\beta^*$ *is a minimum basis of* $\mathcal{M}$. (ii) *For any minimum basis* B^* *of* $\mathcal{M}$, $B^* \cap U_j^*$ *is a basis of* $\mathcal{M}_j^*$, $j = 1, 2, \ldots, \beta$.

Proof. (i) For $j \in [1, \beta]$, $B_j^* \in \mathcal{I}^*$ holds, and there is a minimum basis of $\mathcal{M}$, say $\hat{B}_j^*$, that contains B_j^* as a subset. For $k \in [1, \beta]$, let $\hat{B}_{j,k}^* = \hat{B}_j^* \cap U_k^*$, where $\hat{B}_{j,j}^* = B_j^*$ holds. Let us take up $\hat{B}_1^* = B_1^* \cup \hat{B}_{1,2}^* \cup \hat{B}_{1,3}^* \cup \cdots \cup \hat{B}_{1,\beta}^*$ and $\hat{B}_2^* = \hat{B}_{2,1}^* \cup B_2^* \cup \hat{B}_{2,3}^* \cup \cdots \cup \hat{B}_{2,\beta}^*$. Using Lemma 2, we perform (x, y)-exchanges on $\hat{B}_2^*$ such that $x \in \hat{B}_{2,1}^*$ and $y \in B_1^*$ to obtain a minimum basis $B_1^* \cup B_2^* \cup \hat{B}_{2,3}^* \cup \cdots \cup \hat{B}_{2,\beta}^*$ of $\mathcal{M}$ that contains $B_1^* \cup B_2^*$. Repeating this to $\hat{B}_3^*, \hat{B}_4^*, \ldots, \hat{B}_\beta^*$, we have a minimum basis $B_1^* \cup B_2^* \cup \cdots \cup B_\beta^*$ of $\mathcal{M}$. (ii) is immediate by the definition of $\mathcal{M}_j^*$. $\qquad\square$

Generating a minimum basis B^* of $\mathcal{M}$ by the MinB-oracle, we first enumerate all bases of $\mathcal{M}_1^*, \mathcal{M}_2^*, \ldots, \mathcal{M}_\beta^*$. By Lemma 5, once all bases of $\mathcal{M}_1^*, \mathcal{M}_2^*, \ldots, \mathcal{M}_\beta^*$ are obtained, we can enumerate the remaining minimum bases of $\mathcal{M}$ by taking the union of any β bases of $\mathcal{M}_1^*, \mathcal{M}_2^*, \ldots, \mathcal{M}_\beta^*$, which can be done in $O(r)$ delay. We can enumerate bases of a matroid $\mathcal{M}_j^*$, $j \in [1, \beta]$ by running Algorithm 1 on $\mathcal{M}_j^*$. To use the algorithm, we can easily confirm that the three oracles are available to $\mathcal{M}_j^*$ so that the time complexity remains the same under the big-O notation. (**MinB-oracle**) For the minimum basis B^* generated by the MinB-oracle of the given $\mathcal{M}$, it suffices to take its subset of the j-th smallest elements, say B_j^*. (**IND-oracle**) A subset $B_j \subseteq U_j^*$ is independent for $\mathcal{M}_j^*$ if and only if $(B^* - B_j^*) \cup B_j$ is independent for $\mathcal{M}$. (**REL-oracle**) The REL-oracle of $\mathcal{M}$ lists relevant elements in U^* in the non-decreasing order of weight. We can use it as the REL-oracle of $\mathcal{M}_j^*$ as long as we deal with $\mathcal{M}_1^*, \mathcal{M}_2^*, \ldots, \mathcal{M}_\beta^*$ in this order.

We have the following theorem immediately.

Theorem 2. *Suppose that a matroid $\mathcal{M} = (U, \mathcal{I})$ with a weight function $w : U \to \mathbb{R}$ is given. There is an integer L such that we can generate the ℓ-th minimum basis of $\mathcal{M}$ in $\mathcal{O}((r + \ell)\tau_{\mathrm{REL}} + \ell^2 r\tau_{\mathrm{IND}})$ time for $\ell \leq L$ and the remaining minimum bases in $\mathcal{O}(r)$ delay after $\mathcal{O}(\tau_{\mathrm{MinB}} + \tau_{\mathrm{REL}}^{\mathrm{pre}})$-time preprocessing.*

Let $\mathcal{B}_j^*$, $j \in [1, \beta]$ denote the set of all bases of the matroid $\mathcal{M}_j^*$. Any integer no less than $1 + \sum_{j=1}^{\beta}(|\mathcal{B}_j^*| - 1)$ suffices as L in the theorem.

4 Application to Binary Matroids with An Exponentially Large Ground Set

In this section, we describe polynomial-time implementations of the oracles for binary matroids that are well-known in the literature. The matroids that we consider include the binary matroids from cycle spaces and cut spaces. These matroids arise from a given graph and have a ground set of an exponential size with respect to the graph size. Using our algorithm, we can enumerate minimum bases of these matroids efficiently. A highlight of this section is a poly-delay algorithm for enumerating all relevant cuts in a given edge-weighted undirected graph in non-decreasing order of weight, which is used as the REL-oracle for a binary matroid from the cut space.

Throughout this section, we assume that a graph G is connected and undirected. We denote its vertex (resp., edge) set by $V(G)$ (resp., $E(G)$), where we let $n = |V(G)|$ and $m = |E(G)|$. An undirected edge $\{s, t\} \in E(G)$ is written as st (or ts) for simplicity. Let $E(G) = \{e_1, e_2, \ldots, e_m\}$. We represent a subset of edges by the incidence vector $\boldsymbol{u} = (u_1, u_2, \ldots, u_m) \in GF(2)^m$, that is, $u_i = 1$ (resp., 0) if e_i belongs (resp., does not belong) to the subset.

Cycle Space. We assume an edge-weight to be positive (i.e., $w : E(G) \to \mathbb{R}_+$). A subset $F \subseteq E(G)$ of edges is called a *cycle* if all vertex degrees in the spanning subgraph $(V(G), F)$ are even. The weight of a cycle F is the summation of the weight of edges in F (i.e., $W(F) := \sum_{e \in F} w(e)$). The vector space over $GF(2)$ spanned by the incidence vectors of all cycles in G is called the *cycle space of G*. Let γ denote the number of connected components in G. The number $\mu := m - n + \gamma$ is called the *cyclomatic number of G*, and the dimension of the cycle space is equal to μ [2].

We say that cycles are independent if their incidence vectors are linearly independent in $GF(2)$. Let U denote the set of all cycles and $\mathcal{I}$ denote the family of all subsets of independent cycles in U. We consider a matroid $\mathcal{M} = (U, \mathcal{I})$. We call a relevant element of this matroid a *relevant cycle*. Obviously, the cardinality $|U|$ of the ground set can be up to an exponential number with respect to n and m.

The rank of this matroid is the cyclomatic number $\mu = \mathcal{O}(m)$. We can decide whether a set $I \subseteq U$ of cycles is independent or not by using Gaussian elimination on an $m \times |I|$ matrix. We can assume that $|I| = \mathcal{O}(m)$ when we run the IND-oracle in Algorithm 1. Then $\tau_{\mathrm{IND}} = \mathcal{O}(m^3)$ holds. We can construct a minimum basis in this matroid (i.e., minimum cycle basis) in $\mathcal{O}(m^3 n)$ time by Horton's algorithm [9].[1] We can implement the oracle so that $\tau_{\mathrm{MinB}} = \mathcal{O}(m^3 n)$ holds.

We can use Vismara's algorithms [22] for the REL-oracle. Vismara introduced a partition of the set U^* of relevant cycles, say $U^* = U_1^* \sqcup U_2^* \sqcup \cdots \sqcup U_q^*$, such that all relevant cycles in U_p^*, $p = 1, 2, \ldots, q$ have the same weight and that $q = O(m^2)$, where there is a unique cycle $x_p^* \in U_p^*$ for each p that is called the *prototype of U_p^**. Vismara presented two algorithms, one for enumerating all prototypes in $O(\mu m^3)$ time and the other for enumerating all relevant cycles in U_p^* in $O(n|U_p^*|)$ time. To be more precise, the latter algorithm attains $O(n)$ delay. We can scan all relevant cycles in the non-decreasing order of weight as follows; first, we enumerate all prototypes by the first algorithm and sort them in the non-decreasing order with respect to weight. This can be done in $O(\mu m^3 + m^2 \log m) = O(\mu m^3)$ time. Let $w(x_1^*) \leq w(x_2^*) \leq \cdots \leq w(x_q^*)$ without loss of generality. Next, for each $p = 1, 2, \ldots, q$, we enumerate all relevant cycles in U_p^* by the second algorithm. This can be done in $O(n)$ delay. To realize the REL-oracle, we use this enumeration algorithm as a co-routine. We have that $\tau_{\mathrm{REL}}^{\mathrm{pre}} = O(\mu m^3)$ and $\tau_{\mathrm{REL}} = O(n)$.

By Theorem 2, we have the following corollary immediately.

Corollary 1. *Given a connected graph G with an edge-weight function $w : E(G) \to \mathbb{R}_+$, there is an integer L such that we can generate the ℓ-th minimum cycle basis ($\ell \leq L$) in $\mathcal{O}(\ell^2 m^4)$ time, where $m = |E(G)|$, and the remaining ones in $\mathcal{O}(m)$ delay after $\mathcal{O}(m^4)$-time preprocessing.*

Cut Space. We assume an edge-weight to be nonnegative (i.e., $w : E(G) \to \mathbb{R}_{\geq 0}$). Recall that we assume G to be connected. A subset $F \subseteq E(G)$ of edges is a *cut* (*of G*) if $G - F$ is disconnected. Equivalently, F is a cut if there is a vertex subset $W \subseteq V(G)$ such that $F = E(W; G) := \{st \in E(G) \mid s \in W, \, t \in V(G) - W\}$. The weight of a cut F is the total weight of edges in F. Let $\mathcal{C}_G$ denote the set of all cuts in G. Let $F \in \mathcal{C}_G$ be a cut that satisfies $F = E(W; G)$ for some $W \subseteq V$. For two distinct vertices $s, t \in V$, we call F an *s, t-cut* if $s \in W$ and $t \notin W$. Let $\mathcal{C}_G(s, t)$ denote the set of all s, t-cuts and $\lambda(s, t; G)$ denote the weight of minimum s, t-cut of G. For vertex subsets $S, T \subseteq V$, let $\lambda_{\max}(S, T; G) := \max_{a \in S, \, b \in T : \, a \neq b} \lambda(a, b; G)$. For simplicity, we write $\lambda_{\max}(V, V; G)$ as $\lambda_{\max}(G)$.

The vector space over $GF(2)$ spanned by the incidence vectors of all cuts in G is called the *cut space of G*. We say that cuts are independent if their incidence

---

[1] A faster algorithm is proposed in [17], but it does not improve the preprocessing time of the enumeration algorithm in Corollary 1.

vectors are linearly independent in $GF(2)$. We consider a matroid $\mathcal{M} = (U, \mathcal{I})$ such that the ground set $U = \mathcal{C}_G$ is the set of all cuts in G and $\mathcal{I}$ is the family of all independent cuts in U. We call a relevant element of this matroid a *relevant cut*. The cardinality $|U|$ of the ground set can be up to an exponential number with respect to n and m.

It is known that the rank of this matroid is $n - \gamma$ [6]. Similarly to the previous matroid, we can implement the IND-oracle by using Gaussian elimination so that $\tau_{\mathrm{IND}} = \mathcal{O}(m^3)$ holds. A minimum basis in the matroid is called a *minimum cut basis*. We can construct a minimum cut basis from a Gomory-Hu tree [7] of G; the set of $n - 1$ edges in a Gomory-Hu tree corresponds to a minimum cut basis [6]. We can construct a Gomory-Hu tree in $\mathcal{O}(n\varphi)$ time, where φ denotes the time to compute a minimum s, t-cut in an edge-weighted undirected graph G for arbitrarily chosen vertices $s, t \in V(G)$ [7]; e.g., $\varphi = \mathcal{O}(mn)$ [12,19]. Then we can implement the MinB-oracle so that $\tau_{\mathrm{MinB}} = \mathcal{O}(n\varphi)$.

Now we describe how to list all relevant cuts in G in non-decreasing order of weight. This is different from the problems addressed in Yeh et al. [23]. They studied the problem of enumerating all cuts or s, t-cuts for specified vertices s, t in non-decreasing order of weight and did not take the notion of relevant cuts into account.

Recall that a cut is relevant if it belongs to at least one minimum cut basis. Let $\mathcal{C}_G^* \subseteq \mathcal{C}_G$ denote the set of all relevant cuts of G. A necessary and sufficient condition that a cut is relevant is shown in the following lemma.

Lemma 6 ([5]). *For an edge-weighted graph G, a cut $C \in \mathcal{C}_G$ is relevant if and only if there exists a pair of distinct vertices $s, t \in V(G)$ such that C is a minimum s, t-cut.*

Let us overview our algorithm. We build graphs $G_0 := G, G_1, \ldots, G_\tau$ for some integer τ iteratively, where G_{i+1} is obtained by contracting two certain vertices s_i, t_i in G_i. We will show that relevant cuts of G appear as minimum s_i, t_i-cuts of G_i and that the weights of minimum cuts are in non-increasing order over $G_0, G_1, \ldots, G_\tau$. We can achieve the goal by enumerating minimum s_i, t_i-cuts of G_i for $i = \tau, \tau - 1, \ldots, 0$, where we employ Yeh et al.'s algorithm [23] to enumerate minimum s_i, t_i-cuts of G_i.

For $X \subseteq V(G)$, let G/X denote the multigraph that is obtained by contracting all vertices in X into a single vertex, where we denote by the low-ercase (i.e., x) the new vertex. Multiedges may appear, and we do not take self-loops into account. Formally, $V(G/X)$ and $E(G/X)$ are defined as follows; $V(G/X) = (V(G) - X) \cup \{x\}$ and $E(G/X) = \{uv \in E(G) \mid u, v \notin X\} \cup \left(\bigcup_{uv \in E(G):\ u \notin X,\ v \in X} \{ux\} \right)$. For $v \in V(G)$, we regard $G/\{v\} = G$. We inherit the edge-weight function on G/X from G and hence regard w as the edge-weight function on G/X.

Let $\mathcal{X} = \{X_1, X_2, \ldots, X_k\}$ be a partition of V. We denote by $G/\mathcal{X}$ the multigraph that is obtained by contracting all vertices in $X_i \in \mathcal{X}$ into a single vertex x_i for each $i = 1, 2, \ldots, k$, respectively. Let $G' = G/\mathcal{X}$. For $1 \leq i < j \leq k$, every x_i, x_j-cut of G' is an s, t-cut of G for all $s \in X_i$ and $t \in X_j$ and hence we have $\lambda(x_i, x_j; G') \geq \lambda_{\max}(X_i, X_j; G)$.

Let $s, t \in V(G)$. The set $\mathcal{C}_G$ of all cuts is partitioned into two disjoint subsets, one for the set of s, t-cuts and the other for the set of any other cuts. The latter subset is equal to the set $\mathcal{C}_{G/\{s,t\}}$ of all cuts in the contracted graph $G/\{s,t\}$, as shown in the following lemma.

Lemma 7. *For a graph G, let $s, t \in V(G)$ be two distinct vertices. Then it holds that $\mathcal{C}_G = \mathcal{C}_G(s,t) \sqcup \mathcal{C}_{G/\{s,t\}}$.*

For $s, t \in V(G)$, let $\mathcal{C}_G^*(s,t)$ denote the set of all minimum s, t-cuts. By Lemma 6, any cut in $\mathcal{C}_G^*(s,t)$ is relevant (i.e., $\mathcal{C}_G^*(s,t) \subseteq \mathcal{C}_G^*$). For a partition $\mathcal{X} = \{X_1, X_2, \ldots, X_k\}$ of $V(G)$, we define $\Lambda(\mathcal{X}) := \{(i,j) \mid 1 \leq i < j \leq k, \ \lambda(x_i, x_j; G/\mathcal{X}) = \lambda_{\max}(X_i, X_j; G)\}$.

Lemma 8. *For an edge-weighted graph G, let $\mathcal{X} = \{X_1, X_2, \ldots, X_k\}$ be a partition of $V(G)$. If $\Lambda(\mathcal{X}) = \emptyset$, then no cut of $G/\mathcal{X}$ is relevant for G. Otherwise, for any $(i,j) \in \Lambda(\mathcal{X})$, all minimum x_i, x_j-cuts in $G/\mathcal{X}$ are relevant for G.*

Proof. For the former, we show that the contraposition is true. Let $C \subseteq E(G/\mathcal{X})$ be a cut of $G/\mathcal{X}$ that is relevant for G. By Lemma 6, there are $s, t \in V(G)$ such that C is a minimum s, t-cut in G. Let X_i and X_j denote the subsets that contain s and t, respectively. We see $i \neq j$ since C separates x_i and x_j in $G/\mathcal{X}$. Then we have $w(C) \geq \lambda(x_i, x_j; G/\mathcal{X}) \geq \lambda_{\max}(X_i, X_j; G) \geq \lambda(s, t; G)$, where all the inequalities hold by equalities by $w(C) = \lambda(s, t; G)$, showing that $(i,j) \in \Lambda(\mathcal{X})$.

For the latter, there exist two vertices $s \in X_i$ and $t \in X_j$ such that $\lambda(x_i, x_j; G/\mathcal{X}) = \lambda_{\max}(X_i, X_j; G) = \lambda(s, t; G)$. Any minimum x_i, x_j-cut C of $G/\mathcal{X}$ is an s, t-cut of G that satisfies $w(C) = \lambda(x_i, x_j; G/\mathcal{X}) = \lambda(s, t; G)$, and thus is relevant for G by Lemma 6. $\square$

Let us call a cut C a *solution* if $C \in \mathcal{C}_G^*$. We outline our algorithm to enumerate all solutions in non-decreasing order of weight. Let $V(G) = \{v_1, v_2, \ldots, v_n\}$. Initializing $\mathcal{X} := \{\{v_1\}, \{v_2\}, \ldots, \{v_n\}\}$, we execute the following: (i) If $\Lambda(\mathcal{X}) \neq \emptyset$, then let $(i,j) := \arg\max_{(x,y) \in \Lambda(\mathcal{X})}\{\lambda(x, y; G/\mathcal{X})\}$. All minimum x_i, x_j-cuts in $G/\mathcal{X}$ are solutions by the latter half of Lemma 8. Before enumerating them, we make a recursive call for the partition that is obtained by merging X_i and X_j (i.e., $(\mathcal{X} - \{X_i, X_j\}) \cup \{X_i \cup X_j\}$) to enumerate solutions that have no more weights than minimum x_i, x_j-cuts. After the recursive call is complete, we enumerate all minimum x_i, x_j-cuts in $G/\mathcal{X}$ as solutions, using the algorithm in [23]. (ii) Otherwise (i.e., if $\Lambda(\mathcal{X}) = \emptyset$), we do nothing since no cut of $G/\mathcal{X}$ is a solution by the former half of Lemma 8. Although all outputs by the above procedure are solutions, it is not guaranteed that they are all solutions. We utilize the following two lemmas to prove it.

Lemma 9. *For an edge-weighted graph G, let $\mathcal{X} = \{X_1, X_2, \ldots, X_k\}$ be a partition of $V(G)$. For a cut C of $G/\mathcal{X}$, it holds that $C \in \mathcal{C}_G^* \cap \mathcal{C}_{G/\mathcal{X}}^*$ if and only if there is $(i,j) \in \Lambda(\mathcal{X})$ such that C is a minimum x_i, x_j-cut in $G/\mathcal{X}$.*

Algorithm 2. An algorithm to enumerate all relevant cuts of a given edge-weighted connected undirected graph G

Require: An edge-weighted connected undirected graph G
Ensure: All relevant cuts of G in non-decreasing order of weights
1: $\mathcal{X} \leftarrow \{\{v_1\}, \{v_2\}, \ldots, \{v_n\}\};$ $\triangleright\ V(G) = \{v_1, v_2, \ldots, v_n\}$
2: **for** each $u, v \in V(G)$ $(u \neq v)$ **do**
3:      $W[\{u\}, \{v\}] \leftarrow \lambda(u, v; G)$
4: **end for**;
5: call ENUMRELEVANTCUTS($\mathcal{X}$)

6: **procedure** ENUMRELEVANTCUTS($\mathcal{X} = \{X_1, X_2, \ldots, X_k\}$)
7:      Compute $\Lambda(\mathcal{X})$;
8:      **if** $\Lambda(\mathcal{X}) \neq \emptyset$ **then**
9:          $(i, j) \leftarrow \arg\max_{(x,y) \in \Lambda(\mathcal{X})} \{W[X_x, X_y]\};$
10:         **for** each $X_r \in \mathcal{X} - \{X_i, X_j\}$ **do**
11:            $W[X_r, X_i \cup X_j] \leftarrow \max\{W[X_r, X_i], W[X_r, X_j]\}$
12:         **end for**;
13:         call ENUMRELEVANTCUTS$((\mathcal{X} - \{X_i, X_j\}) \cup \{X_i \cup X_j\})$;
14:         output all minimum x_i, x_j-cuts of $G/\mathcal{X}$
15:      **end if**;
16: **end procedure**

Lemma 10. *For an edge-weighted graph $G = (V, E)$, let $\mathcal{X} = \{X_1, X_2, \ldots, X_k\}$ be a partition of $V(G)$. Let $(i, j) = (x, y)$ denote a maximizer of $\lambda(x, y; G/\mathcal{X})$ among all $(x, y) \in \Lambda(\mathcal{X})$. Then it holds that $\mathcal{C}_G^* \cap \mathcal{C}_{G/\mathcal{X}}^* \subseteq \mathcal{C}_{G/\mathcal{X}}^*(x_i, x_j) \sqcup \mathcal{C}_{(G/\mathcal{X})/\{x_i, x_j\}}^*.$*

We show the enumeration algorithm in Algorithm 2. In (i) of the above outline, to compute $\Lambda(\mathcal{X})$, we identify whether $\lambda(x_i, x_j; G/\mathcal{X}) = \lambda_{\max}(X_i, X_j; G)$ holds for all $x_i, x_j \in V(G/\mathcal{X})$. During the algorithm, we store $\lambda_{\max}(X_i, X_j; G)$ by using an array $W[X_i, X_j]$, where we assume that the key is unordered; i.e., $W[X_i, X_j] = W[X_j, X_i]$. First, we initialize $W[\{u\}, \{v\}] := \lambda(u, v; G)$ for any two $u, v \in V(G)$. When we contract $x_i, x_j \in G/\mathcal{X}$ into a new vertex, for all $x_r \in V(G/\mathcal{X}) - \{x_i, x_j\}$, we set $\max\{W[X_r, X_i], W[X_r, X_j]\}$ to the new entry $W[X_r, X_i \cup X_j]$ since $\lambda_{\max}(X_r, X_i \cup X_j; G) = \max\{\lambda_{\max}(X_r, X_i; G), \lambda_{\max}(X_r, X_j; G)\}$.

Theorem 3. *Given a connected undirected graph G with an edge-weight function $w : E(G) \to \mathbb{R}_{\geq 0}$, Algorithm 2 outputs all relevant cuts of G in non-decreasing order of weight in $\mathcal{O}(nm \log(n^2/m))$ delay except that the computation time for the first output is $\mathcal{O}(n^2 \varphi)$.*

Regarding that the computation until the first output as preprocessing of the REL-oracle, we have $\tau_{\mathrm{REL}}^{\mathrm{pre}} = \mathcal{O}(n^2 \varphi)$ and $\tau_{\mathrm{REL}} = \mathcal{O}(nm \log(n^2/m))$.

Corollary 2. *Given a connected graph $G = (V, E)$ with an edge-weight function $w : E \to \mathbb{R}_{\geq 0}$, there is an integer L such that we can generate the ℓ-th minimum* *cut basis in $\mathcal{O}((n + \ell)nm \log(n^2/m) + \ell^2 nm^3)$ time for $\ell \leq L$ and the remaining minimum cut bases in $\mathcal{O}(n)$ delay after $\mathcal{O}(n^2\varphi)$-time preprocessing.*

5 Concluding Remarks

In this paper, we have studied problems of enumerating bases in a matroid with an exponentially large ground set. Examples of such a matroid includes binary matroids from cycle spaces and cut spaces. Our result is an incremental polynomial algorithm that enumerates all minimum-weight bases. We have shown that it is applicable to the above binary matroids so that the time complexity is polynomially bounded by the graph size.

We also developed a poly-delay algorithm for enumerating all (unweighted) bases of a binary matroid $\mathcal{M} = (U, \mathcal{I})$ such that the summation of the vectors corresponding to one or more independent elements is a vector corresponding to an element in U. This algorithm is more efficient than Algorithm 1. We assume only the IND-oracle, which can be realized by any algorithm that computes a matrix determinant over $GF(2)$. The algorithm is based on Avis and Fukuda's well-known *reverse search* [1]. We show that the delay of the algorithm is $\mathcal{O}(r^3 + r^2\tau_{\mathrm{DET}}(r))$, where r denotes the rank of $\mathcal{M}$ and $\tau_{\mathrm{DET}}(r)$ denotes the computation time for obtaining the determinant of a given $r \times r$ matrix over $GF(2)$. Again, the polynomial does not depend on the ground set size $|U|$. The algorithm immediately yields poly-delay enumeration of bases in binary matroids from cycle spaces and cut spaces. See the preprint [18] for details.

For future work, it would be interesting to explore whether or not it is possible to enumerate all minimum bases in poly-delay in our context (i.e., the bounding polynomial does not contain the ground set size) and/or in poly-space. Among the remaining issues are to study the k-best version, to improve the complexities, to look for other applications, and so on.

References

1. Avis, D., Fukuda, K.: Reverse search for enumeration. Discret. Appl. Math. **65**(1), 21–46 (1996). https://doi.org/10.1016/0166-218X(95)00026-N
2. Berger, F., Gritzmann, P., de Vries, S.: Minimum cycle bases for network graphs. Algorithmica **40**(1), 51–62 (2004). https://doi.org/10.1007/s00453-004-1098-x
3. Cao, Y.: Enumerating maximal induced subgraphs. In: Gørtz, I.L., Farach-Colton, M., Puglisi, S.J., Herman, G. (eds.) 31st Annual European Symposium on Algorithms (ESA 2023). Leibniz International Proceedings in Informatics (LIPIcs), vol. 274, pp. 31:1–31:13. Schloss Dagstuhl – Leibniz-Zentrum für Informatik, Dagstuhl, Germany (2023). https://doi.org/10.4230/LIPIcs.ESA.2023.31
4. Diestel, R.: Graph theory. Springer Berlin Heidelberg, Berlin, Heidelberg (2017). https://doi.org/10.1007/978-3-662-53622-3_1
5. Domschke, N., Gatter, T., Stadler, P.F.: A short note on relevant cuts. The Art of Discrete and Applied Mathematics (2025). https://doi.org/10.26493/2590-9770.1845.02e, the preprint is available at arXiv: 2410.20257

6. Golynski, A., Horton, J.D.: A Polynomial Time Algorithm to Find the Minimum Cycle Basis of a Regular Matroid. In: Penttonen, M., Schmidt, E.M. (eds.) SWAT 2002. LNCS, vol. 2368, pp. 200–209. Springer, Heidelberg (2002). https://doi.org/10.1007/3-540-45471-3_21
7. Gomory, R.E., Hu, T.C.: Multi-terminal network flows. J. Soc. Ind. Appl. Math. **9**(4), 551–570 (1961)
8. Hamacher, H.W., Queyranne, M.: K best solutions to combinatorial optimization problems. Ann. Oper. Res. **4**(1), 123–143 (1985). https://doi.org/10.1007/BF02022039
9. Horton, J.D.: A polynomial-time algorithm to find the shortest cycle basis of a graph. SIAM J. Comput. **16**(2), 358–366 (1987). https://doi.org/10.1137/0216026
10. Johnson, D.S., Yannakakis, M., Papadimitriou, C.H.: On generating all maximal independent sets. Inf. Process. Lett. **27**(3), 119–123 (1988). https://doi.org/10.1016/0020-0190(88)90065-8
11. Khachiyan, L., Boros, E., Elbassioni, K., Gurvich, V., Makino, K.: On the complexity of some enumeration problems for matroids. SIAM J. Discret. Math. **19**(4), 966–984 (2005). https://doi.org/10.1137/S0895480103428338
12. King, V., Rao, S., Tarjan, R.: A faster deterministic maximum flow algorithm. J. Algorithms **17**(3), 447–474 (1994). https://doi.org/10.1006/jagm.1994.1044
13. Kobayashi, Y., Kurita, K., Wasa, K.: Polynomial-delay enumeration of large maximal common independent sets in two matroids and beyond. Inf. Comput. **304**, 105282 (2025). https://doi.org/10.1016/j.ic.2025.105282
14. Korte, B., Vygen, J.: Combinatorial optimization: theory and algorithms. Springer, sixth edn. (2018)
15. Madan, V., Nikolov, A., Singh, M., Tantipongpipat, U.: Maximizing determinants under matroid constraints. In: 2020 IEEE 61st Annual Symposium on Foundations of Computer Science (FOCS), pp. 565–576 (2020).https://doi.org/10.1109/FOCS46700.2020.00059
16. Mary, A., Strozecki, Y.: Efficient enumeration of solutions produced by closure operations. Discrete Math. Theor. Comput. Sci. **21**(3), 22 (2019). https://doi.org/10.23638/DMTCS-21-3-22
17. Mehlhorn, K., Michail, D.: Minimum cycle bases: faster and simpler **6**(1) (2010). https://doi.org/10.1145/1644015.1644023
18. Nishimura, Y., Haraguchi, K.: Enumeration of bases in matroid with exponentially large ground set (2025). https://arxiv.org/abs/2504.11728
19. Orlin, J.B.: Max flows in $O(nm)$ time, or better. In: Proceedings of the Forty-Fifth Annual ACM Symposium on Theory of Computing, pp. 765–774. STOC '13, Association for Computing Machinery, New York, NY, USA (2013). https://doi.org/10.1145/2488608.2488705
20. Oxley, J.: Matroid theory. oxford university press, 2nd edn. (2011)
21. Vasiliauskaite, V., Evans, T.S., Expert, P.: Cycle analysis of directed acyclic graphs. Physica A: Stat. Mech. Appl. **596**, 127097 (2022). https://doi.org/10.1016/j.physa.2022.127097
22. Vismara, P.: Union of all the minimum cycle bases of a graph. Electron. J. Comb. **4**(1), R9 (1997)
23. Yeh, L.P., Wang, B.F., Su, H.H.: Efficient algorithms for the problems of enumerating cuts by non-decreasing weights. Algorithmica **56**(3), 297–312 (2010). https://doi.org/10.1007/s00453-009-9284-5

Author Index

E. Di Giacomo and D. Mondal (Eds.): WALCOM 2026, LNCS 16444, pp. 591–592, 2026.
https://doi.org/10.1007/978-981-95-7127-7

GPSR Compliance
The European Union's (EU) General Product Safety Regulation (GPSR) is a set
of rules that requires consumer products to be safe and our obligations to
ensure this.

If you have any concerns about our products, you can contact us on

ProductSafety@springernature.com

In case Publisher is established outside the EU, the EU authorized
representative is:

Springer Nature Customer Service Center GmbH
Europaplatz 3
69115 Heidelberg, Germany